GERARD J. TORTORA

Biology Coordinator, Bergen Community College, Paramus, New Jersey

ROBERT B. TALLITSCH

Augustana College, Rock Island, Illinois

Laboratory Exercises in ANATOMY AND PHYSIOLOGY with Cat Dissections

FIFTH EDITION

Prentice Hall
Upper Saddle River, NJ 07458

Acquisitions Editor: David Kendric Brake
Managing Editor: Linda Schreiber
Assistant Vice President and Director of Production: David W. Riccardi
Special Projects Manager: Barbara A. Murray
Total Concept Coordinator: Kimberly P. Karpovich
Editorial Production/Design: Publication Services
Text Illustrations: Mary Dersh; Publication Services
Cover Design: Joseph Sengotta
Cover Illustrations: Mary Dersh; Publication Services
Photo Researcher: Jan Fisher

Printed in the United States of America

ISBN 0-13-237579-6

10 9 8 7 6 5 4 3 2 1

ISBN   0-13-237579-6

Prentice-Hall International (UK) Limited, *London*
Prentice-Hall of Australia Pty. Limited, *Sydney*
Prentice-Hall Canada Inc., *Toronto*
Prentice-Hall Hispanoamericana, S.A., *Mexico*
Prentice-Hall of India Private Limited, *New Delhi*
Prentice-Hall of Japan, Inc., *Tokyo*
Simon & Schuster Asia Private Limited., *Singapore*
Editora Prentice Hall do Brazil, Ltda., *Rio de Janeiro*

Preface

Laboratory Exercises in Anatomy and Physiology with Cat Dissections, Fifth Edition, has been written to guide students in the laboratory study of introductory anatomy and physiology. The manual was written to accompany most of the leading anatomy and physiology textbooks.

COMPREHENSIVENESS

This manual examines virtually every structure and function of the human body that is typically studied in an introductory anatomy and physiology course. Because of its detail, the need for supplemental handouts is minimized; the manual is a strong teaching device in itself.

USE OF THE SCIENTIFIC METHOD

Anatomy (the science of structure) and physiology (the science of function) cannot be understood without the practical experience of laboratory work. The exercises in this manual challenge students to understand the way scientists work by asking them to make microscopic examinations and evaluations of cells and tissues, to observe and interpret chemical reactions, to record data, to make gross examinations of organs and systems, to dissect, and to conduct physiological laboratory work and interpret and apply the results of this work.

ILLUSTRATIONS

The manual contains a large number and variety of illustrations. The illustrations of the body systems of the human have been carefully drawn to exhibit structures that are essential to students' understanding of anatomy and physiology. Numerous photographs, photomicrographs, and scanning electron micrographs are presented to show stu-dents how the structures of the body actually look. We feel that this laboratory manual has better and more complete illustrations than any other anatomy and physiology manual.

IMPORTANT FEATURES

Among the key features of this manual are (1) dissection of the white rat, and selected mammalian organs; (2) numerous physiological experiments; (3) emphasis on the study of anatomy through histology; (4) lists of appropriate terms accompanying drawings and photographs to be labeled; (5) inclusion of numerous scanning electron micrographs and specimen photos; (6) phonetic pronunciations and derivations for the vast majority of anatomical and physiological terms; (7) diagrams of commonly used laboratory equipment; (8) laboratory report questions and reports at the end of each exercise that can be filled in, removed, and turned in for grading if the instructor so desires; (9) three appendixes dealing with units of measurement, a periodic table of elements, and eponyms used in the laboratory manual; and (10) emphasis on laboratory safety throughout the manual.

NEW TO FIFTH EDITION

Numerous changes have been made in the fifth edition of this manual in response to suggestions from instructors and students. A significant change is the addition of full color throughout the manual. We have also added some new physiology experiments, line drawings, photomicrographs, and phonetic pronunciations and derivations. Also new to this edition is the inclusion of many computer programs that simulate various physiology experiments. These are presented as alternative procedures for instructors who prefer not to use live animals or do not have access to specific physiological equipment.

The principal additions to various exercises are as follows:

Exercise 2, "Introduction to the Human Body," now has sections dealing with life processes and homeostasis.

Exercise 3, "Cells," contains revised sections on the movement of substances across and through plasma membranes, especially active transport.

In Exercise 4, "Tissues," new color photomicrographs of epithelial and connective tissues have been added.

A new discussion on the homeostasis of body temperature has been added to Exercise 5, "Integumentary System."

In Exercise 6, "Bone Tissue," the chemistry of bone tissue is now discussed before the structure of a long bone. Also, there are new sections on ossification, bone growth, and fracture.

In Exercise 7, "Bones," the black and white bone photos have been replaced by new, clearer illustrations that are colored.

In Exercise 8, "Articulations," several photographs have been replaced.

Exercise 9, "Muscle Tissue," has revised sections on the physiology of muscle contraction, effect of fatigue on a single muscle twitch, refractory period muscle response to variation of stimulus frequency, and electromyography. There is a new section on unit summation.

In Exercise 10, "Skeletal Muscles," there is a new section on the arrangement of muscle fascicles and a new exercise on naming skeletal muscles.

Exercise 12, "Nervous Tissue and Physiology," has a revised introduction, along with spinal reflexes of the frog. There is also a new section on neuronal circuits.

Exercise 13, "Nervous System," now contains new illustrations on plexuses and a new section on somatic sensory and motor pathways. Phonetic pronunciations have been added to *all* the nerves of the plexuses. The sections on reflex experiments and cranial nerve functions have been reorganized.

In Exercise 14, "Sensory Receptors and Sensory and Motor Pathways," the discussion of somatic sensory and motor pathways has been expanded and the details of sensory and motor tracts have been moved to Exercise 13.

In Exercise 15, "The Endocrine System," all hormone functions have been updated and the histology of the testes and ovaries has been revised.

In Exercise 16, "Blood," there are no longer any tests that involve drawing human blood or working with drawn human blood. The various tests involving blood may be conducted with blood from a clinical laboratory that has been tested and certified as noninfectious, mammalian blood (other than a human), and simulated blood.

Several new illustrations have been added to Exercise 17, "The Heart."

In Exercise 18, "Blood Vessels," phonetic pronunciations have been added to *all* blood vessel names and the discussions of hepatic portal, pulmonary, and fetal circulation have been expanded.

The discussions of the cardiac conduction system, cardiac cycle, and heart sounds have been rewritten in Exercise 19, "Cardiovascular Physiology."

In Exercise 21, "Respiratory System," the introduction is new. The sections dealing with the nose, inspiration, and pulmonary volumes and capacities have been revised. New art has also been added.

In Exercise 22, "Digestive System," the sections dealing with the salivary glands and stomach histology have been revised and a new illustration has been added.

Exercise 24, "pH and Acid-Base Balance," is new to this edition.

In Exercise 25, "Reproductive Systems," a new introduction has been added and the discussions of the testes, ovaries, and female reproductive cycle have been revised.

The discussions of spermatogenesis, oogenesis, and the embryonic period have been expanded and revised in Exercise 26, "Development."

CHANGES IN TERMINOLOGY

In recent years, the use of eponyms for anatomical terms has been minimized or eliminated. Anatomical eponyms are terms named after various individuals. Examples include Fallopian tube (after Gabriello Fallopio) and Eustachian tube (after Bartolommeo Eustachio).

Anatomical eponyms are often vague and nondescriptive and do not necessarily mean that the person whose name is applied contributed anything very original. For these reasons, we have also decided to minimize their use. However, because some still prevail, we have provided eponyms, in parentheses, after the first reference in each chapter to the more acceptable synonym. Thus, you will expect to see terms such as *uterine (Fallopian) tube* or *auditory (Eustachian) tube.* See Appendix C.

INSTRUCTOR'S GUIDE

A complimentary instructor's guide by Gerard J. Tortora to accompany the manual is available from the publisher. This comprehensive guide contains: (1) a listing of materials needed to complete each exercise, (2) suggested audiovisual materials, (3) answers to illustrations and questions within the exercises, and (4) answers to laboratory report questions.

Gerard J. Tortora
Biology Coordinator
Natural Sciences and Mathematics, S229
Bergen Community College
400 Paramus Road
Paramus, NJ 07652

Contents

Laboratory Safety   xi

Commonly Used Laboratory Equipment   xv

Pronunciation Key   xvii

1. Microscopy   1

A. Compound Light Microscope   1
B. Electron Microscope   6

2. Introduction to the Human Body   9

A. Anatomy and Physiology   9
B. Levels of Structural Organization   10
C. Systems of the Body   10
D. Life Processes   12
E. Homeostasis   13
F. Anatomical Position and Regional Names   14
G. External Features of the Body   15
H. Directional Terms   15
I. Planes of the Body   18
J. Body Cavities   19
K. Abdominopelvic Regions   19
L. Abdominopelvic Quadrants   21
M. Dissection of White Rat   21

3. Cells   31

A. Cell Parts   31
B. Diversity of Cells   33
C. Movement of Substances Across and Through Plasma Membranes   33
D. Cell Inclusions   39
E. Extracellular Materials   40
F. Cell Division   40

4. Tissues   51

A. Epithelial Tissue   51
B. Connective Tissue   58
C. Membranes   65

5. Integumentary System   71

A. Skin   71
B. Hair   73
C. Glands   75
D. Nails   76
E. Homeostasis of Body Temperature   76

6. Bone Tissue   81

A. Functions of Bone   81
B. Chemistry of Bone   84
C. Gross Structure of a Long Bone   84
D. Histology of Bone   85
E. Bone Formation: Ossification   85
F. Bone Growth   85
G. Fractures   87
H. Types of Bones   87
I. Bone Surface Markings   89

7. Bones   93

A. Bones of Adult Skull   93
B. Sutures of Skull   100
C. Fontanels of Skull   100
D. Paranasal Sinuses of Skull   100
E. Vertebral Column   102
F. Vertebrae   102
G. Sternum and Ribs   106
H. Pectoral (Shoulder) Girdles   108
I. Upper Limbs   110
J. Pelvic (Hip) Girdles   113
K. Lower Limbs   114
L. Articulated Skeleton   118
M. Skeletal System of Cat   118

8. Articulations   123

A. Kinds of Joints   123
B. Synarthroses (Immovable Joints)   123
C. Amphiarthroses (Slightly Movable Joints)   123
D. Diarthroses (Freely Movable Joints)   124
E. Knee Joint   127

9. Muscle Tissue   135

A. Kinds of Muscle Tissue   135
B. Skeletal Muscle Tissue   135
C. Cardiac Muscle Tissue   136
D. Smooth (Visceral) Muscle Tissue   137
E. Physiology of Skeletal Muscle Contraction   137
F. Laboratory Tests on Skeletal Muscle Contraction   140
G. Biochemistry of Skeletal Muscle Contraction   148
H. Electromyography   148

10. Skeletal Muscles   157

A. How Skeletal Muscles Produce Movement   157
B. Arrangement of Fascicles   158
C. Naming Skeletal Muscles   158
D. Connective Tissue Components   158
E. Principal Muscles   160
F. Composite Muscular System   202
G. Dissection of Cat Muscular System   202

11. Surface Anatomy   225

A. Head   225
B. Neck   225
C. Trunk   226
D. Upper Limb (Extremity)   229
E. Lower Limb (Extremity)   233

12. Nervous Tissue and Physiology   245

A. Nervous System Divisions   245
B. Histology of Nervous Tissue   246
C. Histology of Neuroglia   248
D. Neuronal Circuits   249
E. Reflex Arc   250
F. Demonstration of Reflex Arc   251
G. Spinal Reflexes of the Frog   251

13. Nervous System   259

A. Spinal Cord and Spinal Nerves   259
B. Brain   273
C. Electroencephalogram (EEG)   281
D. Cranial Nerves: Names and Components   284
E. Tests of Cranial Nerve Function   285
F. Dissection of Nervous System   288
G. Autonomic Nervous System   296

14. Sensory Receptors and Sensory and Motor Pathways   307

A. Characteristics of Sensations   307
B. Classification of Receptors   308
C. Receptors for General Senses   309
D. Tests for General Senses   311
E. Somatic Sensory Pathways   314
F. Olfactory Sensations   314
G. Gustatory Sensations   317
H. Visual Sensations   321
I. Auditory Sensations and Equilibrium   331
J. Sensory-Motor Integration   339
K. Somatic Motor Pathways   340

15. Endocrine System   351

A. Endocrine Glands   351
B. Pituitary Gland (Hypophysis)   351
C. Thyroid Gland   354
D. Parathyroid Glands   355
E. Adrenal (Suprarenal) Glands   355
F. Pancreas   357
G. Testes   358
H. Ovaries   359
I. Pineal Gland (Epiphysis Cerebri)   360
J. Thymus Gland   360
K. Other Endocrine Tissues   361
L. Physiology of the Endocrine System   361

16. Blood   369

A. Components of Blood   369
B. Plasma   369
C. Erythrocytes   371
D. Red Blood Cell Tests   372
E. Leukocytes   382
F. White Blood Cell Tests   382
G. Platelets   385
H. Drawings of Blood Cells   386
I. Blood Grouping (Typing)   386

17. Heart   395

A. Pericardium   395
B. Heart Wall   395
C. Chambers of Heart   396
D. Great Vessels and Valves of Heart   396
E. Blood Supply of Heart   396
F. Dissection of Cat Heart   401
G. Dissection of Sheep Heart   402

18. Blood Vessels   409

A. Arteries and Arterioles   409
B. Capillaries   409

C. Venules and Veins 409
D. Circulatory Routes 409
E. Blood Vessel Exercise 440
F. Dissection of Cat Cardiovascular System 440

19. Cardiovascular Physiology   453

A. Cardiac Conduction System and Electrocardiogram (ECG or EKG) 453
B. Cardiac Cycle 459
C. Cardiac Cycle Experiments 461
D. Heart Sounds 464
E. Pulse Rate 465
F. Blood Pressure (Auscultation Method) 466
G. Observing Blood Flow 467

20. Lymphatic System   477

A. Lymphatic Vessels 477
B. Lymphatic Tissue 478
C. Lymph Circulation 479
D. Dissection of Cat Lymphatic System 481

21. Respiratory System   485

A. Organs of the Respiratory System 485
B. Dissection of Cat Respiratory System 492
C. Dissection of Sheep Pluck 497
D. Laboratory Tests on Respiration 497
E. Laboratory Tests Combining Respiratory and Cardiovascular Interactions 509

22. Digestive System   519

A. General Organization of Digestive System 519
B. Organs of Digestive System 519
C. Dissection of Cat Digestive System 532
D. Deglutition 538
E. Observation of Movements of the Gastrointestinal Tract 539
F. Physiology of Intestinal Smooth Muscle 541
G. Chemistry of Digestion 545

23. Urinary System   557

A. Organs of Urinary System 557
B. Dissection of Cat Urinary System 564
C. Dissection of Sheep (or Pig) Kidney 567
D. Renal Physiology Experiments 569

E. Urine 570
F. Urinalysis 571

24. pH and Acid-Base Balance   583

A. The Concept of pH 583
B. Measuring pH 583
C. Acid-Base Balance 584
D. Acid-Base Imbalances 588
E. Renal Regulation of Hydrogen Ion Concentration 589

25. Reproductive Systems   595

A. Organs of Male Reproductive System 595
B. Organs of Female Reproductive System 601
C. Dissection of Cat Reproductive Systems 612
D. Dissection of Fetus-Containing Pig Uterus 616

26. Development   621

A. Spermatogenesis 621
B. Oogenesis 623
C. Embryonic Period 624
D. Fetal Period 632

27. Genetics   637

A. Genotype and Phenotype 637
B. Punnett Squares 637
C. Sex Inheritance 638
D. Sex-Linked Inheritance 640
E. Mendelian Laws 641
F. Multiple Alleles 643
G. Genetics Exercises 643

Appendix A: Some Important Units of Measurement   653

Appendix B: Periodic Table of the Elements   655

Appendix C: Eponyms Used in This Laboratory Manual   657

Figure Credits 659

Index 661

Laboratory Safety*

In 1989, The Centers for Disease Control and Prevention (CDC) published "Guidelines for Prevention of Transmission of Human Immunodeficiency Virus and Hepatitis B Virus to Health-Care and Public-Safety Workers" (*MMWR*, vol. 36, No. 6S). The CDC guidelines recommend precautions to protect health care and public safety workers from exposure to human immunodeficiency virus (HIV), the causative agent of acquired immunodeficiency syndrome (AIDS), and hepatitis B virus (HBV), the causative agent of hepatitis B. These guidelines are presented to reaffirm the basic principles involved in the transmission of not only the AIDS and hepatitis B viruses, but also any disease-producing organism.

Based on the CDC guidelines for health care workers, as well as on other standard additional laboratory precautions and procedures, the following list has been developed for your safety in the laboratory. Although specific cautions and warnings concerning laboratory safety are indicated throughout the manual, read the following *before* performing any experiments.

A. General Safety Precautions and Procedures

1. Arrive on time. Laboratory directions and procedures are given at the beginning of the laboratory period.
2. Read all experiments before you come to class to be sure that you understand all the procedures and safety precautions. Ask the instructor about any procedure you do not understand exactly. Do not improvise any procedure.
3. Protective eyewear and laboratory coats or aprons must be worn by all students performing or observing experiments.

*The authors and publisher urge consultation with each instructor's institutional policies concerning laboratory safety and first-aid procedures.

4. Do not perform any unauthorized experiments.
5. Do not bring any unnecessary items to the laboratory and do not place any personal items (pocketbooks, bookbags, coats, umbrellas, etc.) on the laboratory table or at your feet.
6. Make sure each apparatus is supported and squarely on the table.
7. Tie back long hair to prevent it from becoming a laboratory fire hazard.
8. Never remove equipment, chemicals, biological materials, or any other materials from the laboratory.
9. Do not operate any equipment until you are instructed in its proper use. If you are unsure of the procedures, ask the instructor.
10. Dispose of chemicals, biological materials, used apparatus, and waste materials according to your instructor's directions. Not all liquids are to be disposed of in the sink.
11. Some exercises in the laboratory manual are designed to induce some degree of cardiovascular stress. Students should not participate in these exercises if they are pregnant or have hypertension or any other known or suspected condition that might compromise health. Before you perform any of these exercises, check with your physician.
12. Do not put anything in your mouth while in the laboratory. Never eat, drink, taste chemicals, lick labels, smoke, or store food in the laboratory.
13. Your instructor will show you the location of emergency equipment such as fire extinguishers, fire blankets, and first-aid kits as well as eyewash stations. Memorize their locations and know how to use them.
14. Wash your hands before leaving the laboratory. Because bar soaps can become contaminated, liquid or powdered soaps should be used. Before leaving the laboratory, remove any protective clothing, such as laboratory coats or aprons, gloves, and eyewear.

B. PRECAUTIONS RELATED TO WORKING WITH BLOOD, BLOOD PRODUCTS, OR OTHER BODY FLUIDS

1. Work only with *your own* body fluids, such as blood, saliva, urine, tears, and other secretions and excretions; blood from a clinical laboratory that has been tested and certified as noninfectious; or blood from a mammal (other than a human).

2. Wear gloves when touching another person's blood or other body fluids.

3. Wear safety goggles when working with another person's blood.

4. Wear a mask and protective eyewear or a face shield during procedures that are likely to generate droplets of blood or other body fluids.

5. Wear a gown or an apron during procedures that are likely to generate splashes of blood or other body fluids.

6. Wash your hands immediately and thoroughly if contaminated with blood or other body fluids. Hands can be rapidly disinfected by using (1) a phenol disinfectant-detergent for 20 to 30 seconds (sec) and then rinsing with water, or (2) alcohol (50 to 70%) for 20 to 30 sec, followed by a soap scrub of 10 to 15 sec and rinsing with water.

7. Spills of blood, urine, or other body fluids onto bench tops can be disinfected by flooding them with a disinfectant-detergent. The spill should be covered with disinfectant for 20 minutes (min) before being cleaned up.

8. Potentially infectious wastes, including human body secretions and fluids, and objects such as slides, syringes, bandages, gloves, and cotton balls contaminated with those substances, should be placed in an autoclave container. Sharp objects (including broken glass) should be placed in a puncture-proof sharps container. Contaminated glassware should be placed in a container of disinfectant and autoclaved before it is washed.

9. Use only single-use, disposable lancets, and needles. Never recap, bend, or break the lancet once it has been used. Place used lancets, needles, and other sharp instruments in a *fresh* 1 : 10 dilution of household bleach (sodium hypochlorite) or other disinfectant such as phenols (Amphyl), aldehydes (glutaraldehyde, 1%), and 70% ethyl alcohol and then dispose of the instruments in a puncture-proof container. These disinfectants disrupt the envelope of HIV and HBV. The fresh household bleach solution or other disinfectant should be prepared for *each* laboratory session.

10. All reusable instruments, such as hemocytometers, well slides, and resuable pipettes, should be disinfected with a *fresh* 1 : 10 solution of household bleach or other disinfectant and thoroughly washed with soap and hot water. The fresh household bleach solution or other disinfectant should be prepared for *each* laboratory session.

11. A laboratory disinfectant should be used to clean laboratory surfaces *before* and after procedures, and should be available for quick cleanup of any blood spills.

12. Mouth pipetting should never be done. Use mechanical pipetting devices for manipulating all liquids in the laboratory.

13. All procedures and manipulations that have a high potential for creating aerosols or infectious droplets (such as centrifuging, sonicating, and blending) should be performed carefully. In such instances, a biological safety cabinet or other primary containment device is required.

C. PRECAUTIONS RELATED TO WORKING WITH REAGENTS

1. Use extreme care when working with reagents. Should any reagents make contact with your eyes, flush with water for 15 min; or, if they make contact with your skin, flush with water for 5 min. Notify your instructor immediately should a reagent make contact with your eyes or skin, and seek immediate medical attention.

2. Report all accidents to your instructor, no matter how minor they may appear.

3. When you are working with chemicals or preserved specimens, the room should be well ventilated. Avoid breathing fumes for any extended period of time.

4. Never point the opening of a test tube containing a reacting mixture (especially when heating it) toward yourself or another person.

5. Exercise care in noting the odor of fumes. Use "wafting" if you are directed to note an odor. Your instructor will demonstrate this procedure.

6. Do not force glass tubing or a thermometer into rubber stoppers. Lubricate the tubing and introduce it gradually and gently into the stopper. Protect your hands with toweling when inserting the tubing or thermometer into the stopper.

7. Never heat a flammable liquid over or near an open flame.

8. Use only glassware marked Pyrex or Kimax. Other glassware may shatter when heated. Handle hot glassware with test-tube holders.

9. If you have to dilute an acid, always add acid (AAA) to water.

10. When shaking a test tube or bottle to mix its contents, do not use your fingers as a stopper.

11. Read the label on a chemical twice before using it.

12. Replace caps or stoppers on bottles immediately after using them. Return spatulas to their correct place immediately after using them and do not mix them up.

13. Mouth pipetting should never be done. Use mechanical pipetting devices for manipulating all liquids in the laboratory.

D. PRECAUTIONS RELATED TO DISSECTION

1. When you are working with chemicals or preserved specimens, the room should be well ventilated. Avoid breathing fumes for any extended period of time.

2. Wear rubber gloves when dissecting.

3. To reduce the irritating effects of chemical preservatives to your skin, eyes, and nose, soak or wrap your specimen in a substance such as "Biostat." If this is not available, hold your specimen under running water for several minutes to wash away excess preservative and dilute what remains.

4. When dissecting, there is always the possibility of skin cuts or punctures from dissecting equipment or the specimens themselves, such as the teeth or claws of an animal. Should you sustain a cut or puncture in this manner, wash your hands with disinfectant soap, notify your instructor, and seek immediate medical attention to decrease the possibility of infection. A first-aid kit should be readily available for your use.

5. When cleaning dissecting instruments, always hold the sharp edges away from you.

6. Dispose of any damaged or worn-out dissecting equipment in an appropriate container supplied by your instructor.

SELECTED LABORATORY SAFETY SIGNS/LABELS

⚠ CAUTION

Protect eyes. Wear goggles at all times.

⚠ CAUTION

Hot surface. Do not touch.

⚠ CAUTION

Cancer suspect agent. Trained personnel only.

⚠ CAUTION

Radiation area. Authorized personnel only.

⚠ CAUTION

Biological hazard. Authorized personnel only.

⚠ DANGER

Highly toxic. Handle with care.

⚠ DANGER

Do not smoke in this area.

⚠ DANGER

Do not smoke, eat or drink in this area.

⚠ DANGER

Do not pipet liquids by mouth.

⚠ DANGER

Corrosive. Avoid contact with eyes and skin.

⚠ DANGER

Flammable material. Keep fire away.

EMERGENCY

Eye Wash Station. Keep area clear.

EMERGENCY

Safety Shower. Keep area clear.

EMERGENCY

First Aid Station.

Fire extinguisher. Remove pin and squeeze trigger.

COMMONLY USED LABORATORY EQUIPMENT

Beaker

Erlenmeyer flask

Florence flask

funnel

Graduated cylinder

Pipet

Mortar and pestle

Watch glass

Stirring rod

Test tube

Test tube brush

Test tube holder

Test tube rack

Ring stand and ring

Pinch clamp

Utility clamp

Tripod

Clay triangle

Wire gauze

Crucible tongs

Beaker tongs

Forceps

Medicine dropper

Nichrome wire

Spatula

Pronunciation Key

A unique feature of this revised manual is the phonetic pronunciations given for many anatomical and physiological terms. The pronunciations are given in parentheses immediately after the particular term is introduced. The following key explains the essential features of the pronunciations.

1. The syllable with the strongest accent appears in capital letters; for example, bilateral (bī-LAT-er-al) and diagnosis (dī-ag-NŌ-sis).
2. A secondary accent is denoted by a single quote mark ('); for example, constitution (kon'-sti-TOO-shun) and physiology (fiz'-ē-OL-ō-jē). Additional secondary accents are also noted by a single quotation mark; for example, decarboxylation (dē'-kar-bok'-si-LĀ-shun).

3. Vowels marked with a line above the letter are pronounced with the long sound, as in the following common words:

ā as in *māke*	ī as in *īvy*
ē as in *bē*	ō as in *pōle*

4. Unmarked vowels are pronounced with the short sound, as in the following words:

e as in *bet*	o as in *not*
i as in *sip*	u as in *bud*

5. Other phonetic symbols are used to indicate the following sounds:

a as in *above*	yoo as in *cute*
oo as in *soon*	oy as in *oil*

Microscopy

Note: *Before you begin any laboratory exercises in this manual, please read the section on LABORATORY SAFETY on page xi.*

One of the most important instruments that you will use in your anatomy and physiology course is a compound light microscope. In this instrument, the lenses are arranged so that images of objects too small to be seen with the naked eye can become highly magnified; that is, apparent size can be increased, and their minute details can be revealed. Before you actually learn the parts of a compound light microscope and how to use it properly, discussion of some of the principles employed in light microscopy (mī-KROS-kō-pē) will be helpful. Later in this exercise, some of the principles used in electron microscopy will also be discussed.

A. COMPOUND LIGHT MICROSCOPE

A *compound light microscope* uses two sets of lenses, ocular and objective, and employs light as its source of illumination. Magnification is achieved as follows. Light rays from an illuminator are passed through a condenser, which directs the light rays through the specimen under observation; from here, light rays pass into the objective lens, the magnifying lens that is closest to the specimen; the image of the specimen then forms on a prism and is magnified again by the ocular lens.

A general principle of microscopy is that the shorter the wavelength of light used in the instrument, the greater the resolution. *Resolution*, or *resolving power*, is the ability of the lenses to distinguish fine detail and structure, that is, to distinguish between two points as separate objects. As an example, a microscope with a resolving power of 0.3 micrometers (mī-KROM-e-ters), symbolized μm, is capable of distinguishing two points as separate objects if they are at least 0.3 μm apart. 1 μm = 0.000001 or 10^{-6} m. (See Appendix A.) The light used in a compound light microscope has a rela-tively long wavelength and cannot resolve structures smaller than 0.3 μm. This fact, as well as practical considerations, means that even the best compound light microscopes can magnify images only about 2000 times.

A *photomicrograph* (fō-tō-MĪ-krō'-graf), a photograph of a specimen taken through a compound light microscope, is shown in Figure 4.1. In later exercises you will be asked to examine photomicrographs of various specimens of the body before you actually view them yourself through the microscope.

1. Parts of the Microscope

Carefully carry the microscope from the cabinet to your desk by placing one hand around the arm and the other hand firmly under the base. Gently place it on your desk, directly in front of you, with the arm facing you. Locate the following parts of the microscope and, as you read about each part, label Figure 1.1 by placing the correct numbers in the spaces next to the list of terms that accompanies the figure.

1. *Base* The bottom portion on which the microscope rests.
2. *Body tube* The portion that receives the ocular.
3. *Arm* The angular or curved part of the frame.
4. *Inclination joint* A movable hinge in some microscopes that allows the instrument to be tilted to a comfortable viewing position.
5. *Stage* A platform on which slides or other objects to be studied are placed. The opening in the center, called the *stage opening*, allows light to pass from below through the specimen being examined. Some microscopes have a *mechanical stage*. An adjustor knob below the stage moves the stage forward and backward and from side to side. With a mechanical stage, the slide and the stage move simultaneously. A mechanical stage permits a smooth, precise movement of a slide. Sometimes a mechanical stage is fitted with calibrations that permit the numerical "mapping" of a specimen on a slide.

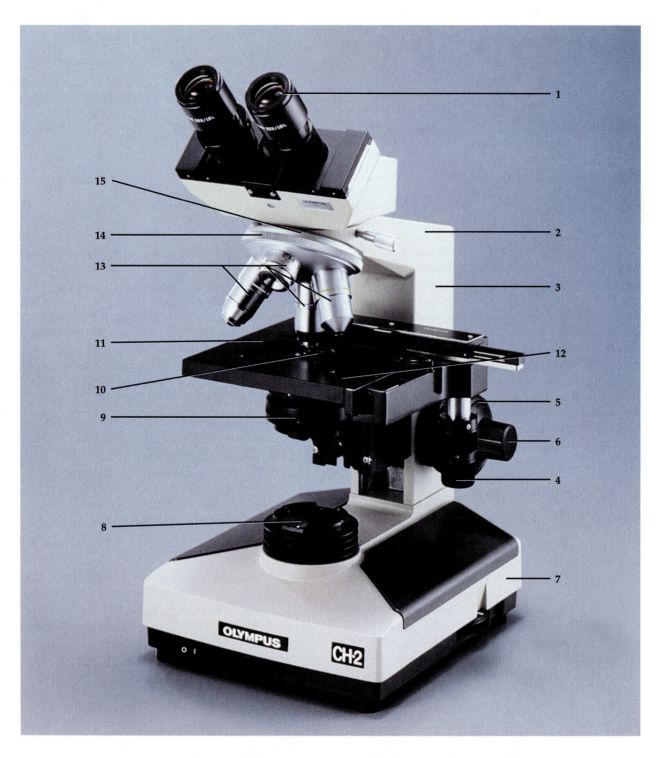

FIGURE 1.1 Olympus CH-2 Microscope

___ Arm
___ Base
___ Body tube
___ Coarse adjustment knob
___ Condenser

___ Diaphragm
___ Fine adjustment knob
___ Mechanical stage knob
___ Nosepiece
___ Objectives

___ Ocular
___ Revolving nosepiece
___ Stage
___ Stage clip of mechanical stage
___ Substage lamp

6. *Stage (spring) clips* Two clips mounted on the stage that hold slides securely in place.

7. *Substage lamp* The source of illumination for some light microscopes with a built-in lamp.

8. *Mirror* A feature found in some microscopes below the stage. The mirror directs light from its source through the stage opening and through the lenses. If the light source is built-in, a mirror is not necessary.

9. *Condenser* A lens located beneath the stage opening that concentrates the light beam on the specimen.

10. *Condenser adjustment knob* A knob that functions to raise and lower the condenser. In its highest position, it allows full illumination and thus can be used to adjust illumination.

11. *Diaphragm* (DĪ-a-fram) A device located below the condenser that regulates light intensity passing through the condenser and lenses to the observer's eyes. Such regulation is needed because transparent or very thin specimens cannot be seen in bright light. One of two types of diaphragms is usually used. An ***iris diaphragm***, as found in cameras, is a series of sliding leaves that vary the size of the opening and thus the amount of light entering the lenses. The leaves are moved by a ***diaphragm lever*** to regulate the diameter of a central opening. A ***disc diaphragm*** consists of a plate with a graded series of holes, any of which can be rotated into position.

12. *Coarse adjustment knob* A usually larger knob that raises and lowers the body tube (or stage) to bring a specimen into general view.

13. *Fine adjustment knob* A usually smaller knob found below or external to the coarse adjustment knob and used for fine or final focusing. Some microscopes have both coarse and fine adjustment knobs combined into one.

14. *Nosepiece* A plate, usually circular, at the bottom of the body tube.

15. *Revolving nosepiece* The lower, movable part of the nosepiece that contains the various objective lenses.

16. *Scanning objective* A lens, marked 5× on most microscopes (× means the same as "times"); it is the shortest objective and is not present on all microscopes.

17. *Low-power objective* A lens, marked 10× on most microscopes; it is the next shortest objective.

18. *High-power objective* A lens, marked 43× or 45× on most microscopes; also called a ***high-dry objective***; it is the second longest objective.

19. *Oil-immersion objective* A lens, marked 100× on most microscopes and distinguished by an etched colored circle (special instructions for this objective are discussed later); it is the longest objective.

20. *Ocular (eyepiece)* A removable lens at the top of the body tube, marked 10× on most microscopes. An ocular is sometimes fitted with a pointer or measuring scale.

2. Rules of Microscopy

You must observe certain basic rules at all times to obtain maximum efficiency and provide proper care for your microscope.

1. Keep all parts of the microscope clean, especially the lenses of the ocular, objectives, condenser, and also the mirror. *You should use the special lens paper that is provided and never use paper towels or cloths, because these tend to scratch the delicate glass surfaces.* When using lens paper, use the same area on the paper only once. As you wipe the lens, change the position of the paper as you go.

2. Do not permit the objectives to get wet, especially when observing a ***wet mount***. You must use a ***cover slip*** when you examine a wet mount or the image becomes distorted.

3. Consult your instructor if any mechanical or optical difficulties rise. *Do not try to solve these problems yourself.*

4. Keep *both* eyes open at all times while observing objects through the microscope. This is difficult at first, but with practice becomes natural. This important technique will help you to draw and observe microscopic specimens without moving your head. Only your eyes will move.

5. Always use either the scanning or low-power objective first to locate an object; then, if necessary, switch to a higher power.

6. If you are using the high-power or oil-immersion objectives, *never focus using the coarse adjustment knob.* The distance between these objectives and the slide, called ***working distance***, is very small and you may break the cover slip and the slide and scratch the lens.

7. Some microscopes have a stage that moves while focusing, others have a body tube that moves while focusing. Be sure you are familiar with which type you are using. Look at your microscope from the side (place objective on scanning or low power) and gently turn the coarse adjustment knob. Which moves? The stage or the body tube? *Never focus downward if*

the microscope's body tube moves when focusing. *Never focus upward* if the microscope's stage moves when focusing. By observing from one side you can see that the objectives do not make contact with the cover slip or slide.

8. Make sure that you raise the body tube before placing a slide on the stage or before removing a slide.

3. Setting Up the Microscope

PROCEDURE

1. Place the microscope on the table with the ocular toward you and with the back of the base at least 1 inch (in.) from the edge of the table.
2. Position yourself and the microscope so that you can look into the ocular comfortably.
3. Wipe the objectives, the top lens of the ocular, the condenser, and the mirror with lens paper. Clean the most delicate and the least dirty lens first. Apply xylol or ethanol to the lens paper only to remove grease and oil from the lenses and microscope slides.
4. Position the low-power objective in line with the body tube. When it is in its proper place, it will click. Lower the body tube using the coarse adjustment knob until the bottom of the lens is approximately 1/4 in. from the stage.
5. Admit the maximum amount of light by opening the diaphragm, if it is an iris diaphragm, or turning the disc to its largest opening if it is a disc diaphragm.
6. Place your eye to the ocular, and adjust the light. When a uniform circle (the *microscopic field*) appears without any shadows, the microscope is ready for use.

4. Using the Microscope

PROCEDURE

1. Using the coarse adjustment knob, raise the body tube to its highest fixed position.
2. Make a temporary mount using a single letter of newsprint, or use a slide that has been specially prepared with a letter, usually the letter "e." If you prepare such a slide, cut a single letter— "a," "b," or "e"—from the smallest print available and place this letter in the correct position to be read with the naked eye. Your instructor will provide directions for preparing the slide.
3. Place the slide on the stage, making sure that the letter is centered over the opening in the stage, directly over the condenser. Secure the slide in place with the stage clips.

4. Align the low-power objective with the body tube.
5. Lower the body tube or raise the stage as far as it will go *while you watch it from the side*, taking care not to touch the slide. The tube should reach an automatic stop that prevents the low-power objective from hitting the slide.
6. While looking through the ocular, turn the coarse adjustment knob counterclockwise, raising the body tube. Or, turn the coarse adjustment knob clockwise, lowering the stage. When focusing, always *raise* the body tube or *lower* the stage. Watch for the object to suddenly appear in the microscopic field. If it is in proper focus, the low-power objective is about 1/2 in. above the slide. When focusing, always *raise* the body tube.
7. Use the fine adjustment knob to complete the focusing; you will usually use a counterclockwise motion once again.
8. Compare the position of the letter as originally seen with the naked eye to its appearance under the microscope.
   Has the position of the letter been changed?

   _____

9. While looking at the slide through the ocular, move the slide by using your thumbs, or, if the microscope is equipped with them, the mechanical stage knobs. This exercise teaches you to move your specimen in various directions quickly and efficiently.
   In which direction does the letter move when you move the slide to the left?

   _____

This procedure, "scanning" a slide, will be useful for examining living objects and for centering specimens so you can observe them easily.

   Make a drawing of the letter as it appears under low power in the microscopic field in the space below on the left side.

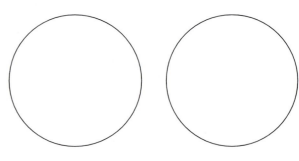

Drawing of letter   Drawing of letter
as seen under   as seen under
low power   high power

10. Change your magnification from low to high power by carrying out the following steps:

   a. Place the letter in the center of the field under low power. Centering is important because you are now focusing on a smaller area of the microscopic field. As you will see, *microscopic field size decreases with higher magnifications.*
   b. Make sure the illumination is at its maximum. Illumination must be increased at higher magnifications because the amount of light entering the lens decreases as the size of the objective lens increases.
   c. The letter should be in focus, and if the microscope is *parfocal* (meaning that when clear focus has been attained using any objective at random, revolving the nosepiece results in a change in magnification but leaves the specimen still in focus), the high-power objective can be switched into line with the body tube without changing focus. If it is not completely in focus after switching the lens, a slight turn of the fine adjustment knob will focus it.
   d. If your microscope is not parfocal, observe the stage from one side and carefully switch the high-power objective in line with the body tube.
   e. While still observing from the side and using the coarse adjustment knob, *carefully* lower the objective or raise the stage until the objective almost touches the slide.
   f. Look through the ocular and focus up slowly. Finish focusing by turning the fine adjustment knob.
   g. If your microscope has an oil-immersion objective, you must follow special procedures. Place a drop of special *immersion oil* directly over the letter on the microscope slide, and lower the oil-immersion objective until it just contacts the oil. If your microscope is parfocal, you do not have to raise or lower the objectives. For example, if you are using the high-power objective and the specimen is in focus, just switch the high-power objective out of line with the body tube. Then add the oil and switch the oil-immersion objective into position; the specimen should be in focus. The same holds true when you switch from low power to high power. The special light-transmitting properties of the oil are such that light is refracted (bent) toward the specimen, permitting the use of powerful objectives in a

relatively narrow field of vision. This objective is extremely close to the slide being examined, so when it is in position, take precautions *never to focus downward* while you are looking through the ocular. Whenever you finish using immersion oil, be sure to saturate a piece of lens paper with xylol or alcohol and clean the oil-immersion objective and the slide if it is to be used again.

Is as much of the letter visible under high power as under low power? Explain. _____

_____

Make a drawing of the letter as it appears under high power in the microscopic field in the space next to the drawing you made for low power following step 9 on page 4.

11. Now select a prepared slide of three different-colored threads. Examination will show that a specimen mounted on a slide has depth as well as length and width. At lower magnification the amount of depth of the specimen that is clearly in focus, the depth of field, is greater than that at higher magnification. You must focus at different depths to determine the position (depth) of each thread.

   After you make your observation under low power and high power, answer the following questions about the location of the different threads:

   What color is at the bottom, closest to the

   slide? _____

   On top, closest to the cover slip? _____

   _____

   In the middle? _____

   _____

12. Your instructor might want you to prepare a wet mount as part of your introduction to microscopy. If so, the directions are given in Exercise 4, A.2.f, on page 56.
13. When you are finished using the microscope

   a. Remove the slide from the stage.
   b. Clean all lenses with lens paper.
   c. Align the mechanical stage so that it does not protrude.
   d. Leave the scanning or low-power objective in place.
   e. Lower the body tube or raise the stage as far as it will go.

f. Wrap the cord according to your instructor's directions.

g. Replace the dust cover or place the microscope in a cabinet.

5. Magnification

The total magnification of your microscope is calculated by multiplying the magnification of the ocular by the magnification of the objective used. Example: An ocular of 10× used with an objective of 5× gives a total magnification of 50× . Calculate the total magnification of each of the objectives on your microscope:

1. Ocular _____ × _____ Objective = _____

2. Ocular _____ × _____ Objective = _____

3. Ocular _____ × _____ Objective = _____

4. Ocular _____ × _____ Objective = _____

B. ELECTRON MICROSCOPE

Examination of specimens smaller than 0.3 μm requires an *electron microscope*, an instrument with a much greater resolving power than a compound light microscope. An electron microscope uses a beam of electrons instead of light. Electrons travel in waves just as light does. Instead of glass lenses, however, magnets are used in an electron microscope to focus a beam of electrons through a vacuum tube onto a specimen. Since the wavelength of electrons is about 1/100,000 that of visible light, the resolving power of very sophisticated transmission electron microscopes (described shortly) is close to 10 nanometers (nm). 1 nm = 0.000,000,001 m or 10^{-9} m. (See Appendix A.) Most electron microscopes have a working resolving power of about 100 nm and can magnify images up to 200,000×.

Two types of electron microscopes are available. In a *transmission electron microscope*, a finely focused beam of electrons passes through a specimen, usually ultrathin sections of material. The beam is then refocused, and the image reflects what the specimen has done to the transmitted electron beam. The image may be used to produce what is called a *transmission electron micrograph*. The method is extremely valuable in providing details of the interior of specimens at different layers but does not give a three-dimensional effect. Such an effect can be obtained with a *scanning electron microscope*. With this instrument, a finely focused beam of electrons is directed over the specimen and then reflected from the surface of the specimen onto a television-like screen or photographic plate. The photograph produced from an image generated in this manner is called a *scanning electron micrograph* (see Figures 5.2 and 21.10). Scanning electron micrographs commonly magnify specimens up to 10,000× and are especially useful in studying surface features of specimens.

Note: A section of *LABORATORY REPORT QUESTIONS* is located at the end of each exercise. These questions can be answered by the student and handed in for grading at the discretion of the instructor. Even if the instructor does not require you to answer these questions, we recommend that you do so anyway to check your understanding.

Some exercises also have a section of *LABORATORY REPORT RESULTS*, in which students can record results of laboratory exercises, in addition to laboratory report questions. As with the laboratory report questions, the laboratory report results are located at the end of selected exercises and can be handed in as the instructor directs. Instructions in the manual tell students when and where to record laboratory results.

ANSWER THE LABORATORY REPORT QUESTIONS AT THE END OF THE EXERCISE.

Microscopy 1

Student _____   **Date** _____

Laboratory Section _____   **Score/Grade** _____

PART 1. Multiple Choice

_____ 1. The amount of light entering a microscope may be adjusted by regulating the (a) ocular (b) diaphragm (c) fine adjustment knob (d) nosepiece

_____ 2. If the ocular on a microscope is marked 10× and the low-power objective is marked 15×, the total magnification is (a) 50× (b) 25× (c) 150× (d) 1500×

_____ 3. The size of the light beam that passes through a microscope is regulated by the (a) revolving nosepiece (b) coarse adjustment knob (c) ocular (d) condenser

_____ 4. Parfocal means that (a) the microscope employs only one lens (b) final focusing can be done only with the fine adjustment knob (c) changing objectives by revolving the nosepiece will still keep the specimen in focus (d) the highest magnification attainable is 1000×.

_____ 5. Which of these is *not* true when changing magnification from low power to high power? (a) the specimen should be centered (b) illumination should be decreased (c) the specimen should be in clear focus (d) the high-power objective should be in line with the body tube

_____ 6. The ability of a microscope to distinguish between two points as separate objects is called (a) parfocal focusing (b) working distance (c) diffraction (d) resolution

PART 2. Completion

7. The advantage of using immersion oil is that it has special _____ properties that permit the use of a powerful objective in a narrow field of vision.

8. The uniform circle of light that appears when one looks into the ocular is called the

_____ .

9. In determining the position (depth) of the colored threads, the _____ (red, blue, yellow) colored thread was in the middle.

10. If you move your slide to the right, the specimen moves to the _____ as you are viewing it microscopically.

11. After switching from low power to high power, _____ (more or less) of the specimen will be visible.

12. Microscopic field size _____ (increases or decreases) with higher magnifications.

13. The wavelength of electrons is about _____ (what proportion?) that of visible white light.

14. A photograph of a specimen taken through a compound light microscope is called a(n)

_____ .

15. The type of microscope that provides three-dimensional images of nonliving specimens is the

_____ microscope.

PART 3. Matching

_____	**16.** Ocular	A. Platform on which slide is placed
_____	**17.** Stage	B. Mounting for objectives
_____	**18.** Arm	C. Lens below stage opening
_____	**19.** Condenser	D. Brings specimen into sharp focus
_____	**20.** Revolving nosepiece	E. Eyepiece
_____	**21.** Low-power objective	F. An objective usually marked 43× or 45×
_____	**22.** Fine adjustment knob	G. An objective usually marked 10×
_____	**23.** Diaphragm	H. Angular or curved part of frame
_____	**24.** Coarse adjustment knob	I. Brings specimen into general focus
_____	**25.** High-power objective	J. Regulates light intensity

Introduction to the Human Body

2

In this exercise, you will be introduced to the organization of the human body through a study of the principal subdivisions of anatomy and physiology, levels of structural organization, principal body systems, anatomical position, regional names, directional terms, planes of the body, body cavities, abdominopelvic regions, and abdominopelvic quadrants.

A. ANATOMY AND PHYSIOLOGY

Whereas *anatomy* (a-NAT-o-mē; *anatome* = to cut up) refers to the study of *structure* and the relationships among structures, *physiology* (fiz-ē-OL-ō-jē) deals with the *functions* of body parts, that is, how they work. Each structure of the body is designed to carry out a particular function.

Following are selected subdivisions of anatomy and physiology. In the spaces provided, define each term.

SUBDIVISIONS OF ANATOMY

Surface anatomy _____

_____

_____

_____

Gross (macroscopic) anatomy _____

_____

_____

_____

Systemic anatomy _____

_____

_____

_____

Regional anatomy _____

_____

_____

_____

Radiographic (rā'-dē-ō-GRAF-ik) *anatomy* ____

_____

_____

_____

Developmental anatomy _____

_____

_____

_____

Embryology (em'-brē-OL-ō-jē) _____

_____

_____

_____

Histology (hiss'-TOL-ō-jē) _____

_____

_____

_____

Cytology (sī-TOL-ō-jē) _____

_____

_____

_____

Pathological (path'-ō-LOJ-i-kal) *anatomy* ____

_____

_____

_____

SUBDIVISIONS OF PHYSIOLOGY

Cell physiology _____

_____

_____

_____

Pathophysiology (PATH-ō-fiz-ē-ol'-ō-jē) _____

_____

_____

_____

Exercise physiology _____

_____

_____

_____

Neurophysiology (NOO-rō-fiz-ē-ol'-ō-jē) _____

_____

_____

_____

Endocrinology (en'-dō-kri-NOL-ō-jē) _____

_____

_____

_____

Cardiovascular (kar-dē-ō-VAS-kyoo-lar) *physiology* _____

_____

_____

_____

Immunology (im'-yoo-NOL-ō-jē) _____

_____

_____

_____

Respiratory (re-SPĪ-ra-tō-rē) *physiology* _____

_____

_____

_____

Renal (RĒ-nal) *physiology* _____

_____

_____

_____

B. LEVELS OF STRUCTURAL ORGANIZATION

The human body is composed of several levels of structural organization associated with one another in various ways:

1. *Chemical level* Composed of all atoms and molecules necessary to maintain life.
2. *Cellular level* Consists of cells, the structural and functional units of the body.
3. *Tissue level* Formed by tissues, groups of cells (and their intercellular material) that usually arise from common ancestor cells and work together to perform a particular function.
4. *Organ level* Consists of organs, structures composed of two or more different tissues, having specific functions and usually having recognizable shapes.
5. *System level* Formed by systems, associations of organs that have a common function. The systems together constitute an *organism.*

C. SYSTEMS OF THE BODY

Using an anatomy and physiology textbook, torso, wall chart, and any other materials that might be available to you, identify the principal organs that compose the following body systems.[1] In the spaces that follow, indicate the organs and functions of the systems.

Integumentary

Organs _____

_____

_____

_____

[1]You will probably need other sources, plus any aids the instructor might provide, to label many of the figures and answer some questions in this manual. You are encouraged to use other sources as you find necessary.

Functions _____

_____

_____

_____

Skeletal

Organs _____

_____

_____

_____

Functions _____

_____

_____

_____

Muscular

Organs _____

_____

_____

_____

Functions _____

_____

_____

_____

Nervous

Organs _____

_____

_____

_____

Functions _____

_____

_____

_____

Endocrine

Organs _____

_____

_____

_____

Functions _____

_____

_____

_____

Cardiovascular

Organs _____

_____

_____

_____

Functions _____

_____

_____

_____

Lymphatic and Immune

Organs _____

_____

_____

_____

Functions _____

_____

_____

_____

Respiratory

Organs _____

_____

_____

_____

Functions _____

_____

_____

_____

Digestive

Organs _____

_____

_____

_____

Functions _____

_____

_____

_____

Urinary

Organs _____

_____

_____

_____

Functions _____

_____

_____

_____

Reproductive

Organs _____

_____

_____

_____

Functions _____

_____

_____

_____

D. LIFE PROCESSES

All living forms carry on certain processes that distinguish them from nonliving things. Using an anatomy and physiology textbook, define the following life processes of humans.

Metabolism _____

_____

_____

_____

Responsiveness _____

_____

_____

_____

Movement _____

_____

_____

_____

Growth _____

_____

_____

_____

Differentiation _____

_____

_____

_____

Reproduction _____

_____

_____

_____

E. HOMEOSTASIS

Homeostasis (hō-mē-ō-STĀ-sis) is a condition in which the body's internal environment (interstitial fluid) remains within certain physiological limits. Homeostasis is achieved through the operation of feedback systems. A *feedback system* involves a cycle of events in which information about the status of a condition is continually monitored and fed back to a central control region.

Using an anatomy and physiology textbook, define the following components of a feedback system.

Stimulus _____

_____

_____

_____

Controlled condition _____

_____

_____

_____

Receptor _____

_____

_____

Input _____

_____

_____

_____

Control center _____

_____

_____

_____

Output _____

_____

_____

_____

Effector _____

_____

_____

_____

Response _____

_____

_____

_____

If the response of the body reverses the original stimulus, the system is a *negative feedback system.* If the response enhances or intensifies the original stimulus, the system is a *positive feedback system.*

Negative feedback systems tend to maintain conditions that require frequent monitoring and adjustments within physiological limits, such as body temperature or blood sugar level. (See Figure 2.1.) Positive feedback systems, on the other hand, are important for conditions that do not require continual fine-tuning. Since positive feedback systems tend to intensify or amplify a controlled condition, they usually are shut off by some mechanism outside the system if they are part of a normal physiological response. Most feedback systems in the body are negative. Positive feedback systems can be destructive and result in various disorders, yet some are normal and beneficial. For example, during blood clotting, which helps stop loss of blood from a cut, the initial signal is amplified until the blood clot forms and bleeding is under control. Then, other substances help turn off the clotting response. Positive feedback mechanisms also contribute during birth of a baby to strengthen labor contractions and during immune responses to provide defense against pathogens.

Label Figure 2.1, the control of blood pressure by a negative feedback system.

F. ANATOMICAL POSITION AND REGIONAL NAMES

Figure 2.2 shows anterior and posterior views of a subject in the *anatomical position*. The subject is standing erect (upright position) and facing the observer when in anterior view, the upper limbs (extremities) are at the sides, the palms are facing forward, and the feet are flat on the floor.

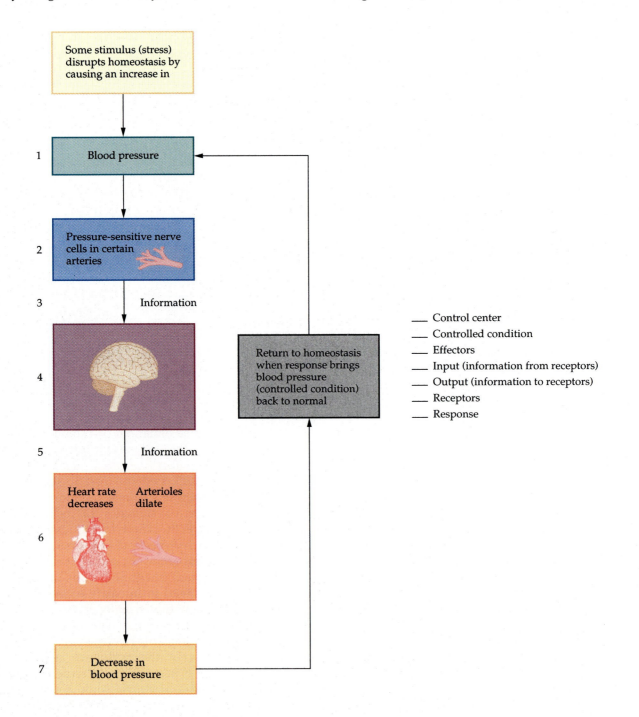

___ Control center
___ Controlled condition
___ Effectors
___ Input (information from receptors)
___ Output (information to receptors)
___ Receptors
___ Response

FIGURE 2.1   The control of blood pressure by a negative feedback system.

The figure also shows the common names for various regions of the body. When you as the observer make reference to the left and right sides of the subject you are studying, this refers to the *subject's* left and right sides. In the spaces next to the list of terms in Figure 2.2 write the number of each common term next to each corresponding anatomical term. For example, the skull (29) is cranial, so write the number *29* next to the term *Cranial*.

G. EXTERNAL FEATURES OF THE BODY

Referring to your textbook and human models, identify the following external features of the body:

1. *Head (cephalic region or caput)* This is divided into the *cranium* (brain case) and *face*.
2. *Neck (collum)* This region consists of an anterior *cervix*, two lateral surfaces, and a posterior *nucha* (NOO-ka), or the nape of the neck.
3. *Trunk* This region is divided into the *back* (dorsum), *chest* (thorax), *abdomen* (venter), and *pelvis*.
4. *Upper limb (extremity)* This consists of the *armpit* (axilla), *shoulder* (acromial region or omos), *arm* (brachium), *elbow* (cubitus), *forearm* (antebrachium), and *hand* (manus). The hand, in turn, consists of the *wrist* (carpus), *palm* (metacarpus), and *fingers* (digits). Individual bones of a digit (finger or toe) are called *phalanges. Phalanx* is singular.
5. *Lower limb (extremity)* This consists of the *buttocks* (gluteal region), *thigh* (femoral region), *knee* (genu), *leg* (crus), and *foot* (pes). The foot includes the *ankle* (tarsus), *sole* (metatarsus), and *toes* (digits).

H. DIRECTIONAL TERMS

To explain exactly where a structure of the body is located, it is a standard procedure to use *directional terms*. Such terms are very precise and avoid the use of unnecessary words. Commonly used directional terms for humans are as follows:

1. *Superior* (soo'-PEER-ē-or) *(cephalic* or *cranial)* Toward the head or the upper part of a structure; generally refers to structures in the trunk.

2. *Inferior* (in'-FEER-ē-or) *(caudal)* Away from the head or toward the lower part of a structure; generally refers to structures in the trunk.
3. *Anterior* (an-TEER-ē-or) *(ventral)* Nearer to or at the front surface of the body. In the *prone position*, the body lies anterior side down; in the *supine position*, the body lies anterior side up.
4. *Posterior* (pos-TEER-ē-or) *(dorsal)* Nearer to or at the back or backbone surface of the body.
5. *Medial* (MĒ-dē-al) Nearer the midline of the body or a structure. The *midline* is an imaginary vertical line that divides the body into equal left and right sides.
6. *Lateral* (LAT-er-al) Farther from the midline of the body or a structure.
7. *Intermediate* (in'-ter-MĒ-dē-at) Between two structures.
8. *Ipsilateral* (ip-si-LAT-er-al) On the same side of the body.
9. *Contralateral* (CON-tra-lat-er-al) On the opposite side of the body.
10. *Proximal* (PROK-si-mal) Nearer the attachment of a limb to the trunk; nearer to the point of origin.
11. *Distal* (DIS-tal) Farther from the attachment of a limb to the trunk; farther from the point of origin.
12. *Superficial* (soo'-per-FISH-al) *(external)* Toward or on the surface of the body.
13. *Deep* (DĒP) *(internal)* Away from the surface of the body.

Using a torso and an articulated skeleton, and consulting with your instructor as necessary, describe the location of the following by inserting the proper directional term.

1. The ulna is on the _____ side of the forearm.
2. The lungs are _____ to the heart.
3. The heart is _____ to the liver.
4. The muscles of the arm are _____ to the skin of the arm.
5. The sternum is _____ to the heart.
6. The humerus is _____ to the radius.
7. The stomach is _____ to the lungs.
8. The muscles of the thoracic wall are _____ _____ to the viscera in the thoracic cavity.

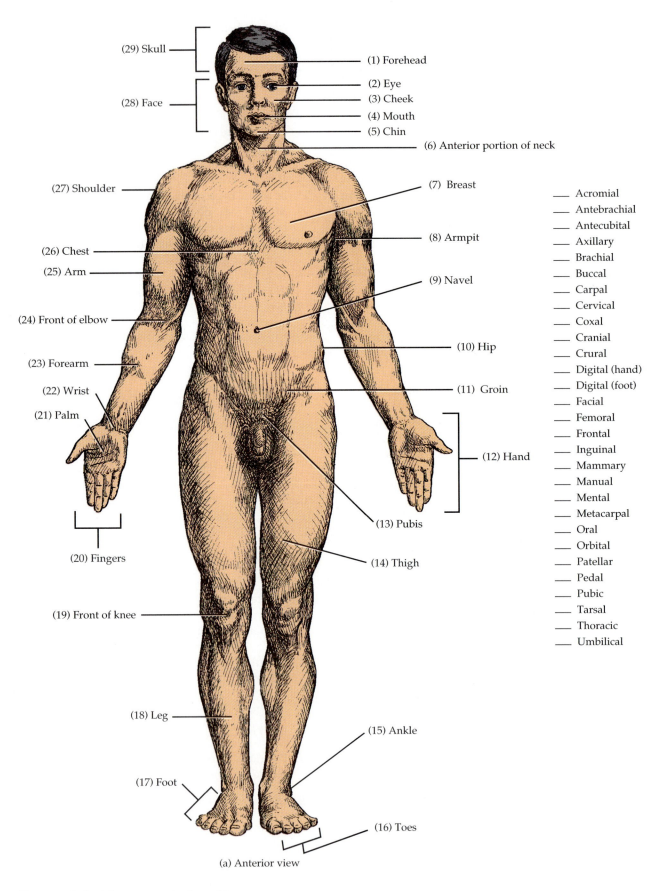

(29) Skull

(28) Face

(1) Forehead

(2) Eye

(3) Cheek

(4) Mouth

(5) Chin

(6) Anterior portion of neck

(27) Shoulder

(7) Breast

(8) Armpit

(26) Chest

(25) Arm

(9) Navel

(24) Front of elbow

(10) Hip

(23) Forearm

(11) Groin

(22) Wrist

(21) Palm

(12) Hand

(20) Fingers

(13) Pubis

(14) Thigh

(19) Front of knee

(18) Leg

(15) Ankle

(17) Foot

(16) Toes

(a) Anterior view

____ Acromial
____ Antebrachial
____ Antecubital
____ Axillary
____ Brachial
____ Buccal
____ Carpal
____ Cervical
____ Coxal
____ Cranial
____ Crural
____ Digital (hand)
____ Digital (foot)
____ Facial
____ Femoral
____ Frontal
____ Inguinal
____ Mammary
____ Manual
____ Mental
____ Metacarpal
____ Oral
____ Orbital
____ Patellar
____ Pedal
____ Pubic
____ Tarsal
____ Thoracic
____ Umbilical

FIGURE 2.2   The anatomical position.

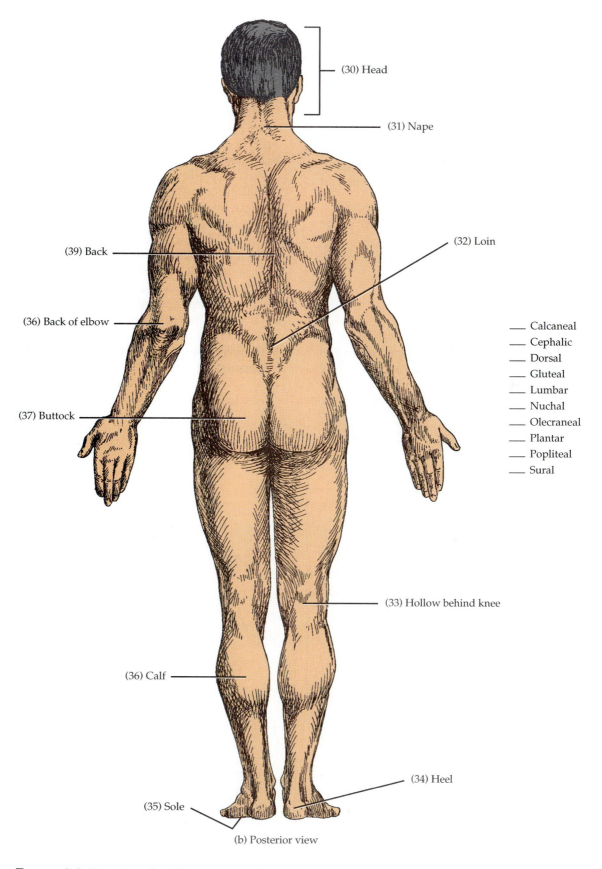

(30) Head

(31) Nape

(39) Back

(32) Loin

(36) Back of elbow

___ Calcaneal
___ Cephalic
___ Dorsal
___ Gluteal
___ Lumbar
___ Nuchal
___ Olecraneal
___ Plantar
___ Popliteal
___ Sural

(37) Buttock

(33) Hollow behind knee

(36) Calf

(34) Heel

(35) Sole

(b) Posterior view

FIGURE 2.2 *(Continued)* The anatomical position.

9. The esophagus is _____ to the trachea.

10. The phalanges are _____ to the carpals.

11. The ring finger is _____ between the little (medial) and middle (lateral) fingers.

12. The ascending colon of the large intestine and the gallblader are _____.

13. The ascending and descending colons of the large intestine are _____.

I. PLANES OF THE BODY

The structural plan of the human body may be described with respect to *planes* (imaginary flat surfaces) passing through it. Planes are frequently used to show the anatomical relationship of several structures in a region to one another.

Commonly used planes are

1. *Midsagittal* (mid-SAJ-it-tal; *sagittalis* = arrow) or *median* A vertical plane that passes through the midline of the body and divides the body or an organ into *equal* right and left sides.
2. *Parasagittal* (par-a-SAJ-it-tal; *para* = near) A vertical plane that does not pass through the midline of the body and divides the body or an organ into *unequal* right and left sides.
3. *Frontal* (*coronal;* kō-RŌ-nal) A vertical plane that divides the body or an organ into anterior (front) and posterior (back) portions.
4. *Transverse* (*horizontal* or *cross-sectional*) A plane that divides the body or an organ into superior (top) and inferior (bottom) portions.
5. *Oblique* (ō-BLĒK) A plane that passes through the body or an organ at an angle between the transverse plane and either the midsagittal, parasagittal, or frontal plane.

Refer to Figure 2.3 and label the planes shown.

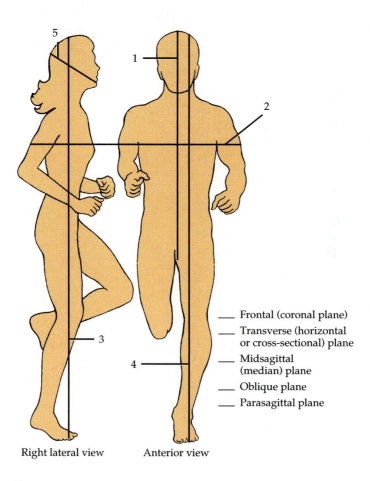

_____ Frontal (coronal plane)

_____ Transverse (horizontal or cross-sectional) plane

_____ Midsagittal (median) plane

_____ Oblique plane

_____ Parasagittal plane

Right lateral view   Anterior view

FIGURE 2.3   Planes of the body.

J. BODY CAVITIES

Confined spaces within the body that contain various internal organs are called *body cavities*. One way of organizing the principal body cavities follows:

> ***Dorsal body cavity***
>   *Cranial* (KRĀ-nē-al) *cavity*
>   *Vertebral* (VER-te-bral) or *spinal cavity*
> ***Ventral body cavity***
>   *Thoracic* (thor-AS-ik) *cavity*
>   Right pleural (PLOOR-al)
>   Left pleural
>   Pericardial (per'-i-KAR-dē-al)
>   *Abdominopelvic cavity*
>   Abdominal
>   Pelvic

The mass of tissue between the pleurae of the lungs and extending from the sternum (breast bone) to the backbone is called the *mediastinum* (mē'-dē-as-TĪ-num). It contains all structures in the thoracic cavity, except the lungs themselves. Included are the heart, thymus gland, esophagus, trachea, and many large blood and lymphatic vessels.

Label the body cavities shown in Figure 2.4. Then examine a torso or wall chart, or both, and determine which organs lie within each cavity.

Using *T* (for thoracic), *A* (for abdominal), and *P* (for pelvic), indicate which organs are found in their respective cavities.

1. ____ Urinary bladder
2. ____ Stomach
3. ____ Spleen
4. ____ Lungs
5. ____ Liver
6. ____ Internal reproductive organs
7. ____ Pancreas
8. ____ Heart
9. ____ Gallbladder
10. ____ Small portion of large intestine

K. ABDOMINOPELVIC REGIONS

To describe the location of viscera more easily, the abdominopelvic cavity may be divided into *nine regions* by using four imaginary lines: (1) an upper horizontal *subcostal* (sub-KOS-tal) *line* that passes just below the bottom of the rib cage through the pylorus (lower portion) of the stomach, (2) a lower

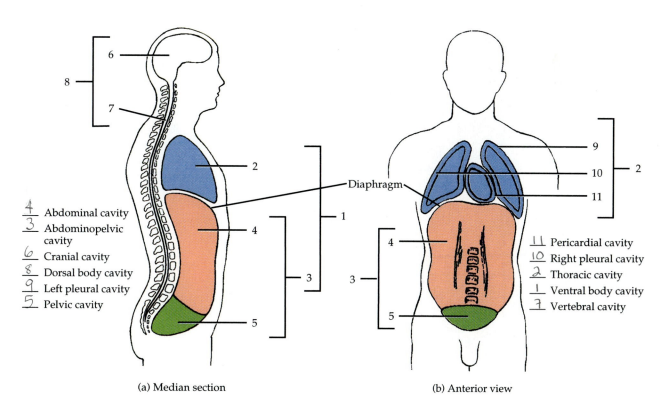

4 Abdominal cavity
3 Abdominopelvic cavity
6 Cranial cavity
8 Dorsal body cavity
9 Left pleural cavity
5 Pelvic cavity

11 Pericardial cavity
10 Right pleural cavity
2 Thoracic cavity
1 Ventral body cavity
7 Vertebral cavity

(a) Median section

(b) Anterior view

FIGURE 2.4 Body cavities.

horizontal line, the ***transtubercular*** (trans-too-BER-kyoo'-lar) *line,* just below the top surfaces of the hipbones, (3) a ***right midclavicular*** (mid-kla-VIK-yoo'-lar) *line* drawn through the midpoint of the right clavicle slightly medial to the right nipple, and (4) a ***left midclavicular line*** drawn through the midpoint of the left clavicle slightly medial to the left nipple.

The four imaginary lines divide the abdominopelvic cavity into the following nine regions: (1) ***umbilical*** (um-BIL-i-kul) *region,* which is centrally located; (2) ***left lumbar*** (*lumbus* = loin) *region,* to the left of the umbilical region; (3) ***right lumbar,*** to the right of the umbilical region; (4) ***epigastric*** (ep-i-GAS-trik; *epi* = above; *gaster* = stomach) ***region,*** directly above the umbilical region; (5) ***left hypochondriac*** (hī'-pō-KON-drē-ak; *hypo* = under;

chondro = cartilage) ***region,*** to the left of the epigastric region; (6) ***right hypochondriac region,*** to the right of the epigastric region; (7) ***hypogastric (pubic) region,*** directly below the umbilical region; (8) ***left iliac*** (IL-ē-ak; *iliacus* = superior part of hipbone) or ***inguinal region,*** to the left of the hypogastric (pubic) region; and (9) ***right iliac (inguinal) region,*** to the right of the hypogastric (pubic) region.

Label Figure 2.5 by indicating the names of the four imaginary lines and the nine abdominopelvic regions.

Examine a torso and determine which organs or parts of organs lie within each of the nine abdominopelvic regions.

In the space provided on page 21, list several organs or parts of organs found in the following abdominopelvic regions:

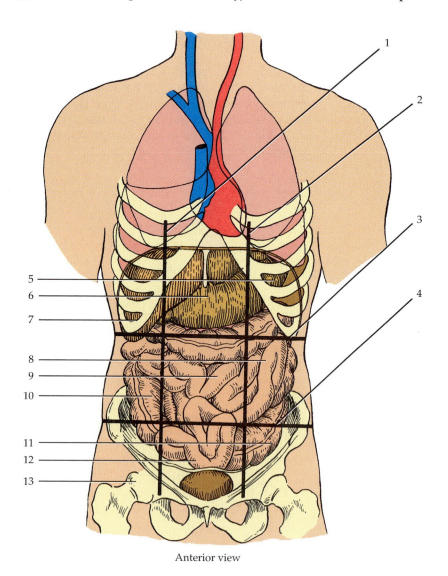

6 Epigastric region
12 Hypogastric region
5 Left hypochondriac region
11 Left iliac region
2 Left midclavicular line
8 Left lumbar region
7 Right hypochondriac region
13 Right iliac region
1 Right midclavicular line
10 Right lumbar region
3 Subcostal line
4 Transtubercular line
9 Umbilical region

Anterior view

FIGURE 2.5   Abdominopelvic regions.

Right hypochondriac _____

_____

_____

Epigastric _____

_____

_____

Left hypochondriac _____

_____

_____

Right lumbar _____

_____

_____

Umbilical _____

_____

_____

Left lumbar _____

_____

_____

Right iliac _____

_____

_____

Hypogastric _____

_____

_____

Left iliac _____

_____

_____

L. ABDOMINOPELVIC QUADRANTS

Another way to divide the abdominopelvic cavity is into *quadrants* by passing one horizontal line and one vertical line through the umbilicus (navel). The two lines thus divide the abdominopelvic cavity into a *right upper quadrant (RUQ)*, *left upper quadrant (LUQ)*, *right lower quadrant (RLQ)*, and *left lower quadrant (LLQ)*. Quadrant names are frequently used by health care professionals for locating the site of an abdominopelvic pain, tumor, or other abnormality.

Examine a torso or wall chart, or both, and determine which organs or parts of organs lie within each of the abdominopelvic quadrants.

In the space provided, list several organs or parts of organs found in the following abdominopelvic quadrants:

Right upper quadrant _____

_____

_____

_____

Left upper quadrant _____

_____

_____

_____

Right lower quadrant _____

_____

_____

_____

Left lower quadrant _____

_____

_____

_____

M. DISSECTION OF WHITE RAT

Now that you have some idea of the names of the various body systems and the principal organs that comprise each, you can actually observe some of these organs by dissecting a white rat. *Dissect* means "to separate." This dissection gives you an excellent opportunity to see the different sizes, shapes, locations, and relationships of organs and to compare the different textures and external features of organs. In addition, this exercise will introduce you to the general procedure for dissection before you dissect in later exercises.

CAUTION! *Please reread Section D, "Precautions Related to Dissection" at the beginning of the laboratory manual on page xiii before you begin your dissection.*

PROCEDURE

1. Place the rat on its backbone on a wax dissecting pan (tray). Using dissecting pins, anchor each of the four limbs to the wax (Figure 2.6a).

2. To expose the contents of the thoracic, abdominal, and pelvic cavities, you will have to first make a midline incision. This is done by lifting the abdominal skin with a forceps to separate the skin from the underlying connective tissue and muscles. While lifting the abdominal skin, cut through it with scissors and make an incision that extends from the lower jaw to the anus (Figure 2.6a).

3. Now make four lateral incisions that extend from the midline incision into the four limbs (Figure 2.6a).

4. Peel the skin back and pin the flaps to the wax to expose the superficial muscles (Figure 2.6b).

5. Next, lift the abdominal muscles with a forceps and cut through the muscle layer, being careful not to damage any underlying organs. Keep the scissors parallel to the rat's backbone. Extend this incision from the anus to a point just below the bottom of the rib cage (Figure 2.6b). Make two lateral incisions just below the rib cage and fold back the muscle flaps to expose the abdominal and pelvic viscera (Figure 2.6b).

6. To expose the thoracic viscera, cut through the ribs on either side of the sternum. This incision should extend from the diaphragm to the neck (Figure 2.6b). The *diaphragm* is the thin muscular partition that separates the thoracic from the abdominal cavity. Again make lateral incisions in the chest wall so that you can lift the ribs to view the thoracic contents.

1. Examination of Thoracic Viscera

You will first examine the thoracic viscera (large internal organs). As you dissect and observe the various structures, palpate (feel with the hand) them so that you can compare their texture. Use Figure 2.7 as a guide.

a. *Thymus gland* An irregular mass of glandular tissue superior to the heart and superficial to the trachea. Push the thymus gland aside or remove it.

b. *Heart* A structure located in the midline, deep to the thymus gland and between the

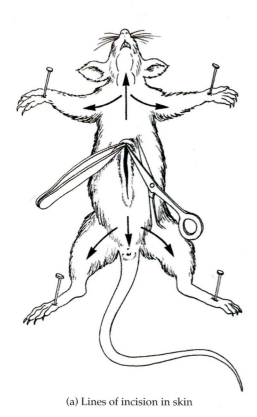

(a) Lines of incision in skin   (b) Peeling back skin and lines of incision in muscles

FIGURE 2.6   Dissection procedure for exposing thoracic and abdominopelvic viscera of the white rat for examination.

lungs. The sac covering the heart is the *pericardium*, which may be removed. The large vein that returns blood from the lower regions of the body is the *inferior vena cava*; the large vein that returns blood to the heart from the upper regions of the body is the *superior vena cava*. The large artery that car-

ries blood from the heart to most parts of the body is the *aorta.*
c. *Lungs* Reddish, spongy structures on either side of the heart. Note that the lungs are divided into regions called *lobes.*
d. *Trachea* A tubelike passageway superior to the heart and deep to the thymus gland. Note that

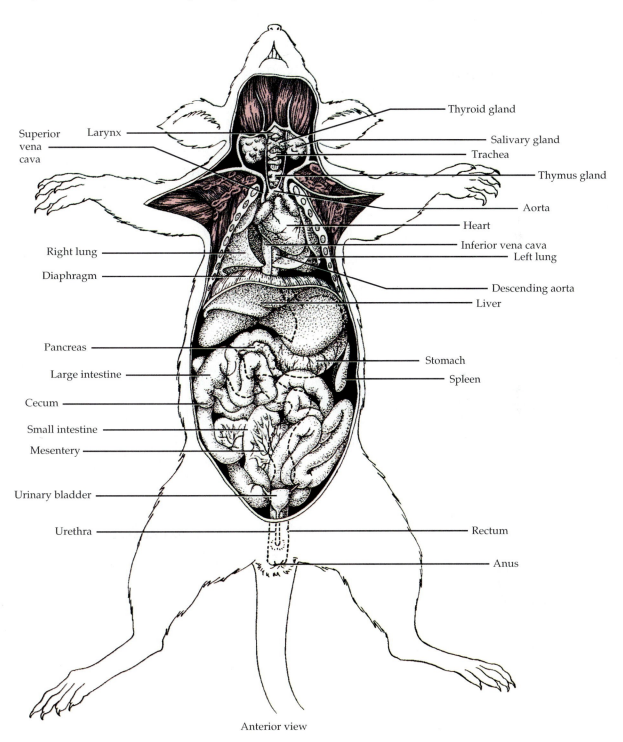

Anterior view

FIGURE 2.7 Superficial structures of the thoracic and abdominopelvic cavities of the white rat.

the wall of the trachea consists of rings of cartilage. Identify the *larynx* (voice box) at the superior end of the trachea and the *thyroid gland*, a bilobed structure on either side of the larynx. The lobes of the thyroid gland are connected by a band of thyroid tissue, the isthmus.

e. *Bronchial tubes* Trace the trachea inferiorly and note that it divides into bronchial tubes that enter the lungs and continue to divide within them.

f. *Esophagus* A muscular tube posterior to the trachea that transports food from the throat into the stomach. Trace the esophagus inferiorly to see where it passes through the diaphragm to join the stomach.

2. Examination of Abdominopelvic Viscera

You will now examine the principal viscera of the abdomen and pelvis. As you do so, again refer to Figure 2.7.

a. *Stomach* An organ located on the left side of the abdomen and in contact with the liver. The digestive organs are attached to the posterior abdominal wall by a membrane called the *mesentery*. Note the blood vessels in the mesentery.

b. *Small intestine* An extensively coiled tube that extends from the stomach to the first portion of the large intestine called the *cecum*.

c. *Large intestine* A wider tube than the small intestine that begins at the cecum and ends at the rectum. The cecum is a large, saclike structure. In humans, the appendix arises from the cecum.

d. *Rectum* A muscular passageway, located on the midline in the pelvic cavity, that terminates in the anus.

e. *Anus* Terminal opening of the digestive system to the exterior.

f. *Pancreas* A pale gray, glandular organ posterior and inferior to the stomach.

g. *Spleen* A small, dark red organ lateral to the stomach.

h. *Liver* A large, brownish-red organ directly inferior to the diaphragm. The rat does not have a gallbladder, a structure associated with the liver. To locate the remaining viscera, either move the superficial viscera aside or remove them. Use Figure 2.8 as a guide.

i. *Kidneys* Bean-shaped organs embedded in fat and attached to the posterior abdominal wall on either side of the backbone. As will be explained later, the kidneys and a few other structures are behind the membrane that lines the abdomen (*peritoneum*). Such structures are referred to as *retroperitoneal* and are not actually within the abdominal cavity. See if you can find the *abdominal aorta*, the large artery located along the midline behind the inferior vena cava. Also, locate the *renal arteries* branching off the abdominal aorta to enter the kidneys.

j. *Adrenal (suprarenal) glands* Glandular structures. One is located on top of each kidney.

k. *Ureters* Tubes that extend from the medial surface of the kidney inferiorly to the urinary bladder.

l. *Urinary bladder* A saclike structure in the pelvic cavity that stores urine.

m. *Urethra* A tube that extends from the urinary bladder to the exterior. Its opening to the exterior is called the *urethral orifice*. In male rats, the urethra extends through the penis; in female rats, the tube is separate from the reproductive tract.

If your specimen is female (no visible scrotum anterior to the anus), identify the following:

n. *Ovaries* Small, dark structures inferior to the kidneys.

o. *Uterus* An organ located near the urinary bladder consisting of two sides (horns) that join separately into the vagina.

p. *Vagina* A tube that leads from the uterus to the external vaginal opening, the *vaginal orifice*. This orifice is in front of the anus and behind the urethral orifice.

If your specimen is male, identify the following:

q. *Scrotum* Large sac anterior to the anus that contains the testes.

r. *Testes* Egg-shaped glands in the scrotum. Make a slit into the scrotum and carefully remove one testis. See if you can find a coiled duct attached to the testis (*epididymis*) and a duct that leads from the epididymis into the abdominal cavity (*vas deferens*).

s. *Penis* Organ of copulation medial to the testes.

When you have finished your dissection, store or dispose of your specimen according to your instructor's directions. Wash your dissecting pan and dissecting instruments with laboratory detergent, dry them, and return them to their storage areas.

ANSWER THE LABORATORY REPORT QUESTIONS AT THE END OF THE EXERCISE.

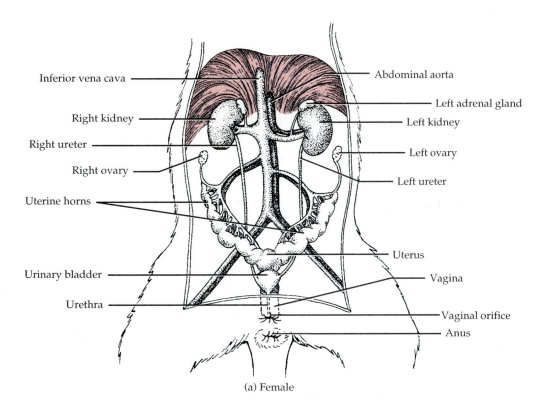

Inferior vena cava

Right kidney

Right ureter

Right ovary

Uterine horns

Urinary bladder

Urethra

Abdominal aorta

Left adrenal gland

Left kidney

Left ovary

Left ureter

Uterus

Vagina

Vaginal orifice

Anus

(a) Female

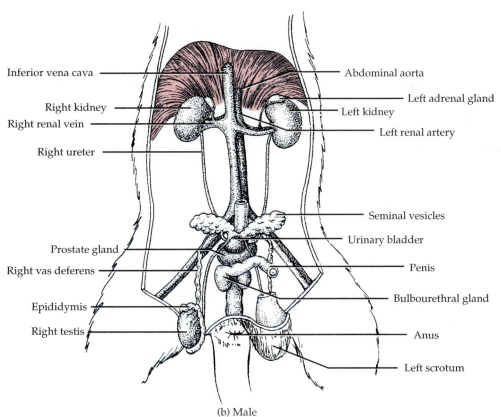

Inferior vena cava

Right kidney

Right renal vein

Right ureter

Prostate gland

Right vas deferens

Epididymis

Right testis

Abdominal aorta

Left adrenal gland

Left kidney

Left renal artery

Seminal vesicles

Urinary bladder

Penis

Bulbourethral gland

Anus

Left scrotum

(b) Male

FIGURE 2.8   Deep structures of the abdominopelvic cavity of the white rat.

Introduction to the Human Body 2

Student _____   Date _____

Laboratory Section _____   Score/Grade _____

PART 1. Multiple Choice

_d_ 1. The directional term that best describes the eyes in relation to the nose is (a) distal (b) superficial (c) anterior (d) lateral

_c_ 2. Which does *not* belong with the others? (a) right pleural cavity (b) pericardial cavity (c) vertebral cavity (d) left pleural cavity

_a_ 3. Which plane divides the brain into an anterior and a posterior portion? (a) frontal (b) median (c) sagittal (d) transverse

_b_ 4. The urinary bladder lies in which region? (a) umbilical (b) hypogastric (c) epigastric (d) left iliac

_c_ 5. Which is *not* a characteristic of the anatomical position? (a) the subject is erect (b) the subject faces the observer (c) the palms face backward (d) the upper limbs are at the sides

_b_ 6. The abdominopelvic region that is bordered by all four imaginary lines is the (a) hypogastric (b) epigastric (c) left hypochondriac (d) umbilical

_b_ 7. Which directional term best describes the position of the phalanges with respect to the carpals? (a) lateral (b) distal (c) anterior (d) proximal

_a_ 8. The pancreas is found in which body cavity? (a) abdominal (b) pericardial (c) pelvic (d) vertebral

_c_ 9. The anatomical term for the leg is (a) brachial (b) tarsal (c) crural (d) sural

_d_ 10. In which abdominopelvic region is the spleen located? (a) left lumbar (b) right lumbar (c) epigastric (d) left hypochondriac

_a_ 11. Which of the following represents the most complex level of structural organization? (a) organ (b) cellular (c) tissue (d) chemical

_c_ 12. Which body system is concerned with support, protection, leverage, blood-cell production, and mineral storage? (a) cardiovascular (b) lymphatic (c) skeletal (d) digestive

_b_ 13. The skin and structures derived from it, such as nails, hair, sweat glands, and oil glands, are components of which system? (a) respiratory (b) integumentary (c) muscular (d) lymphatic

_c_ 14. Hormone-producing glands belong to which body system? (a) cardiovascular (b) lymphatic (c) endocrine (d) digestive

_d_ 15. Which body system brings about movement, maintains posture, and produces heat? (a) skeletal (b) respiratory (c) reproductive (d) muscular

_a_ 16. Which abdominopelvic quadrant contains most of the liver? (a) RUQ (b) RLQ (c) LUQ (d) LLQ

_____d_____ **17.** The physical and chemical breakdown of food for use by body cells and the elimination of solid wastes are accomplished by which body system? (a) respiratory (b) urinary (c) cardio-vascular (d) digestive

_____c_____ **18.** The ability of an organism to detect and respond to environmental changes is called (a) metabolism (b) differentiation (c) responsiveness (d) respiration

_____a_____ **19.** In a feedback system, the component that produces a response is the (a) effector (b) receptor (c) input (d) output

PART 2. Completion

20. The tibia is _medial_ to the fibula.

21. The ovaries are found in the _abdominopelvic_ body cavity.

22. The upper horizontal line that helps divide the abdominopelvic cavity into nine regions is the _subcostal_ line.

23. The anatomical term for the hollow behind the knee is _popliteal_.

24. A plane that divides the stomach into a superior and an inferior portion is a(n) _transverse_ plane.

25. The wrist is described as _distal_ to the elbow.

26. The heart is located in the _mediastinal_ cavity within the thoracic cavity.

27. The abdominopelvic region that contains the rectum is the _hypogastric_ region.

28. A plane that divides the body into unequal left and right sides is the _parasaggital_ plane.

29. The spinal cord is located within the _spinal_ cavity.

30. The body system that removes carbon dioxide from body cells, delivers oxygen to body cells, helps maintain acid-base balance, helps protect against disease, helps regulate body temperature, and prevents hemorrhage by forming clots is the _cardiovascular_ system.

31. The _left lower_ abdominopelvic quadrant contains the descending colon of the large intestine.

32. Which body system returns proteins and plasma to the cardiovascular system, transports fats from the digestive system to the cardiovascular system, filters blood, protects against disease, and produces white blood cells? _lymphatic_

33. The sum of all chemical processes that occur in the body is called _metabolism_.

34. A structure that monitors changes in a controlled condition and sends the information to the control center is the _receptor_.

PART 3. Matching

___D___ 35. Right hypochondriac region

___I___ 36. Hypogastric region

___A___ 37. Left iliac region

___F___ 38. Right lumbar region

___H___ 39. Epigastric region

___C___ 40. Left hypochondriac region

___E___ 41. Right iliac region

___G___ 42. Umbilical region

___B___ 43. Left lumbar region

A. Junction of descending and sigmoid colons of large intestine

B. Descending colon of large intestine

C. Spleen

D. Most of right lobe of liver

E. Appendix

F. Ascending colon of large intestine

G. Middle of transverse colon of large intestine

H. Adrenal (suprarenal) glands

I. Sigmoid colon of large intestine

PART 4. Matching

___R___ 44. Anterior

___H___ 45. Skull

___A___ 46. Transtubercular line

___F___ 47. Armpit

___M___ 48. Umbilical region

___C___ 49. Medial

___T___ 50. Cranial cavity

___P___ 51. Front of knee

___K___ 52. Breast

___D___ 53. Chest

___J___ 54. Buttock

___Q___ 55. Superior

___U___ 56. Groin

___B___ 57. Vertebral cavity

___N___ 58. Cheek

___E___ 59. Front of neck

___O___ 60. Distal

___G___ 61. Pericardial cavity

___I___ 62. Forearm

___L___ 63. Plantar

___S___ 64. Mouth

A. Passes through iliac crests

B. Contains spinal cord

C. Nearer the midline

D. Thoracic

E. Cervical

F. Axillary

G. Contains the heart

H. Cranial

I. Antebrachial

J. Gluteal

K. Mammary

L. Sole

M. Contains navel

N. Buccal

O. Farther from the attachment of a limb

P. Patellar

Q. Toward the head

R. Nearer to or at the front of the body

S. Oral

T. Contains brain

U. Inguinal

Cells

A *cell* is the basic living structural and functional unit of the body. The study of cells is called *cytology* (sī-TOL-ō-jē; *cyto* = cell; *logos* = study of). The different kinds of cells—blood, nerve, bone, muscle, epithelial, and others—perform specific functions and differ from one another in shape, size, and structure. You will start your study of cytology by learning the important components of a theoretical, generalized cell.

A. CELL PARTS

Refer to Figure 3.1, a generalized cell based on electron micrograph studies. With the aid of your textbook and any other items made available by your instructor, label the parts of the cell indicated. In the spaces that follow, describe the function of the cellular structures indicated:

1. *Plasma (cell) membrane* separates internal components from outside environment.

2. *Cytoplasm* (SĪ-tō-plazm') Everything inside the cell wall and outside the nucleus.

3. *Nucleus* (NOO-klē-us) The control center for the cell. Contains the DNA that regulates all cell activities

4. *Endoplasmic reticulum* (en'-dō-PLAS-mik re-TIK-yoo-lum) or *ER* Rough ER attached to nuclear envelope - site of protein synthesis. Smooth ER detoxifies and produces hormones,

5. *Ribosome* (RĪ-bo-sōm) Protein producing unit located freely + attached to rough endo plasmic reticulum

6. *Golgi* (GOL-jē) *complex* Located close to the endoplasmic reticulum. Packages protein for export out of the cell.

7. *Mitochondrion* (mī'-tō-KON-drē-on) Power house of the cell. Splits ATP into ADP + P

8. *Lysosome* (LĪ-sō-sōm) An phagocytic vessicle that has digestive enzymes. It engulfs solid particles.

9. *Peroxisome* (pe-ROKS-i-sōm) A pinocytic vessicle that uses H2O2 to digest matter.

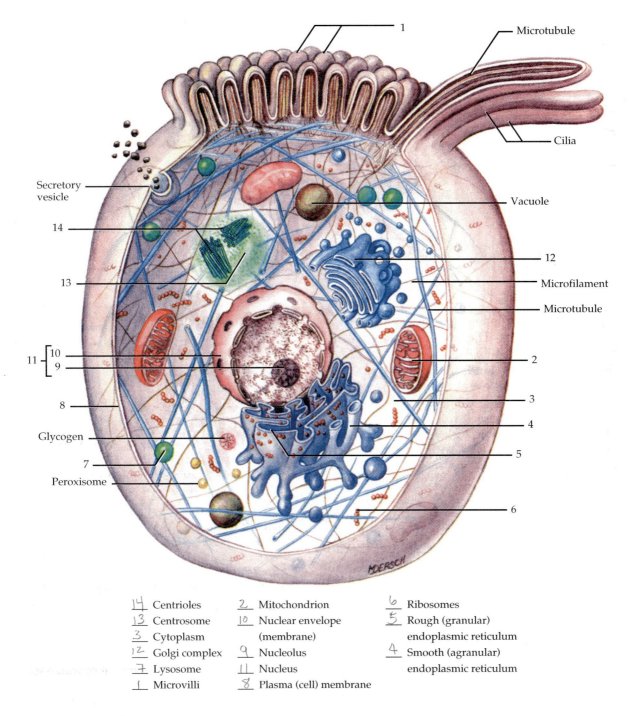

1

Microtubule

Cilia

Secretory
vesicle

Vacuole

14

13

12

Microfilament

Microtubule

11 ⎡ 10
   ⎣ 9

2

8

3

4

Glycogen

5

7

Peroxisome

6

MOERSCH

14 Centrioles	2 Mitochondrion	6 Ribosomes
13 Centrosome	10 Nuclear envelope	5 Rough (granular)
3 Cytoplasm	(membrane)	endoplasmic reticulum
12 Golgi complex	9 Nucleolus	4 Smooth (agranular)
7 Lysosome	11 Nucleus	endoplasmic reticulum
1 Microvilli	8 Plasma (cell) membrane	

FIGURE 3.1   Generalized animal cell.

10. *Cytoskeleton* _____

_____

_____

_____

11. *Centrosome* (SEN-trō-sōm') _____

_____

_____

_____

12. *Cilium* (SIL-ē-um) Attached to cell wall. Responsible for movement/passage of matter.

13. *Flagellum* (fla-JEL-um) Comprised of microtubules. Responsible for cell locomotion.

B. DIVERSITY OF CELLS

Now obtain prepared slides of the following types of cells and examine them under the magnifications suggested:

1. Ciliated columnar epithelial cells (high power)
2. Sperm cells (oil immersion)
3. Nerve cells (high power)
4. Muscle cells (high power)

After you have made your examination, draw an example of each of these kinds of cells in the spaces provided and under each cell indicate how each is adapted to its particular function.

Ciliated columnar epithelial cell

Sperm cell

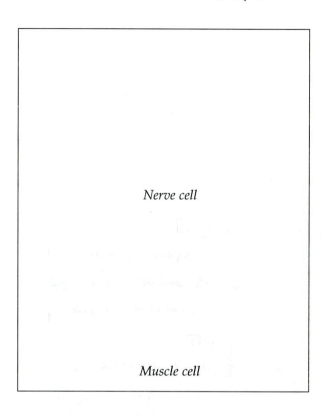

Nerve cell

Muscle cell

C. MOVEMENT OF SUBSTANCES ACROSS AND THROUGH PLASMA MEMBRANES

Biological membranes serve as permeability barriers. They maintain both the internal integrity of the cell and the solute concentrations within the cell that are considerably different from those found in the extracellular environment. In general, most substances in the body have a high solubility in polar liquids, such as water, and low solubility in nonpolar liquids, such as alcohol. Because cellular membranes are composed of a combination of polar and nonpolar components, they present a formidable barrier to the movement of water-soluble substances into and out of a cell. In general, substances move across plasma membranes by two principal kinds of transport processes—*passive transport processes*, which do not require the expenditure of cellular energy, and *active transport processes*, which do require the expenditure of energy, usually via the chemical breakdown of adenosine triphosphate (ATP) into adenosine diphosphate (ADP), inorganic phosphate, and energy. In passive transport processes, which are physical processes, substances move because of differences in concentration (or pressure) from areas of higher concentration (or pressure) to areas of lower concentration (or pressure). The movement continues until the concentration of substances (or pressure) reaches equilibrium. Passive transport processes are the result of the kinetic energy (energy of motion) of the substances themselves and the cell does not expend energy to move the substances. Examples of passive transport processes are simple diffusion, facilitated diffusion, osmosis, filtration, and dialysis.

By contrast, in active transport processes, which are physiological processes, substances move against their concentration gradient, from areas of lower concentration to areas of higher concentration. Moreover, cells must expend energy (ATP) to carry on active transport processes. Examples of active transport processes are active transport and endocytosis (phagocytosis and pinocytosis).

CAUTION! *Please reread Section A, "General Safety Precautions and Procedures," on page xi, and Section C, "Precautions Related to Working with Reagents," on page xii, at the beginning of the laboratory manual before you begin any of the following experiments. Read the experiments before you perform them to be sure that you understand all the procedures and safety cautions.*

1. Passive Transport Processes

Before looking at examples of passive transport processes, we will first examine how molecules move.

a. BROWNIAN MOVEMENT
At temperatures above absolute zero (–273°C or –460°F), all molecules are in constant random motion because of their inherent kinetic energy. This phenomenon is called *Brownian movement*. Less energy is required to move small molecules or particles than large ones. Because all molecules are constantly bombarded by the molecules surrounding them, the smaller the particle, the greater its random motion in terms of speed and distance.

PROCEDURE
1. With a medicine dropper, place a drop of dilute detergent solution in a depression (concave) slide.
2. Using another medicine dropper, add a drop of dilute India ink to the first drop in the depression slide.
3. Stir the two solutions with a toothpick and cover the depression with a cover glass.

4. Let the slide stand for 10 min and then place it on your microscope stage and observe it under high power.
5. Using forceps, place the slide on a hot plate and warm it for 15 sec. Then remove the slide with forceps and observe it again under high power.
6. Record your observations in Section C.1.a of the LABORATORY REPORT RESULTS at the end of the exercise.

b. SIMPLE DIFFUSION
Simple diffusion is the net (greater) movement of molecules or ions from a region of higher concentration to a region of lesser concentration until they are evenly distributed (in equilibrium). An example in the human body is the movement of oxygen and carbon dioxide between body cells and blood.

The following two experiments illustrate simple diffusion. Either or both may be performed.

PROCEDURE
1. To demonstrate simple diffusion of a solid in a liquid, *using forceps, carefully* place a large crystal of potassium permanganate ($KMnO_4$) into a test tube filled with water.

CAUTION! *Avoid contact of $KMnO_4$ with your skin by using acid- or caustic-resistant gloves.*

2. Place the tube in a rack against a white background where it will not be disturbed.
3. Note the diffusion of the crystal material through the water at 15-min intervals for 2 hr.
4. Record the diffusion of the crystal in millimeters (mm) per minute at 15-min intervals in Section C.1.b of the LABORATORY REPORT RESULTS at the end of the exercise. Simply measure the distance of diffusion using a millimeter ruler.

PROCEDURE
1. To demonstrate simple diffusion of a solid in a solid, *using forceps, carefully* place a large crystal of methylene blue on the surface of agar in the center of a petri plate.
2. Note the diffusion of the crystal through the agar at 15-min intervals for 2 hr.
3. Record the diffusion of the crystal in millimeters (mm) per minute at 15-min intervals, using a millimeter ruler, in Section C.1.b of the LABORATORY REPORT RESULTS at the end of the exercise.

c. FACILITATED DIFFUSION

Facilitated diffusion is the movement of a substance across a selectively permeable membrane from a region of higher concentration to a region of lower concentration with the assistance of integral proteins that serve as transporters (carriers) in the plasma membrane for each type of substance. This process does not require the expenditure of cellular energy. Different sugars, especially glucose, cross plasma membranes by facilitated diffusion.

In this experiment, the substance undergoing facilitated diffusion is neutral red, which is red at a pH just below 7 but yellow at a pH between 7 and 8.

PROCEDURE

1. In a 125-ml flask, combine one gram (1 g) of bakers' yeast with 25 ml 0.75% Na_2CO_3 (sodium carbonate) solution. Swirl until the yeast is evenly suspended in the solution.
2. Divide the suspension into two large test tubes marked "U1" (unboiled) and "B1" (boiled) with a wax pencil.
3. Place tube "B1" in a boiling water bath for 2 to 3 minutes. Be sure that all the suspension is below water level so that all the yeast cells will be killed.

CAUTION! *Make sure that the mouth of the test tube is pointed away from you and all other persons in the area.*

4. Using a test tube holder, place both test tubes into a rack and add 7.5 ml of 0.02% neutral red to each.
5. Record the color of the solution in both test tubes in Section C.1.c of the LABORATORY REPORT RESULTS at the end of the exercise.
6. After 15 minutes, place 1 ml from each test tube into test tubes marked "U2" and "B2."
7. Add enough drops of 0.75% acetic acid to tube "B2" to make its color identical to that of tube "U2."
8. Mark two microscope slides "U" and "B"; examine a sample from both tubes under high power (using a cover slip) and note which cells are stained. _____

_____

9. Filter half of the solution remaining in tube "B1" and examine the filtrate. Repeat for tube "U1."

10. Observe and record the color of the filtrate and that of the yeast cells on the filter paper.

_____

_____

11. To the suspension remaining in tubes "U1" and "B1," add about 9 ml 0.75% acetic acid. Filter again according to the instructions in step 9.
12. Record the color of these "U1" and "B1" filtrates and cells in Section C.1.c of the LABORATORY REPORT RESULTS at the end of the exercise.
13. How does boiling affect the facilitated diffusion of neutral red by the cells? _____

_____

Describe the membrane's permeability to neutral red. _____

_____

Did acetic acid enter the living cells? _____

_____

Did sodium carbonate enter the living cells?

_____

d. OSMOSIS

Osmosis (oz-MŌ-sis) is the net movement of *water* through a selectively permeable membrane from a region of higher water (lower solute) concentration to a region of lower water (higher solute) concentration. In the body, fluids move between cells as a result of osmosis.

PROCEDURE

1. Refer to the osmosis apparatus in Figure 3.2.
2. Tie a knot very tightly at one end of a 4-in. piece of cellophane dialysis tubing that has been soaking in water for a few minutes. Fill the dialysis tubing with a 10% sugar (sucrose) solution that has been colored with Congo red (red food coloring can also be used).
3. Close the open end of the dialysis tubing with a one-hole rubber stopper into which a glass tube has already been inserted by a laboratory assistant or your instructor.

CAUTION! *If you are unfamiliar with the procedure, do not attempt to insert the glass tube into the stopper yourself because it may break and result in serious injury.*

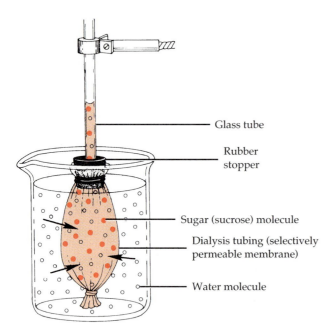

Glass tube

Rubber stopper

Sugar (sucrose) molecule

Dialysis tubing (selectively permeable membrane)

Water molecule

FIGURE 3.2   Osmosis apparatus.

4. Tie a piece of string tightly around the dialysis tubing to secure it to the stopper.
5. Secure and suspend the glass tube and dialysis tubing by means of a clamp attached to a ring stand.
6. Insert the dialysis tubing into a beaker or flask of water until the water comes up to the bottom of the rubber stopper.
7. As soon as the sugar solution becomes visible in the glass tube, mark its height with a wax pencil and note the time.
8. Mark the height of liquid in the glass tube after 10-, 20-, and 30-min intervals by using your millimeter (mm) ruler and record your results in Section C.1.d of the LABORATORY REPORT RESULTS at the end of the exercise.

e. HEMOLYSIS AND CRENATION

Osmosis can be demonstrated by noting the effects of different water concentrations on red blood cells. Red blood cells maintain their normal shape when placed in an *isotonic* (*iso* = same) *solution.* An isotonic solution has the same salt concentration (0.90% NaCl) as that found in red blood cells. If, however, red blood cells are placed in a *hypotonic* (*hypo* = lower) *solution* (a salt solution with less than 0.90% NaCl), a net movement of water into the cells occurs, causing the cells to swell and possibly burst. The red blood cells swell because the difference in osmotic pressure inside the cell compared to outside the cell results in a net movement of water

into the cell. The rupture of blood cells in this manner and the resulting loss of hemoglobin into the surrounding liquid is termed *hemolysis* (hē-MOL-i-sis). If, instead, red blood cells are placed in a *hypertonic* (*hyper* = higher) *solution* (a salt solution with more than 0.90% NaCl), a net movement of water out of the cells occurs, causing the cells to shrink. The red blood cells undergo this shrinkage, known as *crenation* (kri-NĀ-shun) because the difference in osmotic pressure inside the cell compared to outside the cell results in the net movement of water out of the cell.

PROCEDURE

CAUTION! *Please reread Section B, "Precautions Related to Working with Blood, Blood Products, or Other Body Fluids," on page xii, at the beginning of the laboratory manual, before you begin any of the following experiments. Read the experiments before you perform them to be sure that you understand all the procedures and safety precautions. When working with whole blood, take care to avoid any kind of contact with an open sore, cut, or wound. Wear tight-fitting surgical gloves and safety goggles.*

When you finish this part of the exercise, place the reusable items in a fresh bleach solution and the discarded items in a biohazard container.

1. With a wax marking pencil, mark three microscope slides as follows: 0.90%, DW (distilled water), and 3%.
2. Using a medicine dropper, place three drops of fresh (uncoagulated) ox blood on a microscope slide that contains 2 milliliters (ml) of a 0.90% NaCl solution (isotonic solution). Mix gently and thoroughly with a clean toothpick.
3. Using a medicine dropper, place three drops of fresh ox blood on a microscope slide that contains 2 ml of distilled water (hypotonic solution). Mix gently and thoroughly with a clean toothpick.
4. Now, using a medicine dropper, add three drops of fresh ox blood to a microscope slide that contains 2 ml of a 3% NaCl solution (hypertonic solution). Mix gently and thoroughly with a clean toothpick.
5. Using a medicine dropper, place two drops of the red blood cells in the isotonic solution on another microscope slide, cover with a cover slip, and examine the red blood cells under high power. Reduce your illumination.

What is the shape of the cells?_____

_____

Explain their shape._____

_____

6. Using a medicine dropper, place two drops of the red blood cells in the hypotonic solution on another microscope slide, cover with a cover slip, and examine the red blood cells under high power. Reduce your illumination.

    What is the shape of the cells?_____

    _____

    Explain their shape._____

    _____

7. Using a medicine dropper, place two drops of the red blood cells in the hypertonic solution on another microscope slide, cover with a cover slip, and examine the red blood cells under high power. Reduce your illumination.

    What is the shape of the cells?_____

    _____

    Explain their shape._____

    _____

ALTERNATE PROCEDURE

1. Obtain three pieces of raw potato that have an *identical* weight.
2. Immerse one piece in a beaker that contains an isotonic solution; immerse a second piece in a hypotonic solution; immerse the third piece in a hypertonic solution.
3. Continue the experiment for 1 hr. Record the time.
4. At the end of 1 hr, remove the pieces of potato and weigh them separately.

    Weight of potato in isotonic solution_____

    _____

    Weight of potato in hypotonic solution_____

    _____

    Weight of potato in hypertonic solution_____

    _____

    Explain the differences in the weights of the

    three pieces of potato._____

    _____

    _____

_____

_____

_____

_____

_____

f. FILTRATION

Filtration is the movement of solvents and dissolved substances across a selectively permeable membrane from regions of higher pressure to regions of lower pressure. Movement of the solvents and dissolved substances occurs under the influence of gravity and the pressure exerted by the solvent, which is termed *hydrostatic pressure.* The selectively permeable membrane prevents molecules with higher molecular weights from passing through the membrane, while the solvent and substances with lower molecular weights easily pass through the selectively permeable membrane. In general, any substance having a molecular weight of less than 100 is filtered, because the pores in the filter paper are larger than the molecules of the substance. Filtration is one mechanism by which the kidneys regulate the chemical composition of the blood.

PROCEDURE

1. Refer to Figure 3.3, which shows the filtration apparatus. In this apparatus, the filter paper represents the selectively permeable membrane of a cell.

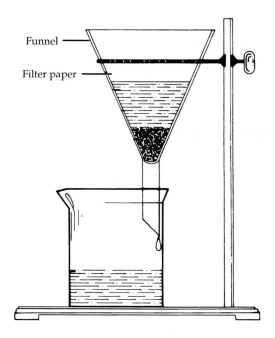

Funnel

Filter paper

FIGURE 3.3 Filtration apparatus.

2. Fold a piece of filter paper in half and then in half again.

3. Open it into a cone, place it in a funnel, and place the funnel over the beaker.

4. Shake a mixture of a few particles of powdered wood charcoal (black), 1% copper sulfate (blue), boiled starch (white), and water, and slowly pour it into the funnel until the mixture almost reaches the top of the filter paper. Gravity will pull the particles through the pores of the filter paper.

5. Count the number of drops passing through the funnel for the following time intervals: 10, 30, 60, 90, and 120 seconds. Record your observations in Section C.1.f of the LABORATORY REPORT RESULTS at the end of the exercise.

6. Observe which substances passed through the filter paper by noting their color in the filtered fluid in the beaker.

7. Examine the filter paper to determine whether any colored particles were not filtered.

8. To determine if any starch is in the liquid (termed the *filtrate*) in the beaker, add several drops of 0.01 M IKI solution. A blue-black color reaction indicates the presence of starch.

g. DIALYSIS

Dialysis (dī-AL-i-sis) is the separation of smaller molecules from larger ones by a selectively permeable membrane. Such a membrane permits diffusion of the small molecules but not the large ones. The principle of dialysis is employed in artificial kidneys.

PROCEDURE

1. Refer to the dialysis apparatus in Figure 3.4.

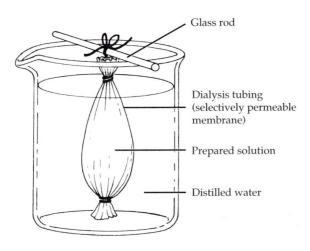

— Glass rod

— Dialysis tubing (selectively permeable membrane)

— Prepared solution

— Distilled water

FIGURE 3.4   Dialysis apparatus.

2. Tie off one end of a piece of dialysis tubing that has been soaking in water. Place a prepared solution containing starch, sodium chloride, 5% glucose, and albumin into the dialysis tubing.

3. Tie off the other end of the dialysis tubing and immerse it in a beaker of distilled water.

4. After 1 hr, test the solution in the beaker for the presence of each of the substances in the tubing, as follows, and record your observations in Section C.1.g of the LABORATORY REPORT RESULTS at the end of the exercise.

CAUTION! *Be extremely careful using nitric acid. It can severely damage your skin. Use acid- or caustic-resistant gloves.*

    a. *Albumin*—Carefully add several drops of concentrated nitric acid to a test tube containing 2 ml of the solution in the beaker. Positive reaction = white coagulate.

    b. *Sugar*—Test 5 ml of the solution in the beaker in a test tube with 5 ml of Benedict's solution. Place the test tube in a boiling water bath for three minutes.

CAUTION! *Make sure that the mouth of the test tube is pointed away from you and everyone else in the area.*

    Using a test tube holder, remove the test tube from the water bath. Note the color. Positive reaction = green, yellow, orange, or red precipitate.

    c. *Starch*—Add several drops of IKI solution to 2 ml of the solution in the beaker in a test tube. Note the color. Positive reaction = blue-black color.

    d. *Sodium chloride*—Place 2 ml of the solution in the beaker in a test tube and add several drops of 1% silver nitrate. Note the color. Positive reaction = white precipitate.

2. Active Transport Processes

a. ACTIVE TRANSPORT

There are two types of *active transport.* In *primary active transport,* energy derived from splitting ATP *directly* moves, or "pumps," a substance across a plasma membrane. The cell uses energy from ATP to change the shape of transport proteins in the membrane. An example of primary active transport is the sodium pump, which maintains a low concentration of sodium ions (Na^+) in the cytosol, the fluid portion of the cytoplasm, by pumping sodium ions out against their concentration gradient. The pump also moves potassium ions (K$^+$) into cells against their concentration gradient. In *secondary active transport,* the energy stored in ion concentration gradients (differences) drives substances across a plasma membrane. Since ion gradients are established by primary active transport, secondary active transport *indirectly* uses energy obtained from splitting ATP. An example of secondary active transport is the movement of an amino acid and Na$^+$ in the same direction across a plasma membrane with the assistance of an integral membrane protein. Another example is the movement of calcium ions (Ca^{2+}) and Na$^+$ in opposite directions across a membrane with the assistance of an integral membrane protein.

Because reliable results are difficult to demonstrate simply, you will not be asked to demonstrate active transport.

b. ENDOCYTOSIS

Endocytosis refers to the passage of large molecules and particles across a plasma membrane, in which a segment of the membrane surrounds the substance, encloses it, and brings it into the cell. Here we will consider two types of endocytosis—phagocytosis and pinocytosis.

(1) Phagocytosis

Phagocytosis (fag'-ō-sī-TŌ-sis; *phagein* = to eat), or cell "eating," is the engulfment of solid particles or organisms by *pseudopods,* temporary fingerlike projections of the cytoplasm of the cell. Once the particle is surrounded by the membrane, the membrane folds inward, pinches off from the rest of the plasma membrane, and forms a *phagocytic vesicle* around the particle. The particle within the vesicle is digested either by the secretion of enzymes into the vesicle or by the combining of the vesicle with an enzyme-containing lysosome.

Phagocytosis can be demonstrated by observing the feeding of an amoeba, a unicellular organism whose movement and ingestion are similar to those of human leukocytes (white blood cells).

PROCEDURE

1. Using a medicine dropper, place a drop of culture containing amoebas that have been starved for 48 hr into the well of a depression slide and cover the well with a cover slip. Cultures containing *Chaos chaos* or *Amoeba proteus* should be used. (Your instructor may wish to use the hanging-drop method instead. If so, she or he will give you verbal instructions.)

2. Examine the amoebas under low power, and be sure that your light is reduced considerably.
3. Observe the locomotion of an amoeba for several minutes. Pay particular attention to the pseudopods that appear to flow out of the cell.
4. To observe phagocytosis, use a medicine dropper and add a drop containing small unicellular animals called *Tetrahymena pyriformis* to the culture containing the amoebas.
5. Examine under low power, and observe the ingestion of *Tetrahymena pyriformis* by an amoeba. Note the action of the pseudopods and the formation of the phagocytic vesicle around the ingested organism.

(2) Pinocytosis

Pinocytosis (pi-nō-sī-TŌ-sis; *pinein* = to drink), or cell "drinking," is the engulfment of a liquid. The liquid is attracted to the surface of the membrane; the membrane folds inward and surrounds the liquid and detaches from the rest of the intact membrane forming a *pinocytic vesicle.*

D. CELL INCLUSIONS

Cell inclusions are a large and diverse group of mostly organic substances produced by cells that may appear or disappear at various times in the life of a cell. Using your textbook as a reference, indicate the function of the following cell inclusions:

1. *Melanin* _____

_____

_____

_____

2. *Glycogen* _____

_____

_____

_____

3. *Triglycerides* _____

_____

_____

_____

E. EXTRACELLULAR MATERIALS

Substances that lie outside the plasma membranes of body cells are referred to as *extracellular materials*. They include body fluids, such as interstitial fluid and plasma, which provide a medium for dissolving, mixing, and transporting substances. Extracellular materials also include special substances in which some cells are embedded.

Some extracellular materials are produced by certain cells and deposited outside their plasma membranes where they support cells, bind them together, and provide strength and elasticity. They have no definite shape and are referred to as *amorphous*. These include hyaluronic (hī-a-loo-RON-ik) acid and chondroitin (kon-DROY-tin) sulfate. Others are *fibrous* (threadlike). Examples include collagen, reticular, and elastic fibers.

Using your textbook as a reference, indicate the location and function for each of the following extracellular materials:

Hyaluronic acid

Location _____

_____

_____

_____

Function _____

_____

_____

Chondroitin sulfate

Location _____

_____

_____

_____

Function _____

_____

_____

_____

Collagen fibers

Location _____

_____

_____

_____

Function _____

_____

_____

Reticular fibers

Location _____

_____

_____

_____

Function _____

_____

_____

Elastic fibers

Location _____

_____

_____

_____

Function _____

_____

_____

_____

F. CELL DIVISION

Cell division is the basic mechanism by which cells reproduce themselves. It consists of a nuclear division and a cytoplasmic division. Because nuclear division can be of two types, two kinds of cell division are recognized. In the first type, called *somatic cell division*, a single starting cell called a *parent cell* duplicates itself and the result is two identical cells called *daughter cells*. Somatic cell division consists of a nuclear division called *mitosis* (mī-TŌ-sis) and a cytoplasmic division called *cytokinesis* (sī-tō-ki-NĒ-sis; *cyto* = cell; *kinesis* = motion). It provides the body with a means of growth and of replacement of diseased or damaged cells (Figure 3.5). The second type of cell division is called *reproductive cell division* and is the mechanism by which sperm and eggs are produced (Exercise 25). Reproductive cell division consists of a nuclear division called *meiosis* and two cytoplasmic divisions (cytokinesis), and it results in the development of four nonidentical daughter cells.

In order to study somatic cell division, obtain a prepared slide of a whitefish blastula and examine it under high power.

A cell between divisions is said to be in the *interphase* of the cell cycle. Interphase is the longest part of the cell cycle and is the period of time during which a cell carries on its physiological activities. One of the most important activities of interphase is the replication of DNA so that the two daughter cells that eventually form will each have the same kind and amount of DNA as the parent cell. In addition, the proteins needed to produce structures required for doubling all cellular components are manufactured. Scan your slide and find a cell in interphase. Such a parent cell is characterized by a clearly defined nuclear envelope. Within the nucleus, look for the nucleolus (or nucleoli) and *chromatin,* DNA that is associated with protein in the form of a granular substance. Also locate the centrioles.

Draw a labeled diagram of an interphase cell in the space provided.

Once a cell completes its interphase activities, mitosis begins. Mitosis is the distribution of two sets of chromosomes into two separate and equal nuclei after replication of the chromosomes of the parent cell, an event that takes place in the interphase preceding mitosis. Although a continuous process, mitosis is divided into four stages for pur-

poses of study: prophase, metaphase, anaphase, and telophase.

1. *Prophase*—The first stage of mitosis is called *prophase* (*pro* = before). During early prophase, the chromatin condenses and shortens into chromosomes. Because DNA replication took place during interphase, each prophase chromosome contains a pair of identical double-stranded DNA molecules called *chromatids*. Each chromatid pair is held together by a small spherical body called a *centromere* that is required for the proper segregation of chromosomes. Attached to the outside of each centromere is a protein complex known as the *kinetochore* (ki-NET-ō-kor), whose function will be described shortly. Later in prophase, the nucleolus (or nucleoli) disperses, and the nuclear envelope breaks up. In addition, the centrosomes each with its pair of centrioles, move to opposite poles (ends) of the cell. As they do so, the centrosomes start to form the *mitotic spindle*, a football-shaped assembly of microtubules that are responsible for the movement of chromosomes.

The lengthening of microtubules between centrosomes pushes the centrosomes to the poles of the cell so that the spindle extends from pole to pole. As the mitotic spindle continues to develop, three types of microtubules form: (a) *nonkinetochore microtubules*, which grow from centrosomes and extend inward, but do not bind to kinetochores; (b) *kinetochore microtubules*, which grow from centrosomes, extend inward, and attach to kinetochores; and (c) *aster microtubules*, which grow out of centrosomes but radiate outward from the mitotic spindle. Overall, the spindle is an attachment site for chromosomes. It also distributes chromosomes to opposite poles of the cell. Draw and label a cell in prophase in the space provided.

2. *Metaphase*—During *metaphase* (*meta* = after), the second stage of mitosis, the kinetochore microtubules line up the centromeres of the chromatid pairs at the exact center of the mitotic spindle. This midpoint region is called

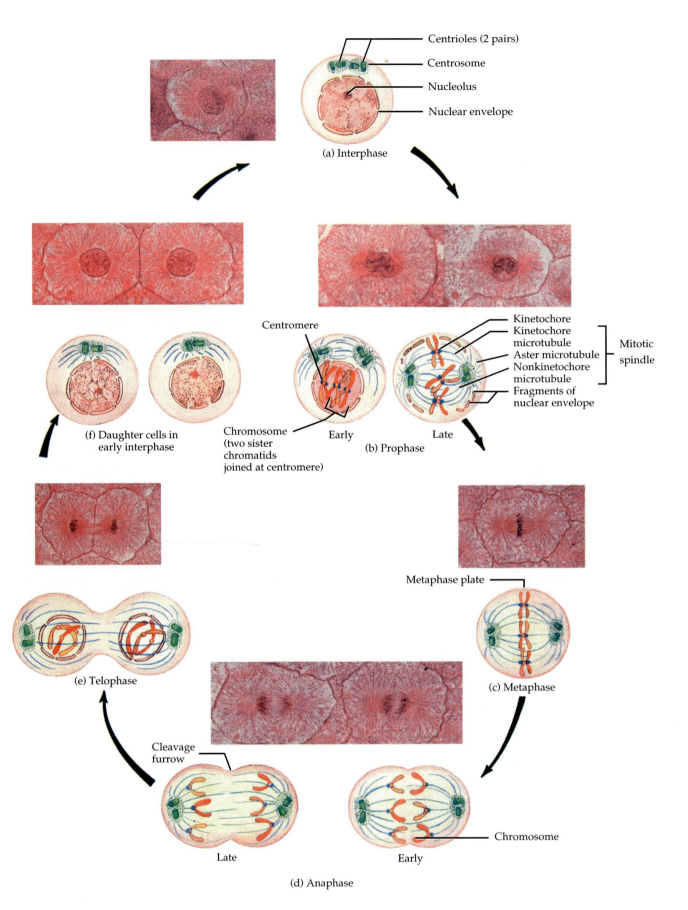

Centrioles (2 pairs)
Centrosome
Nucleolus
Nuclear envelope

(a) Interphase

Kinetochore
Kinetochore microtubule
Aster microtubule
Nonkinetochore microtubule
Fragments of nuclear envelope

Mitotic spindle

Centromere

Chromosome (two sister chromatids joined at centromere)

Early   Late

(b) Prophase

(f) Daughter cells in early interphase

Metaphase plate

(c) Metaphase

(e) Telophase

Cleavage furrow

Chromosome

Late   Early

(d) Anaphase

FIGURE 3.5   Cell division: mitosis and cytokinesis. Diagrams and photomicrographs of the various stages of cell division in whitefish eggs.

the *metaphase plate*, or *equatorial plane region.* Draw and label a cell in metaphase in the space provided.

the mitotic spindle breaks up. Draw and label a cell in telophase in the space provided.

3. *Anaphase*—The third stage of mitosis, *anaphase* (*ana* = upward), is characterized by the splitting and separation of the centromeres (and kinetochores) and the movement of the two sister chromatids of each pair toward opposite poles of the cell. Once separated, the sister chromatids are referred to as *chromosomes.* The movement of chromosomes is the result of the shortening of kinetochore microtubules and elongation of the nonkinetochore microtubules, processes that increase the distance between separated chromosomes. As the chromosomes move during anaphase, they appear V-shaped. Draw and label a cell in anaphase in the space provided.

4. *Telophase*—The final stage of mitosis, *telophase* (*telo* = far or end), begins as soon as chromosomal movement stops. Telophase is essentially the opposite of prophase. During telophase, the identical sets of chromosomes at opposite poles of the cell uncoil and revert to their threadlike chromatin form; kinetochore microtubules disappear; nonkinetochore microtubules elongate even more; a new nuclear envelope reforms around each chromatin mass; new nucleoli reappear in the daughter nuclei; and eventually

Cytokinesis begins during late anaphase and terminates during telophase with the formation of a *cleavage furrow*, a slight indentation of the plasma membrane that extends around the center of the cell. The furrow gradually deepens until opposite surfaces of the cell make contact and the cell is split in two. The result is two separated daughter cells, each with separate portions of cytoplasm and organelles and its own set of identical chromosomes.

Following cytokinesis, each daughter cell returns to interphase. Each cell in most tissues of the body eventually grows and undergoes mitosis and cytokinesis, and a new divisional cycle begins. Examine your telophase cell again and be sure that it contains a cleavage furrow.

Using high power and starting at 12 o'clock, move around the blastula and count the number of cells in interphase and in each mitotic phase. It will be easier to do this if you imagine lines dividing the blastula into quadrants. Count the interphase cells in each quadrant, then assign each dividing cell to a specific mitotic stage. It will be hard to assign some cells to a phase—e.g., to distinguish late anaphase from early telophase. If you cannot make a decision, assign one cell to the earlier phase in question and the next cell to the later phase.

Divide the number of cells in each stage by the total number of cells counted and multiply by 100 to determine the percent of the cells in each mitotic stage at a given point in time. Record your results in Section F of the LABORATORY REPORT RESULTS at the end of the exercise.

ANSWER THE LABORATORY REPORT QUESTIONS AT THE END OF THE EXERCISE.

Cells 3

Student _____ Date _____

Laboratory Section _____ Score/Grade _____

SECTION C. MOVEMENT OF SUBSTANCES ACROSS PLASMA MEMBRANES

1. Passive Transport Processes

a. BROWNIAN MOVEMENT

Describe the movement of the India ink particles on the unheated slide. _____

_____

How does this movement differ from that on the heated slide? _____

_____

b. SIMPLE DIFFUSION

Solid in Liquid		Solid in Solid	
Time, min	Distance, mm	Time, min	Distance, mm
15	_____	15	_____
30	_____	30	_____
45	_____	45	_____
60	_____	60	_____
75	_____	75	_____
90	_____	90	_____
105	_____	105	_____
120	_____	120	_____

c. FACILITATED DIFFUSION

Neutral Red		Acetic Acid	
Tube "U1"	_____	Tube "U1"	_____
Tube "B1"	_____	Tube "B1"	_____

d. OSMOSIS

Time, min   Height of liquid, mm

    10   _____

    20   _____

    30   _____

Explain what happened. _____

_____

f. FILTRATION

    10 sec _____

    30 sec _____

    60 sec _____

    90 sec _____

  120 sec _____

g. DIALYSIS
Place a check in the appropriate place to indicate if the following tests are positive (+) or negative (–).

	(+)	(–)
Albumin	_____	_____
Sugar	_____	_____
Starch	_____	_____
Sodium chloride	_____	_____

SECTION F. Cell Division

Percent of cells in interphase   _____

Percent of cells in prophase   _____

Percent of cells in metaphase   _____

Percent of cells in anaphase   _____

Percent of cells in telophase   _____

Cells 3

Student _____ **Date** _____

Laboratory Section _____ **Score/Grade** _____

PART 1. Multiple Choice

_d_   1. The portion of the cell that forms part of the mitotic spindle during division is the (a) endoplasmic reticulum (b) Golgi complex (c) cytoplasm (d) centrosome

_b_   2. Movement of molecules or ions from a region of higher concentration to a region of lower concentration via a process that does not require energy is called (a) phagocytosis (b) simple diffusion (c) active transport (d) pinocytosis

_c_   3. If red blood cells are placed in a hypertonic solution of sodium chloride, they will (a) swell (b) burst (c) shrink (d) remain the same

_d ?_   4. The reagent used to test for the presence of sugar is (a) silver nitrate (b) nitric acid (c) IKI (d) Benedict's solution

_a_   5. A cell that carries on a great deal of digestion also contains a large number of (a) lysosomes (b) centrioles (c) mitochondria (d) nuclei

_a_   6. Which process does *not* belong with the others? (a) active transport (b) dialysis (c) phagocytosis (d) pinocytosis

_d_   7. Movement of oxygen and carbon dioxide between blood and body cells is an example of (a) osmosis (b) active transport (c) simple diffusion (d) facilitated diffusion

_b_   8. Which type of solution will cause hemolysis? (a) isotonic (b) hypotonic (c) isometric (d) hypertonic

_c_   9. In addition to active transport and simple diffusion, the kidneys regulate the chemical composition of blood by utilizing the process of (a) phagocytosis (b) osmosis (c) filtration (d) pinocytosis

_c_   10. Engulfment of solid particles or organisms by pseudopods is called (a) active transport (b) dialysis (c) phagocytosis (d) filtration

_a_   11. Rupture of red blood cells with subsequent loss of hemoglobin into the surrounding medium is called (a) hemolysis (b) plasmolysis (c) plasmoptysis (d) hemoglobinuria

_d_   12. The area of the cell between the plasma membrane and nuclear envelope where chemical reactions occur is the (a) centrosome (b) vacuole (c) peroxisome (d) cytoplasm

_b_   13. The "powerhouses" of the cell where ATP is produced are the (a) ribosomes (b) mitochondria (c) centrioles (d) lysosomes

_c_   14. The sites of protein synthesis in the cell are (a) peroxisomes (b) flagella (c) ribosomes (d) centrosomes

_a_   15. Which process does *not* belong with the others? (a) simple diffusion (b) phagocytosis (c) active transport (d) pinocytosis

___b___   **16.** A cell inclusion that is a pigment in skin and hair is (a) glycogen (b) melanin (c) mucus (d) collagen

___a___   **17.** Which extracellular material is found in ligaments and tendons? (a) elastic fibers (b) chondroitin sulfate (c) collagen fibers (d) mucus

___d___   **18.** The organelles that contain enzymes for the metabolism of hydrogen peroxide are (a) lysosomes (b) mitochondria (c) Golgi complexes (d) peroxisomes

___d___   **19.** The framework of cilia, flagella, centrioles, and the mitotic spindle is formed by (a) endoplasmic reticulum (b) collagen fibers (c) chondroitin sulfate (d) microtubules

___a___   **20.** A viscous fluidlike substance that binds cells together, lubricates joints, and maintains the shape of the eyeballs is (a) elastin (b) hyaluronic acid (c) mucus (d) plasmin

PART 2. Completion

21. The external boundary of the cell through which substances enter and exit is called the

_____.

22. The cytoskeleton is formed by microtubules, intermediate filaments, and _____.

23. The portion of the cell that contains hereditary information is the _____.

24. The tail of a sperm cell is a long whiplash structure called a(n) _____.

25. Cells placed in a _____ solution will undergo hemolysis.

26. Division of cytoplasm is referred to as _____.

27. Lipid and protein secretion, carbohydrate synthesis, and assembly of glycoproteins are functions of the _____.

28. Storage of digestive enzymes is accomplished by the _____ of a cell.

29. Projections of cells that move substances along their surfaces are called _____.

30. The _____ provides a surface area for chemical reactions, a pathway for transporting molecules, and a storage area for synthesized molecules.

31. _____ is a cell inclusion that represents stored glucose in the liver and skeletal muscles.

32. A jellylike substance that supports cartilage, bone, heart valves, and the umbilical cord is

_____.

33. The framework of many soft organs is formed by _____ fibers.

34. The net movement of water through a selectively permeable membrane from a region of higher concentration of water to a region of lower concentration of water is known as _____.

35. The principle of _____ is employed in the operation of an artificial kidney.

36. In an interphase cell, DNA is in the form of a granular substance called _____.

37. Distribution of chromosomes into separate and equal nuclei is referred to as _____.

38. The constant random motion of molecules caused by their inherent kinetic energy is called

_____.

PART 3. Matching

_____ **39.** Anaphase

_____ **40.** Metaphase

_____ **41.** Interphase

_____ **42.** Telophase

_____ **43.** Prophase

A. Mitotic spindle appears

B. Movement of chromosome sets to opposite poles of cell

C. Centromeres line up on metaphase plate

D. Formation of two identical nuclei

E. Phase between divisions

Tissues

4

A *tissue* is a group of similar cells that usually have the same embryological origin and function together to perform a specific function. The study of tissues is called *histology* (hiss-TOL-ō-jē; *histio* = tissue; *logos* = study of). The various body tissues can be categorized into four principal kinds: (1) epithelial, (2) connective, (3) muscular, and (4) nervous. In this exercise you will examine the structure and functions of epithelial and connective tissues, except for bone or blood. Other tissues will be studied later as parts of the systems to which they belong.

A. EPITHELIAL TISSUE

Epithelial (ep'-i-THĒ-lē-al) *tissue,* or *epithelium,* may be divided into two types: (1) covering and lining and (2) glandular. Covering and lining epithelium forms the outer layer of the skin and some internal organs; forms the inner lining of blood vessels, ducts, body cavities, and many internal organs; and helps make up special sense organs for smell, hearing, vision, and touch. Glandular epithelium constitutes the secreting portion of glands.

1. Characteristics

Following are the general characteristics of epithelial tissue:

a. Epithelium consists largely or entirely of closely packed cells with little extracellular material between adjacent cells.
b. Epithelial cells are arranged in continuous sheets, in either single or multiple layers.
c. Epithelial cells have an *apical* (free) *surface* that is exposed to a body cavity, lining of an internal organ, or the exterior of the body and a *basal surface* that is attached to the basement membrane (described shortly).
d. Cell junctions are plentiful, providing secure attachments among the cells.
e. Epithelia are *avascular* (*a* = without; *vascular* = blood vessels). The vessels that supply nutrients

and remove wastes are located in the adjacent connective tissue. The exchange of materials between epithelium and connective tissue is by diffusion.
f. Epithelia adhere firmly to nearby connective tissue, which holds the epithelium in position and prevents it from being torn. The attachment between the epithelium and the connective tissue is a thin extracellular layer called the *basement membrane*. It consists of two layers. The *basal lamina* contains collagen, laminin, and proteoglycans secreted by the epithelium. Cells in the connective tissue secrete the second layer, the *reticular lamina*, which contains reticular fibers, fibronectin, and glycoproteins. The basement membrane provides physical support for epithelium, provides for cell attachment, serves as a filter in the kidneys, and guides cell migration during development and tissue repair.
g. Epithelial tissue has a nerve supply.
h. Epithelial tissue is the only tissue that makes direct contact with the external environment.
i. Since epithelium is subject to a certain amount of wear and tear and injury, it has a high capacity for renewal (high mitotic rate).
j. Epithelia are diverse in origin. They are derived from all three primary germ layers (ectoderm, mesoderm, and endoderm).
k. Functions of epithelia include protection, filtration, lubrication, secretion, digestion, absorption, transportation, excretion, sensory reception, and reproduction.

2. Covering and Lining Epithelium

Before you start your microscopic examination of epithelial tissues, refer to Figure 4.1. Study the tissues carefully to familiarize yourself with their general structural characteristics. For each of the types of epithelium listed, obtain a prepared slide and, unless otherwise specified by your instructor, examine each under high power. In conjunction with your examination, consult a textbook of anatomy.

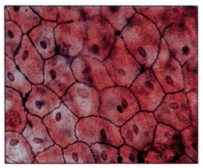

Surface view of mesothelial lining of
peritoneal cavity (240×)

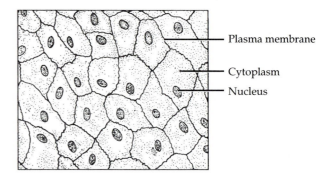

Plasma membrane

Cytoplasm

Nucleus

(a) Simple squamous epithelium

FIGURE 4.1 Epithelial tissues. Photomicrographs are unlabeled; the line drawings of the same tissues are labeled.

a. *Simple squamous* (SKWĀ-mus; *squama* = flat) ***epithelium*** This tissue consists of a single layer of flat cells and is highly adapted for diffusion, osmosis, and filtration because of its thinness. Simple squamous epithelium lines the air sacs (alveoli) of the lungs, glomerular (Bowman's) capsules (filtering units) of the kidneys, and inner surface of the tympanic membrane (eardrum) of the ear. Simple squamous epithelium that lines the heart, blood vessels, and lymphatic vessels and forms capillary walls is called ***endothelium*** (*endo* = within; *thelium* = covering). Simple squamous epithelium that forms the epithelial layer of a serous membrane is called ***mesothelium*** (*meso* = middle). Serous membranes line the thoracic and abdominopelvic cavities and cover viscera within the cavities. After you make your microscopic examination, draw several cells in the space that follows and label plasma membrane, cytoplasm, and nucleus.

Simple squamous epithelium

b. *Simple cuboidal epithelium* This tissue consists of a single layer of cube-shaped cells. When the tissue is sectioned at right angles to the surface, its cuboidal nature is obvious. Highly adapted for secretion and absorption, cuboidal tissue covers the surface of the ovaries; lines the smaller ducts of some glands and the anterior surface of the lens capsule of the eye; and forms the pigmented epithelium of the retina of the eye, part of the tubules of the kidneys, and the secreting units of other glands, such as the thyroid gland.

After you make your microscopic examination, draw several cells in the space that follows and label plasma membrane, cytoplasm, nucleus, basement membrane, and connective tissue layer.

Simple cuboidal epithelium

c. *Simple columnar (nonciliated) epithelium*
This tissue consists of a single layer of columnar cells and, when sectioned at right angles,

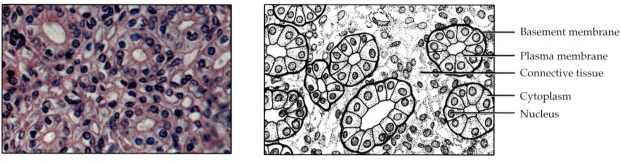

Sectional view of kidney tubules (400×)

(b) Simple cuboidal epithelium

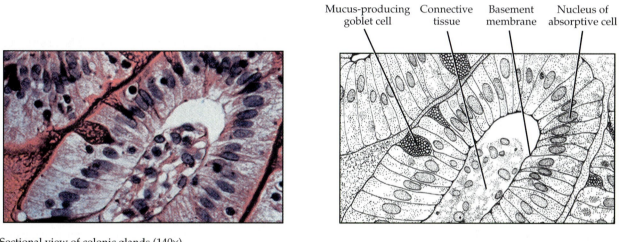

Sectional view of colonic glands (140×)

(c) Simple columnar (nonciliated) epithelium

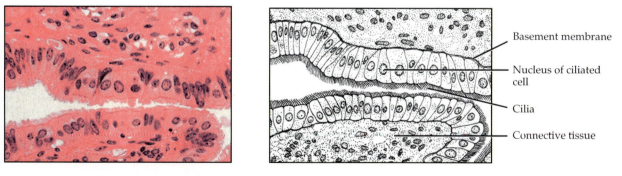

Sectional view of uterine (Fallopian) tube (100×)

(d) Simple columnar (ciliated) epithelium

Figure 4.1 *(Continued)* Epithelial tissues.

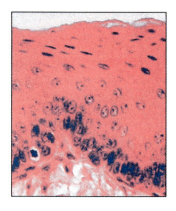
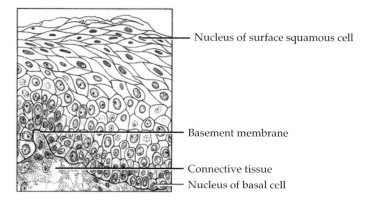

(e) Stratified squamous epithelium

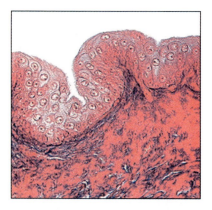

Sectional view of urinary bladder in
relaxed state (100×)

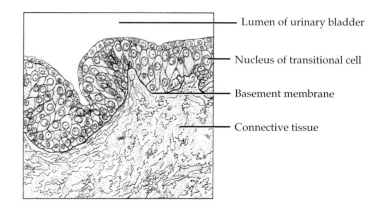

(f) Transitional epithelium

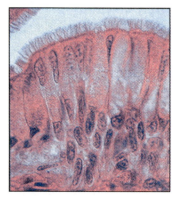

Sectional view of trachea (250×)

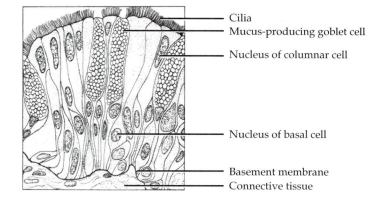

(g) Pseudostratified columnar epithelium

FIGURE 4.1 *(Continued)* Epithelial tissues.

these cells appear as rectangles. Adapted for secretion and absorption, this tissue lines the gastrointestinal tract from the stomach to the anus, gallbladder, and ducts of many glands. Some columnar cells are modified in that the plasma membranes are folded into microscopic fingerlike cytoplasmic projections called *microvilli* (*micro* = small; *villus* = tuft of hair) that increase the surface area for absorption. Other cells are *goblet cells*, modified columnar cells that secrete and store mucus to protect the lining of the gastrointestinal tract. After you make your microscopic examination, draw several cells in the space that follows and label plasma membrane, cytoplasm, nucleus, goblet cell, absorptive cell, basement membrane, and connective tissue layer.

Simple columnar (nonciliated) epithelium

d. Simple columnar (ciliated) epithelium This type of epithelium consists of a single layer of columnar absorptive, goblet, and ciliated cells. *Cilia* (*cilia* = eyelashes) are hairlike processes that move substances over the surfaces of cells. Simple columnar (ciliated) epithelium lines some portions of the upper respiratory tract, uterine (Fallopian) tubes, uterus, some paranasal sinuses, and the central canal of the spinal cord. Mucus produced by goblet cells forms a thin film over the surface of the tissue, and movements of the cilia propel the mucus and the trapped substances over the surface of the tissue. After you make your microscopic examination, draw several cells in the space that follows and label plasma membrane, cytoplasm, nucleus, cilia, goblet cell, basement membrane, and connective tissue layer.

Simple columnar (ciliated) epithelium

e. Stratified squamous epithelium This tissue consists of several layers of cells and affords considerable protection against friction. The superficial cells are flat whereas cells of the deep layers vary in shape from cuboidal to columnar. The basal (bottom) cells continually multiply by cell division. As surface cells are sloughed off, new cells replace them from the basal layer. The surface cells of *keratinized stratified squamous* contain a waterproofing protein called *keratin* (*kerato* = horny) that also resists friction and bacterial invasion. The keratinized variety forms the outer layer of the skin. Surface cells of *nonkeratinized stratified squamous* do not contain keratin. The nonkeratinized variety lines wet surfaces such as the mouth, esophagus, vagina, and part of the epiglottis, and it covers the tongue. After you make your microscopic examination, draw several cells in the space that follows and label plasma membrane, cytoplasm, nucleus, squamous surface cells, basal cells, basement membrane, and connective tissue layer.

Stratified squamous epithelium

**f. *Stratified squamous epithelium (student pre-
pared)*** Before examining the next slide, prepare
a smear of cheek cells from the epithelial lining
of the mouth. As noted previously, epithelium
that lines the mouth is nonkeratinized stratified
squamous epithelium. However, you will be
examining surface cells only, and these will
appear similar to simple squamous epithelium.

PROCEDURE

CAUTION! *Please reread Section B, "Precautions
Related to Working with Blood, Blood Products, or
Other Body Fluids" on page xii at the beginning of the
laboratory manual before you begin any of the following
experiments. You should also read the experiments
before you perform them to be sure that you understand
all the procedures and safety precautions. When you
finish this part of the exercise, place the reusable items
in a fresh bleach solution and the discarded items in a
biohazard container.*

    a. Using the blunt end of a toothpick, *gently*
scrape the lining of your cheek several times
to collect some surface cells of the stratified
squamous epithelium.

    b. Now move the toothpick across a clean glass
microscope slide until a thin layer of scrap-
ings is left on the slide.

    c. Allow the preparation to air dry.

    d. Next, cover the smear with several drops of
1% methylene blue stain. After about 1 min,
gently rinse the slide in cold tap water or
distilled water to remove excess stain.

    e. *Gently* blot the slide dry using a paper towel.

    f. Examine the slide under low and high
power. See if you can identify the plasma
membrane, cytoplasm, nuclear membrane,
and nucleoli. Some bacteria are commonly
found on the slide and usually appear as
very small rods or spheres.

g. *Transitional epithelium* This tissue resembles
nonkeratinized stratified squamous, except
that the superficial cells are larger and more
rounded. When stretched, the surface cells are
drawn out into squamouslike cells. This drawing out permits the tissue to stretch without the
outer cells breaking apart from one another.
The tissue lines parts of the urinary system that
are subject to expansion from within, such as
the urinary bladder, parts of the ureters, and
urethra. After you have made your microscopic
examination, draw several cells in the space
that follows and label plasma membrane, cytoplasm, nucleus, surface cells, basement membrane, and connective tissue layer.

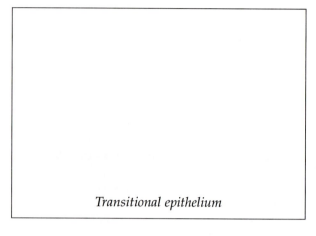

Transitional epithelium

h. *Pseudostratified columnar epithelium* Nuclei
of cells in this tissue are at varying depths,
and, although all the cells are attached to the
basement membrane in a single layer, some do
not reach the surface. This arrangement gives
the impression of a multilayered tissue when
sectioned, thus the name *pseudostratified*. This
tissue lines large ducts of many glands, the
epididymis, parts of the male urethra, and
parts of the auditory (Eustachian) tubes; and a
special variety that lines most of the upper respiratory tract is called pseudostratified ciliated
columnar epithelium. After you have made
your microscopic examination, draw and label
basement membrane, a cell that reaches the
surface, a cell that does not reach the surface,
and nuclei of each cell.

Pseudostratified columnar epithelium

3. Glandular Epithelium

A ***gland*** may consist of a single epithelial cell or a
group of highly specialized epithelial cells that
secrete various substances. Glands that have no
ducts (ductless), secrete hormones, and release their
secretions into the blood are called ***endocrine glands***.

Examples include the pituitary gland and thyroid gland (Exercise 15). Glands that secrete their products into ducts are called *exocrine glands*. Examples include sweat glands and salivary glands.

a. STRUCTURAL CLASSIFICATION OF EXOCRINE GLANDS

Based on the shape of the secretory portion and the degree of branching of the duct, exocrine glands can be structurally classified as follows:

1. *Unicellular* One-celled glands that secrete mucus. An example is the goblet cell (see Figure 4.1c). These cells line portions of the respiratory and digestive systems.
2. *Multicellular* Many-celled glands that occur in several different forms (Figure 4.2).
   Simple—Single, nonbranched duct.
   Tubular—Secretory portion is straight and tubular (intestinal glands).
   Branched tubular—Secretory portion is branched and tubular (gastric and uterine glands).

Coiled tubular—Secretory portion is coiled (sudoriferous [soo-dor-IF-er-us], or sweat, glands).
Acinar (AS-i-nar)—Secretory portion is flasklike (seminal vesicle glands).
Branched acinar—Secretory portion is branched and flasklike (sebaceous [se-BĀ-shus], or oil, glands).
Compound—Branched duct.
Tubular—Secretory portion is tubular (bulbourethral or Cowper's glands, testes, liver).
Acinar—Secretory portion is flasklike (sublingual and submandibular salivary glands).
Tubuloacinar—Secretory portion is both tubular and flasklike (parotid salivary glands, pancreas).

Obtain a prepared slide of a representative of each of the types of multicellular exocrine glands just described. As you examine each slide, compare your observations to the diagrams of the glands in Figure 4.2.

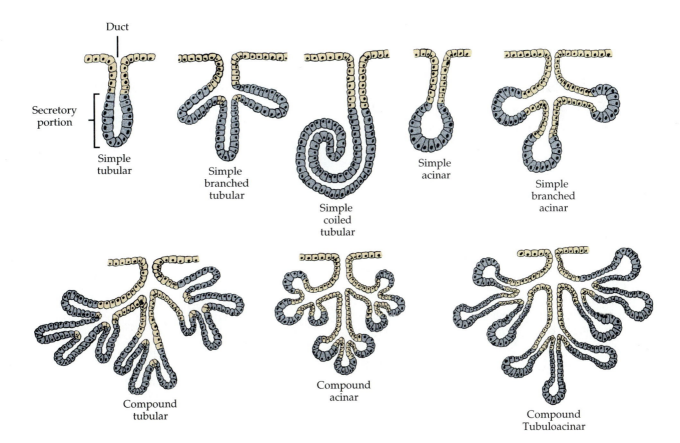

FIGURE 4.2 Structural classification of multicellular exocrine glands.

b. FUNCTIONAL CLASSIFICATION OF EXOCRINE GLANDS

Functional classification is based on whether a secretion is a product of a cell or consists of entire or partial glandular cells themselves. *Holocrine glands*, such as sebaceous (oil) glands, accumulate their secretory product in their cytoplasm. The cell then dies and is discharged with its contents as the glandular secretion; the discharged cell is replaced by a new one. *Apocrine glands*, such as the sudoriferous (sweat) glands in the axilla, accumulate secretory products at the outer margins of the secreting cells. The margins pinch off as the secretion and the remaining portions of the cells are repaired so that the process can be repeated. *Merocrine (eccrine) glands*, such as the pancreas and salivary glands, produce secretions that are simply formed by the secretory cells and then discharged into a duct.

B. CONNECTIVE TISSUE

Connective tissue, the most abundant tissue in the body, functions by protecting, supporting, and separating structures (e.g., skeletal muscles), and binding structures together.

1. Characteristics

Following are the general characteristics of connective tissue.

a. Connective tissue consists of three basic elements: cells, ground substance, and fibers. Together, the ground substance (fluid, gel, or solid) and fibers, both of which are outside the cells, form the *matrix*. Unlike epithelial cells, connective tissue cells rarely touch one another; they are separated by a considerable amount of matrix.

b. In contrast to epithelia, connective tissues do not usually occur on free surfaces, such as the surfaces of a body cavity or the external surface of the body.

c. Except for cartilage, connective tissue, like epithelium, has a nerve supply.

d. Unlike epithelium, connective tissue usually is highly vascular (has a rich blood supply). Exceptions include cartilage, which is avascular, and tendons, which have a scanty blood supply.

e. The matrix of a connective tissue, which may be fluid, semifluid, gelatinous, fibrous, or calcified, is usually secreted by the connective tissue cells and adjacent cells and determines the tissue's qualities. In blood the matrix, which is not secreted by blood cells, is fluid. In cartilage it is firm but pliable. In bone it is considerably harder and not pliable.

2. Connective Tissue Cells

Following are some of the cells contained in various types of connective tissue. The specific tissues to which they belong will be described shortly.

a. *Fibroblasts* (FĪ-brō-blasts; *fibro* = fiber) are large, flat, spindle-shaped cells with branching processes; they secrete the molecules that become the connective tissue fibers and ground substance of the matrix.

b. *Macrophages* (MAK-rō-fā-jez; *macro* = large; *phagein* = to eat), or *histiocytes*, develop from *monocytes*, a type of white blood cell. Macrophages have an irregular shape with short branching projections and are capable of engulfing bacteria and cellular debris by phagocytosis. Thus, they provide a vital defense for the body.

c. *Plasma cells* are small and either round or irregular in shape. They develop from a type of white blood cell called a *B lymphocyte (B cell)*. Plasma cells secrete specific antibodies and, accordingly, provide a defense mechanism through immunity.

d. *Mast cells* are abundant alongside blood vessels. They produce histamine, a chemical that dilates small blood vessels and increases their permeability during inflammation.

e. Other cells in connective tissue include *adipocytes (fat cells)* and *white blood cells (leukocytes)*.

3. Connective Tissue Ground Substance

The *ground substance* is amorphous, meaning that it has no specific shape. Connective tissue cells usually produce the ground substance and deposit it in the space between the cells.

Several examples of ground substance are as follows. *Hyaluronic* (hī-a-loo-RON-ik) acid is a viscous, slippery substance that binds cells together, lubricates joints, and helps maintain the shape of the eyeballs. It also appears to play a role in helping phagocytes migrate through connective tissue during development and wound repair. ***Chondroitin*** (kon-DROY-tin) ***sulfate*** is a jellylike substance that provides support and adhesiveness in cartilage, bone, the skin, and blood vessels. The skin, tendons, blood vessels, and heart valves contain ***dermatan sulfate,*** while bone, cartilage, and the cornea of the eye contain ***keratan sulfate. Adhesion proteins*** (fibronectin, laminin, collagen, and fibrinogen) interact with plasma membrane receptors to anchor cells in position.

The ground substance supports cells and binds them together and provides a medium through which substances are exchanged between the blood and cells. Until recently, the ground substance was thought to function mainly as an inert scaffolding to support tissues. Now it is clear that the ground substance is quite active in functions such as influencing development, migration, proliferation, shape, and even metabolic functions.

4. Connective Tissue Fibers

Fibers in the matrix are secreted by fibroblasts and provide strength and support for tissues. Three types of fibers are embedded in the matrix between the cells of connective tissue: collagen, elastic, and reticular fibers.

a. *Collagen* (*kolla* = glue) *fibers*, of which there are at least five different types, are very tough and resistant to a pulling force, yet allow some flexibility in the tissue because they are not taut. These fibers often occur in bundles made up of many minute fibrils lying parallel to one another. The bundle arrangement affords great strength. Chemically, collagen fibers consist of the protein *collagen*. This is the most abundant protein in your body, representing about 25% of the total protein. Collagen fibers are found in most types of connective tissues, especially bone, cartilage, tendons, and ligaments.

b. *Elastic fibers* are smaller than collagen fibers and freely branch and rejoin one another. They consist of a protein called *elastin*. Like collagen fibers, elastic fibers provide strength. In addition, they can be stretched 150% of their relaxed length without breaking. Elastic fibers are plentiful in the skin, blood vessels, and lungs.

c. *Reticular* (*rete* = net) *fibers* consisting of the protein collagen and a coating of glycoprotein,

provide support in the walls of blood vessels and form a network around fat cells, nerve fibers, and skeletal and smooth-muscle cells. They are much thinner than collagen fibers and form branching networks. Like collagen fibers, reticular fibers provide support and strength and also form the *stroma* (framework) of many soft-tissue organs, such as the spleen and lymph nodes. These fibers also help form the basement membrane.

5. Types

Before you start your microscopic examination of connective tissues, refer to Figure 4.3. Study the tissues carefully to familiarize yourself with their general structural characteristics. For each type of connective tissue listed, obtain a prepared slide and, unless otherwise specified by your instructor, examine each under high power.

Here, we will concentrate on various types of *mature connective tissues*, meaning connective tissues that are present in the newborn and that do not change afterward. The types of mature connective tissue are *loose connective tissue, dense connective tissue, cartilage, bone,* and *blood.*

a. LOOSE CONNECTIVE TISSUE
In this general type of connective tissue, the fibers are *loosely* woven and there are many cells.

1. *Areolar* (a-RĒ-ō-lar; *areola* = small space) *connective tissue* One of the most widely distributed connective tissues in the body. It contains at one time or another all cells normally found in connective tissue, including fibroblasts, macrophages, plasma cells, mast cells, adipocytes, and a few white blood cells. All three types of fibers—collagen, elastic, and reticular—are present and randomly arranged. The fluid, semifluid, or gelatinous ground substance contains hyaluronic acid, chondroitin sulfate, dermatan sulfate, and keratan sulfate. Areolar connective tissue is present in many mucous membranes, superficial region of the dermis of the skin, around blood vessels, nerves, and organs, and, together with adipose tissue, forms the *subcutaneous* (sub'-kyoo-TĀ-nē-us) *layer* or *superficial fascia* (FASH-ē-a). This layer is located between the skin and underlying tissues. After you make your microscopic examination, draw a small area of

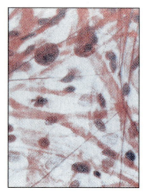

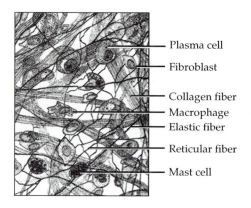

Plasma cell
Fibroblast
Collagen fiber
Macrophage
Elastic fiber
Reticular fiber
Mast cell

Surface view of subcutaneous
tissue (160×)

(a) Areolar connective tissue

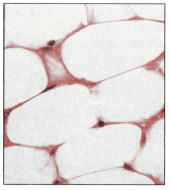

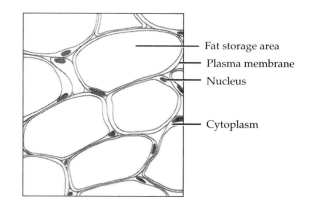

Fat storage area
Plasma membrane
Nucleus

Cytoplasm

Sectional view of white fat of
pancreas (1,600×)

(b) Adipose tissue

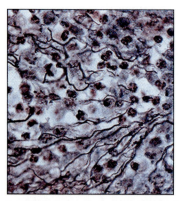

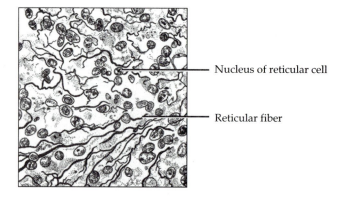

Nucleus of reticular cell

Reticular fiber

Sectional view of lymph node (250×)

(c) Reticular connective tissue

FIGURE 4.3   Connective tissues. Photomicrographs are unlabeled; the line drawings of the same tissues are labeled.

Sectional view of capsule of adrenal gland (250×)

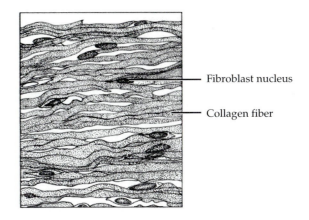

Fibroblast nucleus

Collagen fiber

(d) Dense regular connective tissue

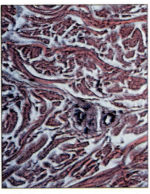

Sectional view of dermis of skin (275×)

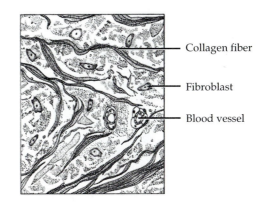

Collagen fiber

Fibroblast

Blood vessel

(e) Dense irregular connective tissue

Sectional view of ligamentum nuchae (400×)

Elastic fibers

Fibro-blast

(f) Elastic connective tissue

FIGURE 4.3 *(Continued)*   Connective tissues.

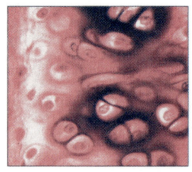

Sectional view of hyaline cartilage
from trachea (160×)

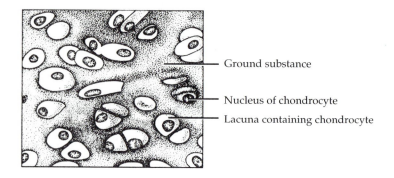

Ground substance

Nucleus of chondrocyte

Lacuna containing chondrocyte

(g) Hyaline cartilage

Sectional view of fibrocartilage from medial meniscus
of knee (315×)

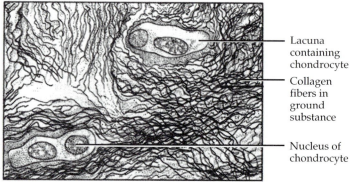

Lacuna
containing
chondrocyte

Collagen
fibers in
ground
substance

Nucleus of
chondrocyte

(h) Fibrocartilage

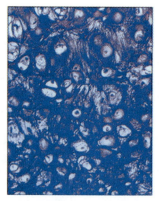

Sectional view of elastic cartilage
from auricle (pinna) of external ear
(175×)

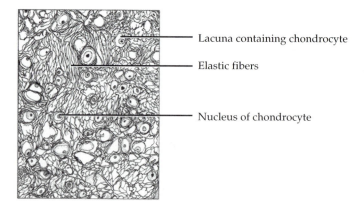

Lacuna containing chondrocyte

Elastic fibers

Nucleus of chondrocyte

(i) Elastic cartilage

FIGURE 4.3 (*Continued*)   Connective tissues.

the tissue in the space that follows and label the elastic fibers, collagen fibers, fibroblasts, and mast cells.

Areolar connective tissue

2. *Adipose tissue* This is fat tissue in which cells derived from fibroblasts, called *adipocytes* (*adeps* = fat), are modified for triglyceride (fat) storage. The cytoplasm and nuclei of the cells are pushed to the edge. The tissue is found wherever areolar connective tissue is located and around the kidneys and heart, in the yellow bone marrow of long bones, and behind the eyeball. It provides insulation, energy reserve, support, and protection. After your microscopic examination, draw several cells in the space that follows and label the fat storage area, cytoplasm, nucleus, and plasma membrane.

Adipose tissue

3. *Reticular connective tissue* This tissue consists of fine interlacing reticular fibers in which reticular cells are interspersed between the fibers. It provides the stroma (framework) of certain organs and helps bind certain cells together. It is found in the liver, spleen, and lymph nodes; in a portion of the basement membrane; and in red bone marrow. After you make your microscopic

examination, draw a sample of the tissue in the space that follows and label the reticular fibers and cells of the organ.

Reticular connective tissue

b. DENSE CONNECTIVE TISSUE
In this general type of connective tissue, the fibers are more numerous, *thicker,* and *densely* packed and there are fewer cells than loose connective tissue.

1. *Dense regular connective tissue* In this tissue, bundles of collagen fibers have an orderly, parallel arrangement that confers great strength. The tissue structure withstands pulling in one direction. Fibroblasts, which produce the fibers and ground substance, appear in rows between the fibers. The tissue is silvery white, tough, yet somewhat pliable. Because of its great strength, it is the principal component of *tendons*, which attach muscles to bones; *aponeuroses* (ap'-ō-noo-RŌ-sēz), which are sheetlike tendons connecting one muscle with another or with bone; and most *ligaments* (collagen ligaments), which hold bones together at joints. After you make your microscopic examination, draw a sample of the tissue in the space that follows and label the collagen fibers and fibroblasts.

Dense regular connective tissue

2. *Dense irregular connective tissue* This tissue contains collagen fibers that are irregularly arranged (without regular orientation) and is found in parts of the body where tensions are exerted in various directions. The tissue usually occurs in sheets. It forms some fasciae, the reticular (deeper) region of the dermis of the skin, the pericardium of the heart, the periosteum of bone, the perichondrium of cartilage, joint capsules, heart valves, and the membrane (fibrous) capsules around organs, such as the kidneys, liver, testes, and lymph nodes. After you make your microscopic examination, draw a sample of the tissue in the space that follows and label the collagen fibers.

Dense irregular connective tissue

3. *Elastic connective tissue* This tissue has a predominance of freely branching elastic fibers. These fibers give the unstained tissue a yellowish color. Fibroblasts are present in the spaces between fibers. Elastic connective tissue can be stretched and will snap back into shape (elasticity). It is a component of the walls of elastic arteries, the trachea, bronchial tubes of the lungs, and the lungs themselves. Elastic connective tissue provides stretch and strength, allowing structures to perform their functions efficiently. Yellow elastic ligaments, as contrasted with collagen ligaments, are composed mostly of elastic fibers; they form the ligamenta flava of the vertebrae (ligaments between successive vertebrae), the suspensory ligament of the penis, and the true vocal cords. After you make your microscopic examination, draw a sample of the tissue in the space that follows and label the elastic fibers and fibroblasts.

Elastic connective tissue

C. CARTILAGE

Cartilage is capable of enduring considerably more stress than the tissues just discussed. Unlike other connective tissues, cartilage has no blood vessels or nerves, except for those in the perichondrium (membranous covering). Cartilage consists of a dense network of collagen fibers and elastic fibers firmly embedded in chondroitin sulfate, a rubbery component of the ground substance. Whereas the strength of cartilage is due to its collagen fibers, its resilience (ability to assume its original shape after deformation) is due to chondroitin sulfate.

The cells of mature cartilage, called ***chondrocytes*** (KON-drō-sīts; *chondros* = cartilage), occur singly or in groups within spaces called ***lacunae*** (la-KOO-nē; *lacuna* = little lake) in the matrix. The surface of cartilage, except for fibrocartilage, is surrounded by dense irregular connective tissue called the ***perichondrium*** (per'-i-KON-drē-um; *peri* = around). Three kinds of cartilage are recognized: hyaline cartilage, fibrocartilage, and elastic cartilage.

1. *Hyaline cartilage* This cartilage, also called *gristle*, contains a resilient gel as its ground substance and appears in the body as a bluish-white, shiny substance. The fine collagen fibers, although present, are not visible with ordinary staining techniques, and the prominent chondrocytes are found in lacunae. Hyaline cartilage is the most abundant cartilage in the body. It is found at joints over the ends of the long bones (articular cartilage) and at the anterior ends of the ribs (costal cartilage). Hyaline cartilage also helps to support the nose, larynx, trachea, bronchi, and bronchial tubes leading to the lungs. Most of the embryonic skeleton consists of hyaline cartilage, which gradually becomes calcified and develops into bone. Hyaline cartilage affords flexibility and support and, at joints, reduces friction and absorbs shock. After you make your microscopic examination, draw a sample of the tissue in the space that follows and label the perichondrium, chondrocytes, lacunae, and ground substance.

Hyaline cartilage

2. *Fibrocartilage* Chondrocytes are scattered among clearly visible bundles of collagen fibers within the matrix of this type of cartilage. Fibrocartilage forms the pubic symphysis, the point where the hipbones fuse anteriorly at the midline. It is also found in the intervertebral discs between vertebrae, and the menisci (cartilage pads) of the knee. This tissue combines strength and rigidity. After you make your microscopic examination, draw a sample of the tissue in the space that follows and label the chondrocytes, lacunae, ground substance, and collagen fibers.

Fibrocartilage

3. *Elastic cartilage* In this tissue, chondrocytes are located in a threadlike network of elastic

fibers within the matrix. Elastic cartilage provides strength and elasticity and maintains the shape of organs—the epiglottis of the larynx, the external part of the ear (auricle), and the auditory (Eustachian) tubes. After you make your microscopic examination, draw a sample of the tissue in the space that follows and label the perichondrium, chondrocytes, lacunae, ground substance, and elastic fibers.

Elastic cartilage

C. MEMBRANES

The combination of an epithelial layer and an underlying layer of connective tissue constitutes an *epithelial membrane*. Examples are mucous, serous, and cutaneous membranes (skin). Another kind of membrane, a synovial membrane, has no epithelium. It contains only connective tissue. *Mucous membranes*, also called the *mucosa*, line body cavities that open directly to the exterior, such as the gastrointestinal, respiratory, urinary, and reproductive tracts. The surface tissue of a mucous membrane consists of epithelium and has a variety of functions, depending on location. Accordingly, the epithelial layer secretes mucus but may also secrete enzymes, filter dust, and have a protective and absorbent action. The underlying connective tissue layer of a mucous membrane, called the *lamina propria* (LAM-i-na PRŌ-prē-a), binds the epithelial layer in place, protects underlying tissues, provides the epithelium with nutrients and oxygen and removes wastes, and holds blood vessels in place.

Serous (*serous* = watery) *membranes*, also called the *serosa*, line body cavities that do not open to the exterior and cover organs that lie within the cavities. Serous membranes consist of a surface layer of mesothelium (simple squamous epithelium) and an underlying layer of areolar connective tissue. The mesothelium secretes a lubricating fluid. Serous membranes consist of two layers. The layer attached to the cavity wall is called the *parietal* (pa-RĪ-e-tal; *paries* = wall) *layer*; the layer that covers the organs in the cavity is called the *visceral* (*viscus* = body organ) *layer*. Examples of serous membranes are the pleurae, pericardium, and peritoneum.

The *cutaneous membrane,* or skin, is the principal component of the integumentary system, which will be considered in the next exercise.

Synovial (sin-Ō-vē-al) *membranes* line joint cavities. They do not contain epithelium but rather consist of areolar connective tissue, adipose tissue, and elastic fibers. Synovial membranes produce synovial fluid, which lubricates the ends of bones as they move at joints and nourishes the articular cartilage around the ends of bones.

ANSWER THE LABORATORY REPORT QUESTIONS AT THE END OF THE EXERCISE.

Tissues 4

Student _____   Date _____

Laboratory Section _____   Score/Grade _____

PART 1. Multiple Choice

_____ 1. In parts of the body such as the urinary bladder, where considerable distension (stretching) occurs, you can expect to find which epithelial tissue? (a) pseudostratified columnar (b) cuboidal (c) columnar (d) transitional

_____ 2. Stratified epithelium is usually found in areas of the body where the principal activity is (a) filtration (b) absorption (c) protection (d) diffusion

_____ 3. Ciliated epithelium destroyed by disease would cause malfunction in which system? (a) digestive (b) respiratory (c) skeletal (d) cardiovascular

_____ 4. The tissue that provides the skin with resistance to wear and tear and serves to waterproof it is (a) keratinized stratified squamous (b) pseudostratified columnar (c) transitional (d) simple columnar

_____ 5. The connective tissue cell that would most likely increase its activity during an infection is the (a) melanocyte (b) macrophage (c) adipocyte (d) fibroblast

_____ 6. Torn ligaments would involve damage to which tissue? (a) dense regular (b) reticular (c) elastic (d) areolar

_____ 7. Simple squamous tissue that lines the heart, blood vessels, and lymphatic vessels is called (a) transitional (b) adipose (c) endothelium (d) mesothelium

_____ 8. Microvilli and goblet cells are associated with which tissue? (a) hyaline cartilage (b) simple columnar nonciliated (c) transitional (d) stratified squamous

_____ 9. Superficial fascia contains which tissue? (a) elastic (b) reticular (c) fibrocartilage (d) areolar connective tissue

_____ 10. Which tissue forms articular cartilage and costal cartilage? (a) fibrocartilage (b) elastic cartilage (c) adipose (d) hyaline cartilage

_____ 11. Because the sublingual gland contains a branched duct and flasklike secretory portions, it is classified as (a) simple coiled tubular (b) compound acinar (c) simple acinar (d) compound tubular

_____ 12. Which glands discharge an entire dead cell and its contents as their secretory products? (a) merocrine (b) apocrine (c) endocrine (d) holocrine

_____ 13. Membranes that line cavities that open directly to the exterior are called (a) synovial (b) serous (c) mucous (d) cutaneous

_____ 14. Which statement about connective tissue is false? (a) Cells are always very closely packed together. (b) Connective tissue always has an abundant blood supply. (c) Matrix is always present in large amounts. (d) It is the most abundant tissue in the body.

_____ 15. A group of similar cells that has a similar embryological origin and operates together to perform a specialized activity is called a(n) (a) organ (b) tissue (c) system (d) organ system

_____ 16. Which statement best describes covering and lining epithelium? (a) It is always arranged in a single layer of cells. (b) It contains large amounts of intercellular substance. (c) It has an abundant blood supply. (d) Its free surface is exposed to the exterior of the body or to the interior of a hollow structure.

_____ 17. Which statement best describes connective tissue? (a) usually contains a large amount of matrix (b) always arranged in a single layer of cells (c) primarily concerned with secretion (d) usually lines a body cavity

_____ 18. A gland (a) is either exocrine or endocrine (b) may be single celled or multicellular (c) consists of epithclial tissue (d) is described by all of the preceding statements.

_____ 19. Which of the following statements is not correct? (a) Simple squamous epithelium lines blood vessels. (b) Endothelium is composed of cuboidal cells. (c) Ciliated squamous epithelium is found in the respiratory system. (d) Transitional epithelium is found in the urinary bladder.

PART 2. Completion

20. Single cells found in epithelium that secrete mucus are called_____cells.

21. A type of epithelium that appears to consist of several layers but actually contains only one layer of cells is_____.

22. The cell in connective tissue that forms new fibers is the_____.

23. Histamine, a substance that dilates small blood vessels during inflammation, is secreted by_____cells.

24. Cartilage cells found in lacunae are called _____.

25. The simple squamous epithelium of a serous membrane that covers viscera is called_____.

26. The tissue that provides insulation, support, protection, and serves as a food reserve is_____.

27. _____tissue forms the stroma of organs such as the liver and spleen.

28. The cartilage that provides support for the larynx and external ear is_____.

29. The ground substance that helps lubricate joints and binds cells together is_____.

30. Ductless glands that secrete hormones are called_____glands.

31. Multicellular exocrine glands that contain branching ducts are classified as_____glands.

32. The mammary glands are classified as_____glands because their secretory products are the pinched-off margins of cells.

33. _____membranes consist of parietal and visceral layers and line cavities that do not open to the exterior.

34. If the secretory portion of a gland is flasklike, it is classified as a(n)_____gland.

35. Membranes that line joint cavities are called_____membranes.

36. An example of a simple branched acinar gland is a(n)_____gland.

37. The structure that attaches epithelium to underlying connective tissue is called the

_____.

PART 3. Matching

_____ **38.** Lines the inner surface of the stomach and intestine

_____ **39.** Lines urinary tract, as in bladder, permitting distention

_____ **40.** Lines mouth; present on outer surface of skin

_____ **41.** Single layer of cube-shaped cells; found in kidney tubules and ducts of some glands

_____ **42.** Lines air sacs of lungs where thin cells are required for diffusion of gases into blood

_____ **43.** Not a true stratified tissue; all cells on basement membrane, but some do not reach surface

_____ **44.** Derived from lymphocyte, gives rise to antibodies and so is helpful in defense

_____ **45.** Phagocytic cell; engulfs bacteria and cleans up debris; important during infection

_____ **46.** Believed to form collagen and elastic fibers in injured tissue

_____ **47.** Abundant along walls of blood vessels; produces histamine, which dilates blood vessels

_____ **48.** Contains lacunae and chondrocytes

_____ **49.** Forms fasciae and dermis of skin

_____ **50.** Stores fat and provides insulation

A. Transitional epithelium

B. Fibroblast

C. Pseudostratified columnar epithelium

D. Dense irregular connective tissue

E. Simple columnar epithelium

F. Macrophage

G. Stratified squamous epithelium

H. Adipose

 I. Simple cuboidal epithelium

J. Plasma cell

K. Simple squamous epithelium

L. Mast cell

M. Cartilage

Integumentary System

5

The skin and the organs derived from it (hair, nails, and glands), and several specialized receptors constitute the *integumentary* (in-teg-yoo-MEN-tar-ē; *integumentum* = covering) *system*, which you will study in this exercise. An *organ* is an aggregation of tissues of definite form and usually recognizable shape that performs a specific function; a *system* is a group of organs that operate together to perform specialized functions.

A. SKIN

The *skin* is one of the largest organs of the body in terms of surface area, occupying a surface area of about 2 square meters (2 m^2) (22 ft^2). Among the functions performed by the skin are regulation of body temperature; protection of underlying tissues from physical abrasion, microorganisms, dehydration, and ultraviolet (UV) radiation; excretion of water and salts and several organic compounds; synthesis of vitamin D in the presence of sunlight; reception of stimuli for touch, pressure, pain, and temperature change sensations; serving as a blood reservoir; and immunity.

The skin consists of an outer, thinner *epidermis* (*epi* = above), which is avascular, and an inner, thicker *dermis* (*derm* = skin), which is vascular. Below the dermis is the *subcutaneous layer* (*superficial fascia* or *hypodermis*) that attaches the skin to underlying tissues and organs.

1. Epidermis

The epidermis consists of four principal kinds of cells. *Keratinocytes* (ker-a-TIN-ō-sīts; *kerato* = horny) are the most numerous cells and they undergo keratinization, that is, newly formed cells produced in the basal layers are pushed up to the surface and in the process synthesize a waterproofing chemical, keratin. *Melanocytes* (MEL-a-nō-sīts; *melan* = black) are pigment cells that impart color to the skin. The third type of cell in the epidermis is called a *Langerhans* (LANG-er-

hans) *cell*. These cells are a small population of cells that arise from red bone marrow and migrate to the epidermis and other stratified squamous epithelial tissue in the body. They are sensitive to UV radiation and lie above the basal layer of keratinocytes. Langerhans cells interact with white blood cells called *helper T cells* to assist in the immune response. *Merkel cells* are found in the bottom layer of the epidermis. Their bases are in contact with flattened portions of the terminations of sensory nerves (*Merkel discs*) and function as receptors for touch. At this point, we will concentrate only on keratinocytes.

Obtain a prepared slide of human skin and carefully examine the epidermis. Identify the following layers from the outside inward:

a. *Stratum corneum* (*corneum* = horny) 25 to 30 rows of flat, dead cells that are filled with *keratin*, a waterproofing protein; these cells are continuously shed and replaced by cells from deeper strata.
b. *Stratum lucidum* (*lucidus* = clear) Several rows of clear, flat cells that contain *eleidin* (el-Ē-i-din), a precursor of keratin; more apparent in the palms and soles.
c. *Stratum granulosum* (*granulum* = little grain) 3 to 5 rows of flat cells that contain *keratohyalin* (ker'-a-tō-HĪ-a-lin), a precursor of eleidin.
d. *Stratum spinosum* (*spinosum* = thornlike) 8 to 10 rows of polyhedral (many-sided) cells.
e. *Stratum basale* (*basale* = base) Single layer of cuboidal to columnar cells that constantly undergo division. In hairless skin, this layer contains tactile (Merkel) discs, receptors sensitive to touch. Also called the *stratum germinativum.*

Label the epidermal layers in Figures 5.1 and 5.2.

2. Dermis

The dermis is divided into two regions and is composed of connective tissue containing collagen and elastic fibers and a number of other struc-

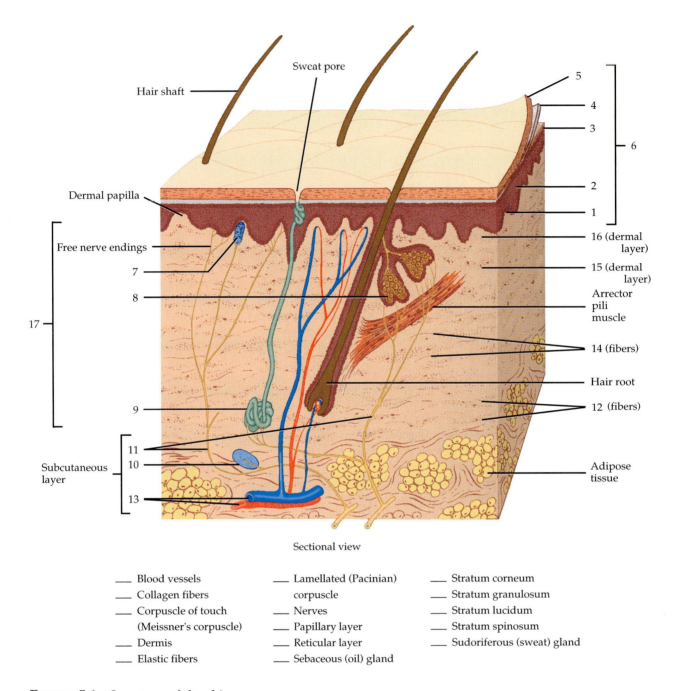

FIGURE 5.1 Structure of the skin.

Labels in figure:

Hair shaft
Sweat pore
Dermal papilla
Free nerve endings
7
8
17
9
11
10
Subcutaneous layer
13

5
4
3
6
2
1
16 (dermal layer)
15 (dermal layer)
Arrector pili muscle
14 (fibers)
Hair root
12 (fibers)
Adipose tissue

Sectional view

___ Blood vessels
___ Collagen fibers
___ Corpuscle of touch (Meissner's corpuscle)
___ Dermis
___ Elastic fibers

___ Lamellated (Pacinian) corpuscle
___ Nerves
___ Papillary layer
___ Reticular layer
___ Sebaceous (oil) gland

___ Stratum corneum
___ Stratum granulosum
___ Stratum lucidum
___ Stratum spinosum
___ Sudoriferous (sweat) gland

tures. The superficial region of the dermis *(papillary layer)* is areolar connective tissue containing fine elastic fibers. This layer contains fingerlike projections, the *dermal papillae* (pa-PIL-ē; *papilla* = nipple). Some papillae enclose blood capillaries; others contain *corpuscles of touch (Meissner's corpuscles)*, nerve endings sensitive to touch. The deeper region of the dermis *(reticular layer)* consists of dense, irregular connective tissue with interlacing bundles of larger collagen and some coarse elastic fibers. Spaces between the fibers may be occupied by *hair follicles, sebaceous (oil) glands, bundles of smooth muscle (arrector pili muscle), sudoriferous (sweat) glands, blood vessels,* and *nerves.*

The reticular layer of the dermis is attached to the underlying structures (bones and muscles) by the subcutaneous layer. This layer also contains nerve endings sensitive to pressure called *lamellated* or *Pacinian* (pa-SIN-ē-an) *corpuscles.*

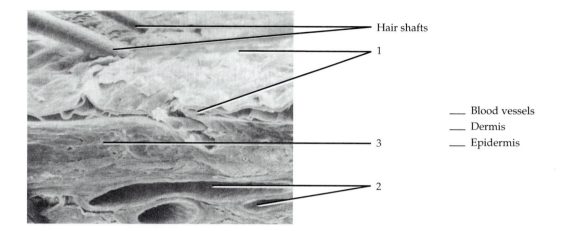

Hair shafts
1
Blood vessels
Dermis
Epidermis
3
2

FIGURE 5.2   Scanning electron micrograph of the skin and several hairs at a magnification of 260×. (Reproduced by permission from R. G. Kessel and R. H. Kardon, *Tissues and Organs: A Text Atlas of Scanning Electron Microscopy*, W. H. Freeman and 1979.)

Carefully examine the dermis and subcutaneous layer on your microscope slide. Label the following structures in Figure 5.1: papillary layer, reticular layer, corpuscle of touch, blood vessels, nerves, elastic fibers, collagen fibers, sebaceous gland, sudoriferous gland, and lamellated corpuscle. Also label the dermis and blood vessels in Figure 5.2.

If a model of the skin and subcutaneous layer is available, examine it to see the three-dimensional relationship of the structures to one another.

3. Skin Color

The color of skin results from (1) *hemoglobin in red blood cells in capillaries* of the dermis (beneath the epidermis); (2) *carotene* (KAR-o-tēn; *keraton* = carrot), a yellow-orange pigment in the stratum corneum of the epidermis and fatty areas of the dermis and subcutaneous layer; and (3) *melanin* (MEL-a-nin), a pale yellow to black pigment found primarily in the melanocytes in the stratum basale and spinosum of the epidermis. Whereas hemoglobin in red blood cells in capillaries imparts a pink color to Caucasian skin, carotene imparts a yellowish color to skin. Because the number of *melanocytes* (MEL-a-nō-sīts), or melanin-producing cells, is about the same in all races, most differences in skin color are due to the amount of melanin that the melanocytes synthesize and disperse. Exposure to ultraviolet (UV) radiation increases melanin synthesis, resulting in darkening (tanning) of the skin to protect the body against further UV radiation.

An inherited inability of a person of any race to produce melanin results in *albinism* (AL-bi-nizm).

The pigment is absent from the hair and eyes as well as from the skin, and the individual is referred to as an *albino*. In some people, melanin tends to form in patches called *freckles*. Others inherit patches of skin that lack pigment, a condition called *vitiligo* (vit-i-LĪ-gō).

B. HAIR

Hairs (pili) develop from the epidermis and are variously distributed over the body. Each hair is composed of columns of dead, keratinized cells and consists of a *shaft*, most of which is visible above the surface of the skin, and a *root*, the portion below the surface that penetrates deep into the dermis and even into the subcutaneous layer.

The shaft of a coarse hair consists of the following parts:

1. *Medulla* Inner region composed of several rows of polyhedral cells containing pigment and air spaces. Poorly developed or not present in fine hairs.
2. *Cortex* Middle layer; contains several rows of dark cells surrounding the medulla; contains pigment in dark hair and mostly air in white hair.
3. *Cuticle of the hair* Outermost layer; consists of a single layer of flat, keratinized cells arranged like shingles on a house.

The root of a hair also contains a medulla, cortex, and cuticle of the hair along with the following associated parts:

1. *Hair follicle* Structure surrounding the root that consists of an external root sheath and an internal

root sheath. These epidermally derived layers are surrounded by a dermal layer of connective tissue.

2. *External root sheath* Downward continuation of the epidermis.

3. *Internal root sheath* Cellular tubular sheath that separates the hair from the external root sheath; consists of (a) the *cuticle of the internal root sheath*, an inner single layer of flattened cells with atrophied nuclei, (b) *granular (Huxley's) layer*, a middle layer of one to three rows of cells with flattened nuclei, and (c) *pallid (Henle's) layer*, an outer single layer of cuboidal cells with flattened nuclei.

4. *Bulb* Enlarged, onion-shaped structure at the base of the hair follicle.

5. *Papilla of the hair* Dermal indentation into the bulb; contains areolar connective tissue and blood vessels to nourish the hair.

6. *Matrix* Region of cells at the base of the bulb derived from the stratum basale and that divides to produce new hair.

7. *Arrector* (*arrector* = to raise) *pili muscle* Bundle of smooth muscle extending from the superficial dermis of the skin to the side of the hair follicle; its contraction, under the influence of fright or cold, causes the hair to move into a vertical position, producing "goose bumps."

8. *Hair root plexus* Nerve endings around each hair follicle that are sensitive to touch and respond when the hair shaft is moved.

Obtain a prepared slide of a cross section and a longitudinal section of a hair root and identify as many parts as you can. Using your textbook as a reference, label Figure 5.3. Also label the parts of a hair shown in Figure 5.4.

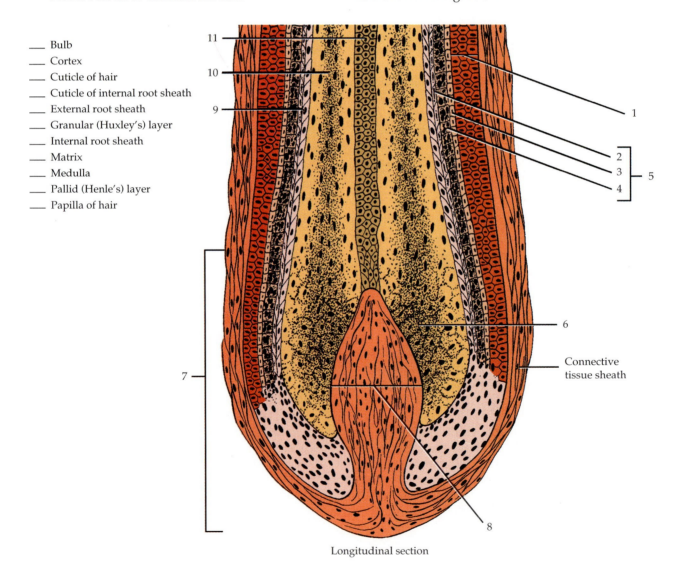

____ Bulb
____ Cortex
____ Cuticle of hair
____ Cuticle of internal root sheath
____ External root sheath
____ Granular (Huxley's) layer
____ Internal root sheath
____ Matrix
____ Medulla
____ Pallid (Henle's) layer
____ Papilla of hair

Connective tissue sheath

Longitudinal section

FIGURE 5.3 Hair root.

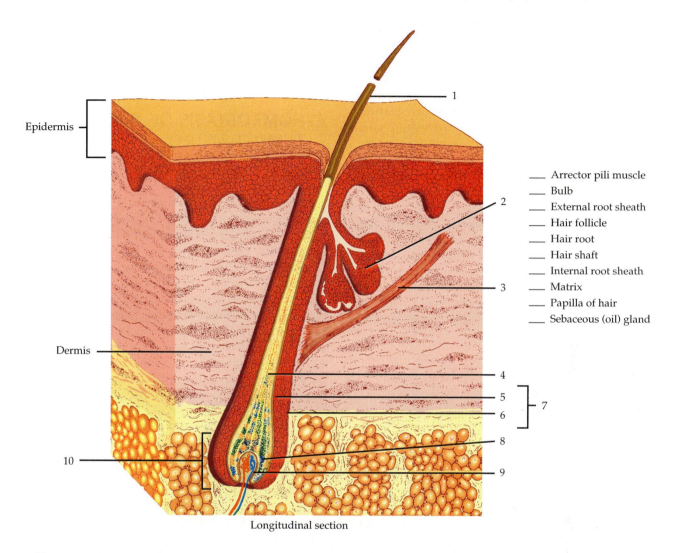

Epidermis

Dermis

___ Arrector pili muscle
___ Bulb
___ External root sheath
___ Hair follicle
___ Hair root
___ Hair shaft
___ Internal root sheath
___ Matrix
___ Papilla of hair
___ Sebaceous (oil) gland

Longitudinal section

FIGURE 5.4 Parts of a hair and associated structures.

C. GLANDS

Sebaceous (se-BĀ-shus; *sebo* = grease) or *oil glands*, with few exceptions, are connected to hair follicles (see Figures 5.1 and 5.4). They are single, branched acinar glands and secrete an oily substance called *sebum* (SĒ-bum), a mixture of fats, cholesterol, proteins, and salts. Sebaceous glands are absent in the skin of the palms and soles but are numerous in the skin of the face, neck, upper chest, and breasts. Sebum helps prevent hair from drying and forms a protective film over the skin that prevents excessive evaporation and keeps the skin soft and pliable.

   Sudoriferous (soo'-dor-IF-er-us; *sudor* = sweat; *ferre* = to bear) or *sweat glands* are separated into two principal types on the basis of structure, location, and secretion. *Apocrine sweat glands* are simple, branched tubular glands found primarily in the skin of the axilla (armpit), pubic region, and areolae (pigmented areas) of the breasts. Their secretory portion is located in the dermis or subcutaneous layer; the excretory duct opens into hair follicles. Apocrine sweat glands begin to function at puberty and produce a more viscous secretion than the other type of sweat gland. *Eccrine sweat glands* are simple, coiled tubular glands found throughout the skin, except for the margins of the lips, nail beds of the fingers and toes, glans penis, glans clitoris, and eardrums. They are most numerous in the skin of the palms and soles. The secretory portion of these glands is in the subcutaneous layer; the excretory duct projects upward and terminates at a pore at the surface of the epidermis (see Figure 5.1). Eccrine sweat glands function throughout life and produce a more watery secretion than the apocrine glands. Sudoriferous glands produce *perspiration*, a mixture of water, salt, urea, uric acid, amino acids, ammonia, sugar, lactic acid, and ascorbic acid. The evaporation of perspiration helps to maintain normal body temperature.

Ceruminous (se-ROO-mi-nus; *cera* = wax) *glands* are modified sudoriferous glands in the external auditory meatus. They are simple, coiled tubular glands. The combined secretion of ceruminous and sudoriferous glands is called *cerumen* (earwax). Cerumen, together with hairs in the external auditory meatus, provides a sticky barrier that prevents foreign bodies from reaching the eardrum.

D. NAILS

Nails are plates of tightly packed, hard, keratinized epidermal cells that form a clear covering over the dorsal surfaces of the terminal portions of the fingers and toes. Each nail consists of the following parts:

1. *Nail body* Portion that is visible.
2. *Free edge* Part that may project beyond the distal end of the digit.
3. *Nail root* Portion hidden in nail groove (see item 7).
4. *Lunula* (LOO-nyoo-la; *lunula* = little moon) Whitish semilunar area at proximal end of body.
5. *Nail fold* Fold of skin that extends around the proximal end and lateral borders of the nail.
6. *Nail bed* Strata basale and spinosum of the epidermis beneath the nail.
7. *Nail groove* Furrow between the nail fold and nail bed.
8. *Eponychium* (ep'-ō-NIK-ē-um) Cuticle; a narrow band of epidermis.
9. *Hyponychium* Thickened area of stratum corneum below the free edge of the nail.
10. *Nail matrix* Epithelium of the proximal part of the nail bed; division of the cells brings about growth of nails.

Using your textbook as a reference, label the parts of a nail shown in Figure 5.5. Also, identify the parts that are visible on your own nails.

E. HOMEOSTASIS OF BODY TEMPERATURE

One of the best examples of homeostasis in humans is the regulation of body temperature by the skin. As warm-blooded animals, we are able to maintain a remarkably constant body temperature of 37°C (98.6°F) even though the environmental temperature varies greatly.

Suppose you are in an environment where the temperature is 38°C (101°F). Heat (the stimulus) continually flows from the environment to your body, raising body temperature. To counteract these changes in a controlled condition, a sequence of events is set into operation. Temperature-sensitive receptors (nerve endings) in the skin called *thermoreceptors* detect the stimulus and send nerve impulses (input) to your brain (control center). A temperature control region of the brain (called the hypothalamus) then sends nerve impulses (output) to the sudoriferous glands (effectors), which produce perspiration more rapidly. As the sweat evaporates from the surface of your skin, heat is lost and your body temperature decreases (response). This cycle continues until body temperature drops to normal (returns to homeostasis). When environmental temperature is low, sweat glands produce less perspiration.

Your brain also sends output to blood vessels (a second set of effectors), dilating (widening) those in the dermis so that skin blood flow increases. As more warm blood flows through capillaries close to the body surface, more heat can be lost to the environment, which lowers body temperature. Thus heat is lost from the body, and body temperature falls to the normal value to restore homeostasis. In response to low environmental temperature, blood vessels in the dermis constrict, blood flow decreases, and heat is lost by radiation.

This temperature regulation involves a *negative feedback system* because the response (cooling) is opposite to the stimulus (heating) that started the cycle. Also, the thermoreceptors continually monitor body temperature and feed this information back to the brain. The brain, in turn, continues to send impulses to the sweat glands and blood vessels until the temperature returns to 37°C (98.6°F).

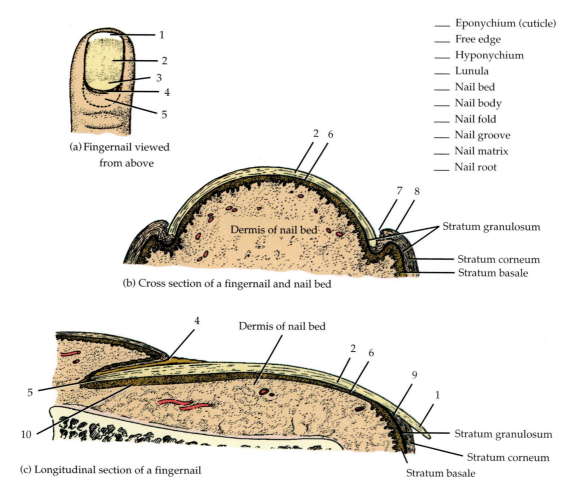

(a) Fingernail viewed from above

____ Eponychium (cuticle)
____ Free edge
____ Hyponychium
____ Lunula
____ Nail bed
____ Nail body
____ Nail fold
____ Nail groove
____ Nail matrix
____ Nail root

Dermis of nail bed

Stratum granulosum
Stratum corneum
Stratum basale

(b) Cross section of a fingernail and nail bed

Dermis of nail bed

Stratum granulosum
Stratum corneum
Stratum basale

(c) Longitudinal section of a fingernail

FIGURE 5.5 Structure of nails.

Regulating the rate of sweating and changing dermal blood flow are only two mechanisms by which body temperature can be adjusted. Other mechanisms include regulating metabolic rate (a slower metabolic rate reduces heat production) and regulating skeletal muscle contractions (decreased muscle tone results in less heat production).

Label Figure 5.6, the role of the skin in regulating the homeostasis of body temperature.

ANSWER THE LABORATORY REPORT QUESTIONS AT THE END OF THE EXERCISE.

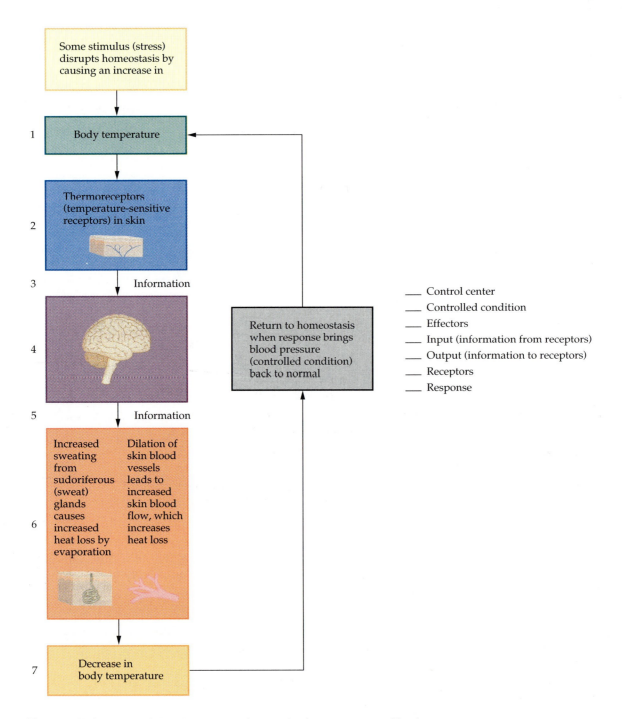

Some stimulus (stress) disrupts homeostasis by causing an increase in

1   Body temperature

2   Thermoreceptors (temperature-sensitive receptors) in skin

3   Information

4

5   Information

6   Increased sweating from sudoriferous (sweat) glands causes increased heat loss by evaporation   Dilation of skin blood vessels leads to increased skin blood flow, which increases heat loss

7   Decrease in body temperature

Return to homeostasis when response brings blood pressure (controlled condition) back to normal

___ Control center
___ Controlled condition
___ Effectors
___ Input (information from receptors)
___ Output (information to receptors)
___ Receptors
___ Response

FIGURE 5.6   Role of the skin in regulating the homeostasis of body temperature.

Integumentary System 5

Student _____   Date _____

Laboratory Section _____   Score/Grade _____

PART 1. Multiple Choice

_____ **1.** The waterproofing quality of skin is due to the presence of (a) melanin (b) carotene (c) keratin (d) receptors

_____ **2.** Sebaceous glands (a) produce a watery solution called sweat (b) produce an oily substance that prevents excessive water evaporation from the skin (c) are associated with mucous membranes (d) are part of the subcutaneous layer

_____ **3.** Which of the following is the proper sequence of layering of the epidermis, going from the free surface toward the underlying tissues? (a) basale, spinosum, granulosum, corneum (b) spinosum, basale, granulosum, corneum (c) corneum, lucidum, granulosum, spinosum, basale (d) corneum, granulosum, lucidum, spinosum

_____ **4.** Skin color is *not* determined by the presence or absence of (a) melanin (b) carotene (c) keratin (d) hemoglobin in red blood cells in capillaries in the dermis

_____ **5.** Destruction of what part of a single hair would result in its inability to grow? (a) sebaceous gland (b) arrector pili muscle (c) matrix (d) bulb

_____ **6.** One would expect to find relatively few, if any, sebaceous glands in the skin of the (a) palms (b) face (c) neck (d) upper chest

_____ **7.** Which of the following sequences, from outside to inside, is correct? (a) epidermis, reticular layer, papillary layer, subcutaneous layer (b) epidermis, subcutaneous layer, reticular layer, papillary layer (c) epidermis, reticular layer, subcutaneous layer, papillary layer (d) epidermis, papillary layer, reticular layer, subcutaneous layer

_____ **8.** The attached visible portion of a nail is called the (a) nail bed (b) nail root (c) nail fold (d) nail body

_____ **9.** Nerve endings sensitive to touch are called (a) corpuscles of touch (Meissner's corpuscles) (b) papillae (c) lamellated (Pacinian) corpuscles (d) follicles

_____ **10.** The cuticle of a nail is referred to as the (a) matrix (b) eponychium (c) hyponychium (d) fold

_____ **11.** One would *not* expect to find sudoriferous glands associated with the (a) forehead (b) axilla (c) palms (d) nail beds

_____ **12.** Fingerlike projections of the dermis that contain loops of capillaries and receptors are called (a) dermal papillae (b) nodules (c) polyps (d) pili

_____ **13.** Which of the following statements about the function of skin is *not* true? (a) it helps control body temperature (b) it prevents excessive water loss (c) it synthesizes several compounds (d) it absorbs water and salts

_____ **14.** Which is *not* part of the internal root sheath? (a) granular (Huxley's) layer (b) cortex (c) pallid (Henle's) layer (d) cuticle of the internal root sheath

_____ **15.** Growth in the length of nails is the result of the activity of the (a) eponychium (b) nail matrix (c) hyponychium (d) nail fold

PART 2. Completion

16. A group of tissues that performs a definite function is called a(n) _____.

17. The outer, thinner layer of the skin is known as the _____.

18. The skin is attached to underlying tissues and organs by the _____.

19. A group of organs that operate together to perform a specialized function is called a(n)

_____.

20. The epidermal layer that contains eleidin is the stratum _____.

21. The epidermal layers that produce new cells are the stratum spinosum and stratum

_____.

22. The smooth muscle attached to a hair follicle is called the _____ muscle.

23. An inherited inability to produce melanin is called _____.

24. Nerve endings sensitive to deep pressure are referred to as _____ corpuscles.

25. The inner region of a hair shaft and root is the _____.

26. The portion of a hair containing areolar connective tissue and blood vessels is the

_____.

27. Modified sweat glands that line the external auditory meatus are called _____ glands.

28. The whitish semilunar area at the proximal end of the nail body is referred to as the

_____.

29. The secretory product of sudoriferous glands is called _____.

30. Melanin is synthesized in cells called _____.

31. In the control of body temperature, the effectors are blood vessels in the dermis and

_____ glands.

Bone Tissue

6

Structurally, the *skeletal system* consists of two types of connective tissue: cartilage and bone. The microscopic structure of cartilage has been discussed in Exercise 4. In this exercise the gross structure of a typical bone and the histology of *bone (osseous) tissue* will be studied. *Osteology* (os-tē-OL-ō-jē; *osteo*=bone; *logos*=study of) is the study of bone structure and the treatment of bone disorders.

A. FUNCTIONS OF BONE

The skeletal system has the following basic functions:

1. *Support* It provides a supporting framework for the soft tissues, maintaining the body's shape and posture and provides points of attachment for many skeletal muscles.
2. *Protection* It protects delicate structures such as the brain, spinal cord, heart, lungs, major blood vessels in the chest, and pelvic viscera.
3. *Assistance in movement* When skeletal muscles contract, they pull on bones to help produce body movements.
4. *Mineral homeostasis* Bone tissue stores several minerals, especially calcium and phosphorus, which are important in muscle contraction and nerve activity, among other functions. On demand, bone releases minerals into the blood to maintain critical mineral balances and for distribution to other parts of the body.
5. *Blood cell production* or *hemopoiesis* (hē'-mō-poy-Ē-sis) *Red bone marrow* in certain parts of bones consists of primitive blood cells in immature stages, fat cells, and macrophages. It is responsible for producing red blood cells, white blood cells, and platelets.
6. *Storage of energy* Lipids stored in cells of yellow bone marrow are an important source of a chemical energy reserve. *Yellow bone marrow* consists of mostly adipose cells and a few scattered blood cells.

Depending on the size and distribution of spaces between its hard components, bone tissue may be categorized as spongy (cancellous) or compact (dense). *Spongy (cancellous) bone tissue,* which contains many large spaces that store mainly red bone marrow, composes most of the bone tissue of short, flat, and irregularly shaped bones and most of the epiphyses of long bones. The spaces within the spongy bone tissue of some bones contain red bone marrow. *Compact (dense) bone tissue,* which contains few spaces, forms the external layer of all bones of the body and the bulk of the diaphyses of long bones.

Label the spongy and compact bone tissue in Figure 6.1.

Spongy bone tissue is composed of concentric layers of hardened matrix, called *lamellae* (la-MEL-ē), that are arranged in an irregular latticework of thin plates of bone called *trabeculae* (tra-BEK-yoo-lē). See Figure 6.2a.

Compact bone tissue is composed of microscopic units called *osteons* (*Haversian systems*).

Obtain a prepared slide of compact bone tissue in which several osteons (Haversian systems) are shown in cross section. Observe under high power. Look for the following structures:

1. *Central (Haversian) canal* Circular canal in the center of an osteon (Haversian system) that runs longitudinally through the bone; the canal contains blood vessels, lymphatic vessels, and nerves.
2. *Concentric lamellae* Concentric layers of calcified matrix.
3. *Lacunae* (la-KOO-nē; *lacuna* = little lake) Spaces or cavities between lamellae that contain osteocytes.
4. *Canaliculi* (kan'-a-LIK-yoo-lē; *canaliculi* = small canal) Minute canals that radiate in all directions from the lacunae and interconnect with each other; contain slender processes of osteocytes; canaliculi provide routes so that nutrients can reach osteocytes and wastes can be removed from them.

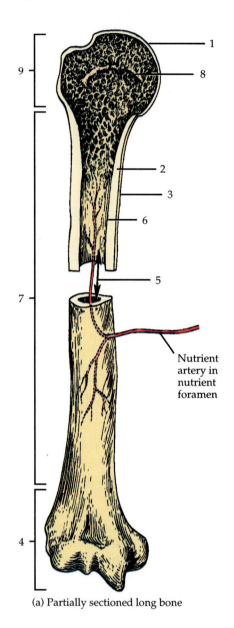

Nutrient
artery in
nutrient
foramen

_____ Articular cartilage
_____ Compact bone tissue
_____ Diaphysis
_____ Distal epiphysis
_____ Endosteum
_____ Medullary (marrow) cavity
_____ Periosteum
_____ Proximal epiphysis
_____ Spongy bone tissue

(a) Partially sectioned long bone

FIGURE 6.1   Parts of a long bone.

5. *Osteocyte* Mature bone cell located within a lacuna.
6. *Osteon (Haversian system)* Microscopic structural unit of compact bone made up of central (Haversian) canal plus its surrounding lamellae, lacunae, canaliculi, and osteocytes.

Label the parts of the osteon (Haversian system) shown in Figure 6.2.

Now obtain a prepared slide of a longitudinal section of compact bone tissue and examine under high power. Locate the following:

1. *Perforating (Volkmann's) canals* Canals that extend obliquely or horizontally inward from the periosteum and contain blood vessels, lymphatic vessels, and nerves; extend into central (Haversian) canals and medullary cavity.
2. *Endosteum*
3. *Medullary (marrow) cavity*
4. *Concentric lamellae*
5. *Lacunae*
6. *Canaliculi*
7. *Osteocytes*

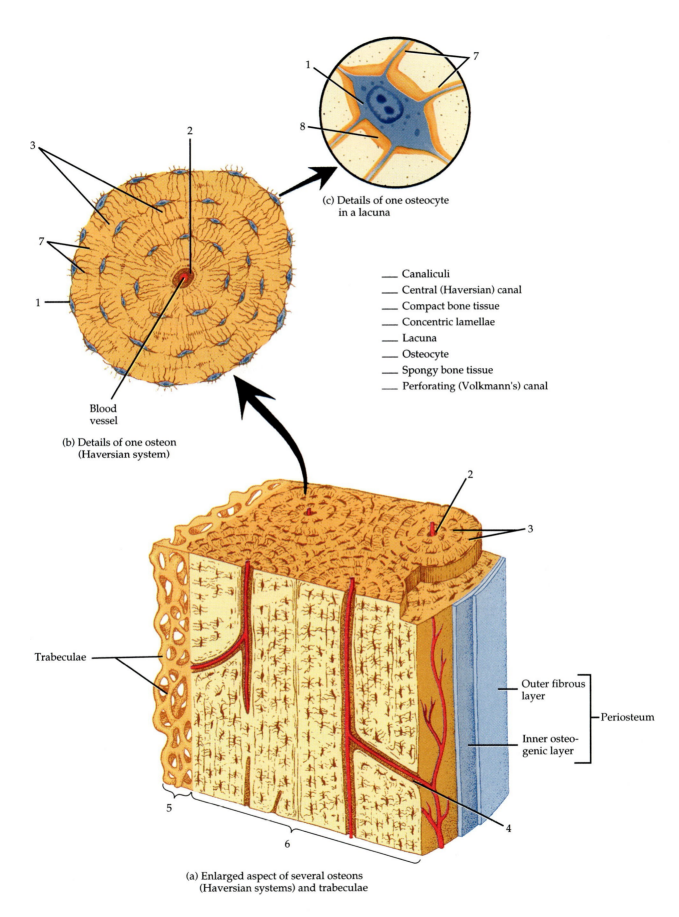

7

1

7

8

(c) Details of one osteocyte
in a lacuna

3

2

7

1

Blood
vessel

(b) Details of one osteon
(Haversian system)

___ Canaliculi
___ Central (Haversian) canal
___ Compact bone tissue
___ Concentric lamellae
___ Lacuna
___ Osteocyte
___ Spongy bone tissue
___ Perforating (Volkmann's) canal

2

3

Trabeculae

Outer fibrous
layer

Inner osteo-
genic layer

Periosteum

5

6

4

(a) Enlarged aspect of several osteons
(Haversian systems) and trabeculae

FIGURE 6.2   Histology of bone.

Label the parts indicated in the microscopic view of bone in Figure 6.2.

B. CHEMISTRY OF BONE

Unlike other connective tissues, the matrix of bone is very hard. This hardness results from the presence of mineral salts, mainly *tricalcium phosphate* ($Ca_3(PO_4)_2 \cdot [OH]_2$), called *hydroxyapatite*, and some calcium carbonate ($CaCO_3$). Mineral salts compose about 50% of the weight of bone. Despite its hardness, bone is also flexible, a characteristic that enables it to resist various forces. The flexibility of bone comes from organic substances in its matrix, especially collagen fibers. Organic materials compose about 25% of the weight of bone. The remaining 25% of the bone matrix is water.

PROCEDURE

1. Obtain a bone that has been baked. How does

   this compare to an untreated one? _____

   _____

   _____

   _____

   What substances does baking remove from the

   bone (inorganic or organic)? _____

   _____

   _____

   _____

2. Now obtain a bone that has already been soaked in nitric acid by your instructor. How does this bone compare to an untreated one?

   _____

   _____

   _____

   _____

   What substances does nitric acid treatment remove from the bone (inorganic or organic)?

   _____

   _____

   _____

   _____

C. GROSS STRUCTURE OF A LONG BONE

Examine the external features of a fresh long bone and locate the following structures:

1. *Diaphysis* (dī-AF-i-sis; *dia* = through) Elongated shaft of a bone between the epiphyses.
2. *Epiphysis* (e-PIF-i-sis; *epi* = above; *physis* = growth) End or extremity of a bone; the epiphyses are referred to as proximal and distal.
3. *Metaphysis* (me-TAF-i-sis; *meta* = after or beyond) In mature bone, the region where the diaphysis joins the epiphysis; in growing bone, the region that includes the epiphyseal plate where calcified cartilage is replaced by bone as the bone lengthens.
4. *Articular cartilage* Thin layer of hyaline cartilage covering the ends of the bone where joints are formed. It reduces friction and absorbs shock at freely movable joints.
5. *Periosteum* (per'-ē-OS-tē-um; *peri* = around; *osteo* = bone) Connective tissue membrane covering the surface of the bone, except for areas covered by articular cartilage. The outer *fibrous layer* is composed of dense, irregular connective tissue and contains blood vessels, lymphatic vessels, nerves that pass into the bone; the inner *osteogenic* (os'-te-ō-JEN-ik) *layer* contains elastic fibers, blood vessels, *osteoprogenitor* (os'-tē-ō-prō-JEN-i-tor, *pro* = precursor, *gen* = to produce) *cells* (stem cells derived from mesenchyme that differentiate into osteoblasts), and *osteoblasts* (OS-tē-ō-blasts; *blast* = germ or bud), cells that secrete the organic components and mineral salts involved in bone formation. The periosteum functions in bone growth, nutrition, and repair, and as an attachment site for tendons and ligaments.
6. *Medullary* (MED-yoo-lar-ē; *medulla* = central part) or *marrow cavity* Cavity within the diaphysis that contains yellow bone marrow in adults.
7. *Endosteum* (end-OS-tē-um; *endo* = within) Membrane that lines the medullary cavity and contains osteoprogenitor cells and *osteoclasts* (OS-tē-ō-clasts; *clast* = to break), cells that remove bone by destroying the matrix, a process called *resorption*.

Label the parts of a long bone indicated in Figure 6.1.

D. HISTOLOGY OF BONE

Bone tissue, like other connective tissues, contains a large amount of matrix that surrounds widely separated cells. As noted earlier, this matrix consists of abundant mineral salts, collagen fibers, and water. The cells include osteoprogenitor cells, osteoblasts, osteoclasts, and *osteocytes* (OS-tē-ō-sīts; *cyte* = cell), mature bone cells that maintain daily activities of bone tissue.

E. BONE FORMATION: OSSIFICATION

The process by which bone forms is called *ossification* (os'-i-fi-KĀ-shun; *facere* = to make). The "skeleton" of a human embryo is composed of fibrous connective tissue membranes formed by embryonic connective tissue (mesenchyme) or hyaline cartilage that are loosely shaped like bones. They provide the supporting structures for ossification. Ossification begins around the sixth or seventh week of embryonic life and continues throughout adulthood. Bone formation follows one of two patterns.

1. *Intramembranous* (in'-tra-MEM-bra-nus; *intra* = within; *membranous* = membrane) *ossification* refers to the formation of bone directly on or within loose fibrous connective tissue membranes. Such bones form *directly* from mesenchyme without first going through a cartilage stage. The flat bones of the skull, mandible (lower jawbone), and clavicles (collarbones) form by this process. As you will see later, the fontanels ("soft spots") of an infant's skull, which are composed of loose fibrous connective tissue membranes, are also eventually replaced by bone through intramembranous ossification.
2. *Endochondral* (en'-dō-KON-dral; *endo* = within; *chondro* = cartilage) *ossification* refers to the formation of bone in hyaline cartilage. In this process, mesenchyme is transformed into chondroblasts which produce a hyaline cartilage matrix that is gradually replaced by bone. Most bones of the body form by this process.

These two kinds of ossification do *not* lead to differences in the structure of mature bones. They are simply different methods of bone formation. Both mechanisms involve the replacement of a preexisting connective tissue with bone.

The first stage in the development of bone is the migration of embryonic mesenchymal cells into the area where bone formation is about to begin. These cells increase in number and size and become osteoprogenitor cells. In some skeletal structures where capillaries are lacking, they become chondroblasts; in others where capillaries are present, they become osteoblasts. The **chondroblasts** are responsible for cartilage formation. Osteoblasts form bone tissue by intramembranous or endochondral ossification.

F. BONE GROWTH

During childhood, bones throughout the body grow in diameter by appositional growth (deposition of matrix on the surface) and long bones lengthen by interstitial growth (the addition of bone material at the epiphyseal plate). Growth in length of bones normally ceases by age 25, although bones may continue to thicken.

1. Growth in Length

To understand how a bone grows in length, you will need to know some of the details of the structure of the epiphyseal plate.

The *epiphyseal* (ep'-i-FIZ-ē-al; *epiphyein* = to grow upon) *plate* is a layer of hyaline cartilage in a growing bone that consists of four zones. The *zone of resting cartilage* is near the epiphysis and consists of small, scattered chondrocytes. The cells do not function in bone growth (thus the term "resting"); they anchor the epiphyseal plate to the bone of the epiphysis.

The *zone of proliferating cartilage* consists of slightly larger chondrocytes arranged like stacks of coins. Chondrocytes divide to replace those that die at the diaphyseal surface of the epiphyseal plate.

The zone of *hypertrophic* (hī-per-TROF-ik) or *maturing cartilage* consists of even larger chondrocytes that are also arranged in columns. The lengthwise expansion of the epiphyseal plate is the result of cell divisions in the zone of proliferating cartilage and maturation of the cells in the zone of hypertrophic cartilage.

The *zone of calcified cartilage* is only a few cells thick and consists mostly of dead cells because the matrix around them has calcified. The calcified matrix is taken up by osteoclasts, and the area is invaded by osteoblasts and capillaries from the bone in the diaphysis. These cells lay down bone on the calcified cartilage that persists. As a result, the diaphyseal border of the epiphyseal plate is firmly cemented to the bone of the diaphysis.

Label the various zones in the epiphyseal plate in Figure 6.3.

The activity of the epiphyseal plate is the only mechanism by which the diaphysis can increase in length. Unlike cartilage, which can grow by both interstitial and appositional growth, bone can grow in diameter only by appositional growth. Eventually, the epiphyseal cartilage cells stop dividing and bone replaces the cartilage. The newly formed bony structure is called the *epiphyseal line*, a remnant of the once active epiphyseal plate. With the appearance of the epiphyseal line, bone stops growing in length. The clavicle is the last bone to stop growing. In general, lengthwise growth in bones in females is completed before that in males.

2. Growth in Thickness

Enlargement of bone thickness or diameter is by appositional growth and occurs as follows. First, the bone lining the medullary cavity is destroyed by osteoclasts in the endosteum so that the cavity increases in diameter. At the same time, osteoblasts from the periosteum add new bone tissue to the outer surface. Initially, diaphyseal and epiphyseal ossification produce only spongy bone. Later, the outer region of spongy bone is reorganized into compact bone.

Epiphyseal side

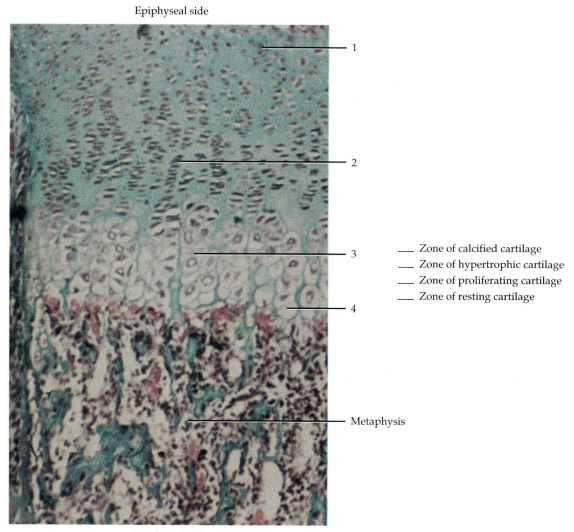

1

2

3

4

____ Zone of calcified cartilage
____ Zone of hypertrophic cartilage
____ Zone of proliferating cartilage
____ Zone of resting cartilage

Metaphysis

Diaphyseal side

Photomicrograph of epiphyseal plate (200×)

FIGURE 6.3   Photomicrograph of epiphyseal plate. (*Source:* Biophoto Associates Photo Researchers.)

G. FRACTURES

A *fracture* is any break in a bone. Usually, the fractured ends of a bone can be reduced (aligned to their normal positions) by manipulation without surgery. This procedure of setting a fracture is called *closed reduction.* In other cases, the fracture must be exposed by surgery before the break is rejoined. This procedure is known as *open reduction*.

Using your textbook as a reference, define the fractures listed in Table 6.1 and label the fractures indicated in Figure 6.4.

H. TYPES OF BONES

The 206 named bones of the body may be classified by shape into four principal types:

TABLE 6.1
Summary of Selected Fractures

Type of fracture	Definition
Partial	
Complete	
Closed (simple)	
Open (compound)	
Comminuted (KOM-i-nyoo'-ted)	
Greenstick	
Spiral	
Transverse	
Impacted	
Displaced	
Nondisplaced	
Stress	
Pathologic	
Pott's	
Colles' (KOL-ez)	

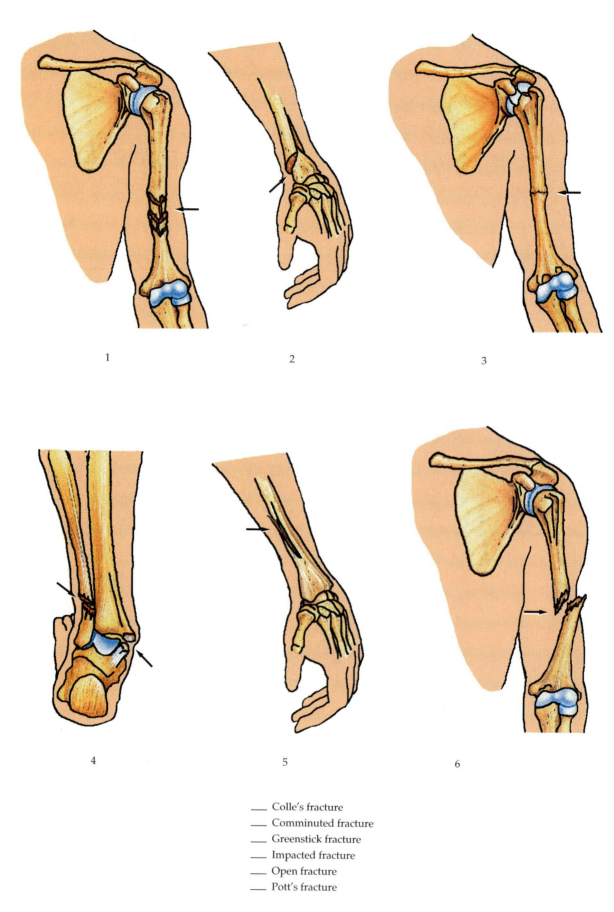

1

2

3

4

5

6

___ Colle's fracture

___ Comminuted fracture

___ Greenstick fracture

___ Impacted fracture

___ Open fracture

___ Pott's fracture

FIGURE 6.4 Types of fractures.

1. *Long* Have greater length than width, consist of a diaphysis and a variable number of epiphyses, and have a marrow cavity; contain more compact than spongy bone tissue. Example: humerus.
2. *Short* Somewhat cube-shaped, and contain more spongy than compact bone tissue. Example: wrist bones.
3. *Flat* Generally thin and flat and composed of two more-or-less parallel plates of compact bone tissue enclosing a layer of spongy bone tissue. Example: sternum.
4. *Irregular* Very complex shapes; cannot be grouped into any of the three categories just described. Example: vertebrae.

Other bones that are recognized, although not considered in the structural classification, include the following:

1. *Sutural* (SOO-chur-al) Small bones between certain cranial bones; variable in number.
2. *Sesamoid* *Sesamoid* means "resembling a grain of sesame." Small bones found in tendons; variable in number; the only constant sesamoid bones are the paired kneecaps.

Examine the disarticulated skeleton, Beauchene (disarticulated) skull, and articulated skeleton and find several examples of long, short, flat, and irregular bones. List examples of each type you find.

1. *Long* _____

_____

_____

_____

2. *Short* _____

_____

_____

_____

3. *Flat* _____

_____

_____

_____

4. *Irregular* _____

_____

_____

_____

I. BONE SURFACE MARKINGS

The surfaces of bones contain various structural features that have specific functions. These features are called *bone surface markings* and are listed in Table 6.2. Knowledge of the bone surface markings will be very useful when you learn the bones of the body in Exercise 7.

Next to each marking listed in Table 6.2, write its definition, using your textbook as a reference.

ANSWER THE LABORATORY REPORT QUESTIONS AT THE END OF THE EXERCISE.

TABLE 6.2
Bone Surface Markings

Marking	Description
DEPRESSIONS AND OPENINGS	
Fissure (FISH-ur)	
Foramen (fō-R-Ā-men; *foramen* = hole)	
Meatus (mē-Ā-tus; *meatus* = canal)	
Paranasal sinus (*sin* = cavity)	
Sulcus (*sulcus* = ditchlike groove)	
Fossa (*fossa* = basinlike depression)	
PROCESSES (PROJECTIONS) THAT FORM JOINTS	
Condyle (KON-dīl; *condulus* = knucklelike process)	
Head	
Facet	
PROCESSES (PROJECTIONS) TO WHICH TENDONS, LIGAMENTS, AND OTHER CONNECTIVE TISSUES ATTACH	
Tubercle (TOO-ber-kul; *tube* = knob)	
Tuberosity	
Trochanter (trō-KAN-ter)	
Crest	
Line	
Spinous process (spine)	
Epicondyle (*epi* = above)	

Bone Tissue 6

Student _____   **Date** _____

Laboratory Section _____   **Score/Grade** _____

PART 1. Completion

1. Small clusters of bones between certain cranial bones are referred to as _____ bones.

2. The technical name for a mature bone cell is a(n) _____.

3. Canals that extend obliquely inward or horizontally from the bone surface and contain blood vessels and lymphatic vessels are _____ canals.

4. The end, or extremity, of a bone is referred to as the _____.

5. Cube-shaped bones that contain more spongy bone tissue than compact bone tissue are known as _____ bones.

6. The cavity within the shaft of a bone that contains yellow bone marrow in the adult is the _____ cavity.

7. The thin layer of hyaline cartilage covering the end of a bone where joints are formed is called _____ cartilage.

8. Minute canals that connect lacunae are called _____.

9. The covering around the surface of a bone, except for the areas covered by cartilage, is the _____.

10. The shaft of a bone is referred to as the _____.

11. The _____ are concentric layers of calcified matrix.

12. The membrane that lines the medullary cavity and contains osteoblasts and a few osteoclasts is the _____.

13. The technical name for bone tissue is _____ tissue.

14. In a mature bone, the region where the shaft joins the extremity is called the _____.

15. The microscopic structural unit of compact bone tissue is called a(n) _____.

16. The hardness of bone is primarily due to the mineral salt _____.

17. The process by which bone is formed is called _____.

18. The zone of _____ is closest to the diaphysis of the bone.

19. Growth in diameter of bones occurs by _____ growth.

20. The destruction of matrix by osteoclasts is called _____.

21. Most bones of the body form by which type of ossification? _____

22. The lengthwise expansion of the epiphyseal plate is partly the result of cell divisions in the zone of

_____.

23. Any break in a bone is called a _____.

24. The term _____ refers to the irregular latticework of thin plates of spongy bone.

25. On the basis of shape, vertebrae are classified as _____ bones.

Bones

The 206 named bones of the adult skeleton are grouped into two divisions: axial and appendicular. The *axial skeleton* consists of bones that compose the axis of the body. The longitudinal axis is an imaginary straight line that runs through the center of gravity of the body, through the head, and down to the space between the feet. The *appendicular skeleton* consists of the bones of the limbs or extremities (upper and lower) and the girdles (pectoral and pelvic), which connect the limbs to the axial skeleton.

In this exercise you will study the names and locations of bones and their markings by examining various regions of the skeleton:

Region	Number of bones
Axial skeleton	
Skull	
Cranium	8
Face	14
Hyoid (above the larynx)	1
*Auditory ossicles, 3 in each ear	6
Vertebral column	26
Thorax	
Sternum	1
Ribs	24
	80
Appendicular Skeleton	
Pectoral (shoulder) girdles	
Clavicle	2
Scapula	2
Upper limbs (extremities)	
Humerus	2
Ulna	2
Radius	2
Carpals	16
Metacarpals	10
Phalanges	28
Pelvic (hip) girdles	
Hipbone (pelvic or coxal bone)	2

Region	Number of bones
Appendicular Skeleton (Continued)	
Lower limbs (extremities)	
Femur	2
Fibula	2
Tibia	2
Patella	2
Tarsals	14
Metatarsals	10
Phalanges	28
	126

*Not actually part of either the axial or appendicular skeleton, but are placed in the axial skeleton for convenience.

A. BONES OF ADULT SKULL

The *skull* is composed of two sets of bones—cranial and facial. The 8 *cranial* (*cranium* = brain case) *bones* form the cranial cavity and enclose and protect the brain. They are 1 *frontal*, 2 *parietals* (pa-RĪ-i-tals), 2 *temporals*, 1 *occipital* (ok-SIP-i-tal), 1 *sphenoid* (SFĒ-noyd), and 1 *ethmoid*. The 14 *facial bones* form the face and include 2 *nasals*, 2 *maxillae* (mak-SIL-ē), 2 *zygomatics*, 1 *mandible*, 2 *lacrimals* (LAK-ri-mals), 2 *palatines* (PAL-a-tīns), 2 *inferior conchae* (KONG-kē), or *turbinates*, and 1 *vomer*. These bones are indicated by *arrows* in Figure 7.1. Using the series of photographs in Figure 7.1 for reference, locate the cranial and facial bones on both a Beauchene (disarticulated) and an articulated skull.

Obtain a Beauchene skull and an articulated skull and observe them in anterior view. Using Figure 7.1a for reference, locate the parts indicated in the figure.

Turn the Beauchene skull and articulated skull so that you are looking at the right side. Using Figure 7.1b for reference, locate the parts indicated in the figure. Note also the *hyoid bone* below the mandible. This is not a bone of the skull; it is noted here because of its proximity to the skull.

If a skull in median section is available, use Figure 7.1c for reference and locate the parts indicated in the figure.

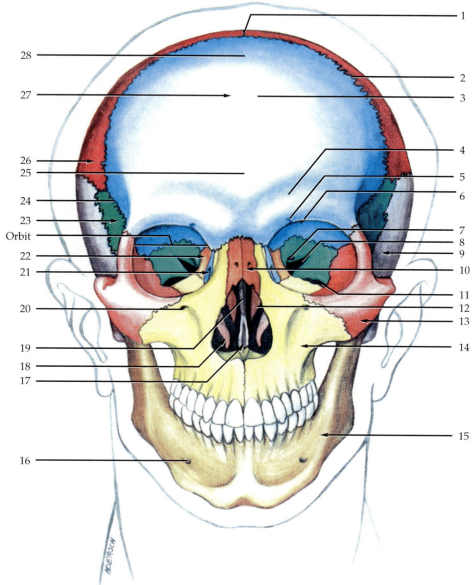

(a) Anterior view

1. Sagittal suture
2. Coronal suture
3. Frontal squama
4. Superciliary arch
5. Supraorbital foramen
6. Supraorbital margin
7. Optic foramen
8. Superior orbital fissure
9. Temporal bone
10. Nasal bone
11. Inferior orbital fissure
12. Middle nasal concha (turbinate)
13. Zygomatic bone
14. Maxilla
15. Mandible
16. Mental foramen
17. Vomer
18. Inferior nasal concha (turbinate)
19. Perpendicular plate
20. Infraorbital foramen
21. Lacrimal bone
22. Ethmoid bone
23. Sphenoid bone
24. Squamous suture
25. Glabella
26. Parietal bone
27. Frontal bone
28. Frontal eminence

FIGURE 7.1   Skull.

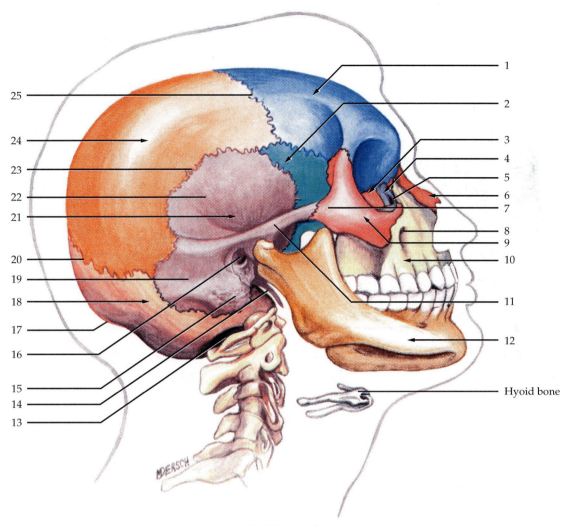

(b) Right lateral view

1. Frontal bone	11. Zygomatic process	18. Occipital bone
2. Sphenoid bone	12. Mandible	19. Mastoid portion
3. Ethmoid bone	13. Foramen magnum	20. Lambdoid suture
4. Lacrimal bone	14. Styloid process	21. Temporal bone
5. Lacrimal foramen	15. Mastoid process	22. Squamous suture
6. Nasal bone	16. External auditory	23. Temporal squama
7. Temporal process	(acoustic) meatus	24. Parietal bone
8. Infraorbital foramen	17. External occipital	25. Coronal suture
9. Zygomatic bone	protuberance	
10. Maxilla		

FIGURE 7.1 (*Continued*) Skull.

Take an articulated skull and turn it upside down so that you are looking at the inferior surface. Using Figure 7.1d for reference, locate the parts indicated in the figure.

Obtain an articulated skull with a removable crown. Using Figure 7.1e for reference, locate the parts indicated in the figure.

Examine the right orbit of an articulated skull. Using Figure 7.2 for reference, locate the parts indicated in the figure.

Obtain a mandible and, using Figure 7.3 for reference, identify the parts indicated in the figure.

Before you move on, refer to Table 7.1, "Summary of Foramina of the Skull" on page 100. Complete the

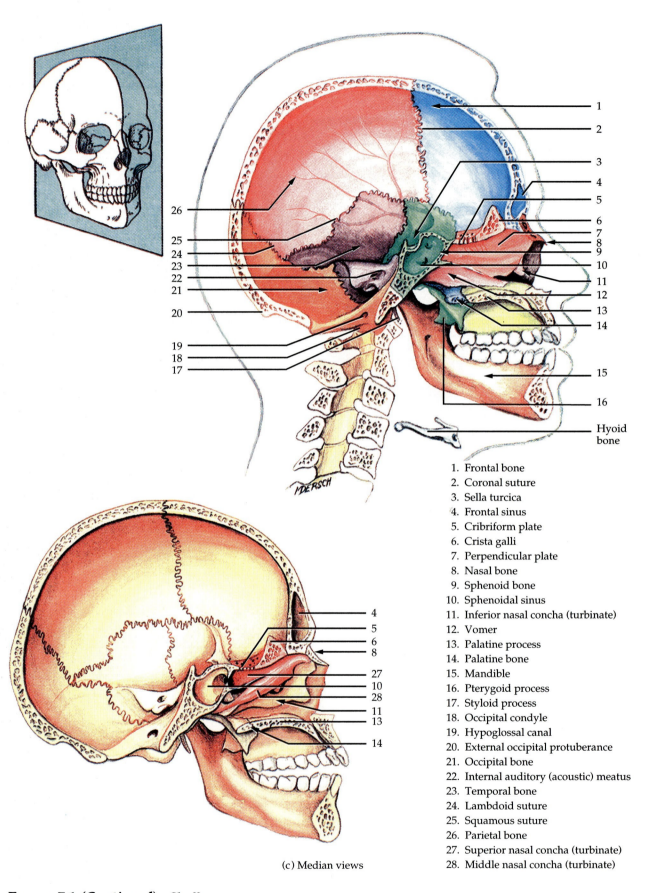

Hyoid
bone

1. Frontal bone
2. Coronal suture
3. Sella turcica
4. Frontal sinus
5. Cribriform plate
6. Crista galli
7. Perpendicular plate
8. Nasal bone
9. Sphenoid bone
10. Sphenoidal sinus
11. Inferior nasal concha (turbinate)
12. Vomer
13. Palatine process
14. Palatine bone
15. Mandible
16. Pterygoid process
17. Styloid process
18. Occipital condyle
19. Hypoglossal canal
20. External occipital protuberance
21. Occipital bone
22. Internal auditory (acoustic) meatus
23. Temporal bone
24. Lambdoid suture
25. Squamous suture
26. Parietal bone
27. Superior nasal concha (turbinate)
28. Middle nasal concha (turbinate)

(c) Median views

FIGURE 7.1 (Continued) Skull.

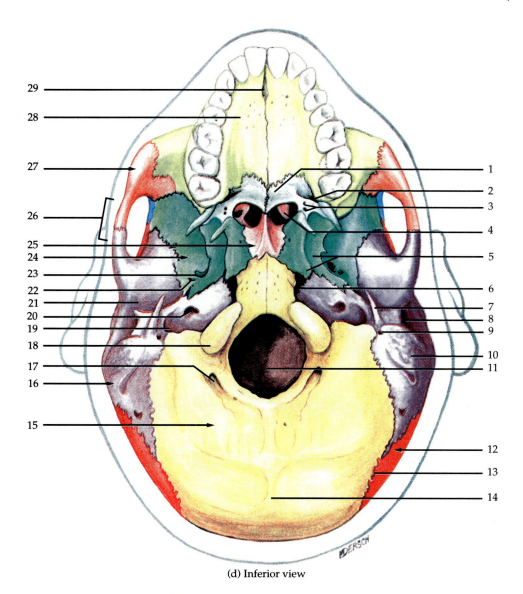

(d) Inferior view

1. Horizontal plate
2. Greater palatine foramen
3. Lesser palatine foramina
4. Middle nasal concha (turbinate)
5. Pterygoid process
6. Foramen lacerum
7. Styloid process
8. External auditory (acoustic) meatus
9. Stylomastoid foramen
10. Mastoid process
11. Foramen magnum
12. Parietal bone
13. Lambdoid suture
14. External occipital protuberance
15. Occipital bone

16. Temporal bone
17. Condylar canal
18. Occipital condyle
19. Jugular foramen
20. Carotid foramen
21. Mandibular fossa
22. Foramen spinosum
23. Foramen ovale
24. Sphenoid bone
25. Vomer
26. Zygomatic arch
27. Zygomatic bone
28. Palatine process
29. Incisive foramen

FIGURE 7.1 *(Continued)* Skull.

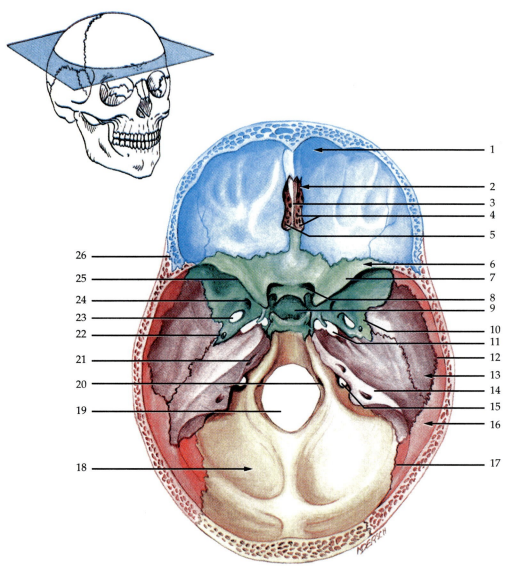

(e) Superior view of floor of cranium

1. Frontal bone	14. Petrous portion
2. Ethmoid bone	15. Jugular foramen
3. Crista galli	16. Parietal bone
4. Olfactory foramina	17. Lambdoid suture
5. Cribriform plate	18. Occipital bone
6. Sphenoid bone	19. Foramen magnum
7. Lesser wing	20. Hypoglossal canal
8. Optic foramen	21. Internal auditory (acoustic) meatus
9. Sella turcica	22. Foramen spinosum
10. Greater wing	23. Foramen ovale
11. Foramen lacerum	24. Foramen rotundum
12. Squamous suture	25. Superior orbital fissure
13. Temporal bone	26. Coronal suture

FIGURE 7.1 (*Continued*)   Skull.

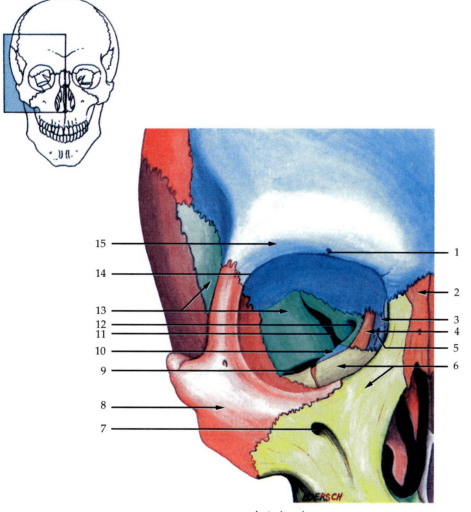

1. Supraorbital foramen
2. Nasal bone
3. Lacrimal bone
4. Ethmoid bone
5. Lacrimal foramen
6. Maxilla
7. Infraorbital foramen
8. Zygomatic bone
9. Inferior orbital fissure
10. Palatine bone
11. Superior orbital fissure
12. Optic foramen
13. Sphenoid bone
14. Supraorbital margin
15. Frontal bone

Anterior view

FIGURE 7.2   Right orbit.

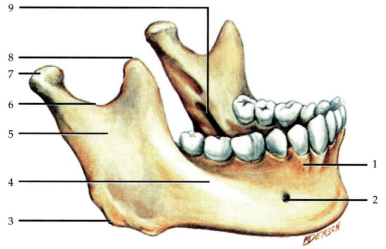

1. Alveolar process
2. Mental foramen
3. Angle
4. Body
5. Ramus
6. Mandibular notch
7. Condylar process
8. Coronoid process
9. Mandibular foramen

Right lateral view

FIGURE 7.3   Mandible.

TABLE 7.1
Summary of Foramina of the Skull

Foramen	Structures Passing Through
Carotid (relating to carotid artery in neck)	
Greater palatine (*palatum* = palate)	
Hypoglossal (*hypo* = under; *glossus* = tongue)	
Incisive (*incisive* = pertaining to incisor teeth)	
Inferior orbital (*inferior* = below; *orbital* = orbit)	
Infraorbital (*infra* = below)	
Jugular (*jugular* = pertaining to jugular vein)	
Lacerum (*lacerum* = lacerated)	
Lacrimal (*lacrima* = pertaining to tears)	
Lesser palatine (*palatum* = palate)	
Magnum (*magnum* = large)	
Mandibular (*mandere* = to chew)	
Mastoid (*mastoid* = breast-shaped)	
Mental (*mentum* = chin)	
Olfactory (*olfacere* = to smell)	
Optic (*optikas* = eye)	
Ovale (*ovale* = oval)	
Rotundum (*rotundum* = round opening)	
Spinosum (*spinosum* = resembling a spine)	
Stylomastoid (*stylo* = stake or pole)	
Superior orbital (*superior* = above)	
Supraorbital (*supra* = above)	
Zygomaticofacial (*zygoma* = cheekbone)	

table by indicating the structures that pass through the foramina listed.

B. SUTURES OF SKULL

A *suture* (SOO-chur; *sutura* = seam) is an immovable joint found only between skull bones. The four prominent sutures are the *coronal* (corona = crown), *sagittal* (sagitta = arrow), *lambdoid* (LAM-doyd), and *squamous* (SKWĀ-mos; *squama* = flat). Using Figures 7.1 and 7.4 for reference, locate these sutures on an articulated skull.

C. FONTANELS OF SKULL

At birth, the skull bones are separated by fibrous connective-tissue membrane-filled spaces called *fontanels* (fon'-ta-NELZ; *fontanelle* = little fountain).

They (1) enable the fetal skull to compress as it passes through the birth canal, (2) permit rapid growth of the brain during infancy, (3) facilitate determination of the degree of brain development by their state of closure, (4) serve as landmarks (anterior fontanel) for withdrawal of blood from the superior sagittal sinus, and (5) aid in determining the position of the fetal head prior to birth. The principal fontanels are *anterior (frontal), posterior (occipital), anterolateral (sphenoidal),* and *posterolateral (mastoid).* Using Figure 7.4 for reference, locate the fontanels on the skull of a newborn infant.

D. PARANASAL SINUSES OF SKULL

A *paranasal* (para = beside) *sinus,* or simply *sinus,* is a cavity in a bone located near the nasal cavity.

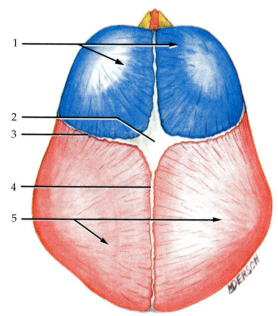

(a) Superior view

1. Frontal bone
2. Anterior (frontal)
   fontanel
3. Coronal suture
4. Sagittal suture
5. Parietal bones

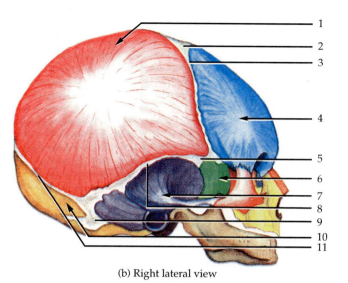

(b) Right lateral view

1. Parietal bones
2. Anterior (frontal) fontanel
3. Coronal suture
4. Frontal bone
5. Anterolateral (sphenoid) frontanel
6. Sphenoid bone
7. Temporal bone
8. Squamous suture
9. Posterolateral (mastoid) fontanel
10. Occipital bone
11. Lambdoid suture

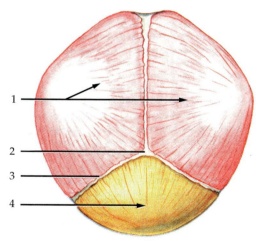

(c) Posterior view

1. Parietal bones
2. Posterior (occipital)
   fontanel
3. Lambdoid suture
4. Occipital bone

FIGURE 7.4   Fontanels.

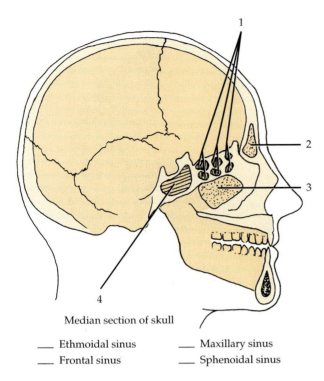

Median section of skull

___ Ethmoidal sinus   ___ Maxillary sinus
___ Frontal sinus   ___ Sphenoidal sinus

FIGURE 7.5 Paranasal sinuses.

Paired paranasal sinuses are found in the maxillae and the frontal, sphenoid, and ethmoid bones. Locate the paranasal sinuses on the Beauchene skull or other demonstration models that may be available. Label the paranasal sinuses shown in Figure 7.5.

E. VERTEBRAL COLUMN

The *vertebral column* (*backbone* or *spine*) is composed of a series of bones called *vertebrae*. The vertebrae of the adult column are distributed as follows: 7 *cervical* (SER-vi-kal) (neck), 12 *thoracic* (thō-RAS-ik) (chest), 5 *lumbar* (lower back), 5 *sacral* (fused into one bone), the *sacrum* (SĀ-krum) (between the hipbones), and usually 4 *coccygeal* (kok-SIJ-ē-al) (fused into one bone, the *coccyx* [KOK-six] forming the tail of the column). Locate each of these regions on the articulated skeleton. Label the same regions in Figure 7.6, the anterior view.

Examine the vertebral column on the articulated skeleton and identify the cervical, thoracic, lumbar, and sacral (sacrococcygeal) curves. Label the curves in Figure 7.6, the right lateral view.

F. VERTEBRAE

A typical *vertebra* consists of the following portions:

1. *Body* Thick, disc-shaped anterior portion.
2. *Vertebral (neural) arch* Posterior extension from the body that surrounds the spinal cord and consists of the following parts:
   a. *Pedicles* (PED-i-kuls; *pediculus* = little feet) Two short, thick processes that project posteriorly; each has a superior and inferior notch (*vertebral notch*) and, when successive vertebrae are fitted together, the adjoining notches form an *intervertebral foramen* through which a spinal nerve and blood vessels pass.
   b. *Laminae* (LAM-i-nē; *lamina* = thin layer) Flat portions that form the posterior wall of the vertebral arch.
   c. *Vertebral foramen* Opening through which the spinal cord passes; when all the vertebrae are fitted together, the foramina form a canal, the *vertebral canal*.
3. *Processes* Seven processes arise from the vertebral arch:
   a. Two *transverse processes* Lateral extensions where the laminae and pedicles join.
   b. One *spinous process* (*spine*) Posterior projection of the lamina.
   c. Two *superior articular processes* Articulate with the vertebra above. Their top surfaces, called *superior articular facets* (*facet* = little face), articulate with the vertebra above.
   d. Two *inferior articular processes* Articulate with the vertebra below. Their bottom surfaces, called *inferior articular facets*, articulate with the vertebra below.

Obtain a thoracic vertebra and locate each part just described. Now label the vertebra in Figure 7.7a. You should be able to distinguish the general parts on all the different vertebrae that contain them.

Although vertebrae have the same basic design, those of a given region have special distinguishing features. Obtain examples of the following vertebrae and identify their distinguishing features.

1. *Cervical vertebrae*
   a. *Atlas (C1)* First cervical vertebra (Figure 7.7b)
      Transverse foramen Opening in transverse process through which an artery, a vein, and a branch of a spinal nerve pass.

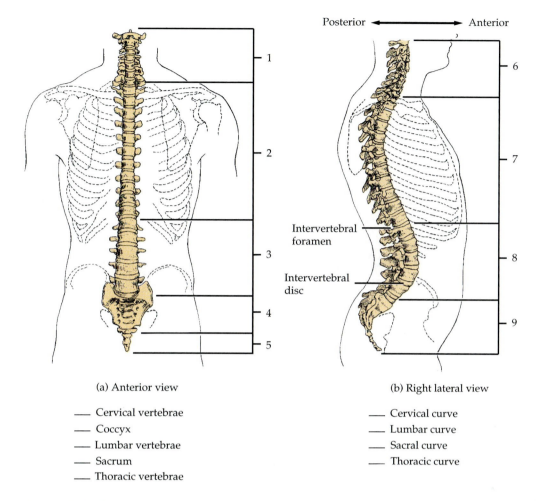

(a) Anterior view

—— Cervical vertebrae
—— Coccyx
—— Lumbar vertebrae
—— Sacrum
—— Thoracic vertebrae

(b) Right lateral view

—— Cervical curve
—— Lumbar curve
—— Sacral curve
—— Thoracic curve

FIGURE 7.6 Vertebral column.

Anterior arch Anterior wall of vertebral foramen.
Posterior arch Posterior wall of vertebral foramen.
Lateral mass Side wall of vertebral foramen.

Label the other indicated parts.

b. *Axis (C2)* Second cervical vertebra (Figure 7.7c)
Dens (*dens* = tooth) Superior projection of body that articulates with atlas.

Label the other indicated parts.

c. *Cervicals 3 through 6* (Figure 7.7d)
Bifid spinous process (C3–C6) Cleft in spinous processes of cervical vertebrae 2 through 6.

Label the other indicated parts.

d. *Vertebra prominens (C7)* Seventh cervical vertebra; contains a nonbifid and long spinous process.
2. *Thoracic vertebrae (T1–T12)* (Figure 7.7e)
a. *Facets* For articulation with the tubercle and head of a rib; found on body and transverse processes.
b. *Spinous processes* Usually long, pointed, downward projections.

Label the other indicated parts.

3. *Lumbar vertebrae (L1–L5)* (Figure 7.7f)
a. *Spinous processes* Broad, blunt.
b. *Superior articular processes* Directed medially, not superiorly.
c. *Inferior articular processes* Directed laterally, not inferiorly.

Posterior

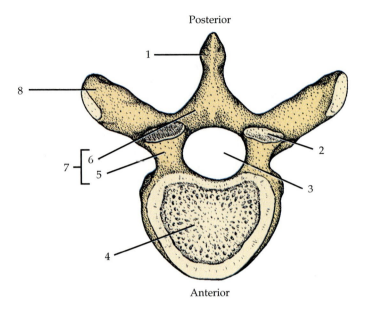

Anterior

(a) Superior view of a typical vertebra

___ Body
___ Lamina
___ Pedicle
___ Spinous process
___ Superior articular facet
___ Transverse process
___ Vertebral arch
___ Vertebral foramen

Posterior

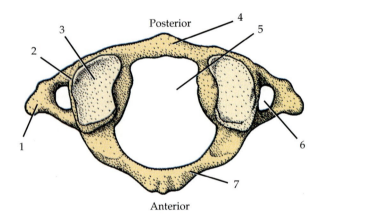

Anterior

(b) Superior view of atlas

___ Anterior arch
___ Lateral mass
___ Posterior arch
___ Superior articular facet
___ Transverse foramen
___ Transverse process
___ Vertebral foramen

Superior

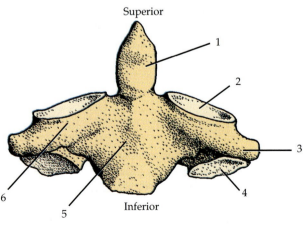

Inferior

(c) Anterior view of the axis

___ Body
___ Dens
___ Inferior articular facet
___ Lateral mass
___ Superior articular facet
___ Transverse process

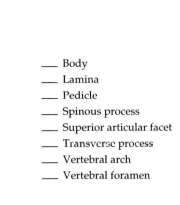

FIGURE 7.7 Vertebrae.

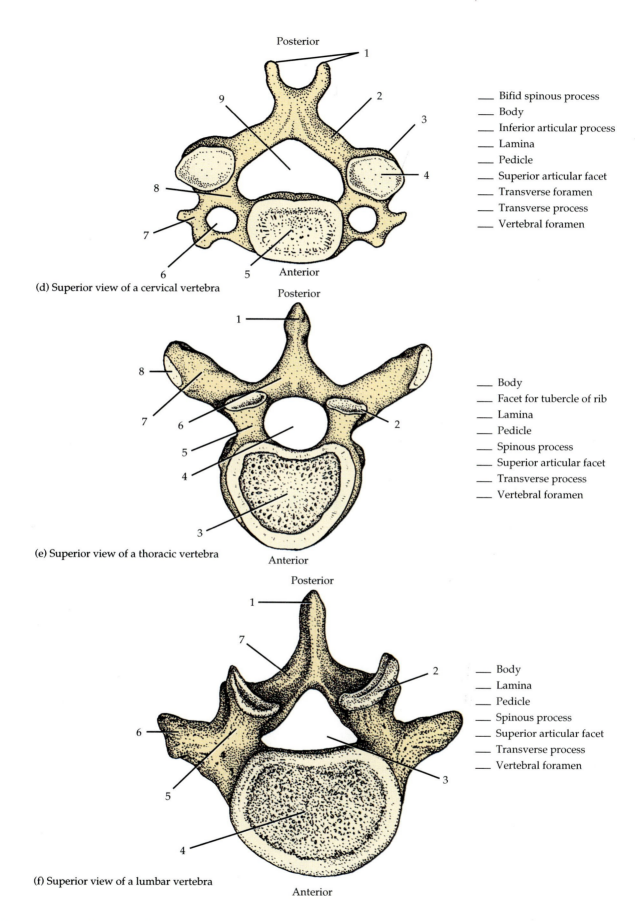

Posterior

Anterior

— Bifid spinous process
— Body
— Inferior articular process
— Lamina
— Pedicle
— Superior articular facet
— Transverse foramen
— Transverse process
— Vertebral foramen

(d) Superior view of a cervical vertebra

Posterior

Anterior

— Body
— Facet for tubercle of rib
— Lamina
— Pedicle
— Spinous process
— Superior articular facet
— Transverse process
— Vertebral foramen

(e) Superior view of a thoracic vertebra

Posterior

Anterior

— Body
— Lamina
— Pedicle
— Spinous process
— Superior articular facet
— Transverse process
— Vertebral foramen

(f) Superior view of a lumbar vertebra

FIGURE 7.7 (Continued) Vertebrae.

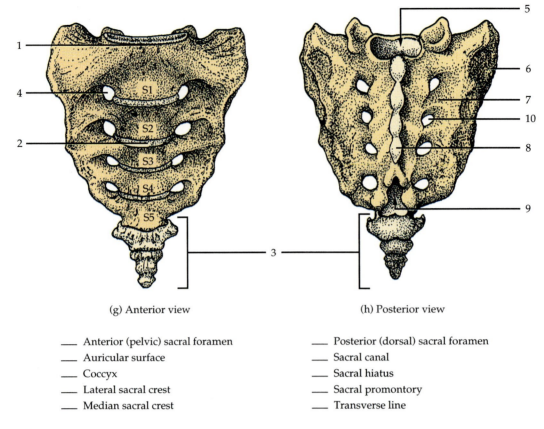

(g) Anterior view

(h) Posterior view

___ Anterior (pelvic) sacral foramen
___ Auricular surface
___ Coccyx
___ Lateral sacral crest
___ Median sacral crest

___ Posterior (dorsal) sacral foramen
___ Sacral canal
___ Sacral hiatus
___ Sacral promontory
___ Transverse line

FIGURE 7.7 (Continued) Vertebrae. Sacrum and coccyx.

Label the other indicated parts.

4. *Sacrum* (Figure 7.7g and h). Formed by the fusion of five sacral vertebrae.
   a. *Transverse lines* Areas where bodies of sacral vertebrae are joined.
   b. *Anterior (pelvic) sacral foramina* Four pairs of foramina that communicate with dorsal sacral foramina; passages for blood vessels and nerves.
   c. *Median sacral crest* Spinous processes of fused sacral vertebrae.
   d. *Lateral sacral crest* Transverse processes of fused sacral vertebrae.
   e. *Posterior (dorsal) sacral foramina* Four pairs of foramina that communicate with pelvic foramina; passages for blood vessels and nerves.
   f. *Sacral canal* Continuation of vertebral canal.
   g. *Sacral hiatus* (hi-Ā-tus) Inferior entrance to sacral canal where laminae of S5, and sometimes S4, fail to meet.

h. *Sacral promontory* Superior, anterior projecting border.
   i. *Auricular surface* Articulates with ilium of hipbone.
5. *Coccyx* (Figure 7.7g and h) Formed by the fusion of usually four coccygeal vertebrae.

Label the coccyx and the parts of the sacrum in Figure 7.7g and h.

G. STERNUM AND RIBS

The skeleton of the *thorax* consists of the *sternum, costal cartilages, ribs,* and bodies of the *thoracic vertebrae.*

Examine the articulated skeleton and disarticulated bones and identify the following:

1. *Sternum (breastbone)* Flat bone in midline of anterior thorax.

a. *Manubrium* (ma-NOO-brē-um; *manubrium* = handle-like) Superior portion.
b. *Body* Middle, largest portion.
c. *Sternal angle* Junction of the manubrium and body.
d. *Xiphoid* (ZI-foyd; *xipho* = sword-like) *process* Inferior, smallest portion.
e. *Suprasternal notch* Depression on the superior surface of the manubrium that may be felt.
f. *Clavicular notches* Articular surfaces lateral to suprasternal notches for articulation with the clavicles.

Label the parts of the sternum in Figure 7.8a.

2. *Costal* (*costa* = rib) *cartilage* Strip of hyaline cartilage that attaches a rib to the sternum.
3. *Ribs* Parts of a typical rib (third through ninth) include
  a. *Body* Shaft, main part of rib.
  b. *Head* Posterior projection.
  c. *Neck* Constricted portion behind head.
  d. *Tubercle* (TOO-ber-kul) Knoblike elevation just below neck; consists of a *nonarticular part* that affords attachment for a ligament and an *articular part* that articulates with an inferior vertebra.
  e. *Costal groove* Depression on the inner surface containing blood vessels and a nerve.

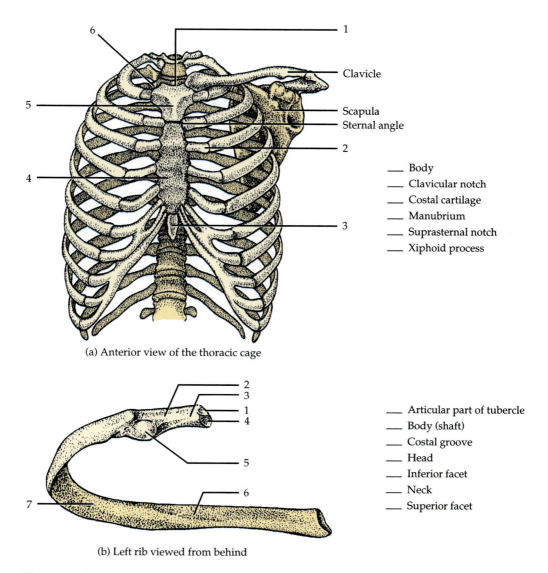

(a) Anterior view of the thoracic cage

___ Body
___ Clavicular notch
___ Costal cartilage
___ Manubrium
___ Suprasternal notch
___ Xiphoid process

___ Articular part of tubercle
___ Body (shaft)
___ Costal groove
___ Head
___ Inferior facet
___ Neck
___ Superior facet

(b) Left rib viewed from behind

FIGURE 7.8 Bones of the thorax.

f. *Superior facet* Articulates with facet on superior vertebra.

g. *Inferior facet* Articulates with facet on inferior vertebra.

Label the parts of a rib in Figure 7.8b.

H. PECTORAL (SHOULDER) GIRDLES

Each *pectoral* (PEK-tō-ral) or *shoulder girdle* consists of two bones—*clavicle* (collar bone) and *scapula* (shoulder blade). Its purpose is to attach the bones of the upper limb to the axial skeleton.

Examine the articulated skeleton and disarticulated bones and identify the following:

1. *Clavicle* (KLAV-i-kul; *clavus* = key) Slender bone with a double curvature; lies horizontally in superior and anterior part of the thorax.
   a. *Sternal extremity* Rounded, medial end that articulates with manubrium of sternum.
   b. *Acromial* (a-KRŌ-mē-al) *extremity* Broad, flat, lateral end that articulates with the acromion of the scapula.
   c. *Conoid tubercle* (TOO-ber-kul; *konos* = cone) Projection on the inferior, lateral surface for attachment of a ligament.

Label the parts of the clavicle in Figure 7.9a.

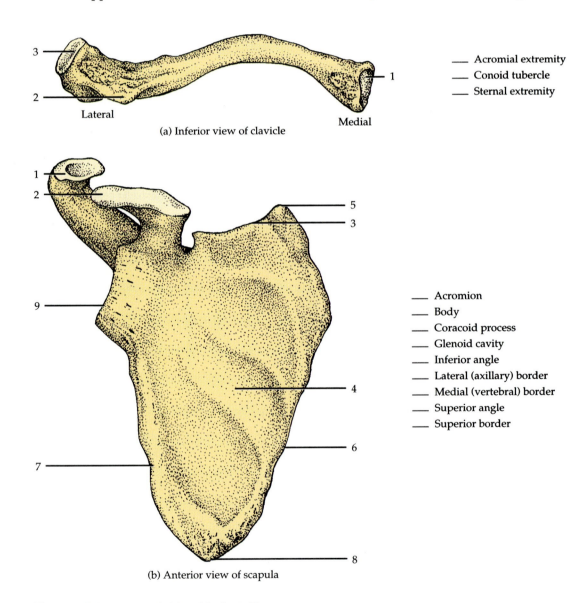

(a) Inferior view of clavicle

___ Acromial extremity
___ Conoid tubercle
___ Sternal extremity

(b) Anterior view of scapula

___ Acromion
___ Body
___ Coracoid process
___ Glenoid cavity
___ Inferior angle
___ Lateral (axillary) border
___ Medial (vertebral) border
___ Superior angle
___ Superior border

FIGURE 7.9   Pectoral (shoulder) girdle.

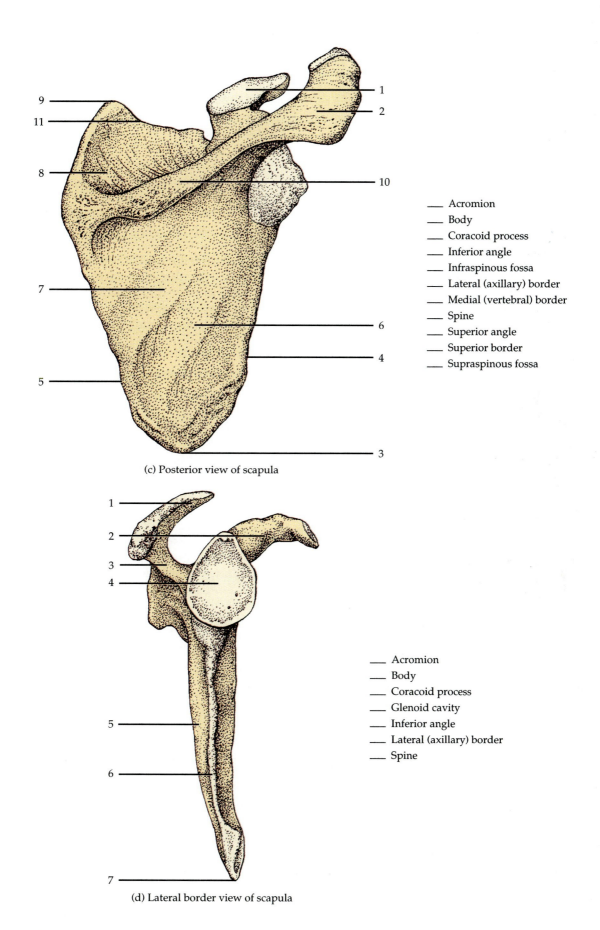

___ Acromion
___ Body
___ Coracoid process
___ Inferior angle
___ Infraspinous fossa
___ Lateral (axillary) border
___ Medial (vertebral) border
___ Spine
___ Superior angle
___ Superior border
___ Supraspinous fossa

(c) Posterior view of scapula

___ Acromion
___ Body
___ Coracoid process
___ Glenoid cavity
___ Inferior angle
___ Lateral (axillary) border
___ Spine

(d) Lateral border view of scapula

FIGURE 7.9 (Continued) Pectoral (shoulder) girdle.

2. *Scapula* (SCAP-yoo-la) Large, flat triangular bone in dorsal thorax between the levels of ribs 2 through 7.

   a. ***Body*** Flattened, triangular portion.

   b. ***Spine*** Ridge across posterior surface.

   c. ***Acromion*** (a-KRŌ-me-on; *acro* = top or summit) Flattened, expanded process of spine.

   d. ***Medial (vertebral) border*** Edge of body near vertebral column.

   e. ***Lateral (axillary) border*** Edge of body near arm.

   f. ***Inferior angle*** Bottom of body where medial and lateral borders join.

   g. ***Glenoid cavity*** Depression below acromion that articulates with head of humerus to form the shoulder joint.

   h. ***Coracoid*** (KOR-a-koyd; *korakodes* = like a crow's beak) ***process*** Projection at lateral end of superior border.

   i. ***Supraspinous*** (soo'-pra-SPĪ-nus) ***fossa*** Surface for muscle attachment above spine.

   j. ***Infraspinous fossa*** Surface for muscle attachment below spine.

   k. ***Superior border*** Superior edge of the body.

   l. ***Superior angle*** Top of body where superior and medial borders join.

Label the parts of the scapula in Figure 7.9b, c, and d.

I. UPPER LIMBS

The skeleton of the ***upper limbs*** consists of a humerus in each arm, an ulna and radius in each forearm, carpals in each wrist, metacarpals in each palm, and phalanges in the fingers.

   Examine the articulated skeleton and disarticulated bones and identify the following:

1. *Humerus* (HYOO-mer-us) Arm bone; longest and largest bone of the upper limb.

   a. ***Head*** Articulates with the glenoid cavity of scapula.

   b. ***Anatomical neck*** Oblique groove below the head; the former site of the epiphyseal plate.

   c. ***Greater tubercle*** Lateral projection below the anatomical neck.

   d. ***Lesser tubercle*** Anterior projection.

   e. ***Intertubercular sulcus (bicipital groove)*** Between the tubercles.

   f. ***Surgical neck*** Constricted portion below the tubercles.

   g. ***Body*** Shaft.

   h. ***Deltoid tuberosity*** Roughened, V-shaped area about midway down the lateral surface of shaft.

   i. ***Capitulum*** (ka-PIT-yoo-lum) Rounded knob that articulates with head of radius.

   j. ***Radial fossa*** Anterior lateral depression that receives head of radius when forearm is flexed.

   k. ***Trochlea*** (TRŌK-le-a) Projection that articulates with the ulna.

   l. ***Coronoid*** (KOR-ō-noyd; *korne* = crown-shaped) ***fossa*** Anterior medial depression that receives part of the ulna when the forearm is flexed.

   m. ***Olecranon*** (ō-LEK-ra-non) ***fossa*** Posterior depression that receives the olecranon of the ulna when the forearm is extended.

   n. ***Medial epicondyle*** Projection on medial side of distal end.

   o. ***Lateral epicondyle*** Projection on lateral side of distal end.

Label the parts of the humerus in Figure 7.10a and b.

2. *Ulna* Medial bone of forearm.

   a. ***Olecranon (olecranon process)*** Prominence of elbow at proximal end.

   b. ***Coronoid process*** Anterior projection that, with olecranon, receives trochlea of humerus.

   c. ***Trochlear (semilunar) notch*** Curved area between olecranon and coronoid process into which trochlea of humerus fits.

   d. ***Radial notch*** Depression lateral and inferior to trochlear notch that receives the head of the radius.

   e. ***Head*** Rounded portion at distal end.

   f. ***Styloid*** (*stylo* = pillar) ***process*** Projection on posterior side of distal end.

Label the parts of the ulna in Figure 7.10c and d.

3. *Radius* Lateral bone of forearm.

   a. ***Head*** Disc-shaped process at proximal end.

   b. ***Radial tuberosity*** Medial projection for insertion of the biceps brachii muscle.

   c. ***Styloid process*** Projection on lateral side of distal end.

   d. ***Ulnar notch*** Medial, concave depression for articulation with ulna.

Label the parts of the radius in Figure 7.10c and d.

4. *Carpus* Wrist, consists of eight small bones, called ***carpals***, united by ligaments.

a. *Proximal row* From medial to lateral are called *pisiform, triquetrum, lunate,* and *scaphoid.*
b. *Distal row* From medial to lateral are called *hamate, capitate, trapezoid,* and *trapezium.*
5. *Metacarpus* (*meta* = after or beyond) Five bones in the palm, numbered as follows beginning with thumb side: I, II, III, IV, and V metacarpals.

6. *Phalanges* (fa-LAN-jēz; *phalanx* = closely knit row) Bones of the fingers; two in each thumb (proximal and distal) and three in each finger (proximal, middle, and distal). The singular of phalanges is *phalanx.*

Label the parts of the carpus, metacarpus, and phalanges in Figure 7.10f.

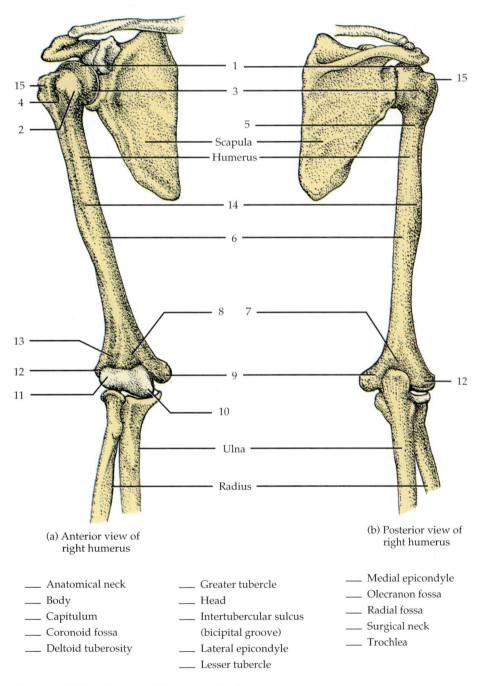

(a) Anterior view of right humerus

(b) Posterior view of right humerus

___ Anatomical neck
___ Body
___ Capitulum
___ Coronoid fossa
___ Deltoid tuberosity

___ Greater tubercle
___ Head
___ Intertubercular sulcus (bicipital groove)
___ Lateral epicondyle
___ Lesser tubercle

___ Medial epicondyle
___ Olecranon fossa
___ Radial fossa
___ Surgical neck
___ Trochlea

FIGURE 7.10 Bones of the upper limb.

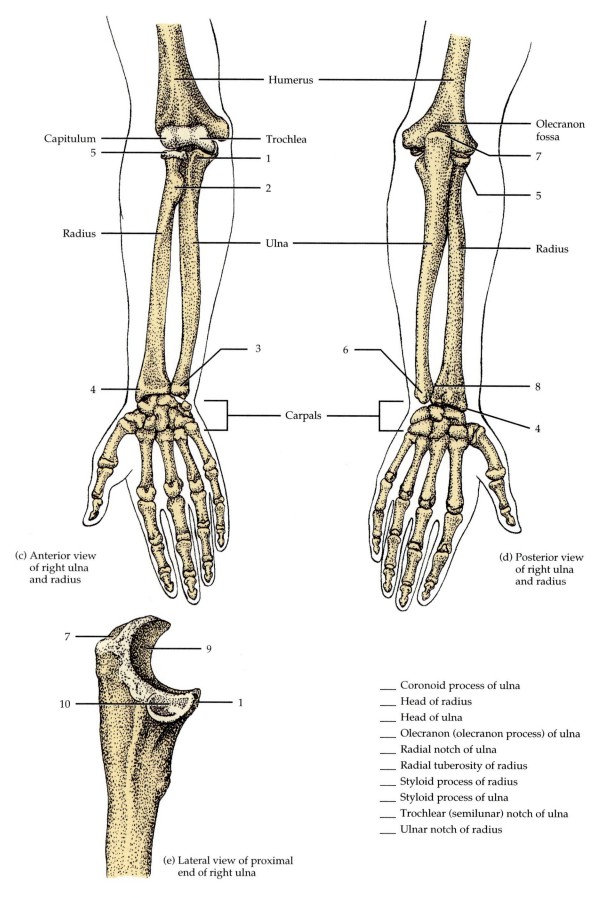

Capitulum

Trochlea

5

1

2

Humerus

Radius

Ulna

(c) Anterior view
of right ulna
and radius

3

4

Carpals

Olecranon
fossa

7

5

Radius

6

8

4

(d) Posterior view
of right ulna
and radius

7

9

10

1

(e) Lateral view of proximal
end of right ulna

___ Coronoid process of ulna
___ Head of radius
___ Head of ulna
___ Olecranon (olecranon process) of ulna
___ Radial notch of ulna
___ Radial tuberosity of radius
___ Styloid process of radius
___ Styloid process of ulna
___ Trochlear (semilunar) notch of ulna
___ Ulnar notch of radius

FIGURE 7.10 (Continued) Bones of the upper limb.

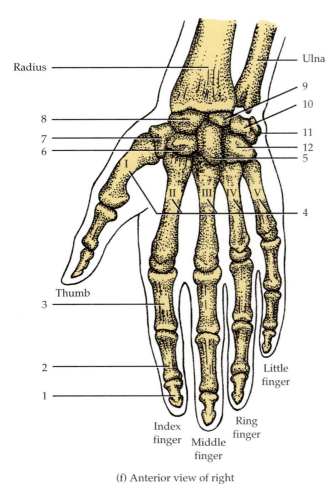

Radius

Ulna

9
10
8
7
6
11
12
5

I
II III IV V
4

Thumb
3

Index
finger

Middle
finger

Ring
finger

Little
finger

2
1

___ Capitate
___ Distal phalanx
___ Hamate
___ Lunate
___ Metacarpal
___ Middle phalanx
___ Pisiform
___ Proximal phalanx
___ Scaphoid
___ Trapezium
___ Trapezoid
___ Triquetrum

(f) Anterior view of right
wrist and hand

FIGURE 7.10 *(Continued)*   Bones of the upper limb.

J. PELVIC (HIP) GIRDLES

The *pelvic (hip) girdles* consist of the two hip-bones or *coxal* (KOK-sal) *bones.* They provide a strong and stable support for the lower limbs on which the weight of the body is carried and attach the lower limbs to the axial skeleton.

Examine the articulated skeleton and disarticulated bones and identify the following parts of the hipbone:

1. *Ilium* Superior flattened portion.
   a. *Iliac crest* Superior border of ilium.
   b. *Anterior superior iliac spine* Anterior projection of iliac crest.
   c. *Anterior inferior iliac spine* Projection under anterior superior iliac spine.
   d. *Posterior superior iliac spine* Posterior projection of iliac crest.

   e. *Posterior inferior iliac spine* Projection below posterior superior iliac spine.
   f. *Greater sciatic* (sī-AT-ik) *notch* Concavity under posterior inferior iliac spine.
   g. *Iliac fossa* Medial concavity for attachment of iliacus muscle.
   h. *Iliac tuberosity* Point of attachment for sacroiliac ligament posterior to iliac fossa.
   i. *Auricular* (*auricula* = little ear) *surface* Point of articulation with sacrum.
2. *Ischium* (IS-kē-um) Lower, posterior portion.
   a. *Ischial spine* Posterior projection of ischium.
   b. *Lesser sciatic notch* Concavity under ischial spine.
   c. *Ischial tuberosity* Roughened projection.
   d. *Ramus* Portion of ischium that joins the pubus and surrounds the *obturator* (OB-too-rā-ter) *foramen.*

3. *Pubis* Anterior, inferior portion.
   a. *Superior ramus* Upper portion of pubis.
   b. *Inferior ramus* Lower portion of pubis.
   c. *Pubic symphysis* (SIM-fi-sis) Joint between left and right hipbones.
4. *Acetabulum* (as'-e-TAB-yoo-lum) Socket that receives the head of the femur to form the hip joint; two-fifths of the acetabulum is formed by the ilium, two-fifths is formed by the ischium, and one-fifth is formed by the pubis.

Label Figure 7.11.

Again, examine the articulated skeleton. This time compare the male and female pelvis. The *pelvis* consists of the two hipbones, sacrum, and coccyx. Identify the following:

1. *Greater (false) pelvis* Expanded portion situated above the brim of the pelvis; bounded laterally by the ilia (plural of ilium) and posteriorly by the upper sacrum.
2. *Lesser (true) pelvis* Below and behind the brim of the pelvis; constructed of parts of the ilium, pubis, sacrum, and coccyx; contains an opening above, the *pelvic inlet,* and an opening below, the *pelvic outlet.*

K. LOWER LIMBS

The bones of the *lower limbs* consist of a femur in each thigh, a patella in front of each knee joint, a tibia and fibula in each leg, tarsals in each ankle, metatarsals in each foot, and phalanges in the toes.

Examine the articulated skeleton and disarticulated bones and identify the following:

1. *Femur* Thigh bone; longest and heaviest bone in the body.
   a. *Head* Rounded projection at proximal end that articulates with acetabulum of hipbone.
   b. *Neck* Constricted portion below head.
   c. *Greater trochanter* (trō-KAN-ter) Prominence on lateral side.
   d. *Lesser trochanter* Prominence on dorsomedial side.
   e. *Intertrochanteric line* Ridge on anterior surface.
   f. *Intertrochanteric crest* Ridge on posterior surface.
   g. *Linea aspera* (LIN-ē-a AS-per-a) Vertical ridge on posterior surface.
   h. *Medial condyle* Medial posterior projection on distal end that articulates with tibia.

   i. *Lateral condyle* Lateral posterior projection on distal end that articulates with tibia.
   j. *Intercondylar* (in'-ter-KON-di-lar) *fossa* Depressed area between condyles on posterior surface.
   k. *Medial epicondyle* Projection above medial condyle.
   l. *Lateral epicondyle* Projection above lateral condyle.
   m. *Patellar surface* Between condyles on the anterior surface.

Label the parts of the femur in Figure 7.12a and b.

2. *Patella* (*patella* = small plate) Kneecap in front of knee joint; develops in tendon of quadriceps femoris muscle.
   a. *Base* Broad superior portion.
   b. *Apex* Pointed inferior portion.
   c. *Articular facets* Articulating surfaces on posterior surface for medial and lateral condyles of femur.

Label the parts of the patella in Figure 7.12c and d.

3. *Tibia* Shinbone; medial bone of leg.
   a. *Lateral condyle* Articulates with lateral condyle of femur.
   b. *Medial condyle* Articulates with medial condyle of femur.
   c. *Intercondylar eminence* Upward projection between condyles.
   d. *Tibial tuberosity* Anterior projection for attachment to patellar ligament.
   e. *Medial malleolus* (mal-LĒ-ō-lus; *malleus* = little hammer) Distal projection that articulates with talus bone of ankle.
   f. *Fibular notch* Distal depression that articulates with the fibula.
4. *Fibula* Lateral bone of leg.
   a. *Head* Proximal projection that articulates with tibia.
   b. *Lateral malleolus* Projection at distal end that articulates with the talus bone of ankle.

Label the parts of the tibia and fibula in Figure 7.12e.

5. *Tarsus* Seven bones of the ankle called *tarsals.*
   a. *Posterior bones—talus* (TĀ-lus) and *calcaneus* (kal-KĀ-nē-us) (heel bone).
   b. *Anterior bones—cuboid, navicular (scaphoid),* and three *cuneiforms* called the *first (medial), second (intermediate),* and *third (lateral) cuneiforms.*

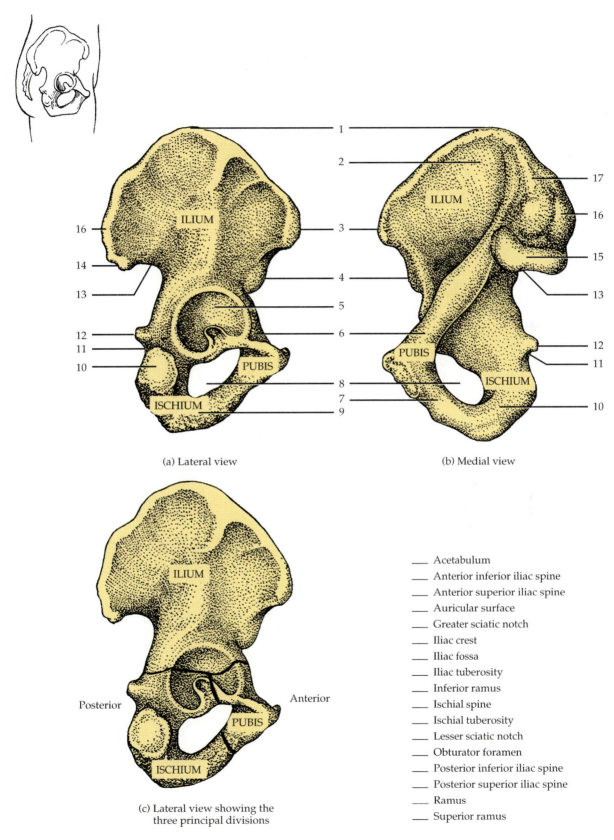

(a) Lateral view

(b) Medial view

(c) Lateral view showing the
three principal divisions

_____ Acetabulum
_____ Anterior inferior iliac spine
_____ Anterior superior iliac spine
_____ Auricular surface
_____ Greater sciatic notch
_____ Iliac crest
_____ Iliac fossa
_____ Iliac tuberosity
_____ Inferior ramus
_____ Ischial spine
_____ Ischial tuberosity
_____ Lesser sciatic notch
_____ Obturator foramen
_____ Posterior inferior iliac spine
_____ Posterior superior iliac spine
_____ Ramus
_____ Superior ramus

FIGURE 7.11   Right hipbone.

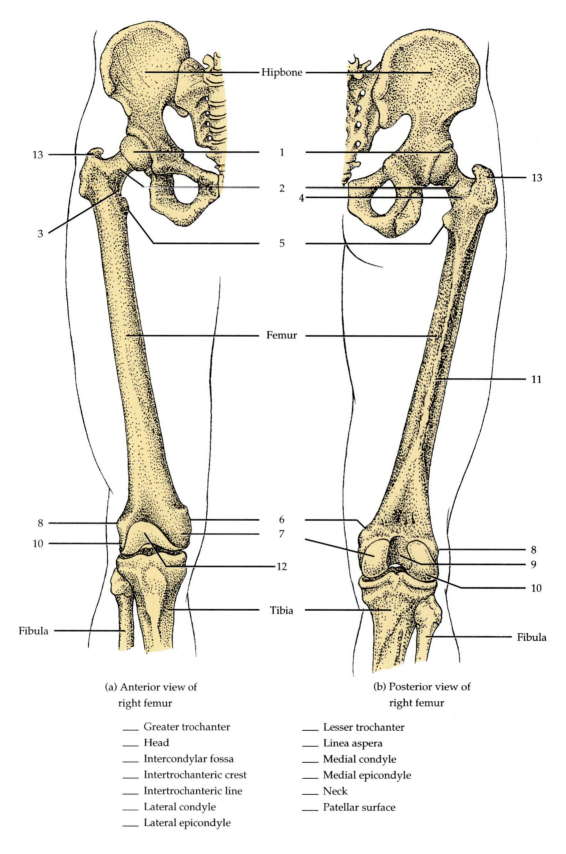

Hipbone

13
3

1
2

5

Femur

11

8
10

6
7

12

Tibia

8
9
10

Fibula

13

4

Fibula

(a) Anterior view of
right femur

(b) Posterior view of
right femur

____ Greater trochanter
____ Head
____ Intercondylar fossa
____ Intertrochanteric crest
____ Intertrochanteric line
____ Lateral condyle
____ Lateral epicondyle

____ Lesser trochanter
____ Linea aspera
____ Medial condyle
____ Medial epicondyle
____ Neck
____ Patellar surface

FIGURE 7.12   Bones of the lower limb.

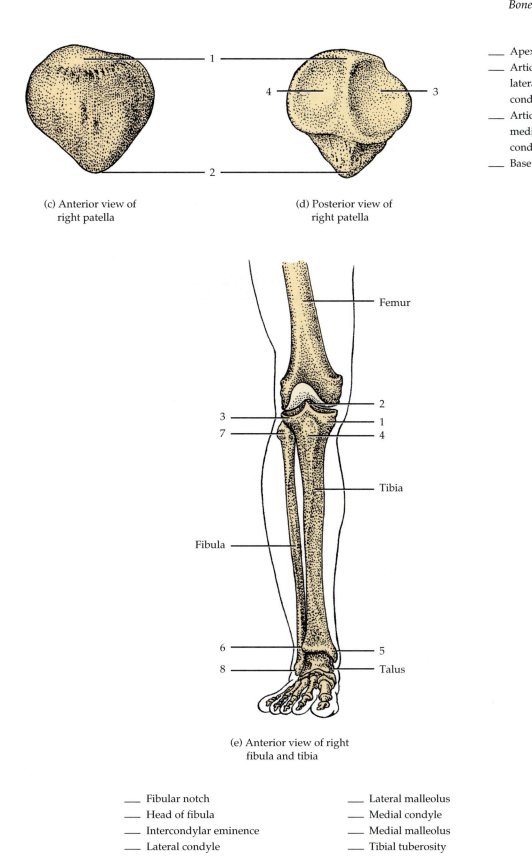

(c) Anterior view of
right patella

(d) Posterior view of
right patella

____ Apex
____ Articular facet for
lateral femoral
condyle
____ Articular facet for
medial femoral
condyle
____ Base

(e) Anterior view of right
fibula and tibia

____ Fibular notch   ____ Lateral malleolus
____ Head of fibula   ____ Medial condyle
____ Intercondylar eminence ____ Medial malleolus
____ Lateral condyle   ____ Tibial tuberosity

FIGURE 7.12 (Continued)   Bones of the lower limb.

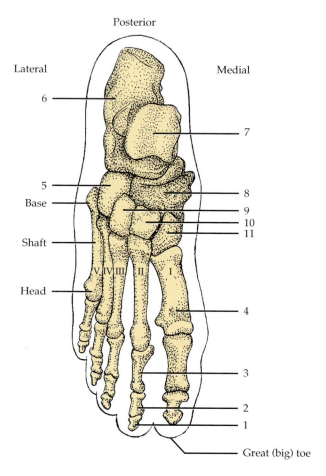

Posterior

Lateral   Medial

6

7

5
Base

8
9
10
11

Shaft

V IV III II I

Head

4

3

2

1

Great (big) toe

(f) Superior view of right foot

___ Calcaneus
___ Cuboid
___ Distal phalanx
        First (medial) cuneiform
___ Metatarsal
___ Middle phalanx
___ Navicular
___ Proximal phalanx
___ Second (intermediate) cuneiform
___ Talus
___ Third (lateral) cuneiform

FIGURE 7.12 (Continued)   Bones of the lower limb.

6. *Metatarsus* Consists of five bones of the foot called metatarsals, numbered as follows, beginning on the medial (large toe) side: I, II, III, IV, and V metatarsals.
7. *Phalanges* Bones of the toes, comparable to phalanges of fingers; two in each large toe (proximal and distal) and three in each small toe (proximal, middle, and distal).

Label the parts of the tarsus, metatarsus, and phalanges in Figure 7.12f.

L. ARTICULATED SKELETON

Now that you have studied all the bones of the body, label the entire articulated skeleton in Figure 7.13.

M. SKELETAL SYSTEM OF CAT

The *skeletal system* of the cat, like that of the human, is conveniently divided into two principal divisions: axial skeleton and appendicular skeleton. The *axial skeleton* of the cat consists of the bones of the skull, vertebral column, ribs, sternum, and hyoid apparatus. The *appendicular skeleton* of the cat consists of the bones of the pectoral girdle, pelvic girdle, forelimbs, and hindlimbs.

Before you start your examination of the bones of the cat skeleton, it should be noted that the cat walks on its digits with the remainder of the palm and foot elevated. This is referred to as *digitigrade locomotion*. Humans, by contrast, walk on the entire sole of the foot. This is called *plantigrade locomotion*.

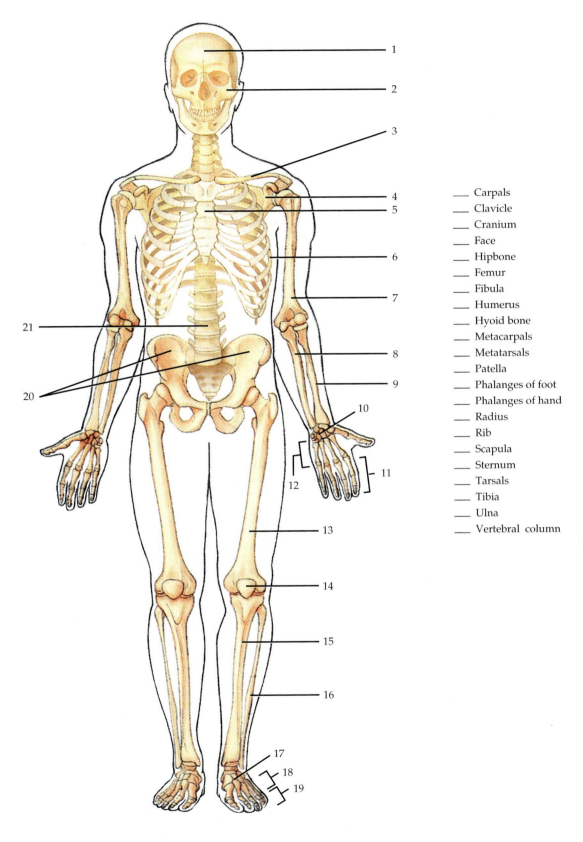

___ Carpals
___ Clavicle
___ Cranium
___ Face
___ Hipbone
___ Femur
___ Fibula
___ Humerus
___ Hyoid bone
___ Metacarpals
___ Metatarsals
___ Patella
___ Phalanges of foot
___ Phalanges of hand
___ Radius
___ Rib
___ Scapula
___ Sternum
___ Tarsals
___ Tibia
___ Ulna
___ Vertebral column

Anterior view

FIGURE 7.13   Entire skeleton.

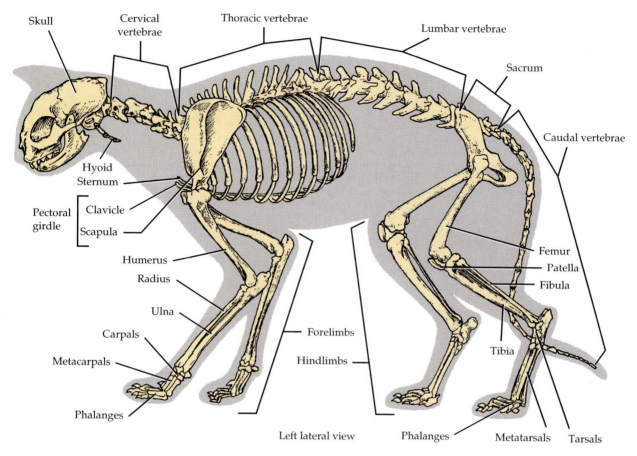

FIGURE 7.14 Skeleton of cat.

Refer to Figure 7.14, which shows the skeleton of the cat in lateral view. Be sure that you study the figure carefully so that you can identify the principal bones.

Obtain a mounted articulated skeleton and identify all the bones indicated in Figure 7.14.

Compare the cat skeleton to a human articulated skeleton.

ANSWER THE LABORATORY REPORT QUESTIONS AT THE END OF THE EXERCISE.

Bones 7

Student _____   **Date** _____

Laboratory Section _____   **Score/Grade** _____

PART 1. Multiple Choice

_____ 1. The suture between the parietal and temporal bone is the (a) lambdoid (b) coronal (c) squamous (d) sagittal

_____ 2. Which bone does *not* contain a paranasal sinus? (a) ethmoid (b) maxilla (c) sphenoid (d) sacral

_____ 3. Which is the superior, concave curve in the vertebral column? (a) thoracic (b) lumbar (c) cervical (d) sacral

_____ 4. The fontanel between the parietal and occipital bones is the (a) anterolateral (b) anterior (c) posterior (d) posterolateral

_____ 5. Which is *not* a component of the upper limb? (a) radius (b) femur (c) carpus (d) humerus

_____ 6. All are components of the appendicular skeleton *except* the (a) humerus (b) occipital bone (c) calcaneus (d) triquetral

_____ 7. Which bone does *not* belong with the others? (a) occipital (b) frontal (c) parietal (d) mandible

_____ 8. Which region of the vertebral column is closer to the skull? (a) thoracic (b) lumbar (c) cervical (d) sacral

_____ 9. Of the following bones, the one that does *not* help form part of the orbit is the (a) sphenoid (b) frontal (c) occipital (d) lacrimal

_____ 10. Which bone does *not* form a border for a fontanel? (a) maxilla (b) temporal (c) occipital (d) parietal

PART 2. Identification

For each surface marking listed, identify the skull bone to which it belongs:

11. Glabella _____

12. Mastoid process _____

13. Sella turcica _____

14. Cribriform plate _____

15. Foramen magnum _____

16. Mental foramen _____

17. Infraorbital foramen _____

18. Crista galli _____

19. Foramen ovale _____

20. Horizontal plate _____

21. Optic foramen _____

22. Superior nasal concha _____

23. Zygomatic process _____

24. Styloid process _____

25. Mandibular fossa _____

PART 3. Matching

_____	**26.** Iliac crest	A. Inferior portion of sternum
_____	**27.** Capitulum	B. Medial bone of distal carpals
_____	**28.** Medial malleolus	C. Distal projection of tibia
_____	**29.** Laminae	D. Lateral end of clavicle
_____	**30.** Vertebral foramen	E. Points where bodies of sacral vertebrae join
_____	**31.** Talus	F. Portion of rib that contains blood vessels
_____	**32.** Olecranon	G. Prominence of elbow
_____	**33.** Pisiform	H. Lateral projection of humerus
_____	**34.** Acromial extremity	I. Medial projection for insertion of biceps brachii muscle
_____	**35.** Pubic symphysis	J. Lower posterior portion of hipbone
_____	**36.** Hamate	K. Articulates with head of radius
_____	**37.** Costal groove	L. Prominence on lateral side of femur
_____	**38.** Xiphoid process	M. Distal projection of fibula
_____	**39.** Greater trochanter	N. Superior border of ilium
_____	**40.** Transverse lines	O. Articulates with head of humerus
_____	**41.** Radial tuberosity	P. Anterior joint between hipbones
_____	**42.** Greater tubercle	Q. Opening through which spinal cord passes
_____	**43.** Ischium	R. Form posterior wall of vertebral arch
_____	**44.** Glenoid cavity	S. Medial bone of proximal carpals
_____	**45.** Lateral malleolus	T. Component of tarsus

Articulations

An *articulation* (ar-tik'-yoo-LĀ-shun), or *joint*, is a point of contact between bones, cartilage and bones, or teeth and bones. When we say that one bone articulates with another, we mean that one bone forms a joint with another. The scientific study of joints is called *arthrology* (ar-THROL-ō-jē; *arthro* = joint; *logos* = study of).

Some joints permit no movement, others permit a slight degree of movement, and still others permit free movement. In this exercise you will study the structure and action of joints.

A. KINDS OF JOINTS

The joints of the body may be classified into several principal kinds on the basis of their structure and function (degree of movement they permit).

The structural classification of joints is based on the presence or absence of a space between articulating bones called a *synovial* or *joint cavity* and the type of connective tissue that binds the bones together. Structurally, a joint is classified as

1. *Fibrous* (FĪ-brus) There is no synovial cavity and the bones are held together by fibrous connective tissue.
2. *Cartilaginous* (kar-ti-LAJ-i-nus) There is no synovial cavity and the bones are held together by cartilage.
3. *Synovial* (si-NŌ-vē-al) There is a synovial cavity and the bones forming the joint are united by a surrounding articular capsule and frequently by accessory ligaments (described in detail later).

The functional classification of joints is as follows:

1. *Synarthroses* (sin'-ar-THRŌ-sēz; *syn* = together; *arthros* = joint) Immovable joints.
2. *Amphiarthroses* (am'-fē-ar-THRŌ-sēz; *amphi* = on both sides) Slightly movable joints.

3. *Diarthroses* (dī-ar-THRŌ-sēz; *diarthros* = movable joint) Freely movable joints.

We will discuss the joints of the body on the basis of their functional classification, referring to their structural classification as well.

B. SYNARTHROSES (IMMOVABLE JOINTS)

A synarthrosis may be of three types: sutures, synchondroses, and gomphoses.

1. *Sutures* (SOO-cherz; *sutura* = seam) A fibrous joint composed of a thin layer of dense fibrous connective tissue that unites the bones of the skull. Example: coronal suture between the frontal and parietal bones.
2. *Gomphoses* (gom-FŌ-sēz; *gomphosis* = to bolt together) A type of fibrous joint in which a cone-shaped peg fits into a socket. The substance between the bones is the periodontal ligament. Example: articulations of the roots of the teeth with the alveoli (sockets) of the maxillae and mandible.
3. *Synchondroses* (sin'-kon-DRŌ-sēz; *syn* = together; *chondro* = cartilage) A cartilaginous joint in which the connecting material is hyaline cartilage. Example: epiphyseal plate between the epiphysis and diaphysis of a growing bone.

C. AMPHIARTHROSES (SLIGHTLY MOVABLE JOINTS)

Amphiarthroses (slightly movable joints) may be of two types: syndesmoses and symphyses.

1. *Syndesmoses* (sin'-dez-MŌ-sēz; *syndesmo* = band or ligament) A fibrous joint in which

there is considerably more fibrous connective tissue than in a suture; the fibrous connective tissue forms an interosseous membrane or ligament that permits some flexibility and movement. Example: the distal articulation of the tibia and fibula.

2. *Symphyses* (SIM-fi-sēz; *symphysis* = growing together) A cartilaginous joint in which the connecting material is a broad, flat disc of fibrocartilage. Example: intervertebral discs between the bodies of vertebrae and the pubic symphysis between the anterior surfaces of the hipbones.

D. DIARTHROSES (FREELY MOVABLE JOINTS)

Diarthroses, or freely movable joints, have a variety of shapes and permit several different types of movements. First, we will discuss the general structure of a diarthrosis and then consider the different types.

1. Structure of a Diarthrosis

A distinguishing anatomical feature of a diarthrosis is a space, called a *synovial* (si-NŌ-vē-al) or *joint cavity* (see Figure 8.1), that separates the articulating bones. Thus diarthroses are also called *synovial joints*. Another characteristic of such joints is the presence of *articular cartilage*. Articular cartilage (hyaline) covers the surfaces of the articulating bones but does not bind the bones together.

A sleevelike *articular capsule* surrounds a synovial joint, encloses the synovial cavity, and unites the articulating bones. The articular capsule is composed of two layers. The outer layer, the *fibrous capsule*, usually consists of dense, irregular connective tissue. It attaches to the periosteum of the articulating bones at a variable distance from the edge of the articular cartilage. The flexibility of the fibrous capsule permits movement at a joint, whereas its great tensile strength resists dislocation. The fibers of some fibrous capsules are arranged in parallel bundles and are therefore highly adapted to resist recurrent strain. Such fibers are called *ligaments* (*ligare* = to bind) and are given special names. The strength of the ligaments is one of the principal factors in holding bone to bone. Diarthroses are freely movable joints because of the synovial cavity and the arrangement of the articular capsule and accessory ligaments. The inner layer of the articular capsule is

formed by a *synovial membrane*. The synovial membrane is composed of areolar connective tissue with elastic fibers and a variable amount of adipose tissue. It secretes *synovial fluid (SF)*, which lubricates the joint and provides nourishment for the articular cartilage.

Many diarthroses also contain *accessory ligaments*, called extracapsular ligaments and intracapsular ligaments. *Extracapsular ligaments* lie outside the articular capsule. An example is the fibular (lateral) collateral ligament of the knee joint (see Figure 8.4b). *Intracapsular ligaments* occur within the articular capsule but are excluded from the synovial cavity by folds of the synovial membrane. Examples are the cruciate ligaments of the knee joint (see Figure 8.4b).

Inside some synovial joints are pads of fibrocartilage that lie between the articular surfaces of the bones and are attached by their margins to the fibrous capsule. These pads are called *articular discs (menisci)*. Singular is *meniscus*. See Figure 8.4b. The discs usually subdivide the synovial cavity into two separate spaces. Articular discs allow two bones of different shapes to fit tightly, modify the shape of the joint surfaces of the articulating bones, help to maintain the stability of the joint, and direct the flow of synovial fluid to areas of greatest friction.

Saclike structures called *bursae* (*bursa* = pouch) are strategically situated to alleviate friction in some spots (see Figure 8.4f). Bursae resemble joint capsules in that their walls consist of connective tissue lined by a synovial membrane. They are also filled with a fluid similar to synovial fluid. Bursae are located between the skin and bone in places where skin rubs over bone. They are also found between tendons and bones, muscles and bones, and ligaments and bones. Such fluid-filled sacs cushion the movement of one part of the body over another. Inflammation of a bursa is called *bursitis*.

Label Figure 8.1, the principal parts of a diarthrosis.

2. Movements at Diarthroses

Movements at diarthroses may be classified as follows:

a. *Gliding* One surface moves back and forth and from side to side without any angular or rotary movement.

b. *Angular* Increase or decrease the angle between bones.

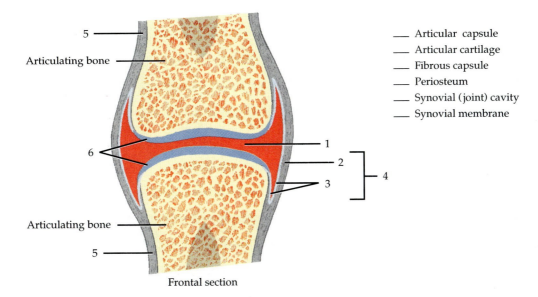

5

Articulating bone

6

1

2

3

4

Articulating bone

5

Frontal section

___ Articular capsule
___ Articular cartilage
___ Fibrous capsule
___ Periosteum
___ Synovial (joint) cavity
___ Synovial membrane

Figure 8.1   Parts of a diarthrosis (synovial joint).

1. *Flexion* Decrease in angle between articulating bones.
2. *Extension* Increase in angle between articulating bones.
3. *Hyperextension* Continuation of extension beyond the anatomical position.
4. *Abduction* (= taking away) Movement of a bone away from midline.
5. *Adduction* (= to move toward) Movement of a bone toward midline.
   c. *Rotation* Movement of a bone in a plane around its longitudinal axis.
   d. *Circumduction* Movement in which the distal end of a bone moves in a circle while the proximal end remains relatively stable; bone outlines a cone in the air.
   e. *Special* Found only at the joints indicated:
      1. *Inversion* Movement of sole of foot inward (medially).
      2. *Eversion* Movement of sole of foot outward (laterally).
      3. *Dorsiflexion* Bending the foot in the direction of the dorsum (upper surface).
      4. *Plantar flexion* Bending the foot in the direction of the plantar surface (sole).
      5. *Protraction* Movement of mandible or shoulder girdle forward on plane parallel to ground.
      6. *Retraction* Movement of protracted part of the body backward on plane parallel to ground.
      7. *Supination* Movement of forearm in which palm is turned anteriorly or superiorly.

8. *Pronation* Movement of forearm in which palm is turned posteriorly or inferiorly.
9. *Depression* Movement in which part of body, such as the mandible or scapula, moves inferiorly.
10. *Elevation* Movement in which part of body moves superiorly.

Label the various movements illustrated in Figure 8.2.

3. Types of Diarthroses

The principal types of diarthroses are as follows:

a. *Gliding* Articulating surfaces usually flat; permits gliding movement in two planes, side-to-side and back-and-forth. Example: between carpals, tarsals, sacrum and ilium, sternum and clavicle, scapula and clavicle, and articular processes of vertebrae.
b. *Hinge* Convex surface of one bone fits into the concave surface of another; movement in single plane (monoaxial or uniaxial movement), usually flexion and extension. Example: elbow, knee, ankle, interphalangeal joints.
c. *Pivot* Rounded or pointed surface of one bone articulates within a ring formed partly by another bone and partly by a ligament; primary movement is rotation; joint is monoaxial. Example: between atlas and axis and between proximal ends of radius and ulna.

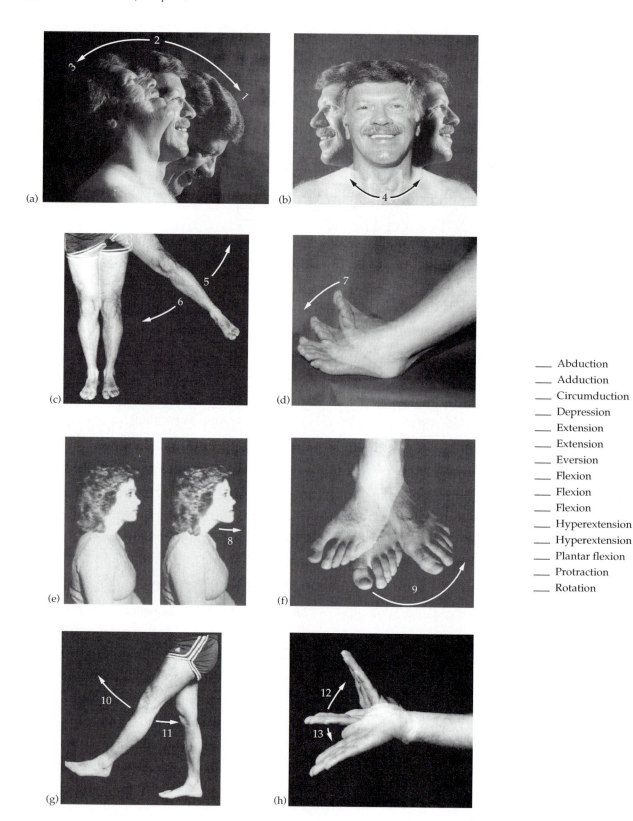

___ Abduction
___ Adduction
___ Circumduction
___ Depression
___ Extension
___ Extension
___ Eversion
___ Flexion
___ Flexion
___ Flexion
___ Hyperextension
___ Hyperextension
___ Plantar flexion
___ Protraction
___ Rotation

Figure 8.2   Movements at diarthroses (synovial joints).

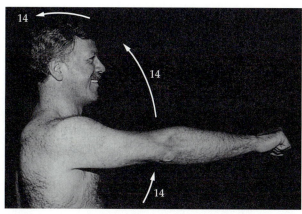

(i)

(j)

Figure 8.2 *(Continued)*   Movements at diarthroses.

d. *Condyloid* Oval-shaped condyle of one bone fits into an elliptical cavity of another bone; movement is biaxial, side-to-side, and back-and-forth. Example: between radius and carpals.

e. *Saddle* Articular surface of one bone is saddle-shaped and the articular surface of the other bone is shaped like a rider sitting in the saddle; movement similar to that of a condyloid joint. Example: between trapezium of carpus and metacarpal of thumb.

f. *Ball-and-socket* Ball-like surface of one bone fits into cuplike depression of another bone;

movement is in three planes (triaxial movement), flexion-extension, abduction-adduction, and rotation. Example: shoulder and hip joints.

Examine the articulated skeleton and find as many examples as you can of the joints just described. As part of your examination, be sure to note the shapes of the articular surfaces and the movements possible at each joint.

Label Figure 8.3.

E. KNEE JOINT

The knee (tibiofemoral) joint is one of the largest joints in the body and illustrates the basic structure of a diarthrosis and the limitations on its movement. Some of the structures associated with the knee joint are as follows:

1. *Tendon of quadriceps femoris muscle* Strengthens joint anteriorly and externally.
2. *Gastrocnemius muscle* Strengthens joint posteriorly and externally.
3. *Patellar ligament* Continuation of the tendon of the quadriceps femoris tendon below the patella that strengthens anterior portion of joint and prevents leg from being flexed too far backward.
4. *Fibular (lateral) collateral ligament* Between femur and fibula; strengthens the lateral side of the joint and prohibits side-to-side movement at the joint.
5. *Tibial (medial) collateral ligament* Between femur and tibia; strengthens the medial side of the joint and prohibits side-to-side movement at the joint.
6. *Oblique popliteal ligament* Starts in a tendon that lies over the tibia and runs upward and laterally to the lateral side of the femur; supports the posterior surface of the knee.
7. *Anterior cruciate ligament (ACL)* Passes posteriorly and laterally from the tibia and attaches to the femur; strengthens the joint internally and may help stabilize the knee during its movements.
8. *Posterior cruciate ligament (PCL)* Passes anteriorly and medially from the tibia and attaches to the femur; strengthens the joint internally and may help stabilize the knee during its movements.
9. *Articular discs (menisci)* Concentric wedge-shaped pieces of fibrocartilage between the

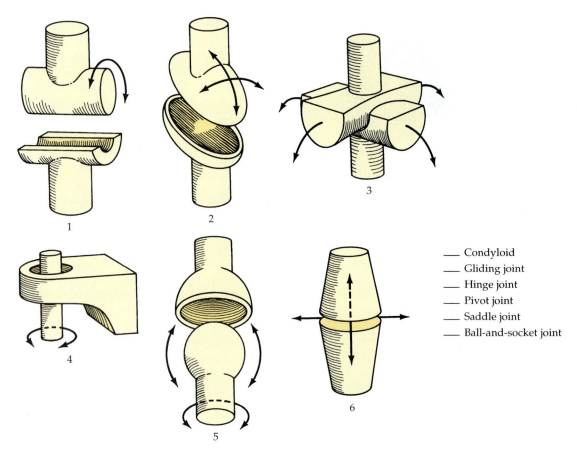

Figure 8.3 Types of synovial joints.

_____ Condyloid
_____ Gliding joint
_____ Hinge joint
_____ Pivot joint
_____ Saddle joint
_____ Ball-and-socket joint

femur and tibia; called *lateral meniscus* and the *medial meniscus;* help compensate for the irregular shapes of articulating bones and circulate synovial fluid.

Label the structures associated with the knee joint in Figure 8.4.

If a longitudinally sectioned knee joint of a cow or lamb is available, examine it and see how many structures you can identify.

ANSWER THE LABORATORY REPORT QUESTIONS AT THE END OF THE EXERCISE.

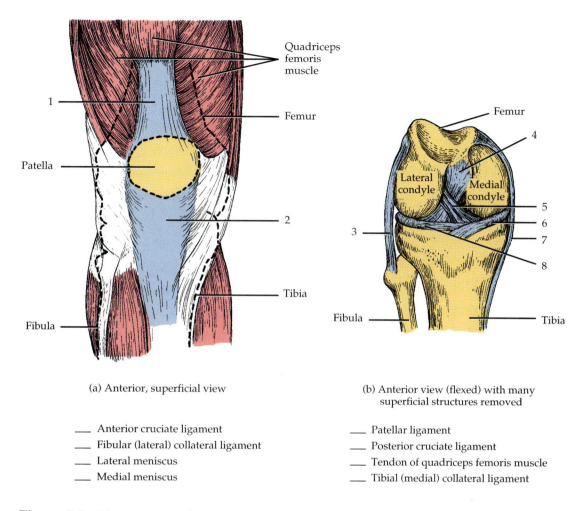

(a) Anterior, superficial view

(b) Anterior view (flexed) with many superficial structures removed

___ Anterior cruciate ligament
___ Fibular (lateral) collateral ligament
___ Lateral meniscus
___ Medial meniscus

___ Patellar ligament
___ Posterior cruciate ligament
___ Tendon of quadriceps femoris muscle
___ Tibial (medial) collateral ligament

Figure 8.4   Ligaments, tendons, bursae, and menisci of right knee joint.

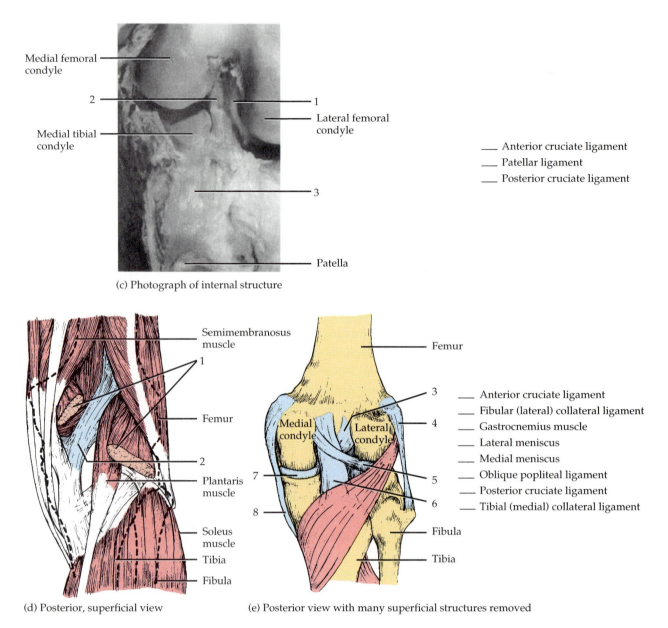

Medial femoral condyle

2

Medial tibial condyle

1

Lateral femoral condyle

3

Patella

(c) Photograph of internal structure

___ Anterior cruciate ligament
___ Patellar ligament
___ Posterior cruciate ligament

Semimembranosus muscle

1

Femur

2

Plantaris muscle

Soleus muscle

Tibia

Fibula

(d) Posterior, superficial view

Femur

3

Medial condyle

Lateral condyle

4

7

5

8

6

Fibula

Tibia

(e) Posterior view with many superficial structures removed

___ Anterior cruciate ligament
___ Fibular (lateral) collateral ligament
___ Gastrocnemius muscle
___ Lateral meniscus
___ Medial meniscus
___ Oblique popliteal ligament
___ Posterior cruciate ligament
___ Tibial (medial) collateral ligament

Figure 8.4 *(Continued)* Ligaments, tendons, bursae, and menisci of right knee joint.

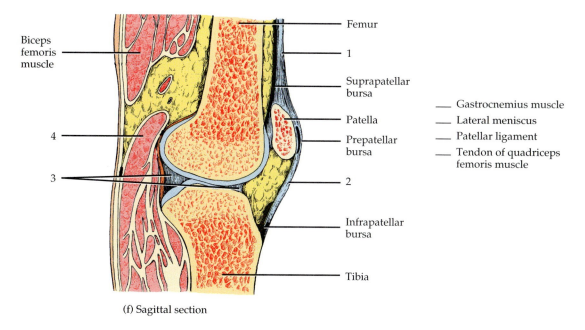

Biceps femoris muscle

Femur

1

Suprapatellar bursa

Patella

Prepatellar bursa

2

Infrapatellar bursa

Tibia

4

3

___ Gastrocnemius muscle
___ Lateral meniscus
___ Patellar ligament
___ Tendon of quadriceps femoris muscle

(f) Sagittal section

Figure 8.4 *(Continued)*   Ligaments, tendons, bursae, and menisci of right knee joint.

Articulations 8

Student _____   Date _____

Laboratory Section _____   Score/Grade _____

PART 1. Multiple Choice

_____ 1. A joint united by dense fibrous tissue that permits a slight degree of movement is a (a) suture (b) syndesmosis (c) symphysis (d) synchondrosis

_____ 2. A joint that contains a broad flat disc of fibrocartilage is classified as a (a) ball-and-socket joint (b) suture (c) symphysis (d) gliding joint

_____ 3. The following characteristics define what type of joint? Presence of a synovial cavity, articular cartilage, synovial membrane, and ligaments. (a) suture (b) synchondrosis (c) syndesmosis (d) hinge

_____ 4. Which joints are slightly movable? (a) diarthroses (b) amphiarthroses (c) synovial (d) synarthroses

_____ 5. Which type of joint is immovable? (a) synarthrosis (b) syndesmosis (c) symphysis (d) diarthrosis

_____ 6. What type of joint provides triaxial movement? (a) hinge (b) ball-and-socket (c) saddle (d) condyloid

_____ 7. Which ligament provides strength on the medial side of the knee joint? (a) oblique popliteal (b) posterior cruciate (c) fibular collateral (d) tibial collateral

_____ 8. On the basis of structure, which joint is fibrous? (a) symphysis (b) synchondrosis (c) pivot (d) syndesmosis

_____ 9. The elbow, knee, and interphalangeal joints are examples of which type of joint? (a) pivot (b) hinge (c) gliding (d) saddle

_____ 10. Functionally, which joint provides the greatest degree of movement? (a) diarthrosis (b) synarthrosis (c) amphiarthrosis (d) syndesmosis

PART 2. Completion

11. The thin layer of hyaline cartilage on articulating surfaces of bones is called _____ cartilage.

12. The synovial membrane and fibrous capsule together form the _____ capsule.

13. Pads of fibrocartilage between the articular surfaces of bones that maintain stability of the joint are

called _____.

14. Fluid-filled connective tissue sacs that cushion movements of one body part over another are

referred to as _____.

15. The _____ ligament supports the back of the knee and helps to prevent hyperextension.

PART 3. Matching

_____	**16.** Circumduction	A. Decrease in the angle between articulating bones
_____	**17.** Adduction	B. Moving a part upward
_____	**18.** Flexion	C. Bending the foot in the direction of the upper surface
_____	**19.** Pronation	D. Forward movement parallel to the ground
_____	**20.** Elevation	E. Movement of the sole inward at the ankle joint
_____	**21.** Protraction	F. Movement toward the midline
_____	**22.** Rotation	G. Bending the foot in the direction of the sole
_____	**23.** Plantar flexion	H. Turning the palm posterior
_____	**24.** Dorsiflexion	I. Movement of a bone around its own axis
_____	**25.** Inversion	J. Distal end of a bone moves in a circle while the proximal end remains relatively stable

Muscle Tissue

9

Muscle tissue constitutes 40 to 50% of total body weight and is composed of fibers (cells) that are highly specialized with respect to four characteristics: (1) *excitability (irritability),* or ability to receive and respond to certain stimuli by producing electrical signals called action potentials (impulses); (2) *contractility,* or ability to contract (shorten and thicken); (3) *extensibility (extension),* or ability to stretch when pulled; and (4) *elasticity,* or ability to return to original shape after contraction or extension. Through contraction, muscle performs three basic functions: motion, maintenance of posture, and heat production. In this exercise you will examine the histological structure of muscle tissue and conduct exercises on the physiology of frog muscle.

A. KINDS OF MUSCLE TISSUE

Histologically, three kinds of muscle tissue are recognized:

1. *Skeletal muscle tissue* Usually attached to bones; contains conspicuous striations (light and dark bands) when viewed microscopically; voluntary because it contracts under conscious control.
2. *Cardiac muscle tissue* Found only in the wall of the heart; striated when viewed microscopically; involuntary because it contracts usually without conscious control.
3. *Smooth (visceral) muscle tissue* Located in walls of viscera and blood vessels; referred to as nonstriated because it lacks striations when viewed microscopically; involuntary because it contracts usually without conscious control.

B. SKELETAL MUSCLE TISSUE

Examine a prepared slide of skeletal muscle tissue in longitudinal and transverse section under high power. Look for the following:

1. *Sarcolemma* (*sarco* = flesh; *lemma* = sheath) Plasma membrane of the muscle fiber.
2. *Sarcoplasm* Cytoplasm of the muscle fiber.
3. *Nuclei* Several in each muscle fiber lying close to sarcolemma.
4. *Striations* Alternating light and dark bands in each muscle fiber (described on page 136).
5. *Epimysium* (ep'-i-MĪZ-ē-um; *epi* = upon) Fibrous connective tissue that surrounds the entire skeletal muscle.
6. *Perimysium* (per'-i-MĪZ-ē-um; *peri* = around) Fibrous connective tissue surrounding a bundle (fascicle) of muscle fibers.
7. *Endomysium* (en'-dō-MĪZ-ē-um; *endo* = within) Fibrous connective tissue surrounding individual muscle fibers.

Refer to Figure 9.1 and label the structures indicated.

With the use of an electron microscope, additional details of skeletal muscle tissue may be noted. Among these are the following:

1. *Mitochondria* Organelles that have smooth outer membrane and folded inner membrane in which ATP is generated.
2. *Sarcoplasmic reticulum* (sar'-kō-PLAZ-mik re-TIK-yoo-lum), or *SR* Network of fluid-filled cisterns similar to the endoplasmic reticulum of nonmuscle cells; stores calcium ions in relaxed muscle fibers.
3. *Transverse (T) tubules* Tunnel-like infoldings of sarcolemma that run perpendicular to and connect with sarcoplasmic reticulum; open to outside of muscle fiber.
4. *Triad* Transverse tubule and the segments of sarcoplasmic reticulum on either side of it.
5. *Myofibrils* Threadlike structures that run lengthwise through a fiber and consist of *thin filaments* composed of the protein actin and *thick filaments* composed of the protein myosin.

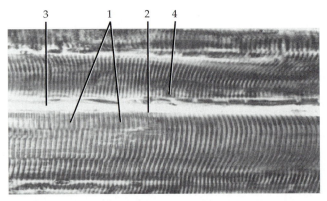

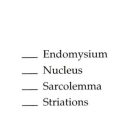

_____ Endomysium
_____ Nucleus
_____ Sarcolemma
_____ Striations

(a) Photomicrograph of several muscle fibers in longitudinal section

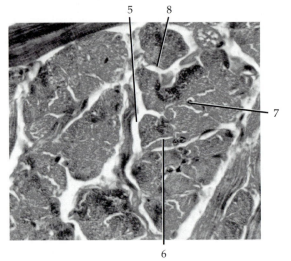

_____ Endomysium
_____ Nucleus
_____ Perimysium
_____ Sarcolemma

(b) Photomicrograph of several muscle fibers in transverse section

FIGURE 9.1   Histology of skeletal muscle tissue.

6. *Sarcomere* (*meros* = part) Contractile unit of a muscle fiber; compartment within a muscle fiber separated from other sarcomeres by dense material called *Z discs (lines)*.

7. *A (anisotropic) band* Dark region in a sarcomere that consists mostly of thick filaments and portions of thin filaments where they overlap the thick filaments.

8. *I (isotropic) band* Light region in a sarcomere composed of the rest of the thin filaments but no thick filaments. The combination of alternating dark A bands and light I bands gives the muscle fiber the striated (striped) appearance.

9. *H zone* Region in the center of the A band of a sarcomere consisting of thick filaments only.

10. *M line* Series of fine threads in the center of the H zone formed by proteins that connect adjacent thick filaments.

Figure 9.2 is a diagram of skeletal muscle tissue based on electron micrographic studies. Label the structures shown.

C. CARDIAC MUSCLE TISSUE

Examine a prepared slide of cardiac muscle tissue in longitudinal and transverse section under high power. Locate the following structures: *sarcolemma, endomysium, nuclei, striations,* and *intercalated*

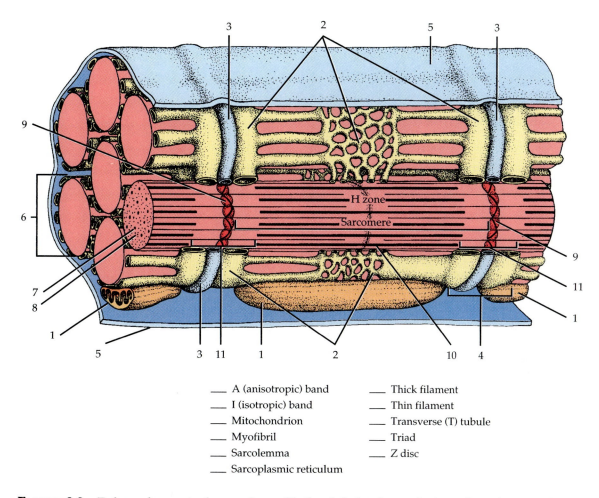

H zone
Sarcomere

___ A (anisotropic) band
___ I (isotropic) band
___ Mitochondrion
___ Myofibril
___ Sarcolemma
___ Sarcoplasmic reticulum
___ Thick filament
___ Thin filament
___ Transverse (T) tubule
___ Triad
___ Z disc

FIGURE 9.2 Enlarged aspect of several myofibrils of skeletal muscle tissue based on an electron micrograph.

discs (transverse thickenings of the sarcolemma that separate individual fibers). Label Figure 9.3.

D. SMOOTH (VISCERAL) MUSCLE TISSUE

Examine a prepared slide of smooth muscle tissue in longitudinal and transverse section under high power. Locate and label the following structures in Figure 9.4: *sarcolemma, sarcoplasm, nucleus,* and *muscle fiber.*

E. PHYSIOLOGY OF SKELETAL MUSCLE CONTRACTION

The process of skeletal muscle contraction involves a series of electrical, chemical, and mechanical events. The linkage between the production of the electrical and chemical events to the actual mechanical contraction of skeletal muscle is termed *excitation-contraction coupling.* Muscle contraction occurs only if the muscle is stimulated with a stimulus of threshold or suprathreshold intensity. A *threshold stimulus* is defined as the minimum strength stimulus necessary to initiate a contraction. Under normal physiological conditions, this is accomplished by a nerve impulse's being transmitted to the skeletal muscle fiber via a nerve cell called a *motor neuron.* A motor neuron and all of the muscle fibers it innervates is termed a *motor unit.*

The portion of the motor neuron that extends from the nerve cell body to a muscle fiber is termed an *axon.* Upon entering the connective tissue surrounding each individual muscle fiber (termed the endomysium), an axon branches into *axon terminals.* These axon terminals approach

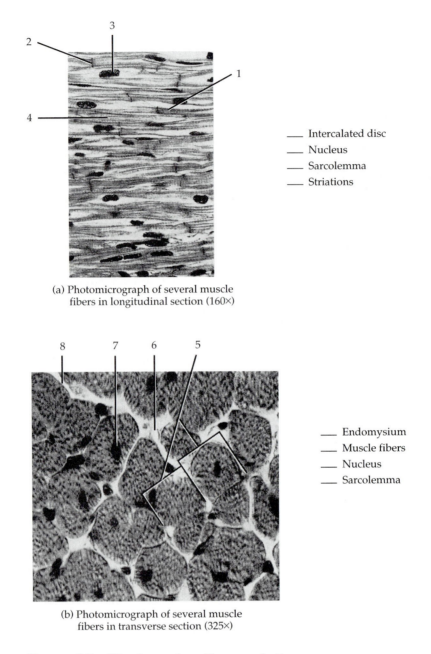

(a) Photomicrograph of several muscle fibers in longitudinal section (160×)

___ Intercalated disc
___ Nucleus
___ Sarcolemma
___ Striations

(b) Photomicrograph of several muscle fibers in transverse section (325×)

___ Endomysium
___ Muscle fibers
___ Nucleus
___ Sarcolemma

FIGURE 9.3 Histology of cardiac muscle tissue.

the sarcolemma of a muscle fiber, but do not come into actual contact with the muscle fiber. The structure formed by the axon terminal and the nearby skeletal muscle sarcolemma is termed a *neuromuscular (myoneural) junction* (Figure 9.5). The portion of the skeletal muscle fiber immediately deep to the axon terminal demonstrates a large number of anatomical specializations and is termed the *motor end plate.*

Close examination of a neuromuscular junction reveals that the distal ends of the axon terminals are expanded into bulblike structures termed *synaptic end bulbs.* These synaptic end bulbs contain membrane-enclosed sacs, called *synaptic vesicles,* that store chemicals termed *neurotransmitters.* The invaginated area of the sarcolemma under the axon terminal is termed the *synaptic gutter (trough);* the space between the axon terminal and sarcolemma is termed the *synaptic cleft.* The neurotransmitters diffuse across the synaptic cleft and bind to receptors on the motor end plate, initiating the excitation-contraction coupling process.

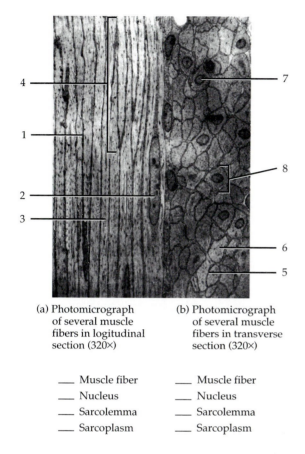

(a) Photomicrograph of several muscle fibers in logitudinal section (320×)

(b) Photomicrograph of several muscle fibers in transverse section (320×)

____ Muscle fiber   ____ Muscle fiber

____ Nucleus   ____ Nucleus

____ Sarcolemma   ____ Sarcolemma

____ Sarcoplasm   ____ Sarcoplasm

FIGURE 9.4 Histology of smooth (visceral) muscle tissue.

In a skeletal muscle neuromuscular junction, the nerve impulse reaches the end of an axon, causing it to depolarize. This depolarization suddenly and temporarily increases the permeability of the synaptic end bulb to calcium ions (Ca^{2+}). This increase in Ca^{2+} permeability causes the synaptic vesicles to fuse with the synaptic end bulb membrane, thereby releasing the neurotransmitter called *acetylcholine* (as'-ē-til-KŌ-lēn), or *ACh.* ACh diffuses across the synaptic cleft and binds to receptors on the motor end plate. This binding of Ach to the motor end plate initiates a localized depolarization called the *end-plate potential,* which achieves threshold intensity and initiates an action potential along the sarcolemma of the muscle fiber. The action potential travels along the sarcolemma and into the interior of the muscle fiber via the transverse tubules (T tubules). The action potential then travels from the T tubules to the sarcoplasmic reticulum (SR), depolarizes the SR, thereby causing the release of Ca^{2+} among the actin and myosin filaments. Ca^{2+} then binds to a regulatory protein, termed *troponin,* found in association with the actin filament. This Ca^{2+}-troponin complex causes a conformational change (change in shape) in the actin filament, thereby allowing the myosin cross bridges to attach to receptors on actin. The attachment between the myosin cross bridges and the receptors on the

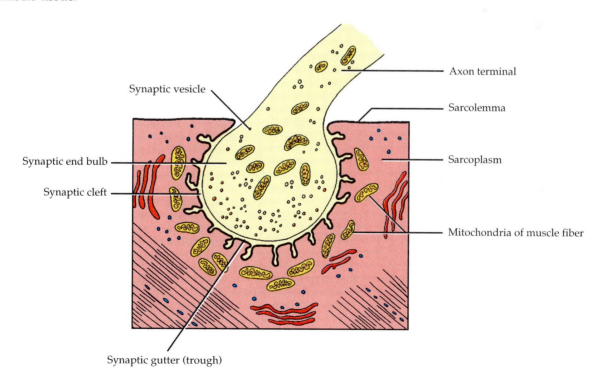

FIGURE 9.5 Neuromuscular junction.

actin causes ATP to be split by myosin ATPase, an enzyme found on the myosin filament. The energy released by the splitting of ATP into ADP and inorganic phosphate causes the myosin cross bridges to move, and the actin filaments slide inward toward the H zone, causing the muscle fiber to shorten (Figure 9.6). As this is occurring, newly synthesized ATP displaces ADP from the myosin molecule. If Ca^{2+} has been taken back into the sarcoplasmic reticulum by the calcium pump, the myosin cross bridges release from the actin receptors and relaxation occurs. However, if Ca^{2+} is still present within the sarcoplasm, the entire process repeats and further contraction occurs. This theory of skeletal muscle contraction is termed the *sliding filament theory of skeletal muscle contraction.*

A contracting skeletal muscle fiber follows the *all-or-none principle.* Simply stated, this principle states that a skeletal muscle fiber will respond maximally or not at all to a stimulus. If the stimulus is of threshold or greater intensity the fiber will respond maximally; if the stimulus is subthreshold in intensity the fiber will not respond. This principle, however, does not imply that the entire skeletal muscle must be either fully relaxed or fully contracted, because individual fibers within a muscle possess varying thresholds for stimulation. Therefore, the muscle as a whole can have graded contractions.

With these facts in mind, you will now perform a few simple lab tests that will illustrate the physiology of skeletal muscle contraction.

F. LABORATORY TESTS ON SKELETAL MUSCLE CONTRACTION[1]

Skeletal muscles produce different kinds of contractions depending on the intensity and frequency of the stimulus applied. Twitch contractions do not occur in the body but are worth demonstrating because they show the different phases of muscle

[1]These tests involve skeletal muscle tissue only. Tests on cardiac muscle tissue are included in Exercise 19 ("Cardiovascular Physiology") and tests on smooth muscle in Exercise 22 ("Digestive System"). At the discretion of the instructor, muscle tissue tests may be done in individual exercises or combined into a single exercise.

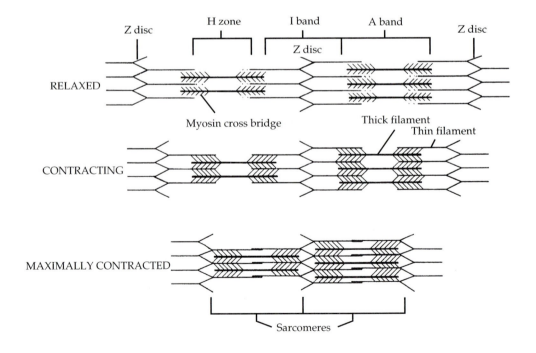

FIGURE 9.6 Sliding filament theory of a skeletal muscle contraction. The positions of the various parts of two sarcomeres in relaxed, contracting, and maximally contracted states are shown. Note the movement of thin filaments and relative size of H zone.

contraction quite clearly. Any record of a muscle contraction is called a *myogram.*

Note: *Depending on the availability of live animals and the type of laboratory equipment available, you may select from the following procedures related to skeletal muscle contraction.*

a. PROCEDURE USING PHYSIOFINDER

Physiofinder is an interactive computer program that permits laboratory simulations which do not involve the use of animals or advanced or expensive laboratory equipment. It permits students to perform experiments, analyze data, draw conclusions, and form hypotheses on the basis of collected data. The program is available from HarperCollins Publishers (1-800-8HEART1).

For activities related to skeletal muscle contraction, select the appropriate experiments from Physiofinder.

Module 1—The Action Potential
Module 2—Synaptic Integration
Module 3—The Neuromuscular Junction

b. PROCEDURE USING PHYSIOGRIP™

Physiogrip™ is a computer hardware and software system designed to demonstrate the contractile characteristics of human skeletal muscle. Through the use of a specially designed displacement transducer, Physiogrip™ provides an alternative to vivisection for studying classic muscle physiology by allowing students to experience their own flexor digitorum superficialis muscle responding to motor point stimulation. The program is available from INTELITOOL® (1-800-227-3805).

c. PROCEDURE USING POLYGRAPH AND PITHED FROGS

You may test muscle contraction through a team exercise or by observing a demonstration prepared by your instructor. Read the sections that follow on pithing a frog and muscle or nerve-muscle preparations *before* you do the exercise or observe the demonstration. Depending on the size of the class and the equipment available, you should work in teams of three to five students: one or two students should be assigned to assist the instructor to set up the physiologic apparatus, one or two other students should be assigned as "surgeons" to isolate, remove, and suspend the frog muscle, and another student should act as the recorder and coordinate the work.

1. Polygraph

The polygraph (Figure 9.7) records physiological events and works by receiving a signal from a sensing device, amplifying it, and changing it into a perceptible form. For example, a sensing device called a *myograph transducer* measures the force exerted by a contracting muscle and converts this mechanical force into an electrical signal. This signal is magnified by the amplifier and then changed into perceptible form by a writing pen that makes a permanent, visible record of the contraction.

A polygraph usually has a number of recording channels. Each is used to record a particular physiological event as it occurs. Different physiological events may be recorded simultaneously by using different channels. Each channel consists of a sensing device to detect the event, a coupler to join the sensing device to the amplifier, an amplifier, and a writing pen to make a record of what happened. A fifth pen, marked TIME AND EVENTS, is used to keep track of time and to mark the points at which stimuli are delivered. This time pen is the lowest pen on the instrument. The other pens are numbered to correspond to the recording channel with which they work.

Either electrodes or a transducer can serve as the sensing device for the polygraph. *Electrodes* are made of metal plates or small rods that can detect electrophysiological phenomena such as changes in electrical potential of the skin or nerves. Electrodes transmit the original electrical signal without change to the amplifier. In many cases a transducer must be used as the sensing device. The *transducer* converts nonelectrical signals into electrical signals: for example, the myograph transducer already mentioned converts mechanical force from a muscle contraction into an electrical signal; a heat transducer converts heat into an electrical signal; and a pressure transducer converts pressure changes into an electrical signal.

PROCEDURE

a. ORIENTATION TO THE POLYGRAPH AND RELATED ACCESSORIES

Your instructor will give you an orientation to the procedures to be followed for the particular brand of polygraph to be utilized in your laboratory. *Do not turn the instrument on or utilize it in any way prior to this orientation lecture.* Following the orientation lecture you should feel comfortable regulating all of the components of the polygraph, including the

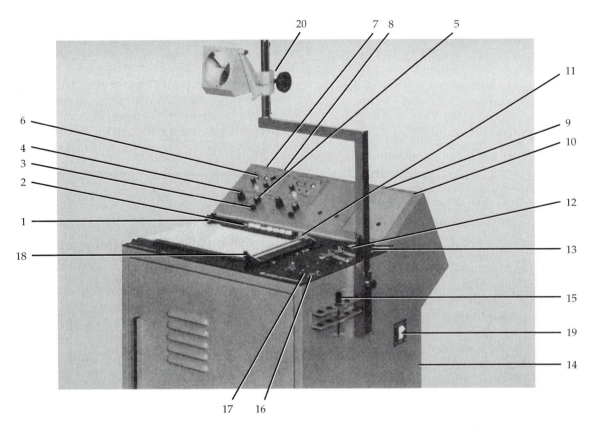

1. Chart drive tension wheel
2. Paper speed switches
3. Record
4. Sensitivity
5. Position
6. Coupler compartment
7. Auxiliary input

8. Monitor output
9. Event input
10. Event output
11. Timer switch
12. Power switch
13. Ground terminal
14. Power connector

15. Inkwell
16. External marker
17. Event marker
18. Pen lifter assembly
19. Projector power switch
20. Projector lens assembly

FIGURE 9.7 A Physiograph polygraph.

regulation of the writing pens, paper drive and paper controls, time and event marker, amplifiers, preamplifiers, couplers, and transducers.

b. CALIBRATION OF THE SYSTEM

1. Suspend a 100-g weight from the muscle transducer and adjust the sensitivity of the amplifier so that a 5-cm pen deflection is obtained.
2. If at any time this calibration proves unsatisfactory, change the sensitivity to best fit the situation. *Record the calibration factor* for future reference during this experiment.

2. Pithing of Frogs

By definition, *pithing* is destruction of the central nervous system by piercing the brain or spinal cord. This procedure is used in animal experimentation to render the animal unconscious so that it feels no pain. A single-pithed frog is one in which only the brain is destroyed, whereas a double-pithed frog has its spinal cord destroyed too. Usually the double-pithing procedure is used.

Note: It is strongly suggested that either your instructor or a laboratory assistant perform the following procedure for pithing a frog. The procedure requires some practice and may prove difficult for inexperienced individuals.

PROCEDURE

1. There are two accepted methods for pithing a frog.
2. One method of single pithing a frog entails holding the animal in a paper towel with its

dorsal side up and with the index finger pressing the nose down so that the head makes a right angle with the trunk (Figure 9.8).

3. Locate the slight depression formed by the first vertebra and skull about 3 mm behind the line joining the posterior borders of the tympanic membrane. This groove represents the area of the foramen magnum.

4. Carefully insert a long sharp-tipped needle or probe into the foramen magnum and direct it forward and a little downward.

5. Exert a steady pressure and rotate it, moving it from side to side in the cranial cavity to destroy the brain.

6. To double pith the frog, insert the needle into the vertebral canal, directing it downward until it has reached the end of the canal. Move the needle from side to side as you go.

7. A second acceptable method to single pith a frog involves holding the animal in a paper towel with its dorsal side up and inserting one blade of a pair of scissors into the animal's mouth and cutting off the top of the head just posterior to the eyes.

8. To double pith the animal insert a long sharp-tipped needle or probe into the exposed vertebral column and direct it downward until it has reached the end of the canal, moving the needle from side to side as you go.

3. Experimental Preparation[2]

PROCEDURE

a. ISOLATED MUSCLE PREPARATION

The gastrocnemius muscle of the frog is commonly used in the laboratory to demonstrate muscle contraction.

1. After double-pithing the frog, cut the leg off two-thirds of the way up the thigh. Remove the skin by pulling it down over the ankle.

2. Cut the calcaneal (Achilles) tendon as far down over the ankle as possible and then gently pull the gastrocnemius free from the lower part of the leg.

3. Cut through the lower part of the leg so as to leave the gastrocnemius muscle attached to the femur, and carefully cut away the severed muscle of the upper part of the leg.

4. The muscle preparation consists of the gastrocnemius muscle, with its calcaneal tendon, attached to the bare lower half of the femur.

[2]You will need to follow all eight steps in the preparation sequence *only* if your laboratory instructor informs you that you will be doing the procedure outlined in Section 8, entitled "Demonstration of Refractory Period." If you will not be doing that particular experimental procedure *do not remove the second leg from the frog at this time.*

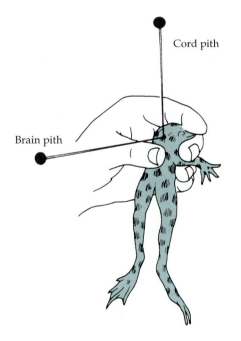

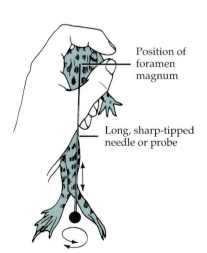

FIGURE 9.8   Procedure for pithing a frog.

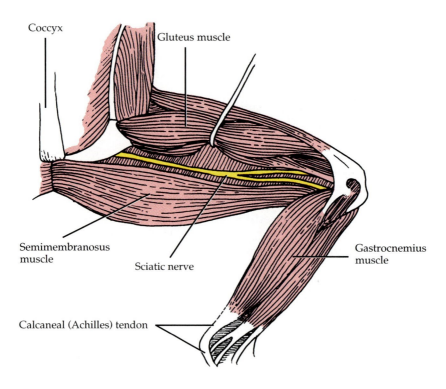

Coccyx

Gluteus muscle

Semimembranosus
muscle

Sciatic nerve

Gastrocnemius
muscle

Calcaneal (Achilles) tendon

FIGURE 9.9 Structures in the lower limb of the frog seen in posterior view.

The other gastrocnemius muscle should not be removed until needed (number 8). The gastrocnemius muscle is shown in Figure 9.9.

5. Place the muscle preparation in the apparatus with the bone held firmly in a femur clamp and the tendon connected to a muscle transducer by means of an S-hook stuck through the tendon (Figure 9.10).

6. Remember that an isolated skeletal muscle has no blood supply. Therefore moisten the muscle with Ringer's solution frequently and follow procedures carefully.

7. Adjust the height of the femur clamp so that the muscle has a slight amount of tension on it.

8. After the first preparation has been completed remove the second leg from the animal and immerse the entire leg, with the skin intact, in the iced Ringer's solution that has been prepared for use in Section 7 later in the laboratory period.

b. NERVE-MUSCLE PREPARATION

Another preparation that may be substituted for the isolated muscle preparation procedure is the nerve-muscle preparation. In this preparation the nerve innervating the gastrocnemius muscle is stimulated rather than directly stimulating the muscle.

Note: It is strongly suggested that either your instructor or a laboratory assistant perform the following procedure for pithing a frog. The procedure requires some practice and may prove difficult for inexperienced individuals.

1. Double-pith the frog and pin it down onto a frog board in the prone position. Remove the skin from the entire lower limb by first making a cut around the upper part of the leg and pulling the skin down over the ankle.

2. Cut the calcaneal tendon as far down over the ankle as possible and then gently pull the gastrocnemius free from the lower part of the leg.

3. Cut through the lower leg so as to leave the gastrocnemius muscle attached to the femur. Gently spread the gluteus muscle from the semimembranosus muscle and expose the sciatic nerve (see Figure 9.9).

4. *Do not stretch the nerve, and avoid any unnecessary contact between the nerve, metal instruments, and other tissues.*

5. Gently place a ligature under the nerve. *Do not tie.* Utilize this ligature to place the stimulating electrodes under the nerve. Gently lower the nerve to lie over the electrode tips.

6. Attach the muscle to a transducer via an S-hook stuck through the calcaneal tendon.

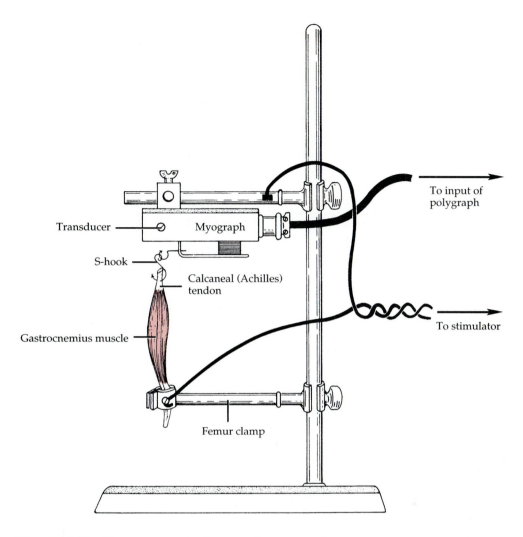

FIGURE 9.10   Frog gastrocnemius muscle preparation.

Periodically moisten the gastrocnemius muscle and sciatic nerve with Ringer's solution (Figure 9.10).

7. Adjust the height of the muscle transducer so that a slight amount of tension is placed upon the muscle.

4. Demonstration of Threshold Stimulus

PROCEDURE

1. With the polygraph set at its lowest paper speed and the stimulus intensity knob placed at its minimum setting, apply a single stimulus to the muscle by utilizing the single stimulus key.
2. Record the stimulus strength on the polygraph paper by the pen mark made by the event marker.

3. Soak the muscle preparation with Ringer's solution between each stimulus.
4. In a stepwise manner increase the stimulus intensity and determine when a threshold strength stimulus has been reached.

5. Demonstration of Unit Summation

The process of recruiting more muscle fibers to increase the strength of a muscular contraction is called *unit summation.*

PROCEDURE

1. Starting with a threshold strength stimulus continue to increase the stimulus strength in a step-wise manner, always increasing the stimulus strength by the same amount. Record the

stimulus strength on the polygraph paper by the pen mark made by the event marker.

2. Soak the muscle preparation with Ringer's solution between each stimulus.

3. Continue to increase the stimulus strength until you have recorded three voltages where the strength does not increase further.

4. If skeletal muscle obeys the all-or-none principle, are the results obtained in this procedure a contradiction of this principle? _____

_____

5. Under what circumstances would unit summation occur in the body? _____

_____

By what mechanism would unit summation occur in the body? _____

_____

Define the term **motor unit recruitment** and describe how it pertains to the application of unit summation under normal physiological conditions in Section F.1 of the LABORATORY REPORT RESULTS at the end of the exercise.

_____

_____

6. Include your results and response in Section F.1 of the LABORATORY REPORT RESULTS at the end of the exercise.

6. Contraction Curve

PROCEDURE

1. With the polygraph set at its fastest paper speed, record a single muscle contraction from a suprathreshold stimulus.

2. Calculate the latent period, contraction period, and relaxation period for this contraction curve.

3. After rinsing the muscle with Ringer's solution, repeat this procedure several times. Make sure to allow adequate time between each individual contraction, and to rinse the muscle with Ringer's solution after each contraction.

4. Calculate the average for the latent period, contraction period and relaxation period.

5. Place one tracing in Section F.3 of the LABORATORY REPORT RESULTS at the end of the exercise and record the average values.

7. Effect of Fatigue on a Single Muscle Twitch

PROCEDURE

1. *Familiarize yourself with the entire procedure before starting.* Set the polygraph at its lowest paper speed. Using a frequency of 60 stimuli per second and the multiple stimulus setting, stimulate the muscle until it fatigues. (Fatigue is demonstrated when the contraction strength is reduced by 50% or more as compared to the initial contraction strength.)

2. *Do not* rinse the muscle with Ringer's solution *anytime* during this procedure once you have started to stimulate the muscle. Rinsing the muscle at any time during this exercise will produce faulty results.

3. Upon onset of fatigue, rapidly turn the stimulator off, set the paper speed at its highest level, and stimulate the muscle with a single suprathreshold stimulus and record a contraction curve for the fatigued muscle. Obtain several additional contraction curves and calculate the average latent period, contraction period, and relaxation period for the fatigued muscle.

4. Place one tracing in Section F.4 of the LABORATORY REPORT RESULTS at the end of the exercise and record the average values.

5. Compare these values to those obtained in Section 6, "Contraction Curve." Explain any differences seen between the two tracings in the space provided in Section F.5 of the LABORATORY REPORT RESULTS at the end of the exercise.

6. Before proceeding to the next section discard the muscle utilized up to this point and replace it with a fresh preparation *that has been soaking in the iced Ringer's solution.* Follow the procedures outlined earlier for removing the skin and attaching the preparation to the muscle clamp and transducer. *Continue to bathe the preparation with iced Ringer's solution.*

8. Demonstration of Refractory Period

PROCEDURE

1. *Make sure to bathe the preparation with iced Ringer's solution throughout this entire procedure.*

2. Quickly determine the threshold for the new preparation.

3. Set the stimulator on the *twin pulse* setting. Set the delay switch for the longest time period possible on the stimulator.

4. With the polygraph set at its fastest paper speed, stimulate the muscle with a single application of a suprathreshold-strength twin pulse stimulus.

5. Rinse the muscle with iced Ringer's solution. Reduce the time period by 10 milliseconds (msec) on the delay dial. Stimulate the muscle again with a single application of a suprathreshold-strength twin pulse stimulus. Continue to rinse the preparation with iced Ringer's solution between each application of stimulus, and continue to reduce the time period on the delay dial by 10 msec increments until the two contractions have merged into a single contraction curve on the polygraph record. Record the interval indicated on the delay dial when the two contractions have merged into a single contraction curve as the refractory period for your preparation.

Refractory period _____

6. At the completion of this section rinse the preparation repeatedly with *room temperature* Ringer's solution prior to proceeding on to the next section. The temperature of the preparation needs to be elevated to room temperature prior to initiating the next procedure in order to obtain the proper results.

9. Muscle Response to Variation of Stimulus Frequency

When a single threshold or suprathreshold stimulus is applied to either a single skeletal muscle fiber or a complete muscle, the contraction response follows the all-or-none principle. However, if two stimuli are applied in rapid succession so that the second stimulus is applied before complete relaxation from the first, one notices that the second contraction is greater than the first. The extent of the increased response to the second stimulus is dependent upon the interval between the successive stimuli, as long as the interval between the two stimuli is greater than the refractory period for that fiber or muscle. This response to increased stimulus frequency is termed *wave summation,* and does not follow the all-or-none principle. Wave summation is believed to be due to the increased amount of calcium ions accumulating within the muscle fiber after successive

stimuli. If you were to continue to increase the stimulus frequency you would note that the amount of relaxation between successive stimuli would progressively decrease until further stimuli would not increase the strength of contraction and no relaxation would be seen. This type of sustained contraction without any relaxation as a result of a high stimulus frequency is termed *tetanus.* During tetanus the contraction strength may be three or four times that seen with a single contraction.

PROCEDURE

1. Set the paper speed at a medium value.
2. Turn the stimulus frequency knob to 1 stimulus per second and set the stimulator mode selector switch to repetitive stimuli. *Be sure to bathe the muscle with Ringer's solution frequently throughout this procedure.*
3. Progressively increase the frequency in a stepwise manner until wave summation and tetany are observed. Record the frequencies at which wave summation and tetany occur on the polygraph recording and in Section F.6 of the LABORATORY REPORT RESULTS at the end of this exercise.
4. Turn the stimulator off and bathe the muscle with Ringer's solution frequently.
5. Allow 2 or 3 minutes of recovery time before proceeding to the next section.

10. Effect of Resting Length upon Contraction Strength

PROCEDURE

1. Adjust either the femur clamp or muscle transducer height (depending upon the preparation used) so that a minimal amount of resting tension is applied to the muscle.
2. With the paper drive set at its lowest speed, apply a single suprathreshold stimulus to the muscle. Observe the contraction strength.
3. Rinse the muscle with Ringer's solution after each stimulation.
4. Adjust the level of the femur clamp or transducer so that the resting tension is increased slightly. Stimulate again with a single stimulus and rinse. Observe the contraction strength at this new resting tension.
5. Repeat until the contraction strength has reached a maximum and started to decline.
6. Graph your results in Section F.7 of the LABORATORY REPORT RESULTS at the end of the

exercise and explain your findings in the space provided.

G. BIOCHEMISTRY OF SKELETAL MUSCLE CONTRACTION

You will examine the effect of the following solutions on the contraction of glycerinated skeletal muscle fibers:[3] (1) ATP solution, (2) mineral ion solution, and (3) ATP plus mineral ion solution.

PROCEDURE

1. Using 7× or 10× magnification and glass needles or clean stainless steel forceps, gently tease the muscle into very thin groups of myofibers. Single fibers or thin groups must be used because strands thicker than a silk thread curl when they contract.
2. Using a Pasteur pipette or medicine dropper, transfer one strand into a drop of glycerol on a clean glass slide and cover the preparation with a cover slip. Examine the strand under low and high power and note the striations. Also note that each fiber has several nuclei.
3. Transfer one of the thinnest strands to a drop of glycerol on a second microscope slide. Do not add a cover slip. If the amount of glycerol on the slide is more than a small drop, soak the excess into a piece of lens paper held at the edge of the glycerol farthest from the fibers. Using a dissecting microscope and a millimeter ruler held beneath the slide, measure the length of one of the fibers.
4. Now flood the fibers with the solution containing only ATP and observe their reaction. After 30 sec or more, remeasure the same fiber.

   How much did the fiber contract? _____mm
5. Using clean slides and medicine droppers, and being especially sure to use clean teasing needles or forceps, transfer other fibers to a drop of glycerol on a slide. Again measure the length of one fiber. Next flood the fibers with a solution containing mineral ions, observe their reaction, and remeasure the fiber.

   How much did the fiber contract? _____ mm
6. Repeat the exercise, this time using a solution containing a combination of ATP and ions.

[3]Glycerinated muscle preparation and solutions are supplied by the Carolina Biological Supply Company, Burlington, North Carolina 27215.

7. Observe a contracted fiber under low and high power, and look for differences in appearance between muscle in a contracted state and muscle in a relaxed state (see Figure 9.6).

H. ELECTROMYOGRAPHY

During the contraction of a single muscle fiber, an action potential is generated that lasts between 1 and 4 msec. This electrical activity is dissipated throughout the surrounding tissues. In order to produce a smooth muscle contraction, motor units fire asynchronously, thereby causing muscle fibers to contract at different times. This asynchronous firing of motor units also prolongs the electrical activity resulting from skeletal muscle contraction. By placing two electrodes on the skin or directly within the muscle an electrical recording, termed an *electromyograph*, or *EMG*, of this muscular activity may be obtained when the muscle is stimulated (Figure 9.11). EMGs are utilized clinically to distinguish between peripheral neurological and muscular diseases, and for differentiating abnormalities resulting in reductions of either muscular strength or sensation that may have been caused by either dysfunction or peripheral nervous tissue or central nervous system (CNS) centers. Typically EMGs are recorded under three different activity levels: complete inactivity, slight muscular activity, and extreme (maximal) muscular activity. Abnormal differences in records

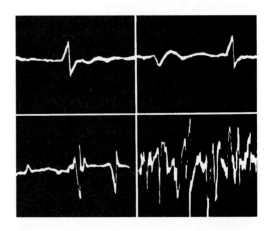

FIGURE 9.11 Diagrammatic representation of normal electromyograms. The single potential in the upper left corner has a measured amplitude of 0.8 mV and a duration of 7 msec.

obtained under these three conditions may also indicate an abnormality in motor unit recruitment.

1. Subject Preparation

PROCEDURE

1. Locate the *flexor digitorum superficialis* muscle (see Figure 10.14a) on the anterior surface of the forearm and the *triceps brachii* muscle (see Figure 10.13b) on the posterior surface of the subject's arm. EMG electrodes (or ECG electrodes) will be placed on the skin superficial to these muscles.
2. Prior to placing the electrodes on the skin, *gently* scrub the area with a dishwashing pad in order to remove dead epithelial cells and to facilitate recording. *Make sure that the capillaries are not damaged and no bleeding occurs.*
3. Apply a small amount of electrode gel to each electrode and apply them to the proper locations.
4. Connect the two recording electrodes (placed on the anterior surface of the forearm) and the ground electrodes (placed on the posterior surface of the arm) to the preamplifier (high gain coupler) and set the gain to × 100, sensitivity at an appropriate setting between 20 and 100, time constant to 0.03, and paper speed at high.

2. Recording of Spontaneous Muscle Activity

PROCEDURE

1. With the subject completely relaxed and his or her arm placed horizontally on a laboratory table, record any spontaneously occurring electrical activity within the flexor digitorum superficialis over a time period of 1 to 3 minutes.

2. Record your observations in Section H.1 of the LABORATORY REPORT RESULTS at the end of the exercise.

3. Recruitment of Motor Units

PROCEDURE

1. Have the subject gently flex her or his digits while recording, and note the electrical activity.
2. Have the subject relax his or her digits completely, followed by a more forceful contraction. Note any change in the EMG.
3. Place a tennis ball in your subject's hand and ask him or her to squeeze the ball once or twice as strongly as possible. Note the EMG recording, and then let your subject rest thoroughly for at least five minutes.
4. Record your observations in Section H.2 of the LABORATORY REPORT RESULTS at the end of the exercise.

4. Effect of Fatigue

PROCEDURE

1. *Without recording* an EMG on the polygraph, have your subject repeatedly squeeze the tennis ball as strongly as possible until no longer able to squeeze the ball.
2. When fatigue has been achieved *quickly* start recording and ask the subject to attempt to squeeze the ball as strongly as possible four or five times in quick succession.
3. Note the EMG recording and record your observations in Section H.3 of the LABORATORY REPORT RESULTS at the end of the exercise.

ANSWER THE LABORATORY REPORT QUESTIONS AT THE END OF THE EXERCISE.

Muscle Tissue 9

Student _____   Date _____

Laboratory Section _____   Score/Grade _____

SECTION F. LABORATORY TESTS ON SKELETAL MUSCLE CONTRACTION

1. Complete the following table with the values obtained.

Stimulus Strength	Contraction Strength
_____	_____
_____	_____
_____	_____
_____	_____
_____	_____
_____	_____
_____	_____

2. The all-or-none principle states that, once threshold is reached, a muscle cell will contract maximally, and that a subthreshold stimulus will not elicit a response. Based upon the data you collected in Section

   F.4, did the muscle obey the all-or-none principle? _____

   Give a physiological explanation for your answer._____

   _____

3. Attach and label the myogram obtained with a single stimulus. Label latent period, contraction period, and relaxation period. Include the average values obtained.

4. Attach and label the myogram obtained with a single stimulus in a fatigued muscle. Label the latent period, contraction period, and relaxation period, and include the average values obtained.

5. Explain the physiological reason for the differences observed between the two myograms.

_____

_____

_____

_____

6. Draw and label the first myogram obtained in Section F.8. Then draw two more myograms obtained in the same section, with the last myogram being the record indicating the refractory period for your preparation. Indicate on the diagram what the time value for the refractory period was with your preparation.

7. Why was the muscle preparation utilized in Section F.8 soaked in iced Ringer's solution prior to the start of the experiment?

8. Draw and label the myogram obtained as the stimulus frequency was increased. Label tetanus in your diagram.

9. Complete the following graph based upon your results obtained in Section F.10.

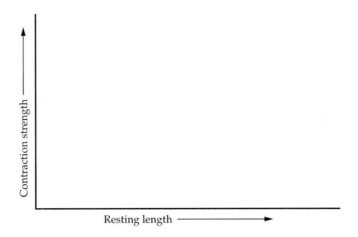

SECTION H. ELECTROMYOGRAPHY

1. Include your record of spontaneous electrical activity recorded while the subject was resting.

Give a physiological explanation. _____

_____

2. Include your record of the EMG recorded during the recruitment exercise. Give a physiological explanation and possible function of recruitment. _____

_____

3. Include your record of the EMG recorded following fatigue, and give a physiological explanation of your observations. _____

_____

Muscle Tissue 9

Student _____   Date _____

Laboratory Section _____   Score/Grade _____

PART 1. Multiple Choice

_____ 1. The ability of muscle tissue to return to its original shape after contraction or extension is called (a) excitability (b) elasticity (c) extension (d) tetanus

_____ 2. Which of the following is striated and voluntary? (a) skeletal muscle tissue (b) cardiac muscle tissue (c) visceral muscle tissue (d) smooth muscle tissue

_____ 3. The portion of a sarcomere composed of thin filaments only is the (a) H zone (b) A band (c) I band (d) Z disc

_____ 4. The area of contact between a motor axon terminal and a muscle fiber sarcolemma (motor end plate) is called the (a) synapse (b) filament (c) transverse tubule (d) neuromuscular junction

_____ 5. The connective tissue layer surrounding bundles of muscle fibers is called the (a) perimysium (b) endomysium (c) ectomysium (d) myomysium

_____ 6. Which of the following is striated and involuntary? (a) smooth muscle tissue (b) skeletal muscle tissue (c) cardiac muscle tissue (d) visceral muscle tissue

_____ 7. The portion of a sarcomere that consists mostly of thick filaments and portions of filaments where thin and thick filaments overlap is called the (a) H zone (b) triad (c) A band (d) I band

_____ 8. The portion of a polygraph that converts nonelectrical signals into electrical signals is the (a) transducer (b) electrode (c) amplifier (d) coupler

_____ 9. In a skeletal muscle twitch, which period of the contraction curve requires the expenditure of energy? (a) latent period (b) contraction period (c) relaxation period (d) extension

_____ 10. A state of sustained muscle contraction is known as (a) tetanus (b) treppe (c) myopathy (d) hypertrophy

PART 2. Completion

11. The ability of muscle tissue to receive and respond to stimuli is called _____.

12. Fibrous connective tissue located between muscle fibers is known as _____.

13. The sections of a muscle fiber separated by Z discs are called _____.

14. The plasma membrane surrounding a muscle fiber is called the _____.

15. The region in a sarcomere consisting of thick filaments only is known as the _____.

16. Muscle tissue that is nonstriated and involuntary is _____.

17. The ability of muscle tissue to stretch when pulled is called _____.

18. The phenomenon by which a muscle fiber contracts to its fullest or not at all is known as the

_____.

19. A record of muscle contraction is referred to as a(n) _____.

20. A stimulus too weak to initiate a nerve impulse is called a(n) _____ stimulus.

21. The linkage between the production of the electrical and chemical events and the contraction of

skeletal muscle is called _____.

22. A localized depolarization of a motor end plate that initiates an action potential is called an

_____.

23. In relaxed muscle fibers, calcium ions are stored in the _____.

24. Whereas thin filaments are composed of the protein actin, thick filaments are composed of the pro-

tein _____.

25. The combination of alternating dark A bands and light _____ bands gives a muscle fiber its striated appearance.

26. A motor neuron and all the muscle fibers it innervates is called a(n) _____.

27. Neurotransmitters are stored in _____, which are located within synaptic end
bulbs.

Skeletal Muscles

10

In this exercise you will learn the names, locations, and actions of the principal skeletal muscles of the body.

A. HOW SKELETAL MUSCLES PRODUCE MOVEMENT

1. Origin and Insertion

Skeletal muscles produce movements by exerting force on tendons, which in turn pull on bones or other structures, such as the skin. Most muscles cross at least one joint and are usually attached to the articulating bones that form the joint. When such a muscle contracts, it draws one articulating bone toward the other. The two articulating bones usually do not move equally in response to the contraction. One is held nearly in its original position because other muscles contract to pull it in the opposite direction or because its structure makes it less movable. Ordinarily, the attachment of a muscle tendon to the stationary bone is called the *origin*. The attachment of the other muscle tendon to the movable bone is the *insertion*. A good analogy is a spring on a door. The part of the spring attached to the door represents the insertion; the part attached to the frame is the origin. The fleshy portion of the muscle between the tendons of the origin and insertion is called the *belly (gaster)*. The origin is usually proximal and the insertion distal, especially in the limbs. In addition, muscles that move a body part generally do not cover the moving part. For example, although contraction of the biceps brachii muscle moves the forearm, the belly of the muscle lies over the humerus.

2. Group Actions

Most movements require several skeletal muscles acting in groups rather than individually. Also, most skeletal muscles are arranged in opposing pairs at joints, that is, flexors-extensors, abductors-

adductors, and so on. Consider flexing the forearm at the elbow, for example. A muscle that causes a desired action is referred to as the *prime mover* or *agonist* (*agogos* = leader). In this instance, the biceps brachii is the prime mover (see Figure 10.13a). Simultaneously with the contraction of the biceps brachii, another muscle, called the *antagonist* (*antagonistes* = opponent), is stretching. In this movement, the triceps brachii serves as the antagonist (see Figure 10.13b). The antagonist has an action that is opposite to that of the prime mover; that is, the antagonist stretches and yields to the movement of the prime mover. You should not assume, however, that the biceps brachii is always the prime mover and the triceps brachii is always the antagonist. For example, when extending the forearm at the elbow, the triceps brachii serves as the prime mover, and the biceps brachii functions as the antagonist; their roles are reversed. Note that if the prime mover and antagonist contracted simultaneously with equal force, there would be no movement.

In addition to prime movers and antagonists, most movements also involve muscles called *synergists* (SIN-er-jists; *syn* = together; *ergon* = work), which serve to steady a movement, thus preventing unwanted movements and helping the prime mover function more efficiently. For example, flex your hand at the wrist and then make a fist. Note how difficult this is to do. Now extend your hand at the wrist and then make a fist. Note how much easier it is to clench your fist. In this case, the extensor muscles of the wrist act as synergists in cooperation with the flexor muscles of the fingers acting as prime movers. The extensor muscles of the fingers serve as antagonists (see Figure 10.14c, d).

Some synergist muscles in a group also act as *fixators*, which stabilize the origin of the prime mover so that the prime mover can act more efficiently. For example, the scapula is a freely movable bone in the pectoral (shoulder) girdle that serves as an origin for several muscles that move the arm. However, for the scapula to serve as a firm origin for muscles that move the arm, it must be held steady. This is accomplished by fixator muscles that hold the scapula firmly against the back of the chest. In abduction of the arm, the deltoid muscle serves as the prime mover, whereas fixators (pectoralis minor, rhomboideus major, rhomboideus minor, trapezius, subclavius, and serratus anterior muscles) hold the scapula firmly (see Figure 10.11). These fixators stabilize the scapula that serves as the attachment site for the origin of the deltoid muscle while the insertion of the muscle pulls on the humerus to abduct the arm. Under different conditions and depending on the movement and which point is fixed, many muscles act, at various times, as prime movers, antagonists, synergists, or fixators.

B. ARRANGEMENT OF FASCICLES

Recall from Exercise 9 that skeletal muscle fibers (cells) are arranged within the muscle in bundles called *fascicles (fasciculi)*. The muscle fibers are arranged in a parallel fashion within each bundle,

but the arrangement of the fascicles with respect to the tendons may take several characteristic patterns.

Table 10.1 describes the major patterns of fascicles. Using your textbook as a guide, provide an example of each.

C. NAMING SKELETAL MUSCLES

Most of the almost 700 skeletal muscles of the body are named on the basis of one or more distinctive characteristics. If you understand these characteristics, you will find it much easier to learn and remember the names of individual muscles.

Table 10.2 describes the major characteristics that are used to name skeletal muscles. Using your textbook as a guide, provide an example for each.

D. CONNECTIVE TISSUE COMPONENTS

Skeletal muscles are protected, strengthened, and attached to other structures by several connective

TABLE 10.1
Arrangements of Fascicles

Arrangement	Description	Example
PARALLEL	Fascicles are parallel with longitudinal axis of muscle and terminate at either end in flat tendons.	
FUSIFORM	Fascicles are nearly parallel with longitudinal axis of muscle and terminate at either end in flat tendons, but muscle tapers toward tendons where the diameter is less than that of the belly.	
PENNATE	Fascicles are short in relation to muscle length and the tendon extends nearly the entire length of the muscle.	
Unipennate	Fascicles are arranged on only one side of tendon.	
Bipennate	Fascicles are arranged on both sides of a centrally positioned tendon.	
Multipennate	Fascicles attach obliquely from many directions to several tendons.	
CIRCULAR	Fascicles are arranged in a circular pattern and enclose an orifice (opening).	

TABLE 10.2
Characteristics Used for Naming Skeletal Muscles

Characteristic	Description	Example
Direction of muscle fibers	Direction of muscle fibers relative to the midline of the body. **Rectus** means the fibers run parallel to the midline. **Transverse** means the fibers run perpendicular to the midline. **Oblique** means the fibers run diagonally to the midline.	
Location	Structure near which a muscle is found.	
Size	Relative size of the muscle. **Maximus** means largest. **Minimus** means smallest. **Longus** means longest. **Brevis** means short.	
Number of origins	Number of tendons of origin. **Biceps** means two origins. **Triceps** means three origins. **Quadriceps** means four origins.	
Shape	Relative shape of the muscle. **Deltoid** means triangular. **Trapezius** means trapezoidal. **Serratus** means saw-toothed. **Rhomboideus** means rhomboid- or diamond-shaped.	
Origin and insertion	Sites where muscle originates and inserts.	
Action	Principal action of the muscle. **Flexor** (FLEK-sor): decreases the angle at a joint. **Extensor** (eks-TEN-sor): increases the angle at a joint. **Abductor** (ab-DUK-tor): moves a bone away from the midline. **Adductor** (ad-DUK-tor): moves a bone closer to the midline. **Levator** (le-VĀ-tor): produces an upward movement. **Depressor** (de-PRES-or): produces a downward movement. **Supinator** (soo'-pi-NĀ-tor): turns the palm upward or anteriorly. **Pronator** (prō-NĀ-tor): turns the palm downward or posteriorly. **Sphincter** (SFINGK-ter): decreases the size of an opening. **Tensor** (TEN-sor): makes a body part more rigid. **Rotator** (RŌ-tāt-or): moves a bone around its longitudinal axis.	

tissue components. For example, the entire muscle is usually wrapped with a dense, irregular connective tissue called the *epimysium* (ep'-i-MĪZ-ē-um; *epi* = upon). When the muscle is cut in cross section, invaginations of the epimysium divide the muscle into fascicles. These invaginations of the epimysium are called the *perimysium* (per'-i-MĪZ-ē-um; *peri* = around). In turn, invaginations of the perimysium, called ***endomysium*** (en'-dō-MĪZ-ē-um; *endo* = within), penetrate into the interior of each fascicle and separate individual muscle fibers from one another. The epimysium, perimysium, and endomysium are all extensions of deep fascia and are all continuous with the connective tissue that attaches the muscle to another structure, such as bone or other muscle. All three elements may be extended beyond the muscle cells as a ***tendon*** (*tendere* = to stretch out)—a cord of connective tissue that attaches a muscle to the periosteum of bone. The connective tissue may also extend as a broad, flat band of tendons called an ***aponeurosis*** (*apo* = from; *neuron* = a tendon). Aponeuroses also attach to the coverings of a bone or another muscle. When a muscle contracts, the tendon and its corresponding bone or muscle are pulled toward the contracting muscle. In this way skeletal muscles produce movement.

In Figure 10.1, label the epimysium, perimysium, endomysium, fascicles, and muscle fibers.

E. PRINCIPAL MUSCLES

In the pages that follow, a series of tables has been provided for you to learn the principal skeletal muscles by region.[1] Use each table as follows:

[1]A few of the muscles listed are not illustrated in the diagrams. Please consult your textbook to locate these muscles.

1. Take each muscle, in sequence, and study the ***learning key*** that appears in parentheses after the name of the muscle. The learning key is a list of prefixes, suffixes, and definitions that explain the derivations of the muscles' names. It will help you to understand the reason for giving a muscle its name.
2. As you learn the name of each muscle, determine its origin, insertion, and action and write these in the spaces provided in the table. Consult your textbook if necessary.
3. Again, using your textbook as a guide, label the diagram referred to in the table.
4. Try to visualize what happens when the muscle contracts so that you will understand its action.
5. Do steps 1 through 4 for each muscle in the table. Before moving to the next table, examine a torso or chart of the skeletal system so that you can compare and approximate the positions of the muscles.
6. When possible, try to feel each muscle on your own body.

Refer to Tables 10.3 through 10.22 and Figures 10.2 through 10.18.

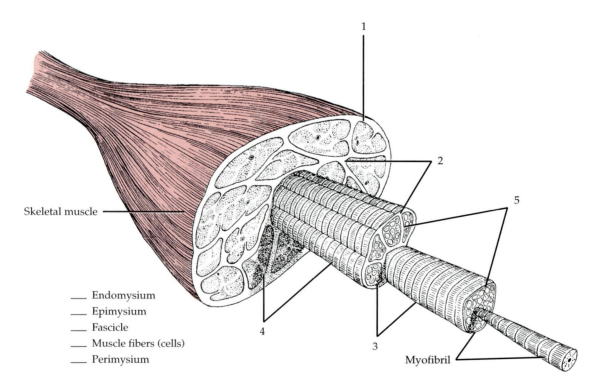

Skeletal muscle

___ Endomysium
___ Epimysium
___ Fascicle
___ Muscle fibers (cells)
___ Perimysium

Myofibril

FIGURE 10.1 Connective tissue components of a skeletal muscle.

TABLE 10.3
Muscles of Facial Expression (After completing the table, label Figure 10.2.)

OVERVIEW: The muscles in this group provide humans with the ability to express a wide variety of emotions, including grief, surprise, fear, and happiness. The muscles themselves lie within the layers of superficial fascia. As a rule, they arise from the fascia or bones of the skull and insert into the skin. Because of their insertions, the muscles of facial expression move the skin rather than a joint when they contract.

Muscle	Origin	Insertion	Action
Epicranius (ep-i-KRĀ-nē-us; *epi* = over; *crani* = skull)	This muscle is divisible into two portions: the frontalis, over the frontal bone, and the occipitalis, over the occipital bone. The two muscles are united by a strong aponeurosis, the galea aponeurotica (epicranial aponeurosis), which covers the superior and lateral surfaces of the skull.		
Frontalis (fron-TA-lis; *front* = forehead)			
Occipitalis (ok-si'-pi-TA-lis; *occipito* = base of skull)			
Orbicularis oris (or-bi'-kyoo-LAR-is OR-is; *orb* = circular; *or* = mouth)			
Zygomaticus (zī-gō-MA-ti-kus) **major** (*zygomatic* = cheek bone; *major* = greater)			
Levator labii superioris (le-VĀ-ter LA-bē-ī soo-per'-ē-OR-is; *levator* = raises or elevates; *labii* = lip; *superioris* = upper)			
Depressor labii inferioris (de-PRE-ser LA-bē-ī in-fer'-ē-OR-is; *depressor* = depresses or lowers; *inferioris* = lower)			
Buccinator (BUK-si-nā'-tor; *bucc* = cheek)			
Mentalis (men-TA-lis; *mentum* = chin)			
Platysma (pla-TIZ-ma; *platy* = flat, broad)			
Risorius (ri-ZOR-ē-us; *risor* = laughter)			
Orbicularis oculi (or-bi'-kyoo-LAR-is O-kyoo-lī; *oculus* = eye)			
Corrugator supercilii (KOR-a-gā'-tor soo-per-SI-lē-ī; *corrugo* = to wrinkle; *supercilium* = eyebrow)			
Levator palpebrae superioris (le-VĀ-tor PAL-pe-brē soo-per'-ē-OR-is; *palpebrae* = eyelids) (See Figure 10.4)			

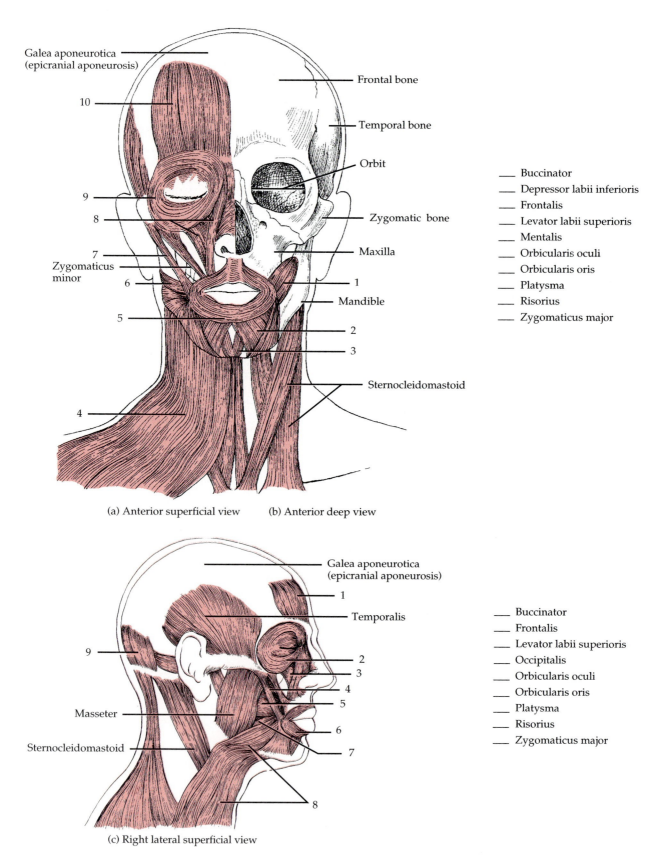

Galea aponeurotica
(epicranial aponeurosis)

10

9

8

7

Zygomaticus
minor

6

5

4

Frontal bone

Temporal bone

Orbit

Zygomatic bone

Maxilla

1

Mandible

2

3

Sternocleidomastoid

(a) Anterior superficial view   (b) Anterior deep view

___ Buccinator
___ Depressor labii inferioris
___ Frontalis
___ Levator labii superioris
___ Mentalis
___ Orbicularis oculi
___ Orbicularis oris
___ Platysma
___ Risorius
___ Zygomaticus major

Galea aponeurotica
(epicranial aponeurosis)

1

Temporalis

9

2

3

4

5

Masseter

6

Sternocleidomastoid

7

8

(c) Right lateral superficial view

___ Buccinator
___ Frontalis
___ Levator labii superioris
___ Occipitalis
___ Orbicularis oculi
___ Orbicularis oris
___ Platysma
___ Risorius
___ Zygomaticus major

FIGURE 10.2 Muscles of facial expression.

TABLE 10.4
Muscles that Move the Mandible (Lower Jaw) (After completing the table, label Figure 10.3.)

OVERVIEW: Muscles that move the mandible (lower jaw) are also known as muscles of mastication because they are involved in biting and chewing. These muscles also assist in speech.

Muscle	Origin	Insertion	Action
Masseter (MA-se-ter; *masseter* = chewer)			
Temporalis (tem'-por-A-lis; *tempora* = temples)			
Medial pterygoid (TER-i-goid; *medial* = closer to midline; *pterygoid* = like a wing)			
Lateral pterygoid (TER-i-goid; *lateral* = farther from midline)			

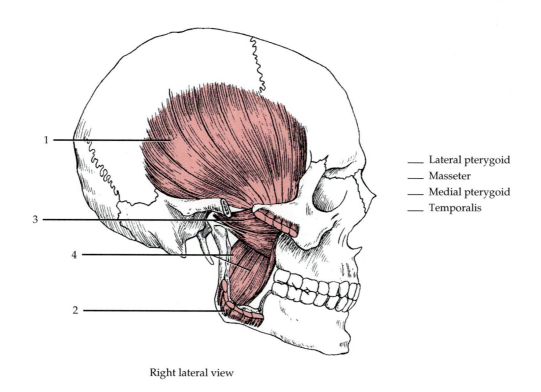

___ Lateral pterygoid
___ Masseter
___ Medial pterygoid
___ Temporalis

Right lateral view

FIGURE 10.3   Muscles that move mandible (lower jaw).

TABLE 10.5
Muscles that Move the Eyeballs—The Extrinsic Muscles* (After completing the table, label Figure 10.4.)

OVERVIEW: Muscles associated with the eyeballs are of two principal types: extrinsic and intrinsic. *Extrinsic muscles* originate outside the eyeballs and are inserted on their outer surfaces (sclera). They move the eyeballs in various directions. *Intrinsic muscles* originate and insert entirely within the eyeballs. They move structures within the eyeballs.

Movements of the eyeballs are controlled by three pairs of extrinsic muscles. The two pairs of rectus muscles move the eyeballs in the direction indicated by their respective names—superior, inferior, lateral, and medial. The pair of oblique muscles—superior and inferior—rotate the eyeballs on their axes. The extrinsic muscles of the eyeballs are among the fastest contracting and precisely controlled skeletal muscles of the body.

Muscle	Origin	Insertion	Action
Superior rectus (REK-tus; *superior* = above; *rectus* = in this case, muscle fibers running parallel to long axis of eyeball) **Inferior rectus** (REK-tus; *inferior* = below)			
Lateral rectus (REK-tus)			
Medial rectus (REK-tus)			
Superior oblique (ō-BLĒK; *oblique* = in this case, muscle fibers running diagonally to long axis of eyeball)			
Inferior oblique (ō-BLĒK)			

*Muscles situated on the outside of the eyeballs.

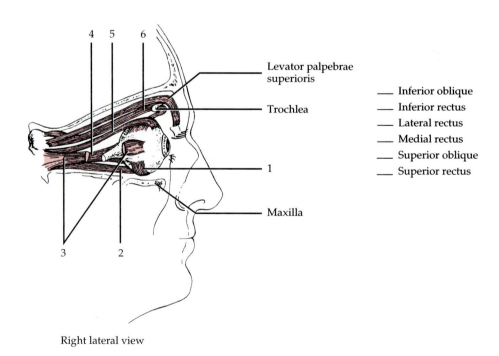

Right lateral view

FIGURE 10.4 Extrinsic muscles of eyeballs.

TABLE 10.6
Muscles that Move the Tongue—The Extrinsic Muscles (After completing the table, label Figure 10.5.)

OVERVIEW: The tongue is divided into lateral halves by a median fibrous septum. The septum extends throughout the length of the tongue and is attached inferiorly to the hyoid bone. Like the muscles of the eyeballs, muscles of the tongue are of two principal types—extrinsic and intrinsic. *Extrinsic muscles* originate outside the tongue and insert into it. They move the entire tongue in various directions, such as anteriorly, posteriorly, and laterally. *Intrinsic muscles* originate and insert within the tongue. They alter the shape of the tongue rather than move the entire tongue. The extrinsic and intrinsic muscles of the tongue are arranged in both lateral halves of the tongue.

Muscle	Origin	Insertion	Action
Genioglossus (jē'-nē-ō-GLOS-us; *geneion* = chin *glossus* = tongue)			
Styloglossus (stī'-lō-GLOS-us; *stylo* = stake or pole)			
Palatoglossus (pal'-a-tō-GLOS-us; *palato* = palate)			
Hyoglossus (hī-ō-GLOS-us)			

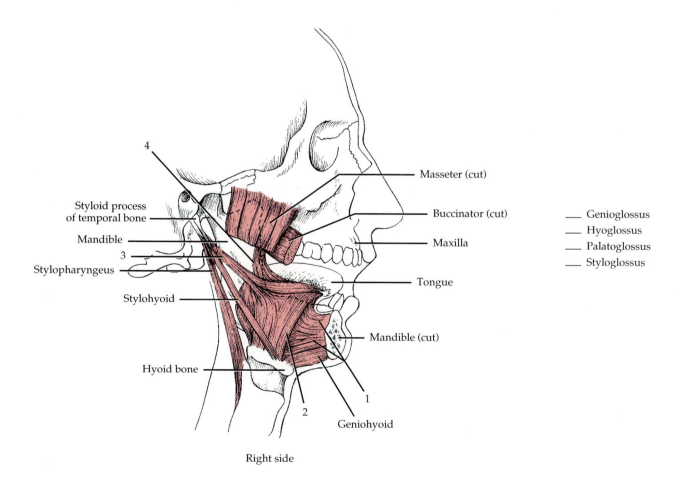

Right side

FIGURE 10.5   Muscles that move the tongue.

TABLE 10.7
Muscles of the Floor of the Oral Cavity (Figure 10.6)

OVERVIEW: As a group, these muscles are referred to as *suprahyoid muscles.* They lie superior to the hyoid bone
and all insert into it. The digastric muscle consists of an anterior belly and a posterior belly united by an intermediate tendon that is held in position by a fibrous group.

Muscle	Origin	Insertion	Action
Digastric (dī'-GAS-trik; *di* = two; *gaster* = belly)			
Stylohyoid (stī'-lō-HĪ-oid; *stylo* = stake or pole, styloid process of temporal bone; *hyoedes* = U-shaped, pertaining to hyoid bone)			
Mylohyoid (mī'-lō-HĪ-oid)			
Geniohyoid (je'-nē-ō-HĪ-oid; *geneion* = chin) (See Figure 10.5)			

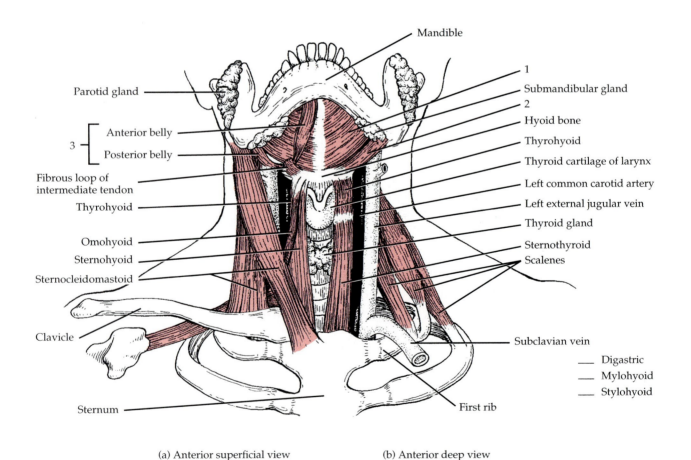

(a) Anterior superficial view   (b) Anterior deep view

FIGURE 10.6   Muscles of the floor of the oral cavity.

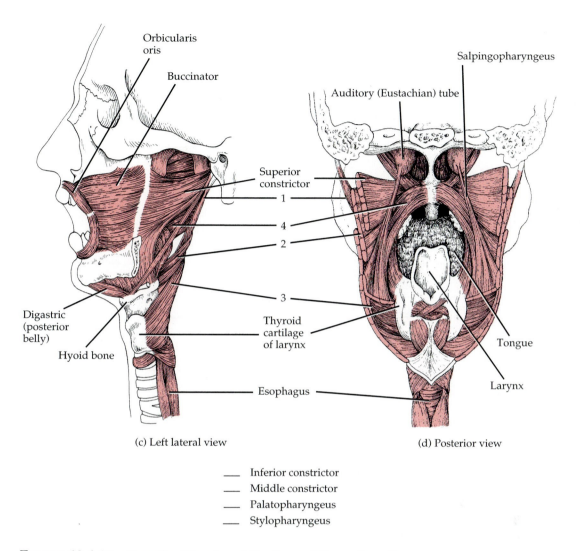

Orbicularis
oris

Buccinator

Salpingopharyngeus

Auditory (Eustachian) tube

Superior
constrictor

1

4

2

3

Digastric
(posterior
belly)

Hyoid bone

Thyroid
cartilage
of larynx

Tongue

Larynx

Esophagus

(c) Left lateral view

(d) Posterior view

____ Inferior constrictor
____ Middle constrictor
____ Palatopharyngeus
____ Stylopharyngeus

FIGURE 10.6 *(Continued)*   Muscles of the floor of the oral cavity.

TABLE 10.8
Muscles of the Larynx (Voice Box) (After completing the table, label Figure 10.7.)

OVERVIEW: The muscles of the larynx (voice box), like those of the eyeballs and tongue, are grouped into *extrinsic* and *intrinsic muscles.* The extrinsic muscles of the larynx marked (*) are together referred to as *infrahyoid (strap) muscles* because they lie inferior to the hyoid bone. The omohyoid muscle, like the digastric muscle, is composed of two bellies and an intermediate tendon. In this case, however, the two bellies are referred to as superior and inferior, rather than anterior and posterior.

Muscle	Origin	Insertion	Action
EXTRINSIC **Omohyoid*** (ō'-mō-HĪ-oid; *omo* = relationship to shoulder; *hyoedes* = U-shaped)			
Sternohyoid* (ster'-nō-HĪ-oid; *sterno* = sternum)			
Sternothyroid* (ster'-nō-THĪ-roid; *thyro* = thyroid gland)			
Thyrohyoid* (thī'-rō-HĪ-oid)			
Stylopharyngeus (stī'-lō-fa-RIN-jē-us)			
Palatopharyngeus (pal'-a-tō-fa-RIN-jē-us)			
Inferior constrictor (kon-STRIK-tor)			
Middle constrictor (kon-STRIK-tor)			
INTRINSIC **Cricothyroid** (kri-kō-THĪ-roid; *crico* = cricoid cartilage of larynx)			
Posterior cricoarytenoid (kri'-kō-ar'-i-TĒ-noid; *arytaina* = shaped like a jug)			
Lateral cricoarytenoid (kri'-kō-ar'-i-TĒ-noid)			
Arytenoid (ar'-i-TĒ-noid)			
Thyroarytenoid (thī'-rō-ar'-i-TĒ-noid)			

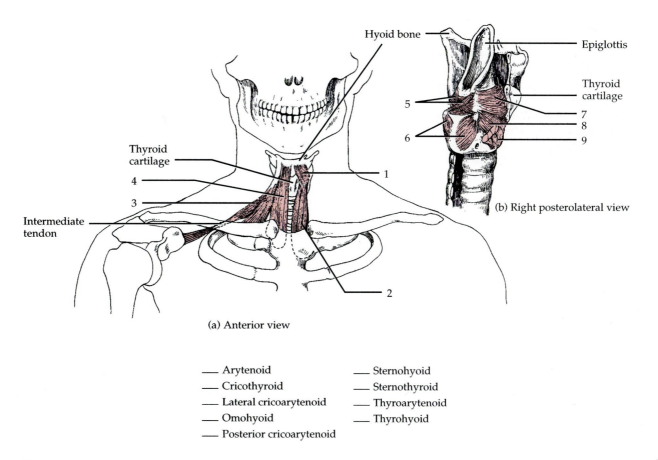

(a) Anterior view

(b) Right posterolateral view

—— Arytenoid   —— Sternohyoid
—— Cricothyroid   —— Sternothyroid
—— Lateral cricoarytenoid   —— Thyroarytenoid
—— Omohyoid   —— Thyrohyoid
—— Posterior cricoarytenoid

FIGURE 10.7   Muscles of larynx (voice box).

TABLE 10.9
Muscles that Move the Head

OVERVIEW: The cervical region is divided by the sternocleidomastoid muscle into two principal triangles—anterior and posterior. The *anterior triangle* is bordered superiorly by the mandible, inferiorly by the sternum, medially by the cervical midline, and laterally by the anterior border of the sternocleidomastoid muscle (see Figure 11.2). The *posterior triangle* is bordered inferiorly by the clavicle, anteriorly by the posterior border of the sternocleidomastoid muscle, and posteriorly by the anterior border of the trapezius muscle (see Figure 11.2). Subsidiary triangles exist within the two principal triangles.

Muscle	Origin	Insertion	Action
Sternocleidomastoid (ster'-nō-klī'-dō-MAS-toid; *sternum* = breastbone; *cleido* = clavicle; *mastoid* = mastoid process of temporal bone) (label this muscle in Figure 10.11)			
Semispinalis capitis (se'-mē-spi-NA-lis KAP-i-tis; *semi* = half; *spine* = spinous process; *caput* = head) (label this muscle in Figure 10.15)			
Splenius capitis (SPLĒ-nē-us KAP-i-tis; *splenion* = bandage) (label this muscle in Figure 10.15)			
Longissimus capitis (lon-JIS-i-mus KAP-i-tis; *longissimus* = longest) (label this muscle in Figure 10.15)			

TABLE 10.10
Muscles that Act on the Anterior Abdominal Wall (After completing the table, label Figure 10.8.)

OVERVIEW: The anterolateral abdominal wall is composed of skin, fascia, and four
  pairs of flat, sheetlike muscles: rectus abdominis, external oblique, internal oblique, and
  transversus abdominis. The anterior surfaces of the rectus abdominis muscles are inter-
  rupted by transverse fibrous bands of tissue called *tendinous intersections,* believed to
  be remnants of septa that separated myotomes during embryonic development. The
  aponeuroses of the external oblique, internal oblique, and transversus abdominis muscles meet at the midline to form the *linea alba* (white line), a tough fibrous band that
  extends from the xiphoid process of the sternum to the pubic symphysis. The inferior
  free border of the external oblique aponeurosis, plus some collagen fibers, forms the
  inguinal ligament, which runs from the anterior superior iliac spine to the pubic tubercle (see Figure 10.16). The ligament demarcates the thigh and body wall.
Just superior to the medial end of the inguinal ligament is a triangular slit in the aponeurosis referred to as the *superficial inguinal ring,* the outer opening of the *inguinal
  canal.* The canal contains the spermatic cord and ilioinguinal nerve in males and round
  ligament of the uterus and ilioinguinal nerve in females.
The posterior abdominal wall is formed by the lumbar vertebrae, parts of the ilia of the
  hipbones, psoas major muscle (described in Table 10.20), quadratus lumborum muscle,
  and iliacus muscle (also described in Table 10.20). Whereas the anterolateral abdominal
  wall is contractile and distensible, the posterior abdominal wall is bulky and stable by
  comparison.

Muscle	Origin	Insertion	Action
Rectus abdominis (REK-tus ab-do-MIN-is; *rectus* = fibers parallel to midline; *abdomino* = abdomen)			
External oblique (ō-BLĒK; *external* = closer to the surface; *oblique* = fibers diagonal to midline)			
Internal oblique (ō-BLĒK; *internal* = farther from the surface)			
Transversus abdominis (tranz-VER-sus ab-do-MIN-is; *transverse* = fibers perpendicular to midline)			
Quadratus lumborum (kwod-RĀ-tus lum-BOR-um; *quad* = four; *lumbo* = lumbar region) (See Figure 10.15a)			

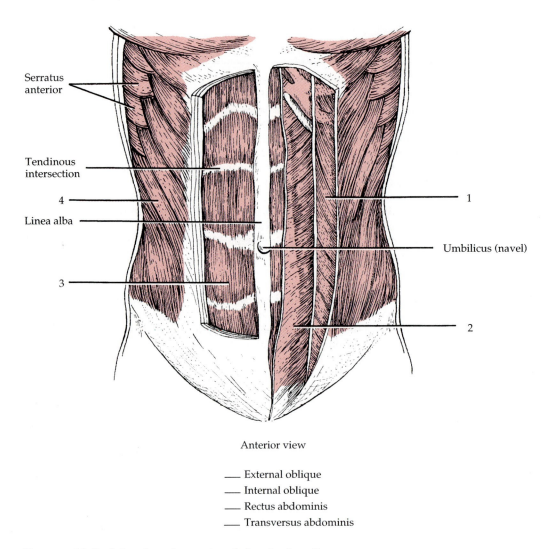

Serratus anterior

Tendinous intersection

4

Linea alba

3

1

Umbilicus (navel)

2

Anterior view

___ External oblique
___ Internal oblique
___ Rectus abdominis
___ Transversus abdominis

FIGURE 10.8 Muscles of anterior abdominal wall.

TABLE 10.11
Muscles Used in Breathing (After completing the table, label Figure 10.9.)

OVERVIEW: The muscles described here are attached to the ribs and by their contraction and relaxation alter the size of the thoracic cavity during breathing. Essentially, inspiration occurs when the thoracic cavity increases in size. Expiration occurs when the thoracic cavity decreases in size. The principal muscles of *inspiration* during normal breathing are the diaphragm and external intercostals. During forced inspiration, accessory muscles, such as the sternocleidomastoid, scalenes, and pectoralis minor are also used. The principal muscles of *expiration* during normal breathing are also the diaphragm and external intercostals. During forced expiration, accessory muscles, such as the internal intercostals and abdominal muscles (external oblique, internal oblique, transversus abdominis, and rectus abdominis) are also used.

The diaphragm, one of the muscles used in breathing, is dome-shaped and has three major openings through which various structures pass between the thorax and abdomen. These structures include the aorta along with the thoracic duct and azygos vein, which pass through the *aortic hiatus;* the esophagus with accompanying vagus (X) nerves, which pass through the *esophageal hiatus;* and the inferior vena cava, which passes through the *foramen for the vena cava.* In a condition called a hiatal hernia, the stomach protrudes superiorly through the esophageal hiatus.

Muscle	Origin	Insertion	Action
Diaphragm (DĪ-a-fram; *dia* = across, between; *phragma* = wall)			
External intercostals (in'-ter-KOS-tals; *inter* = between; *costa* = rib)			
Internal intercostals (in'-ter-KOS-tals; *internal* = farther from surface)			

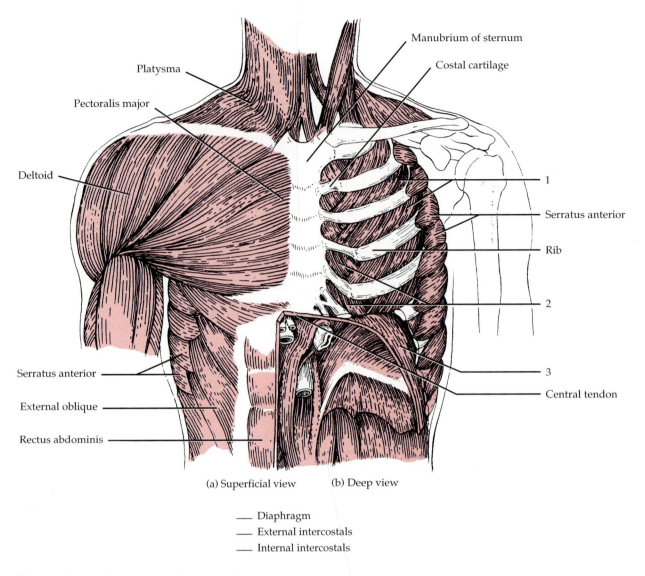

Manubrium of sternum

Costal cartilage

Platysma

Pectoralis major

Deltoid

1

Serratus anterior

Rib

2

Serratus anterior

External oblique

3

Central tendon

Rectus abdominis

(a) Superficial view   (b) Deep view

___ Diaphragm
___ External intercostals
___ Internal intercostals

FIGURE 10.9   Muscles used in breathing.

TABLE 10.12
Muscles of the Pelvic Floor (After completing the table, label Figure 10.10.)

OVERVIEW: The muscles of the pelvic floor, together with the fascia covering their external and internal surfaces, are referred to as the *pelvic diaphragm*. This diaphragm is funnel-shaped and forms the floor of the abdominopelvic cavity. It is pierced by the anal canal and urethra in both sexes and also by the vagina in the female.

Muscle	Origin	Insertion	Action
Levator ani (le-VĀ-tor Ā-nē; *levator* = raises; *ani* = anus)	This muscle is divisible into two parts: the pubococcygeus muscle and the iliococcygeus muscle.		
Pubococcygeus (pu'-bo-kok-SIJ-ē-us; *pubo* = pubis; *coccygeus* = coccyx)			
Iliococcygeus (il'-ē-o-kok-SIJ-ē-us; *ilio* = ilium)			
Coccygeus* (kok-SIJ-ē-us)			

*Not illustrated

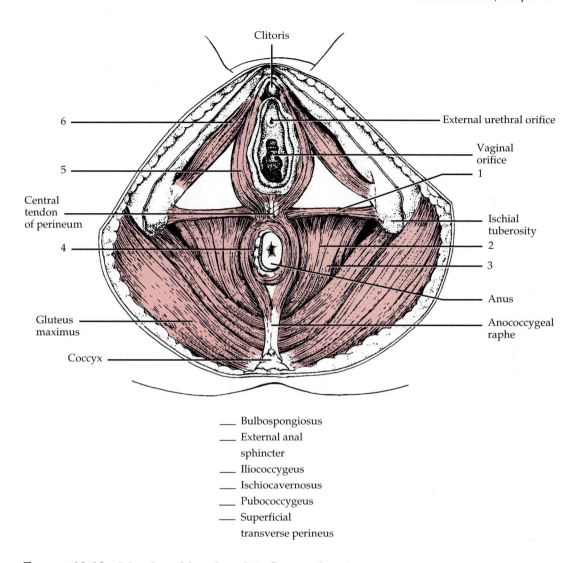

Clitoris

6

External urethral orifice

Vaginal orifice

5

1

Central tendon of perineum

Ischial tuberosity

4

2

3

Anus

Gluteus maximus

Anococcygeal raphe

Coccyx

___ Bulbospongiosus
___ External anal sphincter
___ Iliococcygeus
___ Ischiocavernosus
___ Pubococcygeus
___ Superficial transverse perineus

FIGURE 10.10   Muscles of female pelvic floor and perineum.

TABLE 10.13
Muscles of the Perineum (After completing the table, label Figure 10.10.)

OVERVIEW: The *perineum* is the entire outlet of the pelvis. It is a diamond-shaped area
at the inferior end of the trunk between the thighs and buttocks. It is bordered anteriorly by the pubic symphysis, laterally by the ischial tuberosities, and posteriorly by the
coccyx. A transverse line drawn between the ischial tuberosities divides the perineum
into an anterior *urogenital triangle* that contains the external genitals and a posterior
anal triangle that contains the anus.
The deep transverse perineus muscle, the urethral sphincter, and a fibrous membrane
constitute the *urogenital diaphragm.* It surrounds the urogenital ducts and helps to
strengthen the pelvic floor.

Muscle	Origin	Insertion	Action
Superficial transverse perineus (per-i-NĒ-us; *superficial* = near surface; *transverse* = across; *perineus* = perineum)			
Bulbospongiosus (bul'-bō-spon'-jē-Ō-sus; *bulbus* = bulb; *spongio* = sponge)			
Ischiocavernosus is'-kē-ō-ka'-ver-NŌ-sus; *ischion* = hip)			
Deep transverse perineus* (per-i-NĒ-us; *deep* = farther from surface)			
Urethral sphincter* (yoo-RĒ-thral SFINGK-ter; *urethral* = pertaining to urethra; *sphincter* = circular muscle that decreases the size of an opening)			
External anal (Ā-nal) sphincter			

*Not illustrated

TABLE 10.14
Muscles that Move the Pectoral (Shoulder) Girdle (After completing the table, label Figure 10.11.)

OVERVIEW: Muscles that move the pectoral (shoulder) girdle can be grouped into
anterior and *posterior* muscles. The principal action of the muscles is to stabilize the
scapula so that it can function as a stable point of origin for most of the muscles that
move the humerus (arm).

Muscle	Origin	Insertion	Action
ANTERIOR MUSCLES **Subclavius** (sub-KLĀ-vē-us; *sub* = under; *clavius* = clavicle)			
Pectoralis (pek'-tor-A-lis) **minor** (*pectus* = breast, chest, thorax; *minor* = lesser)			
Serratus (ser-Ā-tus) **anterior** (*serratus* = sawtoothed; *anterior* = front)			
POSTERIOR MUSCLES **Trapezius** (tra-PĒ-zē-us; *trapezoides* = trapezoid-shaped)			
Levator scapulae (le-VĀ-tor SKA-pyoo-lē; *levator* = raises; *scapulae* = scapula)			
Rhomboideus (rom-BOID-ē-us) **major** (*rhomboides* = rhomboid- or diamond-shaped)			
Rhomboideus (rom-BOID-ē-us) **minor**			

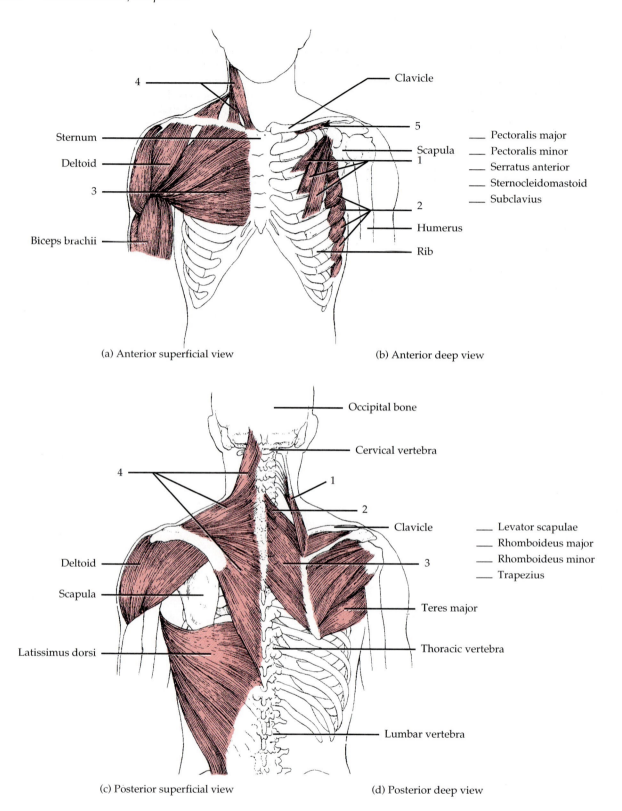

Sternum

Deltoid

3

Biceps brachii

4

Clavicle

5

Scapula

1

2

Humerus

Rib

___ Pectoralis major
___ Pectoralis minor
___ Serratus anterior
___ Sternocleidomastoid
___ Subclavius

(a) Anterior superficial view

(b) Anterior deep view

Occipital bone

Cervical vertebra

4

1

2

Clavicle

3

Deltoid

Scapula

Latissimus dorsi

___ Levator scapulae
___ Rhomboideus major
___ Rhomboideus minor
___ Trapezius

Teres major

Thoracic vertebra

Lumbar vertebra

(c) Posterior superficial view

(d) Posterior deep view

FIGURE 10.11   Muscles that move the pectoral (shoulder) girdle.

TABLE 10.15
Muscles that Move the Humerus (Arm) (After completing the table, label Figure 10.12.)

OVERVIEW: The muscles that move the humerus (arm) cross the shoulder joint. Of the
nine muscles that cross the shoulder joint, only two of them (pectoralis major and latissimus dorsi) do not originate on the scapula. These two muscles are thus designated as
axial muscles, since they originate on the axial skeleton. The remaining seven muscles,
the *scapular muscles,* arise from the scapula.

The strength and stability of the shoulder joint are not provided by the shape of the articulating bones or its ligaments. Instead, four deep muscles of the shoulder—subscapularis, supraspinatus, infraspinatus, and teres minor—strengthen and stabilize the
shoulder joint. The muscles join the scapula to the humerus. Their tendons are
arranged to form a nearly complete circle around the joint. This arrangement is referred
to as the *rotator (musculotendinous) cuff* and is a common site of injury to baseball
pitchers, especially tearing of the supraspinatus muscle tendon. This tendon is especially predisposed to wear-and-tear changes because of its location between the head of
the humerus and acromion of the scapula, which compresses the tendon during shoulder movements.

After you have studied the muscles in this table, arrange them according to the following
actions: flexion, extension, abduction, adduction, medial rotation, and lateral rotation.
(The same muscle can be used more than once.)

Muscle	Origin	Insertion	Action
AXIAL MUSCLES **Pectoralis** (pek'-tor-A-lis) **major** (label this muscle in Figure 10.11)			
Latissimus dorsi (la-TIS-i-mis DOR-sī; *dorsum* = back)			
SCAPULAR MUSCLES **Deltoid** (DEL-toyd; *delta* = triangular)			
Subscapularis (sub-scap'-yoo-LA-ris; *sub* = below; *scapularis* = scapula)			
Supraspinatus (soo'-pra-spi-NĀ-tus; *supra* = above; *spinatus* = spine of scapula)			
Infraspinatus (in'-fra-spi-NĀ-tus; *infra* = below)			
Teres (TE-rēz) **major** (*teres* = long and round)			
Teres (TE-rēz) **minor**			
Coracobrachialis (kor'-a-kō-BRĀ-kē-a'-lis; *coraco* = coracoid process)			

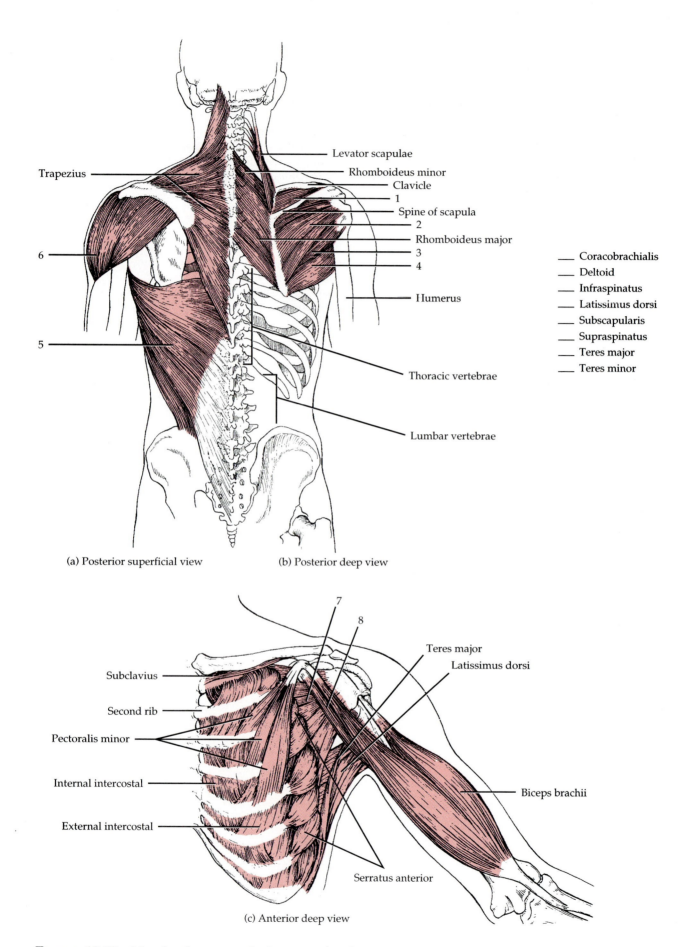

Levator scapulae

Trapezius

Rhomboideus minor

Clavicle

1

Spine of scapula

2

Rhomboideus major

3

4

Humerus

6

5

Thoracic vertebrae

Lumbar vertebrae

___ Coracobrachialis
___ Deltoid
___ Infraspinatus
___ Latissimus dorsi
___ Subscapularis
___ Supraspinatus
___ Teres major
___ Teres minor

(a) Posterior superficial view   (b) Posterior deep view

7

8

Teres major

Latissimus dorsi

Subclavius

Second rib

Pectoralis minor

Internal intercostal

External intercostal

Biceps brachii

Serratus anterior

(c) Anterior deep view

FIGURE 10.12   Muscles that move the humerus (arm).

TABLE 10.16
Muscles that Move the Radius and Ulna (Forearm) (After completing the table, label Figure 10.13.)

OVERVIEW: The muscles that move the radius and ulna (forearm) are divided into *flexors* and *extensors*. Recall that the elbow joint is a hinge joint, capable only of flexion and extension. Whereas the biceps brachii, brachialis, and brachioradialis are flexors of the forearm, the triceps brachii and anconeus are extensors. Other muscles that move the forearm are concerned with pronation and supination. (The biceps brachii muscle also permits supination of the forearm.)

Muscle	Origin	Insertion	Action
FLEXORS			
Biceps brachii (BĪ-ceps BRĀ-kē-ī; *biceps* = two heads of origin; *brachion* = arm)			
Brachialis (brā'-kē-A-lis)			
Brachioradialis (bra'-kē-ō-rā'-dē-A-lis; *radialis* = radius) (See also Figure 10.14a)			
EXTENSORS			
Triceps brachii (TRĪ-ceps BRĀ-kē-ī; *triceps* = three heads of origin)			
Anconeus (an-KŌ-nē-us; *anconeal* = pertaining to the elbow) (label this muscle in Figure 10.14)			
PRONATORS			
Pronator teres (PRŌ-na'-ter TE-rēz; (*pronation* = turning palm downward or posteriorly)			
Pronator quadratus (PRŌ-na'-ter kwod-RĀ-tus; *quadratus* = squared, four-sided) (label the muscle in Figure 10.14)			
SUPINATOR			
Supinator (sup-pi-nā-tor; *supination* = turning palm upward or anteriorly)			

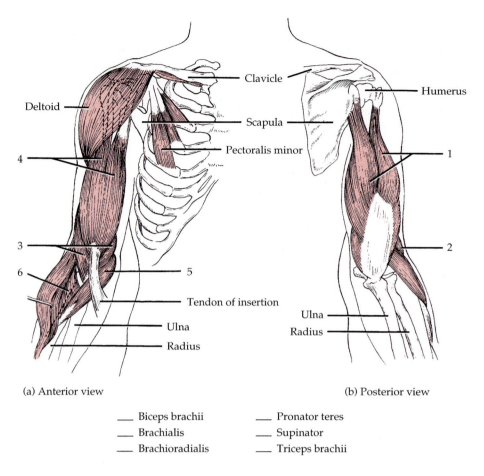

Deltoid

Clavicle

Humerus

Scapula

Pectoralis minor

4

1

3

2

6

5

Tendon of insertion

Ulna

Ulna

Radius

Radius

(a) Anterior view

(b) Posterior view

___ Biceps brachii   ___ Pronator teres
___ Brachialis   ___ Supinator
___ Brachioradialis   ___ Triceps brachii

FIGURE 10.13   Muscles that move the radius and ulna (forearm).

TABLE 10.17
Muscles that Move the Wrist, Hand, and Fingers (Figure 10.14)

OVERVIEW: Muscles that move the wrist, hand, and fingers are many and varied. However, as you will see, their names usually give some indication of their origin, insertion, or action. On the basis of location and function, the muscles are divided into two groups—anterior and posterior compartments. The *anterior compartment muscles* function as flexors. They originate on the humerus and typically insert on the carpals, metacarpals, and phalanges. The bellies of these muscles form the bulk of the proximal forearm. The *posterior compartment muscles* function as extensors. These muscles arise on the humerus and insert on the metacarpals and phalanges. Each of the two principal groups is also divided into superficial and deep muscles.

The tendons of the muscles of the forearm that attach to the wrist or continue into the hand, along with blood vessels and nerves, are held close to bones by strong fascial structures. The tendons are also surrounded by tendon sheaths. At the wrist, the deep fascia is thickened into fibrous bands called *retinacula* (*retinere* = retain). The *flexor retinaculum (transverse carpal ligament)* is located over the palmar surface of the carpal bones. Through it pass the long flexor tendons of the digits and wrist and median nerve. The *extensor retinaculum (dorsal carpal ligament)* is located over the dorsal surface of the carpal bones. Through it pass the extensor tendons of the wrist and digits.

After you have studied the muscles in the table, arrange them according to the following actions: flexion, extension, abduction, adduction, supination, and pronation. (The same muscles can be used more than once.)

Muscle	Origin	Insertion	Action
ANTERIOR GROUP (flexors) Superficial			
Flexor carpi radialis (FLEK-sor KAR-pē rā'-dē-A-lis; *flexor* = decreases angle at a joint; *carpus* = wrist; *radialis* = radius)			
Palmaris longus (pal-MA-ris LON-gus; *palma* = palm *longus* = long)			
Flexor carpi ulnaris (FLEK-sor KAR-pē ul-NAR-is; *ulnaris* = ulna)			
Flexor digitorum superficialis (FLEK-sor di'-ji-TOR-um soo'-per-fish'-ē-A-lis; *digit* = finger or toe; *superficialis* = closer to surface)			

TABLE 10.17 (*Continued*)

Muscle	Origin	Insertion	Action
Deep			
Flexor digitorum profundus (FLEK-sor di'-ji-TOR-um pro-FUN-dus; *profundus* = deep)			
Flexor pollicis longus (FLEK-sor POL-li-kis LON-gus; *pollex* = thumb)			
POSTERIOR GROUP (extensors) Superficial			
Extensor carpi radialis longus (eks-TEN-sor KAR-pē rā'-dē-A-lis LON-gus; *extensor* = increases angle at a joint)			
Extensor carpi radialis brevis (eks-TEN-sor KAR-pē rā'-dē-A-lis BREV-is; *brevis* = short)			
Extensor digitorum (eks-TEN-sor dī'-ji-TOR-um)			
Extensor digiti minimi (eks-TEN-sor DIJ-i-tē MIN-i-mē; *digiti* = digit; *minimi* = finger)			
Extensor carpi ulnaris (eks-TEN-sor KAR-pē ul-NAR-is)			
Abductor pollicis longus (ab-DUK-tor POL-li-kis LON-gus; *abductor* = moves a part away from midline)			
Extensor pollicis brevis (eks-TEN-sor POL-li-kis BREV-is)			
Extensor pollicis longus (eks-TEN-sor POL-li-kis LON-gus)			
Extensor indicis (eks-TEN-sor IN-di-kis; *indicis* = index)			

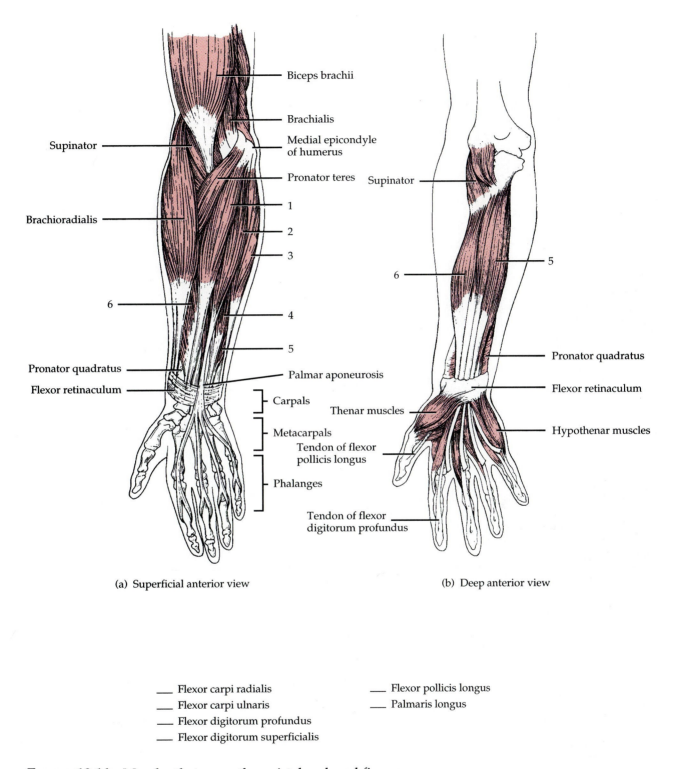

Biceps brachii

Brachialis

Medial epicondyle of humerus

Supinator

Pronator teres

Brachioradialis

1

2

3

Supinator

6

5

6

4

5

Pronator quadratus

Pronator quadratus

Flexor retinaculum

Palmar aponeurosis

Flexor retinaculum

Carpals

Thenar muscles

Hypothenar muscles

Metacarpals

Tendon of flexor pollicis longus

Phalanges

Tendon of flexor digitorum profundus

(a) Superficial anterior view

(b) Deep anterior view

___ Flexor carpi radialis

___ Flexor carpi ulnaris

___ Flexor digitorum profundus

___ Flexor digitorum superficialis

___ Flexor pollicis longus

___ Palmaris longus

FIGURE 10.14 Muscles that move the wrist, hand, and fingers.

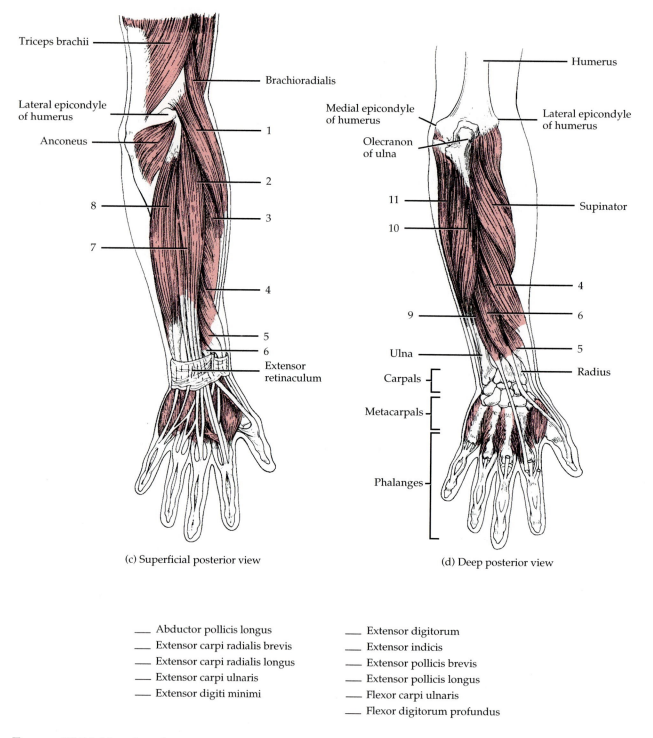

Triceps brachii

Brachioradialis

Lateral epicondyle of humerus

1

Anconeus

2

8

3

7

4

5

6

Extensor retinaculum

(c) Superficial posterior view

Humerus

Medial epicondyle of humerus

Lateral epicondyle of humerus

Olecranon of ulna

11

Supinator

10

4

9

6

Ulna

5

Carpals

Radius

Metacarpals

Phalanges

(d) Deep posterior view

____ Abductor pollicis longus
____ Extensor carpi radialis brevis
____ Extensor carpi radialis longus
____ Extensor carpi ulnaris
____ Extensor digiti minimi

____ Extensor digitorum
____ Extensor indicis
____ Extensor pollicis brevis
____ Extensor pollicis longus
____ Flexor carpi ulnaris
____ Flexor digitorum profundus

FIGURE 10.14 (*Continued*)   Muscles that move the wrist, hand, and fingers.

TABLE 10.18
Intrinsic Muscles of the Hand

OVERVIEW: Several of the muscles discussed in Table 10.17 help to move the digits in various ways. In addition, there are muscles in the palmar surface of the hand called *intrinsic muscles* that also help to move the digits. Such muscles are so named because their origins and insertions are both within the hands. These muscles assist in the intricate and precise movements that are characteristic of the human hand.

The intrinsic muscles of the hand are divided into three principal groups—thenar, hypothenar, and intermediate. The four *thenar* (THĒ-nar) *muscles*, shown in Figure 10.14b, act on the thumb and form the *thenar eminence* (see Figure 11.8). The four *hypothenar* (HĪ-pō-thē-nar) *muscles*, shown in Figure 10.14b, act on the little finger and form the *hypothenar eminence* (see Figure 11.8). The eleven *intermediate* (midpalmar) *muscles* act on all the digits, except the thumb.

The functional importance of the hand is readily apparent when one considers that certain hand injuries can result in permanent disability. In fact, most of the dexterity of the hand depends on the movements of the thumb. The general activities of the hand are free motion, power grip (forcible movement of the fingers and thumb against the palm, as in squeezing), precision handling (a change in position of a handled object that requires exact control of finger and thumb positions, as in winding a watch or threading a needle), and pinch (compression between the thumb and index finger or between the thumb and first two fingers).

Movement of the thumb is very important in the precise activities of the hand. The five principal movements of the thumb, illustrated below, are flexion (movement of the thumb medially across the palm), extension (movement of the thumb laterally away from the palm), abduction (movement of the thumb in an anteroposterior plane away from the palm), adduction (movement of the thumb in an anteroposterior plane toward the palm), and opposition (movement of the thumb across the palm so that the tip of the thumb meets the tips of the fingers). Opposition is the single most distinctive digital movement that gives humans and other primates the ability to precisely grasp and manipulate objects.

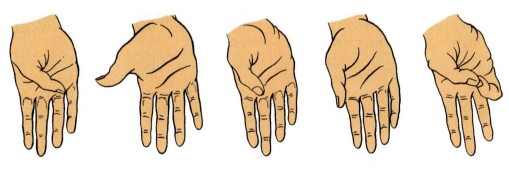

| Flexion | Extension | Abduction | Adduction | Opposition |

TABLE 10.19
Muscles that Move the Vertebral Column (Backbone) (Figure 10.15)

OVERVIEW: The muscles that move the vertebral column (backbone) are quite complex because they have multiple origins and insertions and there is considerable overlapping among them. One way to group the muscles is on the basis of the general direction of the muscle bundles and their approximate lengths. For example, the *splenius muscles* arise from the midline and run laterally and superiorly to their insertions. The *erector spinae (sacrospinalis) muscle* arises from either the midline or more laterally, but usually runs almost longitudinally, with neither a marked lateral nor medial direction as it is traced superiorly. The *transversospinalis muscles* arise laterally, but run toward the midline as they are traced superiorly. Deep to these three muscle groups are small *segmental muscles* that run between spinous processes or transverse processes of vertebrae. Since the scalene muscles also assist in moving the vertebral column, they are included in this table.

Note in Table 10.10 that the rectus abdominis and quadratus lumborum muscles also assume a role in moving the vertebral column.

Muscle	Origin	Insertion	Action
SPLENIUS (SPLĒ-nē-us) MUSCLES **Splenius capitis** (SPLĒ-nē-us KAP-i-tis; *splenium* = bandage; *caput* = head)			
Splenius cervicis (SPLĒ-nē-us SER-vi-kis; *cervix* = neck)			
ERECTOR SPINAE (SACROSPINALIS) This is the largest muscular mass of the back and consists of three groupings—iliocostalis, longissimus, and spinalis. These groups, in turn, consist of a series of overlapping muscles. The iliocostalis group is laterally placed, the longissimus group is intermediate in placement, and the spinalis group is medially placed.			
Iliocostalis (Lateral) Group			
Iliocostalis cervicis (il'-ē-ō-kos-TAL-is SER-vi-kis; *ilium* = flank; *costa* = rib)			
Iliocostalis thoracis (il'-ē-ō-kos-TAL-is thō-RA-kis; *thorax* = chest			
Iliocostalis lumborum (il'-ē-ō-kos-TAL-is lum-BOR-um)			
Longissimus (Intermediate) Group			
Longissimus capitis (lon-JIS-i-mus KAP-i-tis; *longissimus* = longest)			
Longissimus cervicis (lon-JIS-i-mus SER-vi-kis)			
Longissimus thoracis (lon-JIS-i-mus thō-RA-kis)			

TABLE 10.19 (*Continued*)

Muscle	Origin	Insertion	Action
Spinalis (Medial) Group			
Spinalis capitis (spi-NA-lis KAP-i-tis; *spinalis =* vertebral column)			
Spinalis cervicis (spi-NA-lis SER-vi-kis)			
Spinalis thoracis (spi-NA-lis tho-RA-kis)			
TRANSVERSOSPINALIS (trans-ver'-sō-spi-NA-lis) MUSCLES			
Semispinalis capitis (sem'-ē-spi-NA-lis KAP-i-tis; *semi* = partially or one half)			
Semispinalis cervicis (sem'-ē-spi-NA-lis SER-vi-kis)			
Semispinalis thoracis (sem'-ē-spi-NA-lis tho-RA-kis)			
Multifidus (mul-TIF-i-dus; *multi* = many; *fidere* = to split)			
Rotatores (rō'-ta-TO-rēz; *rotate* = to turn)			
SEGMENTAL MUSCLES **Interspinales** (in-ter-SPĪ-na-lēz; *inter* = between)			
Intertransversarii (in'-ter-trans-vers-AR-ē-ī)			
SCALENE (SKĀ-lēn) MUSCLES **Anterior scalene** (SKĀ-lēn; *anterior* = front *skalenos* = uneven)			
Middle scalene (SKĀ-lēn)			
Posterior scalene (SKĀ-lēn; *posterior* = back)			

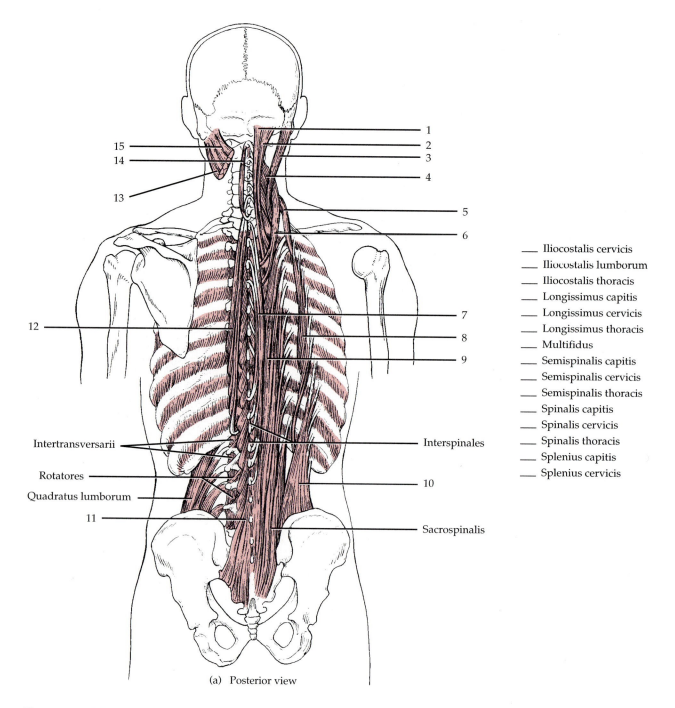

(a) Posterior view

___ Iliocostalis cervicis
___ Iliocostalis lumborum
___ Iliocostalis thoracis
___ Longissimus capitis
___ Longissimus cervicis
___ Longissimus thoracis
___ Multifidus
___ Semispinalis capitis
___ Semispinalis cervicis
___ Semispinalis thoracis
___ Spinalis capitis
___ Spinalis cervicis
___ Spinalis thoracis
___ Splenius capitis
___ Splenius cervicis

Intertransversarii

Rotatores
Quadratus lumborum
11

Interspinales

Sacrospinalis

FIGURE 10.15   Muscles that move the vertebral column (backbone). Many superficial muscles have been removed or cut to show deep muscles.

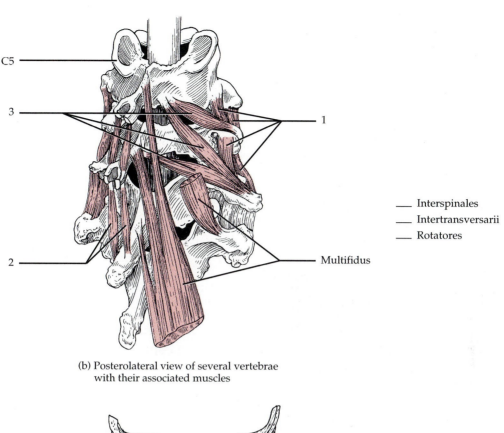

C5

3

1

___ Interspinales
___ Intertransversarii
___ Rotatores

2

Multifidus

(b) Posterolateral view of several vertebrae
    with their associated muscles

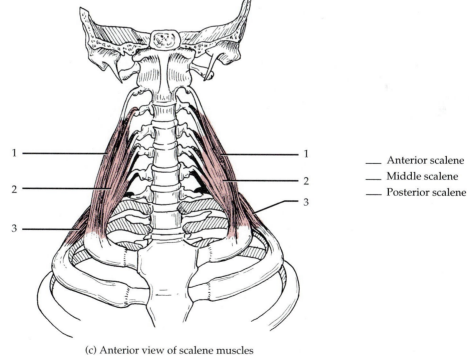

1   1

2   2

                     3

3

___ Anterior scalene
___ Middle scalene
___ Posterior scalene

(c) Anterior view of scalene muscles

FIGURE 10.15 (*Continued*)   Muscles that move the vertebral column (backbone).

TABLE 10.20
Muscles that Move the Femur (Thigh) (After completing the table, label Figure 10.16.)

OVERVIEW: As you will see, muscles of the lower limbs are larger and more power-
    ful than those of the upper limbs since lower limb muscles function in stability, loco-
    motion, and maintenance of posture. Upper limb muscles are characterized by
    versatility of movement. In addition, muscles of the lower limbs frequently cross two
    joints and act equally on both.
The anterior muscles are the psoas major and iliacus, together referred to as the iliop-
    soas muscle. The remaining muscles (except for the pectineus, adductors, and tensor
    fasciae latae) are posterior muscles. Technically, the pectineus and adductors are
    components of the medial compartment of the thigh, but they are included in this
    table because they act on the thigh. The tensor fasciae latae muscle is laterally placed.
    The *fascia lata* is a deep fascia of the thigh that encircles the entire thigh. It is well
    developed laterally where, together with the tendons of the gluteus maximus and
    tensor fasciae latae
    muscles, it forms a structure called the *iliotibial tract*. The tract inserts into the lat-
    eral condyle of the tibia.
After you have studied the muscles in this table, arrange them according to the follow-
    ing actions: flexion, extension, abduction, adduction, medial rotation, and lateral
    rotation. (The same muscles can be used more than once.)

Muscle	Origin	Insertion	Action
Psoas (SŌ-as) major (*psoa* = muscle of loin)			
Iliacus (il'-ē-AK-us; *iliac* = ilium)			
Gluteus maximus (GLOO-tē-us MAK-si-mus; *glutos* = buttock; *maximus* = largest)			
Gluteus medius (GLOO-tē-us MĒ-dē-us; *media* = middle)			
Gluteus minimus (GLOO-tē-us MIN-i-mus; *minimus* = smallest)			
Tensor fasciae latae (TEN-sor FA-shē-ē LĀ-tē; *tensor* = makes tense; *fascia* = band; *latus* = wide)			

TABLE 10.20 (*Continued*)

Muscle	Origin	Insertion	Action
Piriformis (pir-i-FOR-mis; *pirum* = pear; *forma* = shape)			
Obturator internus* (OB-too-rā'-tor in-TER-nus; *obturator* = obturator foramen; *internus* = inside)			
Obturator externus (OB-too-rā'-tor ex-TER-nus; *externus* = outside)			
Superior gemellus (jem-EL-lus; *superior* = above; *gemellus* = twins)			
Inferior gemellus (jem-EL-lus; *inferior* = below)			
Quadratus femoris (kwod-RĀ-tus FEM-or-is; *quad* = four; *femoris* = femur)			
Adductor longus (LONG-us; *adductor* = moves part closer to midline; *longus* = long)			
Adductor brevis (BREV-is; *brevis* = short)			
Adductor magnus (MAG-nus; *magnus* = large)			
Pectineus (pek-TIN-ē-us; *pecten* = comb-shaped)			

*Not illustrated

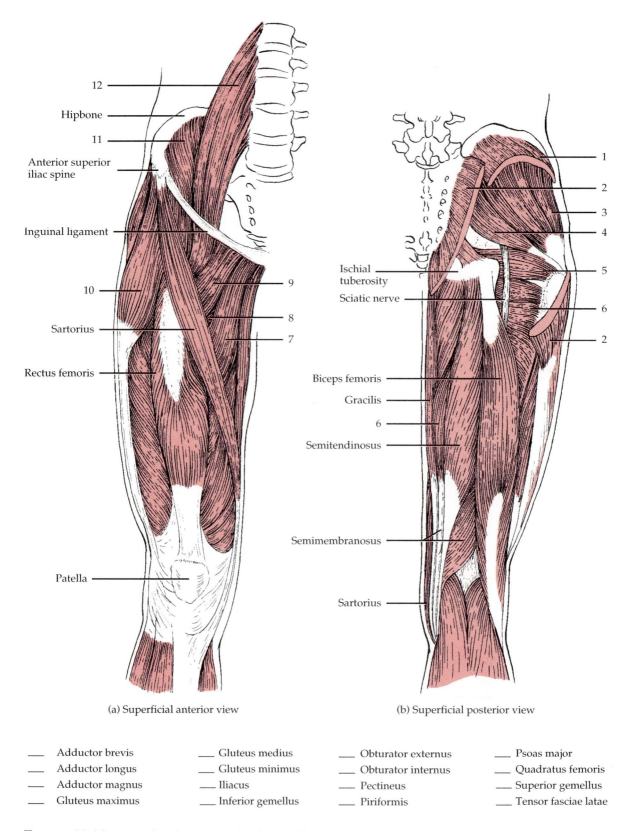

12

Hipbone

11

Anterior superior
iliac spine

Inguinal ligament

10

Sartorius

Rectus femoris

Patella

9

8

7

Ischial
tuberosity

Sciatic nerve

Biceps femoris

Gracilis

6

Semitendinosus

Semimembranosus

Sartorius

1

2

3

4

5

6

2

(a) Superficial anterior view   (b) Superficial posterior view

___ Adductor brevis	___ Gluteus medius	___ Obturator externus	___ Psoas major
___ Adductor longus	___ Gluteus minimus	___ Obturator internus	___ Quadratus femoris
___ Adductor magnus	___ Iliacus	___ Pectineus	___ Superior gemellus
___ Gluteus maximus	___ Inferior gemellus	___ Piriformis	___ Tensor fasciae latae

FIGURE 10.16   Muscles that move the femur (thigh).

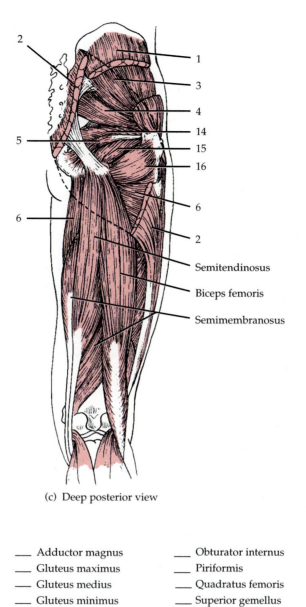

(c) Deep posterior view

___ Adductor magnus   ___ Obturator internus
___ Gluteus maximus   ___ Piriformis
___ Gluteus medius   ___ Quadratus femoris
___ Gluteus minimus   ___ Superior gemellus
___ Inferior gemellus

FIGURE 10.16 (*Continued*)   Muscles that move the femur (thigh).

TABLE 10.21
Muscles that Act on the Tibia and Fibula (Leg) (After completing the table, label Figure 10.17.)

OVERVIEW: The muscles that act on the tibia and fibula (leg) are separated into compartments by deep fascia. The *medial (adductor) compartment* is so named because its muscles adduct the thigh. As noted earlier, the adductor magnus, adductor longus, adductor brevis, and pectineus muscles, components of the medial compartment, are included in Table 10.20 because they act on the femur. The gracilis, the other muscle in the medial compartment, not only adducts the thigh but also flexes the leg. For this reason, it is included in this table.

The *anterior (extensor) compartment* is so designated because its muscles act to extend the leg (some also flex the thigh). It is composed of the quadriceps femoris and sartorius muscles. The quadriceps femoris muscle is a composite muscle that includes four distinct parts, usually described as four separate muscles (rectus femoris, vastus lateralis, vastus intermedius, and vastus medialis). The common tendon for the four muscles is known as the *quadriceps tendon*, which attaches to the patella. The tendon continues inferior to the patella as the *patellar ligament*, which attaches to the tibial tuberosity.

The *posterior (flexor) compartment* is so named because its muscles flex the leg. Included are the hamstrings (biceps femoris, semitendinosus, and semimembranosus). The hamstrings are so named because their tendons are long and stringlike in the popliteal area. The *popliteal fossa* is a diamond-shaped space on the posterior aspect of the knee bordered laterally by the biceps femoris and medially by the semitendinosus and semimembranosus muscles (see also Figure 11.10b).

Muscle	Origin	Insertion	Action
MEDIAL (ADDUCTOR) COMPARTMENT			
Adductor magnus (MAG-nus)			
Adductor longus (LONG-us)			
Adductor brevis (BREV-is)			
Pectineus (pek-TIN-ē-us)			
Gracilis (gra-SIL-is; *gracilis* = slender)			
ANTERIOR (EXTENSOR) COMPARTMENT			
Quadriceps femoris (KWOD-ri-ceps FEM-or-is; *quadriceps* = four heads of origin; *femoris* = femur)			

TABLE 10.21 (*Continued*)

Muscle	Origin	Insertion	Action
Rectus femoris (REK-tus FEM-or-is; *rectus* = fibers parallel to midline)			
Vastus lateralis (VAS-tus lat'-er-A-lis; *vastus* = large; *lateralis* = lateral)			
Vastus medialis (VAS-tus mē-dē-A-lis; *medialis* = medial)			
Vastus intermedius (VAS-tus in'-ter-MĒ-dē-us; *intermedius* = middle)			
Sartorius (sar-TOR-ē-us; *sartor* = tailor; refers to cross-legged position of tailors)			
POSTERIOR (FLEXOR) COMPARTMENT **Hamstrings** **Biceps femoris** (BĪ-ceps FEM-or-is; *biceps* = two heads of origin)			
Semitendinosus (sem'-ē-TEN-di-nō-sus; *semi* = half; *tendo* = tendon)			
Semimembranosus (sem'-ē-MEM-bra-nō-sus; *membran* = membrane)			

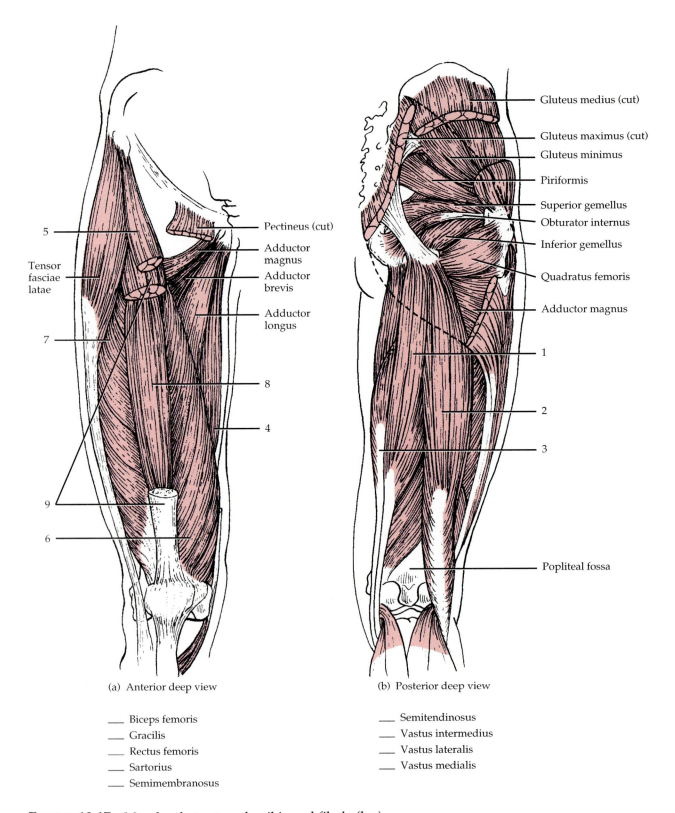

5

Tensor
fasciae
latae

7

9

6

Pectineus (cut)

Adductor
magnus

Adductor
brevis

Adductor
longus

8

4

(a) Anterior deep view

Gluteus medius (cut)

Gluteus maximus (cut)

Gluteus minimus

Piriformis

Superior gemellus

Obturator internus

Inferior gemellus

Quadratus femoris

Adductor magnus

1

2

3

Popliteal fossa

(b) Posterior deep view

___ Biceps femoris

___ Gracilis

___ Rectus femoris

___ Sartorius

___ Semimembranosus

___ Semitendinosus

___ Vastus intermedius

___ Vastus lateralis

___ Vastus medialis

FIGURE 10.17 Muscles that act on the tibia and fibula (leg).

TABLE 10.22
Muscles that Move the Foot and Toes (Figure 10.18)

OVERVIEW: The musculature of the leg, like that of the thigh, can be grouped into
  three compartments by deep fascia. The *anterior compartment* consists of muscles that
  dorsiflex the foot. In a situation analogous to the wrist, the tendons of the muscles of
  the anterior compartment are held firmly to the ankle by thickenings of deep fascia
  called the *superior extensor retinaculum (transverse ligament of the ankle)* and *inferior
  extensor retinaculum (cruciate ligament of the ankle).*
The *lateral (peroneal) compartment* contains two muscles that plantar flex and evert the
  foot.
The *posterior compartment* consists of muscles that are divisible into superficial and deep
  groups. All three superficial muscles share a common tendon of insertion, the calcaneal
  (Achilles) tendon that inserts into the calcaneal bone of the ankle. The superficial muscles are plantar flexors of the foot. Of the four deep muscles, three plantar flex the foot.

Muscle	Origin	Insertion	Action
ANTERIOR COMPARTMENT			
Tibialis (tib'-ē-A-lis) **anterior** (*tibialis* = tibia; *anterior* = front)			
Extensor hallucis longus (HAL-u-kis LON-gus; *extensor* = increases angle at joint; *hallucis* = hallux or great toe; *longus* = long)			
Extensor digitorum longus (di'-ji-TOR-um LON-gus)			
Peroneus tertius (per'-ō-NĒ-us TER-shus; *perone* = fibula; *tertius* = third)			
LATERAL (PERONEAL) COMPARTMENT			
Peroneus longus (per'-ō-NĒ-us LON-gus;)			
Peroneus brevis (per'-ō-NĒ-us BREV-is; *brevis* = short)			

TABLE 10.22 (*Continued*)

Muscle	Origin	Insertion	Action
POSTERIOR COMPARTMENT			
Superficial			
Gastrocnemius (gas'-trok-NĒ-mē-us; *gaster* = belly; *kneme* = leg)			
Soleus (SŌ-lē-us; *soleus* = sole)			
Plantaris (plan-TA-ris; *plantar* = sole)			
Deep			
Popliteus (pop-LIT-ē-us; *poples* = posterior surface of knee)			
Flexor hallucis longus (HAL-u-kis LON-gus; *flexor* = decreases angle at joint)			
Flexor digitorum longus (di'-ji-TOR-um LON-gus; *digitorum* = finger or toe)			
Tibialis (tib'-ē-A-lis) **posterior** (*posterior* = back)			

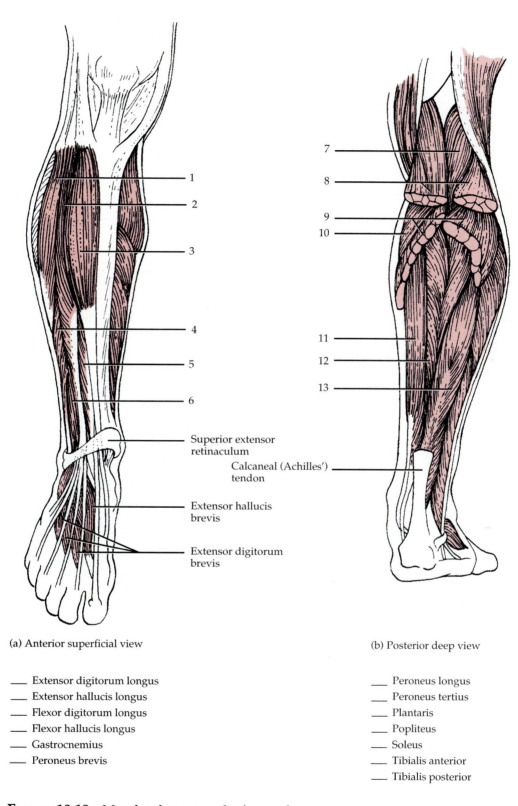

(a) Anterior superficial view

___ Extensor digitorum longus
___ Extensor hallucis longus
___ Flexor digitorum longus
___ Flexor hallucis longus
___ Gastrocnemius
___ Peroneus brevis

(b) Posterior deep view

___ Peroneus longus
___ Peroneus tertius
___ Plantaris
___ Popliteus
___ Soleus
___ Tibialis anterior
___ Tibialis posterior

FIGURE 10.18   Muscles that move the foot and toes.

F. COMPOSITE MUSCULAR SYSTEM

Now that you have studied the muscles of the body by region, label the composite diagram shown in Figure 10.19.

G. DISSECTION OF CAT MUSCULAR SYSTEM

Assuming that you have completed your study of the human muscular system, you can now begin your examination of the cat musculature. The principal objective of dissecting the cat muscular system is to increase your understanding of the human muscular system. Do not be concerned that the cat and human muscular systems differ somewhat. Instead, concentrate on the similarities and, as you perform the dissection, try to relate your knowledge of the human muscular system to that of the cat.

CAUTION! *Please reread Section D, "Precautions Related to Dissection," at the beginning of the laboratory manual on p. xiii before you begin your dissection.*

PROCEDURE

1. After obtaining a cat from your instructor, remove it from the plastic bag and place it on your dissecting tray ventral side up. Keep the liquid preservative in the plastic bag.
2. Determine the gender of your cat. A male has a *scrotum* ventral to the base of the tail. The *prepuce* appears as a slight elevation anterior to the scrotum. In contrast to the human male, the cat penis is usually retracted.
3. A female cat has a *urogenital aperture*, the external opening of the vagina and the urethra, just ventral to the anus. Both male and female cats have four or five *nipples (teats)* on either side of the midline, so their presence does not necessarily mean that your cat is female. Consult your instructor if you have difficulty determining the gender of your cat. Compare your cat with one of the opposite sex and learn to identify both sexes externally.
4. Before skinning your cat, review Section I in Exercise 2, in which you were introduced to the planes of the body. Identify the planes of the cat's body and compare their orientation to

those in the human (Figures 2.3 and 10.20). Also review Section G in Exercise 2 and locate, on your cat, the structures described therein for the human. You will note the following additional structures on the cat:
   a. *Virissae* Tactile whiskers located around the mouth and above the eyes.
   b. *Claws* Derived from the epidermis. They are retractable.
   c. *Tori (foot pads* or *friction pads)* Thickened areas of skin on the ventral surface of the paws. They cushion the body when the cat walks. Count the tori on the front and the back paws.

1. Skinning the Cat

PROCEDURE

Note: Please read this entire section before starting.

1. Grasp the skin at the ventral surface of the neck, pinch it up, and make a small longitudinal incision at the midline with the scissors. The incision, as well as all others you will make in skinning, should not be too deep because the skin is only about ⅛ in. thick and underlying muscles are easily damaged.
2. Now continue the incision upward along the midline to the lower lip and downward along the midline to the genital region. Do not remove the skin from the genital region.
3. If your specimen is female, the mammary glands may be removed. These appear as large masses along the ventral surface of the abdomen. Do not damage the underlying muscles as you remove the glands.
4. Next make incisions from the midline incision on the ventral side to the wrists and from the midline incision to the ankles. Cut all the way around the wrists and ankles.
5. Now you can begin to separate the skin from the underlying muscles to expose the muscles. This can be done by pulling the skin away from the body, using your fingers as much as possible. If you must use a scalpel, make sure that the sharp edge is directed away from the muscle and toward the skin.
6. Continue peeling the skin until only two points of attachment remain, one at the base of the tail and the other at the sides and top of the head.
7. Cut across the skin at the base of the tail to free the skin from the tail. The skin around the tail may be left in position.

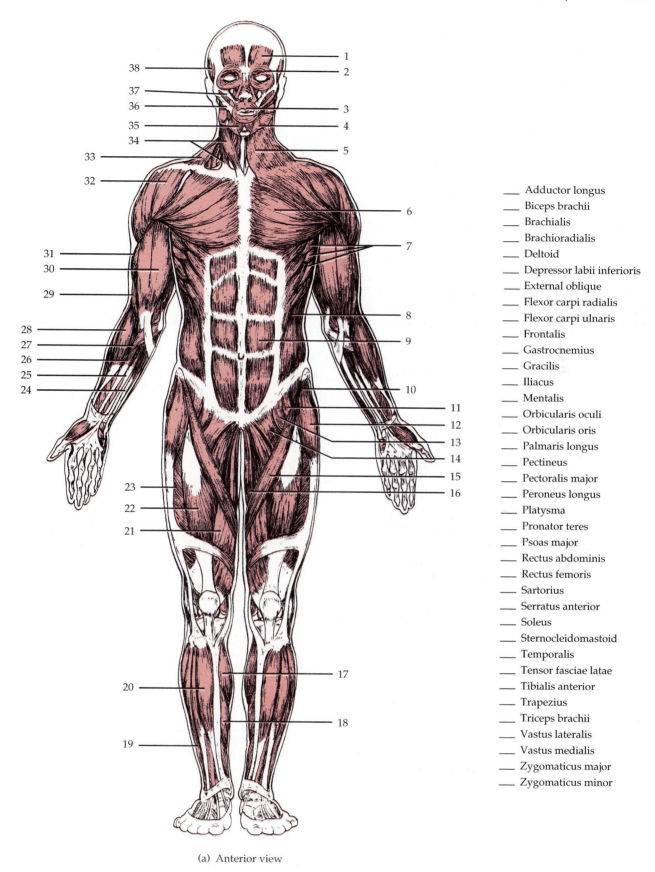

(a) Anterior view

FIGURE 10.19 Principal superficial muscles.

____ Adductor longus
____ Biceps brachii
____ Brachialis
____ Brachioradialis
____ Deltoid
____ Depressor labii inferioris
____ External oblique
____ Flexor carpi radialis
____ Flexor carpi ulnaris
____ Frontalis
____ Gastrocnemius
____ Gracilis
____ Iliacus
____ Mentalis
____ Orbicularis oculi
____ Orbicularis oris
____ Palmaris longus
____ Pectineus
____ Pectoralis major
____ Peroneus longus
____ Platysma
____ Pronator teres
____ Psoas major
____ Rectus abdominis
____ Rectus femoris
____ Sartorius
____ Serratus anterior
____ Soleus
____ Sternocleidomastoid
____ Temporalis
____ Tensor fasciae latae
____ Tibialis anterior
____ Trapezius
____ Triceps brachii
____ Vastus lateralis
____ Vastus medialis
____ Zygomaticus major
____ Zygomaticus minor

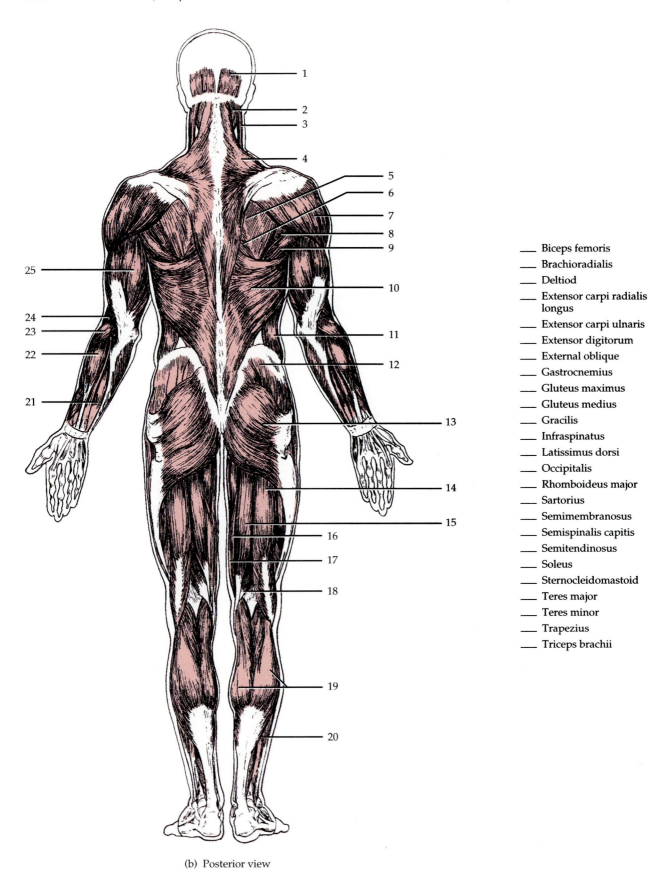

(b) Posterior view

____ Biceps femoris
____ Brachioradialis
____ Deltiod
____ Extensor carpi radialis longus
____ Extensor carpi ulnaris
____ Extensor digitorum
____ External oblique
____ Gastrocnemius
____ Gluteus maximus
____ Gluteus medius
____ Gracilis
____ Infraspinatus
____ Latissimus dorsi
____ Occipitalis
____ Rhomboideus major
____ Sartorius
____ Semimembranosus
____ Semispinalis capitis
____ Semitendinosus
____ Soleus
____ Sternocleidomastoid
____ Teres major
____ Teres minor
____ Trapezius
____ Triceps brachii

FIGURE 10.19 (*Continued*)   Principal superficial muscles.

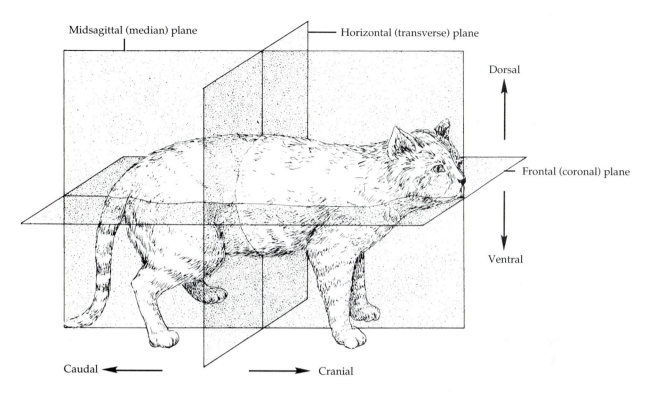

Midsagittal (median) plane   Horizontal (transverse) plane

Dorsal

Frontal (coronal) plane

Ventral

Caudal ◄————   ————► Cranial

FIGURE 10.20 Anatomical planes and directional terms of the cat.

8. Make a final incision from the left side of the lower lip, up to the top of the left eye, across to the top of the right eye, and down to the right side of the lower lip.
9. Remove the skin from the head, cutting around the bases of the ears as you come to them.
10. As part of the skinning procedure, be careful of the following:

    a. While skinning the neck, do not damage the external jugular veins in the neck (see Figure 10.21a).
    b. While skinning the forelimbs, do not damage the cephalic veins (see Figure 18.22).
    c. While skinning the hindlimbs, do not damage the greater saphenous veins (see Figure 18.20).
    d. While skinning the dorsal surface of the abdomen, do not cut through the deep fascia (lumbodorsal fascia) along the dorsal midline (see Figure 10.23a).
    e. The edge of the latissimus dorsi muscle (see Figure 10.23a) has a tendency to tear.

11. Before continuing, find the *cutaneous muscles*, which are attached to the undersurface of the skin and are generally removed with the skin. One of these, the *cutaneous maximus*, is closely bound to the underside of the skin in the trunk area. Another is the *platysma*, which is closely bound to the skin on either side of the neck. The cutaneous muscles move the skin.[2]
12. Clean off as much fat as possible. Large fat deposits are usually found in the neck, axillary regions, and inguinal regions. As you pick away the fat with forceps, do not damage blood vessels or nerves. Also remove as much of the superficial fascia covering muscles as possible. This can be done with your forceps. If your specimen has been properly cleaned, the individual fibers in a muscle will be visible. Usually, the fibers of one muscle run in one direction, while those of another muscle run in another direction.
13. Save the skin and wrap it around the cat at the end of each laboratory period to prevent your specimen from drying out. Moist paper towels or an old cotton towel may also be used.

2. Dissecting Skeletal Muscles

Keep in mind that *dissect* means to separate. Dissection of a skeletal muscle includes separating

---

[2]If preskinned cats are used the instructor may wish to demonstrate the cutaneous maximus and the platysma.

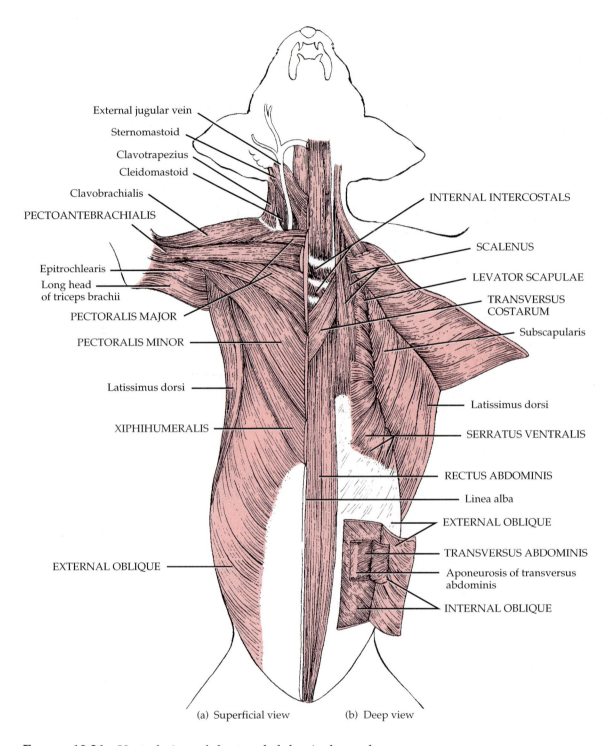

External jugular vein
Sternomastoid
Clavotrapezius
Cleidomastoid
Clavobrachialis
PECTOANTEBRACHIALIS

Epitrochlearis
Long head
of triceps brachii
PECTORALIS MAJOR

PECTORALIS MINOR

Latissimus dorsi

XIPHIHUMERALIS

EXTERNAL OBLIQUE

INTERNAL INTERCOSTALS

SCALENUS

LEVATOR SCAPULAE

TRANSVERSUS
COSTARUM

Subscapularis

Latissimus dorsi

SERRATUS VENTRALIS

RECTUS ABDOMINIS

Linea alba

EXTERNAL OBLIQUE

TRANSVERSUS ABDOMINIS

Aponeurosis of transversus
abdominis

INTERNAL OBLIQUE

(a) Superficial view   (b) Deep view

FIGURE 10.21   Ventral view of chest and abdominal muscles.

the body of the muscle from surrounding tissues, but points of attachment are left intact. These are the *origin* (attachment to a fixed bone) and the *insertion* (attachment to a movable bone). When a muscle contracts, the insertion moves toward the origin. You can pull the insertion toward the origin to observe the action of the muscle. Most muscles

are attached to bones by bands of white fibrous connective tissue called *tendons*. Some muscles are attached by white, flattened, sheetlike tendons called *aponeuroses*.

Skeletal muscles are distinguished from each other by normal lines of cleavage. These lines consist of deep fascia and may be exposed by slightly pulling adjacent muscles apart. The muscles can be separated from each other by picking the fascia out with your forceps. If, after removing the fascia, the muscles separate as distinct units, your procedure is correct. If, instead, the separated muscles have a ragged appearance, you have damaged a muscle by tearing it into two portions.

Sometimes superficial muscles must be severed in order to examine deep muscles. When you cut a muscle, *transect* it (i.e., cut it at right angles to the fibers at about the center of the belly of the muscle) using a scissors. Then *reflect* the cut ends (i.e., pull them back, one toward the origin and one toward the insertion). As you reflect the ends, carefully break the underlying fascia with your scalpel to separate the superficial muscle from the deep muscle. Transection keeps the origin and insertion of the superficial muscle intact and preserves the original shape and position of the superficial muscle.

3. Dissection, Transection, and Reflection of a Single Skeletal Muscle

Before you start to identify skeletal muscles, you should dissect a single muscle to familiarize yourself with the procedure and the terminology. The sample muscle that has been selected is the pectoantebrachialis (see Figure 10.21a).

PROCEDURE
1. Place the cat on its dorsal surface and find the pectoantebrachialis and pectoralis major muscles. Separate the muscles from each other.
2. Dissect the body of the pectoantebrachialis muscle toward its origin and insertion. Its origin is the manubrium of the sternum and it inserts on the fascia on the proximal end of the forearm. Leave the origin and insertion intact.
3. By drawing the insertion toward the origin you can note the action of the muscle—adduction of the arm.
4. Transect the pectoantebrachialis muscle and reflect the cut ends to expose the underlying pectoralis major. As you reflect, break the fascia under the pectoantebrachialis muscle.

4. Identification of Skeletal Muscles

You are now ready to begin your identification of selected skeletal muscles of the cat. The muscles you will study are grouped according to region. Each muscle discussed is shown in the corresponding figure. Locate each muscle discussed and try to determine its origin and insertion. Then dissect it and pull its insertion toward its origin to observe its action. If you are told that a particular muscle is a deep muscle, you will also have to transect and reflect the superficial muscle to locate and study the deep muscle.

As you dissect, transect, and reflect muscles of various regions, work only on either the left or right side of the cat. You will use the other side later on to study blood vessels and nerves.

a. SUPERFICIAL CHEST MUSCLES: PECTORAL GROUP (FIGURE 10.21a)
The *pectoralis* muscle group covers the ventral surface of the chest. The large, triangular muscle group arises from the sternum and inserts primarily on the humerus. On each side, the pectoralis group is divided into four muscles.

1. *Pectoantebrachialis* A narrow ribbonlike muscle superficial to the pectoralis major muscle. It consists of a narrow band (½ in.) of fibers that run from the sternum to the forearm. Dissect, transect, and reflect.
2. *Pectoralis major* A muscle 2 to 3 in. wide that lies immediately deep to the pectoantebrachialis and the clavobrachialis. Its fibers run roughly parallel to those of the pectoantebrachialis. Separate the pectoralis major from the pectoralis minor and clavobrachialis, transect, and reflect.
3. *Pectoralis minor* A muscle lying dorsal and caudal to the pectoralis major. Dissect, transect, and reflect.
4. *Xiphihumeralis* The most caudad of the pectoral group. Its fibers run parallel to pectoralis minor and eventually pass deep to the pectoralis minor. Dissect, transect, and reflect.

In humans, the pectoral group has only two muscles—the pectoralis major and minor.

b. DEEP CHEST MUSCLES (FIGURE 10.21b)
1. *Serratus ventralis* Fan-shaped muscle extending obliquely upward from the ribs to the shoulder blade. Dissect.
2. *Levator scapulae* Continuous with the serratus ventralis at its cranial border. Dissect.
3. *Transversus costarum* Small, thin muscle that covers cranial end of rectus abdominis muscle. Dissect.
4. *Scalenus* Complex muscle that extends longitudinally along the ventrolateral thorax and

neck. Its three main parts are called the scalenus dorsalis, scalenus medius, and scalenus ventralis. The scalenus medius is easiest to recognize. Dissect.

5. *External intercostals* Series of muscles[3] in the intercostal spaces. Their fibers run in the same direction (caudoventrally) as those of the external oblique muscle (next section). Dissect, transect, and reflect.

6. *Internal intercostals* Series of muscles deep to the external intercostals. Their fibers run in the same direction (caudodorsally) as those of the internal oblique muscle (next section).

c. MUSCLES OF THE ABDOMINAL WALL (FIGURE 10.21a AND b)

Before examining the muscles of the abdominal wall, locate the *linea alba,* a midventral fascial line formed by fusion of the aponeuroses of the ventrolateral abdominal muscles. The muscles of the abdominal wall include the following:

1. *External oblique* Most superficial, lateral abdominal muscle. Its dorsal half is fleshy; its ventral half is an aponeurosis. Its fibers run dorsally and caudally. Make a very shallow longitudinal incision in the external oblique from the ribs to the pelvis. Transect and reflect. Find the aponeurosis of the external oblique.

2. *Internal oblique* Deep to the external oblique. Its fibers run ventrally and cranially, opposite those of the external oblique. Make a very shallow longitudinal incision in the internal oblique muscle until you can see transverse fibers. They belong to the transversus abdominis muscle. Reflect the internal oblique and find its aponeurosis.

3. *Transversus abdominis* This is the third and deepest layer of the abdominal wall and is deep to the internal oblique. Its fibers run transversely. Locate the aponeurosis of the transversus abdominis.

4. *Rectus abdominis* Longitudinal band of muscle just lateral to the linea alba. The muscle is visible when the external oblique aponeurosis is reflected.

The abdominal muscles are similar to those in humans.

d. MUSCLES OF THE NECK (FIGURE 10.22a AND b)

1. *Sternomastoid* Band of muscle extending from the manubrium to the mastoid portion of the

[3]Not illustrated.

skull. The muscle is deep to the external jugular vein. Dissect.

2. *Cleidomastoid* Narrow band of muscle that is between the sternomastoid and clavotrapezius. Most of the cleidomastoid is deep to the clavotrapezius. Dissect.

3. *Sternohyoid* Narrow, straight muscle on either side of the midline covering the trachea. Caudal end partially covered by the sternomastoid. Dissect.

4. *Sternothyroid* Narrow band of muscle lateral and dorsal to the sternohyoid. The sternohyoid must be lifted and pushed aside to expose the sternothyroid. The thyroid gland is deep to the sternothyroid.

5. *Thyrohyoid* Small muscle on the lateral side of the thyroid cartilage of the larynx. Its location is just cranial to the sternothyroid.

6. *Geniohyoid* Straplike muscle cranial to the sternohyoid. The mylohyoid must be transected and reflected to locate the geniohyoid. Transect and reflect the geniohyoid.

7. *Mylohyoid* Thin, sheetlike muscle seen after transection and reflection of the digastric muscle. Transect and reflect.

8. *Digastric* Superficial muscle attached to the ventral border of the mandible. Transect and reflect.

9. *Masseter* Large muscle mass at the angle of the jaw behind and below the ear.

10. *Temporalis* A muscle just above and caudal to the eye and medial to the ear (see Figure 10.23a).

11. *Hyoglossus* Short, obliquely directed muscle lateral to the geniohyoid. Dissect.

12. *Styloglossus* Long muscle that runs parallel to the inside of the body and the mandible. Dissect.

13. *Genioglossus* A muscle lying deep to the geniohyoid.

e. SUPERFICIAL SHOULDER MUSCLES (FIGURE 10.23a)

1. *Clavotrapezius* Large, broad muscle on the back and side of the neck. Dissect, transect, and reflect.

2. *Levator scapulae ventralis* A muscle lying caudal to the clavotrapezius and passing underneath the clavotrapezius. Dissect.

3. *Clavobrachialis (clavodeltoid)* Triangular muscle extending onto the shoulder as a continuation of the clavotrapezius. Dissect, transect, and reflect.

4. *Acromiodeltoid* A muscle caudal to the clavobrachialis and lateral to the levator scapulae ventralis. Dissect, transect, and reflect.

5. *Spinodeltoid* Short, thick muscle caudal to the acromiodeltoid. Dissect, transect, and reflect.

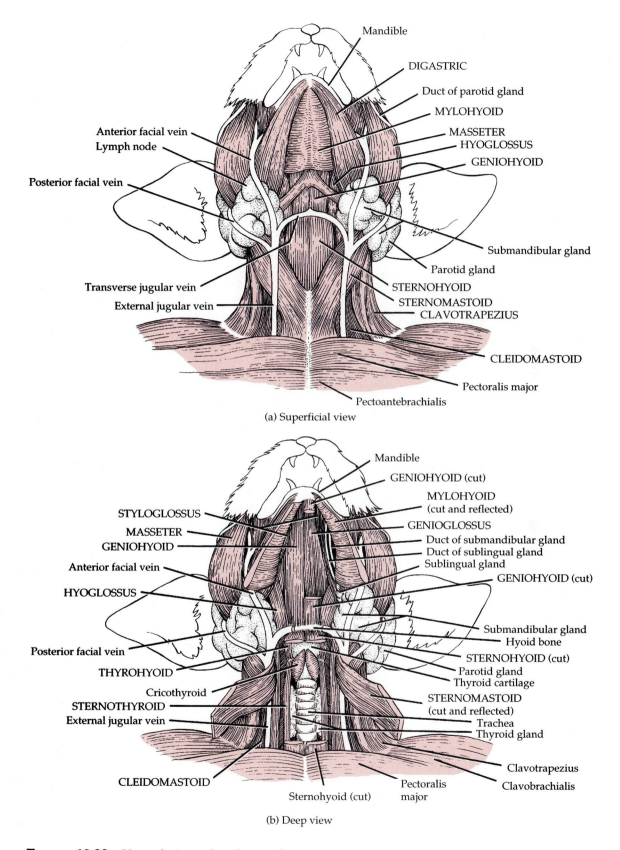

Mandible

DIGASTRIC

Duct of parotid gland

MYLOHYOID

MASSETER

HYOGLOSSUS

GENIOHYOID

Anterior facial vein

Lymph node

Posterior facial vein

Submandibular gland

Parotid gland

STERNOHYOID

STERNOMASTOID

CLAVOTRAPEZIUS

Transverse jugular vein

External jugular vein

CLEIDOMASTOID

Pectoralis major

Pectoantebrachialis

(a) Superficial view

Mandible

GENIOHYOID (cut)

MYLOHYOID
(cut and reflected)

STYLOGLOSSUS

MASSETER

GENIOHYOID

GENIOGLOSSUS

Duct of submandibular gland

Duct of sublingual gland

Sublingual gland

GENIOHYOID (cut)

Anterior facial vein

HYOGLOSSUS

Submandibular gland

Hyoid bone

STERNOHYOID (cut)

Parotid gland

Thyroid cartilage

STERNOMASTOID
(cut and reflected)

Trachea

Thyroid gland

Posterior facial vein

THYROHYOID

Cricothyroid

STERNOTHYROID

External jugular vein

CLEIDOMASTOID

Sternohyoid (cut)

Pectoralis
major

Clavotrapezius

Clavobrachialis

(b) Deep view

FIGURE 10.22   Ventral view of neck muscles.

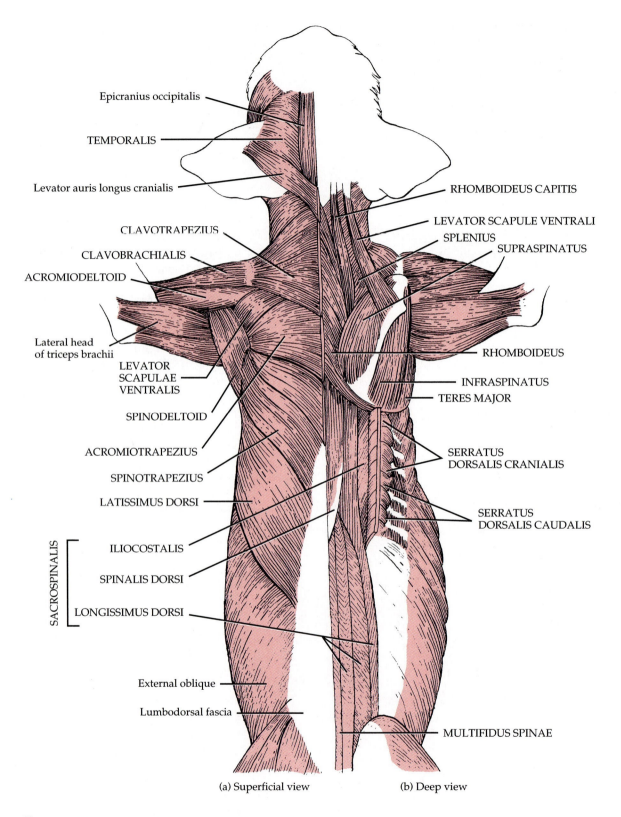

Epicranius occipitalis

TEMPORALIS

Levator auris longus cranialis

CLAVOTRAPEZIUS

CLAVOBRACHIALIS

ACROMIODELTOID

Lateral head of triceps brachii

LEVATOR SCAPULAE VENTRALIS

SPINODELTOID

ACROMIOTRAPEZIUS

SPINOTRAPEZIUS

LATISSIMUS DORSI

SACROSPINALIS

ILIOCOSTALIS

SPINALIS DORSI

LONGISSIMUS DORSI

External oblique

Lumbodorsal fascia

RHOMBOIDEUS CAPITIS

LEVATOR SCAPULE VENTRALI

SPLENIUS

SUPRASPINATUS

RHOMBOIDEUS

INFRASPINATUS

TERES MAJOR

SERRATUS DORSALIS CRANIALIS

SERRATUS DORSALIS CAUDALIS

MULTIFIDUS SPINAE

(a) Superficial view   (b) Deep view

FIGURE 10.23   Dorsal view of shoulder and back muscles.

6. *Acromiotrapezius* Thin, flat muscle extending from the middle of the back to the shoulder. The muscle is caudal to the clavotrapezius. Dissect, transect, and reflect.

7. *Spinotrapezius* A muscle lying caudal to the acromiotrapezius, partly covered by the acromiotrapezius. Dissect, transect, and reflect.

8. *Latissimus dorsi* Large, triangular muscle caudal to the spinotrapezius. This muscle extends from the middle of the back to the humerus. Note its attachment to the lumbodorsal fascia. Dissect, transect, and reflect.

f. DEEP SHOULDER MUSCLES (FIGURE 10.23b)

1. *Rhomboideus capitis* Narrow muscle extending from the back of the skull to the scapula. Dissect.

2. *Rhomboideus* A muscle lying deep to medial portions of the acromiotrapezius and the spinotrapezius, between the scapula and the body wall. Dissect.

3. *Splenius* Broad muscle deep to the clavotrapezius and the rhomboideus capitis. Dissect.

4. *Supraspinatus* A muscle deep to the acromiotrapezius, located in the supraspinous fossa of the scapula. Dissect.

5. *Infraspinatus* A muscle caudal to the supraspinatus, occupying the infraspinous fossa of the scapula. Dissect.

6. *Teres major* Band of thick muscle caudal to the infraspinatus muscle and covering the axillary border of the scapula. Dissect.

7. *Teres minor* Tiny, round muscle caudal and adjacent to the insertion of the infraspinatus.[4] Dissect.

8. *Subscapularis* Large, flat muscle in the subscapular fossa. Dissect (see Figure 10.26b).

g. BACK MUSCLES (FIGURE 10.23a AND b)

1. *Serratus dorsalis cranialis* Four or five short muscle slips above and caudal to the scapula.

2. *Serratus dorsalis caudalis* Four or five short muscle slips caudal to the serratus dorsalis cranialis and medial to the upper part of the latissimus dorsi.

3. *Sacrospinalis* Complex muscle extending along the spine. Its three divisions are the lateral *iliocostalis*, which is deep to the serratus dorsalis cranialis and the serratus dorsalis caudalis; the intermediate *longissimus dorsi*, which fills the spaces between the spines and transverse processes of the thoracic and lumbar vertebrae; and the medial *spinalis dorsi*, next to the spines of the thoracic vertebrae. Dissect these components.

4. *Multifidus spinae* Narrow band of muscle adjacent to the vertebral spines extending the length of the vertebral column and best seen in the sacral region. Its cranial portion is called the *semispinalis cervicis*[5] muscle, which is under the splenius.

h. MUSCLES OF THE ARM: MEDIAL SURFACE (FIGURE 10.24)

1. *Epitrochlearis* Flat band of muscle along the medial surface of the arm between the biceps brachii and the triceps brachii. Dissect, transect, and reflect.

2. *Biceps brachii* Located on the ventromedial surface of the arm. Dissect.

i. MUSCLES OF THE ARM: LATERAL SURFACE (FIGURE 10.25)

1. *Triceps brachii* Muscle with three main heads covering the caudal surface and much of the sides of the arm. The *long head*, located on the caudal surface of the arm, is the largest. The *lateral head* covers much of the lateral surface of the arm. Dissect, transect, and reflect the lateral head. The *medial head*, found on the medial surface of the arm, is located between the long and lateral heads.

2. *Anconeus* A tiny triangular muscle lying deep to the distal end of the lateral head of the triceps brachii. This is a fourth head of the triceps brachii. Dissect.

3. *Brachialis* A muscle located on the ventrolateral surface of the arm, cranial to the lateral head of the triceps brachii. Dissect.

j. MUSCLES OF THE FOREARM: VENTRAL SURFACE (FIGURE 10.26)

1. *Pronator teres* Small, triangular muscle that originates from the medial epicondyle of the humerus. Dissect.

2. *Flexor carpi radialis* Long, spindle-shaped muscle above the pronator teres. Dissect.

3. *Palmaris longus* Large, flat, broad muscle near the center of the medial surface. Dissect, transect, and reflect.

4. *Flexor digitorum sublimis (superficialis)* A muscle deep to the palmaris longus. Dissect.

5. *Flexor carpi ulnaris* Band of muscle above the palmaris longus consisting of two heads. Dissect.

6. *Flexor digitorum profundus* A muscle with five heads of origin, deep to the palmaris longus. Dissect.

[4]Not illustrated.

[5]Not illustrated.

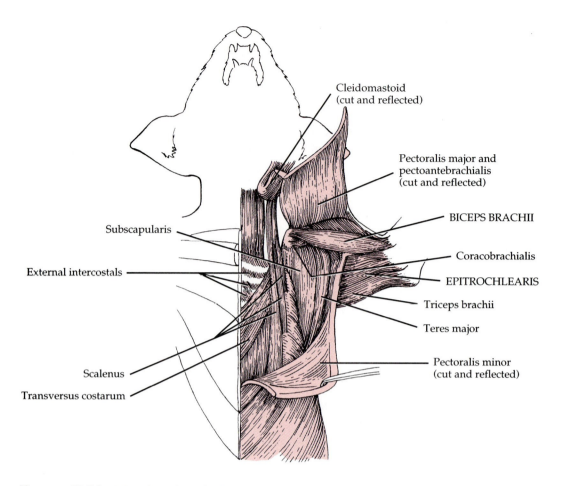

FIGURE 10.24 Muscles of medial surface of arm.

Cleidomastoid (cut and reflected)

Pectoralis major and pectoantebrachialis (cut and reflected)

BICEPS BRACHII

Coracobrachialis

EPITROCHLEARIS

Triceps brachii

Teres major

Pectoralis minor (cut and reflected)

Subscapularis

External intercostals

Scalenus

Transversus costarum

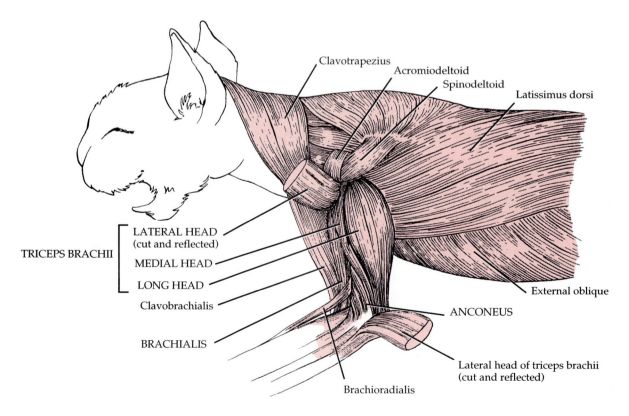

FIGURE 10.25 Muscles of lateral surface of arm.

Clavotrapezius

Acromiodeltoid

Spinodeltoid

Latissimus dorsi

TRICEPS BRACHII

LATERAL HEAD (cut and reflected)

MEDIAL HEAD

LONG HEAD

Clavobrachialis

BRACHIALIS

Brachioradialis

ANCONEUS

External oblique

Lateral head of triceps brachii (cut and reflected)

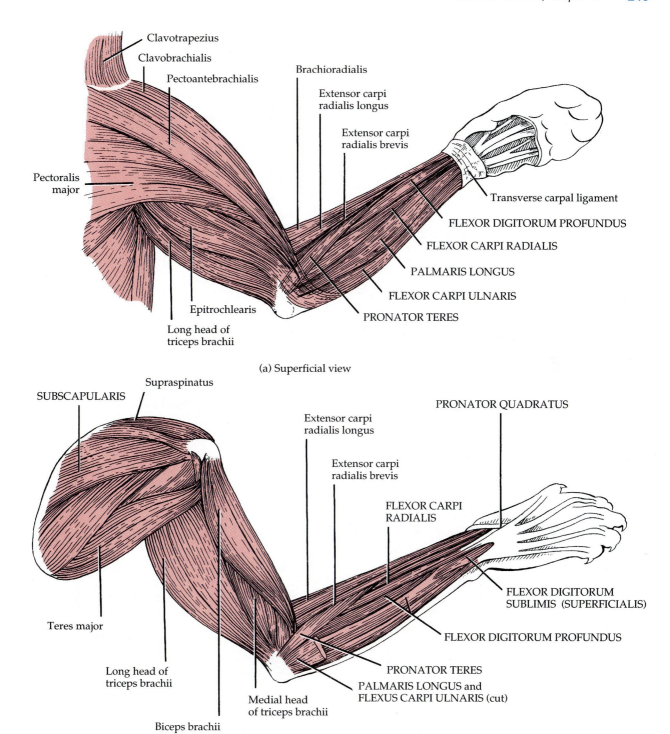

Clavotrapezius

Clavobrachialis

Pectoantebrachialis

Brachioradialis

Extensor carpi radialis longus

Extensor carpi radialis brevis

Pectoralis major

Transverse carpal ligament

FLEXOR DIGITORUM PROFUNDUS

FLEXOR CARPI RADIALIS

PALMARIS LONGUS

FLEXOR CARPI ULNARIS

PRONATOR TERES

Epitrochlearis

Long head of triceps brachii

(a) Superficial view

SUBSCAPULARIS

Supraspinatus

Extensor carpi radialis longus

PRONATOR QUADRATUS

Extensor carpi radialis brevis

FLEXOR CARPI RADIALIS

Teres major

FLEXOR DIGITORUM SUBLIMIS (SUPERFICIALIS)

FLEXOR DIGITORUM PROFUNDUS

PRONATOR TERES

PALMARIS LONGUS and FLEXUS CARPI ULNARIS (cut)

Long head of triceps brachii

Medial head of triceps brachii

Biceps brachii

(b) Deep view

FIGURE 10.26 Ventral view of muscles of forearm.

7. *Pronator quadratus* Flat, quadrilateral muscle extending across the ventral surface of about the distal half of the ulna and radius. Dissect.

k. MUSCLES OF THE FOREARM: DORSAL SURFACE (FIGURE 10.27a AND b)

1. *Brachioradialis* Narrow, ribbonlike muscle on the lateral surface of the forearm. Dissect.
2. *Extensor carpi radialis longus* Larger muscle adjacent to the brachioradialis. Dissect, transect, and reflect.
3. *Extensor carpi radialis brevis* Muscle usually covered by the extensor carpi radialis longus at the proximal end. Dissect.
4. *Extensor digitorum communis* Large muscle running entire length of the lateral surface of forearm. Dissect, transect, and reflect.
5. *Extensor digitorum lateralis* Muscle with about the same structure as the extensor digitorum communis muscle. It lies lateral to the ulna. Dissect, transect, and reflect.
6. *Extensor carpi ulnaris* Muscle adjacent to the extensor digitorum lateralis, extending down the forearm laterally. Dissect, transect, and reflect.
7. *Extensor indicis proprius* Narrow muscle deep to the extensor carpi ulnaris.[6] Dissect.
8. *Extensor pollicis brevis* Muscle extending obliquely across the forearm deep to the extensor carpi ulnaris, extensor digitorum lateralis, and extensor digitorum communis. Dissect.
9. *Supinator* Muscle that passes obliquely across radius deep to the extensor digitorum communis.

l. MUSCLES OF THE THIGH: MEDIAL SURFACE (FIGURE 10.28a AND b)

1. *Sartorius* Straplike muscle on the cranial half of the medial side of the thigh. Dissect, transect, and reflect.
2. *Gracilis* Wide, flat muscle covering most of the caudal portion of the medial side of the thigh. Dissect, transect, and reflect.
3. *Semimembranosus* Thick muscle deep to the gracilis. Dissect.
4. *Adductor femoris* Triangular muscle deep to the gracilis and lateral to the semimembranosus. Dissect.
5. *Adductor longus* Small, triangular muscle cranial to the adductor femoris. Dissect.
6. *Pectineus* Very small triangular muscle cranial to the adductor longus. Dissect.
7. *Iliopsoas* Long, cylindrical muscle that originates on the last two thoracic and lumbar vertebrae and passes along the medial surface of the thigh adjacent to the pectineus. Dissect.
8. *Rectus femoris* Narrow band of muscle deep to the sartorius, one of four components of the *quadriceps femoris* muscle. The others are the vastus medialis, vastus intermedius, and vastus lateralis. Dissect the rectus femoris.
9. *Vastus medialis* Large muscle on the medial surface of the thigh caudal to the rectus femoris and deep to the sartorius. Dissect.
10. *Vastus intermedius* Best seen by separating the rectus femoris and vastus lateralis.[7]

m. MUSCLES OF THE THIGH: LATERAL SURFACE (FIGURE 10.29a AND b)

1. *Tensor fasciae latae* Triangular mass of muscle cranial to the biceps femoris. Dissect, transect, and reflect. Reflect the tough, white fasciae latae.
2. *Biceps femoris* Muscle caudal to the tensor fasciae latae, covers most of the lateral surface of the thigh. Dissect, transect, and reflect.
3. *Semitendinosus* Large band of muscle on the ventral border of the thigh between the biceps femoris and semimembranosus. Dissect.
4. *Caudofemoralis* Band of muscle cranial and dorsal to the biceps femoris. Dissect.
5. *Tenuissimus* Very slender band of muscle deep to the biceps femoris. The sciatic nerve runs parallel to it. Dissect.
6. *Gluteus maximus* Triangular mass of muscle just cranial to the caudofemoralis. Dissect.
7. *Gluteus medius* Relatively large, somewhat triangular muscle cranial to the gluteus maximus. Dissect.
8. *Vastus lateralis* Large muscle deep to the tensor fasciae latae that occupies the craniolateral surface of the thigh. Dissect.

n. MUSCLES OF THE SHANK: MEDIAL SURFACE (FIGURE 10.28a AND b)

1. *Gastrocnemius (medial head)* Muscle originating on medial epicondyle of femur. The medial and lateral heads unite to form a large muscle mass that inserts into the calcaneus by the *calcaneal (Achilles) tendon*. Separate the medial head from the lateral head and reflect the medial head.
2. *Flexor digitorum longus* Muscle consisting of two heads. The first is partly covered by the soleus muscle. The second is between the first head and the medial head of the gastrocnemius. Dissect, transect, and reflect.

---

[6]Not illustrated.

[7]Not illustrated.

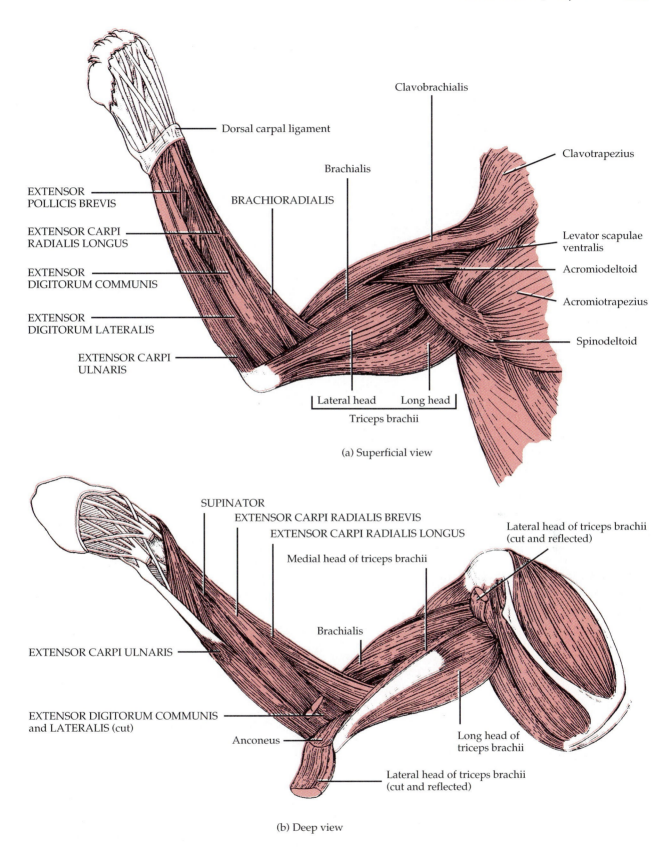

Dorsal carpal ligament

Clavobrachialis

Clavotrapezius

Brachialis

BRACHIORADIALIS

EXTENSOR
POLLICIS BREVIS

EXTENSOR CARPI
RADIALIS LONGUS

EXTENSOR
DIGITORUM COMMUNIS

EXTENSOR
DIGITORUM LATERALIS

EXTENSOR CARPI
ULNARIS

Levator scapulae
ventralis

Acromiodeltoid

Acromiotrapezius

Spinodeltoid

Lateral head   Long head
Triceps brachii

(a) Superficial view

SUPINATOR

EXTENSOR CARPI RADIALIS BREVIS

EXTENSOR CARPI RADIALIS LONGUS

Medial head of triceps brachii

Lateral head of triceps brachii
(cut and reflected)

Brachialis

EXTENSOR CARPI ULNARIS

EXTENSOR DIGITORUM COMMUNIS
and LATERALIS (cut)

Anconeus

Long head of
triceps brachii

Lateral head of triceps brachii
(cut and reflected)

(b) Deep view

FIGURE 10.27   Dorsal view of muscles of forearm.

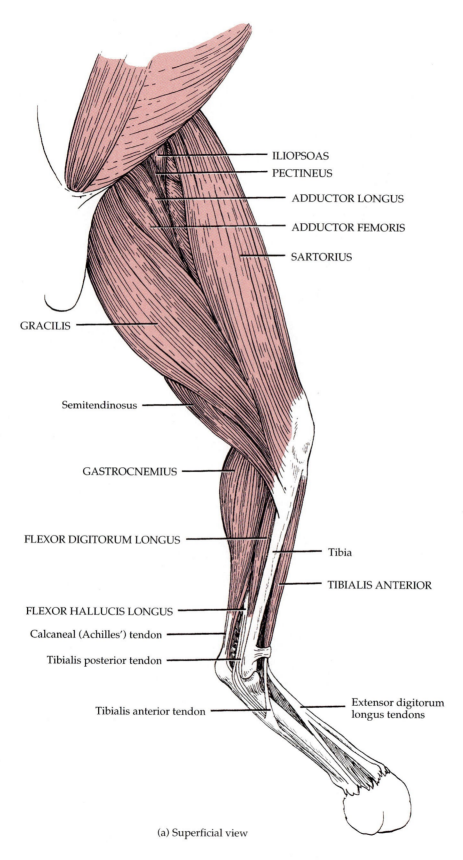

ILIOPSOAS

PECTINEUS

ADDUCTOR LONGUS

ADDUCTOR FEMORIS

SARTORIUS

GRACILIS

Semitendinosus

GASTROCNEMIUS

FLEXOR DIGITORUM LONGUS

Tibia

TIBIALIS ANTERIOR

FLEXOR HALLUCIS LONGUS

Calcaneal (Achilles') tendon

Tibialis posterior tendon

Tibialis anterior tendon

Extensor digitorum
longus tendons

(a) Superficial view

FIGURE 10.28   Muscles of medial surface of thigh and shank.

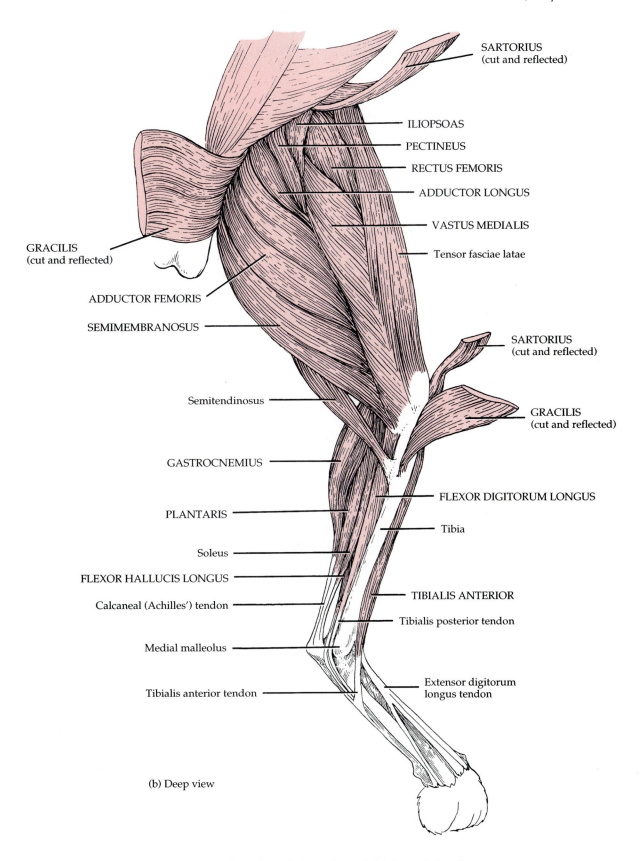

SARTORIUS
(cut and reflected)

ILIOPSOAS

PECTINEUS

RECTUS FEMORIS

ADDUCTOR LONGUS

VASTUS MEDIALIS

Tensor fasciae latae

GRACILIS
(cut and reflected)

ADDUCTOR FEMORIS

SEMIMEMBRANOSUS

SARTORIUS
(cut and reflected)

GRACILIS
(cut and reflected)

Semitendinosus

GASTROCNEMIUS

FLEXOR DIGITORUM LONGUS

PLANTARIS

Tibia

Soleus

FLEXOR HALLUCIS LONGUS

TIBIALIS ANTERIOR

Calcaneal (Achilles') tendon

Tibialis posterior tendon

Medial malleolus

Extensor digitorum
longus tendon

Tibialis anterior tendon

(b) Deep view

FIGURE 10.28 (*Continued*)   Muscles of medial surface of thigh and shank.

3. *Flexor hallucis longus* Long, heavy muscle deep to the flexor digitorum longus. Dissect.
4. *Tibialis anterior* Tapered band of muscle on the anterior aspect of the tibia. Dissect.
5. *Plantaris* Strong, round muscle between the two heads of the gastrocnemius. Also inserts into the calcaneal tendon. Dissect.

O. MUSCLES OF THE SHANK: LATERAL SURFACE (FIGURE 10.29a AND b)

1. *Gastrocnemius (lateral head)* Muscle originating on the lateral epicondyle of the femur. Dissect, transect, and reflect.
2. *Soleus* Small band of muscle next to the lateral head of the gastrocnemius and inserting into the calcaneal (Achilles) tendon. Dissect.

3. *Extensor digitorum longus* Muscle posterior to the tibialis anterior. Dissect.
4. *Peroneus longus* Slender, superficial muscle caudal to the extensor digitorum longus. Dissect, transect, and reflect.
5. *Peroneus brevis* Short, thick muscle with the lateral surface covered by the peroneus longus. Dissect.
6. *Peroneus tertius* Slender muscle adjacent to the extensor digitorum longus. Dissect.

ANSWER THE LABORATORY REPORT QUESTIONS AT THE END OF THE EXERCISE.

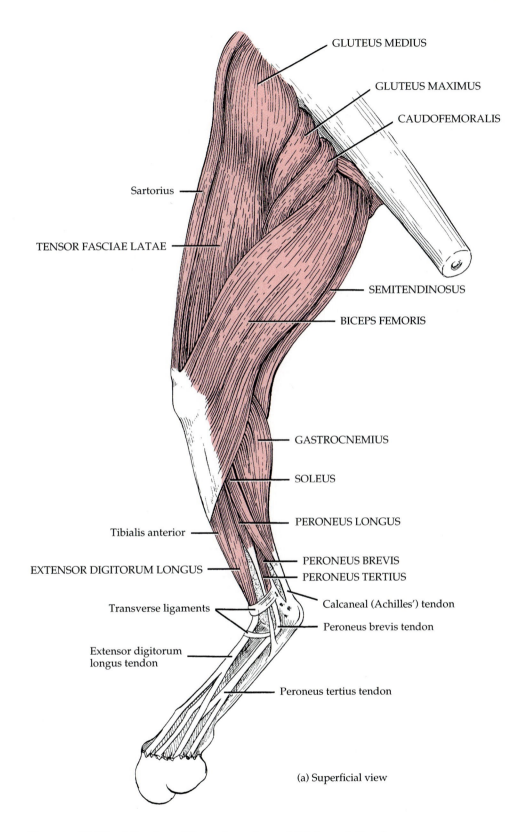

GLUTEUS MEDIUS

GLUTEUS MAXIMUS

CAUDOFEMORALIS

Sartorius

TENSOR FASCIAE LATAE

SEMITENDINOSUS

BICEPS FEMORIS

GASTROCNEMIUS

SOLEUS

PERONEUS LONGUS

Tibialis anterior

EXTENSOR DIGITORUM LONGUS

PERONEUS BREVIS

PERONEUS TERTIUS

Transverse ligaments

Calcaneal (Achilles') tendon

Peroneus brevis tendon

Extensor digitorum longus tendon

Peroneus tertius tendon

(a) Superficial view

FIGURE 10.29   Muscles of lateral surface of thigh and shank.

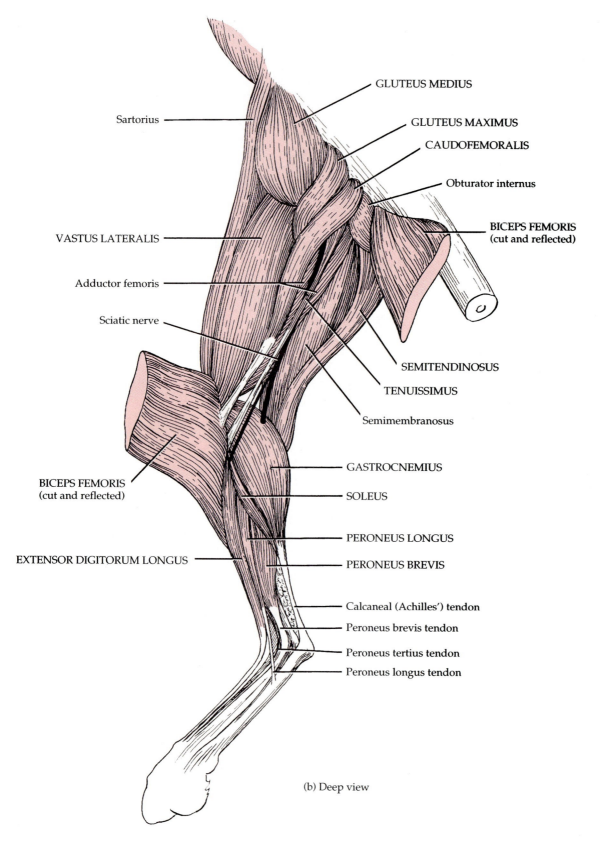

GLUTEUS MEDIUS

Sartorius

GLUTEUS MAXIMUS

CAUDOFEMORALIS

Obturator internus

**BICEPS FEMORIS
(cut and reflected)**

VASTUS LATERALIS

Adductor femoris

Sciatic nerve

SEMITENDINOSUS

TENUISSIMUS

Semimembranosus

GASTROCNEMIUS

**BICEPS FEMORIS
(cut and reflected)**

SOLEUS

PERONEUS LONGUS

EXTENSOR DIGITORUM LONGUS

PERONEUS BREVIS

Calcaneal (Achilles') tendon

Peroneus brevis tendon

Peroneus tertius tendon

Peroneus longus tendon

(b) Deep view

FIGURE 10.29 (*Continued*)   Muscles of lateral surface of thigh and shank.

Skeletal Muscles 10

Student _____   Date _____

Laboratory Section _____   Score/Grade _____

PART 1. Multiple Choice

_____ 1. The connective tissue covering that encloses the entire skeletal muscle is the (a) perimysium (b) endomysium (c) epimysium (d) mesomysium

_____ 2. A cord of connective tissue that attaches a skeletal muscle to the periosteum of bone is called a(n) (a) ligament (b) aponeurosis (c) perichondrium (d) tendon

_____ 3. A skeletal muscle that decreases the angle at a joint is referred to as a(n) (a) flexor (b) abductor (c) pronator (d) evertor

_____ 4. The name *abductor* means that a muscle (a) produces a downward movement (b) moves a part away from the midline (c) elevates a body part (d) increases the angle at a joint

_____ 5. Which connective tissue layer directly encircles the fascicles of skeletal muscles? (a) epimysium (b) endomysium (c) perimysium (d) mesomysium

_____ 6. Which muscle is *not* associated with a movement of the eyeball? (a) superior rectus (b) superior oblique (c) medial rectus (d) genioglossus

_____ 7. Of the following, which muscle is involved in compression of the abdomen? (a) external oblique (b) superior oblique (c) medial rectus (d) genioglossus

_____ 8. A muscle directly concerned with breathing is the (a) sternocleidomastoid (b) mentalis (c) brachialis (d) external intercostal

_____ 9. Which muscle is *not* related to movement of the wrist? (a) extensor carpi ulnaris (b) flexor carpi radialis (c) supinator (d) flexor carpi ulnaris

_____ 10. A muscle that helps move the thigh is the (a) piriformis (b) triceps brachii (c) hypoglossus (d) peroneus tertius

_____ 11. Which muscle is *not* related to mastication? (a) temporalis (b) masseter (c) lateral rectus (d) medial pterygoid

_____ 12. Which muscle elevates the tongue? (a) genioglossus (b) styloglossus (c) hyoglossus (d) omohyoid

_____ 13. Which muscle does *not* belong with the others because of its relationship to the hyoid bone? (a) digastric (b) mylohyoid (c) geniohyoid (d) thyrohyoid

_____ 14. Which muscle is *not* a component of the anterolateral abdominal wall? (a) external oblique (b) psoas major (c) rectus abdominis (d) internal oblique

_____ 15. All are components of the pelvic diaphragm *except* the (a) anconeus (b) coccygeus (c) iliococcygeus (d) pubococcygeus

———   16. Which is *not* a flexor of the forearm? (a) biceps brachii (b) brachialis (c) brachioradialis (d) triceps brachii

———   17. The abductor pollicis brevis, opponeus pollicis, flexor pollicis brevis, and adductor pollicis are all components of the (a) thenar eminence (b) midpalmar muscles (c) hypothenar eminence (d) erector spinae

———   18. Which muscle is *not* involved in moving the vertebral column? (a) splenius (b) longissimus (c) spinalis (d) sartorius

———   19. Which muscle flexes the wrists? (a) palmaris longus (b) extensor carpi radialis longus (c) supinator (d) extensor indicis

———   20. Which muscle is *not* involved in flexion of the thigh? (a) rectus femoris (b) sartorius (c) biceps femoris (d) vastus intermedius

———   21. Of the muscles that move the foot and toes, the anterior compartment muscles are involved in (a) plantar flexion (b) dorsiflexion (c) abduction (d) adduction

———   22. Which muscle is deepest? (a) plantar interosseous (b) adductor hallucis (c) quadratus plantae (d) abductor hallucis

PART 2. Matching

Identify the characteristic(s) used to name the following muscles:

———   23. Supinator   A. Location
———   24. Deltoid   B. Shape
———   25. Stylohyoid   C. Size
———   26. Flexor carpi radialis   D. Direction of fibers
———   27. Gluteus maximus   E. Action
———   28. External oblique   F. Number of origins
———   29. Triceps brachii   G. Insertion and origin
———   30. Adductor longus
———   31. Temporalis
———   32. Trapezius

PART 3. Completion

33. The principal muscle used in compression of the cheek is the ———————————.

34. The muscle that protracts the tongue is the ———————————.

35. The eye muscle that rolls the eyeball downward is the ———————————.

36. The ——————————— muscle flexes the neck on the chest.

37. The abdominal muscle that flexes the vertebral column is the ———————————.

38. The muscle of the pectoral (shoulder) girdle that depresses the clavicle is the

———————————.

39. Flexion, adduction, and medial rotation of the arm are accomplished by the ——————————— muscle.

40. The ——————————— muscle is the most important extensor of the forearm.

41. The muscle that flexes and abducts the wrist is the _____.

42. The four muscles that extend the legs are the vastus lateralis, vastus medialis, vastus intermedius,

    and _____.

43. Muscles that lie inferior to the hyoid bone are referred to as _____ muscles.

44. The neck region is divided into two principal triangles by the _____ muscle.

45. Muscles that move the pectoral (shoulder) girdle originate on the axial skeleton and insert on the

    clavicle or _____.

46. The muscles that move the humerus (arm) and do not originate on the scapula are called

    _____ muscles.

47. Together, the subscapularis, supraspinatus, infraspinatus, and teres major muscles form the

    _____.

48. The posterior muscles involved in moving the wrist, hand, and fingers function in extension and

    _____.

49. The posterior muscles of the thigh are involved in _____ of the leg.

50. Together, the biceps femoris, semitendinosus, and semimembranosus are referred to as the

    _____ muscles.

51. _____ fascicles attach obliquely from many directions to several tendons.

52. Movement of the thumb medially across the palm is called _____.

Surface Anatomy

11

Now that you have studied the skeletal and muscular systems, you will be introduced to the study of **surface anatomy**, in which you will study the form and markings of the surface of the body.[1] A knowledge of surface anatomy will help you identify certain superficial structures by visual inspection or palpation through the skin. **Palpation** means to feel with the hand. Knowledge of surface anatomy is important in health-related activities such as taking a pulse, listening to internal organs, drawing blood, and inserting needles and tubes.

A convenient way to study surface anatomy is first to divide the body into its principal regions: head, neck, trunk, and upper and lower limbs. These may be reviewed in Figure 2.2.

A. HEAD

The **head** (cephalic region or caput) is divisible into the cranium and face. Several surface features of the head are

1. **Cranium (skull, or brain case)**

   a. **Frontal region** Front of skull (sinciput) that includes forehead.
   b. **Parietal region** Crown of skull (vertex).
   c. **Temporal region** Side of skull (tempora).
   d. **Occipital region** Base of skull (occiput).

2. **Face (facies)**

   a. **Orbital** or **ocular region** Includes eyeballs (bulbi oculorum), eyebrows (supercilia), and eyelids (palpebrae).
   b. **Nasal region** Nose (nasus).
   c. **Infraorbital region** Inferior to orbit.

   d. **Oral region** Mouth.
   e. **Mental region** Anterior part of mandible.
   f. **Buccal region** Cheek.
   g. **Parotid-masseteric region** External to parotid gland and masseter muscle.
   h. **Zygomatic region** Inferolateral to orbit.
   i. **Auricular region** Ear.

Using your textbook as an aid, label Figure 11.1.

Using a mirror, examine the various features of the head just described. Working with a partner, be sure that you can identify the regions by both common *and* anatomical names.

The surface anatomy features of the eyeball and accessory structures are presented in Figure 14.10, of the ear in Figure 14.19, and of the nose in Figure 21.3.

B. NECK

The **neck** (collum) can be divided into an **anterior cervical region**, two **lateral cervical regions**, and a **posterior (nuchal) region**. Among the surface features of the neck are

1. **Thyroid cartilage (Adam's apple)** Triangular laryngeal cartilage in the midline of the anterior cervical region.
2. **Hyoid bone** First resistant structure palpated in the midline inferior to the chin, lying just superior to the thyroid cartilage opposite the superior border of C4.
3. **Cricoid cartilage** Inferior laryngeal cartilage that attaches larynx to trachea. This structure can be palpated by running your fingertip down from your chin over the thyroid cartilage. (After you pass the cricoid cartilage, your fingertip sinks in.) This cartilage is used as a landmark in locating the rings of cartilage in the trachea (windpipe) when performing a tracheostomy. The incision is made through the second, third, or fourth tracheal rings and a tube is inserted to assist breathing.

---

[1]At the discretion of your instructor, surface anatomy may be studied either before or in conjunction with your study of various body systems. This exercise can be used as an excellent review of many topics already studied.

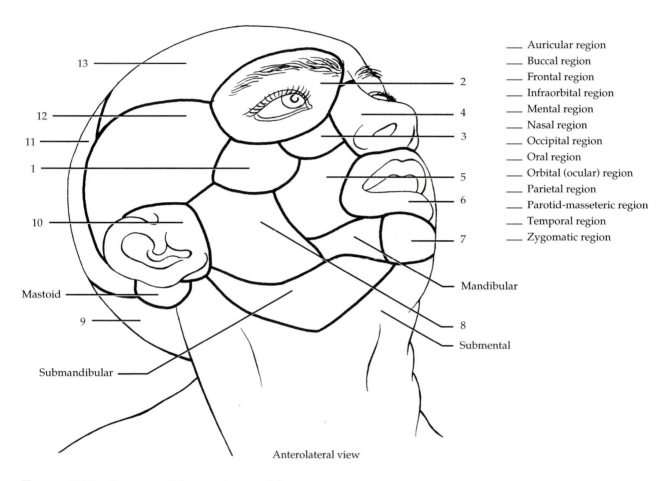

Anterolateral view

___ Auricular region
___ Buccal region
___ Frontal region
___ Infraorbital region
___ Mental region
___ Nasal region
___ Occipital region
___ Oral region
___ Orbital (ocular) region
___ Parietal region
___ Parotid-masseteric region
___ Temporal region
___ Zygomatic region

Mandibular

Submental

FIGURE 11.1   Regions of the cranium and face.

4. ***Sternocleidomastoid muscles*** Form major portion of lateral cervical regions, extending from mastoid process of temporal bone (felt as bump behind auricle of ear) to sternum and clavicle. Each muscle divides the neck into an anterior and posterior triangle. Carotid (neck) pulse is felt along the anterior border of the sternocleidomastoid muscle.

5. ***Trapezius muscles*** Form portion of lateral cervical region, extending inferiorly and laterally from base of skull. "Stiff neck" is frequently associated with inflammation of these muscles.

6. ***Anterior triangle of neck*** Bordered superiorly by mandible, inferiorly by sternum, medially by cervical midline, and laterally by anterior border of sternocleidomastoid muscle.

7. ***Posterior triangle of neck*** Bordered inferiorly by clavicle, anteriorly by posterior border of sternocleidomastoid muscle, and posteriorly by anterior border of trapezius muscle.

8. ***External jugular veins*** Prominent veins along lateral cervical regions, readily seen when a person is angry or a collar fits too tightly.

Using a mirror and working with a partner, use your textbook as an aid in identifying the surface features of the neck just described. Then label Figure 11.2.

C. TRUNK

The ***trunk*** is divided into the back, chest, abdomen, and pelvis. Its surface features include

1. ***Back (dorsum)***

   a. ***Spinous processes (spines)*** Posteriorly pointed projections of vertebrae. The spinous process of C2 is the first bony prominence encountered when the finger is drawn inferiorly in the midline; the spinous process of C7 ***(vertebra prominens)*** is the superior of the two prominences found at the base of the neck; the spinous process of T1 is the lower prominence at the base of the neck; the spinous process of T3 is at about the same level as the spinous process of the scapula; the spinous process of T7 is about opposite the inferior

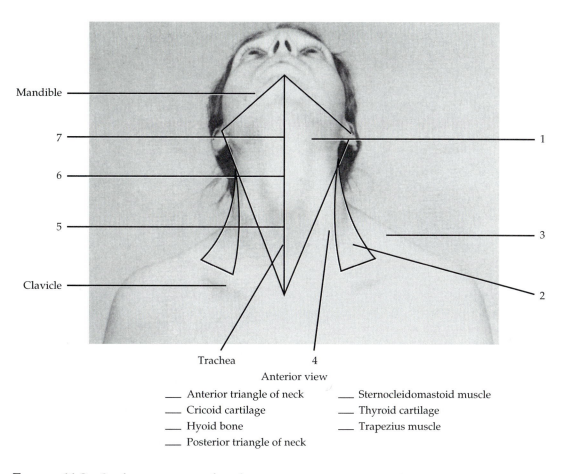

Mandible

7

6

5

Clavicle

1

3

2

Trachea   4

Anterior view

_____ Anterior triangle of neck   _____ Sternocleidomastoid muscle
_____ Cricoid cartilage   _____ Thyroid cartilage
_____ Hyoid bone   _____ Trapezius muscle
_____ Posterior triangle of neck

FIGURE 11.2   Surface anatomy of neck.

angle of the scapula; a line passing through the highest points of the iliac crests, called the supracristal line, passes through the spinous process of L4.

b. ***Scapula*** Shoulder blade. Several parts of the scapula, such as the medial (axillary) border, lateral (vertebral) border, inferior angle, spine, and acromion, may be observed or palpated.

c. ***Ribs*** These may be seen in thin individuals.

d. ***Muscles*** Among the visible superficial back muscles are the ***latissimus dorsi*** (covers lower half of back), ***erector spinae*** (on either side of vertebral column), ***infraspinatus*** (inferior to spine of scapula), ***trapezius***, and ***teres major*** (inferior to infraspinatus).

e. ***Posterior axillary fold*** Formed by the latissimus dorsi and teres major muscles; can be palpated between the finger and thumb.

f. ***Triangle of auscultation*** Triangle formed by latissimus dorsi muscle, trapezius muscle, and vertebral border of scapula. The space between the muscles in the region permits respiratory sounds to be heard clearly with a stethoscope.

Using your textbook as an aid, label Figure 11.3.

2. Chest (thorax)

a. ***Clavicles*** Collarbones. These lie in superior region of thorax and can be palpated along their entire length.

b. ***Sternum*** Breastbone. Lies in midline of chest. The following parts of the sternum are important surface features:

i. ***Suprasternal notch*** Depression on superior surface of manubrium of sternum between medial ends of clavicles. The trachea can be palpated in the notch.

ii. ***Manubrium of sternum*** Superior portion of sternum at the same levels as the bodies of the third and fourth thoracic vertebrae and anterior to the arch of the aorta.

iii. ***Body of sternum*** Midportion of sternum anterior to heart and the vertebral bodies of T5–T8.

iv. ***Sternal angle*** Formed by junction of manubrium and body of sternum, about 4 cm (1½ in.) inferior to suprasternal

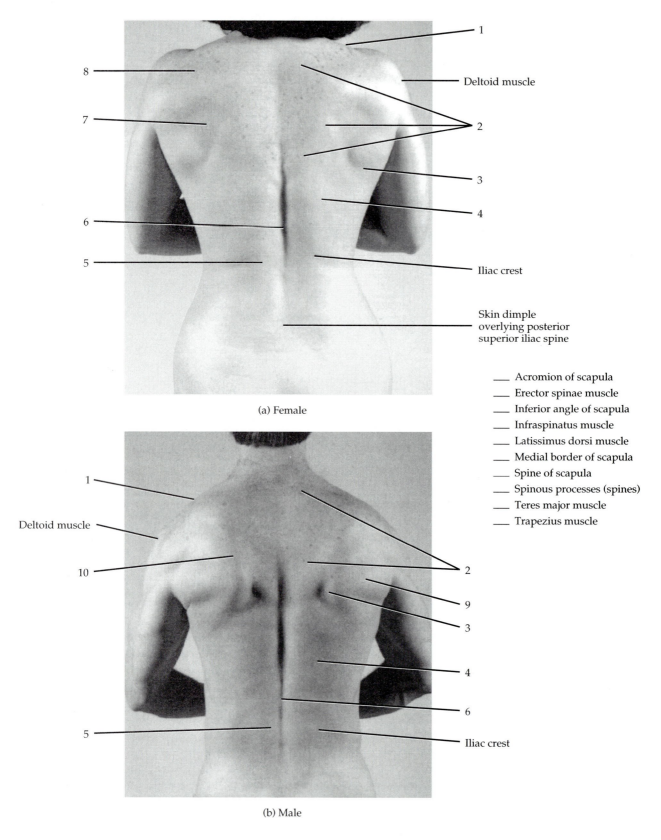

8

7

1

Deltoid muscle

2

3

4

6

5

Iliac crest

Skin dimple
overlying posterior
superior iliac spine

(a) Female

____ Acromion of scapula
____ Erector spinae muscle
____ Inferior angle of scapula
____ Infraspinatus muscle
____ Latissimus dorsi muscle
____ Medial border of scapula
____ Spine of scapula
____ Spinous processes (spines)
____ Teres major muscle
____ Trapezius muscle

1

Deltoid muscle

10

2

9

3

4

6

5

Iliac crest

(b) Male

FIGURE 11.3 Surface anatomy of back.

notch. This is palpable under the skin, locates the costal cartilage of the second rib, and is the starting point from which the ribs are counted.

    v. *Xiphoid process of sternum* Inferior portion of sternum medial to the seventh costal cartilages.

c. *Ribs* Form bony cage of thoracic cavity. The apex beat of the heart in adults is heard in the left fifth intercostal space, just medial to the left midclavicular line.

d. *Costal margins* Inferior edges of costal cartilages of ribs 7 through 10. The first costal cartilage lies inferior to the medial end of the clavicle; the seventh costal cartilage is the most inferior to articulate directly with the sternum; the tenth costal cartilage forms the most inferior part of the costal margin, when viewed anteriorly.

e. *Muscles* Among the superficial chest muscles that can be seen are the *pectoralis major* (principal upper chest muscle) and *serratus anterior* (inferior and lateral to pectoralis major).

f. *Mammary glands* Accessory organs of the female reproductive system located inside the breasts. They overlie the pectoralis major muscle (two-thirds) and serratus anterior muscle (one-third). After puberty, they enlarge to their hemispherical shape, and in young adult females, they extend from the second through sixth ribs and from the lateral margin of the sternum to the midaxillary line.

g. *Nipples* Superficial to fourth intercostal space or fifth rib about 10 cm (4 in.) from the midline in males and most females. The position of the nipples in females is variable, depending on the size and pendulousness of the breasts. The right dome of the diaphragm is just inferior to the right nipple, the left dome is about 2–3 cm (1 in.) inferior to the left nipple, and the central tendon is at the level of the junction of the body and xiphoid process of the sternum.

h. *Anterior axillary fold* Formed by the lateral border of the pectoralis major muscle; can be palpated between the fingers and thumb.

Using your textbook as an aid, label Figure 11.4.

3. *Abdomen* and *Pelvis*

a. *Umbilicus* Also called *navel*; previous site of attachment of umbilical cord to fetus. It is level with the intervertebral disc between the bodies of L3 and L4 and is the most obvious surface marking on the abdomen of

most individuals. The abdominal aorta bifurcates into the right and left common iliac arteries anterior to the body of vertebra L4. The inferior vena cava lies to the right of the abdominal aorta and is wider; it rises anterior to the body of vertebra L5.

b. *Linea alba* Flat, tendonous raphe forming a furrow along midline between rectus abdominis muscles. The furrow extends from the xiphoid process to pubic symphysis. It is particularly obvious in thin, muscular individuals. It is broad superior to the umbilicus and narrow inferior to it. The linea alba is a frequently selected site for abdominal surgery since an incision through it severs no muscles and only a few blood vessels and nerves.

c. *Tendinous intersections* Fibrous bands that run transversely across the rectus abdominis muscle. Three or more are visible in muscular individuals. One intersection is at the level of the umbilicus, one at the level of the xiphoid process, and one midway between.

d. *Muscles* Among the superficial abdominal muscles are the *external oblique* (inferior to serratus anterior) and *rectus abdominis* (just lateral to midline of abdomen).

e. *Pubic symphysis* Anterior joint of hipbones. This structure is palpated as a firm resistance in the midline at the inferior portion of the anterior abdominal wall.

Using your textbook as an aid, label Figure 11.5.

D. UPPER LIMB (EXTREMITY)

The *upper limb (extremity)* consists of the armpit, shoulder, arm, elbow, forearm, wrist, and hand (palm and fingers).

1. Major surface features of the *shoulder (acromial)* region are

a. *Acromioclavicular joint* Slight elevation at lateral end of clavicle.

b. *Acromion* Expanded lateral end of spine of scapula. This is clearly visible in some individuals and can be palpated about 2.5 cm (1 in.) distal to acromioclavicular joint (see Figure 11.3).

c. *Deltoid muscle* Triangular muscle that forms rounded prominence of shoulder. This is a frequent site for intramuscular injections (see Figure 11.3).

2. Major surface features of the *arm (brachium)* and *elbow (cubitus)* are

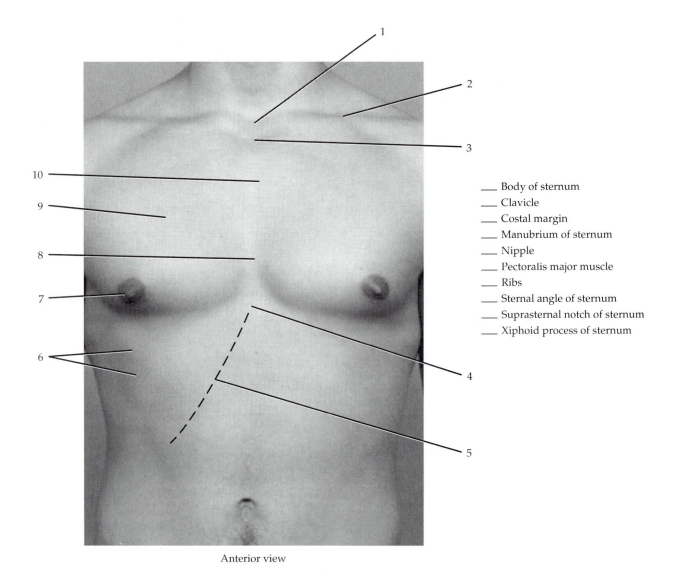

Anterior view

____ Body of sternum
____ Clavicle
____ Costal margin
____ Manubrium of sternum
____ Nipple
____ Pectoralis major muscle
____ Ribs
____ Sternal angle of sternum
____ Suprasternal notch of sternum
____ Xiphoid process of sternum

FIGURE 11.4   Surface anatomy of chest. (The serratus anterior muscle is shown in Figure 11.5.)

a. ***Biceps brachii muscle*** Forms bulk of anterior surface of arm.

b. ***Triceps brachii muscle*** Forms bulk of posterior surface of arm.

c. ***Medial epicondyle*** Medial projection at distal end of humerus.

d. ***Ulnar nerve*** Can be palpated as a rounded cord in a groove posterior to the medial epicondyle.

e. ***Lateral epicondyle*** Lateral projection at distal end of humerus.

f. ***Olecranon*** Projection of proximal end of ulna that lies between and slightly superior to epicondyles when forearm is extended; forms elbow.

g. ***Cubital fossa*** Triangular space in anterior region of elbow bounded proximally by an imaginary line between humeral epicondyles, laterally by the medial border of the brachioradialis muscle, and medially by the lateral border of the pronator teres muscle; contains tendon of biceps brachii muscle, brachial artery and its terminal branches (radial and ulnar arteries), and parts of median and radial nerves.

h. ***Median cubital vein*** Crosses cubital fossa obliquely. This vein is frequently selected for removal of blood.

i. ***Brachial artery*** Continuation of axillary artery that passes posterior to coracobrachialis

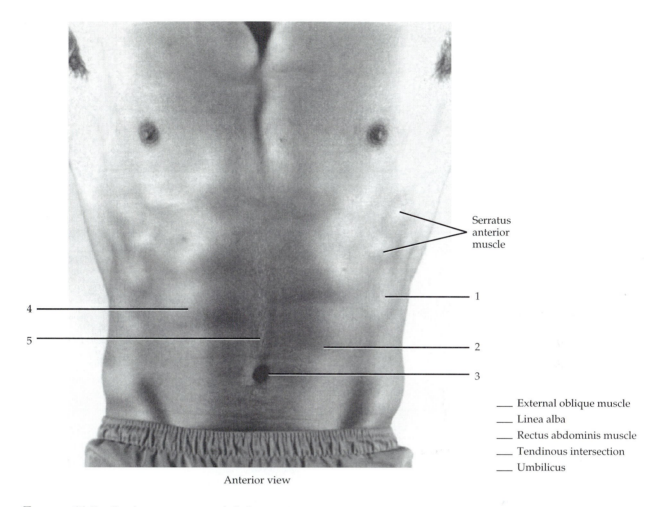

Serratus
anterior
muscle

1

4

5

2

3

___ External oblique muscle
___ Linea alba
___ Rectus abdominis muscle
___ Tendinous intersection
___ Umbilicus

Anterior view

FIGURE 11.5   Surface anatomy of abdomen.

muscle and then medial to biceps brachii muscle. It enters the middle of the cubital fossa and passes inferior to the bicipital aponeurosis, which separates it from the median cubital vein. The artery is frequently used to take blood pressure. Pressure may be applied to it in cases of severe hemorrhage in the forearm and hand.

j. ***Bicipital aponeurosis*** An aponeurotic band that inserts the biceps brachii muscle into the deep fascia in the medial aspect of the forearm. It can be felt when the muscle contracts.

Using your textbook as an aid, label Figure 11.6.

3. Major surface features of the ***forearm (antebrachium)*** and ***wrist (carpus)*** are

a. ***Styloid process of ulna*** Projection of distal end of ulna at medial side of wrist.

b. ***Styloid process of radius*** Projection of distal end of radius at lateral side of wrist.

c. ***Brachioradialis muscle*** Located at superior and lateral aspect of forearm.

d. ***Flexor carpi radialis muscle*** Located along midportion of forearm.

e. ***Flexor carpi ulnaris muscle*** Located at medial aspect of forearm.

f. ***Tendon of palmaris longus muscle*** Located on anterior surface of wrist near ulna. When you make a fist, you can see this tendon. The palmaris longus muscle is absent in about 13% of the population.

g. ***Tendon of flexor carpi radialis muscle*** Tendon on anterior surface of wrist lateral to tendon of palmaris longus.

h. ***Radial artery*** Can be palpated just medial to styloid process of radius; this artery is frequently used to take pulse.

i. ***Pisiform bone*** Medial bone of proximal carpals. The bone is easily palpated as a

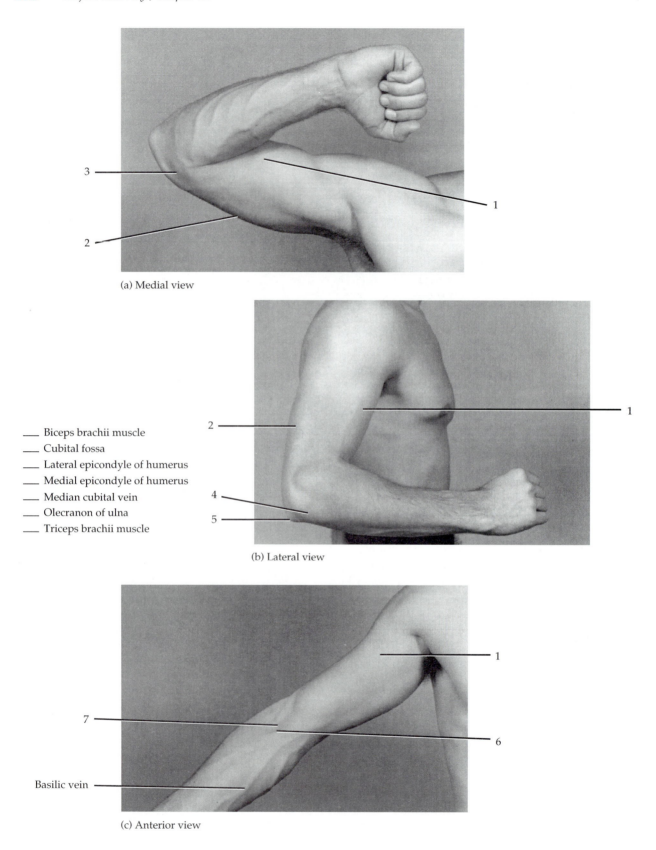

(a) Medial view

___ Biceps brachii muscle
___ Cubital fossa
___ Lateral epicondyle of humerus
___ Medial epicondyle of humerus
___ Median cubital vein
___ Olecranon of ulna
___ Triceps brachii muscle

(b) Lateral view

Basilic vein

(c) Anterior view

FIGURE 11.6 Surface anatomy of arm and elbow.

projection distal and anterior to styloid process of ulna.

   j. ***Tendon of extensor pollicis brevis muscle*** Tendon close to styloid process of radius along posterior surface of wrist, best seen when thumb is bent backward.

   k. ***Tendon of extensor pollicis longus muscle*** Tendon closer to styloid process of ulna along posterior surface of wrist, best seen when thumb is bent backward.

   l. ***"Anatomical snuffbox"*** Depression between tendons of extensor pollicis brevis and extensor pollicis longus muscles. Styloid process of the radius, the base of the first metacarpal, trapezium, scaphoid, and radial artery can all be palpated in the depression.

   m. ***Wrist creases*** Three more or less constant lines on anterior aspect of wrist where skin is firmly attached to underlying deep fascia.

Using your textbook as an aid, label Figure 11.7.

4. Major surface features of the ***hand (manus)*** are

   a. ***"Knuckles"*** Commonly refers to dorsal aspects of distal ends of metacarpals II, III, IV, and V. Term also includes dorsal aspects of metacarpophalangeal and interphalangeal joints.

   b. ***Thenar eminence*** Lateral rounded contour on anterior surface of hand formed by muscles of thumb.

   c. ***Hypothenar eminence*** Medial rounded contour on anterior surface of hand formed by muscles of little finger.

   d. ***Skin creases*** Several more or less constant lines on the anterior aspect of the palm ***(palmar flexion creases)*** and digits ***(digital flexion creases)*** where skin is firmly attached to underlying deep fascia.

   e. ***Extensor tendons*** Besides the tendons of the extensor pollicis brevis and extensor pollicis longus muscles associated with the thumb, the following extensor tendons are also visible on the posterior aspect of the hand: ***extensor digiti minimi tendon*** in line with phalanx V (little finger) and ***extensor digitorum*** in line with phalanges II, III, and IV.

   f. ***Dorsal venous arch*** Superficial veins on posterior surface of hand that form cephalic vein; displayed by compressing the blood vessels at the wrist for a few minutes as the hand is opened and closed.

With the aid of your textbook, label Figure 11.8 on page 235.

E. LOWER LIMB (EXTREMITY)

The ***lower limb (extremity)*** consists of the buttocks, thigh, knee, leg, ankle, and foot.

1. Major surface features of the ***buttocks (gluteal region)*** and ***thigh (femoral region)*** are

   a. ***Iliac crest*** Superior margin of ilium of hipbone, forming outline of superior border of buttock. When you rest your hands on your hips, they rest on the iliac crests.

   b. ***Posterior superior iliac spine*** Posterior termination of iliac crest; lies deep to a dimple (skin depression) about 4 cm (1.5 in.) lateral to midline; dimple forms because skin and underlying fascia are attached to bone. The spine marks the inferior limit of cerebrospinal fluid in the subarachnoid space around the spinal cord.

   c. ***Gluteus maximus muscle*** Forms major portion of prominence of buttock.

   d. ***Gluteus medius muscle*** Superolateral to gluteus maximus. This is a frequent site for intramuscular injections.

   e. ***Gluteal (natal) cleft*** Depression along midline that separates the buttocks; it extends as high as the fourth or third sacral vertebra.

   f. ***Gluteal fold*** Inferior limit of buttock formed by inferior margin of gluteus maximus muscle.

   g. ***Ischial tuberosity*** Bony prominence of ischium of hipbone bears weight of body when seated.

   h. ***Greater trochanter*** Projection of proximal end of femur on lateral surface of thigh felt and seen in front of hollow on side of hip. This can be palpated about 20 cm (8 in.) inferior to iliac crest.

   i. ***Anterior thigh muscles*** Include ***sartorius*** (runs obliquely across thigh) and ***quadriceps femoris***, which consists of ***rectus femoris*** (midportion of thigh), ***vastus lateralis*** (anterolateral surface of thigh), ***vastus medialis*** (medial inferior portion of thigh), and ***vastus intermedius*** (deep to rectus femoris). Vastus lateralis is frequent injection site for diabetics.

   j. ***Posterior thigh muscles*** Include the ***hamstrings (semitendinosus, semimembranosus, and biceps femoris)***.

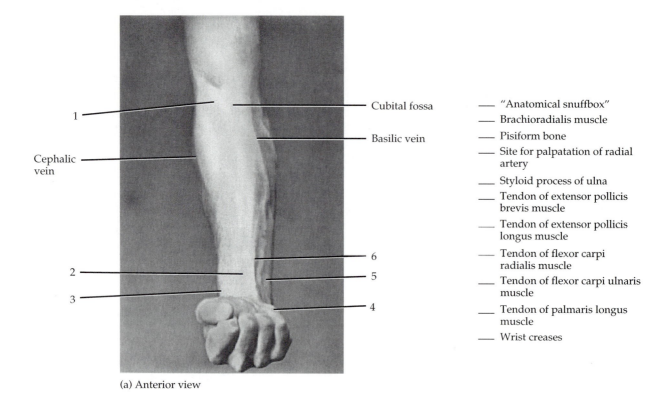

Cubital fossa

Basilic vein

Cephalic vein

____ "Anatomical snuffbox"
____ Brachioradialis muscle
____ Pisiform bone
____ Site for palpatation of radial artery
____ Styloid process of ulna
____ Tendon of extensor pollicis brevis muscle
____ Tendon of extensor pollicis longus muscle
____ Tendon of flexor carpi radialis muscle
____ Tendon of flexor carpi ulnaris muscle
____ Tendon of palmaris longus muscle
____ Wrist creases

(a) Anterior view

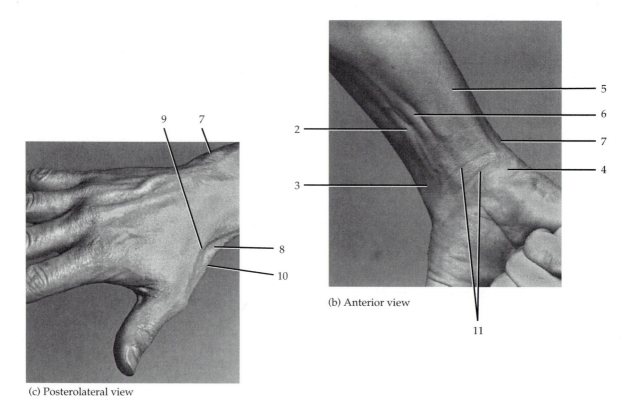

(c) Posterolateral view

(b) Anterior view

FIGURE 11.7   Surface anatomy of forearm and wrist.

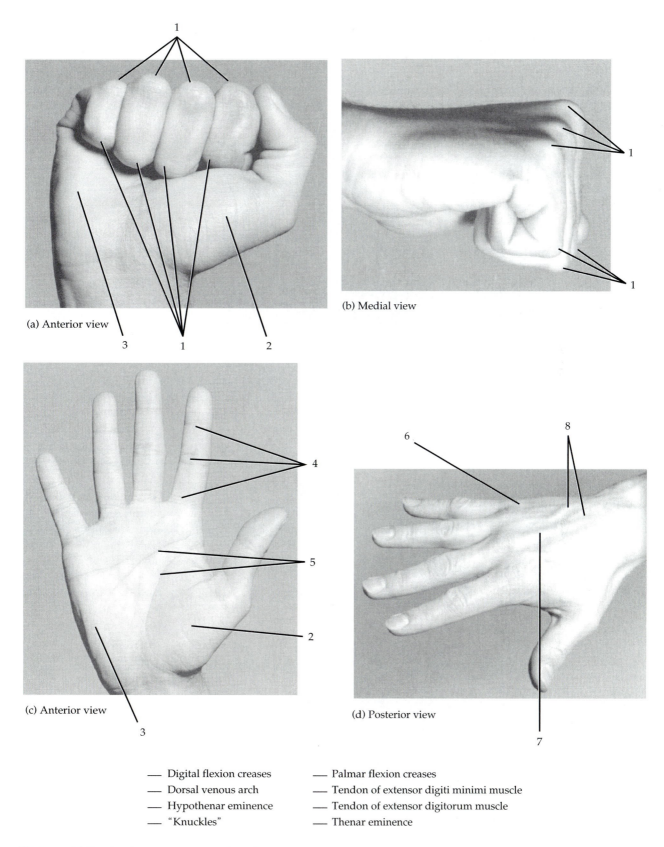

(a) Anterior view

(b) Medial view

(c) Anterior view

(d) Posterior view

— Digital flexion creases   — Palmar flexion creases
— Dorsal venous arch   — Tendon of extensor digiti minimi muscle
— Hypothenar eminence   — Tendon of extensor digitorum muscle
— "Knuckles"   — Thenar eminence

FIGURE 11.8   Surface anatomy of hand.

k. *Medial thigh muscles* Include the *adductor magnus, adductor brevis, adductor longus, gracilis, obturator externus,* and *pectineus.*

With the aid of your textbook, label Figure 11.9.

2. Major surface anatomy features of the **knee** (*genu*) are

a. *Patella* Kneecap, located within quadriceps femoris tendon on anterior surface of **knee**

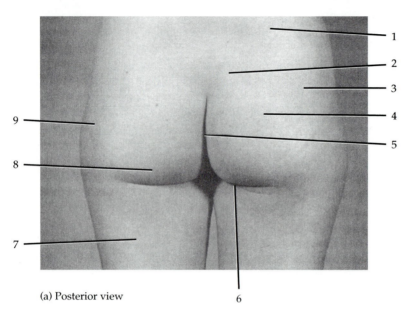

(a) Posterior view

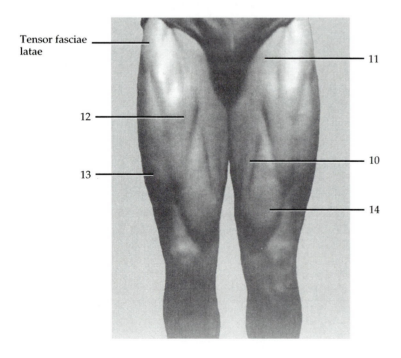

(b) Anterior view

___ Adductor magnus muscle
___ Gluteal (natal) cleft
___ Gluteal fold
___ Gluteus maximus muscle
___ Gluteus medius muscle
___ Greater trochanter of femur
___ Hamstrings
___ Iliac crest
___ Posterior superior iliac spine
___ Rectus femoris muscle
___ Sartorius muscle
___ Site of ischial tuberosity
___ Vastus lateralis muscle
___ Vastus medialis muscle

FIGURE 11.9 Surface anatomy of buttocks and thigh.

along midline. Margins of condyles (described shortly) can be felt on either side of it.

b. *Patellar ligament* Continuation of quadriceps femoris tendon inferior to patella.

c. *Popliteal fossa* Diamond-shaped space on posterior aspect of knee visible when knee is flexed. Fossa is bordered superolaterally by the biceps femoris muscle, superomedially by the semimembranosus and semitendinosus muscles, and inferolaterally and inferomedially by the lateral and medial heads of the gastrocnemius muscle.

d. *Medial condyles of femur and tibia* Medial projections just inferior to patella. Superior part of projection belongs to distal end of femur; inferior part of projection belongs to proximal end of tibia.

e. *Lateral condyles of femur and tibia* Lateral projections just inferior to patella. Superior part of projections belongs to distal end of femur. Inferior part of projections belongs to proximal end of tibia.

With the aid of your textbook, label Figure 11.10.

3. Major surface anatomy features of the *leg (crus)* and *ankle (tarsus)* are

a. *Tibial tuberosity* Bony prominence of tibia inferior to patella into which patellar ligament inserts (Figure 11.10a).

b. *Medial malleolus of tibia* Projection of distal end of tibia that forms medial prominence of ankle.

c. *Lateral malleolus of fibula* Projection of distal end of fibula that forms lateral prominence of ankle.

d. *Calcaneal (Achilles) tendon* Tendon of insertion into calcaneus for gastrocnemius and soleus muscles.

e. *Tibialis anterior muscle* Located on anterior surface of leg.

f. *Gastrocnemius muscle* Forms bulk of middle and superior portions of posterior surface of leg.

g. *Soleus muscle* Mostly deep to gastrocnemius, visible on either side of gastrocnemius below middle of leg.

With the aid of your textbook, label Figure 11.11 on page 239.

4. Major surface anatomy features of the *foot (pes)* are

a. *Calcaneus* Heel bone to which calcaneal (Achilles) tendon inserts.

b. *Tendon of extensor hallucis longus muscle* Visible in line with phalanx of great toe. Pulsations in the dorsalis pedis artery may be felt in most people just lateral to this tendon when the blood vessel passes over the navicular and cuneiform bones.

c. *Tendons of extensor digitorum longus muscles* Visible in line with phalanges II through V.

d. *Dorsal venous arch* Superficial veins on dorsum of foot that unite to form small and great saphenous veins.

With the aid of your textbook, label Figure 11.12 on page 239.

ANSWER THE LABORATORY REPORT QUESTIONS AT THE END OF THE EXERCISE.

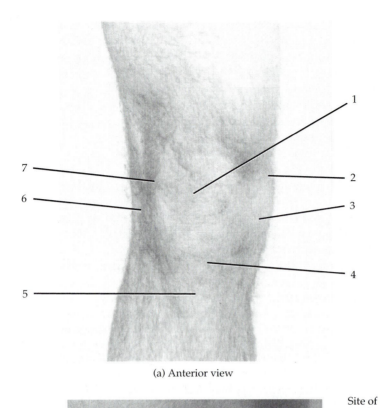

(a) Anterior view

(b) Posterior view

Site of
semitendinosus and
semimembranosus
muscle

Tendon of biceps
femoris muscle

Tendon of semitendinosus
muscle

Gastrocnemius
muscle

—— Lateral condyle of femur   —— Patella
—— Lateral condyle of tibia   —— Patellar ligament
—— Medial condyle of femur   —— Popliteal fossa
—— Medial condyle of tibia   —— Tibial tuberosity

FIGURE 11.10   Surface anatomy of knee.

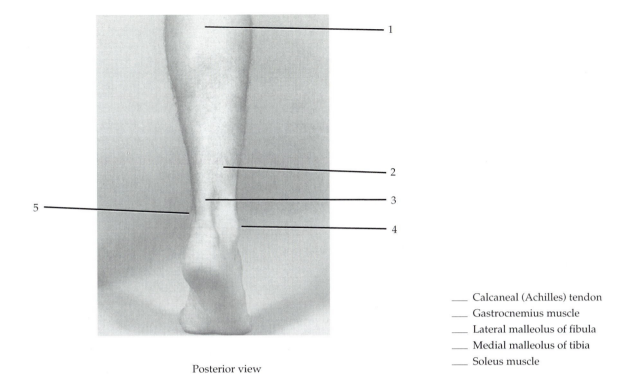

Posterior view

___ Calcaneal (Achilles) tendon
___ Gastrocnemius muscle
___ Lateral malleolus of fibula
___ Medial malleolus of tibia
___ Soleus muscle

FIGURE 11.11   Surface anatomy of leg.

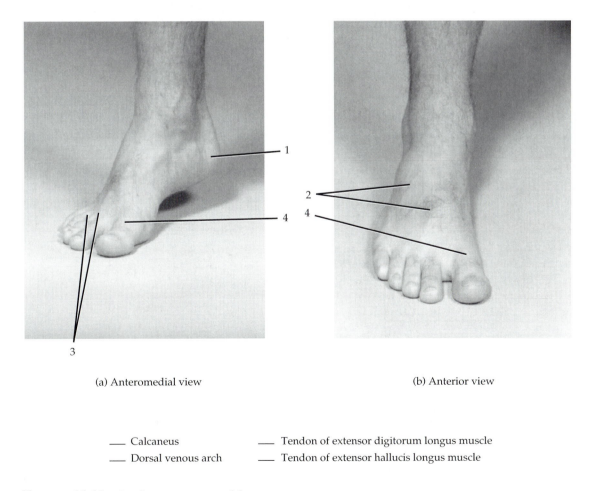

(a) Anteromedial view   (b) Anterior view

___ Calcaneus   ___ Tendon of extensor digitorum longus muscle
___ Dorsal venous arch   ___ Tendon of extensor hallucis longus muscle

FIGURE 11.12   Surface anatomy of foot.

Surface Anatomy 11

Student _____   **Date** _____

Laboratory Section _____   **Score/Grade** _____

PART 1. Multiple Choice

_____ **1.** The term used to refer to the crown of the skull is (a) occipital (b) mental (c) parietal (d) zygomatic

_____ **2.** The laryngeal cartilage in the midline of the anterior cervical region known as the Adam's apple is the (a) cricoid (b) epiglottis (c) arytenoid (d) thyroid

_____ **3.** Inflammation of which muscle is associated with "stiff neck"? (a) teres major (b) trapezius (c) deltoid (d) cervicalis

_____ **4.** The skeletal muscle located directly on either side of the vertebral column is the (a) serratus anterior (b) infraspinatus (c) teres major (d) erector spinae

_____ **5.** The suprasternal notch and xiphoid process are associated with the (a) sternum (b) scapula (c) clavicle (d) ribs

_____ **6.** The expanded end of the spine of the scapula is the (a) acromion (b) linea alba (c) olecranon (d) superior angle

_____ **7.** Which nerve can be palpated as a rounded cord in a groove posterior to the medial epicondyle? (a) median (b) radial (c) ulnar (d) brachial

_____ **8.** Which carpal bone can be palpated as a projection distal to the styloid process of the ulna? (a) trapezoid (b) trapezium (c) hamate (d) pisiform

_____ **9.** The lateral rounded contour on the anterior surface of the hand (at the base of the thumb) formed by the muscles of the thumb is the (a) "anatomical snuffbox" (b) thenar eminence (c) hypothenar eminence (d) dorsal venous arch

_____ **10.** The superior margin of the hipbone is the (a) pubic symphysis (b) iliac spine (c) acetabulum (d) iliac crest

_____ **11.** Which bony structure bears the weight of the body when a person is seated? (a) greater trochanter (b) iliac crest (c) ischial tuberosity (d) gluteal fold

_____ **12.** Which muscle is *not* a component of the quadriceps femoris group? (a) biceps femoris (b) vastus lateralis (c) vastus medialis (d) rectus femoris

_____ **13.** The diamond-shaped space on the posterior aspect of the knee is the (a) cubital fossa (b) posterior triangle (c) popliteal fossa (d) nuchal groove

_____ **14.** The projection of the distal end of the tibia that forms the prominence on one side of the ankle is the (a) medial condyle (b) medial malleolus (c) lateral condyle (d) lateral malleolus

_____ **15.** The tendon that can be seen in line with the great toe belongs to which muscle? (a) extensor digiti minimi (b) extensor digitorum longus (c) extensor hallucis longus (d) extensor carpi radialis

PART 2. Completion

16. The laryngeal cartilage that connects the larynx to the trachea is the _____ cartilage.

17. The _____ triangle is bordered by the mandible, sternum, cervical midline, and sternocleidomastoid muscle.

18. The depression on the superior surface of the sternum between the medial ends of the clavicles is the

    _____ .

19. The principal superficial chest muscle is the _____ .

20. Tendinous intersections are associated with the _____ muscle.

21. The muscle that forms the rounded prominence of the shoulder is the _____ muscle.

22. The triangular space in the anterior aspect of the elbow is the _____ .

23. The "anatomical snuffbox" is bordered by the tendons of the extensor pollicis brevis muscle and the

    _____ muscle.

24. The dorsal aspects of the distal ends of metacarpals II through V are commonly referred to as

    _____ .

25. The tendon of the _____ muscle is in line with phalanx V.

26. The dimple that forms about 4 cm lateral to the midline just above the buttocks lies superficial to the

    _____ .

27. The femoral projection that can be palpated about 20 cm (8 in.) inferior to the iliac crest is the

    _____ .

28. The continuation of the quadriceps femoris tendon inferior to the patella is the

    _____ .

29. The tendon of insertion for the gastrocnemius and soleus muscles is the _____ tendon.

30. Superficial veins on the dorsum of the foot that unite to form the small and great saphenous veins

    belong to the _____ .

31. The prominent veins along the lateral cervical regions are the _____ veins.

32. The pronounced vertebral spine of C7 is the _____ .

33. The most reliable surface anatomy feature of the chest is the _____ .

34. A slight groove extending from the xiphoid process to the pubic symphysis is the

    _____ .

35. The vein that crosses the cubital fossa and is frequently used to remove blood is the

    _____ vein.

PART 3. MATCHING

_____	**36.** Mental region	A. Inferior edges of costal cartilages of ribs 7 through 10
_____	**37.** Xiphoid process	B. Forms elbow
_____	**38.** Arm	C. Inferior portion of sternum
_____	**39.** Nucha	D. Manus
_____	**40.** Shoulder	E. Crus
_____	**41.** Gluteus maximus muscle	F. Brachium
_____	**42.** Costal margin	G. Tarsus
_____	**43.** Olecranon	H. Anterior part of mandible
_____	**44.** Wrist	I. Brain case
_____	**45.** Auricular region	J. Component of hamstrings
_____	**46.** Semitendinosus muscle	K. Forms main part of prominence of buttock
_____	**47.** Leg	L. Carpus
_____	**48.** Cranium	M. Posterior neck region
_____	**49.** Ankle	N. Acromial
_____	**50.** Hand	O. Ear

Nervous Tissue and Physiology

12

The *nervous system* has three basic functions: (1) sensory, (2) integrative, and (3) motor.

1. **Sensory function** It *senses* certain changes (stimuli), both within your body (the internal environment), such as stretching of your stomach or an increase in blood acidity, and outside your body (the external environment), such as a raindrop landing on your arm or the aroma of a rose.
2. **Integrative function** It *analyzes* the sensory information, *stores* some aspects, and *makes decisions* regarding appropriate behaviors.
3. **Motor function** It may *respond* to stimuli by initiating muscular contractions or glandular secretions.

The branch of medical science that deals with the normal functioning and disorders of the nervous system is called **neurology** (noo-ROL-ō-jē; *neuro* = nerve or nervous system; *logos* = study of).

A. NERVOUS SYSTEM DIVISIONS

The two principal divisions of the nervous system are the **central nervous system (CNS)** and the **peripheral** (pe-RIF-er-al) **nervous system (PNS).** The CNS consists of the **brain** and **spinal cord.** Within the CNS, various sorts of incoming sensory information are integrated and correlated, thoughts and emotions are generated, and memories are formed and stored. Most nerve impulses that stimulate muscles to contract and glands to secrete originate in the CNS.

The CNS is connected to sensory receptors, muscles, and glands in peripheral parts of the body by the PNS. The PNS consists of **cranial nerves** that arise from the brain and **spinal nerves** that emerge from the spinal cord. Portions of these nerves carry nerve impulses into the CNS while other portions carry impulses out of the CNS.

The input component of the PNS consists of nerve cells called **sensory** or **afferent** (AF-er-ent; *ad* = toward; *ferre* = to carry) **neurons.** They conduct nerve impulses from sensory receptors in various parts of the body to the CNS and end within the CNS. The output component consists of nerve cells called **motor** or **efferent** (EF-er-ent; *ex* = away from; *ferre* = to carry) **neurons.** They originate within the CNS and conduct nerve impulses from the CNS to muscles and glands.

The PNS may be subdivided further into a **somatic** (*soma* = body) **nervous system (SNS)** and an **autonomic** (*auto* = self; *nomos* = law) **nervous system (ANS).** The major difference between the two is the **effector,** the tissue that receives stimulation or inhibition from the nervous system (skeletal muscle, smooth muscle, cardiac muscle, or a gland). The SNS consists of sensory neurons that convey information from cutaneous and special sense receptors primarily in the head, body wall, and limbs to the CNS and motor neurons from the CNS that conduct impulses to *skeletal muscles* only. Because these motor responses can be consciously controlled, this portion of the SNS is *voluntary.*

The ANS consists of sensory neurons that convey information from receptors primarily in the viscera to the CNS and motor neurons from the CNS that conduct impulses to smooth muscle, cardiac muscle, and glands. Since its motor responses are not normally under conscious control, the ANS is *involuntary.*

The motor portion of the ANS consists of two branches, the **sympathetic division** and the **parasympathetic division.** With few exceptions, the viscera receive instructions from both. Usually, the two divisions have opposing actions. For example, sympathetic neurons speed the heartbeat while parasympathetic neurons slow it down. Processes promoted by sympathetic neurons often involve expenditure of energy while those promoted by parasympathetic neurons restore and conserve body energy.

In this exercise you will identify the parts of a neuron and the components of a reflex arc. You will also perform several experiments on reflexes in the frog.

B. HISTOLOGY OF NERVOUS TISSUE

Despite its complexity, the nervous system consists of only two principal kinds of cells: neurons and neuroglia. *Neurons* (nerve cells) constitute the nervous tissue and are highly specialized for nerve impulse conduction. Mature neurons have only limited capacity for replacement or repair. *Neuroglia* (noo-ROG-lē-a; *neuro* = nerve; *glia* = glue) can divide and multiply and support, nurture, and protect neurons and maintain homeostasis of the fluid that bathes neurons. They do not transmit nerve impulses. Brain tumors are commonly derived from neuroglia. Such tumors, called *gliomas*, are highly malignant and rapidly enlarging.

A neuron consists of the following parts:

1. *Cell body* Contains a nucleus, cytoplasm, lysosomes, mitochondria, Golgi apparatus, chromatophilic substance (Nissl bodies), and neurofibrils.
2. *Dendrites* (*dendro* = tree) Usually short, highly branched extensions of the cell body that conduct nerve impulses toward the cell body.
3. *Axon* (*axon* = axis) Single, usually relatively long process that conducts nerve impulses away from the cell body to another neuron, muscle fiber, or gland cell. An axon, in turn, consists of the following:
   a. *Axon hillock* (*hilloc* = small hill) The origin of an axon from the cell body represented as a small cone-shaped region.
   b. *Initial segment* First portion of a neuron. Except in sensory neurons, nerve impulses arise at the junction of the axon hillock and initial segment, a region called the *trigger zone.*
   c. *Axoplasm* Cytoplasm of an axon.
   d. *Axolemma* (*lemma* = sheath or husk) Plasma membrane around the axoplasm.
   e. *Axon collateral* Side branch of an axon.

   f. *Axon terminals* Fine, branching filaments of an axon or axon collateral.
   g. *Synaptic end bulbs* Bulblike structures at distal end of axon terminals that contain storage sacs *(synaptic vesicles)* for neurotransmitters.
   h. *Myelin sheath* Multilayered, lipid and protein segmented covering of many axons, especially large peripheral ones; the myelin sheath is produced by peripheral nervous system neuroglia called *neurolemmocytes (Schwann cells)* and central nervous system neuroglia called *oligodendrocytes* (described shortly).
   i. *Neurolemma (sheath of Schwann)* Peripheral, nucleated cytoplasmic layer of the neurolemmocyte (Schwann cell) that encloses the myelin sheath. It is found only around axons in the peripheral nervous system.
   j. *Neurofibral node (node of Ranvier)* Unmyelinated gap between segments of the myelin sheath.

Using your textbook and models of neurons as a guide, label Figure 12.1.

Now obtain a prepared slide of an ox spinal cord (cross section), human spinal cord (cross and longitudinal sections), nerve endings in skeletal muscle, and a nerve trunk (cross and longitudinal sections). Examine each under high power and identify as many parts of the neuron as you can.

Neurons may be classified on the basis of structure and function. Structural classification is based on the number of processes extending from the cell body. Structurally, neurons are *multipolar* (several dendrites and one axon), *bipolar* (one dendrite and one axon), and *unipolar* (a single process that branches into an axon and a dendrite). Functional classification is based on the type of information carried and the direction in which the information is carried. Functionally, neurons are classified as *sensory (afferent)*, which carry nerve impulses toward the central nervous system; *motor (efferent)*, which carry nerve impulses away from the central nervous system; and *association (connecting* or *interneurons)*, neurons that are located in the central nervous system and carry nerve impulses between sensory and motor neurons. The neuron you have already labeled in Figure 12.1 is a motor neuron.

Using your textbook and models of neurons as a guide, label Figure 12.2. on page 248.

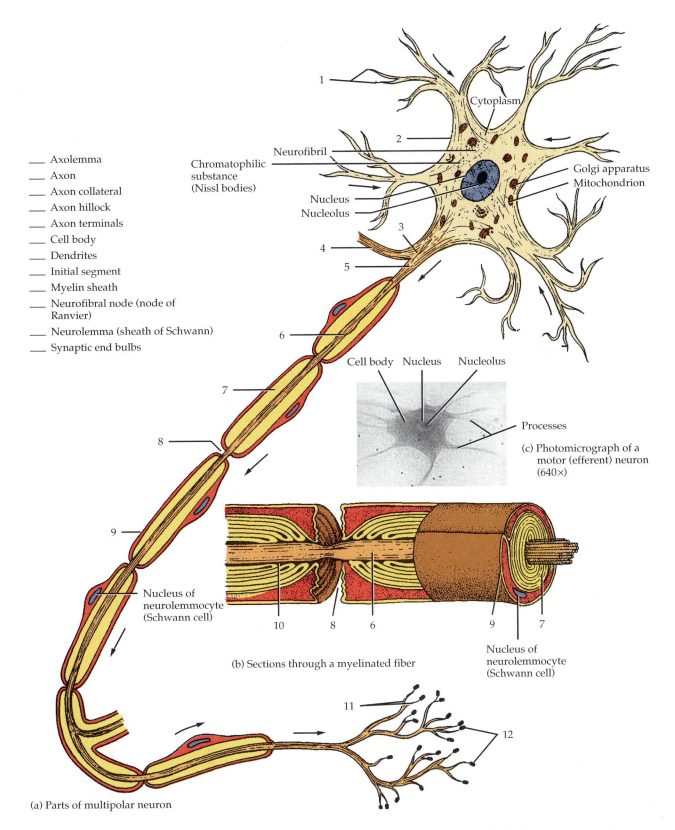

___ Axolemma
___ Axon
___ Axon collateral
___ Axon hillock
___ Axon terminals
___ Cell body
___ Dendrites
___ Initial segment
___ Myelin sheath
___ Neurofibral node (node of Ranvier)
___ Neurolemma (sheath of Schwann)
___ Synaptic end bulbs

Cytoplasm

Neurofibril

Chromatophilic substance (Nissl bodies)

Nucleus
Nucleolus

Golgi apparatus
Mitochondrion

Cell body   Nucleus   Nucleolus

Processes

(c) Photomicrograph of a motor (efferent) neuron (640×)

Nucleus of neurolemmocyte (Schwann cell)

Nucleus of neurolemmocyte (Schwann cell)

(b) Sections through a myelinated fiber

(a) Parts of multipolar neuron

FIGURE 12.1   Structure of a neuron. In (a), arrows indicate the direction in which the nerve impulse travels.

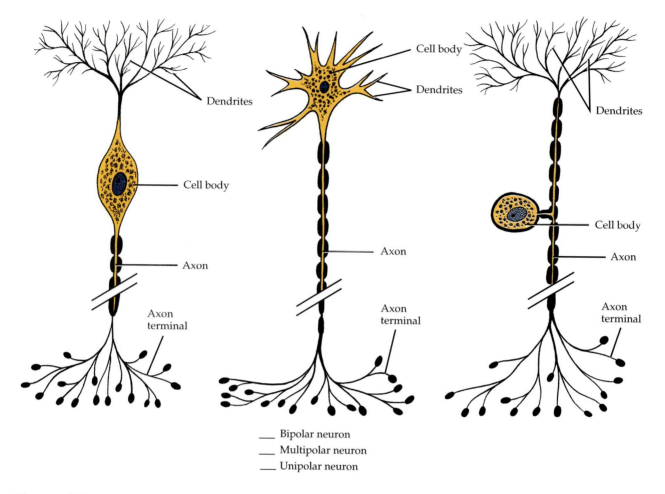

___ Bipolar neuron
___ Multipolar neuron
___ Unipolar neuron

FIGURE 12.2   Structural classification of neurons.

C. HISTOLOGY OF NEUROGLIA

Among the types of neuroglial cells are the following:

1. ***Astrocytes*** (AS-trō-sīts; *astro* = star; *cyte* = cell) Participate in the metabolism of neurotransmitters (glutamate and γ-aminobutyric acid) and maintain the proper balance of potassium ions (K^+) for generation of nerve impulses by CNS neurons; participate in brain development by assisting migration of neurons; help form the blood-brain barrier, which regulates the passage of substances into the brain; twine around neurons to form a supporting network; and provide a link between neurons and blood vessels.

   a. ***Protoplasmic astrocytes*** are found in the gray matter of the central nervous system.
   b. ***Fibrous astrocytes*** are found in the white matter of the central nervous system.

2. ***Oligodendrocytes*** (ol'-i-gō-DEN-drō-sīts; *oligo* = few; *dendro* = tree) Resemble astrocytes but with fewer and shorter processes; they provide support in the CNS and produce a myelin sheath on axons of neurons of the CNS.
3. ***Microglia*** (mī-KROG-lē-a; *micro* = small) Small cells derived from monocytes with few processes; although normally stationary, they can migrate to damaged nervous tissue and there carry on phagocytosis; they are also called ***brain macrophages.***
4. ***Ependymal*** (e-PEN-di-mal) *cells* Epithelial cells arranged in a single layer that range from squamous to columnar in shape; many are ciliated; they form a continuous epithelial lining for the central canal of the spinal cord and for the ventricles of the brain, spaces that contain networks of capillaries that form cerebrospinal fluid; ependymal cells probably assist in circulating cerebrospinal fluid in these areas.
5. ***Neurolemmocytes (Schwann cells)*** Flattened cells arranged around axons. Produce a phos-

pholipid myelin sheath around axon and dendrites of neurons of PNS.

6. *Satellite cells* Flattened cells arranged around the cell bodies of ganglias (collections of neuron cell bodies outside the CNS). Support neurons in ganglia of PNS.

Obtain prepared slides of astrocytes (protoplasmic and fibrous), oligodendrocytes, microglia, ependymal cells, neurolemmocytes, and satellite cells. Using your textbook, Figure 12.3, and models of neuroglia as a guide, identify the various kinds of cells. In the spaces provided, draw each of the cells.

Protoplasmic astrocyte

Fibrous astrocyte

Oligodendrocyte

Microglial cell

Ependymal cell

Neurolemmocyte

Satellite cell

D. NEURONAL CIRCUITS

The CNS contains billions of neurons organized into complicated patterns called *neuronal pools.* Each pool differs from all others and has its own role in regulating homeostasis. A neuronal pool may contain thousands or even millions of neurons.

The functional contact between two neurons or between a neuron and an effector is called a *synapse.* At a synapse, the neuron sending the signal is called a *presynaptic neuron,* and the neuron receiving the message is called a *postsynaptic neuron.*

Neuronal pools in the CNS are arranged in patterns called *circuits* over which the nerve impulses are conducted. In *simple series circuits* a presynaptic neuron stimulates only a single neuron in a pool. The single neuron then stimulates another, and so on. Most circuits, however, are more complex.

A single presynaptic neuron may synapse with several postsynaptic neurons. Such an arrangement, called *divergence,* permits one presynaptic neuron to influence several postsynaptic neurons or several muscle fibers or gland cells at the same time. In a *diverging circuit,* the nerve impulse from a single presynaptic neuron causes the stimulation of increasing numbers of cells along the circuit. For example, a small number of neurons in the brain that govern a particular body movement stimulate a much larger number of neurons in the spinal cord. Sensory signals also feed into diverging circuits and are often relayed to several regions of the brain.

In another arrangement, called *convergence,* several presynaptic neurons synapse with a single postsynaptic neuron. This arrangement permits more effective stimulation or inhibition of the postsynaptic neuron. In one type of *converging circuit,* the postsynaptic neuron receives nerve impulses from several different sources. For example, a single motor neuron that synapses with skeletal muscle fibers at neuromuscular junctions receives input from several pathways that originate in different brain regions.

Some circuits in your body are constructed so that once the presynaptic cell is stimulated, it will cause the postsynaptic cell to transmit a series of nerve impulses. One such circuit is called a *reverberating (oscillatory) circuit.* In this pattern, the incoming impulse stimulates the first neuron, which stimulates the second, which stimulates the third, and so on. Branches from later neurons synapse with earlier ones, however, sending the

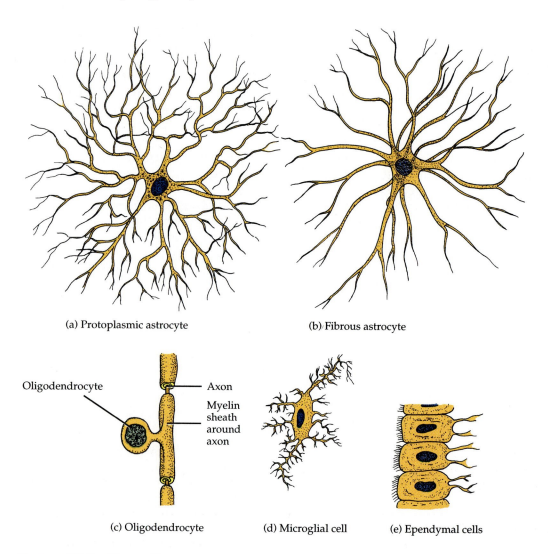

(a) Protoplasmic astrocyte

(b) Fibrous astrocyte

Oligodendrocyte

Axon

Myelin sheath around axon

(c) Oligodendrocyte

(d) Microglial cell

(e) Ependymal cells

Figure 12.3   Neuroglia.

impulse back through the circuit again and again. The output signal may last from a few seconds to many hours, depending on the number of synapses and the arrangement of neurons in the circuit. Inhibitory neurons may turn off a reverberating circuit after a period of time. Among the body responses thought to be the result of output signals from reverberating circuits are breathing, coordinated muscular activities, waking up, sleeping (when reverberation stops), and short-term memory. One form of epilepsy (grand mal) is probably caused by abnormal reverberating circuits.

A fourth type of circuit is the *parallel afterdischarge circuit.* In this circuit, a single presynaptic cell stimulates a group of neurons, each of which synapses with a common postsynaptic

cell. If the input is excitatory, the postsynaptic neuron then can send out a stream of impulses in quick succession. It is thought that parallel afterdischarge circuits may be employed for precise activities such as mathematical calculations.

Using your textbook as a guide, label Figure 12.4.

E. REFLEX ARC

A *reflex* is a fast, predictable, automatic response to a change in the environment (stimulus). The stimulus is applied to the periphery and conducted either to the brain or spinal cord. Reflexes serve to restore functions to homeostasis. For a reflex to occur, a stimulus must elicit a response in

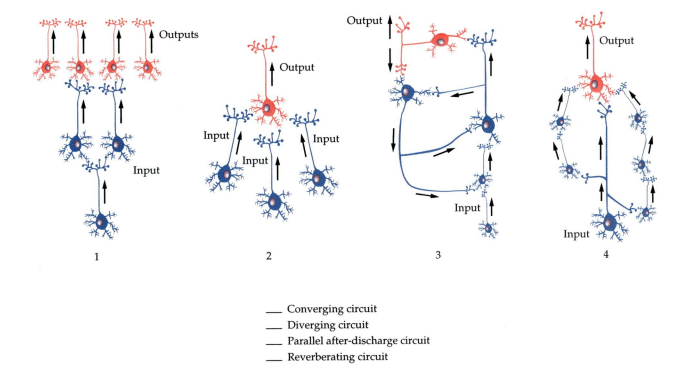

1   2   3   4

_____ Converging circuit
_____ Diverging circuit
_____ Parallel after-discharge circuit
_____ Reverberating circuit

FIGURE 12.4   Neuronal circuit.

one or more structural units within the body termed *reflex arcs.* A reflex arc consists of the following components:

1. *Receptor* The distal end of a sensory neuron (dendrite) or an associated sensory structure that serves as a receptor. A receptor converts a stimulus from its particular form of energy (such as temperature, pressure, or stretch) into the electrical energy utilized by neurons. If this localized depolarization is of threshold value, a nerve impulse will be initiated in a sensory neuron.
2. *Sensory (afferent) neuron* A nerve cell that carries the action potential from the receptor to the central nervous system (CNS).
3. *Integrating center* Region in the CNS where the sensory neuron makes a functional connection with one or more neurons. This CNS synapse may be with an association neuron or a motor neuron. It is here at the CNS synapse where the initial processing of the sensory information occurs: the more complex the reflex involved, the greater the number of CNS synapses involved in a reflex arc.
4. *Motor (efferent) neuron* A nerve cell that transmits the action potential generated by the

sensory neuron or an association neuron away from the CNS to the effector.
5. *Effector* The part of the body, either a muscle or a gland, that responds to the motor neuron impulse and thus the stimulus.

Label the components of a reflex arc in Figure 12.5.

F. DEMONSTRATION OF REFLEX ARC

FLEXICOMP™ is a computer hardware and software system designed to demonstrate characteristics of the reflex arc. It consists of a computer-interfaced goniometer and percussion mallet that permit quantification of the reflex response for various joints of the body and provide visual representation of the reflex arc. The program is available from INTELITOOL® (1-800-227-3805).

G. SPINAL REFLEXES OF THE FROG

The following exercise will demonstrate the varying levels of complexity encountered in several

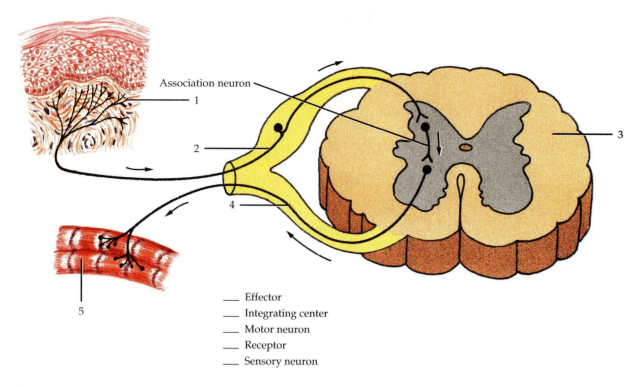

Association neuron

_____ 1

_____ 2

_____ 4

5

3

_____ Effector
_____ Integrating center
_____ Motor neuron
_____ Receptor
_____ Sensory neuron

FIGURE 12.5   Components of a reflex arc.

common reflexes. Through the results obtained in this experiment, you should be able to determine which reflexes involve synapses within the brain and which involve *synapses only within the spinal cord*. You should also be able to determine the relative complexities of the reflexes. Refer to Exercise 9 for pithing instructions and for the nerve-muscle preparation.

All test results should be recorded in the table provided in Section E of the LABORATORY REPORT RESULTS at the end of the exercise.

PROCEDURE

1. Place the frog on the lab bench and observe the position of the head and legs.
2. The rate of respiration can be determined by counting the number of times that the frog raises and lowers the floor of its mouth or opens and closes its nostrils per minute. Determine the

   respiration rate for one minute._____
3. Place the frog on its back and see if it rights itself. Hypothesize what sensory input and peripheral receptors are utilized by the frog in this *righting reflex*.
4. Place the frog on a turtle board and then slowly tilt the board and observe the animal's response, paying particular attention to the

animal's head position and any alteration in limb positioning. Hypothesize **what sensory input and peripheral receptors are utilized by** the frog in this *horizon reflex*.

_____

_____

_____

_____

_____

5. Place the frog in water and observe **whether or** not it stays afloat or swims. Again, hypothesize what sensory input and peripheral receptors are utilized in this reflex.

_____

_____

_____

_____

_____

_____

6. Remove the frog from the water and place it on the laboratory bench. Pinch the toes of a hind leg and observe if the frog withdraws its foot. What sensory input and peripheral receptors would be utilized in this *withdrawal reflex?*

_____

_____

_____

_____

_____

7. Touch the frog's cornea with a fine piece of thread and observe the presence or absence of a corneal reflex. Again, what sensory input and peripheral receptors would be involved in this reflex?

_____

_____

_____

_____

_____

8. Slap the table sharply, close to the frog, and observe its response. What sensory input and peripheral receptors would be involved in this reflex?

_____

_____

_____

_____

_____

Note: *It is strongly suggested that either your instructor or a laboratory assistant perform the following procedure for pithing the frog. The procedure requires some practice and may prove difficult for inexperienced individuals.* Single-pith the frog and repeat the same eight procedures. Upon completion of these procedures, the following questions should be addressed:

1. How were the animal's responses altered in the previously performed tests?

_____

_____

_____

_____

_____

2. Explain how and why the responses were or were not altered following the single-pithing.

_____

_____

_____

_____

_____

_____

3. Which of the previously performed reflexes did and did not require an intact brain?

_____

_____

_____

_____

_____

_____

After answering the above questions for all eight procedures, the following tests should be conducted on a single-pithed frog:

PROCEDURE

1. Insert a small wire through the anterior area of both jaws. Attach the wire so that the frog is suspended over a pan or sink with its legs hanging free.
2. Using a piece of cotton that has been moistened with a 30% acetic acid solution, hold the cotton by forceps and rub the lower portion of one leg. Observe the response. Rinse the affected limb with water. Hypothesize what sensory input

and peripheral receptors are utilized in this reflex.

_____

_____

_____

_____

_____

3. Place the acid on the chest of the frog, and observe the response. Wash the affected region on the frog as before. Hypothesize what sensory input and peripheral receptors are utilized in this reflex. Was the response to the stimulus in this procedure any different from that observed when the acetic acid was applied to the leg? What might account for the difference, or lack thereof, in the responses to these two procedures?

_____

_____

_____

_____

_____

Double-pith the frog and repeat the same 11 procedures. Upon completion of these procedures, the following questions should be addressed:

1. How were the animal's responses altered in the previously performed tests?

_____

_____

_____

_____

2. Explain how and why the responses were or were not altered following the double-pithing.

_____

_____

_____

_____

_____

_____

3. Which of the above reflexes did and did not require an intact CNS?

_____

_____

_____

_____

_____

PROCEDURE

1. Prepare the sciatic nerve and muscle preparation of one leg of the frog as shown in Figure 9.9. Separate the gluteus and semimembranosus muscles to expose the sciatic nerve.
2. Ascertain the reactivity of the sciatic nerve by determining the threshold of the nerve. What response should you observe in order to determine that the threshold, and therefore a nerve action potential, has been obtained in the sciatic nerve?

_____

_____

_____

If the sciatic nerve can be stimulated, what aspects of the reflex arc are functional and what aspects of the reflex arc are not functional in single- and double-pithed frogs?

_____

_____

_____

_____

_____

ANSWER THE LABORATORY REPORT QUESTIONS AT THE END OF THE EXERCISE.

Nervous Tissue and Physiology 12

Student _____   **Date** _____

Laboratory Section _____   **Score/Grade** _____

SECTION G. SPINAL REFLEXES OF THE FROG

Record the results of your observations of the procedure involving differences in the reflexes demonstrated by the control, single-pithed frog, and double-pithed frog in the following table.

Reflex activity	Control frog	Single-pithed frog	Double-pithed frog
1. Head and leg position			
2. Rate of respiration			
3. Righting reflex			
4. Horizontal reflex			
5. Swimming			
6. Withdrawal reflex			
7. Corneal reflex			
8. Response to noise			
9. Response to acetic acid on leg			
10. Response to acetic acid on chest			

1. What conclusions may be drawn regarding the varying levels of complexity of the above reflexes based on the results you have obtained? _____

_____

_____

_____

2. Based upon the results you obtained with the sciatic nerve preparation, what conclusions may be drawn

   as to the reactivity of peripheral nerves in both single- and double-pithed frogs? _____

   _____

   _____

   _____

3. Determine the reactivity of the sciatic nerve by determining the threshold of the nerve. What response
   should you observe in order to determine that threshold, and therefore a nerve impulse, has been

   obtained in the sciatic nerve? _____

   _____

   _____

   _____

4. If the sciatic nerve can be stimulated, what aspects of the reflex arc are functional, and what aspects of

   the reflex arc are not functional in single- and double-pithed frogs? _____

   _____

   _____

   _____

Nervous Tissue and Physiology   12

Student _____   Date _____

Laboratory Section _____   Score/Grade _____

PART 1. Multiple Choice

_____ 1. The portion of a neuron that conducts nerve impulses away from the cell body is the (a) dendrite (b) axon (c) receptor (d) effector

_____ 2. The fine branching filaments of an axon are called (a) myelin sheaths (b) axolemmas (c) axon terminals (d) axon hillocks

_____ 3. The component of a reflex arc that responds to a motor impulse is the (a) integrating center (b) receptor (c) sensory neuron (d) effector

_____ 4. Which type of neuron conducts nerve impulses toward the central nervous system? (a) sensory (b) association (c) connecting (d) motor

_____ 5. In a reflex arc, the nerve impulse is transmitted directly to the effector by the (a) sensory neuron (b) motor neuron (c) integrating center (d) receptor

_____ 6. Bulblike structures at the distal ends of axon terminals that contain storage sacs for neurotransmitters are called (a) dendrites (b) synaptic end bulbs (c) axon collaterals (d) neurofibrils

_____ 7. Which neuroglial cell is phagocytic? (a) oligodendrocyte (b) protoplasmic astrocyte (c) microglial cell (d) fibrous astrocyte

PART 2. Completion

8. A neuron that contains several dendrites and one axon is classified as _____.

9. The portion of a neuron that contains the nucleus and cytoplasm is the _____.

10. The lipid and protein covering around many peripheral axons is called the _____.

11. The two types of cells that compose the nervous system are neurons and _____.

12. The peripheral, nucleated layer of the neurolemmocyte (Schwann cell) that encloses the myelin

    sheath is the _____.

13. The side branch of an axon is referred to as the _____.

14. The part of a neuron that conducts nerve impulses toward the cell body is the

    _____.

15. Neurons that carry nerve impulses between sensory neurons and motor neurons are called

_____ neurons.

16. Neurons with one dendrite and one axon are classified as _____.

17. Unmyelinated gaps between segments of the myelin sheath are known as _____.

18. The neuroglial cell that produces a myelin sheath around axons of neurons of the central nervous

system is called a(n) _____.

19. In a reflex arc, the muscle or gland that responds to a motor impulse is called the

_____.

20. The functional contact between two neurons or between a neuron and an effector is called a(n)

_____.

21. The circuit that probably plays a role in breathing, waking up, sleeping, and short-term memory is

a(n) _____ circuit.

Nervous System

13

In this exercise, you will examine the principal structural features of the spinal cord and spinal nerves, perform several experiments on reflexes, identify the principal structural features of the brain, trace the course of cerebrospinal fluid, identify the cranial nerves, perform several experiments designed to test for cranial nerve function, and examine the structure and function of the autonomic nervous system. You will also dissect and study the cat nervous system and sheep brain.

A. SPINAL CORD AND SPINAL NERVES

1. Meninges

The *meninges* (me-NIN-jēz) are connective tissue coverings that run continuously around the spinal cord and brain (*meninx*, pronounced MĒ-ninks, is singular). They protect the central nervous system. The spinal meninges are:

a. *Dura mater* (DYOO-ra MĀ-ter; *dura* = tough; *mater* = mother) The most superficial meninx composed of dense irregular connective tissue. Between the wall of the vertebral canal and the dura mater is the *epidural space*, which is filled with fat, connective tissue, and blood vessels.

b. *Arachnoid* (a-RAK-noyd; *arachne* = spider) The middle meninx is an avascular covering composed of very delicate collagen and elastic fibers. Between the arachnoid and the dura mater is a space called the *subdural space*, which contains interstitial fluid.

c. *Pia mater* (PĒ-a MĀ-ter; *pia* = delicate) The deep meninx is a thin, transparent connective tissue layer that adheres to the surface of the brain and spinal cord. It consists of interlacing collagen and a few elastic fibers and contains blood vessels. Between the pia mater and the arachnoid is a space called the *subarachnoid*

space where cerebrospinal fluid circulates. Extensions of the pia mater called *denticulate* (den-TIK-yoo-lāt; *denticulus* = a small tooth) *ligaments* are attached laterally to the dura mater along the length of the spinal cord and suspend the spinal cord and afford protection against shock and sudden displacement.

Label the meninges, subarachnoid space, and denticulate ligament in Figure 13.1.

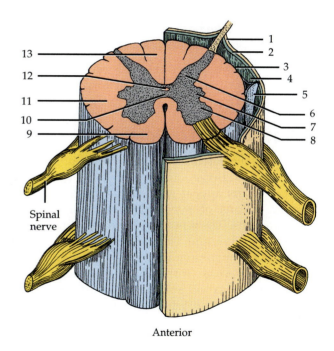

FIGURE 13.1 Cross section of spinal cord showing meninges on right side of figure.

____ Anterior gray horn   ____ Lateral gray horn

____ Anterior white column   ____ Lateral white column

____ Arachnoid   ____ Pia mater

____ Central canal   ____ Posterior gray horn

____ Denticulate ligament   ____ Posterior white column

____ Dura mater   ____ Subarachnoid space

____ Gray commissure

2. General Features

Obtain a model or preserved specimen of the spinal cord and identify the following general features:

a. *Cervical enlargement* Between vertebrae C4 and T1; origin of nerves to upper limbs.

b. *Lumbar enlargement* Between vertebrae T9 and T12; origin of nerves to lower limbs.

c. *Conus medullaris* (KŌ-nus med-yoo-LAR-is; *konos* = cone). Tapered conical portion of spinal cord that ends at the intervertebral disc between L1 and L2.

d. *Filum terminale* (FĪ-lum ter-mi-NAL-ē; *filum* = filament; *terminale* = terminal) Nonnervous fibrous tissue arising from the conus medullaris and extending inferiorly to attach to the coccyx; consists mostly of pia mater.

e. *Cauda equina* (KAW-da ē-KWĪ-na) Spinal nerves that angle inferiorly in the vertebral canal giving the appearance of wisps of coarse hair.

f. *Anterior median fissure* Deep, wide groove on the anterior surface of the spinal cord.

g. *Posterior median sulcus* Shallow, narrow groove on the posterior surface of the spinal cord.

After you have located the parts on a model or preserved specimen of the spinal cord, label the spinal cord in Figure 13.2.

3. Cross Section of Spinal Cord

In a freshly dissected section of the brain or spinal cord, some regions look white and glistening whereas others appear gray. *White matter* refers to aggregations of myelinated processes from many neurons. The whitish color of myelin gives white matter its name. The *gray matter* of the nervous system contains either nerve cell bodies, dendrites, and axon terminals or bundles of unmyelinated axons and neuroglia. They look grayish, rather than white, because there is no myelin in these areas.

Obtain a model or specimen of the spinal cord in cross section and note the gray matter, shaped like a letter H or a butterfly. Identify the following parts:

a. *Gray commissure* (KOM-mi-shur) Cross bar of the letter H.

b. *Central canal* Small space in the center of the gray commissure that contains cerebrospinal fluid.

c. *Anterior gray horn* Anterior region of the upright portion of the H.

d. *Posterior gray horn* Posterior region of the upright portion of the H.

e. *Lateral gray horn* Intermediate region between the anterior and posterior gray horns present in the thoracic, upper lumbar, and sacral segments of the spinal cord.

f. *Anterior white column* Anterior region of white matter.

g. *Posterior white column* Posterior region of white matter.

h. *Lateral white column* Intermediate region of white matter between the anterior and posterior white columns.

Label these parts of the spinal cord in cross section in Figure 13.1.

Examine a prepared slide of a spinal cord in cross section and see how many structures you can identify.

4. Spinal Nerve Attachments

Spinal nerves are the paths of communication between the spinal cord tracts and most of the body. The 31 pairs of spinal nerves are named and numbered according to the region of the spinal cord from which they emerge. The first cervical pair emerges between the atlas and occipital bone; all other spinal nerves leave the vertebral column from intervertebral foramina between adjoining vertebrae. There are 8 pairs of cervical nerves, 12 pairs of thoracic nerves, 5 pairs of lumbar nerves, 5 pairs of sacral nerves, and 1 pair of coccygeal nerves. Label the spinal nerves in Figure 13.2.

Each pair of spinal nerves is connected to the spinal cord by two points of attachment called roots. The *posterior (sensory) root* contains sensory nerve fibers only and conducts nerve impulses from the periphery to the spinal cord. Each posterior root has a swelling, the *posterior (sensory) root ganglion*, which contains the cell bodies of the sensory neurons from the periphery. Fibers extend from the ganglion into the posterior gray horn. The other point of attachment, the *anterior (motor) root*, contains motor nerve fibers only and conducts nerve impulses from the spinal cord to the periphery. The cell bodies of the motor neurons are located in lateral or anterior gray horns.

Label the posterior root, posterior root ganglion, anterior root, spinal nerve, cell body of sensory neuron, axon of sensory neuron, cell body of motor neuron, and axon of motor neuron in Figure 13.3 on page 262.

5. Components and Coverings of Spinal Nerves

The posterior and anterior roots unite to form a spinal nerve at the intervertebral foramen. Because the posterior root contains sensory nerve

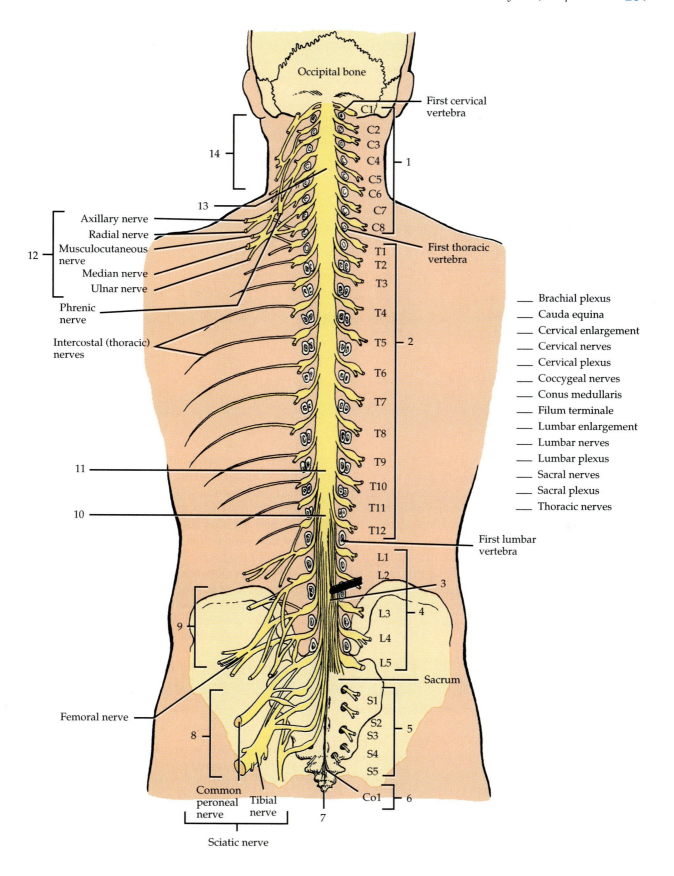

Occipital bone

First cervical vertebra

C1
C2
C3
C4
C5
C6
C7
C8

14

13

1

Axillary nerve
Radial nerve
Musculocutaneous nerve
Median nerve
Ulnar nerve

12

Phrenic nerve

Intercostal (thoracic) nerves

First thoracic vertebra

T1
T2
T3
T4
T5
T6
T7
T8
T9
T10
T11
T12

2

11

10

___ Brachial plexus
___ Cauda equina
___ Cervical enlargement
___ Cervical nerves
___ Cervical plexus
___ Coccygeal nerves
___ Conus medullaris
___ Filum terminale
___ Lumbar enlargement
___ Lumbar nerves
___ Lumbar plexus
___ Sacral nerves
___ Sacral plexus
___ Thoracic nerves

First lumbar vertebra

L1
L2
L3
L4
L5

3
4

9

Sacrum

S1
S2
S3
S4
S5

5

Femoral nerve

8

Common peroneal nerve

Tibial nerve

Co1

6

7

Sciatic nerve

Posterior view

FIGURE 13.2 Spinal cord.

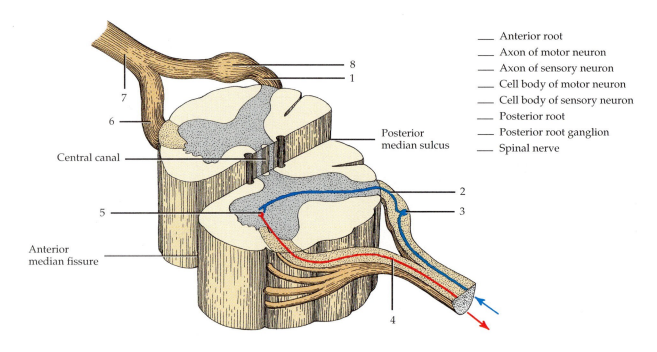

Central canal

Anterior
median fissure

Posterior
median sulcus

___ Anterior root
___ Axon of motor neuron
___ Axon of sensory neuron
___ Cell body of motor neuron
___ Cell body of sensory neuron
___ Posterior root
___ Posterior root ganglion
___ Spinal nerve

Sections through the thoracic spinal cord

FIGURE 13.3   Spinal nerve attachments.

fibers and the anterior root contains motor nerve fibers, all spinal nerves are *mixed nerves.*

Some spinal nerves carry information to or from visceral structures, while others carry information to or from somatic (*soma* = body) structures. The functional components of a spinal nerve may therefore by placed into one of four categories, depending upon the type of information carried within the spinal nerve. The names of these four categories are all prefixed by the word "general." Any one spinal nerve may contain one or more of these functional categories.

a. *General somatic afferent* These fibers are related to receptors for pain, temperature and mechanical stimuli. Receptors attached to these fibers transmit infomation from somatic structures such as skin, muscles, and joints.
b. *General visceral efferent* This type of spinal nerve contains preganglionic autonomic nervous system neurons.
c. *General visceral afferent* These fibers are related to receptors in visceral structures, such as walls of the digestive tract.
d. *General somatic efferent* These neurons innervated skeletal muscle.

Spinal nerves are covered by several connective tissue layers. Individual nerve fibers within a nerve, whether myelinated or unmyelinated, are wrapped in a covering called the *endoneurium* (en'-dō-NOO-rē-um). Groups of fibers with their endoneurium are arranged in bundles called *fascicles*, and each bundle is wrapped in a covering called the *perineurium* (per'-i-NOO-rē-um). All the fascicles, in turn, are wrapped in a covering called the *epineurium* (ep'-i-NOO-rē-um). This is the outermost covering around the entire nerve.

Obtain a prepared slide of a nerve in cross section and identify the fibers, endoneurium, perineurium, epineurium, and fascicles. Now label Figure 13.4.

6. Branches of Spinal Nerves

Shortly after leaving its intervertebral foramen, a spinal nerve divides into several branches called *rami* (singular is *ramus* [RĀ-mus]):

a. *Dorsal ramus* Innervates deep muscles and skin of the dorsal surface of the back.
b. *Ventral ramus* Innervates superficial back muscles and all structures of the limbs and lateral and ventral trunk; except for thoracic nerves T2–T11, the ventral rami of the other spinal nerves form plexuses before innervating their structures.
c. *Meningeal branch* Innervates vertebrae, vertebral ligaments, blood vessels of the spinal cord, and meninges.

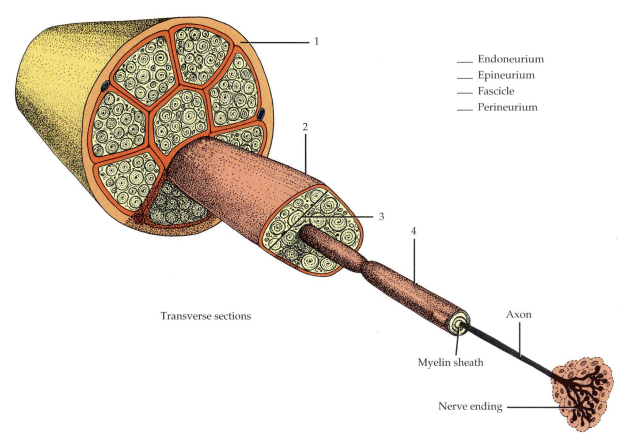

_____ Endoneurium
_____ Epineurium
_____ Fascicle
_____ Perineurium

Transverse sections

Axon

Myelin sheath

Nerve ending

FIGURE 13.4   Coverings of a spinal nerve.

d. *Rami communicantes* (RĀ-mē ko-myoo-nē-KAN-tēz) Gray and white rami communicantes are components of the autonomic nervous system; they connect the ventral rami with sympathetic trunk ganglia.

7. Plexuses

The ventral rami of spinal nerves, except for T2–T11, do not go directly to body structures they supply. Instead, they join with adjacent nerves on either side of the body to form networks called *plexuses* (PLEK-sus-ēz; *plexus* = braid).

a. *Cervical plexus* (SER-vi-kul PLEK-sus) Formed by the ventral rami of the first four cervical nerves (C1–C4) with contributions from C5; one is located on each side of the neck alongside the first four cervical vertebrae; the plexus supplies the skin and muscles of the head, neck, and upper part of shoulders.

Using your textbook as a guide, label the nerves of the cervical plexus in Figure 13.5.

b. *Brachial* (BRĀ-kē-al) *plexus* Formed by the ventral rami of spinal nerves C5–C8 and T1 with contributions from C4 and T2; each is located on either side of the last four cervical and first thoracic vertebrae and extends inferiorly and laterally, superior to the first rib behind the clavicle, and into the axilla; the plexus constitutes the entire nerve supply for the upper limbs and shoulder region.

Using your textbook as a guide, label the nerves of the brachial plexus in Figure 13.6.

c. *Lumbar* (LUM-bar) *plexus* Formed by the ventral rami of spinal nerves L1–L4; each is located on either side of the first four lumbar vertebrae posterior to the psoas major muscle and anterior to the quadratus lumborum muscle; the plexus supplies the anterolateral abdominal wall, external genitals, and part of the lower limbs.

Using your textbook as a guide, label the nerves of the lumbar plexus in Figure 13.7 on page 267.

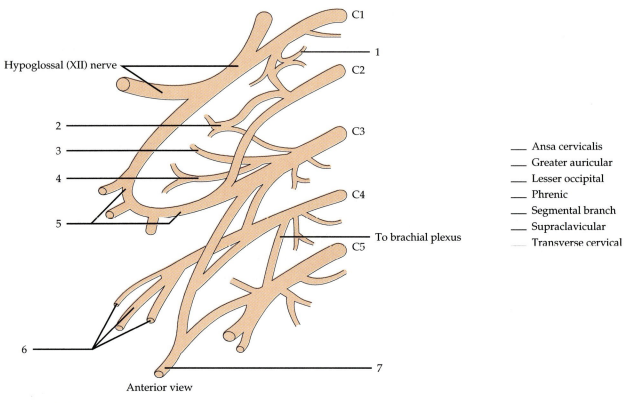

Hypoglossal (XII) nerve

C1
C2
C3
C4
C5

1
2
3
4
5
6
7

To brachial plexus

_____ Ansa cervicalis
_____ Greater auricular
_____ Lesser occipital
_____ Phrenic
_____ Segmental branch
_____ Supraclavicular
_____ Transverse cervical

Anterior view

FIGURE 13.5   Cervical plexus.

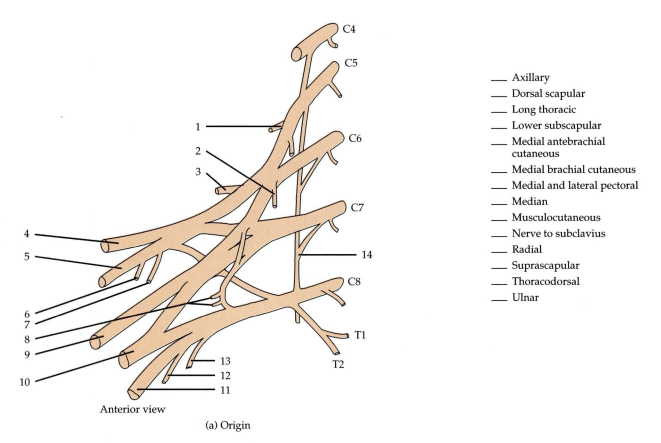

C4
C5
C6
C7
C8
T1
T2

1
2
3
4
5
6
7
8
9
10
11
12
13
14

_____ Axillary
_____ Dorsal scapular
_____ Long thoracic
_____ Lower subscapular
_____ Medial antebrachial cutaneous
_____ Medial brachial cutaneous
_____ Medial and lateral pectoral
_____ Median
_____ Musculocutaneous
_____ Nerve to subclavius
_____ Radial
_____ Suprascapular
_____ Thoracodorsal
_____ Ulnar

Anterior view

(a) Origin

FIGURE 13.6   Brachial plexus.

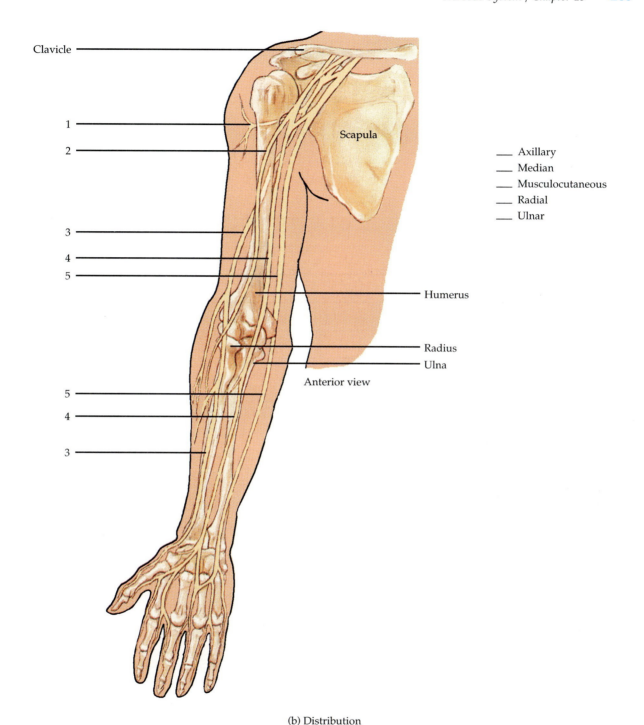

Clavicle

1
2

Scapula

_____ Axillary
_____ Median
_____ Musculocutaneous
_____ Radial
_____ Ulnar

3
4
5

Humerus

Radius
Ulna

Anterior view

5
4

3

(b) Distribution

FIGURE 13.6 *(Continued)* Brachial plexus.

d. *Sacral* (SĀ-kral) *plexus* Formed by the ventral rami of spinal nerves L4–L5 and S1–S4; each is located largely anterior to the sacrum; the plexus supplies the buttocks, perineum, and lower limbs.

Using your textbook as a guide, label the nerves of the sacral plexus in Figure 13.8 on page 267.

Label the cervical, brachial, lumbar, and sacral plexuses in Figure 13.2. Also note the names of some of the major peripheral nerves that arise from the plexuses.

For each nerve listed in the following table, indicate the plexus to which it belongs and the structure(s) it innervates.

Nerve	Plexus	Innervation
Musculocutaneous (mus'-kyoo-lō-kyoo-TAN-ē-us)		
Femoral (FEM-or-al)		
Phrenic (FREN-ik)		
Pudendal (pyoo-DEN-dal)		
Axillary (AK-si-lar-ē)		
Transverse cervical (SER-vi-kul)		
Radial (RĀ-dē-al)		
Obturatur (OB-too-rā-tor)		
Tibial (TIB-ē-al)		
Thoracodorsal (thō-RA-kō-dor-sal)		
Perforating cutaneous (PER-fō-rā-ting kyoo'-TA-nē-us)		
Ulnar (UL-nar)		
Long thoracic (thō-RAS-ik)		
Median (MĒ-dē-an)		
Iliohypogastric (il'-ē-ō-hī-pō-GAS-trik)		
Deep peroneal (per'-ō-NĒ-al)		
Ansa cervicalis (AN-sa ser-vi-KAL-is)		
Sciatic (sī-AT-ik)		

8. Spinal Cord Tracts

The vital function of conveying sensory and motor information to and from the brain is carried out by sensory (ascending) and motor (descending) tracts and pathways in the spinal cord. The names of the tracts and pathways indicate the white column in which the tract travels, where the cell bodies of the tract originate, and where the axons of the tract terminate. For example, the anterior spinothalamic tract is located in the *anterior* white column, it originates in the *spinal cord*, and it terminates in the *thalamus* of the brain. Since it conveys nerve impulses from the spinal cord upward to the brain, it is a sensory (ascending) tract.

a. SOMATIC SENSORY PATHWAYS

Somatic sensory pathways from receptors to the cerebral cortex involve three-neuron sets. Axon collaterals (branches) of somatic sensory neurons simultaneously carry signals into the cerebellum and the reticular formation of the brain stem.

1. *First-order neurons* Carry signals from the somatic receptors into either the brain stem or spinal cord. From the face, mouth, teeth, and eyes, somatic sensory impulses propagate along *cranial nerves* into the brain stem. From the back of the head, neck, and body, somatic sensory impulses propagate along *spinal nerves* into the spinal cord.

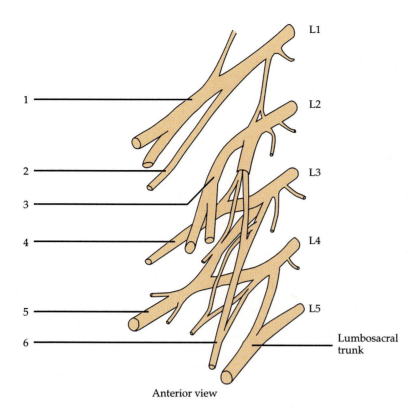

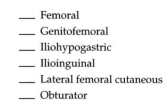

__ Femoral
__ Genitofemoral
__ Iliohypogastric
__ Ilioinguinal
__ Lateral femoral cutaneous
__ Obturator

L1
L2
L3
L4
L5

Lumbosacral trunk

Anterior view

FIGURE 13.7 Lumbar plexus.

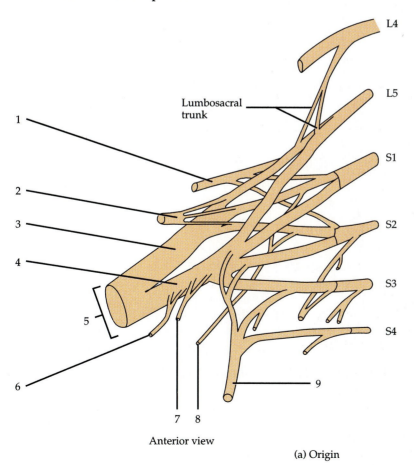

__ Common peroneal
__ Inferior gluteal
__ Nerve to obturator internus and superior gemellus
__ Nerve to quadratus femoris and inferior gemellus
__ Posterior femoral cutaneous
__ Pudendal
__ Sciatic
__ Superior gluteal
__ Tibial

L4
L5

Lumbosacral trunk

S1
S2
S3
S4

Anterior view

(a) Origin

FIGURE 13.8 Sacral plexus.

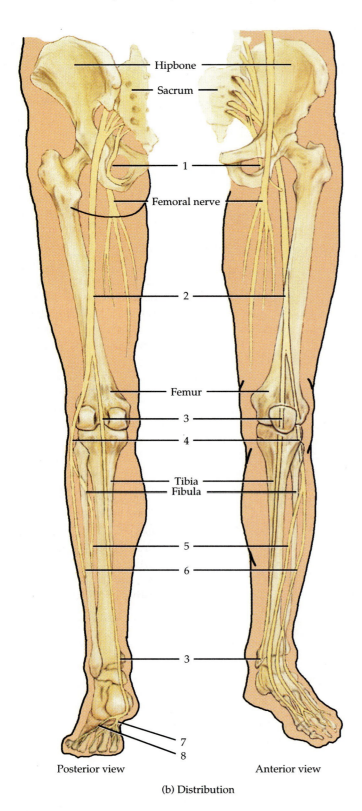

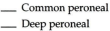

___ Common peroneal
___ Deep peroneal
___ Lateral plantar
___ Medial plantar
___ Pudendal
___ Sciatic
___ Superficial peroneal
___ Tibial

Posterior view   Anterior view

(b) Distribution

FIGURE 13.8 (Continued)   Sacral plexus.

2. *Second-order neurons* Carry signals from the spinal cord and brain stem to the thalamus. Axons of second-order neurons cross over (decussate) to the opposite side in the spinal cord or brain stem before ascending to the thalamus.

3. *Third-order neurons* Project from the thalamus to the primary somatosensory area of the cortex (postcentral gyrus; see Figure 13.14), where conscious perception of the sensations results.

There are two general pathways by which somatic sensory signals entering the spinal cord ascend to the cerebral cortex: the *posterior column–medial lemniscus pathway* and the *anterolateral (spinothalamic) pathways.*

1. Posterior Column–Medial Lemniscus Pathway to the Cortex Nerve impulses for conscious proprioception and most tactile sensations ascend to the cortex along a common pathway formed by three-neuron sets. First-order neurons extend from sensory receptors into the spinal cord and up to the medulla oblongata on the same side of the body. The cell bodies of these first-order neurons are in the posterior (dorsal) root ganglia of spinal nerves. Their axons form the *posterior column* (*fasciculus gracilis;* fa-SIK-yoo-lus gras-I-lus and *fasciculus cuneatus;* kyoo-nē-ĀT-us) in the spinal cord. The axon terminals form synapses with second-order neurons in the medulla. The cell body of a second-order neuron is located in the nucleus cuneatus (which receives input conducted along axons in the fasciculus cuneatus from the neck, upper limbs, and upper chest) or nucleus gracilis (which receives input conducted along axons in the fasciculus gracilis from the trunk and lower limbs). The axon of the second-order neuron crosses to the opposite side of the medulla and enters the *medial lemniscus,* a projection tract that extends from the medulla to the thalamus. In the thalamus, the axon terminals of second-order neurons synapse with third-order neurons, which project their axons to the somatosensory area of the cerebral cortex.

Impulses conducted along the posterior column–medial lemniscus pathway give rise to several highly evolved and refined sensations. These are

Discriminative touch The ability to recognize the exact location of a light touch and to make two-point discriminations.

Stereognosis The ability to recognize by feel the size, shape, and texture of an object. Examples are identifying (with closed eyes) a paper clip put into your hand or reading braille.

Proprioception The awareness of the precise position of body parts, and *kinesthesia,* the awareness of directions of movement.

Weight discrimination The ability to assess the weight of an object.

Vibratory sensations The ability to sense rapidly fluctuating touch.

2. Anterolateral (Spinothalamic) Pathways to the Cortex The *anterolateral (spinothalamic;* spī-nō-THAL-am-ik) *pathways* carry mainly pain and temperature impulses. In addition, they relay the sensations of tickle and itch and some tactile impulses, which give rise to a very crude, not well-localized touch or pressure sensation. Like the posterior column–medial lemniscus pathway, the anterolateral pathways are also composed of three-neuron sets. The first-order neuron connects a receptor of the neck, trunk, or limbs with the spinal cord. The cell body of the first-order neuron is in the posterior root ganglion. The axon of the first-order neuron synapses with the second-order neuron, which is located in the posterior gray horn of the spinal cord. The axon of the second-order neuron continues to the opposite side of the spinal cord and passes superiorly to the brain stem in either the *lateral spinothalamic tract* or the *anterior spinothalamic tract.* The axon from the second-order neuron ends in the thalamus. There, it synapses with the third-order neuron. The axon of the third-order neuron projects to the somatosensory area of the cerebral cortex. The lateral spinothalamic tract conveys sensory impulses for pain and temperature whereas the anterior spinothalamic tract conveys impulses for tickle, itch, crude touch, and pressure.

3. Somatic Sensory Pathways to the Cerebellum Two tracts in the spinal cord, the *posterior spinocerebellar* (spī-nō-ser-e-BEL-ar) *tract* and the *anterior spinocerebellar tract,* are major routes for the subconscious proprioceptive input to reach the cerebellum. Sensory input conveyed to the cerebellum along these two pathways is critical for posture, balance, and coordination of skilled movements.

b. SOMATIC MOTOR PATHWAYS

The most direct somatic motor pathways extend from the cerebral cortex to skeletal muscles. Other pathways are less direct and include synapses in the basal ganglia, thalamus, reticular formation, and cerebellum.

1. Direct Pathways Voluntary motor impulses are propagated from the motor cortex to voluntary motor neurons (somatic motor neurons) that innervate skeletal muscles via the direct or *pyramidal* (pi-RAM-i-dal) *pathways.* The simplest of these pathways consists of sets of two neurons, upper motor neurons and lower motor neurons. About one million pyramidal-shaped cell bodies of direct pathway *upper motor neurons* (**UMNs**) are in the cortex. Their axons descend through the internal capsule of the cerebrum. In the medulla oblongata, the axon bundles form the ventral bulges known as the pyramids. About 90% of these axons also cross to the opposite side in the medulla oblongata. They terminate in nuclei of cranial nerves or in the anterior gray horn of the spinal cord. *Lower motor neurons,* (**LMNs**), extend from the motor nuclei of nine cranial nerves to muscles of the face and head and from the anterior horn of each spinal cord segment to skeletal muscle fibers of the trunk and limbs. Close to their termination point, most upper motor neurons synapse with an association neuron, which, in turn, synapses with a lower motor neuron. A few upper motor neurons synapse directly with lower motor neurons.

The direct pathways convey impulses from the cortex that result in precise, voluntary movements. The main parts of the body governed by the direct pathways are the face, vocal cords (for speech), and hands and feet of the limbs. They channel nerve impulses into three tracts:

a. *Lateral corticospinal* (kor'-ti-kō-SPĪ-nal) *tracts.* These pathways begin in the right and left motor cortex and descend through the *internal capsule* of the cerebrum and through the cerebral peduncle of the midbrain and the pons on the same side. About 90% of the axons of upper motor neurons cross over to the opposite side (decussate) in the medulla oblongata. These axons form the lateral corticospinal tracts in the right and left lateral white columns of the spinal cord. Thus the motor cortex of the right side of the brain controls muscles on the left side of the body, and vice versa. The lower motor neurons then receive input from both upper motor neurons and association neurons. Axons of lower motor neurons (somatic motor neurons) exit all levels of the spinal cord via the anterior roots of spinal nerves and terminate in skeletal muscles. These motor neurons control skilled movements of the distal portions of the limbs.

b. *Corticobulbar* (kor'-ti-kō-BUL-bar) *tracts* The axons of upper motor neurons of these tracts accompany the corticospinal tracts from the motor cortex through the internal capsule to the brain stem. There some cross whereas others remain uncrossed. They terminate in the nuclei of nine pairs of cranial nerves in the pons and medulla: the oculomotor (III), trochlear (IV), trigeminal (V), abducens (VI), facial (VII), glossopharyngeal (IX), vagus (X), accessory (XI), and hypoglossal (XII). The lower motor neurons

of cranial nerves convey impulses that control voluntary movements of the eyes, tongue, and neck; chewing; facial expression; and speech.

c. *Anterior corticospinal tracts* About 10% of the axons of upper motor neurons do not cross in the medulla oblongata. They pass through the medulla oblongata, descend on the same side, and form the anterior corticospinal tracts in the right and left anterior white columns. At several spinal cord levels, some of the axons of these upper motor neurons decussate. After crossing to the opposite side, they synapse with association or lower motor neurons in the anterior gray horn of the spinal cord. Axons of these lower motor neurons exit the cervical and upper thoracic segments of the cord via the anterior roots of spinal nerves. They terminate in skeletal muscles that control movements of the neck and part of the trunk, thus coordinating movements of the axial skeleton.

2. *Indirect Pathways* The *indirect (extrapyramidal) pathways* include all descending (motor) tracts other than the corticospinal and corticobulbar tracts. Nerve impulses conducted along the indirect pathways follow complex, polysynaptic circuits that involve the motor cortex, basal ganglia, limbic system, thalamus, cerebellum, reticular formation, and nuclei in the brain stem. Axons of upper motor neurons that carry motor signals from the indirect pathways descend from various nuclei of the brain stem into five major tracts of the spinal cord and terminate on association neurons or lower motor neurons.

Lower motor neurons receive both excitatory and inhibitory input from many presynaptic neurons in both direct and indirect pathways, an example of convergence. For this reason, lower motor neurons are also called the *final common pathway.* Most nerve impulses from the brain are conveyed to association neurons before being received by lower motor neurons. The sum total of the input from upper motor neurons and association neurons determines the final response of the lower motor neuron. It is not just a simple matter of the brain sending an impulse and the muscle always contracting.

The five major tracts of the indirect pathways are the *rubrospinal* (ROO-brō-spī-nal), *tectospinal* (TEK-tō-spī-nal), *vestibulospinal* (ves-TIB-yoo-lō-spī-nal), *lateral reticulospinal* (re-TIK-yoo-lō-spī-nal), and *medial reticulospinal.*

Label the sensory (ascending) and motor (descending) tracts shown in Figure 13.9.

Indicate the function of each sensory and motor tract listed in the following table:

Sensory (ascending) tracts	Function
Posterior column	
Lateral spinothalamic	
Anterior spinothalamic	
Posterior spinocerebellar	
Anterior spinocerebellar	

Motor (descending) tracts	Function
Direct (pyramidal)	
Lateral corticospinal	
Anterior corticospinal	
Corticobulbar	
Indirect (extrapyramidal)	
Rubrospinal	
Tectospinal	
Vestibulospinal	
Lateral reticulospinal	
Medial reticulospinal	

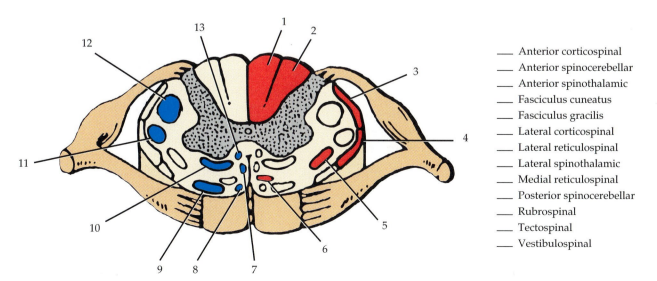

____ Anterior corticospinal
____ Anterior spinocerebellar
____ Anterior spinothalamic
____ Fasciculus cuneatus
____ Fasciculus gracilis
____ Lateral corticospinal
____ Lateral reticulospinal
____ Lateral spinothalamic
____ Medial reticulospinal
____ Posterior spinocerebellar
____ Rubrospinal
____ Tectospinal
____ Vestibulospinal

FIGURE 13.9   Selected sensory (ascending) and motor (descending) tracts of the spinal cord. The sensory tracts are indicated in blue and the motor tracts are indicated in red.

9. Reflex Experiments

In this section, you will determine the responses obtained in various common human reflexes. While performing the procedures it is important for you to consider the various components of a reflex arc, and to try to determine the receptors and effector organs involved.

a. *Patellar reflex* Have your partner sit on a table so that the knee is off of the table and the leg hangs freely. Strike the quadriceps (patellar) tendon just inferior to the kneecap with the reflex hammer. What is the response?

_____

_____

_____

Test the subject while he or she interlocks fingers and pulls one hand against the other. Is there any change in the level of activity (sensitivity) of the reflex?

_____

_____

_____

Formulate a hypothesis regarding the purpose of the patellar reflex.

_____

_____

_____

b. *Achilles reflex* Have a subject kneel on a chair and let the feet hang freely over the edge of the chair. Bend one foot to increase the tension on the gastrocnemius muscle. Tap the calcaneal (Achilles) tendon with the reflex hammer. What is the result?

_____

_____

_____

c. *Plantar reflex* Scratch the sole of your partner's foot by moving a blunt object along the sole toward the toes.

What is the result?

_____

_____

_____

What is the Babinski sign?

_____

_____

_____

Why is it normal in children under the age of 18 months?

_____

_____

_____

What does the Babinski sign indicate in an adult?

_____

_____

_____

_____

10. Measurement of Reaction Times

The time required for an individual to react to a stimulus (*reaction time*) is dependent on several factors. Some of the factors affecting reaction time include (1) the sensitivity and reactive state of the receptors, (2) the amount of convergence or divergence that occurs within the receptive fields, (3) the conduction velocity of the sensory and motor neurons, (4) the length of the neurons involved in the reflex, (5) the number of synapses involved in the reflex arc, and (6) whether the reflex is somatic or visceral.

One of the instruments that can be utilized to determine reaction times is a multichoice reaction timer. This unit is composed of a variable light source, a control box, and a multiple keypad.

1. Divide your group so that one member will serve as subject, another as recorder, and a third as the operator of the reaction timer.
2. Familiarize yourselves with the reaction timer. Note the different key functions, stimulus choices, and selective key choices. Turn the unit on and note the color changes in the light source

and the function of the keypad. Note how the reaction timer works, and how to reset the timer.

3. In order to determine the reaction time of the subject for a predetermined light stimulus, ask the subject to sit comfortably with the right hand positioned close to the keypad. Ask the subject to touch key 2 as quickly as possible when a red light is presented.

4. Turn the unit on and set the stimulus dial on red and the selective key on 1. When the subject is relaxed and attentive, flip the master key switch and then note the reaction time. Record the reaction time below. Repeat the procedure three more times, recording the different reaction times below, and then take an average of the four values, recording the average in the space provided.

Trial 1 _____

Trial 2 _____

Trial 3 _____

Trial 4 _____
Average time for predetermined light stimulus

_____

5. Now determine the reaction time for a variable light stimulus. Tell your subject to press key 1 if an amber light stimulus is presented, key 2 if a red light is presented, and key 3 if a green light is presented. Have the subject sit in a relaxed position with his or her right hand comfortably resting in front of the keypad. The operator should then choose a light stimulus and set the selective key on the corresponding number. (*Note: Because the switches click each time they are changed, the operator needs to be devious and move the switches several times before setting the appropriate stimulus color and selective key to prevent the subject from determining the correct color by counting switch clicks.*) When the subject is relaxed and attentive, flip the master key switch and then note the reaction time. Record the reaction time below. Repeat the procedure three more times, randomly changing the stimulus color. Record the different reaction times below, and then take an average of the four values, recording the average in the space provided.

Trial 1 _____

Trial 2 _____

Trial 3 _____

Trial 4 _____

Average time for variable light stimulus

_____

6. Determine which average stimulus time was greater, and then formulate a hypothesis that would account for the difference in reaction times observed. _____

_____

_____

_____

B. BRAIN

1. Parts

The brain may be divided into four principal parts: (1) *brain stem*, which consists of the medulla oblongata, pons, and midbrain; (2) *diencephalon* (dī-en-SEF-a-lon; *enkephalos* = brain), which consists primarily of the thalamus and hypothalamus; (3) *cerebellum*, (ser'-e-BEL-um) which is posterior to the brain stem; and (4) *cerebrum*, (se-RĒ-brum) which is superior to the brain stem and comprises about seven-eighths of the total weight of the brain.

Examine a model and preserved specimen of the brain and identify the parts just described. Then refer to Figure 13.10 and label the parts of the brain.

2. Meninges

As in the spinal cord, the brain is protected by *meninges*. The cranial meninges are continuous with the spinal meninges. The cranial meninges are the superficial *dura mater*, the middle *arachnoid*, and the deep *pia mater*. The cranial dura mater consists of two layers called the *periosteal layer* (adheres to the cranial bones and serves as a periosteum) and the *meningeal layer* (thinner, inner layer that corresponds to the spinal dura mater).

Refer to Figure 13.11 on page 275 and label all of the meninges.

3. Cerebrospinal Fluid (CSF)

The central nervous system is nourished and protected by *cerebrospinal fluid (CSF)*. The fluid circulates through the subarachnoid space around the brain and spinal cord and through the *ventricles* (VEN-tri-kuls; *ventriculus* = little belly or cavity) of the brain. The ventricles are cavities in the brain that communicate with each other, with the central canal of the spinal cord, and with the subarachnoid space.

Cerebrospinal fluid is formed primarily by filtration and secretion from networks of capillaries and

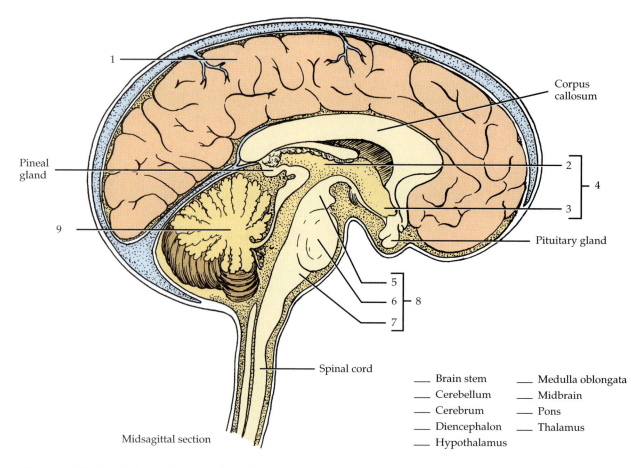

1

Pineal gland

9

Corpus callosum

2

4

3

Pituitary gland

5
6   8
7

Spinal cord

Midsagittal section

___ Brain stem   ___ Medulla oblongata
___ Cerebellum   ___ Midbrain
___ Cerebrum   ___ Pons
___ Diencephalon   ___ Thalamus
___ Hypothalamus

FIGURE 13.10   Principal parts of the brain.

ependymal cells in the ventricles, called *choroid* (KŌ-royd; *chorion* = membrane) *plexuses* (see Figure 13.11). Each of the two *lateral ventricles* is located within a hemisphere (side) of the cerebrum under the corpus callosum. The fluid formed in the choroid plexuses of the lateral ventricles circulates through an opening called the *interventricular foramen* into the third ventricle. The *third ventricle* is a slit between and inferior to the right and left halves of the thalamus and between the lateral ventricles. More fluid is added by the choroid plexus of the third ventricle. Then the fluid circulates through a canal-like structure called the *cerebral aqueduct* (AK-we-dukt) into the fourth ventricle. The *fourth ventricle* lies between the inferior brain stem and the cerebellum. More fluid is added by the choroid plexus of the fourth ventricle. The roof of the fourth ventricle has three openings: one *median aperture* (AP-er-chur) and two *lateral apertures*. The fluid circulates through the apertures into the subarachnoid space around the back of the brain and downward through the central canal of the spinal cord to the subarachnoid space around the posterior surface of the spinal cord, up the anterior surface of the

spinal cord, and around the anterior part of the brain. Most of the cerebrospinal fluid is absorbed into a vein called the superior sagittal sinus through its arachnoid villi. Normally, cerebrospinal fluid is absorbed as rapidly as it is formed.

Refer to Figure 13.11 and label the choroid plexus of the lateral ventricle, lateral ventricle, interventricular foramen, choroid plexus of third ventricle, third ventricle, cerebral aqueduct, choroid plexus of fourth ventricle, fourth ventricle, median aperture, lateral aperture, subarachnoid space of spinal cord, superior sagittal sinus, and arachnoid villus.

Note the arrows in Figure 13.11, which indicate the path taken by cerebrospinal fluid. With the aid of your textbook, starting at the choroid plexus of the lateral ventricle and ending at the superior sagittal sinus, see if you can follow the remaining path of the fluid.

Now complete Figure 13.12 on page 276.

4. Medulla Oblongata

The *medulla oblongata* (me-DULL-la ob'-long-GA-ta), or just simply *medulla*, is a continuation of the superior part of the spinal cord and forms the

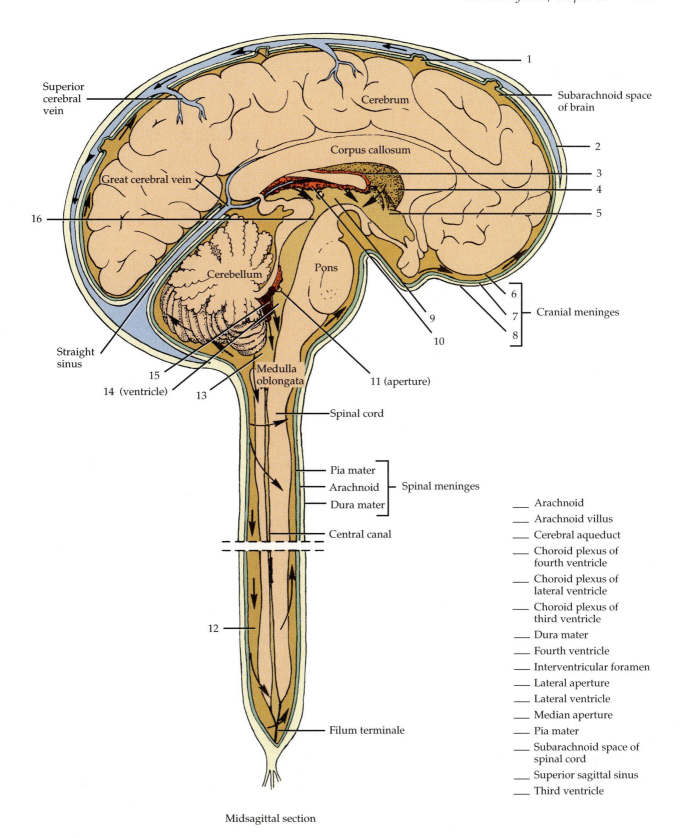

Superior cerebral vein

Cerebrum

Subarachnoid space of brain

1

Corpus callosum

2

3

Great cerebral vein

4

5

16

Cerebellum

Pons

6
7
8

Cranial meninges

9

10

11 (aperture)

Straight sinus

Medulla oblongata

15

14 (ventricle)

13

Spinal cord

Pia mater
Arachnoid
Dura mater

Spinal meninges

Central canal

12

Filum terminale

Midsagittal section

___ Arachnoid
___ Arachnoid villus
___ Cerebral aqueduct
___ Choroid plexus of fourth ventricle
___ Choroid plexus of lateral ventricle
___ Choroid plexus of third ventricle
___ Dura mater
___ Fourth ventricle
___ Interventricular foramen
___ Lateral aperture
___ Lateral ventricle
___ Median aperture
___ Pia mater
___ Subarachnoid space of spinal cord
___ Superior sagittal sinus
___ Third ventricle

FIGURE 13.11   Brain and spinal cord.

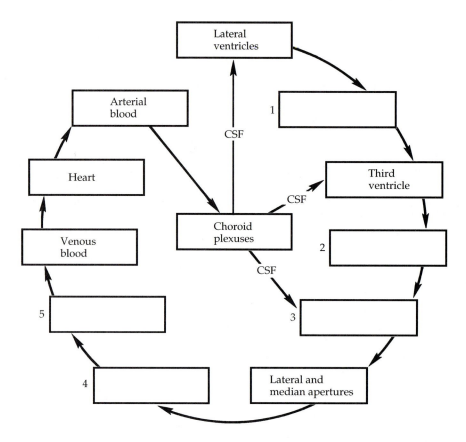

FIGURE 13.12 Formation, circulation, and absorption of cerebrospinal fluid (CSF).

inferior part of the brain stem. The medulla contains all sensory and motor tracts that communicate between the spinal cord and various parts of the brain. On the ventral side of the medulla are two roughly triangular bulges called *pyramids*. They contain the largest motor tracks that run from the cerebral cortex to the spinal cord. Most fibers in the left pyramid cross to the right side of the spinal cord and most fibers in the right pyramid cross to the left side of the spinal cord. This crossing is called the *decussation* (dē'-ku-SĀ-shun) *of pyramids* and explains why motor areas of one side of the cerebral cortex control muscular movements on the opposite side of the body. The medulla contains three vital reflex centers called the *cardiac center*, *respiratory center*, and *vasomotor center*. The dorsal side of the medulla contains two pairs of prominent nuclei: *nucleus gracilis* and *nucleus cuneatus*. Most sensory impulses initiated on one side of the body cross to the thalamus on the opposite side either in the spinal cord or in these medullary nuclei. The medulla also contains the nuclei of origin of several cranial nerves. These are the cochlear and vestibular branches of the vestibulocochlear (VIII) nerves, glossopharyngeal (IX) nerves, vagus

(X) nerves, spinal portions of the accessory (XI) nerves, and hypoglossal (XII) nerves. Examine a model or specimen of the brain and identify the parts of the medulla. Then refer to Figure 13.10 and locate the medulla.

5. Pons

The *pons* (*pons* = bridge) lies directly superior the medulla oblongata and anterior to the cerebellum. The pons contains fibers that connect parts of the cerebellum and medulla with the cerebrum. The pons contains the nuclei or origin of the following pairs of cranial nerves: trigeminal (V) nerves, abducens (VI) nerves, facial (VII) nerves, and vestibular branch of the vestibulocochlear (VIII) nerves. Other important nuclei in the pons are the *pneumotaxic* (noo-mō-TAK-sik) *area* and *apneustic* (ap-NOO-stik) *area* that help control respiration.

Identify the pons on a model or specimen of the brain. Locate the pons in Figure 13.10.

6. Midbrain

The *midbrain* extends from the pons to the inferior portion of the cerebrum. The ventral portion of the midbrain contains the paired *cerebral peduncles* (pe-DUNG-kulz; *pedunculus* = stemlike portion), which connect the upper parts of the brain to inferior parts of the brain and spinal cord. The dorsal part of the midbrain contains four rounded elevations called the *corpora quadrigemina* (KOR-por-ra kwad-ri-JEM-in-a; *corpus* = body; *quadrigeminus* = group of four). Two of the elevations, the *superior colliculi* (ko-LIK-yoo-lī) serve as reflex centers for movements of the eyeballs and the head in response to visual and other stimuli. *Colliculus,* which is singular, means small mound. The other two elevations, the *inferior colliculi,* serve as reflex centers for movements of the head and trunk in response to auditory stimuli. The midbrain contains the nuclei of origin for two pairs of cranial nerves: oculomotor (III) and trochlear (IV).

Identify the parts of the midbrain on a model or specimen of the brain. Locate the midbrain in Figure 13.10.

7. Thalamus

The *thalamus* (THAL-a-mus; *thalamos* = inner chamber) is a large oval structure that comprises about 80% of the diencephalon. It is located superior to the midbrain and consists of two masses of gray matter organized into nuclei and covered by a layer of white matter. The two masses are joined by a bridge of gray matter called the *intermediate mass.* The thalamus contains numerous nuclei that serve as relay stations for all sensory impulses. The most prominent are the *medial geniculate* (je-NIK-yoo-lāt) *nuclei* (hearing), *lateral geniculate nuclei* (vision), *ventral posterior nuclei* (general sensations and taste), *ventral lateral nuclei* (voluntary motor actions), and the *ventral anterior nuclei* (voluntary motor actions and arousal). The thalamus also allows crude appreciation of some sensations, such as pain, temperature, and pressure. Precise localization of such sensations depends on nerve impulses being relayed from the thalamus to the cerebral cortex.

Identify the thalamic nuclei on a model or specimen of the brain. Then refer to Figure 13.13 and label the nuclei.

8. Hypothalamus

The *hypothalamus* (*hypo* = under) is located inferior to the thalamus and forms the floor and part of the wall of the third ventricle. Among the functions served by the hypothalamus are the control and integration of the activities of the autonomic nervous system and parts of the endocrine system (pituitary gland) and the control of body tempera-

ture. The hypothalamus also assumes a role in feelings of rage and aggression, food intake, thirst, and the waking state and sleep patterns.

Identify the hypothalamus on a model or specimen of the brain. Locate the hypothalamus in Figure 13.10.

9. Cerebrum

The *cerebrum* is the largest portion of the brain and is supported on the brain stem. Its outer surface consists of gray matter and is called the *cerebral cortex* (*cortex* = rind or bark). Beneath the cerebral cortex is the cerebral white matter. The upfolds of the cerebral cortex are termed *gyri* (JĪ-rī) or *convolutions*, the deep downfolds are termed *fissures*, and the shallow downfolds are termed *sulci* (SUL-sī).

The *longitudinal fissure* separates the cerebrum into right and left halves called *hemispheres*. Each hemisphere is further divided into lobes by sulci or fissures. The *central sulcus* (SUL-kus) separates the *frontal lobe* from the *parietal lobe*. The *lateral cerebral sulcus* separates the frontal lobe from the *temporal lobe*. The *parietooccipital sulcus* separates the *parietal lobe* from the *occipital lobe*. Another prominent fissure, the *transverse fissure*, separates the cerebrum from the cerebellum. Another lobe of the cerebrum, the *insula*, lies deep within the lateral cerebral fissure under the parietal, frontal, and temporal lobes. It cannot be seen in external view. Two important gyri on either side of the *central sulcus* are the *precentral gyrus* and the *postcentral gyrus*. The olfactory (I) and optic (II) cranial nerves are associated with the cerebrum.

Examine a model of the brain and identify the parts of the cerebrum just described. Refer to Figure 13.14 and label the lobes.

10. Functional Areas of Cerebral Cortex

The functions of the cerebrum are numerous and complex. In a general way, the cerebral cortex can be divided into sensory, motor, and association areas. The *sensory areas* interpret sensory impulses, the *motor areas* control muscular movement, and the *associated areas* are concerned with emotional and intellectual processes.

The principal sensory and motor areas of the cerebral cortex are indicated by numbers based on K. Brodmann's map of the cerebral cortex. His map, first published in 1909, attempts to correlate structure and function. Refer to Figure 13.15 on page 279

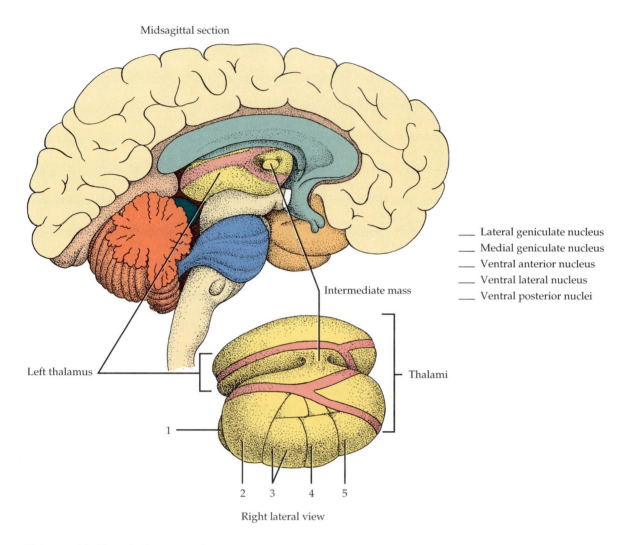

Midsagittal section

_____ Lateral geniculate nucleus
_____ Medial geniculate nucleus
_____ Ventral anterior nucleus
_____ Ventral lateral nucleus
_____ Ventral posterior nuclei

Intermediate mass

Left thalamus

Thalami

1

2   3   4   5

Right lateral view

FIGURE 13.13 Thalamic nuclei.

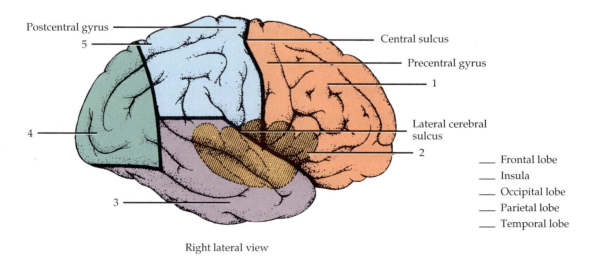

Postcentral gyrus

5

Central sulcus

Precentral gyrus

1

Lateral cerebral sulcus

2

4

3

_____ Frontal lobe
_____ Insula
_____ Occipital lobe
_____ Parietal lobe
_____ Temporal lobe

Right lateral view

FIGURE 13.14 Lobes of cerebrum.

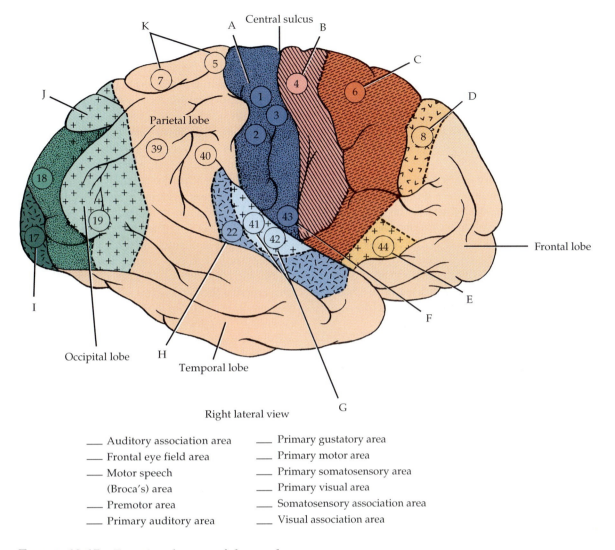

K   A   Central sulcus   B

C

D

J

Parietal lobe

Frontal lobe

E

F

Occipital lobe   H

Temporal lobe

Right lateral view   G

_____ Auditory association area   _____ Primary gustatory area

_____ Frontal eye field area   _____ Primary motor area

_____ Motor speech   _____ Primary somatosensory area

      (Broca's) area   _____ Primary visual area

_____ Premotor area   _____ Somatosensory association area

_____ Primary auditory area   _____ Visual association area

FIGURE 13.15   Functional areas of the cerebrum.

and match the name of the sensory or motor area next to the appropriate letter (A, B, C, etc.).

11. Cerebral White Matter

The white matter underlying the cerebral cortex consists of myelinated axons extending in three principal directions.

  a. *Association fibers* connect and transmit nerve impulses between gyri in the same hemisphere.

  b. *Commissural fibers* transmit impulses from the gyri in one cerebral hemisphere to the corresponding gyri in the opposite cerebral hemisphere. Three important groups of commissural fibers are the *corpus callosum, anterior commissure,* and *posterior commissure.*

  c. *Projection fibers* form motor and sensory tracts that transmit impulses from the cere-

brum and other parts of the brain to the spinal cord or from the spinal cord to the brain.

12. Basal Ganglia

Basal ganglia (GANG-lē-a) or *cerebral nuclei* are groups of gray matter in the cerebral hemispheres. The largest of the basal ganglia is the *corpus striatum* (strī-Ā-tum; *corpus* = body; *striatum* = striped), which consists of the *caudate* (*cauda* = tail) *nucleus* and the *lentiform* (*lenticala* = shaped like a lentil or lens) or *lenticular nucleus*. The lentiform nucleus, in turn, is subdivided into a lateral *putamen* (pu-TĀ-men; *putamen* = shell) and a medial *globus pallidus*; (*globus* = ball; *pallid* = pale).

  The basal ganglia are interconnected by many nerve fibers. They also receive input from and provide output to the cerebral cortex, thalamus, and hypothalamus. The caudate nucleus and the putamen control large automatic movements of skeletal muscles, such as swinging the arms while walking. The globus pallidus is concerned with the regulation of muscle tone required for specific body movements.

Examine a model or specimen of the brain and identify the basal ganglia described. Now refer to Figure 13.16 and label the basal ganglia shown.

13. Cerebellum

The *cerebellum* is inferior to the posterior portion of the cerebrum and separated from it by the transverse fissure. The central constricted area of the cerebellum is called the *vermis* (meaning worm-shaped) and the lateral portions are referred to as *cerebellar hemispheres*. The surface of the cerebellum, called the *cerebellar cortex*, consists of gray matter thrown into a series of slender parallel ridges called *folia* (*folia* = leaf). Deep to the gray matter are tracts of white matter called *arbor vitae* (meaning tree of life).

The cerebellum is attached to the brain stem by three paired bundles of fibers called *cerebellar peduncles*. The *inferior cerebellar peduncles* connect the cerebellum with the medulla oblongata at the base of the brain stem and with the spinal cord. The *middle cerebellar peduncles* connect the cerebellum with the pons. The *superior cerebellar peduncles* connect the cerebellum with the midbrain and thalamus.

The cerebellum functions to compare the intended movement determined by motor areas in the cerebrum with what is actually happening. The cerebellum constantly receives sensory input from proprioceptors in muscles, tendons, and joints, receptors for equilibrium, and visual receptors of the eyes. If the intent of the cerebral motor areas is not being attained by skeletal muscles, the cerebellum detects the variation and sends feedback signals to the motor areas to either stimulate or inhibit the activity of skeletal muscles. This interation helps to smooth and coordinate complex sequences of skeletal muscle contractions. Besides coordinating skilled movements, the cerebellum is the main brain region that regulates posture and balance. These aspects of cerebellar function make possible all skilled motor activities, from catching a baseball to dancing.

Examine a model or specimen of the brain and locate the parts of the cerebellum. Identify the cerebellum in Figure 13.10.

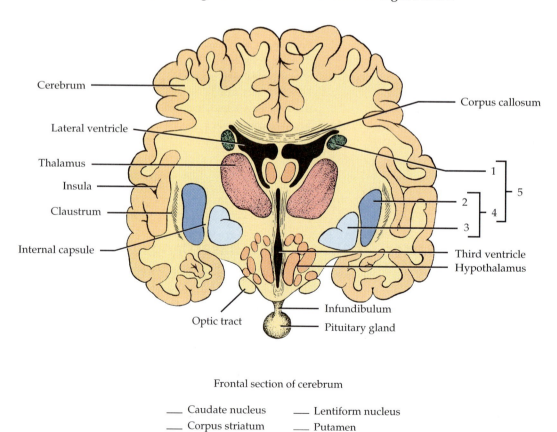

Frontal section of cerebrum

___ Caudate nucleus   ___ Lentiform nucleus
___ Corpus striatum   ___ Putamen
___ Globus pallidus

FIGURE 13.16   Basal ganglia.

14. Determination of Cerebellar Function

Working with your lab partner, perform the following tests to determine cerebellar function:

1. With your upper limbs down straight at your side, walk heel to toe for 20 ft without losing your balance.
2. Stand away from any supporting object and place your feet together and upper limbs down straight at your side; look straight ahead. Now close your eyes and stand for 2 to 3 min. Your lab partner should stand off to one side and then immediately in front of his or her partner and observe any and all body movements.
3. Stand away from any supporting object and place your feet together and upper limbs down straight at your side; look straight ahead. Now, with your eyes open raise one foot off the ground and run the heel of that foot down the shin of the other leg. Your lab partner should observe the smoothness of the movement and whether your heel remains in contact with the shin at all times.
4. Place your left palm in the supinated (palm up) position with the elbow at a 90° angle from the body. Now place the palm of your right hand into the left palm. Alternately pronate (palm down) and supinate the right hand as quickly as possible, while contacting the left palm only when the movement is completed (i.e., supinated right hand gently slaps left palm, and then pronated right hand gently slaps left palm). Do not let the medial surface of the right hand contact the left palm at any time. Note the ease with which your partner completes this activity. Then have your partner repeat the actions with the right and left hands switching roles.
5. Stand away from any supporting object and place your feet together and upper limbs abducted (away from midline) and palms facing anteriorly; look straight ahead. Now, gently close your eyes and touch the tip of your nose with your right index finger and return the right upper limb to its abducted position. Now touch the tip of your nose with your left index finger, and return the left upper limb to its abducted position. Slowly increase the rapidity of the activity.
6. With your eyes closed, touch your nose with the index finger of each hand.

C. ELECTROENCEPHALOGRAM (EEG)

Electrical activity of the brain may be recorded either directly or indirectly. With direct recordings, electrodes are placed directly on the brain surface, while indirect recordings are obtained via electrodes placed on various locations on the forehead and scalp. The pattern and level of activity observed are determined, to a great extent, by the overall excitation level of the brain. Such brain excitation levels are, in turn, largely determined by the level of activity of the *reticular activating system*. The electrical potential variations recorded are called *brain waves* (Figure 13.17), and an entire record of brain waves is termed an *electroencephalogram (EEG)*.

The intensity of brain waves varies from 0 to 300 microvolts (μV), while frequencies range from 0.25 sec to 50 sec or more. Brain waves vary depending on cerebral cortex activity levels, and will vary greatly among coma, sleep, and wakefulness. Although most EEGs are irregular and without any pattern, specific neurological abnormalities such as epilepsy or the results of a cerebral vascular accident (CVA) will present characteristic and repeatable patterns. Normal repeatable patterns also appear during certain conditions in individuals without any neurological abnormalities.

Alpha waves are rhythmic waves with a frequency of 8 to 13 sec and an intensity of 50 μV. They are usually recorded from the occipital region of the brain (or skull), but they may also be recorded from electrodes placed upon the frontal or parietal regions of the brain (or skull). Alpha waves are found in the EEG of almost all normal adults during quiet and restful relaxation, and are replaced by brain waves of lower voltage and higher frequency during directed mental activity. Alpha waves also disappear entirely during sleep.

Beta waves have a frequency between 14 and 25 sec and are most frequently seen during times of tension (termed *beta II waves*), or activation of the central nervous system (termed *beta I waves*). Beta waves are usually obtained by recording from either the parietal or frontal regions of the brain or scalp.

Theta waves are normally seen in children, and may also be seen in normal adults during emotional stress (for example, frustration and/or disappointment), or in adults with certain brain disorders. Theta waves usually exhibit a frequency of 4 to 7 sec. Such waves are usually recorded from the parietal or temporal regions of the brain or scalp.

Delta waves include all waves of the EEG with a frequency of less than 3.5 sec. Such waves occur in

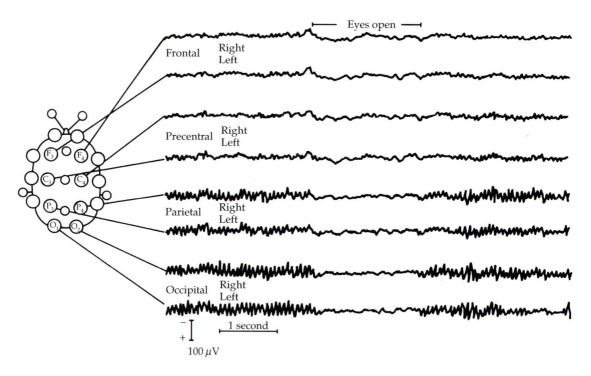

FIGURE 13.17 Diagrammatic representation of a normal EEG of a resting, awake individual. The record represents a simultaneous eight-channel unit recording from the external surface of the skull with a ground electrode placed behind the ear on the temporal lobe. (From Schmidt, R.F., editor: *Fundamentals of Neurophysiology*, ed. 2, New York, 1978, Springer-Verlag.)

infancy, in sleep (irregardless of age), and in serious organic brain disorders. It is believed that delta waves originate in the cerebral cortex independent of lower brain center influence as a result of experimental data demonstrating that delta waves may be seen in preparations in which the cerebral cortex has been functionally separated from lower brain centers.

1. Use of the Oscilloscope

A *cathode-ray oscilloscope* is an electrical instrument that traces a graph of electrical potential difference on a screen in a left-to-right direction. An electron beam is produced by a cathode tube and is aimed at a phosphorescent screen. Electrons emitted from the cathode pass between two pairs of plates en route to the screen (Figure 13.18). Because *bioelectrical potentials*, that is, potentials associated with the activity of living tissues, usually range between a few microvolts (μV) to a few millivolts (mV), some electronic device is necessary to amplify the potentials. The amplified potentials are directed to the **Y** *plates* of the cathode tube, making one plate positive and the other negative, thereby deflecting the electron beam prior to its reaching the screen. The amount of vertical deflection of the spot produced on the screen by the electron beam is determined by

the magnitude of the potential difference across the Y plates and the sensitivity of the instrument.

In order to record an event in time, a *sweep circuit* is present within the instrument, which is able to sweep the electron beam from left to right by applying current to the **X** plates. *Therefore the image produced on the screen of an oscilloscope is actually a plot of voltage (y axis) versus time (x axis).*

Your instructor will give you an orientation to the procedures to be followed for the particular brand of oscilloscope to be utilized in your laboratory. ***Do not turn the instrument on or utilize it in any way prior to this orientation lecture.*** Following the orientation lecture, you should feel comfortable regulating all of the components of the oscilloscope, including the use of the *focus, intensity, scale illumination, sensitivity,* and *position* functions.

2. Preparation of Subject

PROCEDURE

1. For the purpose of recording an EEG in this experiment, place one electrode on the frontal region of the subject's skull immediately below the hair line and place another at the hair line on the temporal region of the skull posterior to the ear.

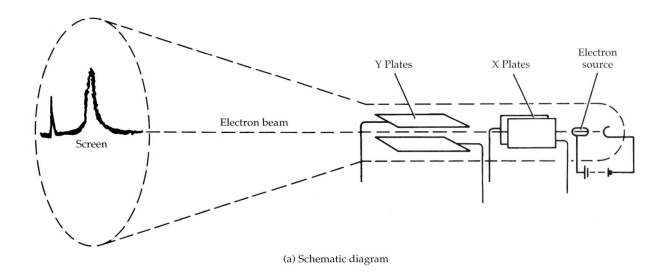

(a) Schematic diagram

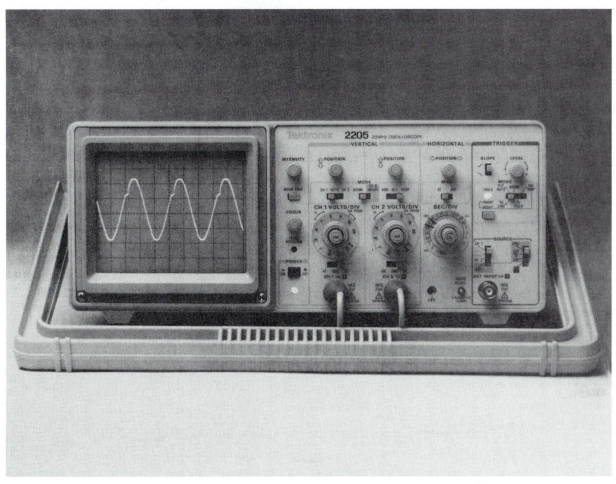

(b) Photograph

FIGURE 13.18   Oscilloscope.

2. Brain waves will be detected from the frontal electrode, while the temporal electrode will serve as the ground.

3. With a dishwashing pad *gently* rub the areas where the electrodes will be placed. This will remove dead epithelial cells that will hamper recording.

CAUTION! *Be sure that the capillaries are not damaged and that bleeding does not occur.*

4. Place electrode jelly on each electrode and fasten into place with a stretchable Band-Aid.

5. Connect the electrode selector box to the "A" and "ground" input jacks of the oscilloscope. Since the EEG activity of each subject will vary, adjust the sweep and sensitivity to a mid-point value and observe the results obtained in the following procedures. If necessary, slowly increase the sensitivity until definitive brain waves are obtained with the following procedures.

3. EEG Recording

a. CONTROL BRAIN WAVE PATTERN

PROCEDURE

1. Have the subject lie down on a cot or laboratory table. The leads should not be binding under the head or shoulders, and the subject must lie perfectly still.

2. Alpha waves will be difficult to demonstrate in a laboratory setting, as common disturbances may disrupt the subject and prevent the recording of alpha waves.

3. With the classroom dimly lit and all observers being absolutely still, ask your subject to close his/her eyes and relax completely.

4. Following five or more minutes of such conditions a control brain wave pattern will be established that may be composed of alpha waves.

5. Record your observations in Section C.1 of the LABORATORY REPORT RESULTS at the end of the exercise.

b. EFFECT OF HYPERVENTILATION

PROCEDURE

1. After the subject has relaxed completely for an adequate amount of time (usually 10–15 min), have the person *slowly* inhale deeply and *slowly but completely* exhale (hyperventilate). (*Note:* It is important that the subject keeps all skeletal muscle activity other than that required for hyperventilation to a minimum!)

2. Have the subject continue to hyperventilate until there is a noticeable change in brain wave pattern.

CAUTION! *The subject should stop hyperventilating at the first sign of dizziness or lightheadedness.*

3. Once a brain wave pattern different from that recorded under control conditions has been observed allow the subject to breathe normally and relax completely until the brain wave pattern again returns to a pattern similar to that seen under control conditions.

4. Record your observations in Section C.2 of the LABORATORY REPORT RESULTS at the end of the exercise.

c. ALPHA BLOCK

PROCEDURE

1. After the subject has relaxed completely following hyperventilation, have the subject open his or her eyes and allow the subject to adjust to the lighting in the room.

2. Quickly increase the lighting in the room and note any changes in the EEG recording. (*Note: Prior to increasing the lighting in the room you will need to increase the intensity of the oscilloscope beam as well as the scale illumination in order to read the EEG easily. If it is impossible or difficult to increase the overall lighting in the room quickly enough to obtain a noticeable change in the EEG recording, you may have to allow the subject to once again adapt to a dimly lit room for 10 min and then quickly shine a bright light into the subject's eyes in order to observe a noticeable change in the EEG recording.*)

3. Following 2 to 3 min of observing the EEG have the subject *mentally* complete a long division problem and observe any changes in the EEG during this process.

4. Record your observations in Section C.3 of the LABORATORY REPORT RESULTS at the end of the exercise.

D. CRANIAL NERVES: NAMES AND COMPONENTS

Of the 12 pairs of *cranial nerves*, 10 originate from the brain stem, but all pass through foramina in the base of the skull. The cranial nerves are designated by Roman numerals and names. The Roman numerals indicate the order in which the nerves arise from the brain, from front to back. The names indicate the distribution or function of the nerves.

As with spinal nerves, cranial nerves may contain different functional components. Those functional components related to visceral and somatic structures are prefixed by the word "general." These four components are the same as those found in spinal nerves and are:

a. *General somatic afferent* These fibers are related to receptors for pain, temperature, and mechanical stimuli. Receptors attached to these fibers transmit information from somatic structures such as skin, muscles, and joints.
b. *General visceral efferent* This type of spinal nerve contains preganglionic autonomic nervous system neurons.
c. *General visceral afferent* These fibers are related to receptors in visceral structures, such as walls of the digestive tract.
d. *General somatic efferent* These neurons innervated skeletal muscle.

All four of the spinal functional components are found in various cranial nerves, where they serve the same functions for the head. Three additional components are also found among the cranial nerves. These new components are prefixed by the word "special," and include:

a. *Special somatic afferent* This function component is related to the special senses of sight, hearing, and equilibrium.
b. *Special visceral afferent* This functional component is related to the special senses of smell and taste.
c. *Special visceral efferent* These fibers innervate branchiometric structures (structures that develop from the branchial arches). They include the muscles of the larynx, pharynx, and face.

No cranial nerve contains all seven functional components. In actuality, there are only three types of cranial nerves based upon functional composition: Cranial nerves III, IV, VI, and XII contain general somatic efferent fibers and little else. Cranial nerves I, II, and VIII contain special sensory fibers and nothing else. Finally, the remaining cranial nerves (V, VII, IX, X and XI) are somewhat more complex and tend to contain several components. In addition, all of the cranial nerves in this last group innervate branchial arch musculature.

Obtain a model of the brain and using your text and any other aids available identify the 12 pairs of cranial nerves. Now refer to Figure 13.19 and label the cranial nerves.

E. TESTS OF CRANIAL NERVE FUNCTION

The following simple tests may be performed to determine cranial nerve function. Although they provide only superficial information, they will help you to understand how the various cranial nerves function. Perform each of the tests with your partner.

1. *Olfactory (I) nerve* Have your partner smell several familiar substances such as spices, first using one nostril and then the other. Your partner should be able to distinguish the odors equally with each nostril.
   What structures could be malfunctioning if the

   ability to smell is lost? _____

   _____

   _____

   _____

2. *Optic (II) nerve* Have your partner read a portion of a printed page using each eye. Do the same while using a Snellen chart at a distance of 6.10 m (20 ft).
   Describe the visual pathway from the optic

   nerve to the cerebral cortex. _____

   _____

   _____

   _____

3. *Oculomotor (III), trochlear* (TROK-lē-ar) *(IV),* and *abducens* (ab-DOO-sens) *(VI) nerves* To test the motor responses of these nerves, have your partner follow your finger with his or her eyes without moving the head. Move your finger up, down, medially, and laterally.
   Which nerves control which movements of the

   eyeball? _____

   _____

   Look for signs of ptosis (drooping of one or both eyelids).
   Which cranial nerve innervates the upper eye

   lid? _____

CAUTION! *Do not make contact with the eyes.*

To test for pupillary light reflex, shine a small flashlight into each eye from the side. Observe the pupil.

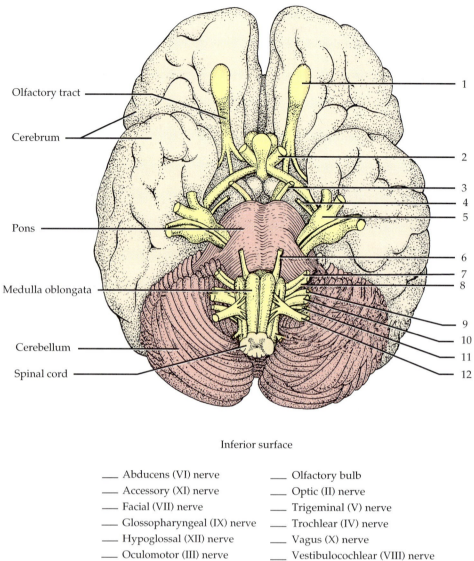

Olfactory tract

Cerebrum

Pons

Medulla oblongata

Cerebellum

Spinal cord

1
2
3
4
5
6
7
8
9
10
11
12

Inferior surface

_____ Abducens (VI) nerve   _____ Olfactory bulb
_____ Accessory (XI) nerve   _____ Optic (II) nerve
_____ Facial (VII) nerve   _____ Trigeminal (V) nerve
_____ Glossopharyngeal (IX) nerve _____ Trochlear (IV) nerve
_____ Hypoglossal (XII) nerve   _____ Vagus (X) nerve
_____ Oculomotor (III) nerve   _____ Vestibulocochlear (VIII) nerve

FIGURE 13.19   Cranial nerves of human brain.

What cranial nerve controls this reflex? _____

4. *Trigeminal* (trī-JEM-i-nal) *(V) nerve* To test the motor responses of this nerve, have your partner close his or her jaws tightly.

What muscles are used? _____

Now while holding your hand under your partner's lower jaw to provide resistance, ask your partner to open his or her mouth. To test the sensory responses of this nerve, have your partner close his or her eyes and lightly whisk a piece of dry cotton over the mandibular, maxillary, and ophthalmic areas on each side of the face. Do the same with cotton that has been moistened with cold water.
What cranial nerves bring about this response?

_____

5. *Facial (VII) nerve* To test the motor responses of this nerve, ask your partner to bring the corners of the mouth straight back, smile while showing his or her teeth, whistle, puff his or her cheeks, frown, raise his or her eyebrows, and wrinkle his or her forehead.

Define Bell's palsy. _____

_____

To test the sensory responses of this nerve, touch the tip of your partner's tongue with an applicator that has been dipped into a salt solution. Rinse the mouth out with water and now place an applicator dipped into a sugar solution along the anterior surface of the tongue. What are the four basic tastes that can be distinguished? _____

_____

6. *Vestibulocochlear* (ves-tib'-yoo-lō-KŌK-lē-ar) *(VIII) nerve*
   a. To test for the functioning of the vestibular portion of this nerve, have your partner sit on a swivel stool with his or her head slightly bent forward toward the chest. Now have a student (the operator) turn your partner *slowly and very carefully* to the right about 10 times (about one turn every 2 sec). Stop the stool suddenly and look for nystagmus (nis-TAG-mus), which is rapid movement or quivering of the eyeballs. Note the direction of the nystagmus.

   CAUTION! *Do not allow the subject to stand or walk until dizziness is completely gone.*

   b. When the nystagmus movements have ceased have the subject sit on the chair and, with his or her eyes open, reach forward and touch the operator's index finger with his or her finger. Now spin the subject to the right about 10 times (as before) and then suddenly stop the stool. *Immediately* after stopping the stool have the subject again, with his or her eyes open, try to reach forward and touch the operator's index finger with his or her finger. Note the degree of ease or difficulty with which this is accomplished.
   c. Again, when the nystagmus movements have ceased, repeat procedure b. This time, however, *immediately* after stopping the chair have the operator extend an index finger and have the subject open his or her eyes, locate the finger visually, and then *close his or her eyes and try to touch the operator's index finger with his or her eyes closed.* Note the degree of ease or difficulty with which this is accomplished, as well as the direction (right or left) in which the subject missed the operator's finger.
   d. Now repeat procedure a with the subject resting his or her head on their right shoulder. Note any changes that might occur with the nystagmus response. _____

_____

Formulate a hypothesis as to the differences in results obtained in procedures a through d.

_____

_____

_____

To test the functioning of the cochlear portion of the nerve, have your subject sit on a chair with his or her eyes closed with six members of the class standing in a wide circle around the individual. Now, one at a time, and in a random order, have the students click two coins together and have the subject point in the direction from which the sound originated. Then, with the students still standing in a circle of equal radius around the chair, and with the subject's eyes still closed, hand a ticking watch to one of the individuals in the circle and have him or her *slowly* move the watch closer to the subject until the subject first hears the sound and can correctly indicate from which direction the sound is originating. Repeat this procedure until the distance at which the sound is first heard and the direction correctly identified has been determined for all six directions. Measure the distances between the subject and each encircling student. Are all six distances equal? _____

_____

7. *Glossopharyngeal* (glos-ō-fa-RIN-jē-al) *(IX)* and *Vagus (X) nerves* The palatal (gag) reflex can be used to test the functioning of both of these nerves. Using a cotton-tipped applicator, *very slowly and gently* touch your partner's uvula.
   Develop a hypothesis concerning the purpose of this reflex. _____

_____

To test the sensory responses of these nerves, have your partner swallow. Does swallowing occur easily? Now *gently* hold your partner's tongue down with a tongue depressor and ask him or her to say "ah." Does the uvula move? Are the movements on both sides of the soft palate the same? The sensory function of the glossopharyngeal nerve may be tested by *lightly* applying a cotton-tipped applicator dipped in quinine to the top, sides, and back of the tongue.

In which area of the tongue was the quinine

tasted? _____

What taste sensation is located there? _____

_____

8. *Accessory (XI) nerve* The strength and muscle tone of the sternocleidomastoid and trapezius muscles indicate the proper functioning of the accessory nerve. To ascertain the strength of the sternocleidomastoid muscle, have your partner turn his or her head from side to side against *slight* resistance that you supply by placing your hands on either side of your partner's head. To ascertain the strength of the trapezius muscle, place your hands on your partner's shoulders and while *gently* pressing down firmly ask him or her to shrug his or her shoulders.

Do both muscles appear to be reasonably

strong? _____

9. *Hypoglossal (XII) nerve* Have your partner protrude his or her tongue. It should protrude without deviation. Now have your partner protrude his or her tongue and move it from side to side while you attempt to *gently* resist the movements with a tongue depressor.

F. DISSECTION OF NERVOUS SYSTEM

CAUTION! *Please reread Section D, "Precautions Related to Dissection" at the beginning of the laboratory manual on page xiii before you begin your dissection.*

The following instructions for dissection of the mammalian brain can be used in cats, fetal pigs, and sheep, because their brains are structurally similar. One main difference is the surface (gyri and sulci) of the cerebrum. Another difference is the proportions of certain parts. For example, the human brain is basically the same as those of sheep, pig, and cat, but the cerebrum, the center of intelligence, is disproportionately larger.

Figures 13.20 through 13.23 are based on the sheep brain, but they can be used as references if you are to dissect the cat brain instead. Because

the only major dissections needed of the brain are a cross-sectional cut through the cerebrum and a midsagittal cut, brains can be returned to their preservative when you finish with them and used again. The instructor should provide brains that show cranial nerves, spinal cord, and meninges. Even if the cat brain is not to be dissected, the instructor might want to demonstrate a newly dissected one to the class.

PROCEDURE

1. If you are to dissect the brain from your cat, start by removing all of the muscles from the dorsal and lateral surfaces of the head.
2. *Very carefully*, using either bone shears or a bone saw, make an opening in the parietal bone. Do not damage the brain below.
3. Slowly chip away all the bone from the dorsal and lateral surfaces of the skull until the brain is exposed.
4. Between the cerebrum and the cerebellum is a transverse bony partition, the tentorium cerebelli, which should be removed carefully.
5. Cut the spinal cord transversely at the foramen magnum.
6. Gently lift the brain out of the floor of the cranium. As you do so, sever each cranial nerve as far from the brain as possible.

Study the following sections and dissect either the cat brain (Section E.2) or sheep brain (Section E.6) as your instructor directs.

1. Meninges

On the brain that you are to use, locate the three layers of the meninges that cover the brain and the spinal cord: outer *dura mater*, middle *arachnoid*, and inner *pia mater*. Refer to pages 259 and 275 as you examine them. Note that the dura mater has two folds, the *falx cerebri* located between the cerebral hemispheres, and the *tentorium cerebelli* between the cerebral hemispheres and the cerebellum (the latter is ossified and fused to the parietal bones in the cat).

2. Dissection of Cat Brain

PROCEDURE

1. First identify on the dorsal surface of the brain the two large, cranial *cerebral hemispheres* (Figure 13.20). Together they constitute the *cerebrum*. The cerebral hemispheres are sepa-

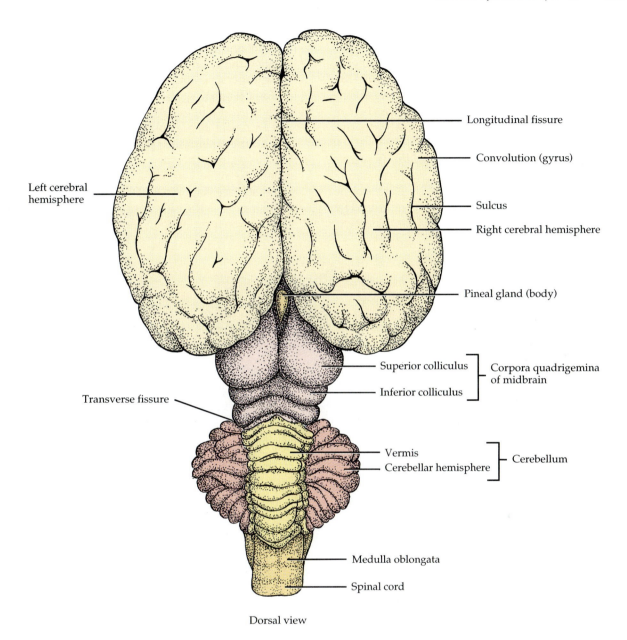

Left cerebral hemisphere

Longitudinal fissure

Convolution (gyrus)

Sulcus

Right cerebral hemisphere

Pineal gland (body)

Superior colliculus
Inferior colliculus
Corpora quadrigemina of midbrain

Transverse fissure

Vermis
Cerebellar hemisphere
Cerebellum

Medulla oblongata

Spinal cord

Dorsal view

FIGURE 13.20   Sheep brain in which cerebellum has been spread apart from the cerebrum.

rated from each other by the deep *longitudinal fissure.* The surface of each hemisphere consists of upward folds called *convolutions (gyri)* and grooves called *sulci.*

2. Spread the hemispheres and look into the longitudinal fissure. Note the *corpus callosum*, a band of transverse fibers that connects the two hemispheres internally.

3. Make a frontal section through a cerebral hemisphere. Note the *gray matter* near the surface of the cerebral cortex and the *white matter* beneath the gray matter.

4. Now locate the *cerebellum* caudal to the cerebral hemisphere and dorsal to the medulla oblongata. It is separated from the cerebral hemispheres by a deep *transverse fissure*. The cerebellum consists of a median *vermis* and two lateral *cerebellar hemispheres*. As does the cerebrum, the cerebellum has folia (convolutions), fissures, and sulci. The treelike arrangement of white matter tracts in the gray matter in the cerebellum is the *arbor vitae* ("tree of life").

5. Section a portion of the cerebellum longitudinally to examine the arbor vitae.

6. The cerebellum is connected to the brain stem by three prominent fiber tracts called *peduncles*. The *superior cerebellar peduncle* connects the cerebellum with the midbrain, the *inferior cerebellar peduncle* connects the cerebellum with the medulla, and the *middle cerebellar peduncle* connects the cerebellum with the pons.

7. The dorsal portion (roof) of the *midbrain* can be examined if you spread the cerebral hemispheres apart from the cerebellum. The root of the midbrain is formed by four rounded bulges, the *corpora quadrigemina*. The two larger cranial bulges are called the *superior colliculi* and the two smaller caudal bulges are called the *inferior colliculi*. If you look between the superior colliculi you should see the *pineal gland (body)*.

8. The *medulla oblongata*, or just simply medulla, is the most caudal portion of the brain and is continuous with the spinal cord.

9. To examine other parts of the brain, you will have to make your observations from its ventral surface (Figure 13.21). At the cranial end of the brain are the *olfactory bulbs.* The olfactory (I) nerves terminate on the bulbs after passing

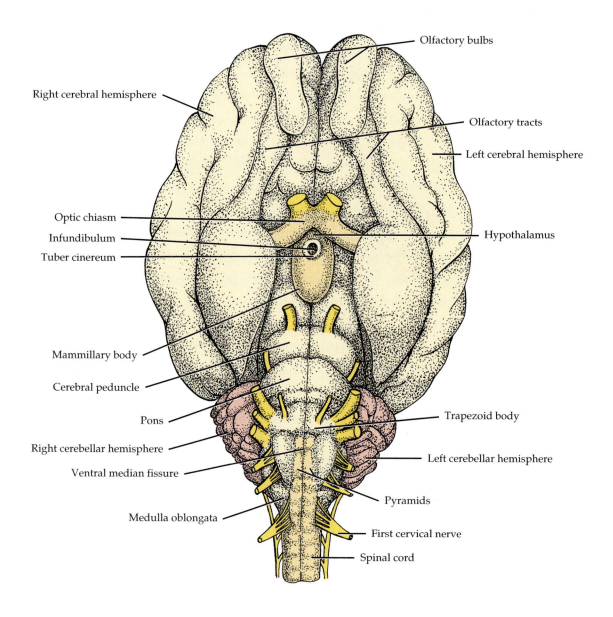

Ventral view

FIGURE 13.21 Sheep brain.

through the cribriform plate of the ethmoid bone. Bands of nerve fibers extending caudally and ventrally from the olfactory bulbs are the *olfactory tracts.* The tracts terminate in the cranial part of the cerebrum.

10. Caudal to the olfactory tracts is the *hypothalamus*. At the cranial border of the hypothalamus, the optic (II) nerves come together and partially cross. This point of crossing is called the *optic chiasm.* The rest of the hypothalamus is caudal to the optic chiasma, covered mostly by the *pituitary gland (hypophysis)*. The pituitary gland is attached to a small round elevation on the hypothalamus called the *tuber cinereum* by a narrow stalk called the *infundibulum*.

11. Now examine the ventral portion (floor) of the midbrain called the *cerebral peduncles*. These bundles of white fibers connect the pons to the cerebrum. They run obliquely and cranially on either side of the hypophysis.

12. *Caudal* to the cerebral peduncles is the *pons*, which consists of transverse fibers that connect the cerebellar hemispheres.

13. Caudal to the pons is the *medulla oblongata.* Identify the *ventral median fissure*, a midline groove. The longitudinal bundles of fibers on either side of the ventral medial fissure are the *pyramids.*

14. Lateral to the cranial end of the pyramids are the *trapezoid bodies*. These are narrow bands of transverse fibers that cross the cranial end of the medulla.

15. The remaining parts of the brain may be located by cutting the brain longitudinally into halves. Cut through the midsagittal plane. A long sharp knife may be used here.

16. Using Figure 13.22 as a guide, identify the *corpus callosum*, a transverse band of white fibers at the bottom of the cerebral hemispheres. The cranial portion of the corpus callosum is called the *genu*, the middle part is known as the *body* or *trunk*, and the caudal portion is called the *splenium*. The ventral deflection of the genu is known as the *rostrum.* Just beneath the splenium identify the *pineal gland (body)*.

17. Ventral to the corpus callosum is the *fornix*, a band of white fibers connecting the cerebral hemispheres. It is cranial to the third ventricle.

18. Between the fornix and corpus callosum is a thin vertical membrane, the *septum pellucidum*, which separates the two lateral ventricles of the brain. If you break the membrane, you can expose the *lateral ventricle*, a cavity in each cerebral hemisphere that contains cerebrospinal fluid in the living animal.

19. Identify the *anterior commissure*, a small group of fibers ventral to the fornix. Extending

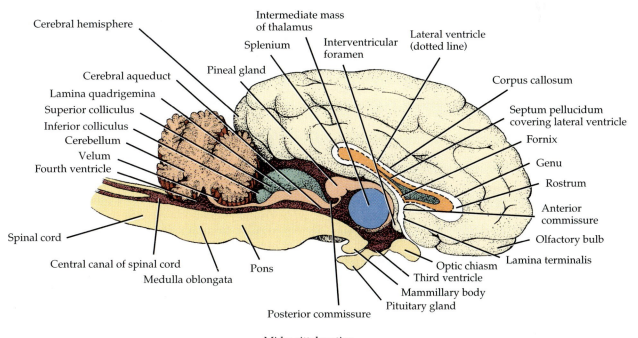

Midsagittal section

FIGURE 13.22 Sheep brain.

from the anterior commissure to the optic chiasma is the *lamina terminalis.* It forms the cranial wall of the third ventricle.

20. The *thalamus* forms the lateral walls of the *third ventricle*, another cavity containing cerebrospinal fluid. The thalamus is caudal to the fornix and extends ventrally to the infundibulum and caudally to the corpora quadrigemina. The third ventricle lies mostly in the thalamus and hypothalamus, and, in the living animal, is filled with cerebrospinal fluid. The lateral ventricles communicate with the third ventricle by an opening called the *interventricular foramen*.

21. The *intermediate mass* is a circular area that extends across the third ventricle, connecting the two sides of the thalamus. The interventricular foramen lies in the depression cranial to the intermediate mass. The intermediate mass is absent in about 30% of human brains and, when present, is proportionately much smaller.

22. Identify the *posterior commissure*, a small round mass of fibers that crosses the midline. It is located just ventral to the pineal gland (body).

23. Now find the *cerebral aqueduct*, a narrow longitudinal channel that passes through the midbrain and connects the third and fourth ventricles. The roof of the cerebral aqueduct is called the *lamina quadrigemina.*

24. The *fourth ventricle* is a cavity superior to the pons and medulla and inferior to the cerebellum. It is continuous with the cerebral aqueduct and contains cerebrospinal fluid in a living animal. A membrane that forms the roof of the fourth ventricle is called the *velum.* The fourth ventricle leads into the *central canal* of the spinal cord.

3. Cranial Nerves

Briefly review Section D, Cranial Nerves (page 284). Using Figure 13.23 as a guide, identify the cranial nerves.

a. *Olfactory (I) nerve* Consists of fibers that pass from the olfactory bulb, through the foramina of the cribriform plate, to the mucous membrane of the nasal cavity. It functions in smell.

b. *Optic (II) nerve* Arises from the optic chiasma and passes through the optic foramen to the retina of the eye. It functions in vision.

c. *Oculomotor (III) nerve* Arises from the ventral surface of the cerebral peduncle, passes through the orbital fissure, and supplies the superior rectus, medial rectus, inferior rectus, inferior oblique, and levator palpebrae eye muscles.

d. *Trochlear (IV) nerve* Arises just caudal to the corpora quadrigemina, curves around the lateral aspect of the cerebral peduncle, passes through the orbital fissure, and supplies the superior oblique muscle of the eyeball.

e. *Trigeminal (V) nerve* Originates from the lateral surface of the pons by a large sensory root and a small motor root. It exits through the orbital fissure and foramen ovale. The three branches of the trigeminal nerve are the *ophthalmic, maxillary,* and *mandibular*. All three branches supply the skin of the head and face, the epithelium of the oral cavity, the anterior two-thirds of the tongue, and the teeth. The mandibular branch also supplies the muscles of mastication.

f. *Abducens (VI) nerve* Arises from between the trapezoid body and pyramid, passes through the orbital fissure, and supplies the lateral rectus muscle of the eyeball.

g. *Facial (VII) nerve* Arises caudal to the trigeminal nerve between the pons and trapezoid body, passes through the middle ear, and emerges at the stylomastoid foramen. It supplies many head and facial muscles and the muscosa of the tongue.

h. *Vestibulocochlear (VIII) nerve* Arises just caudal to the facial nerve, passes into the internal auditory meatus, and divides into two branches. One branch transmits auditory sensations from the cochlea. The other branch transmits sensations for equilibrium from the semicircular canals.

i. *Glossopharyngeal (IX) nerve* Arises just caudal to the vestibulocochlear nerve, emerges from the jugular foramen, and supplies the muscles of the pharynx and the taste buds.

j. *Vagus (X) nerve* Arises just caudal to the glossopharyngeal nerve and exits through the jugular foramen. It supplies most of the thoracic and abdominal viscera.

k. *Accessory (XI) nerve* Arises from the medulla and spinal cord, passes through the jugular foramen, and supplies the neck and shoulder muscles.

l. *Hypoglossal (XII) nerve* Arises from the ventral medulla, passes through the hypoglossal foramen, and supplies tongue muscles.

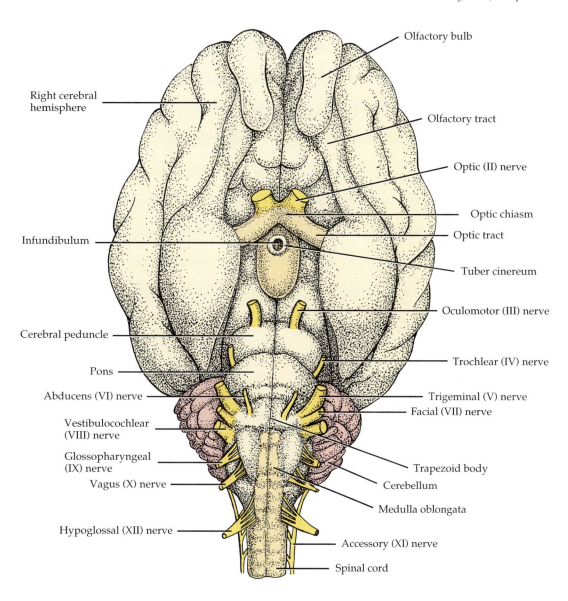

Ventral view

FIGURE 13.23   Cranial nerves of a sheep brain.

4. Dissection of Cat Spinal Cord

PROCEDURE

1. Remove the muscles dorsal to the vertebral column.
2. Using small bone clippers, cut through the laminae of the vertebrae and remove the neural arches of all the vertebrae starting at the foramen magnum.
3. Once the entire spinal cord is exposed, identify the following (Figure 13.24a):
    a. *Cervical enlargement* Between the fourth cervical and first thoracic vertebrae.
    b. *Lumbar enlargement* Between the third and seventh lumbar vertebrae.
    c. *Conus medullaris* Tapered conical tip of the spinal cord caudal to the lumbar enlargement.
    d. *Filum terminale* A slender strand continuous with the conus medullaris, which extends into the caudal region. The filum and caudal spinal nerves are referred to as the *cauda equina.*
    e. *Dorsal median sulcus* Shallow groove on the dorsal surface of the spinal cord along the midline.

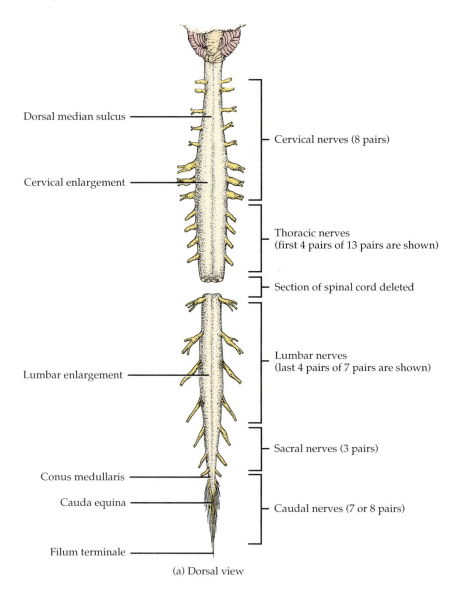

Dorsal median sulcus

Cervical enlargement

Lumbar enlargement

Conus medullaris

Cauda equina

Filum terminale

Cervical nerves (8 pairs)

Thoracic nerves
(first 4 pairs of 13 pairs are shown)

Section of spinal cord deleted

Lumbar nerves
(last 4 pairs of 7 pairs are shown)

Sacral nerves (3 pairs)

Caudal nerves (7 or 8 pairs)

(a) Dorsal view

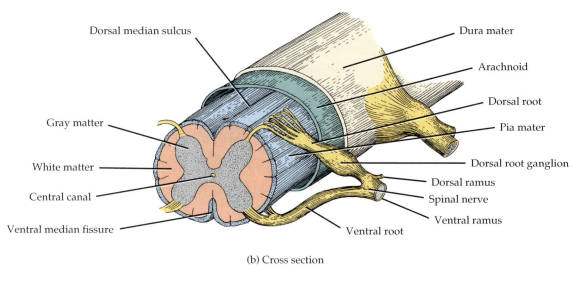

Dorsal median sulcus

Gray matter

White matter

Central canal

Ventral median fissure

Dura mater

Arachnoid

Dorsal root

Pia mater

Dorsal root ganglion

Dorsal ramus

Spinal nerve

Ventral ramus

Ventral root

(b) Cross section

Figure 13.24 The spinal cord.

f. *Ventral median fissure* Deeper groove on the ventral surface of the spinal cord along the midline (Figure 13.24b).

g. *Dorsolateral sulcus* Groove parallel and lateral to the dorsal median sulcus. The dorsal roots of spinal nerves attach to this sulcus (see Figure 13.24b).

h. *Meninges* These are similar in structure and location to those of the brain (see Figure 13.24b).

4. If you examine the cut end of the spinal cord at the foramen magnum with a hand lens, you can see the gray matter arranged like a letter H in the interior and the white matter external to the gray matter. Also identify the dorsal median sulcus and the ventral median fissure.

5. Spinal Nerves of Cat

The cat has 38 or 39 pairs of *spinal nerves.* These include 8 pairs of *cervical*, 13 pairs of *thoracic*, 7 pairs of *lumbar*, and 3 pairs of *sacral*, and 7 or 8 pairs of *caudal (coccygeal)* (see Figure 13.24a).

Each spinal nerve arises by a *dorsal (sensory) root*, which originates from the dorsolateral sulcus, and a *ventral root* (Figure 13.24b). The dorsal root has an oval swelling called the *dorsal root ganglion*. The first cervical nerve exits from the vertebral canal through the atlantal foramen. The second cervical nerve exits between the arches of the atlas and axis. All other spinal nerves exit through intervertebral foramina.

Just outside the intervertebral foramina, a spinal nerve divides into a ventral ramus and a dorsal ramus. The *dorsal rami* are small and are distributed to the skin and muscles of the back. The *ventral rami* are large and supply the skin and muscles of the ventral part of the trunk.

Expose the roots of several spinal nerves and identify the parts just described.

6. Dissection of Sheep Brain

CAUTION! *Please reread Section D, "Precautions Related to Dissection" at the beginning of the laboratory manual on page xiii before you begin your dissection.*

The brains of the fetal pig, sheep, and human show many similarities. They possess the protective membranes called the *meninges*, which can easily be seen as you proceed with the dissections. The outermost layer is the *dura mater*. It is the toughest, protective one, and may be missing in the preserved sheep brain. The middle membrane is the *arachnoid*, and the inner one containing blood vessels and adhering closely to the surface of the brain itself is the *pia mater*.

PROCEDURE

1. Use Figures 13.20 through 13.22 as references for this dissection.

2. The most prominent external parts of the brain are the pair of large *cerebral hemispheres* and the posterior *cerebellum* on the dorsal surface. These large hemispheres are separated from each other by the *longitudinal fissure*. The *transverse fissure* separates the hemispheres from the cerebellum. The surfaces of these hemispheres form many *gyri*, or raised ridges, that are separated by grooves, or *sulci*.

3. If you spread the hemispheres gently apart you can see, deep in the longitudinal fissure, thick bundles of white transverse fibers. These bundles form the *corpus callosum*, which connects the hemispheres.

4. Most of the following structures can be identified by examining a midsagittal section of the sheep brain, or by cutting an intact brain along the longitudinal fissure completely through the corpus callosum.

5. If you break through the thin ventral wall, the *septum pellucidum* of the corpus callosum, you can see part of a large chamber, the *lateral ventricle*, inside the hemisphere.

6. Each hemisphere has one of these ventricles. Ventral to the septum pellucidum, locate a smaller band of white fibers called the *fornix*. Close by where the fornix disappears is a small, round bundle of fibers called the *anterior commissure*.

7. The *third ventricle* and the *thalamus* are located ventral to the fornix. The third ventricle is outlined by its shiny epithelial lining, and the thalamus forms the lateral walls of this ventricle. This ventricle is crossed by a large circular mass of tissue, the *intermediate mass*, which connects the two sides of the thalamus. Each lateral ventricle communicates with the third ventricle through an opening, the *interventricular foramen*, which lies in a depression anterior to the intermediate mass and can be located with a dull probe.

8. Spreading the cerebral hemispheres and the cerebellum apart reveals the roof of the midbrain (mesencephalon), which is seen as two pairs of round swellings collectively called the *corpora quadrigemina*. The larger, more anterior

pair are the *superior colliculi*. The smaller posterior pair are the *inferior colliculi*. The *pineal gland (body)* is seen directly between the superior colliculi. Just posterior to the inferior colliculi, appearing as a thin white strand, is the *trochlear (IV) nerve*.

9. The cerebellum is connected to the brain stem by three prominent fiber tracts called *peduncles*. The *superior cerebellar peduncle* connects the cerebellum with the midbrain, the *inferior cerebellar peduncle* connects the cerebellum with the medulla, and the *cerebellar peduncle* connects the cerebellum with the pons.

10. Most of the following parts can be located on the ventral surface of the intact brain.

11. Just beneath the cerebral hemispheres are two *olfactory bulbs*, which continue posteriorly as two *olfactory tracts*. Posterior to these tracts, the *optic (II) nerves* undergo a crossing (decussation) known as the *optic chiasma*.

12. Locate the *pituitary gland (hypophysis)* just posterior to the chiasma. This gland is connected to the *hypothalamus* portion of the diencephalon by a stalk called the *infundibulum*. The *mammillary body* appears immediately posterior to the infundibulum.

13. Just posterior to this body are the paired *cerebral peduncles*, from which arise the large *oculomotor (III) nerves*. They may be partially covered by the pituitary gland.

14. The *pons* is a posterior extension of the hypothalamus and the *medulla oblongata* is a posterior extension of the pons.

15. The *cerebral aqueduct* dorsal to the peduncles runs posteriorly and connects the third ventricle with the *fourth ventricle*, which is located dorsal to the medulla and ventral to the cerebellum.

16. The medulla merges with the *spinal cord*, and is separated by the *ventral median fissure*. The *pyramids* are the longitudinal bands of tissue on either side of this fissure.

17. Identify the remaining *cranial nerves* on the ventral surface of the brain. They are the trigeminal (V), abducens (VI), facial (VII), vestibulocochlear (VIII), glossopharyngeal (IX), vagus (X), accessory (XI), and hypoglossal (XII). The previously identified cranial nerves are the olfactory (I), optic (II), oculomotor (III), and trochlear (IV), for a total of twelve.

18. A cross section through a cerebral hemisphere reveals *gray matter* near the surface of the *cerebral cortex* and *white matter* beneath this layer.

19. A cross section through the spinal cord reveals the *central canal*, which is connected to the fourth ventricle and contains *cerebrospinal fluid*.

20. A midsagittal section through the cerebellum reveals a treelike arrangement of gray and white matter called the *arbor vitae* (tree of life).

G. AUTONOMIC NERVOUS SYSTEM

The *autonomic nervous system (ANS)* regulates the activities of smooth muscle, cardiac muscle, and certain glands, usually involuntarily. In the somatic nervous system (SNS), which is voluntary, the cell bodies of the motor neurons are in the CNS, and their axons extend all the way to skeletal muscles in spinal nerves. The ANS always has two motor neurons in the pathway. The first motor neuron, the *preganglionic neuron*, has its cell body in the CNS. Its axon leaves the CNS and synapses in an autonomic ganglion with the second neuron called the *postganglionic neuron*. The cell body of the postganglionic neuron is inside an autonomic ganglion, and its axon terminates in a *visceral effector* (muscle or gland).

Label the components of the autonomic pathway shown in Figure 13.25.

The ANS consists of two divisions: sympathetic and parasympathetic (Figure 13.26 on page 298). Most viscera are innervated by both divisions. In general, nerve impulses from one division stimulate a structure, whereas nerve impulses from the other division decrease its activity.

In the *sympathetic division*, the cell bodies of the preganglionic neurons are located in the lateral gray horns of the spinal cord in the thoracic and first two lumbar segments. The axons of preganglionic neurons are myelinated and leave the spinal cord through the ventral (anterior) root of a spinal nerve. Each axon travels briefly in a ventral ramus and then through a small branch called a *white ramus communicans* to enter a sympathetic trunk ganglion. These ganglia lie in a vertical row, on either side of the vertebral column, from the base of the skull to the coccyx. In the ganglion, the axon may synapse with a postganglionic neuron, travel upward or downward through the sympathetic trunk ganglia to synapse with postganglionic neurons at different levels, or pass through the ganglion without synapsing to form part of the

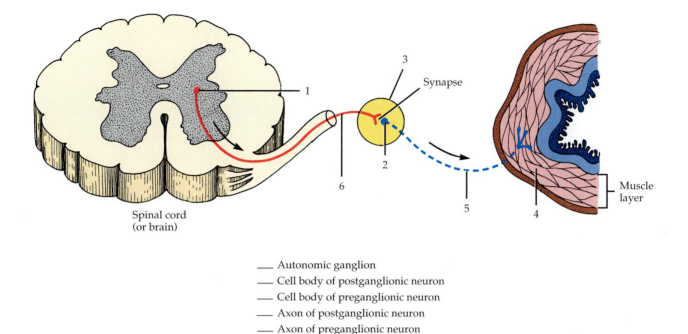

—— Autonomic ganglion

—— Cell body of postganglionic neuron

—— Cell body of preganglionic neuron

—— Axon of postganglionic neuron

—— Axon of preganglionic neuron

—— Visceral effector

FIGURE 13.25   Components of an autonomic pathway.

splanchnic nerves. If the preganglionic axon synapses in a sympathetic trunk ganglion, it reenters the ventral or dorsal ramus of a spinal nerve via a small branch called a *gray ramus communicans*. If the preganglionic axon forms part of the splanchnic nerves, it passes through the sympathetic trunk ganglion but synapses with a postganglionic neuron in a prevertebral (collateral) ganglion. These ganglia are anterior to the vertebral column close to large abdominal arteries from which their names are derived (celiac, superior mesenteric, and inferior mesenteric).

In the *parasympathetic division*, the cell bodies of the preganglionic neurons are located in nuclei in the brain stem and lateral gray horn of the second through fourth sacral segments of the spinal cord. The axons emerge as part of cranial or spinal nerves. The preganglionic axons synapse with postganglionic neurons in terminal ganglia, near or within visceral effectors.

The sympathetic division is primarily concerned with processes that expend energy. During stress, the sympathetic division sets into operation a series of reactions collectively called the *fight-or-flight response*, designed to help the body counteract the stress and return to homeostasis. During the fight-or-flight response, the heart and breathing rates increase and the blood sugar level rises, among other things.

The parasympathetic division is primarily concerned with activities that restore and conserve energy. It is thus called the *rest-repose system*. Under normal conditions, the parasympathetic division dominates the sympathetic division in order to maintain homeostasis.

Autonomic fibers, like other axons of the nervous system, release neurotransmitters at synapses as well as at points of contact with visceral effectors *(neuroeffector junctions)*. On the basis of the neurotransmitter produced, autonomic fibers may be classified as either cholinergic or adrenergic. *Cholinergic* (kō-lin-ER-jik) *fibers* release *acetylcholine (ACh)* and include the following: (1) all sympathetic and parasympathetic preganglionic axons, (2) all parasympathetic postganglionic axons, and (3) a few sympathetic postganglionic axons. *Adrenergic* (ad'-ren-ER-jik) *fibers* produce *norepinephrine (NE)* or *epinephrine (adrenalin)* in the modified postganglionic sympathetic fibers of the adrenal medulla. Most sympathetic postganglionic axons are adrenergic.

The actual effects produced by ACh are determined by the type of receptor with which it interacts. The two types of cholinergic receptors are known as nicotinic receptors and muscarinic receptors. *Nicotinic receptors* are found on both sympathetic and parasympathetic postganglionic neurons. These receptors are so named because

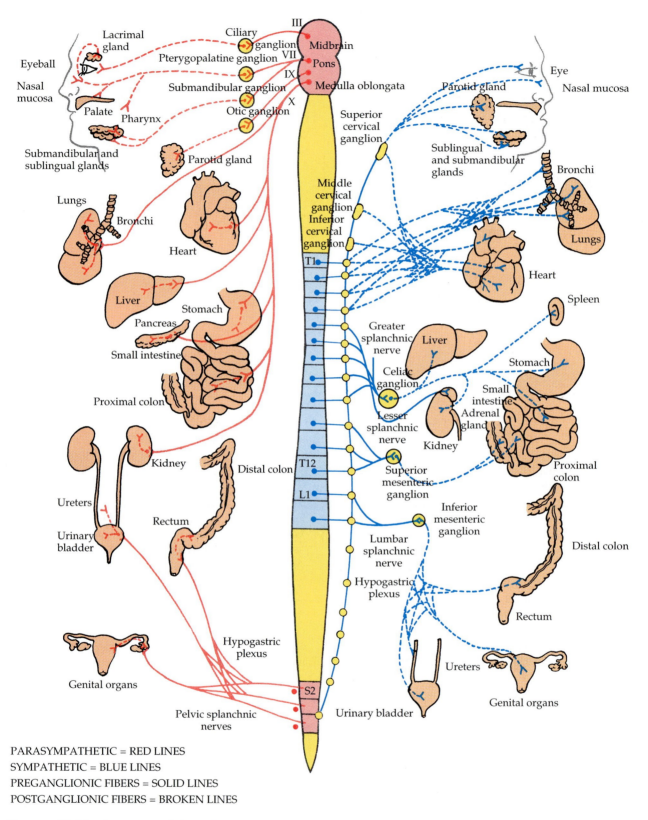

PARASYMPATHETIC = RED LINES
SYMPATHETIC = BLUE LINES
PREGANGLIONIC FIBERS = SOLID LINES
POSTGANGLIONIC FIBERS = BROKEN LINES

FIGURE 13.26   Structure of the autonomic nervous system.

the actions of ACh on them are similar to those produced by nicotine. ***Muscarinic receptors*** are found on effectors (muscles and glands) innervated by parasympathetic postganglionic axons. These receptors are so named because the actions of ACh on them are similar to those produced by muscarine, a toxin produced by a mushroom. The effects of NE and epinephrine, like those of ACh, are also determined by the type of receptor with which they interact. Such receptors are found on visceral effectors innervated by most sympathetic postganglionic axons and are referred to as ***alpha receptors*** and ***beta receptors.*** In general, alpha receptors are excitatory. Some beta receptors are excitatory, while others are inhibitory. Although cells of most effectors contain either alpha or beta receptors, some effector cells contain both. NE, in general, stimulates alpha receptors to a greater extent than beta receptors, and epinephrine, in general, stimulates both alpha and beta receptors about equally.

Using your textbook as a reference, write in the effects of sympathetic and parasympathetic stimulation for the visceral effectors listed in Table 13.1.

ANSWER THE LABORATORY REPORT QUESTIONS AT THE END OF THE EXERCISE.

TABLE 13.1
Activities of the Autonomic Nervous System

Visceral effector	Effect of sympathetic stimulation	Effect of parasympathetic stimulation
GLANDS		
Sweat		
Lacrimal (tear)		
Adrenal medulla		
Liver		
Adipose (fat) cells		
Kidney (juxtaglomerular cells)		
Pancreas		
SMOOTH MUSCLE		
Iris, radial muscle		
Iris, circular muscle		
Ciliary muscle of eye		
Salivary gland arterioles		
Gastric gland arterioles		
Intestinal gland arterioles		
Lungs, bronchial muscle		

TABLE 13.1 (*Continued*)

Visceral effector	Effect of sympathetic stimulation	Effect of parasympathetic stimulation
Heart arterioles		
Skin and mucosal arterioles		
Skeletal muscle arterioles		
Abdominal viscera arterioles		
Brain arterioles		
Systemic veins		
Gallbladder and ducts		
Stomach and intestines		
Kidney		
Ureter		
Spleen		
Urinary bladder		
Uterus		
Sex organs		
Hair follicles (arrector pili muscle)		

CARDIAC MUSCLE (HEART)

Nervous System 13

Student _____   **Date** _____

Laboratory Section _____   **Score/Grade** _____

SECTION C: ELECTROENCEPHALOGRAM (EEG)

1. Did the subject demonstrate an alpha wave pattern during the control conditions? If not, give a physiological explanation for your observations. _____

   _____

   _____

2. Explain the physiological reasons for the observed effect of hyperventilation upon the EEG.

   _____

   _____

   _____

3. Describe the effects of increased lighting and performance of a long division problem upon the EEG.

   _____

   _____

   _____

   _____

Nervous System 13

Student _____   Date _____

Laboratory Section _____   Score/Grade _____

PART 1. Multiple Choice

_____ 1. The tapered, conical portion of the spinal cord is the (a) filum terminale (b) conus medullaris (c) cauda equina (d) lumbar enlargement

_____ 2. The superficial meninx composed of dense fibrous connective tissue is the (a) pia mater (b) arachnoid (c) dura mater (d) denticulate

_____ 3. The portion of a spinal nerve that contains motor nerve fibers only is the (a) posterior root (b) posterior root ganglion (c) lateral root (d) anterior root

_____ 4. The connective tissue covering around individual nerve fibers is the (a) endoneurium (b) epineurium (c) perineurium (d) ectoneurium

_____ 5. On the basis of organization, which does *not* belong with the others? (a) pons (b) medulla oblongata (c) thalamus (d) midbrain

_____ 6. The lateral ventricles are connected to the third ventricle by the (a) interventricular foramen (b) cerebral aqueduct (c) median aperture (d) lateral aperture

_____ 7. The vital centers for heartbeat, respiration, and blood vessel diameter regulation are found in the (a) pons (b) cerebrum (c) cerebellum (d) medulla oblongata

_____ 8. The reflex centers for movements of the head and trunk in response to auditory stimuli are located in the (a) inferior colliculi (b) medial geniculate nucleus (c) superior colliculi (d) ventral posterior nucleus

_____ 9. Which thalamic nucleus controls general sensations and taste? (a) medial geniculate (b) ventral posterior (c) ventral lateral (d) ventral anterior

_____ 10. Integration of the autonomic nervous system, control of body temperature, and the regulation of food intake and thirst are functions of the (a) pons (b) thalamus (c) cerebrum (d) hypothalamus

_____ 11. The left and right cerebral hemispheres are separated from each other by the (a) central sulcus (b) transverse fissure (c) longitudinal fissure (d) insula

_____ 12. Which structure does *not* belong with the others? (a) putamen (b) caudate nucleus (c) insula (d) globus pallidus

_____ 13. Which peduncles connect the cerebellum with the midbrain? (a) superior (b) inferior (c) middle (d) lateral

_____ 14. Which cranial nerve has the most anterior origin? (a) XI (b) IX (c) VII (d) IV

_____ 15. Extensions of the pia mater that suspend the spinal cord and protect against shock are the (a) choroid plexuses (b) pyramids (c) denticulate ligaments (d) superior colliculi

_____   **16.** Which branch of a spinal nerve enters into formation of plexuses? (a) meningeal (b) dorsal (c) rami communicantes (d) ventral

_____   **17.** Which plexus innervates the upper limbs and shoulders? (a) sacral (b) brachial (c) lumbar (d) cervical

_____   **18.** How many pairs of thoracic spinal nerves are there? (a) 1 (b) 5 (c) 7 (d) 12

PART 2. Completion

19. The narrow, shallow groove on the posterior surface of the spinal cord is the _____.

20. The space between the dura mater and wall of the vertebral canal is called the _____.

21. In a spinal nerve, the cell bodies of sensory neurons are found in the _____.

22. The superficial connective tissue covering around a spinal nerve is the _____.

23. The middle meninx is referred to as the _____.

24. The nuclei of origin for cranial nerves IX, X, XI, and XII are found in the _____.

25. The portion of the brain containing the cerebral peduncles is the _____.

26. Cranial nerves V, VI, VII, and VIII have their nuclei of origin in the _____.

27. A shallow downfold of the cerebral cortex is called a(n) _____.

28. The _____ separates the frontal lobe of the cerebrum from the parietal lobe.

29. White matter tracts of the cerebellum are called _____.

30. The space between the dura mater and the arachnoid is referred to as the _____.

31. Together, the thalamus and hypothalamus constitute the _____.

32. Cerebrospinal fluid passes from the third ventricle into the fourth ventricle through the

_____.

33. The cerebrum is separated from the cerebellum by the _____ fissure.

34. The branches of a spinal nerve that are components of the autonomic nervous system are known as

_____.

35. The plexus that innervates the buttocks, perineum, and lower limbs is the _____ plexus.

36. There are _____ pairs of spinal nerves.

37. The part of the brain that coordinates subconscious movements in skeletal muscles is the

_____.

38. The cell bodies of _____ neurons of the ANS are found inside autonomic ganglia.

39. The portion of the ANS concerned with the fight-or-flight response is the _____ division.

40. The autonomic ganglia that are anterior to the vertebral column and close to large abdominal arteries

are called _____ ganglia.

41. The _____ nerve innervates the diaphragm.

42. The _____ tract conveys nerve impulses for touch and pressure.

43. The flexor muscles of the thigh and extensor muscles of the leg are innervated by the

_____ nerve.

44. The _____ tract conveys nerve impulses related to muscle tone and posture.

45. The _____ nerve supplies the extensor muscles of the arm and forearm.

46. The _____ area of the cerebral cortex receives sensations from cutaneous, muscular, and visceral receptors in various parts of the body.

47. The portion of the cerebral cortex that translates thoughts into speech is the _____ area.

48. The term _____ refers to aggregations of myelinated processes from many neurons.

49. A(n) _____ spinal nerve contains preganglionic autonomic nervous system neurons.

50. _____-order neurons carry sensory information from the spinal cord and brain stem to the thalamus.

51. The part of the brain that allows crude appreciation of some sensations, such as pain, temperature, and pressure, is the _____.

52. Autonomic fibers are classified as either cholinergic or _____.

PART 3. Matching

Using B for brachial, C for cervical, L for lumbar, and S for sacral, indicate to which plexus the following nerves belong.

_____ 53. Sciatic

_____ 54. Femoral

_____ 55. Radial

_____ 56. Ansa cervicalis

_____ 57. Median

_____ 58. Obturator

_____ 59. Pudendal

_____ 60. Tibial

_____ 61. Ilioinguinal

_____ 62. Ulnar

_____ 63. Phrenic

PART 4. Matching

_____   **64.** Posterior column–medial
            meniscus pathway

_____   **65.** Spinocerebellar tracts

_____   **66.** Lateral corticospinal tracts

_____   **67.** Lateral spinothalamic tract

_____   **68.** Anterior corticospinal tracts

_____   **69.** Tectospinal tract

_____   **70.** Lateral reticulospinal tract

A. Conduct subconscious proprioceptive input
   to the cerebellum

B. Convey impulses that control movements of the
   neck and part of the trunk

C. Conducts impulses for conscious proprioception
   and most tactile sensations

D. Conveys impulses that facilitate flexon reflexes,
   inhibit extensor reflexes, and decrease muscle tone
   in muscles of the axial skeleton and proximal limbs

E. Control skilled movements of the distal limbs

F. Conduct mainly pain and temperature impulses

G. Conveys impulses that move the head and eyes in
   response to visual stimuli

Sensory Receptors and Sensory and Motor Pathways

14

Now that you have studied the nervous system, you will study sensations. In its broadest sense, the term *sensation* refers to the conscious or subconscious awareness of conditions in the external environment, or within your body. Sensations are detected by *sensory receptors* that change mechanical, thermal, or chemical energy into electrical energy termed *receptor potentials.* If the receptor reaches threshold value, a nervous impulse will be transmitted to the central nervous system for integration and interpretation.

A. CHARACTERISTICS OF SENSATIONS

Conscious sensations may be characterized into three types. *Superficial sensations* include touch, temperature, two-point discrimination, and pain. *Deep sensations* include muscle and joint pain, vibratory sense, and proprioception (muscle and joint position or location). *Combined sensations* are involved in the process termed *stereognosis,* which is the process of recognizing objects by the sense of touch while the eyes are closed, and *topognosis,* which is the ability to localize cutaneous sensation.

For a sensation to be detected (either consciously or subconsciously) four events must occur:

1. *Stimulation* A *stimulus,* or change in the environment, capable of activating certain sensory neurons must be present.
2. *Transduction* A *sensory receptor* or *sense organ* must receive the stimulus and *transduce* (convert) it to a receptor potential. A sensory receptor or sense organ is a specialized peripheral structure or a specialized type of neuron that is sensitive to certain types of changes in internal or external conditions *(stimuli).*
3. *Conduction* The receptor potential elicits nerve impulses that are conducted along a sensory

neural pathway to the CNS. The sensory neuron that conveys such impulses to the CNS is termed a *first-order neuron.* Additional neurons (termed *second-order neurons* and *third-order neurons*) carry the action potential to a particular region within the brain. The number of neurons involved in the transmission of information to the brain (two or three) varies depending upon the sensory tract utilized to transmit the information to the brain.

4. *Integration* A region of the CNS must integrate the information carried via the action potential into a sensation. Most conscious sensations or perceptions occur in the cerebral cortex of the brain after passing through the thalamus. In other words, you see, hear, and feel in the brain. You seem to see with your eyes, hear with your ears, and feel pain in an injured part of your body because sensory impulses from each part of the body arrive in a specific region in the cerebral cortex, which interprets the sensation as coming from the stimulated sensory receptors.

A *sensory unit* consists of a single peripheral neuron and its terminal ending. The neuron's cell body is typically located in either a dorsal spinal ganglion or within a cranial nerve ganglion. The peripheral processes of this neuron may terminate either as free nerve endings or in association with a sensory receptor. Sensory receptors may be either very simple or quite complex, containing highly specialized neurons, epithelium, and connective tissue components. All sensory receptors contain the terminal processes of a neuron, exhibit a high degree of excitability, and possess a specific threshold for a particular type of stimulus, which is termed a *sensory modality.* In addition, a sensory unit possesses a peripheral *receptive field,* which is an area within which a stimulus of appropriate quality and strength will cause a sensory neuron to initiate an action potential. The majority of sensory impulses are conducted to the sensory areas of the cerebral cortex, for this is the region of the brain that initiates the processing of sensory information that ultimately results in the conscious feeling of the sensory stimulus. Different sets of sensory nerve fibers, when activated, will elicit different sensations by virtue of their unique central nervous system connections. Therefore, a particular sensory neuron will provoke an identical sensation *regardless of how it is excited.* This interpretation of sensory perception is termed *Muller's Doctrine of Specific Energies.*

One characteristic of sensations, that of *projection,* describes the process by which the brain refers sensations to their point of *learned origin* of the stimulation. This process accounts for a phenomenon termed *phantom pain* that is quite common in amputees. When phantom pain is experienced the individual feels pain (or one of many other sensations, such as itching or tickling) in the part of the body that was amputated because of the irritation of a nerve ending in the healing wound surface of the amputation. The action potential is carried to the brain and projected to the portion of the limb that is no longer intact.

A second characteristic of many sensations is *adaptation,* that is, a change in sensitivity, usually a decrease, even though a stimulus is still being applied. For example, when you first get into a tub of hot water, you might feel an intense burning sensation. However, after a brief period of time the sensation decreases to one of comfortable warmth, even though the stimulus (hot water) is still present and has not diminished in intensity.

A third characteristic is that of *afterimages,* that is, the persistence of a sensation after the stimulus has been removed. One common example of an afterimage occurs when you look at a bright light and then look away. You will still see the light for several seconds afterward.

A fourth characteristic of sensations is that of *modality.* Modality is the possession of distinct properties by which one sensation may be distinguished from another. For example, pain, pressure, touch, body position, equilibrium, hearing, vision, smell, and taste are all distinctive because the brain perceives each differently.

B. CLASSIFICATION OF RECEPTORS

No single classification format has proved adequate in describing the types of receptors present within the body, nor in describing the method of receptor stimulation or the information conveyed to the central nervous system by the afferent nervous connections. One convenient method of classifying receptors is by their location:

1. *Exteroceptors* (EKS'-ter-ō-sep'-tors), located at or near the body surface, provide information about the *external* environment. They transmit sensations such as hearing, sight, smell, taste, touch, pressure, temperature, and pain.
2. *Interoceptors* or *visceroceptors* (VIS-er-ō-sep'-tors), located in blood vessels and viscera, provide information about the *internal* environment. These stimuli are perceived as pain, taste, fatigue, hunger, thirst, and nausea. Two specific types of interoceptors are *baroreceptors (pressoreceptors),* stimulated by changes in the hydrostatic pressure of blood, and *osmoreceptors,* which are stimulated by changes in the solute concentration of extracellular fluid.
3. *Proprioceptors* (PRŌ-prē-ō-sep'-tors; *proprio* = one's own) or stretch receptors, located in muscles, tendons, joints, and the internal ear, are stimulated by stretching or movement. They provide information about the position and activity of our joints and equilibrium.

Another classification of sensations is based on the type of stimuli they detect:

1. *Mechanoreceptors* detect mechanical pressure or stretching. They provide sensations of touch, pressure, vibration, proprioception (awareness of the location and movement of body parts), hearing, equilibrium, and blood pressure.
2. *Thermoreceptors* detect changes in temperature.
3. *Nociceptors* (NŌ-sē-sep'-tors) detect pain.
4. *Photoreceptors* detect light that strikes the retina of the eye.
5. *Chemoreceptors* detect chemicals in the mouth (taste), nose (smell), and chemicals in body fluids such as water, oxygen, carbon dioxide, hydrogen ions, certain other electrolytes, hormones, and glucose.

Sensations may also be classified into general sensory systems and special sensory systems. *General sensory systems* involve reflex responses through pathways in the spinal cord and the brain stem. Impulses from some of these receptors may even reach the cerebellum. *Special sensory sys-*

tems are related to one of the organs of sight, hearing and equilibrium, smell, and taste.

1. *General senses* involve simple receptors and neural pathways. Taken together, a, b, and c below are referred to as *cutaneous* (*cuta* = skin) *sensations.*

   a. *Tactile* (TAK-tīl; *tact* = touch) *sensations* (touch, pressure, vibration).
   b. *Thermoreceptive sensations* (warmth and coolness).
   c. *Pain sensations.*
   d. *Proprioceptive* (prō'-prē-ō-SEP-tiv) *sensations* (awareness of equilibrium and of the activities of muscles, tendons, and joints).

2. *Special senses* involve complex receptors and neural pathways.

   a. *Olfactory* (ōl-FAK-tō-rē) *sensations* (smell).
   b. *Gustatory* (GUS-ta-tō-rē) *sensations* (taste).
   c. *Visual sensations* (sight).
   d. *Auditory sensations* (hearing).
   e. *Equilibrium* (ē'-kwi-LIB-rē-um) *sensations* (orientation of the body).

C. RECEPTORS FOR GENERAL SENSES

1. Tactile Receptors

Although touch, pressure, and vibration are classified as separate sensations, all are detected by mechanoreceptors.

Touch sensations generally result from stimulation of tactile receptors in the skin or in tissues immediately deep to the skin. *Crude touch* refers to the ability to perceive that something has touched the skin, although the size or texture cannot be determined. *Discriminative touch* refers to the ability to recognize exactly what point on the body is touched. Receptors for touch include corpuscles of touch, hair root plexuses, and type I and type II cutaneous mechanoreceptors. *Corpuscles of touch* or *Meissner's* (MĪS-ners) *corpuscles* are receptors for discriminative touch that are found in dermal papillae. They have already been discussed in Exercise 5. *Hair root plexuses* are dendrites arranged in networks around hair follicles that detect movement when hairs are disturbed. *Type I cutaneous mechanoreceptors,* also called *tactile* or *Merkel* (MER-kel) *discs,* are the flattened portions of dendrites of sensory neurons that make contact

with epidermal cells of the stratum basale called Merkel cells. They are distributed in many of the same locations as corpuscles of touch and also function in discriminative touch. *Type II cutaneous mechanoreceptors,* or *end organs of Ruffini,* are embedded deeply in the dermis and in deeper tissues of the body. They detect heavy and continuous touch sensations.

Receptors for touch, like other cutaneous receptors, are not randomly distributed; some areas contain many receptors, others contain few. Such a clustering of receptors is called *punctate distribution.* Touch receptors are most numerous in the fingertips, palms, and soles. They are also abundant in the eyelids, tip of the tongue, lips, nipples, clitoris, and tip of the penis.

Examine prepared slides of corpuscles of touch (Meissner's corpuscles), hair root plexuses, type I cutaneous mechanoreceptors (tactile or Merkel discs), and type II cutaneous mechanoreceptors (end organs of Ruffini). With the aid of your textbook, draw each of the receptors in the spaces that follow.

Corpuscles of touch (Meissner's corpuscles)

Hair root plexuses

Type I cutaneous mechanoreceptors (tactile or Merkel discs)

Type II cutaneous mechanoreceptors (end organs of Ruffini)

Pressure sensations generally result from stimulation of tactile receptors in deeper tissues. *Pressure* is a sustained sensation that is felt over a larger area than touch. Receptors for pressure sensations include lamellated corpuscles and type II cutaneous mechanoreceptors. *Lamellated (Pacinian) corpuscles* are found in the subcutaneous layer and have already been discussed in Exercise 5.

Pressure receptors are found in the subcutaneous tissue under the skin, in the deep subcutaneous tissues that lie under mucous membranes, around joints and tendons, in the perimysium of muscles, in the mammary glands, in the external genitals of both sexes, and in some viscera.

Examine a prepared slide of lamellated (Pacinian) corpuscles. With the aid of your textbook, draw the receptors in the spaces that follow.

Lamellated (Pacinian) corpuscles

Vibration sensations result from rapidly repetitive sensory signals from tactile receptors. Receptors for vibration include corpuscles of touch (Meissner's corpuscles) that detect low-frequency vibration and lamellated (Pacinian) corpuscles that detect high-frequency vibration.

2. Thermoreceptors

The cutaneous receptors for the sensation of warmth and coolness are free (naked) nerve endings that are widely distributed in the dermis and subcutaneous connective tissue. They are also located in the cornea of the eye, tip of the tongue, and external genitals.

3. Pain Receptors

Receptors for *pain*, called *nociceptors* (NŌ-sē-sep'-tors; *noci* = harmful), are free (naked) nerve endings (see Figure 5.1). Pain receptors are found in practically every tissue of the body and adapt only slightly or not at all. They may be excited by any type of stimulus. Excessive stimulation of any sense organ causes pain. For example, when stimuli for other sensations such as touch, pressure, heat, and cold reach a certain threshold, they stimulate pain receptors as well. Pain receptors, because of their sensitivity to all stimuli, have a general protective function of informing us of changes that could be potentially dangerous to health or life. Adaptation to pain does not readily occur. This low level of adaptation is important, because pain indicates disorder or disease. If we became used to it and ignored it, irreparable damage could result.

4. Proprioceptive Receptors

An awareness of the activities of muscles, tendons, and joints and equilibrium is provided by the *proprioceptive (kinesthetic) sense*. It informs us of the degree to which tendons are tensed and muscles are contracted. The proprioceptive sense enables us to recognize the location and rate of movement of one part of the body in relation to other parts. It also allows us to estimate weight and to determine the muscular effort necessary to perform a task. With the proprioceptive sense, we can judge the position and movements of our limbs without using our eyes when we walk, type, play a musical instrument, or dress in the dark.

Proprioceptive receptors are located in skeletal muscles and tendons, in and around joints, and in the internal ear. Proprioceptors adapt only slightly. This slight adaptation is beneficial because the brain must be apprised of the status of different parts of the body at all times so that adjustments can be made to ensure coordination.

Receptors for proprioception are as follows. The *joint kinesthetic* (kin'-es-THET-ik) *receptors* are located in the articular capsules of joints and ligaments about joints. These receptors provide feedback information on the degree and rate of angulation (change of position) of a joint. *Muscle spindles* are specialized muscle fibers (cells) that consist of endings of sensory neurons. They are located in nearly all skeletal muscles and are more numerous in the muscles of the limbs. Muscle spindles provide feedback information on the degree of muscle stretch. This information is relayed to the central nervous system to assist in the coordination and efficiency of muscle contraction. *Tendon organs (Golgi tendon organs)* are located at the junction of skeletal muscle and tendon. They function by sensing the tension applied to a tendon and the force of contraction of associated muscles. The information is translated by the central nervous system.

Proprioceptors in the internal ear are the maculae and cristae that function in equilibrium. These are discussed at the end of the exercise.

Examine prepared slides of joint kinesthetic receptors, muscle spindles, and tendon organs (Golgi tendon organs). With the aid of your textbook, draw the receptors in the spaces that follow.

Joint kinesthetic receptors

Muscle spindles

Tendon organs (Golgi tendon organs)

D. TESTS FOR GENERAL SENSES

Areas of the body that have few cutaneous receptors are relatively insensitive, whereas those regions that contain large numbers of cutaneous receptors are quite sensitive. This difference can be demonstrated by the *two-point discrimination test* for touch. In the following tests, students can work in pairs, with one acting as subject and the other as experimenter. The subject will keep his or her eyes closed during the experiments.

1. Two-Point Discrimination Test

In this test, the two points of a measuring compass are applied to the skin and the distance in millimeters (mm) between the two points is varied. The subject indicates when he or she feels two points and when he or she feels only one.

PROCEDURE

CAUTION! *Wash the compass in a fresh bleach solution before using it on another subject to prevent transfer of saliva.*

1. *Very gently*, place the compass on the tip of the tongue, an area where receptors are very densely packed.
2. Narrow the distance between the two points to 1.4 mm. At this distance, the points are able to stimulate two different receptors, and the subject feels that he or she is being touched by two objects.
3. Decrease the distance to less than 1.4 mm. The subject feels only one point, even though both points are touching the tongue, because the points are so close together that they reach only one receptor.
4. Now *gently* place the compass on the back of the neck, where receptors are relatively few and far between. Here the subject feels two distinctly different points only if the distance between them is 36.2 mm or more.
5. The two-point discrimination test shows that the more sensitive the area, the closer the compass points can be placed and still be felt separately.
6. The following order, from greatest to least sensitivity, has been established from the test: tip of tongue, tip of finger, side of nose, back of hand, and back of neck.
7. Test the tip of finger, side of nose, and back of hand and record your results in Section D.1 of the LABORATORY REPORT RESULTS at the end of the exercise.

2. Identifying Touch Receptors

PROCEDURE

1. Using a water-soluble colored felt marking pen, draw a 1-in. square on the back of the forearm and divide the square into 16 smaller squares.
2. With the subject's eyes closed, press a Von Frey hair or bristle against the skin, just enough to cause the hair to bend, once in each of the 16 squares. The pressure should be applied in the same manner each time.
3. The subject should indicate when he or she experiences the sensation of touch, and the

experimenter should make dots in square 1 at the places corresponding to the points at which the subject feels the sensations.

4. The subject and the experimenter should switch roles and repeat the test.

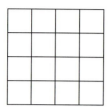

The pair of students working as a team should examine their 1-in. squares after the test is done. They should compare the number of positive and negative responses in each of the 16 small squares, and see how uniformly the touch receptors are distributed throughout the entire 1-in. square. Other general areas used for locating touch receptors are the arm and the back of the hand.

3. Identifying Pressure Receptors

PROCEDURE

1. The experimenter touches the skin of the subject (whose eyes are closed) with the point of a piece of colored chalk.
2. With eyes still closed, the subject then tries to touch the same spot with a piece of differently colored chalk. The distance between the two points is then measured.
3. Proceed using various parts of the body, such as the palm, arm, forearm, and back of neck.
4. Record your results in Section D.3 of the LABORATORY REPORT RESULTS at the end of the exercise.

4. Identifying Thermoreceptors

PROCEDURE

1. Draw a 1-in. square on the back of the wrist.
2. Place a forceps or other metal probe in ice-cold water for a minute, dry it quickly, and, with the *dull* point, explore the area in the square for the presence of cold spots.
3. Keep the probe cold and, using ink, mark the position of each spot that you find.
4. Mark each corresponding place in square 2 with the letter "c."
5. Immerse the forceps in hot water so that it will give a sensation of warmth when removed and

applied to the skin, but *avoid having it so hot that it causes pain.*

6. Proceeding as before, locate the position of the warm spots in the same area of the skin.
7. Mark these spots with ink of a different color, and then mark each corresponding place in square 2 with the letter "h."
8. Repeat the entire procedure, using both cold forceps and warm forceps on the back of the hand and the palm, respectively, and mark squares 3 and 4 as you did square 2.

9. In order to demonstrate adaptation of thermoreceptors, fill three 1-liter beakers with 700 ml of (1) ice water, (2) water at room temperature, and (3) water at 45° C. Immerse your left hand in the ice water and your right hand in the water at 45° C for 1 min. Now move your left hand to the beaker with water at room temperature and record the sensation.

_____

Move your right hand to the beaker with the water at room temperature and record the sensation. _____

How do you explain the experienced difference in the temperature of the water in the beaker with the water at room temperature?

_____

_____

5. Identifying Pain Receptors

PROCEDURE

1. Using the same 1-in. square of the forearm that was previously used for the touch test in Section D.2, perform the following experiment.
2. Apply a piece of absorbent cotton soaked with water to the area of the forearm for 5 min to soften the skin.
3. Add water to the cotton as needed.
4. Place the blunt end of a probe to the surface of the skin and press enough to produce a sensation of pain. Explore the marked area systematically.

5. Using dots, mark the places in square 5 that correspond to the points that give pain sensation when stimulated.
6. Distinguish between sensations of pain and touch. Are the areas for touch and pain identical?

_____

7. At the end of the test, compare your squares as you did in Section D.2.

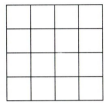

Perform the following test to demonstrate the phenomenon of *referred pain.* Place your elbow in a large shallow pan of ice water, and note the progression of sensation that you experience. At first, you will feel some discomfort in the region of the elbow. Later, pain sensations will be felt elsewhere.

Where do you feel the referred pain? _____

_____

Label the areas of referred pain indicated in Figure 14.1.

6. Identifying Proprioceptors

PROCEDURE

1. Face a blackboard close enough so that you can easily reach to mark it. Mark a small X on the board in front of you and keep the chalk on the X for a moment. Now close your eyes, raise your right hand above your head and then, with your eyes still closed, mark a dot as near as possible to the X. Repeat the procedure by placing your

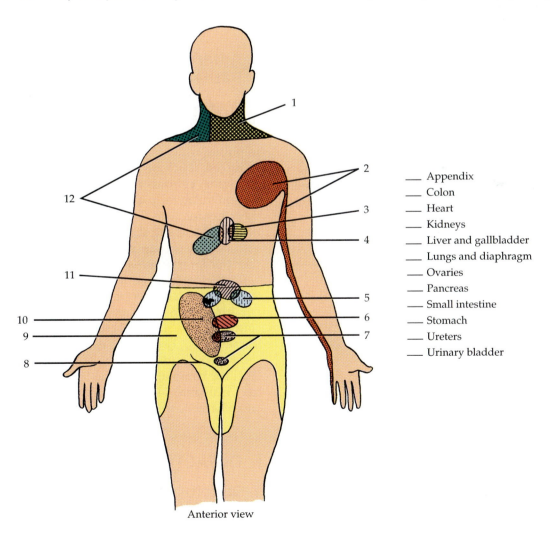

_____ Appendix
_____ Colon
_____ Heart
_____ Kidneys
_____ Liver and gallbladder
_____ Lungs and diaphragm
_____ Ovaries
_____ Pancreas
_____ Small intestine
_____ Stomach
_____ Ureters
_____ Urinary bladder

Anterior view

FIGURE 14.1   Referred pain.

chalk on the X, closing your eyes, raising your arm above your head, and then marking another dot as close as possible to the X. Repeat the procedure a third time. Record your results by estimating or measuring how far you missed the X for each trial in Section D.6 of the LABORATORY REPORT RESULTS at the end of the exercise.

2. Write the word "physiology" on the left line that follows. Now, with your *eyes closed*, write the same word immediately to the right. How do the samples of writing compare?

_____

Explain your results. _____

_____

_____

_____

3. The following experiments demonstrate that kinesthetic sensations facilitate repetition of certain acts involving muscular coordination.

   Students should work in pairs for these experiments.

   a. The experimenter asks the subject to carry out certain movements with his or her eyes closed, for example, point to the middle finger of the subject's left hand with the index finger of the subject's right hand.

   b. With his or her eyes closed, the subject extends the right arm as far as possible behind the body, and then brings the index finger quickly to the tip of his or her nose.
      How accurate is the subject in doing this?

   _____

   c. Ask the subject, with eyes shut, to touch the named fingers of one hand with the index finger of the other hand.
      How well does the subject carry out the directions? _____

   _____

   _____

   _____

E. SOMATIC SENSORY PATHWAYS

Most input from somatic receptors on one side of the body crosses over to the opposite side in the spinal cord or brain stem before ascending to the thalamus. It then projects from the thalamus to the somatosensory area of the cerebral cortex, where conscious sensations result. Axon collaterals (branches) of somatic sensory neurons also carry signals into the cerebellum and the reticular formation of the brain stem. Two general pathways lead from sensory receptors to the cortex: the posterior column–medial lemniscus pathway and the anterolateral (spinothalamic) pathway.

1. Posterior Column–Medial Lemniscus Pathway

Nerve impulses for conscious proprioception and most tactile sensations ascend to the cerebral cortex along a common pathway formed by three-neuron sets.

Based on the description provided on page 269, label the components of the posterior column–medial lemniscus pathway in Figure 14.2.

2. Anterolateral (Spinothalamic) Pathways

The *anterolateral (spinothalamic) pathways* carry mainly pain and temperature impulses. In addition, they relay tickle, itch, and some tactile impulses, which give rise to a very crude, not well-localized touch or pressure sensation. Like the posterior column–medial lemniscus pathway, the anterolateral pathways are also composed of three-neuron sets. The axon of the second-order neuron extends to the opposite side of the spinal cord and passes superiorly to the brain stem in either the *lateral spinothalamic tract* or *anterior spinothalamic tract*. The lateral spinothalamic tract conveys sensory impulses for pain and temperature, whereas the anterior spinothalamic tract conveys tickle, itch, and crude touch and pressure impulses.

Based on the descriptions provided on page 269, label the components of the lateral spinothalamic tract in Figure 14.3 on page 316 and the anterior spinothalamic tract in Figure 14.4 on page 317.

F. OLFACTORY SENSATIONS

1. Olfactory Receptors

The receptors for the *olfactory* (ol-FAK-tō-rē; *olfactus* = smell) *sense* are found in the nasal epithelium

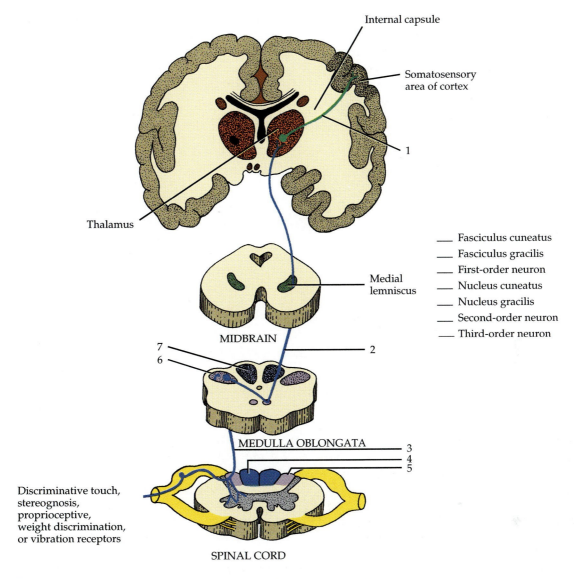

Internal capsule

Somatosensory area of cortex

1

Thalamus

Medial lemniscus

MIDBRAIN

___ Fasciculus cuneatus
___ Fasciculus gracilis
___ First-order neuron
___ Nucleus cuneatus
___ Nucleus gracilis
___ Second-order neuron
___ Third-order neuron

7
6

2

MEDULLA OBLONGATA

3
4
5

Discriminative touch, stereognosis, proprioceptive, weight discrimination, or vibration receptors

SPINAL CORD

FIGURE 14.2   The posterior column–medial lemniscus pathway.

in the superior portion of the nasal cavity on either side of the nasal septum. The nasal epithelium consists of three principal kinds of cells: olfactory receptors, supporting cells, and basal cells. The *olfactory receptors (cells)* are bipolar neurons. Their cell bodies lie between the supporting cells. The distal (free) end of each olfactory cell contains a knob-shaped dendrite from which six to eight cilia, called *olfactory hairs,* protrude. The *supporting (sustentacular) cells* are columnar epithelial cells of the mucous membrane that lines the nose. *Basal cells* lie between the bases of the supporting cells and produce new olfactory receptors. Within the connective tissue deep to the olfactory epithe-

lium are *olfactory (Bowman's) glands* that secrete mucus.

The unmyelinated axons of the olfactory receptors unite to form the *olfactory (I) nerves,* which pass through foramina in the cribriform plate of the ethmoid bone. The olfactory nerves terminate in paired masses of gray matter, the *olfactory bulbs,* which lie inferior to the frontal lobes of the cerebrum on either side of the crista galli of the ethmoid bone. The first synapse of the olfactory neural pathway occurs in the olfactory bulbs between the axons of the olfactory (I) nerves and the dendrites of neurons inside the olfactory bulbs. Axons of these neurons run

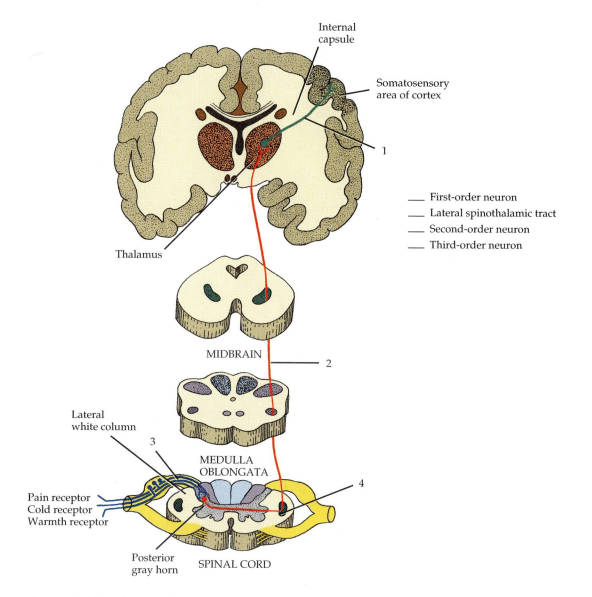

Internal
capsule

Somatosensory
area of cortex

1

_____ First-order neuron
_____ Lateral spinothalamic tract
_____ Second-order neuron
_____ Third-order neuron

Thalamus

MIDBRAIN

2

Lateral
white column

3

MEDULLA
OBLONGATA

4

Pain receptor
Cold receptor
Warmth receptor

Posterior
gray horn   SPINAL CORD

FIGURE 14.3   The lateral spinothalamic pathway.

posteriorly to form the ***olfactory tract.*** From here, nerve impulses are conveyed to the primary olfactory area in the temporal lobe of the cerebral cortex. In the cortex, the nerve impulses are interpreted as odor and give rise to the sensation of smell.

Adaptation happens quickly, especially adaptation to odors. For this reason, we become accustomed to some odors and are also able to endure unpleasant ones. Rapid adaptation also accounts for the failure of a person to detect gas that accumulates slowly in a room.

Label the structures associated with olfaction in Figure 14.5 on page 318.

Now examine a slide of the olfactory epithelium under high power. Identify the olfactory receptors and supporting cells and label Figure 14.6 on page 319.

2. Olfactory Adaptation

PROCEDURE

1. The subject should close his or her eyes after plugging one nostril with cotton.
2. Hold a bottle of oil of cloves, or other substance having a distinct odor, under the open nostril.

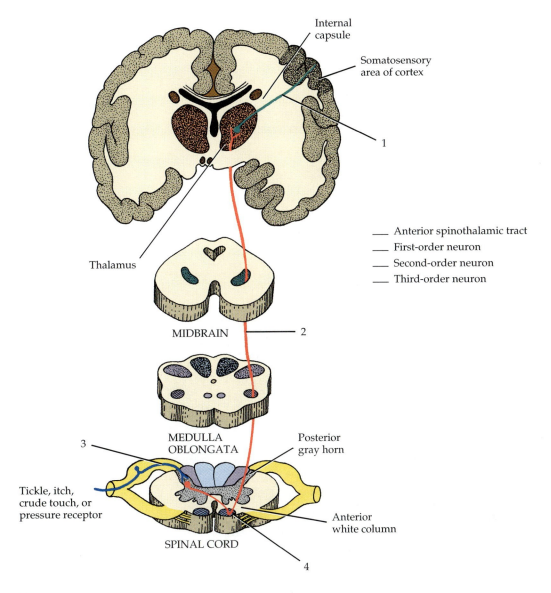

Internal capsule

Somatosensory area of cortex

1

Thalamus

___ Anterior spinothalamic tract
___ First-order neuron
___ Second-order neuron
___ Third-order neuron

MIDBRAIN   2

MEDULLA OBLONGATA

Posterior gray horn

3

Tickle, itch, crude touch, or pressure receptor

Anterior white column

SPINAL CORD

4

FIGURE 14.4 The anterior spinothalamic pathway.

3. The subject breathes in through the open nostril, and exhales through the mouth. Note the time required for the odor to disappear, and repeat with the other nostril.
4. As soon as olfactory adaptation has occurred, test an entirely different substance.
5. Compare results for the various materials tested.

Olfactory stimuli, such as pepper, onions, ammonia, ether, and chloroform, are irritating and may cause tearing because they stimulate the receptors of the trigeminal (V) nerve as well as the olfactory neurons.

G. GUSTATORY SENSATIONS

1. Gustatory Receptors

The receptors for *gustatory* (GUS-ta-tō-rē; *gusto* = taste) *sensations,* or sensations of taste, are located in the taste buds. Although taste buds are most numerous on the tongue, they are also found on the soft palate, pharynx and larynx. *Taste buds* are oval bodies consisting of three kinds of cells: supporting cells, gustatory receptors, and basal cells. The *supporting (sustentacular) cells* are specialized epithelial cells that form a capsule. Inside each capsule are about 50 *gustatory receptors (cells).*

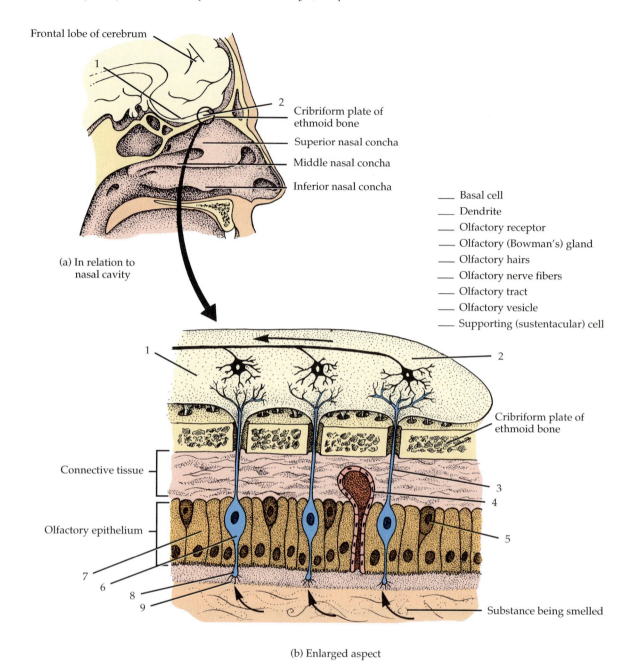

Frontal lobe of cerebrum

Cribriform plate of ethmoid bone

Superior nasal concha

Middle nasal concha

Inferior nasal concha

(a) In relation to nasal cavity

___ Basal cell
___ Dendrite
___ Olfactory receptor
___ Olfactory (Bowman's) gland
___ Olfactory hairs
___ Olfactory nerve fibers
___ Olfactory tract
___ Olfactory vesicle
___ Supporting (sustentacular) cell

Cribriform plate of ethmoid bone

Connective tissue

Olfactory epithelium

Substance being smelled

(b) Enlarged aspect

FIGURE 14.5   Olfactory receptors.

Each gustatory receptor contains a hairlike process *(gustatory hair)* that projects to the surface through an opening in the taste bud called the *taste pore.* Gustatory cells make contact with taste stimuli through the taste pore. **Basal cells** are found at the periphery of the taste bud and produce new supporting cells, which then develop into gustatory receptors.

Examine a slide of taste buds and label the structures associated with gustation in Figure 14.7 on page 320.

Taste buds are located in some connective tissue elevations on the tongue called *papillae* (pa-PILL-ē). They give the upper surface of the tongue its rough texture and appearance. *Circumvallate* (ser-kum-

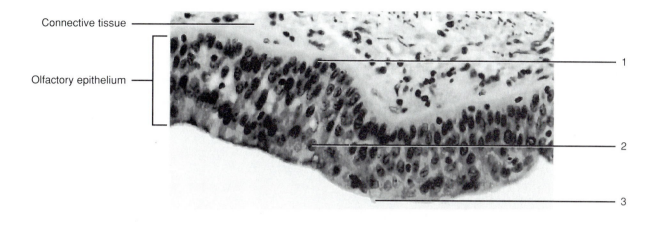

Connective tissue ———

Olfactory epithelium —

—— 1

—— 2

—— 3

___ Basal cell
___ Olfactory receptor
___ Supporting (sustentacular) cell

FIGURE 14.6   Photomicrograph of the olfactory epithelium.

VAL-āt) *papillae* are circular and form an inverted V-shaped row at the posterior portion of the tongue. *Fungiform* (FUN-ji-form) *papillae* are knoblike elevations scattered over the entire surface of the tongue. All circumvallate and most fungiform papillae contain taste buds. *Filiform* (FIL-i-form) *papillae* are threadlike structures that are also distributed over the entire surface of the tongue.

Have your partner protrude his or her tongue and examine its surface with a hand lens to identify the shape and position of the papillae.

2. Identifying Taste Zones

For gustatory cells to be stimulated, substances tasted must be in solution in the saliva in order to enter the taste pores in the taste buds. Despite the many substances tasted, there are basically only four taste sensations: sour, salty, bitter, and sweet. Each taste is due to a different response to different chemicals. Some regions of the tongue react more strongly than others to particular taste sensations.

To identify the taste zones for the four taste sensations, perform the following steps and record the results in Section G.2 of the LABORATORY REPORT RESULTS at the end of the exercise by inserting a plus sign (taste detected) or a minus sign (taste not detected) where appropriate.

PROCEDURE

1. The subject thoroughly dries his or her tongue (use a clean paper towel). The experimenter places some granulated sugar on the tip of the tongue and notes the time. The subject indicates when he or she tastes sugar by raising his or her hand. The experimenter notes the time again and records how long it takes for the subject to taste the sugar.
2. Repeat the experiment, but this time use a drop of sugar solution. Again record how long it takes for the subject to taste the sugar. How do you explain the difference in time periods?
3. The subject rinses his or her mouth again. The experiment is then repeated using the quinine solution (bitter taste), and then the salt solution.

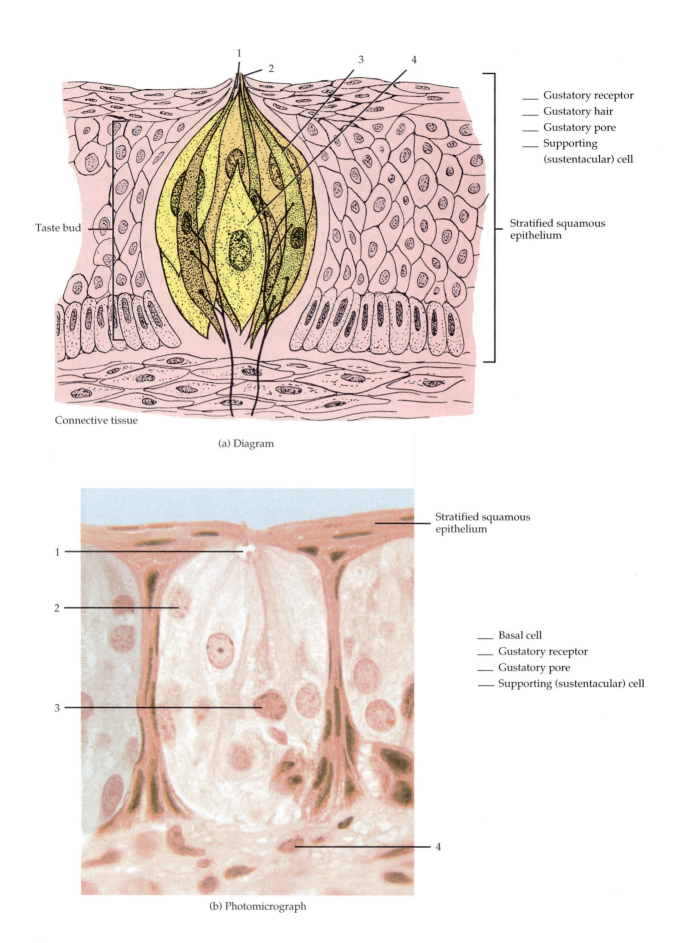

1
2
3
4

_____ Gustatory receptor
_____ Gustatory hair
_____ Gustatory pore
_____ Supporting
    (sustentacular) cell

Stratified squamous
epithelium

Taste bud

Connective tissue

(a) Diagram

Stratified squamous
epithelium

1

2

_____ Basal cell
_____ Gustatory receptor
_____ Gustatory pore
_____ Supporting (sustentacular) cell

3

4

(b) Photomicrograph

FIGURE 14.7 Structure of a taste bud.

4. After rinsing yet again, the experiment is repeated using the acetic acid solution or vinegar (sour) placed on the tip and *sides* of the tongue.

3. Taste and Inheritance

Taste for certain substances is inherited, and geneticists for many years have been using the chemical *phenylthiocarbamide (PTC)* to test taste. To some individuals this substance tastes bitter; to others it is sweet; and some cannot taste it at all.

PROCEDURE

1. Place a few crystals of PTC on the subject's tongue. Does he or she taste it? If so, describe the taste.
2. Special paper that is flavored with this chemical may be chewed and mixed with saliva and tested in the same manner.
3. Record on the blackboard your response to the PTC test. Usually about 70% of the people tested can taste this compound; 30% cannot. Compare this percentage with the class results.

4. Taste and Smell

This test combines the effect of smell on the sense of taste.

PROCEDURE

1. Obtain small cubes of carrot, onion, potato, and apple.
2. The subject dries the tongue with a clean paper towel, closes the eyes, and pinches the nostrils shut. The experimenter places the cubes, one by one, on the subject's tongue.
3. The subject attempts to identify each cube in the following sequences: (1) immediately, (2) after chewing (nostrils closed), and (3) after opening the nostrils.
4. Record your results in Section G.4 of the LABORATORY REPORT RESULTS at the end of the exercise.

H. VISUAL SENSATIONS

Structures related to *vision* are the eyeball (which is the receptor organ for visual sensations), optic (II) nerve, brain, and accessory structures. The extrinsic muscles of the eyeball may be reviewed in Figure 10.4.

1. Accessory Structures

Among the *accessory structures* are the eyebrows, eyelids, eyelashes, lacrimal (tearing) apparatus, and extrinsic eye muscles. *Eyebrows* protect the eyeball from falling objects, prevent perspiration from getting into the eye, and shade the eye from the direct rays of the sun. *Eyelids,* or *palpebrae* (PAL-pe-brē), consist primarily of skeletal muscle covered externally by skin. The underside of the muscle is lined by a mucous membrane called the *palpebral conjunctiva* (kon-junk-TĪ-va). The *bulbar (ocular) conjunctiva* covers the surface of the eyeball. Also within eyelids are *tarsal (Meibomian) glands,* modified sebaceous glands whose oily secretion keeps the eyelids from adhering to each other. Infection of these glands produces a *chalazion* (cyst) in the eyelid. Eyelids shade the eyes during sleep, protect the eyes from light rays and foreign objects, and spread lubricating secretions over the surface of the eyeballs. Projecting from the border of each eyelid is a row of short, thick hairs, the *eyelashes.* Sebaceous glands at the base of the hair follicles of the eyelashes, called *sebaceous ciliary glands (glands of Zeis),* pour a lubricating fluid into the follicles. An infection of these glands is called a *sty.*

The *lacrimal* (LAK-ri-mal; *lacrima* = tear) *apparatus* consists of a group of structures that manufacture and drain tears. Each *lacrimal gland* is located at the superior lateral portion of both orbits. Leading from the lacrimal glands are 6 to 12 *excretory lacrimal ducts* that empty tears onto the surface of the conjunctiva of the upper lid. From here, the tears pass medially and enter two small openings called *lacrimal puncta* that appear as two small pores, one in each papilla of the eyelid, at the medial commissure of the eye. The tears then pass into two ducts, the *lacrimal canals,* and are next conveyed into the lacrimal sac. The *lacrimal sac* is the superior expanded portion of the *nasolacrimal duct,* a canal that carries the tears into the inferior meatus of the nasal cavity. Tears clean, lubricate, and moisten the external surface of the eyeball.

Label the parts of the lacrimal apparatus in Figure 14.8.

2. Structure of the Eyeball

The eyeball can be divided into three principal layers: (1) fibrous tunic, (2) vascular tunic, and (3) retina (nervous tunic). See Figure 14.9.

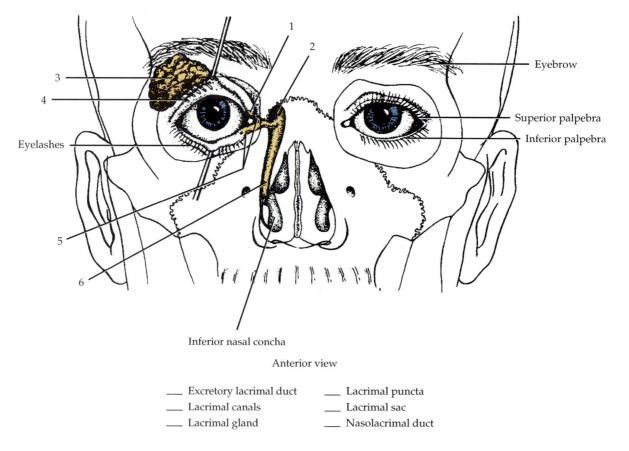

Anterior view

___ Excretory lacrimal duct   ___ Lacrimal puncta
___ Lacrimal canals   ___ Lacrimal sac
___ Lacrimal gland   ___ Nasolacrimal duct

FIGURE 14.8   Lacrimal apparatus.

a. FIBROUS TUNIC

The *fibrous tunic* is the outer coat of the eyeball. It is divided into the posterior sclera and the anterior cornea. The *sclera* (SKLE-ra; *skleros* = hard), called the "white of the eye," is a coat of dense connective tissue that covers all the eyeball except the most anterior portion (cornea). The sclera gives shape to the eyeball and protects its inner parts. The anterior portion of the fibrous tunic is known as the *cornea* (KOR-nē-a). This nonvascular, transparent coat covers the colored iris. Because it is curved, the cornea helps focus light. The outer surface of the cornea contains epithelium that is continuous with the epithelium of the bulbar conjunctiva. At the junction of the sclera and cornea is the *scleral venous sinus (canal of Schlemm).*

b. VASCULAR TUNIC

The *vascular tunic* is the middle layer of the eyeball and consists of three portions: choroid, ciliary body, and iris. The *choroid* (KŌ-royd) is the posterior portion of the vascular tunic. It is a thin, dark brown membrane that lines most of the internal surface of the sclera and contains blood vessels and melanin. The choroid absorbs light rays so they are not reflected back out of the eyeball and maintains the nutrition of the retina. The anterior portion of the choroid is the *ciliary* (SIL-ē-ar'-ē) *body,* the thickest portion of the vascular tunic. It extends from the *ora serrata* (Ō-ra ser-RĀ-ta) of the retina (inner tunic) to a point just behind the sclerocorneal junction. The ora serrata is the jagged margin of the retina. The ciliary body consists of the *ciliary processes* (folds of the ciliary body that secrete aqueous humor) and the *ciliary muscle* (a circular band of smooth muscle that alters the shape of the lens for near or far vision). The *iris* (*irid* = colored circle), the third portion of the vascular tunic, is the colored portion of the eyeball and consists of circular and radial smooth-muscle fibers arranged to form a doughnut-shaped structure. The hole in the center of the iris is the *pupil,* through which light enters the eyeball. One function of the iris is to regulate the amount of light entering the eyeball.

c. RETINA

The third and inner coat of the eyeball, the *retina (nervous tunic),* is found only in the posterior

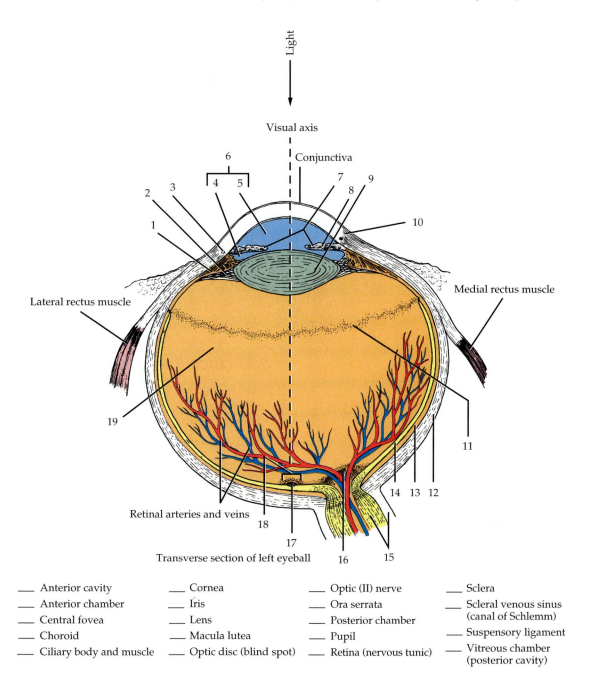

Light

Visual axis

Conjunctiva

6

4 5

3

2

1

7

8

9

10

Lateral rectus muscle

Medial rectus muscle

19

11

Retinal arteries and veins

14 13 12

18

17

16 15

Transverse section of left eyeball

___ Anterior cavity	___ Cornea
___ Anterior chamber	___ Iris
___ Central fovea	___ Lens
___ Choroid	___ Macula lutea
___ Ciliary body and muscle	___ Optic disc (blind spot)

___ Optic (II) nerve	___ Sclera
___ Ora serrata	___ Scleral venous sinus (canal of Schlemm)
___ Posterior chamber	___ Suspensory ligament
___ Pupil	___ Vitreous chamber (posterior cavity)
___ Retina (nervous tunic)	

FIGURE 14.9   Parts of eyeball.

portion of the eye. Its primary function is image formation. It consists of a pigment epithelium and a neural portion. The outer *pigment epithelium* (nonvisual portion) consists of melanin-containing epithelial cells in contact with the choroid. The inner *neural portion* (visual portion) is composed of three zones of neurons. Named in the order in which they conduct nerve impulses, these are the *photoreceptor layer, bipolar cell*

layer, and *ganglion cell layer.* Structurally, the photoreceptor layer is just internal to the pigment epithelium, which lies adjacent to the choroid. The ganglion cell layer is the innermost zone of the neural portion.

The two types of photoreceptors are called rods and cones because of their respective shapes. *Rods* are specialized for vision in dim light. In addition, they allow discrimination between different shades of dark and light and permit discernment of shapes and movement. *Cones* are specialized for color vision and for sharpness of vision, that is, *visual acuity*. Cones are stimulated only by bright light and are most densely concentrated in the *central fovea*, a small depression in the center of the macula lutea. The *macula lutea* (MAK-yoo-la LOO-tē-a), or yellow spot, is situated in the exact center of the posterior portion of the retina and corresponds to the visual axis of the eye. The fovea is the area of sharpest vision because of the high concentration of cones. Rods are absent from the fovea and macula but increase in density toward the periphery of the retina.

When light stimulates photoreceptors, impulses are conducted across synapses to the bipolar neurons in the intermediate zone of the neural portion of the retina. From there, the impulses pass to the ganglion cell layer. Axons of the ganglion neurons extend posteriorly to a small area of the retina called the *optic disc (blind spot)*. This region contains openings through which fibers of the ganglion neurons exit as the *optic (II) nerve*. Because this area contains neither rods nor cones, and only nerve fibers, no image is formed on it. For this reason it is called the blind spot.

d. LENS

The eyeball itself also contains the lens, just behind the pupil and iris. The *lens* is constructed of numerous layers of protein fibers arranged like the layers of an onion. Normally, the lens is perfectly transparent and is enclosed by a clear capsule and held in position by the *suspensory ligaments*. A loss of transparency of the lens is called a *cataract*.

e. INTERIOR

The interior of the eyeball contains a large cavity divided into two smaller cavities. These are called the anterior cavity and the vitreous chamber (posterior cavity) and are separated from each other by the lens. The *anterior cavity*, in turn, has two subdivisions known as the anterior chamber and the posterior chamber. The *anterior chamber* lies posterior to the cornea and anterior to the iris. The *posterior chamber* lies posterior to the iris and anterior to the suspensory ligaments and lens. The anterior cavity is filled with a clear, watery fluid known as the *aqueous* (*aqua* = water) *humor*, which is secreted by the ciliary processes posterior to the iris. From the posterior chamber, the fluid permeates the posterior cavity and then passes anteriorly between the iris and the lens, through the pupil into the anterior chamber. From the anterior chamber, the aqueous humor is drained off into the scleral venous sinus and passes into the blood. Pressure in the eye, called *intraocular pressure (IOP)*, is produced mainly by the aqueous humor. Intraocular pressure keeps the retina smoothly applied to the choroid so that the retina may form clear images. Abnormal elevation of intraocular pressure, called *glaucoma* (glaw-KŌ-ma), results in degeneration of the retina and blindness.

The second, larger cavity of the eyeball is the *vitreous chamber (posterior cavity)*. It is located between the lens and retina and contains a soft, jellylike substance called the *vitreous body*. This substance contributes to intraocular pressure, helps to prevent the eyeball from collapsing, and holds the retina flush against the internal portions of the eyeball.

Label the parts of the eyeball in Figure 14.9.

3. Surface Anatomy

Refer to Figure 14.10 for a summary of several surface anatomy features of the eyeball and accessory structures of the eye.

4. Dissection of Vertebrate Eye (Beef or Sheep)

CAUTION! *Please reread Section D, "Precautions Related to Dissection" at the beginning of the laboratory manual on page xiii before you begin your dissection.*

a. EXTERNAL EXAMINATION

PROCEDURE

1. Note any *fat* on the surface of the eyeball that protects the eyeball from shock in the orbit. Remove the fat.
2. Locate the *sclera*, the tough external white coat, and the *conjunctiva*, a delicate membrane that covers the anterior surface of the eyeball and is attached near the edge of the cornea. The *cornea* is the anterior, transparent portion of the sclera. It is probably opaque in your specimen due to the preservative.
3. Locate the *optic (II) nerve*, a solid, white cord of nerve fibers on the posterior surface of the eyeball.
4. If possible, identify six *extrinsic eye muscles* that appear as flat bands near the posterior part of the eyeball.

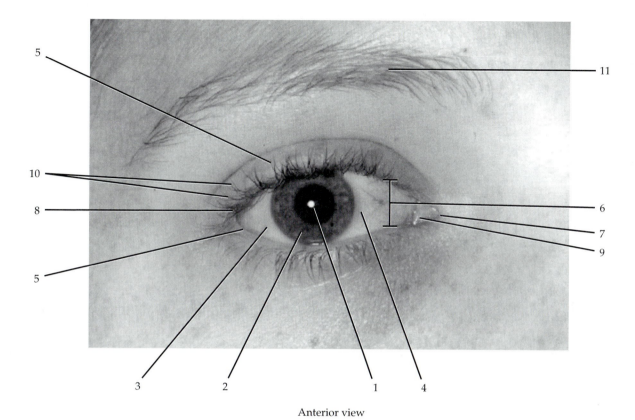

Anterior view

1. **Pupil**. Opening of center of iris of eyeball for light transmission.
2. **Iris**. Circular pigmented muscular membrane behind cornea.
3. **Sclera**. "White" of eye, a coat of fibrous tissue that covers entire eyeball except for cornea.
4. **Conjunctiva**. Membrane that covers exposed surface of eyeball and lines eyelids.
5. **Palpebrae (eyelids)**. Folds of skin and muscle lined by conjunctiva.
6. **Palpebral fissure**. Space between eyelids when they are open.
7. **Medial commissure**. Site of union of upper and lower eyelids near nose.
8. **Lateral commissure**. Site of union of upper and lower eyelids away from nose.
9. **Lacrimal caruncle**. Fleshy, yellowish projection of medial commissure that contains modified sweat and sebaceous glands.
10. **Eyelashes**. Hairs on margins of eyelids, usually arranged in two or three rows.
11. **Eyebrows**. Several rows of hair superior to upper eyelids.

FIGURE 14.10   Surface anatomy of eyeball and accessory structures.

b. INTERNAL EXAMINATION

PROCEDURE

1. With a sharp scalpel, make an incision about 0.6 cm (¼ in.) lateral to the cornea (Figure 14.11).
2. Insert scissors into the incision and carefully and slowly cut all the way around the corneal region. The eyeball contains fluid, so take care that it does not squirt out when you make your first incision. Examine the inside of the anterior part of the eyeball.
3. The *lens* is held in position by **suspensory ligaments**, which are delicate fibers. Around the outer margin of the lens, with a pleated ap-

pearance, is the black **ciliary body**, which also functions to hold the lens in place. Free the lens and notice how hard it is.

4. The *iris* can be seen just anterior to the lens and is also heavily pigmented or black.
5. The *pupil* is the circular opening in the center of the iris.
6. Examine the inside of the posterior part of the eyeball, identifying the thick **vitreous humor** that fills the space between the lens and retina.
7. The *retina* is the white inner coat beneath the choroid coat and is easily separated from it.
8. The **choroid coat** is a dark, iridescent-colored tissue that gets its iridescence from a special structure called the **tapetum lucidum**. The

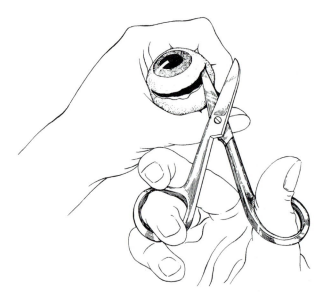

FIGURE 14.11   Procedure for dissecting a vertebrate eye.

tapetum lucidum, which is not present in the human eye, functions to reflect some light back onto the retina.

9. Finally, identify the **blind spot**, the point at which the retina is attached to the back of the eyeball.

5. Ophthalmoscopic Examination of the Eyeball

Examinations of the eye include inspection of the fundus (interior) of the eyeball. This is accomplished by using an instrument called an **ophthalmoscope** (of-THAL-mō-skōp). It consists of a light source, a set of mirrors or a prism arranged to reflect light so that the fundus of the eyeball is illuminated, and a set of lenses arranged on a rotating disc. Ophthalmoscopic examination of the eyeball permits examination of the retina (nervous tunic), optic disc, macula lutea, and blood vessels and is useful in detecting changes associated with conditions such as diabetes mellitus, atherosclerosis, and cataracts. Without using eye drops to dilate the pupil, a physician can see about 15% of the retina; if the pupil is dilated, about half of the retina can be visualized.

The procedure for using the ophthalmoscope is as follows:

CAUTION! *Please limit your examination time to one minute.*

PROCEDURE

1. In a dimly lit or dark room, seat your partner comfortably and have her or him look straight ahead at an object at eye level.
2. To examine your partner's right eye, hold the ophthalmoscope in your right hand and use your right eye. Reverse hands and eyes when viewing your partner's left eye.
3. With your index finger on the edge of the lens selection disc, rotate the disc so that the "O" is in position.
4. Hold the instrument about 15 cm (6 in.) from the subject's eye.

CAUTION! *Direct light toward the edge of the pupil rather than directly in its center.*

(If you direct light toward the center of the pupil, light will be refracted from the subject's cornea back into your eye.) You should now see a red circular area in the fundus of the eyeball.

5. Now, with the red area in view, move the ophthalmoscope to *within 5 cm (2 in.) of the subject*, and, while still directing the beam of light toward the edge of the pupil, rotate the lens selector until the optic disc is in sharp focus (Figure 14.12). Positive (+) settings (may be printed in black) correct for farsightedness and negative (–) settings (may be printed in red) correct for nearsightedness. Reduce the light, if necessary. Examine the optic disc carefully and note its blood vessels. The disc should have a sharp outline.
6. Examine the macula lutea lateral to the optic disc. It lacks blood vessels and should be darker than the optic disc. Locate the central fovea, a slightly lighter area in the center of the macula lutea.

6. Testing for Visual Acuity

The acuteness of vision may be tested by means of a **Snellen Chart.** It consists of letters of different sizes which are read at a distance normally designated at 20 ft. If the subject reads to the line that is marked "50," he or she is said to possess 20/50 vision in that eye, meaning that he or she is reading at 20 ft what a person who has normal vision can read at 50 ft. If he or she reads to the line marked "20," he or she has 20/20 vision in that eye. The normal eye can sufficiently refract light rays from an object 20 ft away to focus a clear object on the retina. Therefore, if you have 20/20 vision, your

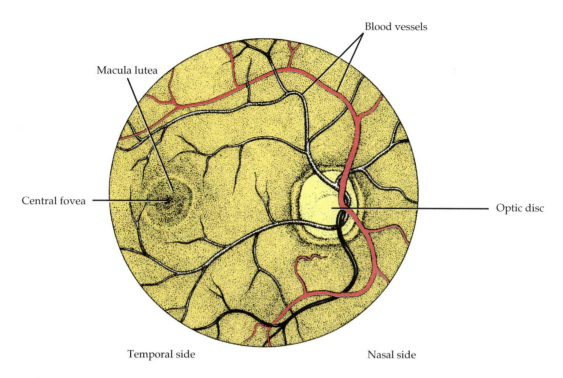

FIGURE 14.12 Diagram of the interior of the right eye as seen by ophthalmoscopy.

eyes are perfectly normal. The higher the bottom number, the larger the letter must be for you to see it clearly, and of course the worse or weaker are your eyes.

PROCEDURE

1. Have the subject stand 20 ft from the Snellen Chart and cover the right eye with a 3″ × 5″ (3-in.-by-5-in.) card.
2. Instruct the subject to slowly read down the chart until he or she can no longer focus the letters.
3. Record the number of the last line (20/20, 20/30, or whichever) that can be successfully read.
4. Repeat this procedure covering the left eye.
5. Now the subject should read the chart using both eyes.
6. Record your results in Section H.6 of the LABORATORY REPORT RESULTS at the end of the exercise and change places.

7. Testing for Astigmatism

The eye, with normal ability to refract light, is referred to as an *emmetropic* (em'-e-TROP-ik) eye. It can sufficiently refract light rays from an object 20 ft away to focus a clear object on the retina. If the lens is normal, objects as far away as the horizon and as close as about 20 ft will form images on the sensitive part of the retina. When objects are closer than 20 ft, however, the lens has to sharpen its focus by using the ciliary muscles. Many individuals, however, have abnormalities related to improper refraction. Among these are *myopia* (mī-Ō-pē-a) (nearsightedness), *hypermetropia* (hī'-per-mē-TRŌ-pē-a) (farsightedness), and *astigmatism* (a-STIG-ma-tizm) (irregularities in the surface of the lens or cornea).

Why do you think nearsightedness can be corrected with glasses containing biconcave lenses?

_____

_____

_____

How would you correct farsightedness? _____

_____

_____

PROCEDURE

1. In order to determine the presence of astigmatism, remove any corrective lenses if you are wearing them and look at the center of the

following astigmatism test chart, first with one eye, then the other:

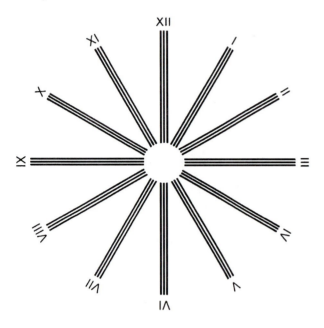

2. If all the radiating lines appear equally sharp and equally black, there is no astigmatism.
3. If some of the lines are blurred or less dark than others, astigmatism is present.
4. If you wear corrective lenses, try the test with them on.

8. Testing for the Blind Spot

PROCEDURE

1. Hold this page about 20 in. from your face with the cross shown below directly in front of your right eye. You should be able to see the cross and the circle when you close your left eye.
2. Now, keeping the left eye closed, slowly bring the page closer to your face while fixing the right eye on the cross.
3. At a certain distance the circle will disappear from your field of vision because its image falls on the blind spot.

9. Image Formation

Formation of an image on the retina requires four basic processes, all concerned with focusing light rays. These are (1) refraction of light rays, (2) ac-

commodation of the lens, (3) constriction of the pupil, and (4) convergence of the eyes.

When light rays traveling through a transparent medium (such as air) pass into a second transparent medium with a different density (such as water), the rays bend at the surface of the two media. This is called *refraction.* As light rays enter the eye, they are refracted at the anterior and posterior surfaces of the cornea. Both surfaces of the lens of the eye further refract the light rays so that they come into exact focus on the retina.

The lens of the eye has the unique ability to change the focusing power of the eye by becoming moderately curved at one moment and greatly curved the next. When the eye is focusing on a close object, the lens curves greatly in order to bend the rays toward the central fovea of the eye. This increase in the curvature of the lens is called *accommodation.*

a. TESTING FOR NEAR-POINT ACCOMMODATION

The following test determines your *near-point accommodation:*

PROCEDURE

1. Using any card that has a letter printed on it, close one eye and focus on the letter.
2. Measure the distance of the card from the eye using a ruler or a meter stick.
3. Now *slowly* bring the card as close as possible to your open eye, and stop when you no longer see a clear, detailed letter.
4. Measure and record this distance. This value is your near-point accommodation.
5. Repeat this procedure three times and then test your other eye.
6. Check Table 14.1 to see whether the near point for your eyes corresponds with that recorded for your age group. (*Note: Use a letter that is the size of typical newsprint.*)

TABLE 14.1
Correlation of Age and Near-point Accommodation

Age	Inches	Centimeters
10	2.95	7.5
20	3.54	9.0
30	4.53	11.5
40	6.77	17.2
50	20.67	52.5
60	32.80	83.3

b. TESTING FOR CONSTRICTION OF THE PUPIL

PROCEDURE

1. Place a 3″ × 5″ (3-in.-by-5-in.) card on the side of the nose so that a light shining on one side of the face will not affect the eye on the other side.
2. Shine the light from a lamp or a flashlight on one eye, 6 in. away (approximately 15 cm), for about 5 sec. Note the change in the size of the pupil of this eye.
3. Remove the light, wait about 3 min, and repeat, but this time observe the pupil of the opposite eye.
4. Wait a few minutes, and repeat the test, observing the pupils of both eyes.

c. TESTING FOR CONVERGENCE

In humans, both eyes focus on only one set of objects—a characteristic called *single binocular vision.* The term *convergence* refers to a medial movement of the two eyeballs so that they are both directed toward the object being viewed. The nearer the object, the greater the degree of convergence necessary to maintain single binocular vision.

PROCEDURE

1. Hold a pencil or pen about 2 ft from your nose and focus on its point. Now slowly bring the pencil toward your nose.
2. At some moment you should suddenly see two pencil points, or a blurring of the point.
3. Observe your partner's eyes when he or she does this test.

Images are actually focused upside down on the retina. They also undergo mirror reversal. That is, light reflected from the right side of an object hits the left side of the retina and vice versa. Reflected light from the top of the object crosses light from the bottom of the object and strikes the retina below the central fovea. Reflected light from the bottom of the object crosses light from the top of the object and strikes the retina above the central fovea.

The reason why we do not see a topsy-turvy world is that the brain learns early in life to coordinate visual images with the exact location of objects. The brain stores memories of reaching and touching objects and automatically turns visual images right-side up and right-side around.

10. Testing for Binocular Vision, Depth Perception, Diplopia, and Dominance

Humans are endowed with binocular vision. Each eye sees a different view, although there is a large degree of overlap. The cerebral cortex uses these discrepancies to produce *stereopsis* or three-dimensional vision, an orientation of the object in space that allows us to perceive its distance from us *(depth perception)*.

Even though there are slight differences in the views from each eye, we see a single image because the cerebral cortex integrates the images from each eye into a single perception. If this integration is not present, *diplopia* (dip-LŌ-pē-a) or double vision results. We do not perceive an "average" or "mean" of both left and right views; rather the view from one of our eyes is *dominant*, i.e., the view we always perceive with both eyes open.

PROCEDURE

1. Place a book at arm's length on a table. Place your left hand at your side and touch the nearest corner of the book with your right index finger. Close the left eye and repeat; close the right eye and repeat. Note the accuracy of your attempts. How accurate was your attempt to touch a

   point with one eye closed? _____
   If the right eye was closed, did you have error

   to the left or the right? _____
2. Use a depth perception tester with both eyes open and with each eye closed. Attempt to align the two arrows and measure monocular perception errors using the scale on the base.

   Length of depth perception error _____
3. Focus on a discrete object in the distance and press *very gently* on your left eyelid. Note how

   many objects you see: _____
4. Stab a pencil through a piece of paper and remove the scrap. Hold the paper at arm's length and focus on a small discrete object which just fills the hole. (You may want to use an X on the blackboard.) *Without moving the paper,* close the left eye and note whether the object is still present. Repeat with right eye

   closed. Note your dominant eye: _____

11. Testing for Afterimages

The rods and cones of the retina are photoreceptors; that is, they contain light-sensitive pigments that absorb light of different wavelengths. *Rhodopsin*, the light-sensitive pigment in rods, consists of opsin and retinal. When light strikes a molecule of rhodopsin, the retinal changes from a curved to a straight shape and breaks away from the opsin. The energy released as a result of the exchange initiates the nerve impulse that causes us to perceive light.

In bright light, all the rhodopsin is decomposed, so that the rods are nonfunctional. Cones function in bright light, in much the same way as rods, but they absorb light of specific wavelengths (red, blue, or green). The particular color perceived depends on which combinations of cones are stimulated. If all three are stimulated, we see white; if none is stimulated, we see black.

If photoreceptors are stimulated by staring at a bright object for a long period of time, they will continue firing briefly after the stimulus has been removed *(positive afterimage)*. However, after prolonged firing, these photoreceptors become "bleached" or fatigued, so that they can no longer fire, resulting in a reversal of the image *(negative afterimage)*.

PROCEDURE

1. Stare at a light source, *but not a bright light*, for 30 sec. Close the eyes briefly and note the positive afterimage. Open the eyes and focus on a piece of white paper. Note the negative afterimage.
2. Draw a red cross on paper about 1 in. high, with lines ½ in. thick. Stare at the cross without moving the eyes for at least 30 sec. Close the eyes very briefly; open them and stare at a piece of white paper. Note the positive and negative afterimages. Record your observations in Section H.11 of the LABORATORY REPORT RESULTS at the end of the exercise.

12. Testing for Color Blindness

Color blindness is an inherited disorder in which certain cones are absent. It is a sex-linked disorder. About 0.5% of all females and about 8% of all males are affected.

PROCEDURE

1. View Ishihara plates in bright light and compare your responses to those in the plate book or use Holmgren's test for matching colored threads.
2. Note the accuracy of your responses: _____

_____

13. Visual Pathway

From the rods and cones, impulses are transmitted through bipolar cells to ganglion cells. The cell bodies of the ganglion cells lie in the retina and their axons leave the eye via the *optic (II) nerve*. The axons pass through the *optic chiasm* (kī-AZ-em), a crossing point of the optic nerves. Fibers from the medial retina cross to the opposite side. Fibers from the lateral retina remain uncrossed. Upon passing through the optic chiasm, the fibers, now part of the *optic tract*, enter the brain and terminate in the lateral geniculate nucleus of the thalamus. Here the fibers synapse with the neurons whose axons pass to the visual centers located in the occipital lobes of the cerebral cortex. Label the visual pathway in Figure 14.13.

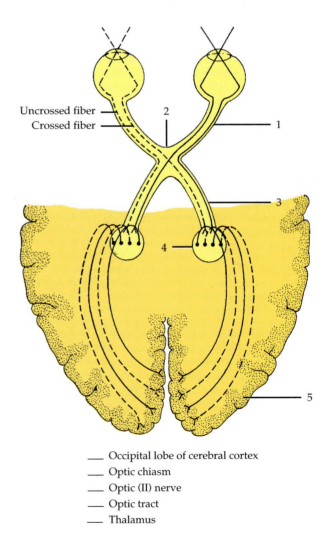

Uncrossed fiber
Crossed fiber

___ Occipital lobe of cerebral cortex
___ Optic chiasm
___ Optic (II) nerve
___ Optic tract
___ Thalamus

FIGURE 14.13 Visual pathway.

I. AUDITORY SENSATIONS AND EQUILIBRIUM

In addition to containing receptors for sound waves, the *ear* also contains receptors for equilibrium. The ear is subdivided into three principal regions: (1) external (outer) ear, (2) middle ear, and (3) internal (inner) ear.

1. Structure of Ear

a. EXTERNAL (OUTER) EAR

The *external (outer) ear* collects sound waves and directs them inward. Its structure consists of the auricle, external auditory canal, and tympanic membrane. The *auricle (pinna)* is a trumpet-shaped flap of elastic cartilage covered by thick skin. The rim of the auricle is called the *helix,* and the inferior portion is referred to as the *lobule.* The auricle is attached to the head by ligaments and muscles. The *external auditory canal (meatus)* is a tube, about 2.5 cm (1 in.) in length that leads from the auricle to the eardrum. The walls of the canal consist of bone lined with cartilage that is continuous with the cartilage of the auricle. Near the exterior opening, the canal contains a few hairs and specialized sebaceous glands called *ceruminous* (se-ROO-me-nus) *glands,* which secrete *cerumen* (earwax). The combination of hairs and cerumen prevents foreign objects from entering the ear. The *eardrum,* or *tympanic* (tim-PAN-ik) *membrane,* is a thin, semitransparent partition of fibrous connective tissue located between the external auditory canal and middle ear.

Examine a model or charts and label the parts of the external ear in Figure 14.14.

b. MIDDLE EAR

The *middle ear (tympanic cavity)* is a small, epithelium-lined, air-filled cavity hollowed out of the temporal bone. The area is separated from the external ear by the eardrum and from the internal ear by a very thin bony partition that contains two small membrane-covered openings, called the oval window and the round window. The posterior wall of the middle ear communicates with the mastoid cells of the temporal bone through a chamber called the *tympanic antrum.*

The anterior wall of the middle ear contains an opening that leads into the *auditory (Eustachian) tube.* The auditory tube connects the middle ear with the nose and nasopharynx. The function of the tube is to equalize air pressure on both sides of the eardrum. Any sudden pressure changes against the eardrum may be equalized by deliberately swallowing.

Extending across the middle ear are three exceedingly small bones called *auditory ossicles* (OS-si-kuls). These are known as the malleus, incus, and stapes. Based on their shape, they are commonly named the hammer, anvil, and stirrup, respectively. The "handle" of the *malleus* is attached to the internal surface of the eardrum. Its head articulates with the base of the *incus,* the intermediate bone in the series, which articulates with the stapes. The base of the *stapes* fits into a small opening between the middle and inner ear called the *oval window.* Directly inferior to the oval window is another opening, the *round window.* This opening, which separates the middle and inner ears, is enclosed by a membrane called the *secondary tympanic membrane.*

Examine a model or charts and label the parts of the middle ear in Figures 14.14 and 14.15 (on page 333).

c. INTERNAL (INNER) EAR

The *internal (inner) ear* is also known as the *labyrinth* (LAB-i-rinth). Structurally, it consists of two main divisions: (1) an outer bony labyrinth and (2) an inner membranous labyrinth that fits within the bony labyrinth. The *bony labyrinth* is a series of cavities within the petrous portion of the temporal bone that can be divided into three regions, named on the basis of shape: vestibule, cochlea, and semicircular canals. The bony labyrinth is lined with periosteum and contains a fluid called the *perilymph.* This fluid surrounds the *membranous labyrinth,* a series of sacs and tubes lying inside and having the same general form as the bony labyrinth. Epithelium lines the membranous labyrinth, which is filled with a fluid called the *endolymph.*

The *vestibule* is the oval, central portion of the bony labyrinth (see Figure 14.15). The membranous labyrinth within the vestibule consists of two sacs called the *utricle* (YOO-tri-kul) and *saccule* (SAK-yool). These sacs are connected to each other by a small duct.

Projecting superiorly and posteriorly from the vestibule are the three bony *semicircular canals* (see Figure 14.15). Each is arranged at approximately right angles to the other two. They are called the anterior, posterior, and lateral canals. One end of each canal enlarges into a swelling called the *ampulla* (am-POOL-la; = little jar). Inside the bony semicircular canals lie portions of the membranous labyrinth, the *semicircular ducts*

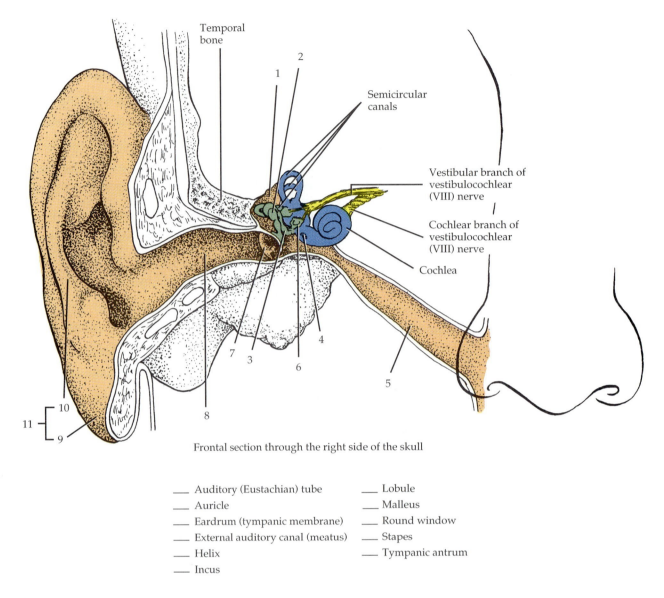

Temporal
bone

Semicircular
canals

Vestibular branch of
vestibulocochlear
(VIII) nerve

Cochlear branch of
vestibulocochlear
(VIII) nerve

Cochlea

Frontal section through the right side of the skull

____ Auditory (Eustachian) tube   ____ Lobule
____ Auricle   ____ Malleus
____ Eardrum (tympanic membrane)   ____ Round window
____ External auditory canal (meatus) ____ Stapes
____ Helix   ____ Tympanic antrum
____ Incus

FIGURE 14.14   Principal subdivisions of ear.

(membranous semicircular canals). These structures communicate with the utricle of the vestibule. Label these structures in Figure 14.16.

Anterior to the vestibule is the *cochlea* (KOK-lē-a; = snail's shell) (label it in Figure 14.16). The cochlea consists of a bony spiral canal that makes about three turns around a central bony core called the *modiolus.* A cross section through the cochlea shows that the canal is divided by partitions into three separate channels resembling the letter Y lying on its side. The stem of the Y is a bony shelf that protrudes into the canal. The wings of the Y are composed of the vestibular and basilar membranes. The channel above the partition is called the *scala vestibuli.* The channel below is known as the *scala tympani.* The cochlea adjoins the wall of the vestibule, into which the scala vestibuli opens. The scala tympani terminates at the round window. The perilymph of the vestibule is continuous with that of the scala vestibuli. The third channel (between the wings of the Y) is the membranous labyrinth, the *cochlear duct (scala media).* This duct is separated from the scala vestibuli by the *vestibular membrane* and from the scala tympani by the *basilar membrane.* Resting on the basilar membrane is the *spiral organ (organ of Corti),* the organ of hearing. Label these structures in Figure 14.17 on page 334.

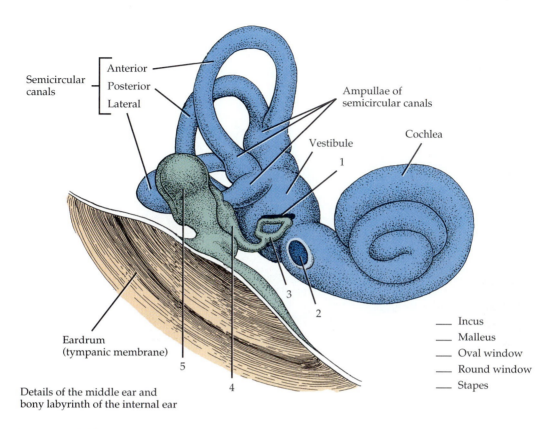

Semicircular canals
- Anterior
- Posterior
- Lateral

Ampullae of semicircular canals

Vestibule

Cochlea

1

3

2

Eardrum (tympanic membrane)

5

4

Details of the middle ear and bony labyrinth of the internal ear

___ Incus
___ Malleus
___ Oval window
___ Round window
___ Stapes

FIGURE 14.15   Ossicles of middle ear.

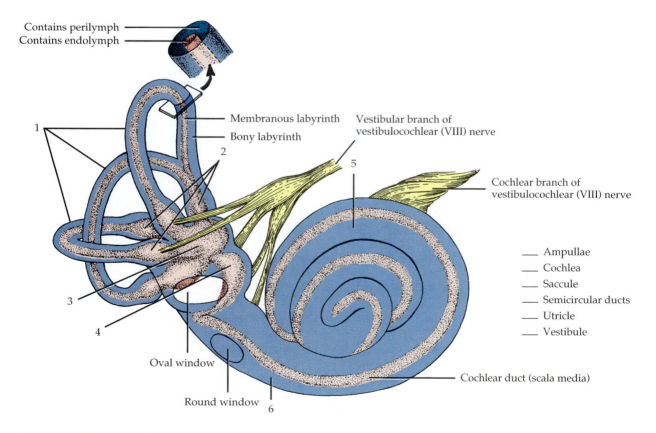

Contains perilymph
Contains endolymph

Membranous labyrinth
Bony labyrinth

Vestibular branch of vestibulocochlear (VIII) nerve

Cochlear branch of vestibulocochlear (VIII) nerve

1

2

5

___ Ampullae
___ Cochlea
___ Saccule
___ Semicircular ducts
___ Utricle
___ Vestibule

3

4

Oval window

Round window

6

Cochlear duct (scala media)

FIGURE 14.16   Details of inner ear.

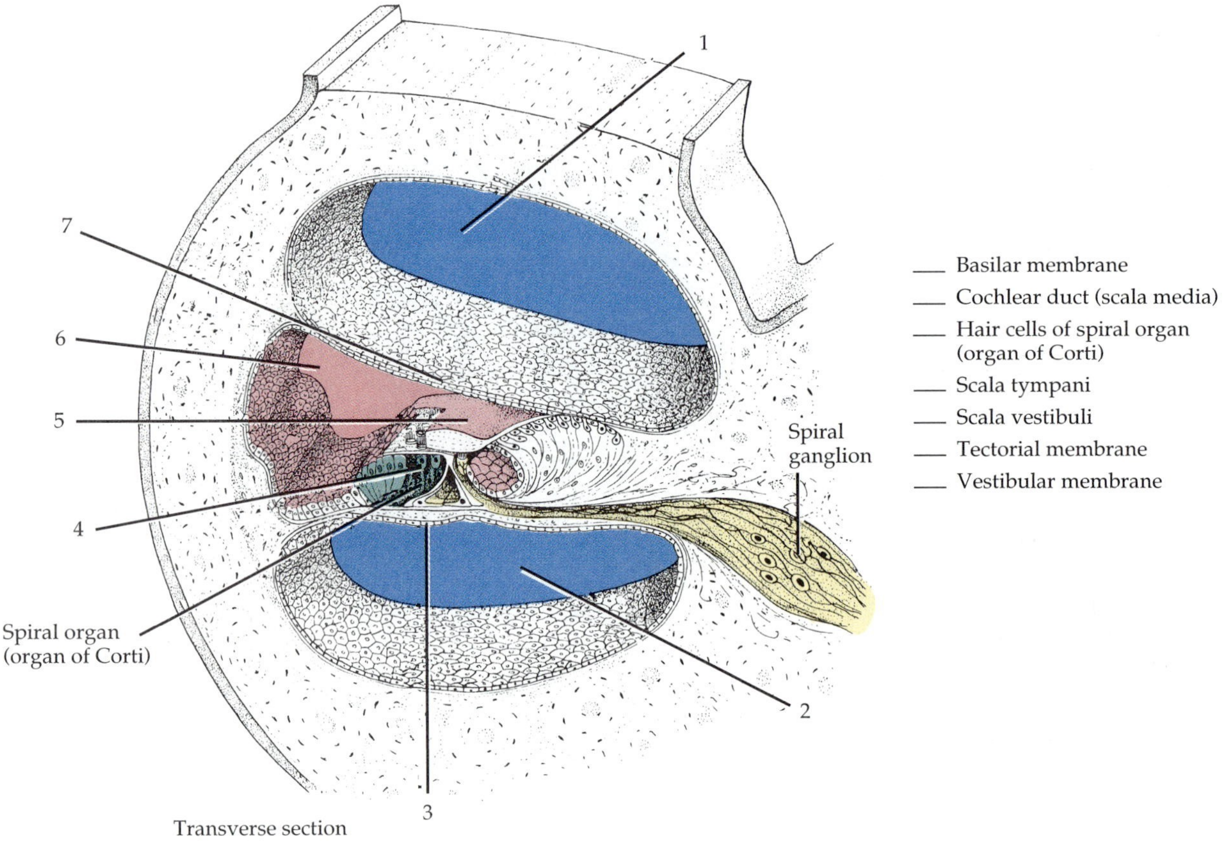

FIGURE 14.17 Cochlea.

The spiral organ (organ of Corti) is a coiled sheet of epithelial cells on the inner surface of the basilar membrane. This structure is composed of supporting cells and hair cells, which are receptors for auditory sensations. The hair cells have long hairlike processes at their free ends that extend into the endolymph of the cochlear duct. The basal ends of the hair cells are in contact with fibers of the cochlear branch of the vestibulocochlear (VIII) nerve. Projecting over and in contact with the hair cells of the spiral organ is the ***tectorial*** (*tectum* = cover) ***membrane,*** a very delicate and flexible gelatinous membrane. Label the tectorial membrane in Figure 14.17.

Obtain a prepared microscope slide of the spiral organ (organ of Corti) and examine under high power. Now label Figure 14.18.

2. Surface Anatomy

Refer to Figure 14.19 on page 336 for a summary of several surface anatomy features of the ear.

3. Otoscopy

An ***otoscope*** (Ō-tō-skōp) is an instrument used to examine the eardrum. Otoscopic visualization of the eardrum can detect infection, lesions, and the orientation of the malleus to the eardrum.

PROCEDURE

1. Clean the speculum (earpiece) of the otoscope with alcohol and turn on the light source.
2. Grasp the pinna of the subject's ear and *gently* pull it upward, backward, and a little laterally.
3. *Gently and slowly* insert the speculum into the external auditory canal (meatus) and angle the speculum downward and forward.
4. Observe the canal and eardrum. The eardrum should be pearly white; tympanic blood vessels should be visible.
5. Record your observations. Describe the structures viewed with the otoscope. ______________

   ____

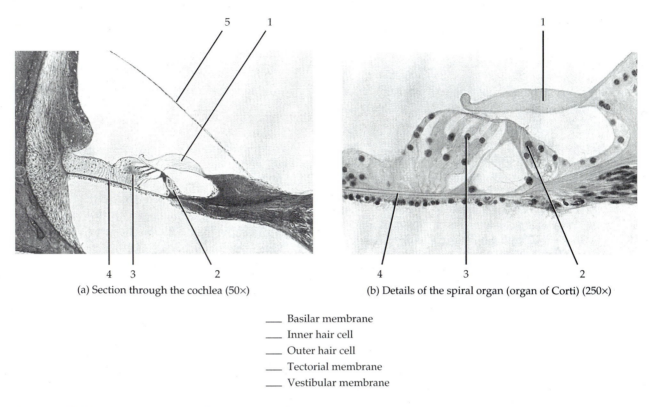

(a) Section through the cochlea (50×)

(b) Details of the spiral organ (organ of Corti) (250×)

___ Basilar membrane
___ Inner hair cell
___ Outer hair cell
___ Tectorial membrane
___ Vestibular membrane

FIGURE 14.18 Photomicrograph of the spiral organ (organ of Corti).

4. Tests for Auditory Acuity

Sound waves result from the alternate compression and decompression of air. The waves have both a frequency and an amplitude. The *frequency* is the distance between crests of a sound wave. It is measured in *Hertz (Hz)*, or *cycles per second*, and is perceived as *pitch*. The higher the frequency of a sound, the higher its pitch. Different frequencies displace different areas of the basilar membrane. The *amplitude* of a sound wave is perceived as *intensity (loudness)*. Intensity is measured in *decibels (dB)*. Population norms have been calculated that measure the intensity of a sound that can just be heard, i.e., the *threshold* of sound. These levels are said to have an intensity of 0 dB. Each 10 dB reflects a tenfold increase in intensity; thus a sound is 10 times louder than threshold at 10 dB, 100 times greater at 20 dB, 1 million times greater at 60 dB, and so on.

a. SOUND LOCALIZATION TEST

The cerebral cortex localizes the source of a sound by evaluating the time lag between the entry of sound into each ear. Without turning the head, it is impossible to distinguish between the origin of

sounds that arise at 30° ahead of the right ear and 30° behind the right ear.

PROCEDURE
1. In a quiet room, have a subject occlude one ear with a cotton plug and close the eyes.
2. Move a ticking watch to various points 6 in. away from the ear (side, front, top, back) and ask the subject to point to the location of the sound.
3. Note in which positions the sound is best localized. In what regions was sound localization most accurate? _____

_____

b. WEBER'S TEST

Weber's test is diagnostic for both conduction and sensorineural deafness. *Conduction deafness* results from interference with sound waves reaching the inner ear (plugged external auditory canal, fusion of the ossicles, etc.); *sensorineural deafness* is caused by damage to the nerve pathway between the cochlea and the brain. This test should not be performed in a quiet room.

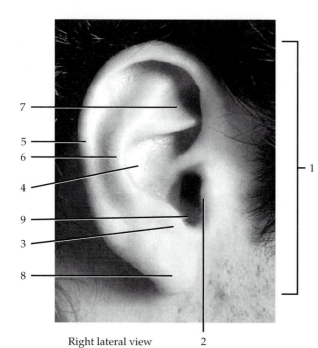

Right lateral view

1. **Auricle.** Portion of external ear not contained in head, also called trumpet.
2. **Tragus.** Cartilaginous projection anterior to external opening of ear.
3. **Antitragus.** Cartilaginous projection opposite tragus.
4. **Concha.** Hollow of auricle.
5. **Helix.** Superior and posterior free margin of auricle.
6. **Antihelix.** Semicircular ridge posterior and superior to concha.
7. **Triangular fossa.** Depression in superior portion of anti-helix.
8. **Lobule.** Inferior portion of auricle devoid of cartilage.
9. **External auditory canal (meatus).** Canal extending from external ear to eardrum.

FIGURE 14.19 Surface anatomy of ear.

PROCEDURE

1. Strike a tuning fork with a mallet and place the tip of the handle on the median line of a subject's forehead.
2. Determine if the tone is equally loud in both ears. _____

3. Have the subject occlude the left ear with cotton and repeat the exercise; occlude the right ear and repeat.

   Describe your results: _____

   _____

If the sound is equally loud in both unoccluded ears, hearing is normal. If the sound is louder in the occluded ear, hearing is normal. This situation par-

allels conduction deafness, when the cochlea is not receiving environmental background noise but only the vibrations of the tuning fork. If there is sensorineural deafness, the sound is louder in the normal (or nonoccluded) ear, because sensorineural loss produces a deficit in transmission to the cerebral cortex.

C. RINNE TEST FOR CONDUCTION DEAFNESS

PROCEDURE

1. In a quiet room, strike a tuning fork with a mallet and place the handle of the tuning fork on the right mastoid process of a subject.
2. When the sound is no longer audible to the subject, move the vibrating fork close to the external auditory canal.
3. Ask the subject to note whether sound is still heard.
4. If the subject hears the fork again when moved near the external auditory canal (by air conduction), hearing is *not* impaired.
5. Repeat the test by testing air conduction hearing first by striking a tuning fork and holding it near the external auditory canal. When the sound is no longer audible, place the handle of the tuning fork on the mastoid process.
6. If the subject now hears the tuning fork again (by bone conduction), this indicates some degree of conduction deafness.
7. Repeat 1–6 using the left ear of the subject.
8. Record your results in Section I.4.c of the LABORATORY REPORT RESULTS at the end of the exercise.

With normal hearing ability, air conduction lasts longer than bone conduction, because the sound waves are not absorbed by the tissues. Therefore, the sound is heard outside the external ear. In conduction deafness, bone conduction is better than air conduction because the sound waves can go to the cochlea without interference by damaged structures. Therefore, the sound will not be heard outside the external ear.

d. AUDIOMETRY

An *audiometer* (aw'-dē-OM-e-ter) is an instrument that tests auditory acuity by delivering stimuli of different frequencies at varying intensities to determine the hearing threshold for each frequency. The audiometer can use either air conduction or bone conduction. The subject reports whether or not a sound is heard. The threshold for sound detection is measured in decibels of loudness. If a threshold is

30 or more decibels (dB) above the population norm for perceiving that sound, the individual is said to have a hearing loss of 30 dB for that specific tone. An *audiogram* plots hearing loss (or normalcy) for each tested frequency in the auditory spectrum.

Children can perceive sound between 30 and 22,000 Hz. A sensorineural loss accompanies aging. It begins after age 20, and the highest frequencies are affected first. By age 40, the upper limit is about 15,000 Hz; by age 80, it may be as low as 8000 Hz.

PROCEDURE

1. In a quiet room, have the subject place the earphones with the red earphone on the right ear, blue on the left.
2. With the intensity control set at audiometric zero (the tone which can just be heard by the normal ear), set the frequency control for 125 Hz.
3. Turn the output switch to red (right ear) and push the bar level to send a continuous tone through the right earphone.
4. Instruct the subject to raise a finger when a tone is heard.
5. Increase the sound intensity gradually until it is heard.
6. Record the results in Section I.4.d of the LABORATORY REPORT RESULTS at the end of the exercise.
7. Repeat this procedure for 500, 1000, 2000, 4000, and 8000 Hz. Repeat for the left ear by turning the output switch to blue.
8. Plot the results in Section I.4.d of the LABORATORY REPORT RESULTS at the end of the exercise.

5. Equilibrium Apparatus

The term *equilibrium* (balance) has two meanings. One kind of equilibrium, called *static equilibrium,* refers to the position of the body (mainly the head) relative to the force of gravity. The second kind of equilibrium, called *dynamic equilibrium,* is the maintenance of the position of the body (mainly the head) in response to sudden movements (rotation, acceleration, and deceleration). Collectively, the receptor organs for equilibrium are called the *vestibular apparatus,* which includes the maculae in the saccule and utricle and the cristae in the semicircular ducts.

The *maculae* (MAK-yoo-lē) in the walls of the *utricle* and *saccule* are the receptors concerned mainly with static equilibrium. The maculae are small, flat regions that resemble the spiral organ (organ of Corti) microscopically. Maculae are located in planes perpendicular to each other and contain two kinds of cells: *hair (receptor) cells* and *supporting cells.* The hair cells project *stereocilia* (microvilli) and a *kinocilium* (conventional cilium). The columnar supporting cells are scattered among the hair cells. Floating over the hair cells is a thick, gelatinous glycoprotein layer, the *otolithic membrane.* A layer of calcium carbonate crystals, called *otoliths* (*oto* = ear; *lithos* = stone), extends over the entire surface of the otolithic membrane. When the head is tilted, the membrane slides over the hair cells in the direction determined by the tilt of the head. This sliding causes the membrane to pull on the stereocilia, thus initiating a nerve impulse that is conveyed via the vestibular branch of the vestibulocochlear (VIII) nerve to the brain (cerebellum). The cerebellum sends continuous nerve impulses to the motor areas of the cerebral cortex in response to input from the maculae in the utricle and saccule, causing the motor system to increase or decrease its nerve impulses to specific skeletal muscles to maintain static equilibrium.

Label the parts of the macula in Figure 14.20a.

Now consider the role of the cristae in the semicircular ducts in maintaining dynamic equilibrium. The three semicircular ducts are positioned at right angles to one another in three planes: the two vertical ones are called the *anterior* and *posterior semicircular ducts* and the horizontal one is called the *lateral semicircular duct.* This positioning permits correction of an imbalance in three planes. In the *ampulla,* the dilated portion of each duct, is a small elevation called the *crista.* Each crista is composed of a group of *hair (receptor) cells* and *supporting cells* covered by a mass of gelatinous material called the *cupula.* When the head moves, endolymph in the semicircular ducts flows over the hairs and bends them as water in a stream bends the plant life growing at its bottom. Movement of the hairs stimulates sensory neurons, and nerve impulses pass over the vestibular branch of the vestibulocochlear (VIII) nerve. The nerve impulses follow the same pathway as those involved in static equilibrium and are eventually sent to the muscles that contract to maintain body balance in the new position.

Label the parts of the crista in Figure 14.20b.

6. Tests for Equilibrium

You can test equilibrium by using a few simple procedures.

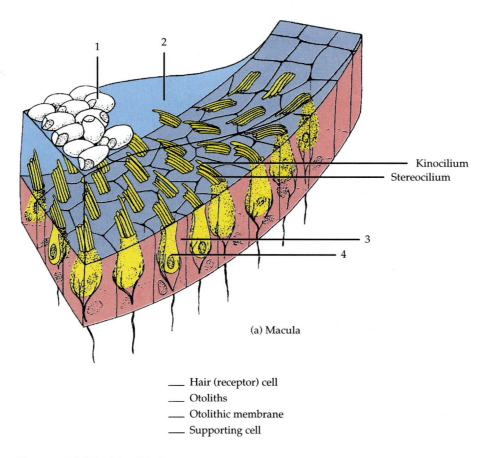

Kinocilium
Stereocilium

(a) Macula

___ Hair (receptor) cell
___ Otoliths
___ Otolithic membrane
___ Supporting cell

FIGURE 14.20   Equilibrium apparatus.

a. BALANCE TEST

This test is used to evaluate static equilibrium.

PROCEDURE

1. Instruct a subject to stand perfectly still with the arms at the sides and the eyes closed.
2. Observe any swaying movements. These are easier to detect if the subject stands in front of a blackboard with a light in front of the subject.
3. A mark is made at the edge of the shadow of the subject's shoulders and movement of the shadow is then observed.

If the static equilibrium system is dysfunctional the subject will sway or fall, although other proprioreceptors may compensate for the defect.

Record your observations in Section I.5.a of the LABORATORY REPORT RESULTS at the end of the exercise.

b. BARANY TEST

This test evaluates the function of each of the semicircular canals. A subject is rotated, and when rotation is stopped, the momentum causes the endolymph in the lateral (horizontal) canal to continue spinning. The spinning bends the cupula in the direction of rotation. Other sensory input informs us that we are no longer rotating. *Nystagmus*, the reflex movement of the eyes rapidly and then slowly, attempts to compensate for the loss of balance by visual fixation on an object. When the head is tipped toward the shoulder, the anterior semicircular canals are stimulated, so nystagmus is vertical; when the head is tipped slightly forward, the lateral canals are stimulated by the rotation, so nystagmus occurs laterally; when the head is bent onto the chest, the posterior canal receptors are stimulated, so nystagmus is rotational.

PROCEDURE

1. Use a person who is not subject to vertigo (dizziness).
2. The subject sits firmly anchored on a stool, legs up on the stool rung, head tilted onto one shoulder. Then the stool is *very carefully*

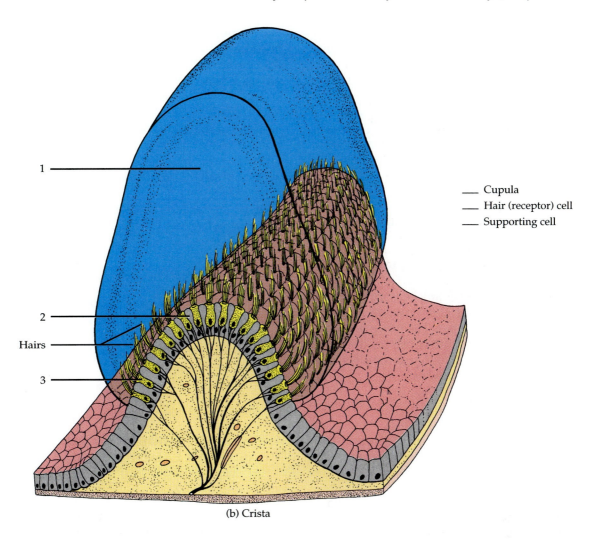

_____ Cupula
_____ Hair (receptor) cell
_____ Supporting cell

(b) Crista

FIGURE 14.20 (Continued)   Equilibrium apparatus.

revolved to the right about 10 times (about one turn every two seconds) and suddenly stopped.

3. Observe the subject's eye movements.

CAUTION! *Watch the subject carefully and be prepared to provide support until the dizziness has passed.*

4. The subject will experience the sensation that the stool is still rotating, which means that the semicircular canals are functioning properly.

5. Repeat on a different subject with the head tipped slightly forward. Use still another subject with the chin touching the chest.

6. Record the direction of nystagmus in all three subjects in Section I.5.b of the LABORATORY REPORT RESULTS at the end of the exercise.

J. SENSORY-MOTOR INTEGRATION

Sensory systems provide the input that keeps the central nervous system informed of changes in the external and internal environment. Responses to this information are conveyed to motor systems, which enable us to move about, alter glandular secretions, and change our relationship to the world around us. As sensory information reaches the CNS, it becomes part of a large pool of sensory input. We do not actively respond to every bit of input the CNS receives. Rather, the incoming information is integrated with other information arriving from all other operating sensory receptors. The integration process occurs not just once, but at many stations along the pathways of the CNS and at both conscious and subconscious levels. It occurs within the spinal cord, brain stem, cerebellum, basal ganglia, and cerebral cortex. As a result, a motor response to make a muscle contract or a gland secrete can be modified at any of these levels. Motor portions of the cerebral motor cortex play the major role in initiating and controlling precise, discrete muscular movements. The basal ganglia largely integrate semivoluntary, automatic movements like walking, swimming, and laughing. The cerebellum assists the motor cortex and basal ganglia by making body movements smooth and coordinated and by contributing significantly to maintaining normal posture and balance.

K. SOMATIC MOTOR PATHWAYS

After receiving and interpreting sensory information, the CNS generates nerve impulses to direct responses to that sensory input. The nerve impulses are sent down the spinal cord in two major motor pathways: the direct (pyramidal) pathways and the indirect (extrapyramidal) pathways.

1. Direct Pathways

Voluntary motor impulses propagate from the motor cortex to voluntary motor neurons (somatic motor neurons) that innervate skeletal muscles via the *direct* or *pyramidal* (pi-RAM-i-dal) *pathways.* The direct pathways convey impulses from the cortex that result in precise, voluntary movements and include three tracts: lateral corticospinal (control muscles in the distal portions of the limbs), anterior corticospinal (control muscles of the neck and trunk), and corticobulbar (control muscles of the eyes, tongue, and neck; chewing, facial expression, and speech) (see page 269).

Based on the description provided on page 270, label the components of the lateral corticospinal tracts in Figure 14.21.

2. Indirect Pathways

The *indirect* or *extrapyramidal pathways* include all motor tracts other than the corticospinal and corticobulbar tracts (see p. 270). These are the rubrospinal, tectospinal, vestibulospinal, and lateral and medial veticulospinal.

ANSWER THE LABORATORY REPORT QUESTIONS AT THE END OF THE EXERCISE.

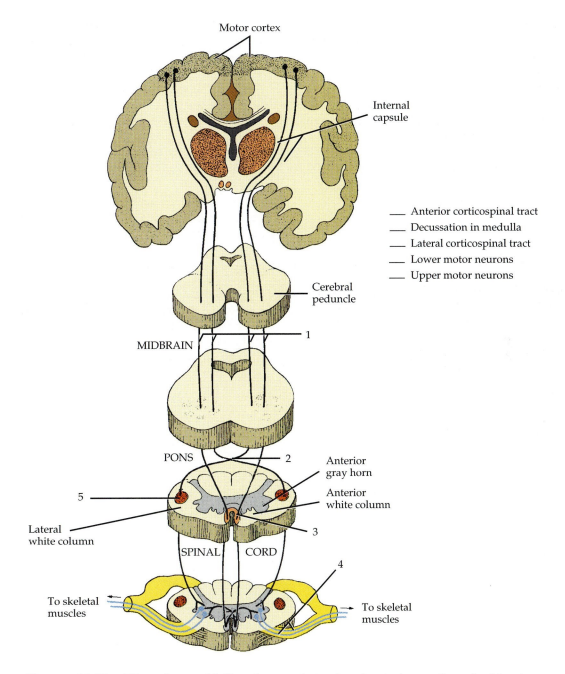

Motor cortex

Internal capsule

___ Anterior corticospinal tract
___ Decussation in medulla
___ Lateral corticospinal tract
___ Lower motor neurons
___ Upper motor neurons

Cerebral peduncle

MIDBRAIN

1

PONS

2

Anterior gray horn

5

Anterior white column

Lateral white column

3

SPINAL CORD

4

To skeletal muscles

To skeletal muscles

FIGURE 14.21   Direct (pyramidal) pathways: lateral and anterior corticospinal tracts.

Sensory Receptors and Sensory and Motor Pathways

14

Student _____   Date _____

Laboratory Section _____   Score/Grade _____

SECTION D. TESTS FOR GENERAL SENSES

1. Two-Point Discrimination Test

Part of body	Least distance at which two points can be detected
Tip of tongue	1.4 mm
Tip of finger	_____
Side of nose	_____
Back of hand	_____
Back of neck	36.2 mm

3. Identifying Pressure Receptors

Part of body	Distances between points touched by chalk
Palm	_____
Arm	_____
Forearm	_____
Back of neck	_____

6. Identifying Proprioceptors

First trial	_____
Second trial	_____
Third trial	_____

SECTION G. GUSTATORY SENSATIONS
2. Identifying Taste Zones

Areas of Tongue in Which Basic Tastes Are Detected

	Sweet	Bitter	Salty	Sour
Tip of tongue				
Back of tongue				
Sides of tongue				

4. Taste and Smell

	Sensations when placed on dry tongue	Sensations while chewing (nostrils closed)	Sensation while chewing with nostrils open
Carrot			
Onion			
Potato			
Apple			

SECTION H. VISUAL SENSATIONS
6. Testing for Visual Acuity

Visual activity, left eye   _____

Visual activity, right eye   _____

Visual activity, both eyes   _____

11. Testing for Afterimages

	Appearance of positive afterimage	Appearance of negative afterimage
Bright light		
Red cross		

SECTION I. AUDITORY SENSATIONS AND EQUILIBRIUM

4. Tests for Auditory Acuity

c. RINNE TEST

	Air conduction	Bone conduction
Right ear		
Left ear		

d. AUDIOMETRY

Frequency in Hz	Threshold: left ear	Threshold: right ear
125		
500		
1000		
2000		
4000		
8000		

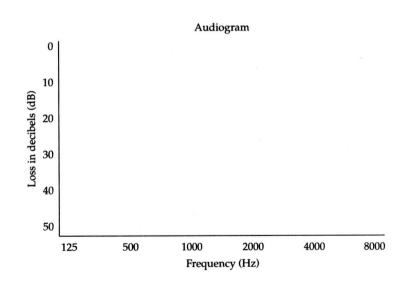

5. Tests For Equilibrium

a. BALANCE TEST

Describe the response of the subject. _____

_____

_____

b. BARANY TEST

Subject 1 _____

Subject 2 _____

Subject 3 _____

Sensory Receptors and Sensory and Motor Pathways **14**

Student _____   Date _____

Laboratory Section _____   Score/Grade _____

PART 1. Multiple Choice

_____ 1. The process by which the brain refers sensations to their point of stimulation is referred to as (a) modality (b) projection (c) accommodation (d) convergence

_____ 2. An awareness of the activities of muscles, tendons, and joints is known as (a) referred pain (b) adaptation (c) refraction (d) proprioception

_____ 3. The papillae located in an inverted V-shaped row at the posterior portion of the tongue are the (a) circumvallate (b) filiform (c) fungiform (d) gustatory

_____ 4. The technical name for the "white of the eye" is the (a) cornea (b) conjunctiva (c) choroid (d) sclera

_____ 5. Which is *not* a component of the vascular tunic? (a) choroid (b) macula lutea (c) iris (d) ciliary body

_____ 6. The amount of light entering the eyeball is regulated by the (a) lens (b) iris (c) cornea (d) conjunctiva

_____ 7. Which region of the eye is concerned primarily with image formation? (a) retina (b) choroid (c) lens (d) ciliary body

_____ 8. The densest concentration of cones is found at the (a) blind spot (b) macula lutea (c) central fovea (d) optic disc

_____ 9. Which region of the eyeball contains the vitreous body? (a) anterior chamber (b) vitreous chamber (c) posterior chamber (d) conjunctiva

_____ 10. Among the structures found in the middle ear are the (a) vestibule (b) auditory ossicles (c) semicircular canals (d) external auditory canal

_____ 11. The receptors for dynamic equilibrium are the (a) saccules (b) utricles (c) cristae in semicircular ducts (d) spiral organs (organs of Corti)

_____ 12. The auditory ossicles are attached to the eardrum, to each other, and to the (a) semicircular ducts (b) semicircular canals (c) oval window (d) labyrinth

_____ 13. Another name for the internal ear is the (a) labyrinth (b) fenestra (c) cochlea (d) vestibule

_____ 14. Orientation of the body relative to the force of gravity is termed (a) postural reflex (b) tonal reflex (c) dynamic equilibrium (d) static equilibrium

_____ 15. The inability to feel a sensation consciously even though a stimulus is still being applied is called (a) modality (b) projection (c) adaptation (d) afterimage formation

———— 16. Which sequence best describes the normal flow of tears from the eyes into the nose? (a) lacrimal canals, lacrimal sacs, nasolacrimal ducts (b) lacrimal sacs, lacrimal canals, nasolacrimal ducts (c) nasolacrimal ducts, lacrimal sacs, lacrimal canals (d) lacrimal sacs, nasolacrimal ducts, lacrimal canals

———— 17. A patient whose lens has lost transparency is suffering from (a) glaucoma (b) conjunctivitis (c) cataract (d) trachoma

———— 18. The portion of the eyeball that contains aqueous humor is the (a) anterior cavity (b) lens (c) posterior chamber (d) macula lutea

———— 19. The organ of hearing located within the inner ear is the (a) vestibule (b) oval window (c) modiolus (d) spiral organ (organ of Corti)

———— 20. The sense organs of static equilibrium are the (a) semicircular ducts (b) membranous labyrinths (c) maculae in the utricle and saccule (d) pinnae

———— 21. Which receptor does *not* belong with the others? (a) muscle spindle (b) tendon organ (Golgi tendon organ) (c) joint kinesthetic receptor (d) lamellated (Pacinian) corpuscle

———— 22. The membrane that is reflected from the eyelids onto the eyeball is the (a) retina (b) bulbar conjunctiva (c) sclera (d) choroid

———— 23. A characteristic of sensations by which one sensation may be distinguished from another is called (a) modality (b) projection (c) adaptation (d) afterimage formation

———— 24. Which are *not* cutaneous receptors? (a) hair root plexuses (b) muscle spindles (c) type I cutaneous mechanoreceptors (tactile or Merkel) discs (d) corpuscles of touch (Meissner's corpuscles)

———— 25. Which region of the tongue reacts strongest to bitter tastes? (a) tip (b) center (c) back (d) sides

———— 26. Which of the following values indicates the best visual acuity? (a) 20/30 (b) 20/40 (c) 20/50 (d) 20/60

———— 27. Nearsightedness is referred to as (a) emmetropia (b) hypermetropia (c) eumetropia (d) myopia

PART 2. Completion

28. Receptors found in blood vessels and viscera are classified as ————————.

29. Structures that collectively produce and drain tears are referred to as the ————————.

30. The three zones of the inner nervous layer of the retina (nervous tunic) are the photoreceptor layer, bipolar cell layer, and ————————.

31. The small area of the retina where no image is formed is referred to as the ————————.

32. Abnormal elevation of intraocular pressure (IOP) is called ————————.

33. In the visual pathway, nerve impulses pass from the optic chiasm to the ———————— before passing to the thalamus.

34. The openings between the middle and inner ears are the oval and ———————— windows.

35. The fluid within the bony labyrinth is called ————————.

36. Tactile sensations include touch, pressure, and ————————.

37. Receptors for pressure are type II cutaneous mechanoreceptors (end organs of Ruffini) and

_____ corpuscles.

38. Proprioceptive receptors that provide information about the degree and rate of angulations of joints

are _____ .

39. The neural pathway for olfaction includes olfactory receptors, olfactory bulbs,

_____ , and cerebral cortex.

40. The posterior wall of the middle ear communicates with the mastoid air cells of the temporal bone

through the _____ .

41. A thin semitransparent partition of fibrous connective tissue that separates the external auditory

meatus from the inner ear is the _____ .

42. The fluid in the membranous labyrinth is called _____ .

43. The cochlear duct is separated from the scala vestibuli by the _____ .

44. The gelatinous glycoprotein layer over the hair cells in the maculae is called

the _____ .

45. _____ is blurred vision caused by an irregular curvature of the surface of the
cornea or lens.

46. Image formation requires refraction, accommodation, constriction of the pupil, and

_____ .

47. The _____ tract conveys sensory impulses for pain and temperature.

48. The _____ tract conveys motor impulses for precise contraction of muscles in the
distal portions of the limbs.

49. Pain receptors are called _____ .

50. The ability to recognize exactly which point of the body is touched is called _____ .

51. _____ neurons extend from cranial nerve motor nuclei or spinal cord anterior
horns to skeletal muscle fibers.

52. _____ tracts convey impulses that control voluntary movements of the head and
neck.

Endocrine System

15

You have learned how the nervous system controls the body through nerve impulses that are delivered over neurons. Another system of the body, the *endocrine system,* is also involved in controlling bodily functions. The endocrine glands affect bodily activities by releasing chemical messengers, called *hormones,* into the blood stream (the term "hormone" means "to urge on"). The nervous and endocrine systems coordinate their regulatory activities via a complex series of interacting activities. Certain parts of the nervous system stimulate or inhibit the release of hormones. The hormones, in turn, are quite capable of stimulating or inhibiting the flow of particular nerve impulses.

The body contains two different kinds of glands: exocrine and endocrine. *Exocrine glands* secrete their products into ducts. The ducts then carry the secretions into body cavities, into the lumens of various organs, or to the external surface of the body. Examples are sudoriferous (sweat), sebaceous (oil), mucous, and digestive glands. *Endocrine glands,* by contrast, secrete their products (hormones) into the extracellular space around the secretory cells. The secretion then diffuses into blood vessels and the blood. Because they have no ducts, endocrine glands are also called *ductless glands.*

A. ENDOCRINE GLANDS

The endocrine glands are the anterior pituitary gland, thyroid gland, parathyroid glands, adrenal cortex, adrenal medulla, pineal gland, and thymus gland. In addition, several organs of the body contain endocrine tissue, but are not endocrine glands exclusively. They include the pancreas, testes, kidney, hypothalamus, and placenta. The stomach, small intestine, skin, and heart also secrete hormone-like substances and are considered to have endocrine tissue with yet to be completely determined functions. The endocrine glands are organs that together form the *endocrine system.*

Locate the endocrine glands on a torso, and, using your textbook or charts for reference, label Figure 15.1.

All hormones maintain homeostasis by changing the physiological activities of cells. A hormone may stimulate changes in the cells of one or more organs. The cells that respond to the effects of a hormone are called *target cells*.

B. PITUITARY GLAND (HYPOPHYSIS)

The hormones of the *pituitary gland*, also called the *hypophysis* (hī-POF-i-sis), regulate so many body activities that the pituitary gland has been nicknamed the "master gland." This gland lies in the sella turcica of the sphenoid bone and is attached to the hypothalamus via a stalklike structure termed the *infundibulum*. Not only is the hypothalamus of the brain an important regulatory center in the nervous system; it is also a crucial endocrine gland. Cells in the hypothalamus synthesize at least nine different hormones, and the pituitary gland secretes seven more. Together, they play important roles in the regulation of virtually all aspects of growth, development, metabolism, and homeostasis.

The pituitary gland is divided structurally and functionally into an *anterior pituitary gland*, also called the *adenohypophysis* (ad'-i-nō-hī-POF-i-sis), and a *posterior pituitary gland*, also called the *neurohypophysis* (noo'-rō-hī-POF-i-sis). The anterior pituitary gland contains many glandular epitheloid (epithelial-like) cells and forms the glandular part of the pituitary gland. The hypothalamus is connected to the anterior pituitary gland by a series of blood vessels, the *hypothalamic–hypophyseal portal system.* The posterior pituitary gland contains the axon terminals of neurons whose cell bodies are in the

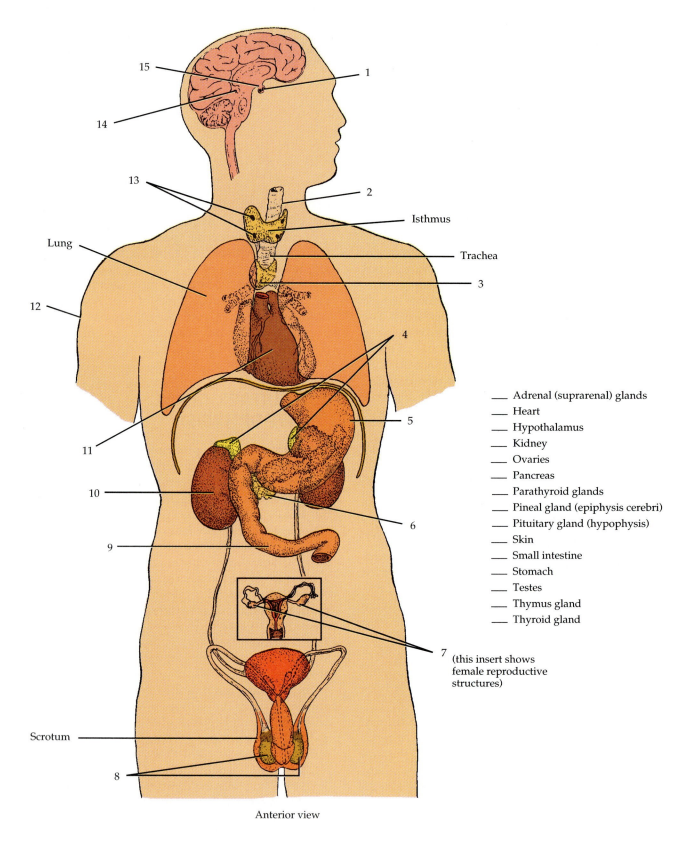

15
1
14
13
2
Isthmus
Lung
Trachea
3
12
4
11
5
10
6
9
7 (this insert shows
female reproductive
structures)
Scrotum
8

___ Adrenal (suprarenal) glands
___ Heart
___ Hypothalamus
___ Kidney
___ Ovaries
___ Pancreas
___ Parathyroid glands
___ Pineal gland (epiphysis cerebri)
___ Pituitary gland (hypophysis)
___ Skin
___ Small intestine
___ Stomach
___ Testes
___ Thymus gland
___ Thyroid gland

Anterior view

FIGURE 15.1   Location of endocrine glands, organs containing endocrine tissue, and associated structures.

hypothalamus. These terminals form the neural part of the pituitary gland. Between the anterior and posterior pituitary glands is a small, relatively avascular rudimentary zone, the **pars intermedia**, whose role in humans is unknown.

1. Histology of the Pituitary Gland

The anterior pituitary gland releases hormones that regulate a whole range of body activities, from growth to reproduction. However, the release of these hormones is stimulated by **releasing hormones** or inhibited by **inhibiting hormones** that are produced by neurosecretory cells in the hypothalamus of the brain. This is one interaction between the nervous system and the endocrine system. The hypothalamic hormones reach the anterior pituitary gland through a network of blood vessels.

When the anterior pituitary gland receives proper stimulation from the hypothalamus via releasing hormones, its glandular cells secrete any one of seven hormones. Special staining techniques have established the division of glandular cells into five principal types:

1. **Somatotrophs** (*soma* = body; *tropos* = changing) produce **human growth hormone (hGH)**, which stimulates general body growth and certain aspects of metabolism.
2. **Lactotrophs** (*lact* = milk) synthesize **prolactin (PRL)**, which initiates milk production by suitably prepared mammary glands.
3. **Corticolipotrophs** (*cortex* = rind or bark) synthesize **adrenocorticotropic** (ad-rē-nō-kor'-ti-kō-TRŌ-pik) **hormone (ACTH)**, which stimulates the adrenal cortex to secrete its hormones and **melanocyte-stimulating hormone (MSH)**, which affects skin pigmentation.
4. **Thyrotrophs** (*thyreos* = shield) manufacture **thyroid-stimulating hormone (TSH)**, which controls the thyroid gland.
5. **Gonadotrophs** (*gonas* = seed) produce **follicle-stimulating hormone (FSH)** and **luteinizing hormone (LH)**. Together these hormones stimulate secretion of estrogens and progesterone by the ovaries and maturation of oocytes and secretion of testosterone and production of sperm in the testes.

Except for human growth hormone (hGH), melanocyte-stimulating hormone (MSH), and prolactin (PRL), all the secretions are referred to as **tropins** or **tropic hormones**, which means that

their target organs are other endocrine glands. Follicle-stimulating hormone (FSH) and luteinizing hormone (LH) are also called **gonadotropic** (gō-nad-ō-TRŌ-pik) **hormones** because they regulate the functions of the gonads (ovaries and testes). The gonads are the endocrine glands that produce sex hormones.

Examine a prepared slide of the anterior pituitary gland and identify as many types of cells as possible with the aid of your textbook or a histology textbook.

The posterior pituitary gland is not really an endocrine gland. Instead of synthesizing hormones, it stores and releases hormones synthesized by cells of the hypothalamus. The posterior pituitary gland consists of (1) cells called **pituicytes** (pi-TOO-i-sītz), which are similar in appearance to the neuroglia of the nervous system, and (2) axon terminations of secretory nerve cells of the hypothalamus. The cell bodies of the neurons, called **neurosecretory cells**, originate in nuclei in the hypothalamus. The fibers project from the hypothalamus, form the **supraopticohypophyseal** (soo'-pra-op'ti-kō-hī'-pō-FIZ-ē-al) or **hypothalamic–hypophyseal tract**, and terminate on blood capillaries in the posterior pituitary gland. The cell bodies of the neurosecretory cells produce the hormones **oxytocin** (ok'-sē-TŌ-sin), or **OT**, and **antidiuretic hormone (ADH)**. These hormones are transported in the neuron fibers into the posterior pituitary gland and are stored in the axon terminals resting on the capillaries. When properly stimulated, the hypothalamus sends impulses over the neurons. The impulses cause release of hormones from the axon terminals into the blood.

Examine a prepared slide of the posterior pituitary gland under high power. Identify the pituicytes and axon terminations of neurosecretory cells with the aid of your textbook.

2. Hormones of the Pituitary Gland

Using your textbook as a reference, give the major functions for the hormones listed below.

a. PITUITARY HORMONES SECRETED BY THE ANTERIOR PITUITARY GLAND

Human growth hormone (hGH) Also called **somatotropin** and **somatotropic hormone (STH)**

Thyroid-stimulating hormone (TSH)   Also called *thyrotropin* _____

_____

_____

Adrenocorticotropic hormone (ACTH) _____

_____

_____

_____

Follicle-stimulating hormone (FSH)

In female: _____

_____

_____

_____

In male: _____

_____

_____

_____

Luteinizing hormone (LH)

In female: _____

_____

_____

_____

In male: _____

_____

_____

_____

Prolactin (PRL)   Also called *lactogenic hormone*

_____

_____

Melanocyte-stimulating hormone (MSH) _____

_____

_____

_____

_____

b. HORMONES STORED AND RELEASED BY THE POSTERIOR PITUITARY GLAND

Oxytocin (OT) _____

_____

_____

_____

Antidiuretic hormone (ADH) _____

_____

_____

_____

C. THYROID GLAND

The *thyroid gland* is the endocrine gland located just inferior to the larynx. The right and left *lateral lobes* lie lateral to the trachea and are connected by a mass of tissue called an *isthmus* (IS-mus) that lies anterior to the trachea just inferior to the cricoid cartilage. The *pyramidal lobe*, when present, extends superiorly from the isthmus.

1. Histology of the Thyroid Gland

Histologically, the thyroid gland consists of spherical sacs called *thyroid follicles*. The walls of each follicle consist of epithelial cells that reach the surface of the lumen of the follicle *(follicular cells)*. In addition to the follicular cells the thyroid gland also contains *parafollicular cells* (also termed *C cells*). These cells may be within a follicle, where they do not reach the surface of the lumen, or they may be found between follicles in groups of three to four cells. Follicular cells synthesize the hormones *thyroxine* (thī-ROX-sēn) (also termed T_4 or *tetraiodothyronine)* and *triiodothyronine* (trī-ī-ōd-ō-THĪ-rō-nēn) *(T_3)*. Together these hormones are referred to as the *thyroid hormones*. Approximately 90% of the hormone secreted by the follicular cells is thyroxine, while the remaining 10% is triiodothyronine. The functions of these two hormones are essentially the same, but they differ in rapidity and intensity of action. T_4 has a significantly longer life within the blood stream, but is also significantly weaker than triiodothyronine. The parafollicular cells produce the hormone *calcitonin* (kal-si-TŌ-nin) *(CT)*. Each thyroid follicle is filled with a glycoprotein called *thyroglobulin (TBG),* which is also called *thyroid colloid.*

Examine a prepared slide of the thyroid gland under high power. Identify the thyroid follicles, epithelial cells forming the follicle, and thyroid colloid. Now label the photomicrograph in Figure 15.2.

2. Hormones of the Thyroid Gland

Using your textbook as a reference, give the major functions of the hormones listed below.

Thyroxine (T_4) and triiodothyronine (T_3)

_____

_____

_____

Calcitonin (CT) _____

_____

_____

_____

D. PARATHYROID GLANDS

Attached to the posterior surfaces of the lateral lobes of the thyroid gland are small, round masses of tissue called the *parathyroid (para = beside) glands*.

Typically two parathyroid glands, superior and inferior, are attached to each lateral thyroid lobe.

1. Histology of the Parathyroid Glands

Histologically, the parathyroid glands contain two kinds of epithelial cells. The first, a larger, more numerous cell called a *principal (chief) cell*, is believed to be the major synthesizer of *parathyroid hormone (PTH)*. The second, a smaller cell called an *oxyphil cell*, may serve as a reserve cell for hormone synthesis.

Examine a prepared slide of the parathyroid glands under high power. Identify the principal and oxyphil cells. Now label the photomicrograph in Figure 15.3.

2. Hormone of the Parathyroid Glands

Using your textbook as a reference, give the major functions of the hormone listed below:

Parathyroid hormone (PTH)   Also called *para-*

thormone _____

_____

_____

_____

E. ADRENAL (SUPRARENAL) GLANDS

The two *adrenal (suprarenal) glands* are superior to each kidney, and each is structurally and functionally differentiated into two separate endocrine

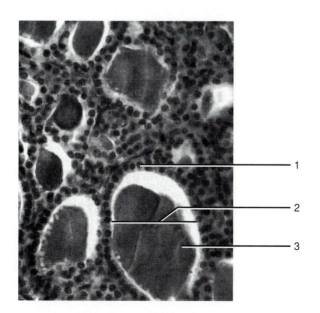

___ Epithelium of follicle
___ Thyroid colloid
___ Thyroid follicle

FIGURE 15.2   Histology of thyroid gland.

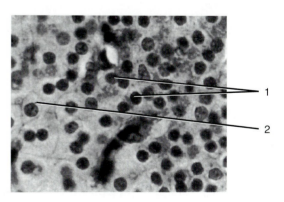

___ Oxyphil cell
___ Principal (chief) cells

FIGURE 15.3   Histology of parathyroid glands.

glands: the superficial *adrenal cortex*, which forms the bulk of the gland in humans, and the deeper *adrenal medulla*. Covering the gland is a thick layer of fatty tissue and an outer thin fibrous connective tissue *capsule*.

1. Histology of the Adrenal Cortex

Histologically, the adrenal cortex is subdivided into three zones, each of which has a different cellular arrangement and secretes different steroid hormones. The outer zone, immediately deep to the capsule, is called the *zona glomerulosa* (*glomerulus* = little ball). Its cells, arranged in arched loops or round balls, primarily secrete a group of hormones called *mineralocorticoids* (min'-er-al-ō-KOR-ti-koyds). The major mineralocorticoid produced by this region is *aldosterone*.

The intermediate zone of the adrenal cortex is the *zona fasciculata* (*fasciculus* = little bundle). This zone, which is the widest of the three, consists of cells arranged in long, relatively straight cords. The zona fasciculata secretes mainly *glucocorticoids* (gloo'-ko-KOR-ti-koyds). The zona fasciculata secretes three glucocorticoids, 95% of which is *cortisol* (also known as *hydrocortisone*). The remaining glucocorticoids synthesized include a small amount of *corticosterone* and a minute amount of *cortisone*.

The deepest zone of the adrenal cortex is termed the *zona reticularis* (*reticular* = net). It is composed of frequently branching cords of cells. This zone synthesizes minute amounts of male hormones (androgens).

Examine a prepared slide of the adrenal cortex under high power. Identify the capsule, zona glomerulosa, zona fasciculata, and zona reticularis. Now label Figure 15.4.

2. Hormones of the Adrenal Cortex

Using your textbook as a reference, give the major functions of the hormones listed below.

Mineralocorticoids (mainly *aldosterone*)

_____

_____

_____

Glucocorticoids (mainly *cortisol*) _____

_____

_____

_____

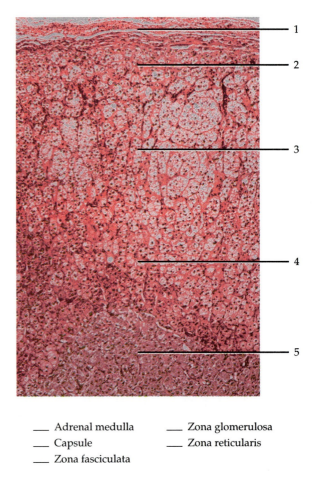

_____ Adrenal medulla _____ Zona glomerulosa
_____ Capsule   _____ Zona reticularis
_____ Zona fasciculata

FIGURE 15.4 Histology of adrenal (suprarenal) glands.

Androgens

In male: _____

_____

In female: _____

_____ .

3. Histology of the Adrenal Medulla

The adrenal medulla consists of hormone-producing cells called *chromaffin* (krō-MAF-in; *chroma* = color; *affinia* = affinity for) *cells*. These cells develop from the same embryonic tissue as the postganglionic cells of the sympathetic division of the nervous system. They are directly innervated by preganglionic cells of the sympathetic division of the autonomic nervous system and may be regarded as postganglionic cells that are specialized to secrete hormones. Secretion of hormones from the chromaffin cells is directly controlled by the sympathetic division of the autonomic nervous system, and innervation by the preganglionic fibers allows the gland to respond extremely rapidly to a stimulus. The adrenal medulla secretes the hormones *epinephrine* and *norepinephrine (NE)*.

Examine a prepared slide of the adrenal medulla under high power. Identify the chromaffin cells. Now locate the cells in Figure 15.4.

4. Hormones of the Adrenal Medulla

Using your textbook as a reference, give the major functions of the hormones listed below.

Epinephrine and *norepinephrine (NE)* _____

_____

_____

_____

F. PANCREAS

The *pancreas* is classified as both an endocrine and an exocrine gland. Thus, it is referred to as a *heterocrine gland*. We shall discuss only its endocrine functions now. The pancreas is a flattened organ located posterior and slightly inferior to the stomach. The adult pancreas consists of a head, body, and tail.

1. Histology of the Pancreas

The endocrine portion of the pancreas consists of clusters of cells called *pancreatic islets (islets of Langerhans)*. They contain four kinds of cells: (1) *alpha cells*, which have more distinguishable plasma membranes, are usually peripheral in the islet, and secrete the hormone *glucagon*; (2) *beta cells*, which generally lie deeper within the islet and secrete the hormone *insulin*; (3) *delta cells*, which secrete *somatostatin*; and (4) *F cells*, which secrete pancreatic polypeptide. The pancreatic islets are surrounded by blood capillaries and by the cells called *acini* that form the exocrine part of the gland.

Examine a prepared slide of the pancreas under high power. Identify the alpha cells, beta cells, and acini (clusters of cells that secrete digestive enzymes) around the pancreatic islets. Now label Figure 15.5.

2. Hormones of the Pancreas

Using your textbook as a reference, give the major functions of the hormones listed below.

Glucagon _____

_____

_____

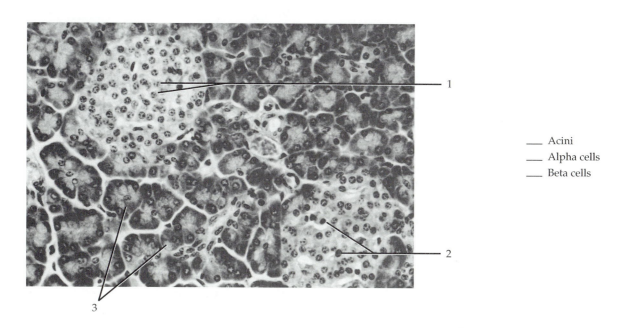

___ Acini
___ Alpha cells
___ Beta cells

FIGURE 15.5 Histology of pancreas.

Insulin _____

_____

_____

Somatostatin _____

_____

_____

Pancreatic polypeptide _____

_____

_____

G. TESTES

The *testes* (male gonads) are paired oval glands enclosed by the scrotum. They are partially covered by a serous membrane called the *tunica* (*tunica* = sheath) *vaginalis*, which is derived from the peritoneum and forms during descent of the testes. Internal to the tunica vaginalis is a dense layer of white fibrous tissue, the *tunica albuginea* (al'-byoo-JIN-ē-a; *albus* = white), which extends inward and divides each testis into a series of internal compartments called *lobules*. Each of the 200–300 lobules contains one to three tightly coiled tubules, the convoluted *seminiferous* (*semen* = seed; *ferre* = to carry) *tubules*, which produce sperm by a process called *spermatogenesis* (sper'-ma-tō-JEN-e-sis).

1. Histology of the Testes

Spermatogenic cells are sperm-forming cells in various stages that undergo mitosis and differentiation to eventually produce sperm. Together with supporting cells, they line the seminiferous tubules. The most immature spermatogenic cells are called *spermatogonia* (sper'-ma-tō-GŌ-nē-a; *sperm* = seed; *gonium* = generation or offspring; the singular is *spermatogonium*). They lie next to the basement membrane. Toward the lumen of the tubules are layers of progressively more mature cells. In order of advancing maturity, these are *primary spermatocytes* (SPER-ma-tō-sīts), *secondary spermatocytes*, and *spermatids*. By the time a *sperm cell* or *spermatozoon* (sper'-ma-tō-ZŌ-on; *zoon* = life; the plural is *sperm* or *spermatozoa*), has nearly reached maturity, it is released into the lumen of the tubule and begins to move out of the rete testis.

Embedded among the spermatogenic cells in the tubules are large *sustentacular* (sus'-ten-TAK-yoo-

lar; *sustentare* = to support) or *Sertoli, cells*, that extend from the basement membrane to the lumen of the tubule. Sustentacular cells support and protect developing spermatogenic cells; nourish spermatocytes, spermatids, and sperm; phagocytize excess spermatid cytoplasm as development proceeds; and mediate the effects of testosterone and follicle-stimulating hormone (FSH). Sustentacular cells also control movements of spermatogenic cells and the release of sperm into the lumen of the seminiferous tubule. They produce fluid for sperm transport and secrete the hormone inhibin, which helps regulate sperm production by inhibiting the secretion of FSH.

In the spaces between adjacent seminiferous tubules are clusters of cells called *interstitial endocrinocytes*, or *Leydig cells*. These cells secrete testosterone, the most important androgen (male sex hormone).

Examine a prepared slide of the testes under high power and identify all of the structures listed. Now label Figure 15.6.

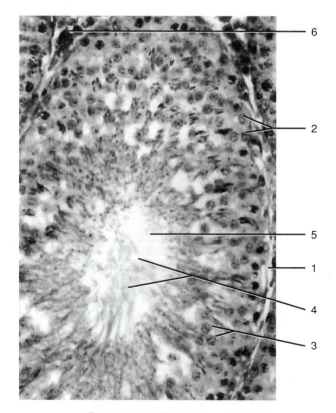

____ Basement membrane
____ Interstitial endocrinocytes (interstitial cells of Leydig)
____ Lumen of seminiferous tubule
____ Spermatids
____ Spermatogonia
____ Spermatozoa

FIGURE 15.6 Histology of testes.

2. Hormones of the Testes

Using your textbook as a reference, give the major functions of the hormones listed below.

Testosterone _____

_____

_____

Inhibin _____

_____

_____

H. OVARIES

The *ovaries* (*ovarium* = egg receptacle), or female gonads, are paired glands resembling unshelled almonds in size and shape. They are positioned in the superior pelvic cavity, one on each side of the uterus, and are maintained in position by a series of ligaments. They are (1) attached to the **broad ligament** of the uterus, which is itself part of the parietal peritoneum, by a fold of peritoneum called the **mesovarium**; (2) anchored to the uterus by the **ovarian ligament**; and (3) attached to the pelvic wall by the **suspensory ligament**. Each ovary also contains a **hilus**, which is the point of entrance for blood vessels and nerves.

1. Histology of the Ovaries

Histologically, each ovary consists of the following parts:

1. **Germinal epithelium** A layer of simple epithelium (low cuboidal or squamous) that covers the free surface of the ovary and is continuous with the mesothelium that covers the mesovarium. The term *germinal epithelium* is a misnomer because it does not give rise to oocytes, although at one time it was believed that it did.
2. **Tunica albuginea** A whitish capsule of dense, irregular connective tissue immediately deep to the germinal epithelium.
3. **Stroma** A region of connective tissue deep to the tunica albuginea and composed of a superficial, dense layer called the **cortex** and a deep, loose layer known as the **medulla**.
4. **Ovarian follicles** (*folliculus* = little bag) Lie in the cortex and consist of **oocytes** (immature ova) in various stages of development and their surrounding cells. When the surrounding cells form a single layer, they are called **follicular cells**. Later in development, when they form

several layers, they are referred to as *granulosa cells*. The surrounding cells nourish the developing oocyte and begin to secrete estrogens as the follicle grows larger. Ovarian follicles undergo a series of changes prior to ovulation, progressing through several distinct stages. The most numerous and peripherally arranged follicles are termed **primordial follicles**. If a primordial follicle progresses to ovulation (release of a mature ova), it will sequentially transform into a **primary (preantral) follicle**, then a **secondary (antral) follicle**, and finally a *mature (Graafian) follicle*.

5. *Mature (Graafian) follicle* A large, fluid-filled follicle that soon will rupture and expel a secondary oocyte, a process called *ovulation*.
6. *Corpus luteum* (= yellow body) Contains the remnants of an ovulated mature follicle. The corpus luteum produces progesterone, estrogens, relaxin, and inhibin until it degenerates and turns into fibrous tissue called a *corpus albicans* (= white body).

Examine a prepared slide of the ovary under high power. (You may need to examine more than one slide to see all of the structures listed.) Label Figure 15.7.

2. Hormones of the Ovaries

Using your textbook as a reference, give the major functions of the hormones listed below.

Estrogens _____

_____

_____

_____

Progesterone _____

_____

_____

_____

Relaxin _____

_____

_____

_____

Inhibin _____

_____

_____

_____

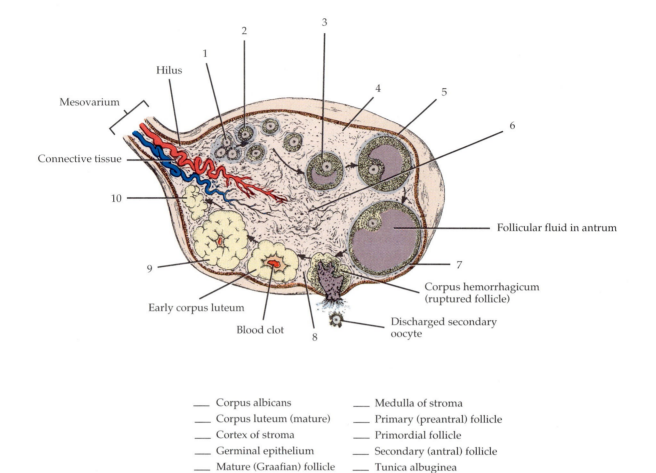

Mesovarium
Hilus
Connective tissue
10
9
Early corpus luteum
Blood clot
1
2
3
4
5
6
Follicular fluid in antrum
7
Corpus hemorrhagicum (ruptured follicle)
8
Discharged secondary oocyte

___ Corpus albicans
___ Corpus luteum (mature)
___ Cortex of stroma
___ Germinal epithelium
___ Mature (Graafian) follicle

___ Medulla of stroma
___ Primary (preantral) follicle
___ Primordial follicle
___ Secondary (antral) follicle
___ Tunica albuginea

FIGURE 15.7   Histology of ovary.

I. PINEAL GLAND (EPIPHYSIS CEREBRI)

The cone-shaped endocrine gland attached to the roof of the third ventricle is known as the *pineal* (PĪN-ē-al) *gland (epiphysis cerebri).*

1. Histology of the Pineal Gland

The gland is covered by a capsule formed by the pia mater and consists of masses of neuroglial cells and secretory cells called *pinealocytes* (pin-ē-AL-ō-sīts). Around the cells are scattered preganglionic sympathetic fibers. The pineal gland starts to calcify at about the time of puberty. Such calcium deposits are called *brain sand*. Contrary to a once widely held belief, no evidence suggests that the pineal gland atrophies with age and that the presence of brain sand indicates atrophy. Rather, brain sand may even denote increased secretory activity.

The physiology of the pineal gland is still somewhat obscure. The gland secretes *melatonin*.

2. Hormone of the Pineal Gland

Using your textbook as a reference, give the major functions of the hormone listed below.

Melatonin _____

_____

_____

J. THYMUS GLAND

The *thymus gland* is a bilobed lymphatic gland located in the upper mediastinum posterior to the sternum and between the lungs. The gland is conspicuous in infants and, during puberty, reaches maximum size.

1. Histology of Thymus Gland

After puberty, thymic tissue, which consists primarily of *lymphocytes*, is replaced by fat. By the time a person reaches maturity, the gland has atrophied. Lymphoid tissue of the body consists primarily of lymphocytes that may be distinguished into two kinds: B cells and T cells. Both are derived originally in the embryo from lymphocytic stem cells in red bone marrow. Before migrating to their positions in lymphoid tissue, the descendants of the system cells follow two distinct pathways. About half of them migrate to the thymus gland, where they are processed to become thymus-dependent lymphocytes, or *T cells*. The thymus gland confers on them the ability to destroy antigens (foreign microbes and substances). The remaining stem cells are processed in some as yet undetermined area of the body, possibly the fetal liver and spleen, and are known as *B cells*. Hormones produced by the thymus gland are *thymosin, thymic humoral factor (THF), thymic factor (TF),* and *thymopoietin*.

2. Hormones of the Thymus Gland

Using your textbook as a reference, write the major functions of the hormones listed below:

Thymosin, thymic humoral factor (THF), thymic

factor (TF), and thymopoietin _____

_____

_____

K. OTHER ENDOCRINE TISSUES

Body tissues other than endocrine glands also secrete hormones. The gastrointestinal tract synthesizes several hormones that regulate digestion in the stomach and small intestine. Among these hormones are *gastrin, secretin, cholecystokinin (CCK),* and *gastric inhibitory peptide (GIP)*.

The placenta produces *human chorionic gonadotropin (HCG), estrogens, progesterone (PROG), relaxin,* and *human chorionic somatomammotrophin (HCS),* all of which are related to pregnancy.

When the kidneys (and liver, to a lesser extent) become hypoxic (subject to below normal levels of oxygen), they release an enzyme called *renal erythropoietic factor*. This is secreted into the blood where it acts on a plasma protein produced in the liver to form a hormone called *erythropoietin* (ē-rith'-rō-POY-ē-tin), or *EPO,* which stimulates red bone marrow to produce more red blood cells and hemoglobin. This ultimately reverses the original stimulus (hypoxia).

Vitamin D, produced by the skin, liver, and kidneys in the presence of sunlight, is converted to its active hormone, *calcitriol,* in the kidneys and liver.

The atria of the heart secrete a hormone called *atrial natriuretic peptide (ANP),* released in response to increased blood volume.

Using your textbook as a reference, write the major functions of the hormones listed below:

Gastrin _____

_____

Secretin _____

_____

Cholecystokinin (CCK) _____

_____

Gastric inhibitory peptide (GIP) _____

_____

Human chorionic gonadotropin (hCG) _____

_____

Human chorionic somatomammotropin (hCS) __

_____

Erythropoietin (EPO) _____

_____

Calcitriol _____

_____

Atrial natriuretic peptide (ANP) _____

_____

L. PHYSIOLOGY OF THE ENDOCRINE SYSTEM

1. Determination of Standard Metabolic Rate in Normal, Hyperthyroid, and Hypothyroid Rats

As was mentioned previously, the thyroid gland produces two hormones related to metabolism, thyroxine (T_4) and triiodothyronine (T_3). This laboratory exercise will help you to determine the effect of a normally functioning, hyperactive, or hypoactive thyroid gland upon a rat's standard metabolic rate.

a. PREPARATION OF RATS

PROCEDURE

1. Hyperthyroid rat: Animals should be injected by your instructor or a laboratory assistant with 25 μg of L-thyroxine interiperitoneally every third day for a two-week period.
2. Hypothyroid rat: Rats should be fed by your instructor or a laboratory assistant *either* 0.5% thiouracil in their rat chow *or* 0.02% propyl-thiouracil *or* 1% potassium perchlorate in their drinking water. (**Note:** *If tetany appears in this group of animals, 1% lactate or gluconate should be given in their drinking water.*) Maintain this dosage for two weeks.

b. CALIBRATION OF SYSTEM

PROCEDURE

1. Before starting the experiment, the system must be calibrated.
2. Place an object of approximately the same volume as the rat in the system (four or five No. 10 stoppers is approximately the same volume as a 3-month-old rat).
3. Completely seal the chamber, *making sure that the syringe is set at 0.*
4. Withdraw 40 ml of air from the system via the syringe and mark the new water level (Figure 15.8).

c. EXPERIMENTAL PROCEDURE

PROCEDURE

1. Remove the syringe from the rubber tubing and clamp the tube.
2. Have your instructor or a laboratory assistant weigh the rat.
3. Remove the stoppers from the system and have your instructor or a laboratory assistant replace them with an experimental animal.
4. Seal the system and record the time and temperature.
5. As the animal utilizes oxygen, it produces carbon dioxide. The carbon dioxide will be removed by the soda lime, causing the fluid to rise in the tubing as a result of decreased air volume within the chamber.
6. Record the time required for the water to rise to the calibration mark.

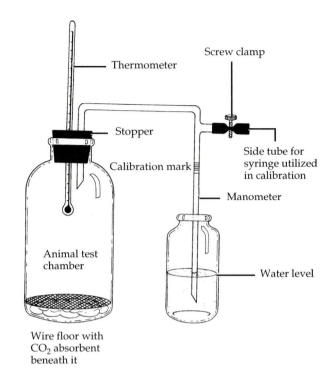

FIGURE 15.8 Experimental setup for measuring basal metabolic rate in rats.

7. Also record the temperature within the system at this time.
8. After recording your calculations, repeat the procedure with another experimental animal until all three classes of animals are utilized.
9. Complete calculations as outlined with the example below. Compute average values for each class of animals.
   Sample calculations:
   Rat weight = 150 g
   40 ml oxygen consumed in 10 min
   Oxygen consumption = 40 ml/10 min
   $\qquad\qquad\qquad\qquad$ = 4 ml/min
   Reduced to standard conditions:
   Temperature = 27°C
   Barometric pressure = 731 mm Hg
   $4 \times {}^{731}\!/_{760} \times {}^{273}\!/_{300}$ = 3.5 ml/min
   (3.5 ml/min)/0.150 kg = 23.33 ml/kg/min
10. Record all calculations in Section L.1 of the LABORATORY REPORT RESULTS at the end of the exercise.

2. Demonstrations on Effects of Estrogens and Testosterone

Review the physiology of the male and female sex hormones upon the function and structure of the accessory sex organs as outlined in your textbook. These accessory sex organs are very sensitive to the effects of these hormones. The rats used in this demonstration have been prepared by your instructor or a laboratory assistant as follows:

Females	Males
Control	Control
Ovariectomized	Castrated
Ovariectomized and injected with estrogens	Castrated and injected with testosterone
Estrogens injection only	Testosterone injection only

Compare the size of uterus in females to the size of seminal vesicles in males.

Record your observations in Section L.2 of the LABORATORY REPORT RESULTS at the end of the exercise.

3. Hyperinsulinemia in Fish

Insulin causes the level of blood glucose to drop (hypoglycemia) by promoting the transport of glucose through the plasma membranes of cells. Insulin also promotes the conversion of glucose into glycogen and fat and increases the rate of synthesis of proteins from amino acids.

The regulation of insulin secretion is a direct effect of the level of sugar in the blood. When blood sugar levels fall below normal limits, the alpha cells of the pancreas secrete *glucagon*. This hormone promotes *glycogenolysis*, the conversion of glycogen to glucose. The release of glucose into the blood counteracts the hypoglycemia through a *negative feedback mechanism*.

Hypoinsulinism (diabetes mellitius) results in *hyperglycemia* and *glucosuria* (urinary excretion of glucose). The high osmotic concentration of glucose in the urine causes the excretion of large urinary volumes *(polyuria)* and consequent thirst *(polydipsia)*.

Hypoinsulinism also depletes the supply of glucose available for the energy requirements for cellular metabolism. Consequently, the body shifts to a higher rate of protein and fat metabolism. The breakdown of high quantities of fats releases ketone (acetone) bodies. These byproducts of fat metabolism have a characteristic acetone odor and lower the blood pH, resulting in acidosis and eventually coma.

Hyperinsulinism can result from a tumor of the beta cells in the pancreatic islets (islets of Langerhans) or from administration of excess insulin by a diabetic. High insulin levels cause hypoglycemia because insulin promotes glucose uptake primarily by liver, muscle, and fat cells. As blood sugar levels drop sharply, the brain is deprived of its major energy source. When blood glucose falls below a critical level, *hypoglycemic (insulin) shock* causes irritability, convulsions, and eventually coma. It is difficult to distinguish the coma resulting from hypoinsulinism from that of hyperinsulinism, even though the causes are radically different.

The effects of hyperinsulinism in fish are comparable to those in humans. The insulin molecule is small enough to diffuse through the gill membranes into the fish blood.

Note: Depending on the availability of live animals and the type of laboratory equipment available, you may select from the following procedures related to hyperinsulinemia in fish.

PROCEDURE USING PHYSIOFINDER

PHYSIOFINDER is an interactive computer program that permits laboratory simulations that do not involve the use of animals or advanced or expensive laboratory equipment. It permits students to perform experiments, analyze data, draw conclusions, and form hypotheses on the basis of collected data.

For activities related to hyperinsulinemia in fish, as well as other activities related to plasma glucose levels, select the appropriate experiment from PHYSIOFINDER Module 11—Control of Plasma Glucose

PROCEDURE USING LIVE FISH

1. Mark two beakers I and G. In beaker I, place 200 ml of water and 10 drops of commercial insulin. Stir the mixture to disperse the insulin. In beaker G, place 200 ml of 5% glucose solution.
2. Place a small (1- to 2-in.) guppy, sunfish, or goldfish in beaker I.
3. Observe the behavior of the fish and note swimming movements, convulsive movements, and coma. If no changes occur within a minute, replace the fish in the aquarium and begin the procedure with another fish.
4. When the fish becomes comatose, note the length of time it was subjected to the insulin-containing environment and quickly move it to beaker G.

Time required for the fish to become comatose:

_____

5. Note the time required for the fish's behavior to return to normal and then replace the fish in the aquarium.

Time required for recovery: _____
How did the glucose solution promote recovery?

_____

If the fish were placed in a glucagon-containing solution after exposure to insulin, do you think the recovery would have been slower or faster than that caused by placement in glucose? Why? (The glucagon molecule is smaller than the insulin molecule.) _____

_____

ANSWER THE LABORATORY REPORT QUESTIONS AT THE END OF THE EXERCISE.

Endocrine System 15

Student _____ **Date** _____

Laboratory Section _____ **Score/Grade** _____

SECTION L. PHYSIOLOGY OF THE ENDOCRINE SYSTEM

1. Complete the following table based upon your calculations.

	Control		Hyperthyroid		Hypothyroid	
	Your value	Class average	Your value	Class average	Your value	Class average
Time to consume 40 ml oxygen						
Rat weight						
Oxygen consumption						

2. Record your observations for each of the demonstrations.

Males
    Normal
    Castrated
    Castrated plus injections
    Injections only
Females
    Normal
    Ovariectomized
    Ovariectomized plus injections
    Injections only

Endocrine System 15

Student _____   Date _____

Laboratory Section _____   Score/Grade _____

PART 1. Multiple Choice

_____ **1.** Somatotrophs, gonadotrophs, and corticolipotrophs are associated with the (a) thyroid gland (b) anterior pituitary gland (c) parathyroid glands (d) adrenal glands

_____ **2.** The posterior pituitary gland is *not* an endocrine gland because it (a) has a rich blood supply (b) is not near the brain (c) does not make hormones (d) contains ducts

_____ **3.** Which hormone assumes a role in the development and discharge of a secondary oocyte? (a) hGH (b) TSH (c) LH (d) PRL

_____ **4.** The endocrine gland that is probably malfunctioning if a person has a high metabolic rate is the (a) thymus gland (b) posterior pituitary gland (c) anterior pituitary gland (d) thyroid gland

_____ **5.** The antagonistic hormones that regulate blood calcium level are (a) hGH-TSH (b) insulin-glucagon (c) aldosterone-cortisone (d) CT-PTH

_____ **6.** The endocrine gland that develops from the sympathetic nervous system is the (a) adrenal medulla (b) pancreas (c) thyroid gland (d) anterior pituitary gland

_____ **7.** The hormone that aids in sodium conservation and potassium excretion is (a) hydrocortisone (b) CT (c) ADH (d) aldosterone

_____ **8.** Which of the following hormones is sympathomimetic? (a) insulin (b) oxytocin (OT) (c) epinephrine (d) testosterone

_____ **9.** Which hormone lowers blood sugar level? (a) glucagon (b) melatonin (c) insulin (d) cortisone

_____ **10.** The endocrine gland that may assume a role in regulation of the menstrual cycle is the (a) pineal gland (b) thymus gland (c) thyroid gland (d) adrenal gland

PART 2. Completion

11. The pituitary gland is attached to the hypothalamus by a stalklike structure called the

_____.

12. A hormone that acts on another endocrine gland and causes that gland to secrete its own hormones

is called a _____.

13. _____ cells of the anterior pituitary gland synthesize ACTH.

14. The hormone that helps cause contraction of the smooth muscle of the pregnant uterus is

_____.

15. Histologically, the spherical sacs that compose the thyroid gland are called thyroid

_____.

16. The thyroid hormones associated with metabolism are triiodothyronine and _____.

17. Principal and oxyphil cells are associated with the _____ gland.

18. The zona glomerulosa of the adrenal cortex secretes a group of hormones called

_____.

19. The hormones that promote normal metabolism, provide resistance to stress, and function as anti-

inflammatories are _____.

20. The pancreatic hormone that raises blood sugar level is _____.

21. In spermatogenesis, the most immature cells near the basement membrane are called

_____.

22. Cells within the testes that secrete testosterone are known as _____.

23. The ovaries are attached to the uterus by means of the _____ ligament.

24. The female hormones that help cause the development of secondary sex characteristics are called

_____.

25. The endocrine gland that assumes a direct function in the proliferation and maturation of T cells is

the _____.

26. Any hormone that regulates the functions of the gonads is classified as a(n) _____
hormone.

27. The hormone that is stored in the posterior pituitary gland that prevents excessive urine production

is _____.

28. The region of the adrenal cortex that synthesizes androgens is the _____.

29. The hormone-producing cells of the adrenal medulla are called _____ cells.

30. Together, alpha cells, beta cells, delta cells, and F cells constitute the _____.

31. Regulating hormones are produced by the _____ and reach the anterior pituitary
gland by a network of blood vessels.

32. FSH and LH are produced by _____ cells of the anterior pituitary gland.

33. The structure in an ovary that produces progesterone, estrogens, relaxin, and inhibin is the

_____.

34. Calcium deposits in the pineal gland are referred to as _____.

35. A gland that is both an exocrine and endocrine gland is known as a _____ gland.

Blood

16

The blood, heart, and blood vessels constitute the **cardiovascular system.** In this exercise you will examine the characteristics of **blood,** a connective tissue also known as **vascular tissue.**

CAUTION! *Please reread Section A, "General Safety Precautions and Procedures," and Section B, "Precautions Related to Working with Blood, Blood Products, or Other Body Fluids," on pages xi–xii, at the beginning of the laboratory manual, before you begin any of the following experiments. Read the experiments before you perform them, to be sure that you understand all the procedures and safety precautions.*

When working with whole blood, wear tight-fitting surgical gloves and safety goggles. Avoid any kind of contact with an open sore, cut, or wound.

After you have completed your experiments, place all glassware in a fresh solution of household bleach or other comparable disinfectant, wash the laboratory tabletop with a fresh solution of household bleach or comparable disinfectant, and dispose of your gloves in the appropriate biohazard container provided by your instructor.

A. COMPONENTS OF BLOOD

Blood is a liquid connective tissue that is composed of two portions: (1) *plasma,* a liquid that contains dissolved substances, and (2) *formed elements,* cells and cell fragments suspended in the plasma. In clinical practice, the most common classification of the formed elements of the blood is the following:

Erythrocytes (e-RITH-rō-sīts), or *red blood cells*
Leukocytes (LOO-kō-sīts), or *white blood cells*
Granular leukocytes (granulocytes)
   Neutrophils
   Eosinophils
   Basophils
Agranular leukocytes (agranulocytes)
   Lymphocytes (T cells, B cells, and natural killer cells)
   Monocytes
Platelets (thrombocytes)

The origin and subsequent development of these formed elements can be seen in Figure 16.1. Blood cells are formed by a process called *hemopoiesis* (hē-mō-poy-Ē-sis; *hemo* = blood; *poiem* = to make), or *hematopoiesis* (hē-ma-tō-poy-Ē-sis), and the process by which erythrocytes are formed is called *erythropoiesis* (e-rith'-rō-poy-Ē-sis). The immature cells that are eventually capable of developing into mature blood cells are called *hematopoietic stem cells,* or *hemocytoblasts* (hē-mō-SĪ-tō-blasts) (see Figure 16.1). Mature blood cells are constantly being replaced, so special cells called *reticuloendothelial cells* (fixed macrophages that line liver sinusoids) have the responsibility of clearing away the dead, disintegrating cell bodies so that small blood vessels are not clogged.

The shapes of the nuclei, staining characteristics, and color of cytoplasmic granules are all useful in differentiation and identification of the various white blood cells. Red blood cells are biconcave discs without nuclei and can be identified easily.

B. PLASMA

When the formed elements are removed from blood, a straw-colored liquid called *plasma* is left. This liquid consists of about 91.5% water and about 8.5% solutes. Among the solutes are proteins (albumins, globulins, and fibrinogen), nonprotein nitrogen (NPN) substances (urea, uric acid, and creatine), foods (amino acids, glucose, fatty acids, glycerides, and glycerol), regulatory substances (enzymes and hormones), gases (oxygen and carbon dioxide), and electrolytes (Na^+, K^+, Ca^{2+}, Mg^{2+}, Cl^-, HCO_3^-, SO_4^{2-}, and HPO_4^{3-}).

1. Physical Characteristics

CAUTION! *Please reread Section A, "General Safety Precautions and Procedures," and Section B, "Precautions Related to Working with Blood, Blood Products, or Other Body Fluids," on pages xi–xii, at*

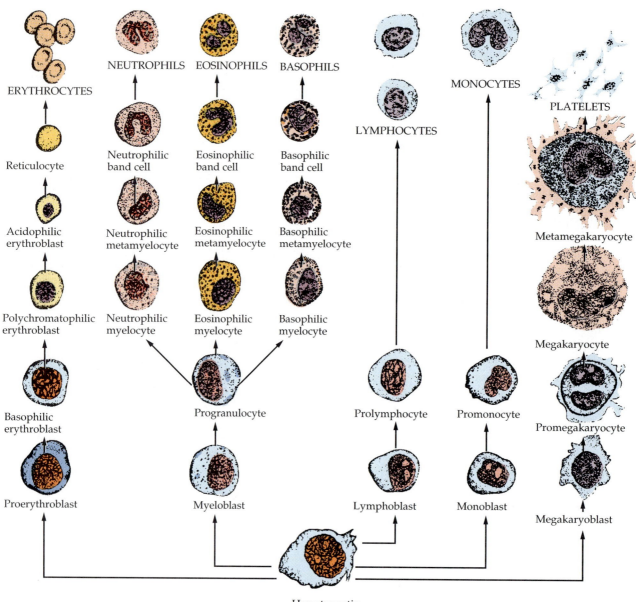

ERYTHROCYTES

Reticulocyte

Acidophilic
erythroblast

Polychromatophilic
erythroblast

Basophilic
erythroblast

Proerythroblast

NEUTROPHILS

Neutrophilic
band cell

Neutrophilic
metamyelocyte

Neutrophilic
myelocyte

Myeloblast

EOSINOPHILS

Eosinophilic
band cell

Eosinophilic
metamyelocyte

Eosinophilic
myelocyte

Progranulocyte

BASOPHILS

Basophilic
band cell

Basophilic
metamyelocyte

Basophilic
myelocyte

LYMPHOCYTES

Prolymphocyte

Lymphoblast

MONOCYTES

Promonocyte

Monoblast

PLATELETS

Metamegakaryocyte

Megakaryocyte

Promegakaryocyte

Megakaryoblast

Hematopoetic
stem cell (hemocytoblast)

FIGURE 16.1   Origin, development, and structure of blood cells.

the beginning of the laboratory manual before you
begin any of the following experiments. Read the
experiments before you perform them, to be sure that
you understand all the procedures and safety
precautions.

When working with whole blood, wear tight-fitting
surgical gloves and safety goggles. Avoid any kind of
contact with an open sore, cut, or wound.

PROCEDURE

Note: Obtain the plasma or whole blood for this procedure from a clinical laboratory where it has been tested

and certified as noninfectious* or from a mammal (other
than a human).

1. Obtain about 5 ml of plasma as follows.
2. Centrifuge the whole blood obtained from a
   clinical laboratory where it has been tested and
   certified as noninfectious or from a mammal
   (other than a human). After centrifugation, the

*Whole blood that has been tested for syphilis; hepatitis A, B,
and C; and HIV is available from Carolina Biological Supply
Company.

bottom of the test tube will contain an anticoagulant (usually EDTA). Right above this will be a layer of erythrocytes, followed by a layer of leukocytes and platelets (buffy coat). The top layer is plasma. Visual inspection of these layers will give an indication of the relative proportions of the various components of blood.

3. Remove 5 ml of plasma using a disposable sterile Pasteur pipette attached to a bulb and place the plasma in a disposable test tube. Note its color. Test the pH of the plasma by dipping the end of a length of pH paper into the plasma and comparing the paper color with the color scale on the pH container. Record the pH in Section B.1 of the LABORATORY REPORT RESULTS at the end of the exercise.

4. Hold the test tube up to a source of natural light and note the color and transparency of the plasma. Record the color and transparency in Section B.1 of the LABORATORY REPORT RESULTS at the end of the exercise.

5. *Dispose of the pipettes and test tubes in the appropriate biohazard container provided by your instructor. Dispose of the pH paper and plasma as per your instructor's directions.*

2. Chemical Constituents

CAUTION! *Please reread Section A, "General Safety Precautions and Procedures," and Section B, "Precautions Related to Working with Blood, Blood Products, or Other Body Fluids," on pages xi–xii, at the beginning of the laboratory manual, before you begin any of the following experiments. Read the experiments before you perform them, to be sure that you understand all the procedures and safety precautions.*

When working with whole blood, wear tight-fitting surgical gloves and safety goggles. Avoid any kind of contact with an open sore, cut, or wound.

PROCEDURE

Note: Obtain the plasma or whole blood for this procedure from a clinical laboratory where it has been tested and certified as noninfectious or from a mammal (other than a human).

1. Using a medicine dropper, add 10 ml of distilled water to 5 ml of plasma in a Pyrex test tube.

CAUTION! *Place the test tube in a test-tube rack because it will become too hot to handle.*

2. With forceps, add one Clinitest tablet.

CAUTION! *The concentrated sodium hydroxide in the tablet generates enough heat to make the liquid in the test tube boil. Make sure that the mouth of the test tube is pointed away from you and all other persons in the area.*

3. The color of the solution is graded as follows:

Color	Results
Blue	Negative
Greenish-yellow	1 + (0.5 g/100 ml)
Olive green	2 + (1 g/100 ml)
Orange-yellow	3 + (1.5 g/100 ml)
Brick red (with precipitate)	4 + (more than 2 g/100 ml)

4. Fifteen seconds after boiling has stopped, shake the test tube *gently* and evaluate the color according to the test table above.

CAUTION! *Make sure that the mouth of the test tube is pointed away from you and others.*

A green, yellow, orange, or red color indicates the presence of glucose. Record your results in Section B.2 of the LABORATORY REPORT RESULTS at the end of the exercise.

5. Test the contents of the filter paper for the presence of protein. If protein is present, it will be coagulated by the acetic acid and filtered out of the plasma. Place the contents of the filter paper in a test tube and add 3 ml of water and 3 ml of Biuret reagent. A violet or purple color indicates the presence of protein. Record your results in Section B.2 of the LABORATORY REPORT RESULTS at the end of the exercise.

C. ERYTHROCYTES

Erythrocytes (red blood cells, or RBCs) are biconcave in appearance, have no nucleus, and can neither reproduce nor carry on extensive metabolic activities (Figure 16.1). The interior of the cell contains a red pigment called *hemoglobin,* which is responsible for the red color of blood. The heme portion of hemoglobin combines with oxygen and, to a lesser extent, carbon dioxide and transports them through the blood vessels. An average red blood cell has a life span of about 120 days. A healthy male has about 5.4 million red blood cells per cubic millimeter (mm^3), or per deciliter (dl), of blood; a healthy female, about 4.8 million. Erythropoiesis and red blood cell destruction normally proceed at the same pace. A diagnostic test that informs the physician about the rate of erythropoiesis is the *reticulocyte* (re-TIK-yoo-lō-sīt) *count. Reticulocytes* (see Figure 16.1) are precursor cells of mature red blood cells. Normally, a reticulocyte count is 0.5 to 1.5%. The reticulocyte count is an important diagnostic tool because it is a relatively accurate predictor of the status of red cell production in red bone marrow. Normally, red bone marrow replaces about 1% of the adult red blood cells each day. A decreased reticulocyte count (reticulocytopenia) is seen in aplastic anemia and in conditions in which the red bone marrow is not producing red blood cells. An increase in reticulocytes (reticulocytosis) is found in acute and chronic blood loss and certain kinds of anemias such as iron-deficiency anemia.

Note: Although many blood tests in this exercise are sufficient for laboratory demonstration, they are not necessarily clinically accurate.

D. RED BLOOD CELL TESTS

CAUTION! *Please reread Section A, "General Safety Precautions and Procedures," and Section B, "Precautions Related to Working with Blood, Blood Products, or Other Body Fluids," on pages xi–xii, at the beginning of the laboratory manual before you begin any of the following experiments. Read the experiments before you perform them, to be sure that you understand all the procedures and safety precautions.*

When working with whole blood, wear tight-fitting surgical gloves and safety goggles. Avoid any kind of contact with an open sore, cut, or wound.

1. Source of Blood

Blood should be obtained from a clinical laboratory where it has been tested and certified as noninfectious or from a mammal (other than a human).

Whole blood that has been tested for syphilis; hepatitis A, B, and C; and HIV is available from Carolina Biological Supply Company. This blood can be used for blood grouping and typing, blood glucose, hematocrit, hemoglobin, red blood cell count, white blood cell count, differential white blood cell count, platelet count, sedimentation rate, osmotic fragility studies, and various plasma chemistries. Carolina Biological Supply Company also provides aseptic red blood cells that can be used for blood grouping and typing, blood glucose,

hematocrit, hemoglobin, red blood cell count, and osmotic fragility studies. These blood cells are suspended in a modified Alsevere's solution, to which several antibiotics have been added. The cells have been tested for syphilis; hepatitis A, B, and C; and HIV.

2. Filling of Hemocytometer (Counting Chamber)

The procedure for filling the hemocytometer is as follows. Please read it carefully *before* doing 3. Red Blood Cell Count.

PROCEDURE

Note: The blood used for this procedure should be obtained from a clinical laboratory where it has been tested and certified as noninfectious or from a mammal (other than a human).

1. Obtain a hemocytometer and cover slip (Figure 16.2).
2. Clean the hemocytometer thoroughly and carefully with alcohol.
3. Place the cover slip on the hemocytometer.
4. Using a Unopette Reservoir System® pipette filled with the proper dilution of blood, place the tip of the pipette on the polished surface of the hemocytometer next to the edge of the cover slip. Although we recommend use of this system, your instructor might wish to use an alternate procedure. *If a different system is used for pipetting and diluting, use rubber suction bulbs or pipette pumps. Do not pipette by mouth.*
5. Deposit a small drop of diluted blood by squeezing the sides of the reservoir, but do not leave the tip of the pipette in contact with the hemocytometer for more than an instant because this will cause the chamber to overfill. The diluted blood must not overflow the moat; overfilling the moat results in an inaccurate cell count. A properly filled hemocytometer has a blood specimen only within the space between the cover glass and counting area.

3. Red Blood Cell Count

The purpose of a red blood cell count is to determine the number of circulating red blood cells per cubic millimeter (mm^3), or per deciliter (dl), of blood. Red blood cells carry oxygen to all tissues; thus a drastic reduction in the red cell count will cause immediate reduction in available oxygen.

A decrease in red blood cells can result from a variety of conditions, including impaired cell production, increased cell destruction, and acute

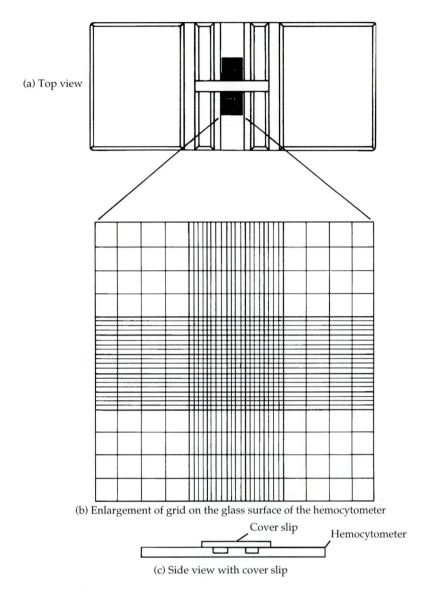

(a) Top view

(b) Enlargement of grid on the glass surface of the hemocytometer

Cover slip

Hemocytometer

(c) Side view with cover slip

FIGURE 16.2   Various parts of a hemocytometer (counting chamber). In (b), the areas used for red blood cell counts are labeled E, F, G, H, and I, and those for white blood cell counts are labeled A, B, C, and D.

blood loss. When the red blood cell count is increased above normal limits, the condition is called *polycythemia* (pol'-ē-sī-THĒ-mē-a).

The procedure we will use for determining the number of red blood cells per cubic millimeter (mm³) of blood is as follows.

PROCEDURE

WARD's Natural Science Establishment, Inc., has made available WARD'S *Simulated Blood*. It is a solution that contains microcomponents that simulate red blood cells, white blood cells, and platelets. The microcomponents are similar in relative pro-

portion to those found in human blood and can be observed under a microscope without staining.

1. Obtain a Simulated Blood Activity kit.
2. Follow the instructions to perform the red blood cell count.
3. Complete the portion of the Data Table related to red blood cells.
4. Answer the questions related to red blood cells.

ALTERNATE PROCEDURE

Note: The blood used for this procedure should be obtained from a clinical laboratory where it has been tested and certified as noninfectious or from a mammal (other than a human).

1. Obtain a Unopette Reservoir System® for red blood cell determination (the color of the reservoir's bottom surface is red). Identify the reservoir chamber, diluent fluid inside the chamber, pipette, and protective shield for the pipette (Figure 16.3). The diluent fluid contains isotonic saline and sodium azide, an antibacterial agent.

CAUTION! *Do not ingest this fluid; it is for in vitro diagnostic use only.*

2. Hold the reservoir on a flat surface in one hand and grasp the pipette assembly in the other hand.
3. Push the tip of the pipette shield firmly through the diaphragm in the neck of the reservoir (Figure 16.4a).
4. Pull out the assembly unit and with a twist remove the protective shield from the pipette assembly.

5. Hold the pipette *almost* horizontally and touch the tip of it to a forming drop of blood that has been placed on a microscope slide (Figure 16.4b). The pipette will fill by capillary action and, when the blood reaches the end of the capillary bore in the neck of the pipette, the filling action will stop.
6. Carefully wipe any excess blood from the tip of the pipette using a Kimwipe or similar wiping tissue, *making certain that no sample is removed from the capillary bore.*
7. Squeeze the reservoir *slightly* to expel a small amount of air.

CAUTION! *Do not expel any liquid. If the reservoir is squeezed too hard, the specimen may be expelled through the overflow chamber, resulting in contamination of the fingers. Reagent contains sodium azide, which is extremely toxic and yields explosive products in metal sinks or pipes. Azide compounds should be diluted with running water before being discarded and disposed of per the directions of your instructor.*

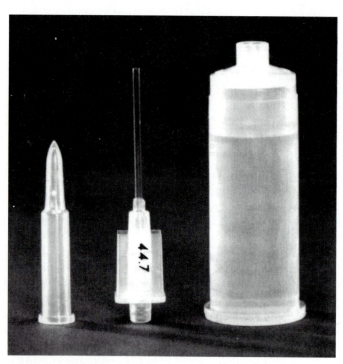

| Pipette shield | Pipette | Reservoir chamber containing diluent fluid |

FIGURE 16.3   Photograph of the Unopette Reservoir System®.

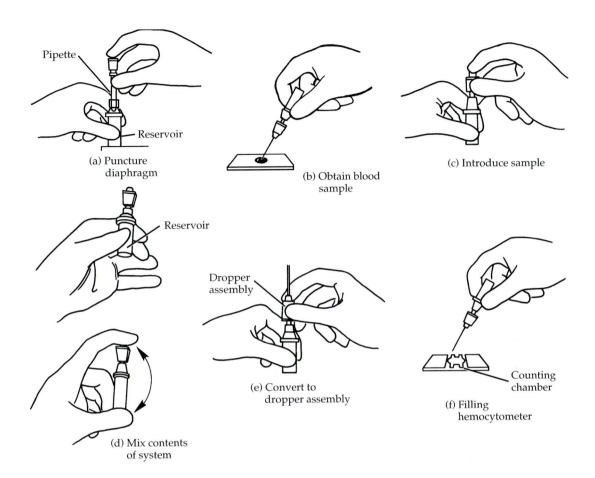

Pipette

Reservoir

(a) Puncture
diaphragm

(b) Obtain blood
sample

(c) Introduce sample

Reservoir

Dropper
assembly

Counting
chamber

(e) Convert to
dropper assembly

(f) Filling
hemocytometer

(d) Mix contents
of system

FIGURE 16.4　Preparation of the Unopette Reservoir System®. (Modified from *RBC Determination for Manual Methods* and *WBC Determination for Manual Methods*, Becton-Dickinson, Division of Becton, Dickinson and Company.)

While still maintaining pressure on the reservoir, cover the opening of the overflow chamber of the pipette with your index finger and push the pipette securely into the reservoir neck (Figure 16.4c).

8. Release the pressure on the reservoir and remove your finger from the pipette opening. This will draw the blood into the diluent fluid.

9. Mix the contents of the reservoir chamber by squeezing the reservoir gently several times. It is important to *squeeze gently* so that the diluent fluid is not forced out of the chamber. In addition to squeezing the reservoir chamber, invert it *gently* several times (Figure 16.4d).

10. Remove the pipette from the reservoir chamber, reverse its position, and replace it on the reservoir chamber. This converts the apparatus into a dropper assembly (Figure 16.4e).

11. Squeeze a few drops out of the reservoir chamber into a disposal container or wipe with

Kimwipe or similar wiping tissue. You are now ready to fill the hemocytometer (Figure 16.4f).

12. Follow the procedure outlined in D.2, filling the hemocytometer.

13. Place the hemocytometer on the microscope stage and allow the red blood cells to settle in the hemocytometer (approximately 1–2 min).

14. When counting red blood cells with the hemocytometer, the loaded (filled) chamber should be first located using the low-power objective of the microscope. Upon finding that, switch to the high-power objective via parfocal procedure. Focus the field using the fine adjustment of the microscope. Each red blood cell counted is equal to 10,000, so the degree of error in manual red blood cell counting is quite high, generally in the range of 15 to 20%.

15. Using the high-power lens, count the cells in each of the five squares (E, F, G, H, I) as shown in Figure 16.5. It is suggested that you use a hand counter. As you count, move up and

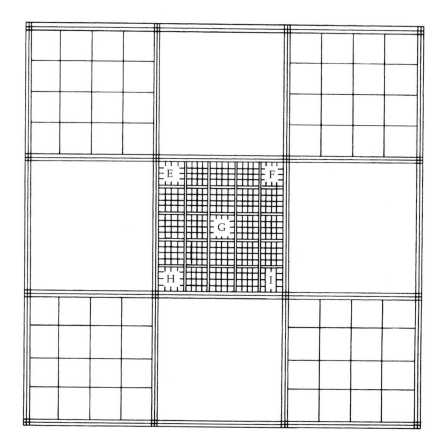

FIGURE 16.5 Hemocytometer (counting chamber). Areas E, F, G, H, and I are all counted in the red blood cell count. The large square containing the smaller squares E, F, G, H, and I is seen in microscopic field under 10× magnification. To count the smaller squares (E, F, G, H, and I), the 45× lens is used, and therefore E, F, G, H, and I each encompass the entire microscopic field.

down in a systematic manner. *Note: To avoid overcounting cells at the boundaries, cells that touch the lines on the left and top sides of the hemocytometer should be counted, but not the ones that touch the boundary lines on the right and bottom sides.*

16. Multiply the total number of red blood cells counted in the five squares by 10,000 to obtain the number of red blood cells in 1 mm³ of blood. Record your value in Section D.1 of the LABORATORY REPORT RESULTS at the end of the exercise.

17. The hemocytometer and cover slip should be saved, not discarded.

CAUTION! *Place the hemocytometer and cover slip in a container of fresh bleach solution and dispose of all other materials as directed by your instructor.*

4. Red Blood Cell Volume (Hematocrit)

The percentage of blood volume occupied by the red blood cells is called the *hematocrit (Hct)* or *packed cell volume (PCV)*. It is an important component of a complete blood count and is a standard test for almost all persons having a physical examination or for diagnostic purposes. When a tube of blood is centrifuged, the red blood cells pack into the bottom part of the tube with the plasma on top. The white blood cells and platelets are found in a thin area, the buffy layer, above the red blood cells.

The Readacrit® centrifuge (Figure 16.6) incorporates a built-in hematocrit scale and tube-holding compartments which, when used with special pre-calibrated capillary tubes, permit direct reading of

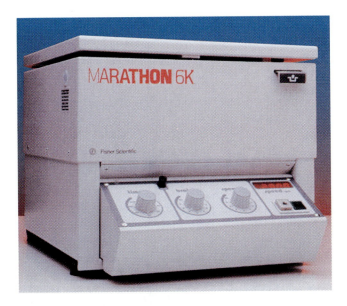

FIGURE 16.6 Centrifuge used for spinning blood to determine hematocrit.

the hematocrit value by measuring the length of the packed red blood cell column. Readacrit® centrifuges present the final hematocrit value without requiring computation by the operator. If the Readacrit® centrifuge or tube reader is not used for direct reading, measure the total length of blood volume, divide this quantity into length of packed red blood cells, and multiply by 100 for percentage.

The calculation is as follows:

$$\frac{\text{length of packed red blood cells}}{\text{total length of blood volume}} \times 100$$

$$= \text{\% volume of whole blood occupied by red blood cells (hematocrit)}$$

In males the normal range is between 40 and 54%, with an average of 47%. In females the normal range is between 38 and 47%, with an average of 42%. Anemic blood may have a hematocrit of 15%; polycythemic blood may have a hematocrit of 65%.

The following procedure for testing red blood cell volume is a micromethod requiring only a drop of blood:

PROCEDURE

Note: The blood used for this procedure should be obtained from a clinical laboratory where it has been tested and certified as noninfectious or from a mammal (other than a human).

1. Place a drop of blood on a microscope slide.
2. Place the unmarked (clear) end of a disposable capillary tube into the drop of blood. Hold the tube slightly below the level of the blood and do not move the tip from the blood or an air bubble will enter the column.

CAUTION! *If an air bubble is present, immediately dispose of the capillary tube in the appropriate biohazard container provided by your instructor and begin again.*

3. Allow blood to flow about two-thirds of the way into the tube. The tube will fill easily if you hold the open end level with or below the blood source.
4. Place your finger over the marked (red) end of the tube and keep it in that position as you seal the blood end of the tube with Seal-ease® or clay or other similar sealing material.
5. According to the directions of your laboratory instructor, place the tube into the centrifuge, making sure that the sealed end is against the rubber ring at the circumference of the centrifuge rotor. Have your laboratory instructor make sure that the tubes are properly balanced. Write the number that indicates the location of your tube. _____
6. When all the tubes from your laboratory section are in place, secure the inside cover of the centrifuge and close the outside cover. Centrifuge on high speed for 4 min.
7. Remove your tube and determine the hematocrit value by reading the length of packed red blood cells directly on the centrifuge scale, or by placing the tube in the tube reader (Figure 16.7) and following instructions on the reader, or by calculation using the equation mentioned previously. When using the tube reader, place the base of the Seal-ease® or clay on the base line and move tube until the top of the plasma is at the line marked 100.

CAUTION! *Immediately dispose of your capillary tube in the appropriate biohazard container provided by your instructor.*

Record your results in Section D.2 of the LABORATORY REPORT RESULTS at the end of the exercise.

FIGURE 16.7   Microhematocrit tube reader.

What effect would dehydration have on hematocrit? _____

_____

5. Sedimentation Rate

If blood is allowed to stand vertically in a tube, the red blood cells fall to the bottom of the tube and leave clear plasma above them. The distance that the cells fall in 1 hr can be measured and is called the **sedimentation rate.** It is a function of the amount of fibrinogen and gamma globulin present in plasma and the tendency of the red blood cells to adhere to one another. This rate is greater than normal during menstruation, pregnancy, malignancy, and most infections. Sedimentation rate may be decreased in liver disease. A high rate may indicate tissue destruction in some part of the body; therefore, sedimentation rate is considered a valuable nonspecific diagnostic tool. The normal rate for adults is 0 to 6 mm per hour, for children 0 to 8 mm per hour.

The following method for determining sedimentation rate requires only one drop of blood and is called the Landau micromethod.

PROCEDURE

Note: The blood used for this procedure should be obtained from a clinical laboratory where it has been tested and certified as noninfectious or from a mammal (other than a human).

1. Place a drop of blood on a microscope slide.
2. Using the mechanical suction device, draw the sodium citrate up to the first line that encircles the pipette.
3. Draw the blood up into a disposable sedimentation pipette to the second line. Do not remove the tip of the pipette from the blood or air bubbles will enter the column.

CAUTION! *If air bubbles are drawn into the blood, carefully expel the mixture onto a clean microscope slide, dispose of the slide and pipette in the appropriate biohazard container provided by your instructor, and begin again.*

4. Draw the mixture of fluids up into the bulb and mix by expelling it into the lumen of the tube. Draw and mix the fluids six times. If any air bubbles appear, use the procedure described in step 3.
5. Adjust the level of the blood as close to zero as possible. (Exactly zero is very difficult to get.)
6. Remove the suction device by placing the lower end of the pipette on the *gloved* index finger of the left hand before removing the device from the other end. The blood will leave the pipette if the end is not held closed.
7. Place the lower end of the pipette on the base of the pipette rack (Figure 16.8) and the opposite end at the top of the rack. The tube must be exactly perpendicular. Record the time at which it is put in the rack.

   Time _____.
8. One hour later, measure the distance from the top of the plasma to the top of the red blood cell layer with an accurate millimeter scale, and record your results in Section D.3 of the LABORATORY REPORT RESULTS at the end of the exercise.

CAUTION! *Dispose of the pipette in the appropriate biohazard container provided by your instructor. Using a medicine dropper, rinse the base of the pipette rack with fresh bleach solution.*

6. Hemoglobin Determination

It is possible for one to have anemia even if one's red blood cell count is normal. For example, the red blood cells present may be deficient in hemoglobin or they may be smaller than normal. Thus,

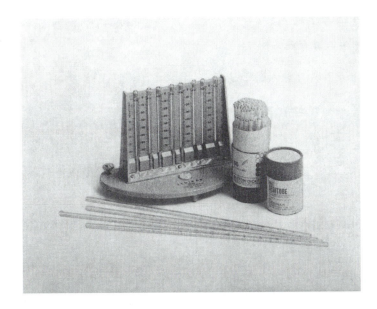

FIGURE 16.8   One type of sedimentation tube rack.

the amount of hemoglobin per unit volume of blood, and not necessarily the number of red blood cells, is the determining factor for anemia. The hemoglobin content of blood is expressed in grams (g) per 100 ml of blood, and the normal ranges vary with the technique used. An average range is 12 to 15 g/100 ml in females and 13 to 16 g/100 ml in males.

One procedure that can be used for determining hemoglobin uses an apparatus called a ***hemoglobinometer*** (see Figure 16.9a). This instrument compares the absorption of light by hemoglobin in a blood sample of known depth to that of a standardized glass plate.[1]

PROCEDURE

Note: *The blood used for this procedure should be obtained from a clinical laboratory where it has been tested and certified as noninfectious or from a mammal (other than a human).*

1. Obtain a hemoglobinometer and examine its parts (Figure 16.9a). Note the four scales on the side of the instrument. The top scale gives hemoglobin content in g/100 ml of blood; the other three scales give comparative readings, that is, the amount of hemoglobin expressed as a percent of 15.6, 14.5, or 13.8 g/100 ml of blood.

2. Remove the two pieces of glass from the blood chamber assembly (Figure 16.9b) and clean them with alcohol and wipe them with lens paper. The piece of glass that contains the moat will receive the blood sample; the other piece of glass serves as a cover glass.

3. Replace the glass pieces half-way onto the clip with the moat plate on the bottom.

4. Obtain a sample of blood.

5. Apply the blood to the moat and completely cover the raised surface of the plate with blood.

6. Hemolyze the blood by agitating it with the tip of a hemolysis applicator for about 30 to 45 sec. Hemolysis is complete when the appearance of the blood changes from cloudy to transparent. This change reflects the lysis of the red blood cell membranes and the liberation of hemoglobin.

7. Push the specimen chamber into the clip and push the clip into the hemocytometer.

8. Holding the hemocytometer in your left hand, look through the eyepiece while depressing the light switch button. You will see a split field.

9. Move the slide button back and forth with the right index finger until the two sides of the field match in color and shading.

10. Read the value indicated on the scale marked 15.6. Determine the grams of hemoglobin per 100 ml of the blood sample by reading the number above the index mark on the hemocytometer. Record your results in Section D.4 of the LABORATORY REPORT RESULTS at the end of the exercise.

[1]If Unopette tests are available, follow the procedure outlined in Unopette test no. 5857, "Cyanmethemoglobin Determination for Manual Methods," for hemoglobin estimation.

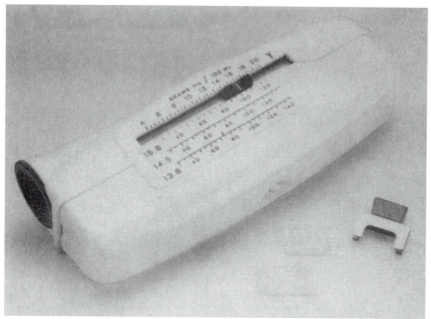

(a) Cambridge Instruments Hemoglobin-Meter

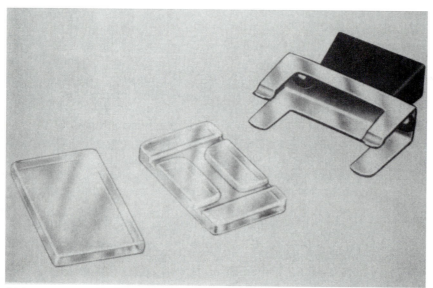

(b) Specimen chamber

FIGURE 16.9   Hemoglobinometer.

CAUTION! *Remove the glass pieces from the chamber and wash in fresh bleach solution. Dry the glass pieces with lens paper.*

11. What type of anemia would reveal a normal or slightly below normal hematocrit, yet a significantly reduced hemoglobin level?

_____

ALTERNATE PROCEDURE

The ***Tallquist measurement*** of hemoglobin is an old and somewhat inaccurate technique, although it is quick and inexpensive. The hemoglobinometer gives more accurate (±5%) results.

Note: The blood used for this procedure should be obtained from a clinical laboratory where it has been tested and certified as noninfectious or from a mammal (other than a human).

1. Obtain a sample of blood.
2. Place one drop of blood on the Tallquist® paper.
3. As soon as the blood no longer appears shiny, match its color with the scale provided. Record your observations in Section D.4 of the LABORATORY REPORT RESULTS at the end of the exercise.

7. Oxyhemoglobin Saturation

The ability of blood to carry oxygen is determined by proper respiratory system function, hematocrit, and the ability of hemoglobin to bind reversibly with oxygen within the alveoli (air sacs) of the lungs and release oxygen at the tissue level. Hemoglobin's interaction with oxygen is measured by an *oxygen-hemoglobin dissociation curve.* Such a curve demonstrates that the number of available oxygen-binding sites on hemoglobin that actually bind to oxygen is proportional to the partial pressure of oxygen (pO_2). Arterial blood usually has a pO_2 of 95, causing 97% of the available hemoglobin to be saturated with oxygen. Venous blood, however, has a pO_2 of 40, so only 75% of the hemoglobin is saturated with oxygen; 25% is given off to tissues. The determination of percent saturation for hemoglobin is a very sensitive test of pulmonary function. However, such a test needs to be interpreted carefully, as an individual with a normally functioning pulmonary system may exhibit abnormally low oxyhemoglobin saturation due to a variety of possible causes, such as *methemoglobinemia* (transformation of normal oxyhemoglobin due to the reduction of normal Fe^{2+} to Fe^{3+} within the hemoglobin molecule) or *carbon monoxide (CO) poisoning.* In order to determine the percent hemoglobin saturation in an unknown sample of blood one would need to compare the absorption spectrum of the blood sample to that obtained from a pure sample of *oxyhemoglobin (HbO2), carboxyhemoglobin (HbCO),* and *reduced hemoglobin (Hb)* (carboxyhemoglobin is included due to the normal presence of carbon monoxide in today's polluted atmosphere). This is shown in Figure 16.10. The spectrum obtained with the unknown sample of blood would be some combination of these three, since it would contain a certain percentage of each form of hemoglobin. Such a test is possible due to the fact that all three of these types of hemoglobin compounds are different colors, and would therefore absorb different portions of the light spectrum. Through a complex determination of the relative contribution of each type of hemoglobin to the resulting spectrum for the unknown sample, the percentage of each hemoglobin form may be determined.

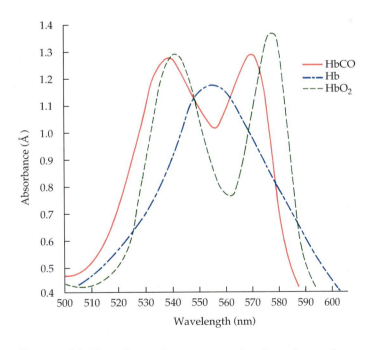

FIGURE 16.10 Absorption spectra of carboxyhemoglobin (HbCO), reduced hemoglobin (Hb), and oxyhemoglobin (HbO$_2$).

8. Spectrum of Oxyhemoglobin and Reduced Hemoglobin

The following procedure details a simplified, although less accurate method for determining the absorption spectra for fully oxygenated and deoxygenated hemoglobin. Exposing blood to air saturates it with oxygen, while sodium dithionate reduces it by removing oxygen.

PROCEDURE

Note: The blood used for this procedure should be obtained from a clinical laboratory where it has been tested and certified as noninfectious or from a mammal (other than a human).

1. Turn on the spectrophotometer and let it warm up for several minutes.
2. After the instrument has warmed up, set the wavelength at 500 nm and adjust the meter needle to 0% transmittance (absorbance at infinity) by using the zero control knob.
3. Place a cuvette containing distilled water into the cuvette holder and adjust the needle to 100% transmittance by using the light control knob.
4. Place 8 ml of distilled water in a clean test tube.
5. Obtain several drops of blood.
6. Mix the distilled water with the blood by placing a stopper in the mouth of the test tube and then inverting the test tube, thereby adding the blood to the distilled water.
7. Transfer half of the test tube contents (4 ml) into a second clean test tube.
8. Add 0.2 ml of 1% sodium dithionite solution to the second test tube and mix thoroughly. (*Note: In order for the procedure to work properly, the sodium dithionite solution must be fresh and made up just prior to use, and a spectral determination must be completed within 5 min of addition of sodium dithionite to the second tube.*)
9. Record the absorbances of solutions 1 (oxyhemoglobin) and 2 (reduced hemoglobin) at 500 nm wavelength.
10. Standardize the spectrophotometer at 510 nm utilizing the cuvette containing only distilled water as outlined in steps 2 and 3 above.
11. Determine the absorbances for solutions 1 and 2 at 510 nm.
12. Repeat procedures 2, 3, and 9 at each of the following wavelengths: 520, 530, 540, 560, 570, 580, 590, 600 nm.
13. Record and graph your results in Section D.5 of the LABORATORY REPORT RESULTS at the end of the exercise. Be sure to compare and give physiological reasons for the observed differences in the two spectra.

E. LEUKOCYTES

Leukocytes (white blood cells, or WBCs) are different from red blood cells in that they have nuclei and do not contain hemoglobin (see Figure 16.1). They are less numerous than red blood cells, ranging from 5000 to 10,000 cells per cubic millimeter (mm^3) or per deciliter (dl) of blood. The ratio, therefore, of red blood cells to white blood cells is about 700:1.

As Figure 16.1 shows, leukocytes can be differentiated by their appearance. They are divided into two major groups, granular leukocytes and agranular leukocytes. *Granular leukocytes,* which are formed from red bone marrow, have large granules in the cytoplasm that can be seen under a microscope and possess lobed nuclei. The three types of granular leukocytes are *neutrophils, eosinophils,* and *basophils. Agranular leukocytes,* which are also formed from red bone marrow, contain small granules that cannot be seen under a light microscope, and usually have spherical nuclei. The two types of agranular leukocytes are *lymphocytes* and *monocytes.*

Leukocytes as a group function in phagocytosis, producing antibodies, and combating allergies. The life span of a leukocyte usually ranges from a few hours to a few months. Some lymphocytes, called T and B memory cells, can live throughout one's life once they are formed.

F. WHITE BLOOD CELL TESTS

CAUTION! *Please reread Section A, "General Safety Precautions and Procedures," and Section B, "Precautions Related to Working with Blood, Blood Products, or Other Body Fluids," on pages xi–xii at the beginning of the laboratory manual, before you begin any of the following experiments. Read the experiments before you perform them, to be sure that you understand all the procedures and safety precautions.*

When working with whole blood, wear tight-fitting surgical gloves and safety goggles. Avoid any kind of contact with an open sore, cut, or wound.

1. White Blood Cell Count

This procedure determines the number of circulating white blood cells in the body. Because white blood cells are a vital part of the body's immune defense system, any abnormalities in the white blood cell count must be carefully noted.

An increase in number (*leukocytosis*) may result from such conditions as bacterial or viral infection, metabolic disorders, chemical and drug poisoning, and acute hemorrhage. A decrease in number (*leukopenia*) may result from typhoid infection, measles, infectious hepatitis, tuberculosis, or cirrhosis of the liver.

PROCEDURE

WARD'S Natural Science Establishment, Inc., has made available WARD'S *Simulated Blood*. It is a solution that contains microcomponents that simulate red blood cells, white blood cells, and platelets. The microcomponents are similar in relative proportion to those found in human blood and can be observed under a microscope without staining.

1. Obtain a Simulated Blood Activity kit.
2. Follow the instructions to perform the white blood cell count.
3. Complete the portion of the Data Table related to counted white blood cells.
4. Answer the questions related to white blood cells.

ALTERNATE PROCEDURE

Note: The blood used for this procedure should be obtained from a clinical laboratory where it has been tested and certified as noninfectious or from a mammal (other than a human).

Whole blood that has been tested for syphilis; hepatitis A, B, and C; and HIV is available from Carolina Biological Supply Company. This blood can be used for white blood cell count and differential white blood cell count.

1. Follow steps 1–14 of D.3, "Red Blood Cell Count," but use the Unopette Reservoir System® for white blood cell determination (the color of the reservoir's bottom surface is white).
2. Using the low-power objective (10×), count the cells in each of the four corner squares (A, B, C, D in Figure 16.11) of the hemocytometer. The direction to follow when counting white blood cells is indicated in square A in Figure 16.11. *Note: To avoid overcounting of cells at the boundaries, the cells that touch the lines on the left and top sides of the hemocytometer should be counted, but not the ones that touch the boundary lines on the right and bottom sides. It is suggested that you use a hand counter.*
3. Multiply the results by 50 to obtain the amount of circulating white blood cells per mm^3 of

blood and record your results in Section F.6 of the LABORATORY REPORT RESULTS at the end of the exercise.

4. The factor of 50 is the dilution factor, or $2.5 \times 20 = 50$. The volume correction factor of 2.5 is arrived at in this manner: each of the corner areas (A, B, C, and D) is exactly 1 mm^3 by 0.1 mm deep (Figure 16.11). Therefore, the volume of each of these corner areas is 0.1 mm^3. Because four of them are counted, the total volume of diluted blood examined is 0.4 mm^3. However, because we want to know the number of cells in 1 mm^3 instead of 0.4 mm^3, we must multiply our count by 2.5 ($0.4 \times 2.5 = 1.0$).

CAUTION! *Place the hemocytometer and cover slip in a container of fresh bleach solution and dispose of all other materials as directed by your instructor.*

5. The hemocytometer should be saved, not discarded.

2. Differential White Blood Cell Count

The purpose of a differential white blood cell count is to determine the relative *percentages* of each of the five normally circulating types of white blood cells in a total count of 100 white blood cells. A normal differential white blood cell count might appear as follows:

Type of white blood cell	Normal
Neutrophils	60–70%
Eosinophils	2–4%
Basophils	0.5–1%
Lymphocytes	20–25%
Monocytes	3–8%

Significant elevations of different types of white blood cells usually indicate specific pathological conditions. For example, a high neutrophil count may indicate acute appendicitis or infection. Lymphocytes predominate in antigen-antibody reactions, specific leukemias, and in infectious mononucleosis. An increase in eosinophils may be seen in allergic reactions and parasitic infections. An elevated percentage of monocytes may result from chronic infections. An increase in basophils is rare and denotes a specific type of leukemia and allergic reactions.

The procedure for making a differential white blood cell count is as follows:

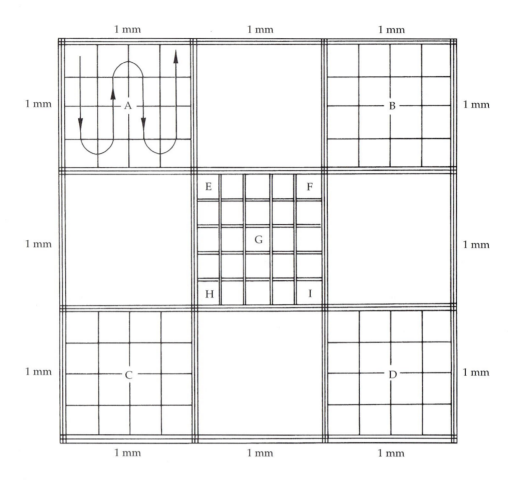

FIGURE 16.11 Hemocytometer (counting chamber) for white blood cell count (as seen through scanning lens). A, B, C, and D are areas counted in white blood cell count (when viewed through 10× lens, 1 sq mm is seen in microscopic field). Areas A, B, C, and D each equal 1 sq mm; therefore, a total of 4 sq mm is counted in the white blood cell count.

PROCEDURE

Note: The blood used for this procedure should be obtained from a clinical laboratory where it has been tested and certified as noninfectious or from a mammal (other than a human).

1. Obtain a drop of blood.
2. Use a second slide as a spreader (Figure 16.12).

3. Draw the spreader toward the drop of blood (in this direction →) until it touches the drop. The blood should fan out to the edges of the spreader slide.
4. Keeping the spreader at a 25° angle, press the edge of the spreader firmly against the slide and push the spreader rapidly over the entire length of the slide (in this direction ←). The

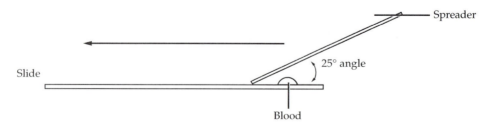

FIGURE 16.12 Procedure to prepare a blood smear.

drop of blood will thin out toward the end of the slide.

5. Let the smear dry.
6. To stain the slide follow this procedure:

   a. Place the slide on a staining rack.
   b. Cover the entire slide with Wright's stain.
   c. Let stain stand for 1 min.
   d. Add 25 drops of buffer solution and mix completely with Wright's stain.
   e. Let the mixture stand for 8 min.
   f. Wash the mixture off completely with distilled water.

7. Let the slide dry completely before counting.
8. To proceed with counting the cells, use an area of slide where blood is thinnest (one cell thick). This is called the *feathering edge*. Count cells under an oil-immersion lens. Count a total of 100 white cells.
9. The actual counting may be done by moving the slide either up and down or from side to side as follows:

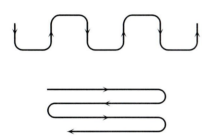

10. Record each white blood cell you observe by making a mark in the following chart until you have recorded 100 cells.

ALTERNATE PROCEDURE

1. Obtain a prepared slide of stained blood.
2. Repeat steps 9 and 10 in the preceding procedure.

G. PLATELETS

Platelets (thrombocytes) are formed from fragments of the cytoplasm of megakaryocytes (see Figure 16.1). The fragments become enclosed in pieces of cell membrane from the megakaryocytes and develop into platelets. Platelets are very small, disc-shaped cell fragments without nuclei. Between 250,000 and 400,000 are found in each cubic millimeter or deciliter (dl) of blood. They function to prevent fluid loss by starting a chain of reactions that results in blood clotting. They have a short life span, probably only one week, because they are expended in clotting and are just too simple to carry on extensive metabolic activity.

PROCEDURE

WARD'S Natural Science Establishment, Inc., has made available WARD'S *Simulated Blood*. It is a solution that contains microcomponents that simulate red blood cells, white blood cells, and platelets. The microcomponents are similar in relative proportion to those found in human blood and can be observed under a microscope without staining.

1. Obtain a Simulated Blood Activity kit.
2. Follow the instructions to perform the platelet count.

Type of white blood cell	Number observed	Percentage
Neutrophils		
Eosinophils		
Basophils		
Lymphocytes		
Monocytes		

3. Complete the portion of the Data Table related to platelets.

4. Answer the questions related to platelets.

H. DRAWINGS OF BLOOD CELLS

Examine a prepared slide of a stained smear of blood cells using the oil-immersion objective. With the aid of your textbook, identify red blood cells, neutrophils, basophils, eosinophils, lymphocytes, monocytes, and platelets. Using colored crayons or pencil crayons, draw and label all of the different blood cells in Section H. of the LABORATORY REPORT RESULTS at the end of the exercise. Be very accurate in drawing and coloring the granules and the nuclear shapes, because both of these are used to identify the various types of cells.

I. BLOOD GROUPING (TYPING)

The plasma membranes of red blood cells contain genetically determined antigens called *isoantigens.* The plasma of blood contains genetically determined antibodies called *isoantibodies.* The antibodies cause the *agglutination (clumping)* of the red blood cells carrying the corresponding antigen.

These proteins, antigens, and antibodies are responsible for the two major classifications of blood groups: the ABO group and the Rh system. In addition to the ABO group and the Rh system, other human blood groups include: MNSs, P, Lutheran, Kell, Lewis, Duffy, Kidd, Diego, and Sutter. Fortunately these different antigenic factors do not exhibit extreme degrees of antigenicity and, therefore, usually cause very weak transfusion reactions or no reaction at all.

1. ABO Group

This major blood grouping is based on two antigens symbolized as *A* and *B* (Figure 16.13). Individuals whose red blood cells produce only antigen *A* have blood type A. Individuals who produce only antigen *B* have blood type B. If both *A* and *B* antigens are produced, the result is type AB, whereas the absence of both *A* and *B* antigens results in the so-called type O.

The antibodies in blood plasma are anti-*A* antibody, which attacks antigen *A*, and anti-*B* antibody, which attacks antigen *B*. The antigens and antibodies formed by each of the four blood types and the various agglutination reactions that occur when whole blood samples are mixed with serum are shown in Figure 16.13. You do not have antibodies that attack the antigens of your own red blood cells. For example, a type A person has antigen *A* but not antibody anti-*A*.

Antigens and antibodies are of critical importance in blood transfusions. In an incompatible blood transfusion, the donated red blood cells are attacked by the recipient antibodies, causing the blood cells to agglutinate. Agglutinated cells become lodged in small capillaries throughout the body and, over a period of hours, the cells swell, rupture, and release hemoglobin into the blood. Such a reaction as it relates to red blood cells is called *hemolysis* (*lysis* = dissolve). The degree of agglutination depends on the titer (strength or amount) of antibody in the blood. Agglutinated cells can block blood vessels and may lead to kidney or brain damage and death, and the liberated hemoglobin may also cause kidney damage.

Because cells of type O blood contain neither of the two antigens (*A* or *B*), moderate amounts of this blood can be transfused into a recipient of any ABO blood type without *immediate* agglutination. For this reason, type O blood is referred to as the *universal donor.* However, transfusing large amounts of type O blood into a recipient of type A, B, or AB can cause *delayed* agglutination of the recipient's red blood cells, because the transfused antibodies (type O blood contains anti-*A* and anti-*B* antibodies) are not sufficiently diluted to prevent the reaction.

Persons with type AB blood are sometimes called *universal recipients* because their blood contains no antibodies to agglutinate donor red blood cells. *Small quantities* of blood from all other ABO types can be transfused into type AB blood without adverse reaction. However, again, if *large quantities* of type A, B, or O blood are transfused into type AB, the antibodies in the donor blood might accumulate in sufficient quantity to clump the recipient's red blood cells.

Table 16.1 summarizes the various interactions of the four blood types. Table 16.2 lists the incidence of these different blood types in the United States, comparing some races.

2. ABO Blood Grouping Test

CAUTION! *Please reread Section A, "General Safety Precautions and Procedures" and Section B, "Precautions Related to Working with Blood, Blood Products, or Other Body Fluids" on pages xi–xii at the beginning of the laboratory manual before you*

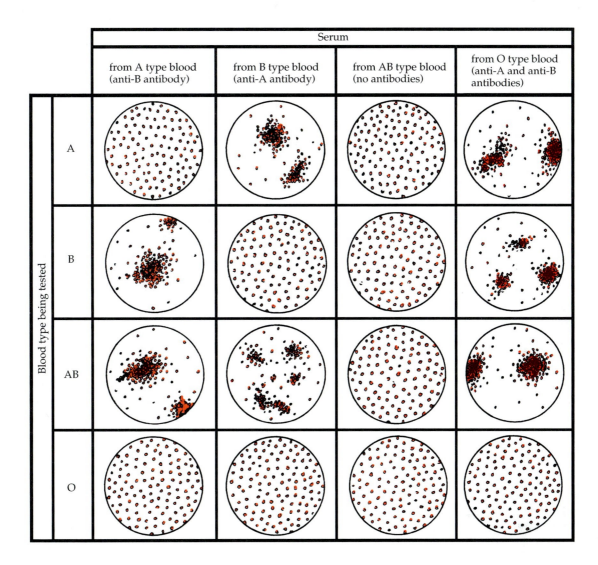

FIGURE 16.13   ABO blood groupings.

TABLE 16.1
Summary of ABO System Interactions

Blood Type	A	B	AB	O
Antigen on red blood cell	*A*	*B*	*A* and *B*	Neither *A* nor *B*
Antibody in plasma	anti-*B*	anti-*A*	Neither anti-*A* nor anti-*B*	*a* and *b*
Compatible donor blood types	A,O	B,O	A, B, AB, O	O
Incompatible donor blood types	B, AB	A, AB	—	A, B, AB
Genotype (genetic make-up)	*AO* and *AA*	*BO* and *BB*	*AB*	*OO*

TABLE 16.2
Incidence of Human Blood Groups in the United States

	Blood Groups (percentages)				
	O	A	B	AB	Rh$^+$
Caucasians	45	41	10	4	85
Blacks	48	27	21	4	88
Japanese	31	38	22	9	~ 100
Chinese	36	28	23	13	~ 100
American Indians	23	76	0	1	~ 100
Hawaiians	37	61	1.5	0.5	~ 100

begin any of the following experiments. You should also read the experiments before you perform them to be sure that you understand all the procedures and safety precautions.

When working with whole blood, wear tight-fitting surgical gloves and safety goggles. Avoid any kind of contact with an open sore, cut, or wound.

Blood should be obtained from a clinical laboratory where it has been tested and certified as noninfectious or from a mammal (other than a human).

Whole blood that has been tested for syphilis; hepatitis A, B, and C; and HIV is available from Carolina Biological Supply Company. This blood can be used for blood grouping and typing. Carolina Biological Supply Company also provides aseptic red blood cells that can be used for blood grouping and typing. These blood cells are suspended in a modified alsevere's solution to which several antibiotics have been added. The cells have been tested for syphilis; hepatitis A, B, and C; and HIV.

Also available for this exercise is WARD'S Simulated ABO and Rh Blood Typing Activity.

The procedure for ABO sampling is as follows:

PROCEDURE
1. Using a wax pencil, divide a glass slide in half and label the left side A and the right side B.
2. On the left side, place one large drop of anti-A serum, and on the right side place one large drop of anti-B serum.
3. Next to the drops of antisera, place one drop of blood, being careful not to mix samples on the left and right sides.
4. Using a mixing stick or a toothpick, mix the blood on the left side with the anti-A serum, and then, using a *different stick or different tooth-*

pick, mix the blood on the right side with the anti-B.

CAUTION! *Immediately dispose of the stick or toothpick in the appropriate biohazard container provided by your instructor.*

5. Gently tilt the slide back and forth and observe it for 1 minute.
6. Record your results in Section I.1 of the LABORATORY REPORT RESULTS at the end of the exercise, using "+" for clumping (agglutination) and "−" for no clumping.

CAUTION! *Dispose of your slide in the appropriate biohazard container provided by your instructor.*

3. Rh System

The Rh factor is the other major classification of blood grouping. This group was designated Rh because the blood of the rhesus monkey was used in the first research and development. As is the ABO grouping, this classification is based on antigens that lie on the surfaces of red blood cells. The designation Rh$^+$ (positive) is given to those that have the antigen, and Rh$^-$ (negative) is for those that lack the antigen. The estimation is that 85% of whites and 88% of blacks in the United States are Rh$^+$, whereas 15% of whites and 12% of blacks are Rh$^-$ (see Table 16.2).

The Rh factor is extremely important in pregnancy and childbirth. Under normal circumstances, human plasma does not contain anti-Rh antibodies. If, however, a woman who is Rh$^-$ becomes pregnant with an Rh$^+$ child, her blood may produce antibodies that will react with the blood of a subsequent child.[2] The first child is unaffected because the mother's body has not yet produced these antibodies. This is a serious reaction and hemolysis may occur in the fetal blood.

The hemolysis produced by this fetal-maternal incompatibility is called **hemolytic disease of newborn,** or **HDN (erythroblastosis fetalis),** and could be fatal for the newborn infant. A drug called Rho-GAM, given to RH⁻ mothers immediately after delivery or abortion, prevents the production of antibodies by the mother so that the fetus of the next pregnancy is protected.

4. Rh Blood Grouping Test

CAUTION! *Please reread Section A, "General Safety Precautions and Procedures," and Section B, "Precautions Related to Working with Blood, Blood Products, or Other Body Fluids," on pages xi–xii, at the beginning of the laboratory manual, before you begin any of the following experiments. Read the experiments before you perform them, to be sure that you understand all the procedures and safety precautions.*

When working with whole blood, wear tight-fitting surgical gloves and safety goggles. Avoid any kind of contact with an open sore, cut, or wound.

Note: *Remember that this procedure is sufficient for laboratory demonstration, but it should not be considered clinically accurate, since the anti-Rh (anti-D) serum deteriorates rapidly at room temperature and the anti-Rh antibodies are far less potent in their agglutinizing capability than the anti-A or anti-B antibodies. This test is less accurate than the ABO determination.*

[2]Be sure to note the important difference in the production of antibodies in the ABO and Rh systems. An Rh⁻ person can *produce antibodies in response to the stimulus of invading Rh antigen;* by contrast, any antibodies of the ABO system that exist in the blood of a person *occur naturally and are present regardless of whether or not ABO antigens are introduced.*

Blood should be obtained from a clinical laboratory where it has been tested and certified as noninfectious or from a mammal (other than a human).

Whole blood that has been tested for syphilis; hepatitis A, B, and C; and HIV is available from Carolina Biological Supply Company. This blood can be used for blood grouping and typing. Carolina Biological Supply Company also provides aseptic red blood cells that can be used for blood grouping and typing. These blood cells are suspended in a modified alsevere's solution to which several antibiotics have been added. The cells have been tested for syphilis; hepatitis A, B, and C; and HIV.

Also available for this exercise is WARD'S Simulated ABO and Rh Blood Typing Activity.

The procedure for Rh grouping is as follows:

PROCEDURE

1. Place one large drop of anti-Rh (anti-D) serum on a glass slide.[3]
2. Add one drop of blood and mix using a mixing stick or toothpick.

CAUTION! *Immediately dispose of the stick or toothpick in the appropriate biohazard container provided by your instructor.*

3. Place the slide on a preheated warming box, and gently rock the box back and forth for 2 minutes. (Unlike ABO typing, Rh typing is better done on a heated warming box.)
4. Record your results, using "+" for clumping and "–" for no clumping. Record whether you are Rh⁺ or Rh⁻ in Section I.4 of the LABORATORY REPORT RESULTS at the end of the exercise.

CAUTION! *Dispose of your slide in the appropriate biohazard container provided by your instructor.*

ANSWER THE LABORATORY REPORT QUESTIONS AT THE END OF THE EXERCISE.

[3]The Rh antigen is more specifically termed the D antigen after the Fisher-Race nomenclature, which is based on genetic concepts or theories of inheritance.

Blood 16

Student _____   **Date** _____

Laboratory Section _____   **Score/Grade** _____

SECTION B. PLASMA

1. *Physical Characteristics*

pH _____

Color _____

Transparency (clear, translucent, opaque) _____

2. *Chemical Constituents*

Is glucose present? _____

Is protein present? _____

SECTION D. RED BLOOD CELL TESTS

1. Red blood cell count results: _____ RBCs per mm^3.

2. Red blood cell volume (hematocrit) results: _____%.

3. Sedimentation rate results: _____ mm per hour.

4. Hemoglobin determination results: hemoglobinometer, _____ g per 100 ml.

Tallquist® paper, _____ g per 100 ml.

5. Complete the following table:

Wavelength

	500	510	520	530	540	550	560	570	580	590	600
Solution 1											
Solution 2											

Graph your results here.

SECTION F. WHITE BLOOD CELL TESTS

6. White blood cell count results: _____ WBCs per mm^3.

SECTION H. DRAWINGS OF BLOOD CELLS

SECTION I. BLOOD GROUPING (TYPING)

1. In determining the ABO blood grouping, did you observe clumping when the blood was mixed with

_____ anti-A serum only

_____ anti-B serum only

_____ both anti-A and anti-B serums

_____ neither anti-A nor anti-B serum

2. Based on your observations, what is the ABO grouping?

_____ A   _____ B   _____ AB   _____ O

3. Based on your observations, briefly explain why you identify your ABO blood grouping as you do.

_____

_____

4. Record the results of the ABO blood grouping test done by your class.

Type	Anti-A (present or absent)	Anti-B (present or absent)	Number of individuals	Class percentage
A				
B				
AB				
O				

5. Based on your observations, are you Rh$^+$ or Rh$^-$?

_____ Rh$^+$ _____ Rh$^-$

6. Record the results of the Rh test done by your class.

Type	Number of individuals	Class percentage
Rh$^+$		
Rh$^-$		

Blood 16

Student _____   **Date** _____

Laboratory Section _____   **Score/Grade** _____

PART 1. Multiple Choice

_____   1. The process by which all blood cells are formed is called (a) hemocytoblastosis (b) erythro-poiesis (c) hemopoiesis (d) leukocytosis

_____   2. An inability of body cells to receive adequate amounts of oxygen may indicate a malfunction of (a) neutrophils (b) leukocytes (c) lymphocytes (d) erythrocytes

_____   3. Special cells of the body that have the responsibility of clearing away dead, disintegrating bodies of red and white blood cells are called (a) agranular leukocytes (b) reticuloendothelial cells (c) erythrocytes (d) thrombocytes

_____   4. The name of the test procedure that informs the physician about the rate of erythropoiesis is called the (a) reticulocyte count (b) sedimentation rate (c) hemoglobin count (d) differential white blood cell count

_____   5. The normal red blood cell count per cubic millimeter in a male is about (a) 5.4 million (b) 7 million (c) 4 million (d) more than 9 million

_____   6. The normal number of leukocytes per cubic millimeter is (a) 5000 to 10,000 (b) 8000 to 12,000 (c) 2000 to 4000 (d) over 15,000

_____   7. Under the microscope, red blood cells appear as (a) circular discs with centrally located nuclei (b) circular discs with lobed nuclei (c) oval discs with many nuclei (d) biconcave discs without nuclei

_____   8. An increase in the number of white blood cells is called (a) leukopenia (b) hematocrit (c) polycythemia (d) leukocytosis

_____   9. Platelets are formed from a special large cell that breaks up into small fragments. This cell is called a(n) (a) eosinophil (b) hemocytoblast (c) megakaryocyte (d) platelet

_____   10. The blood type showing the highest incidence in Caucasians in the United States is (a) A (b) O (c) AB (d) B

PART 2. Completion

11. Another name for red blood cells is _____.

12. Blood gets its red color from the presence of _____.

13. The life span of a red blood cell is approximately _____.

14. A good method for routine testing for anemia is _____.

15. The normal sedimentation rate value for adults is _____.

16. The normal ratio of red blood cells to white blood cells is about _____.

17. The granular leukocytes are formed from _____ tissue.

18. The number of platelets per cubic millimeter found normally in blood is _____.

19. The function of platelets is to prevent blood loss by starting a chain of reactions resulting in

_____.

20. In the ABO blood grouping system, the genetically determined structures on the surface of red blood

cells are called _____.

21. The hemolysis produced by fetal-maternal incompatibility of blood cells is called

_____.

22. The part of the cell where agglutinogens are located is _____.

PART 3. Matching

_____ **23.** Hematocytoblast	A. Response to tissue by invading bacteria.
_____ **24.** Polycythemia	B. An increase in the normal red blood cell count
_____ **25.** A high neutrophil count	C. Leukemia and infectious mononucleosis
_____ **26.** A high monocyte count	D. Immature cells that develop into mature blood cells.
_____ **27.** Leukopenia	E. Chronic infections
_____ **28.** A high lymphocyte count	F. Liquid portion of blood without the formed elements
_____ **29.** Plasma	and clotting substances
_____ **30.** A high eosinophil count	G. An allergic reaction
_____ **31.** Serum	H. A decrease in the normal white blood cell count
	I. Liquid portion of blood without the formed elements

Heart

The **heart** is a hollow, muscular organ that pumps blood through miles and miles of blood vessels. This organ is located in the mediastinum, between the lungs, with two thirds of its mass lying to the left of the body's midline. Its pointed end, the **apex,** projects inferiorly to the left, and its broad end, the **base,** projects superiorly to the right. The main parts of the heart and associated structures to be discussed here are the pericardium, wall, chambers, great vessels, and valves.

A. PERICARDIUM

A loose-fitting serous membrane called the **peri-cardium** (*peri* = around; *cardio* = heart), or **pericardial sac** encloses the heart (Figure 17.1). The membrane is composed of two layers, the fibrous pericardium and the serous pericardium. The **fibrous pericardium** forming the superficial layer is tough, inelastic, dense, irregular connective tissue that adheres to the parietal pleura and anchors the heart in the mediastinum. The deep layer, the **serous pericardium**, is a thinner, delicate membrane that is a double layered structure. The outer

parietal layer of the serous pericardium is directly beneath the fibrous pericardium. The inner **visceral layer** of the serous pericardium is also called the **epicardium**. Between the parietal and visceral layers of the serous pericardium is a potential space, the **pericardial cavity**, which contains pericardial fluid and functions to prevent friction between the layers as the heart beats. Identify these structures using a specimen, model, or chart of a heart and label Figure 17.1.

B. HEART WALL

Three layers of tissue compose the heart: the epicardium (external layer), the myocardium (middle layer), and the endocardium (inner layer). The **epicardium** (*epi* = above), which is also the visceral layer of the serous pericardium, is the thin, transparent superficial layer of the heart wall. The middle **myocardium** (*myo* = muscle), which is composed of cardiac muscle tissue, forms the bulk of the heart and is responsible for contraction. The **endocardium** (*endo* = within) is a thin, deep layer of endothelium and areolar connective tissue that lines the inside of

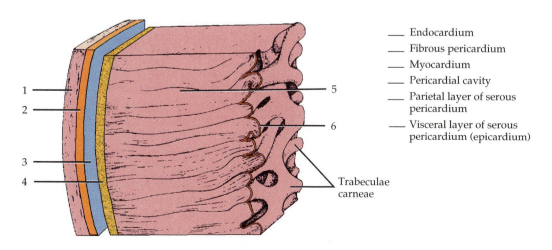

____ Endocardium

____ Fibrous pericardium

____ Myocardium

____ Pericardial cavity

____ Parietal layer of serous pericardium

____ Visceral layer of serous pericardium (epicardium)

Trabeculae carneae

FIGURE 17.1 Structure of pericardium and heart wall.

the myocardium and covers the heart valves and the tendons that hold them open. Label the layers of the heart in Figure 17.1.

C. CHAMBERS OF HEART

The interior of the heart is divided into four cavities called *chambers,* which receive circulating blood. The two superior chambers are known as *right* and *left atria* (*atrium* = entry hall) and are separated internally by a partition called the *interatrial septum* (*septum* = partition). A prominent feature of this septum is an oval depression, the *fossa ovalis,* which corresponds to the site of the *foramen ovale,* an opening in the interatrial septum of the fetal heart that helps blood bypass the nonfunctioning lungs. Each atrium has an appendage called an *auricle* (OR-i-kul; *auris* = ear), so named because its shape resembles a dog's ear. The auricles increase the size of the atria so that they can hold greater volumes of blood. The lining of the atria is smooth, except for the anterior walls and linings of the auricles, which contain projecting muscle bundles called *pectinate* (PEK-ti-nāt) *muscles.*

The two inferior and larger chambers, called the *right* and *left ventricles* (*ventricle* = little belly), are separated internally by a partition called the *interventricular septum.* The irregular surface of ridges and folds of the myocardium in the ventricles is known as the *trabeculae carneae* (tra-BEK-yoo-lē KAR-nē-ē; *trabecula* = little beam; *carneous* = fleshy). Externally, a groove known as the *coronary sulcus* (SUL-kus; plural is *sulci;* SUL-kē) separates the atria from the ventricles. The groove encircles the heart and houses the coronary sinus (a large cardiac vein) and the circumflex branch of the left coronary artery. The *anterior interventricular sulcus* and *posterior interventricular sulcus* separate the right and left ventricles externally. They also contain coronary blood vessels and a variable amount of fat.

Label the chambers and associated structures of the heart in Figures 17.2 and 17.3.

D. GREAT VESSELS AND VALVES OF HEART

The right atrium receives *deoxygenated blood* (blood that gives up some of its oxygen to cells) through three veins: the *superior vena cava (SVC)* brings blood from most parts of the body superior to the

heart, the *inferior vena cava (IVC)* brings blood from all parts of the body inferior to the diaphragm, and the *coronary sinus* brings blood from most of the vessels supplying the heart wall.

Deoxygenated blood is passed from the right atrium into the right ventricle through the atrioventricular valve called the *tricuspid valve,* which consists of three cusps (flaps). The right ventricle then pumps the blood through the *pulmonary semilunar valve* into the *pulmonary trunk.* The pulmonary trunk divides into a *right* and *left pulmonary artery,* each of which carries blood to the lungs where the blood releases its carbon dioxide and takes on oxygen. This blood, called *oxygenated blood,* returns to the heart via four *pulmonary veins* that empty the blood into the left atrium. The blood is then passed into the left ventricle through another atrioventricular valve called the *bicuspid (mitral) valve,* which consists of two cusps. The cusps of the tricuspid and bicuspid valves are connected to cords called *chordae tendineae* (KOR-dē TEN-din-ē-ē), which in turn attach to projections in the ventricular walls called *papillary muscles.* The left ventricle pumps oxygenated blood through the *aortic semilunar valve* into the *ascending aorta.* From this vessel, aortic blood is passed into the *coronary arteries, arch of the aorta,* and *descending aorta.* These blood vessels transport the blood to the body. The function of the heart valves is to permit the blood to flow in only one direction.

Label the great vessels and valves of the heart in Figures 17.2 and 17.3a and b.

E. BLOOD SUPPLY OF HEART

Because of its importance in myocardial infarction (heart attack), the blood supply of the heart will be described briefly at this point. The arterial supply of the heart is provided by the right and left coronary arteries. The *right coronary artery* originates as a branch of the ascending aorta, descends in the coronary sulcus, and gives off a *marginal branch* that supplies the right ventricle. The right coronary artery continues around the posterior surface of the heart in the posterior interventricular sulcus. This portion of the artery is known as the *posterior interventricular branch* and supplies the right and left ventricles and interventricular septum. In its course, the right coronary artery also supplies small branches (*atrial branches*) to the right atrium. The *left coronary artery* also originates as a branch of the ascending aorta. Between the pulmonary

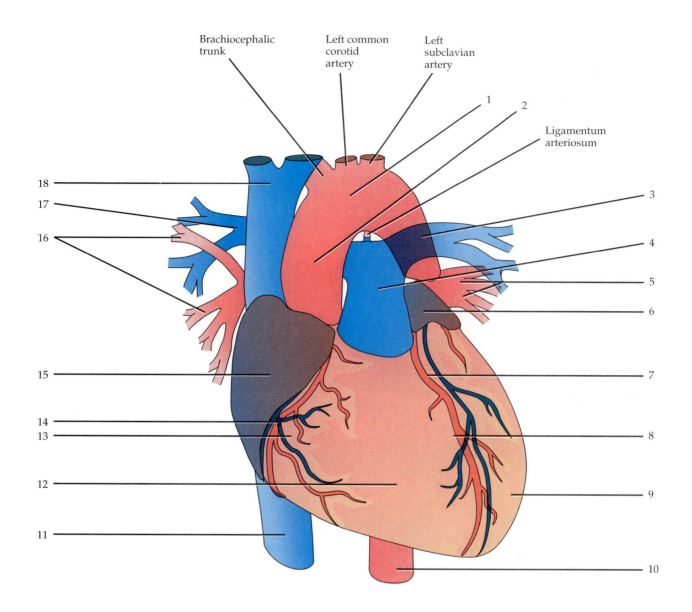

Brachiocephalic trunk

Left common corotid artery

Left subclavian artery

Ligamentum arteriosum

18

17

16

15

14

13

12

11

1

2

3

4

5

6

7

8

9

10

(a) Diagram of anterior external view

___ Anterior interventricular sulcus

___ Arch of aorta

___ Ascending aorta

___ Coronary sulcus

___ Descending aorta

___ Inferior vena cava

___ Left atrium

___ Left coronary artery

___ Left pulmonary artery

___ Left pulmonary veins

___ Left ventricle

___ Pulmonary trunk

___ Right atrium

___ Right coronary artery

___ Right pulmonary artery

___ Right pulmonary veins

___ Right ventricle

___ Superior vena cava

FIGURE 17.2   External surface of the human heart. Red-colored vessels carry oxygenated blood; blue-colored vessels carry deoxygenated blood.

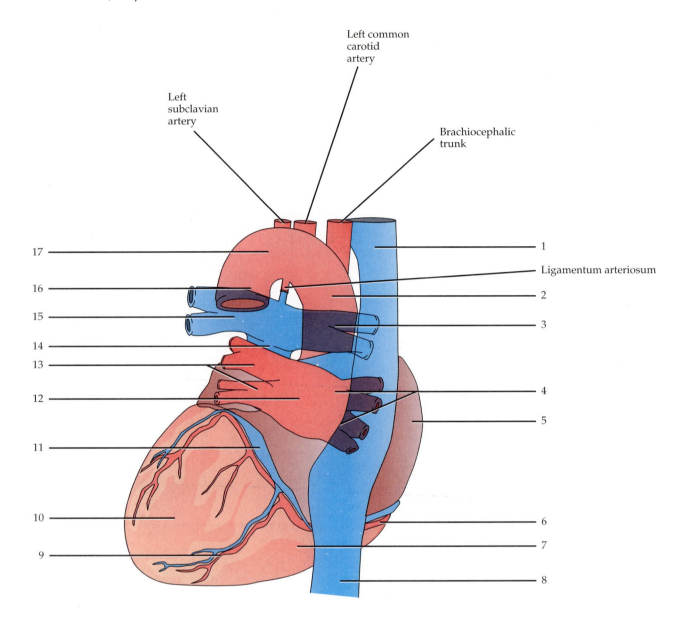

Left common
carotid
artery

Left
subclavian
artery

Brachiocephalic
trunk

17

16

15

14

13

12

11

10

9

1

Ligamentum arteriosum

2

3

4

5

6

7

8

(b) Diagram of posterior external view

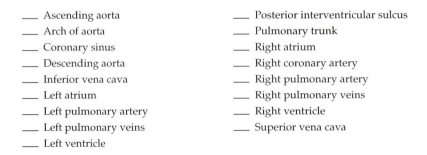

___ Ascending aorta

___ Arch of aorta

___ Coronary sinus

___ Descending aorta

___ Inferior vena cava

___ Left atrium

___ Left pulmonary artery

___ Left pulmonary veins

___ Left ventricle

___ Posterior interventricular sulcus

___ Pulmonary trunk

___ Right atrium

___ Right coronary artery

___ Right pulmonary artery

___ Right pulmonary veins

___ Right ventricle

___ Superior vena cava

FIGURE 17.2 *(Continued)* External surface of the human heart.

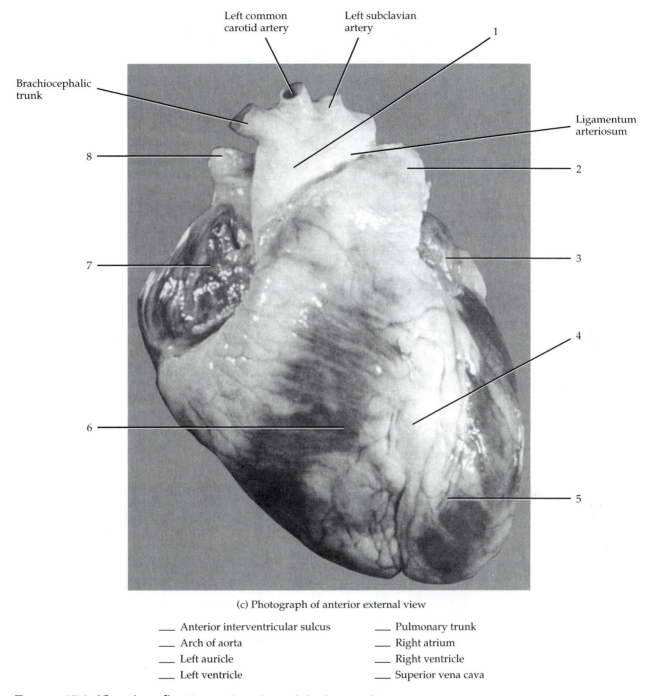

Left common
carotid artery

Left subclavian
artery

1

Brachiocephalic
trunk

Ligamentum
arteriosum

8

2

7

3

4

6

5

(c) Photograph of anterior external view

___ Anterior interventricular sulcus   ___ Pulmonary trunk

___ Arch of aorta   ___ Right atrium

___ Left auricle   ___ Right ventricle

___ Left ventricle   ___ Superior vena cava

FIGURE 17.2 (Continued)   External surface of the human heart.

trunk and left auricle, the left coronary artery divides into two branches: anterior interventricular and circumflex. The ***anterior interventricular branch*** passes in the anterior interventricular sulcus and supplies the right and left ventricles and interventricular septum. The ***circumflex branch*** circles toward the posterior surface of the heart in the coronary sulcus and distributes blood to the left ventricle and left atrium.

Label the arteries in Figure 17.4a on page 401.

Most blood from the heart drains into the ***coronary sinus,*** a venous channel in the posterior portion of the coronary sulcus between the left atrium and left ventricle. The vein empties into the right atrium. The principal tributaries of the coronary sinus are the ***great cardiac vein,*** which drains the anterior aspect of the heart, and the ***middle cardiac vein,*** which drains the posterior aspect of the heart.

Label the veins in Figure 17.4b on page 401.

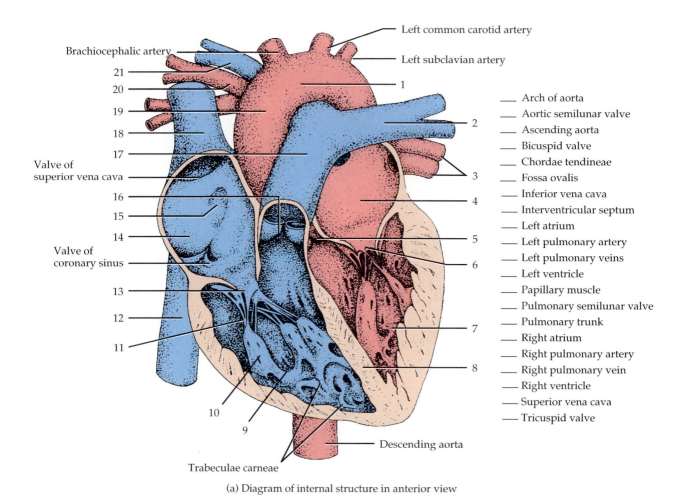

Left common carotid artery

Brachiocephalic artery

Left subclavian artery

21

20

19

18

17

Valve of
superior vena cava

16

15

14

Valve of
coronary sinus

13

12

11

10

9

Trabeculae carneae

1

2

3

4

5

6

7

8

Descending aorta

___ Arch of aorta
___ Aortic semilunar valve
___ Ascending aorta
___ Bicuspid valve
___ Chordae tendineae
___ Fossa ovalis
___ Inferior vena cava
___ Interventricular septum
___ Left atrium
___ Left pulmonary artery
___ Left pulmonary veins
___ Left ventricle
___ Papillary muscle
___ Pulmonary semilunar valve
___ Pulmonary trunk
___ Right atrium
___ Right pulmonary artery
___ Right pulmonary vein
___ Right ventricle
___ Superior vena cava
___ Tricuspid valve

(a) Diagram of internal structure in anterior view

6

5

Trabeculae
carneae

1

2

3

4

___ Bicuspid valve   ___ Left ventricle
___ Chordae tendineae   ___ Papillary muscle
___ Interventricular septum   ___ Right ventricle

(b) Photograph of internal structure in anterior view

FIGURE 17.3   Structure of the human heart.

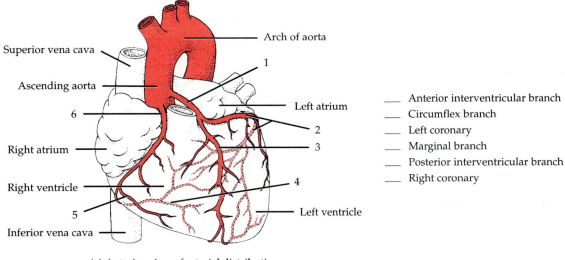

Superior vena cava

Ascending aorta

6

Right atrium

Right ventricle

5

Inferior vena cava

Arch of aorta

1

Left atrium

2

3

4

Left ventricle

___ Anterior interventricular branch
___ Circumflex branch
___ Left coronary
___ Marginal branch
___ Posterior interventricular branch
___ Right coronary

(a) Anterior view of arterial distribution

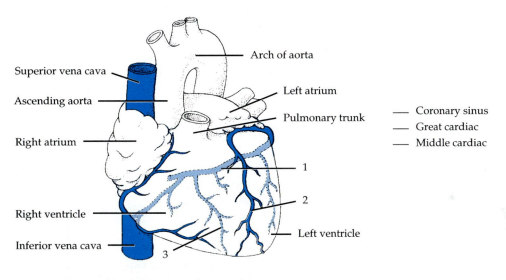

Superior vena cava

Ascending aorta

Right atrium

Right ventricle

Inferior vena cava

3

Arch of aorta

Left atrium

Pulmonary trunk

1

2

Left ventricle

___ Coronary sinus
___ Great cardiac
___ Middle cardiac

(b) Anterior view of venous drainage

FIGURE 17.4 Coronary (cardiac) circulation.

F. DISSECTION OF CAT HEART

CAUTION! *Please reread Section D, "Precautions Related to Dissection," at the beginning of the laboratory manual on page xiii before you begin your dissection.*

PROCEDURE

1. Before examining the cat's heart and major blood vessels, you must open the thoracic cavity.
2. On the ventral surface of the animal, palpate the bottom edge of the rib cage. Using heavy-duty scissors, penetrate the chest cavity immediately below the lowest rib and make a longitudinal incision about ½ in. to the right of the sternum.
3. Cut through the ribs, extending the incision anteriorly to the apex of the thorax. Be careful not to cut into any internal organs.
4. Spread apart the walls of the thorax and locate the diaphragm at the caudal end of the incision. *Be careful not to break any of the ribs,* because the sharp edge of a broken rib could interfere with your examination of the viscera.

5. Make two additional cuts laterally and dorsally toward the spine on both sides, keeping the incisions just anterior to the diaphragm.

6. Note the *right* and *left atria* and the *right* and *left ventricles* (Figure 17.5). Each atrium has an earlike flap on its ventral surface called an *auricle.* Identify the *coronary sulcus* between the right atrium and right ventricle and the *anterior longitudinal sulcus* between the right ventricle and left ventricle.

7. Carefully remove the thymus gland and any fat so that you can clearly see the ventral surface of the heart and its attached vessels (Figure 17.5). Be careful not to damage any nerves in this area.

8. The *pericardium*, the membrane surrounding the heart, consists of two layers. The outer *parietal pericardium* is a tough membrane that surrounds the heart and blood vessels that join the heart. This layer can be removed. The inner *visceral pericardium* invests the heart muscle (myocardium) closely and is very difficult to remove.

9. The *apex* of the heart is directed caudally and to the left and its *base* is directed cranially and to the right.

10. The cat's heart can be examined without removal. Leaving the heart in place will help you examine and identify the major blood vessels later on.

11. If a detailed dissection of the cat heart is to be done, follow the directions of the sheep heart dissection and identify the following structures of the cat heart, which correlate with structures of the sheep heart. Because the cat heart is small, it may be difficult to identify all of them.
  a. Opening of the superior vena cava
  b. Opening of the inferior vena cava
  c. Fossa ovalis
  d. Opening of the coronary sinus
  e. Pectinate muscle
  f. Interatrial septum
  g. Right ventricle
     i. tricuspid valve
     ii. chordae tendineae
     iii. papillary muscles
     iv. interventricular septum
     v. pulmonary semilunar valve
  h. Left atrium: openings of pulmonary veins
  i. Left ventricle
     i. bicuspid (mitral) valve
     ii. chordae tendineae
     iii. papillary muscles
     iv. aortic semilunar valve

G. DISSECTION OF SHEEP HEART

The anatomy of the sheep heart closely resembles that of the human heart. Use Figures 17.5 and 17.6 as references for this dissection. In addition, models of human hearts can also be used as references.

CAUTION! *Please reread Section D, "Precautions Related to Dissection" at the beginning of the laboratory manual on page xiii before you begin your dissection.*

First examine the *pericardium*, a fibroserous membrane that encloses the heart, which may have already been removed in preparing the sheep heart for preservation. The *myocardium* is the middle layer and constitutes the main muscle portion of the heart. The *endocardium* (the third layer) is the inner lining of the heart. Use the figures to determine which is the ventral surface of the heart and then identify the *pulmonary trunk* emerging from the anterior ventral surface, near the midline, and medial to the *left auricle.* A longitudinal depression on the ventral surface, called the *anterior longitudinal sulcus*, separates the right ventricle from the left ventricle. Locate the *coronary blood vessels* lying in this sulcus.

PROCEDURE

1. Remove any fat or pulmonary tissue that is present.

2. In cutting the sheep heart open to examine the chambers, valves, and vessels, the anterior longitudinal sulcus is used as a guide.

3. Carefully make a shallow incision through the ventral wall of the pulmonary trunk and the right ventricle, trying not to cut the dorsal surface of either structure.

4. The incision is best made *less than an inch to the right of, and parallel to*, the previously mentioned anterior longitudinal sulcus.

5. If necessary, the incision can be continued to where the pulmonary trunk branches into a *right pulmonary artery*, which goes to the right lung, and a *left pulmonary artery*, which goes to the left lung. The *pulmonary semilunar valve* of the pulmonary artery can be clearly seen upon opening it. In any of these internal dissections of the heart, any coagulated blood or latex should be immediately removed so that all important structures can be located and identified.

6. Keeping the cut still parallel to the sulcus, extend the incision around and through the

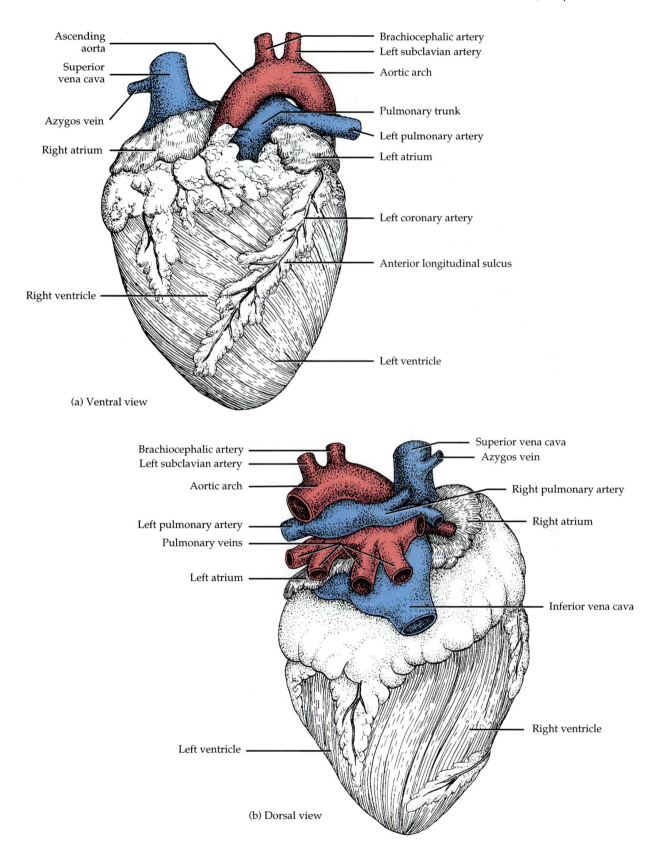

Ascending aorta

Superior vena cava

Azygos vein

Right atrium

Right ventricle

Brachiocephalic artery

Left subclavian artery

Aortic arch

Pulmonary trunk

Left pulmonary artery

Left atrium

Left coronary artery

Anterior longitudinal sulcus

Left ventricle

(a) Ventral view

Brachiocephalic artery

Left subclavian artery

Aortic arch

Left pulmonary artery

Pulmonary veins

Left atrium

Left ventricle

Superior vena cava

Azygos vein

Right pulmonary artery

Right atrium

Inferior vena cava

Right ventricle

(b) Dorsal view

FIGURE 17.5 External structure of a cat or sheep heart.

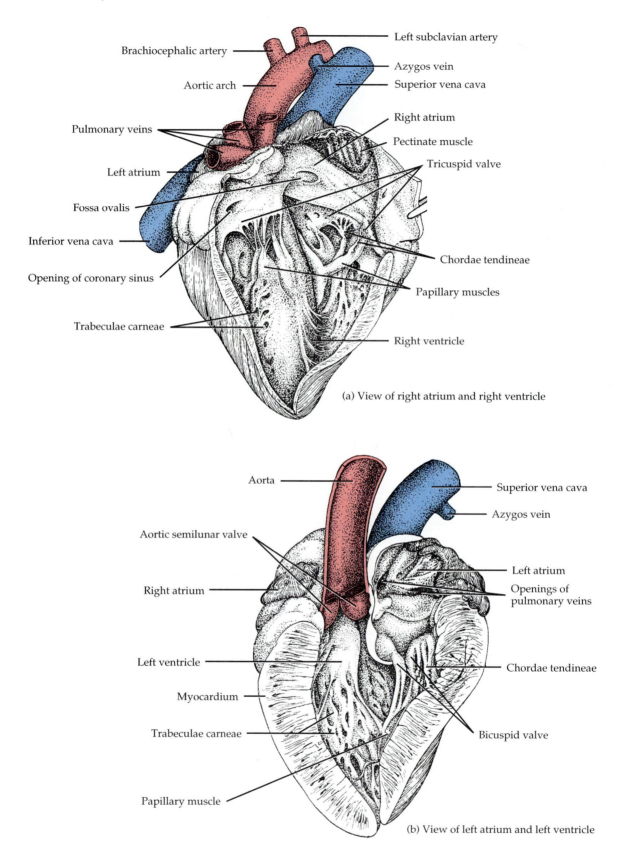

Brachiocephalic artery

Left subclavian artery

Aortic arch

Azygos vein

Superior vena cava

Pulmonary veins

Right atrium

Pectinate muscle

Tricuspid valve

Left atrium

Fossa ovalis

Inferior vena cava

Opening of coronary sinus

Chordae tendineae

Papillary muscles

Trabeculae carneae

Right ventricle

(a) View of right atrium and right ventricle

Aorta

Superior vena cava

Azygos vein

Aortic semilunar valve

Right atrium

Left atrium

Openings of pulmonary veins

Left ventricle

Myocardium

Chordae tendineae

Trabeculae carneae

Bicuspid valve

Papillary muscle

(b) View of left atrium and left ventricle

FIGURE 17.6   Internal structure of a cat or sheep heart.

dorsal ventricular wall until you reach the *interventricular septum.*

7. Now examine the dorsal surface of the heart and locate the thin-walled *superior vena cava* directly above the *right auricle*. This vein proceeds posteriorly straight into the right atrium.

8. Make a second longitudinal cut, this time through the superior vena cava (dorsal wall).

9. Extend the cut posteriorly through the right atrium on the left of the right auricle. Proceed posteriorly to the dorsal right ventricular wall and join your first incision.

10. The entire internal right side of the heart should now be clearly seen when carefully spread apart. The interior of the superior vena cava, right atrium, and right ventricle will now be examined. Start with the right auricle and locate the *pectinate muscle*, the large opening of the *inferior vena cava* on the left side of the right atrium, and the opening of the *coronary sinus* just below the opening of the inferior vena cava. By using a dull probe and gentle pressure, most of the vessels can be traced to the dorsal surface of the heart.

11. Now find the wall that separates the two atria, the *interatrial septum*. Also find the *fossa ovalis*, an oval-shaped depression ventral to the entrance of the inferior vena cava.

12. The *tricuspid valve* between the right atrium and the right ventricle should be examined to locate the three cusps, as its name indicates. From the cusps of the valve itself, and tracing posteriorly, the *chordae tendineae*, which hold the valve in place, should be identified. Still tracing posteriorly, the chordae are seen to originate from the *papillary muscles*, which themselves originate from the wall of the right ventricle itself.

13. Look carefully again at the dorsal surface of the left atrium and locate as many *pulmonary veins* (normally, four) as possible.

14. Make your third longitudinal cut through the most lateral of the pulmonary veins that you have located.

15. Continue posteriorly through the left atrial wall and the left ventricle to the *apex* of the heart.

16. Compare the difference in the thickness of the wall between the right and left ventricles. Explain your answer.

17. Examine the *bicuspid (mitral) valve*, again counting the cusps. Determine if the left side of the heart has basically the same structures as studied on the right side.

18. Probe from the left ventricle to the *aorta* as it emerges from the heart, examining the *aortic semilunar valve*. Find the openings of the right and left main coronary arteries.

19. Locate now the *brachiocephalic artery*, which is one of the first branches from the arch of the aorta. This artery continues branching and terminates by supplying the arms and head as its name indicates.

20. Connecting the aorta with the pulmonary artery is the remnant of the *ductus arteriosus*, called the *ligamentum arteriosum*. It may not be present in your sheep heart.

ANSWER THE LABORATORY REPORT QUESTIONS AT THE END OF THE EXERCISE.

Heart 17

Student _____   Date _____

Laboratory Section _____   Score/Grade _____

PART 1. Multiple Choice

_____ 1. Which of the following veins drains the blood from most of the vessels supplying the heart wall? (a) vasa vasorum (b) superior vena cava (c) coronary sinus (d) inferior vena cava

_____ 2. The atrioventricular valve on the same side of the heart as the origin of the aorta is the (a) aortic semilunar (b) tricuspid (c) bicuspid (d) pulmonary semilunar

_____ 3. Which valve does the blood go through just before entering the pulmonary trunk on the way to the lungs? (a) tricuspid (b) pulmonary semilunar (c) aortic semilunar (d) bicuspid

_____ 4. The pointed end of the heart that projects inferiorly and to the left is the (a) costal surface (b) base (c) apex (d) coronary sulcus

_____ 5. Which of these structures is more internal? (a) fibrous pericardium (b) visceral layer of serous pericardium (c) parietal layer of serous pericardium (d) myocardium

_____ 6. The musculature of the heart is referred to as the (a) endocardium (b) myocardium (c) epicardium (d) pericardium

_____ 7. The depression in the interatrial septum corresponding to the foramen ovale of fetal circulation is the (a) interventricular sulcus (b) pectinate muscle (c) chordae tendineae (d) fossa ovalis

PART 2. COMPLETION

8. Malfunction of the _____ valve would interfere with the flow of blood from the right atrium to the right ventricle.

9. Deoxygenated blood is sent to the lungs through the _____.

10. The loose-fitting serous membrane that encloses the heart is called the _____.

11. The two inferior chambers of the heart are separated by the _____.

12. The earlike flap of tissue on each atrium is called a(n) _____.

13. The large vein that drains blood from most parts of the body superior to the heart and empties into the right atrium is the _____.

14. The cusps of atrioventricular valves are prevented from inverting by the presence of cords called

_____, which are attached to papillary muscle.

15. A groove on the surface of the heart that houses blood vessels and a variable amount of fat is called

    a(n) _____.

16. The branch of the left coronary artery that distributes blood to the left atrium and left ventricle is the

    _____.

PART 3. Special Exercise

Draw a model of the heart and carefully label the four chambers, the four valves in their proper places, and the major blood vessels entering and exiting from the heart.

Blood Vessels

18

Blood vessels are networks of tubes that carry blood throughout the body. Blood vessels are called arteries, arterioles, capillaries, venules, or veins. In this exercise, you will study the histology of blood vessels and identify the principal arteries and veins of the human cardiovascular system.

A. ARTERIES AND ARTERIOLES

Arteries (AR-ter-ēs; *aer* = air; *tereo* = to carry) are blood vessels that carry blood *away* from the heart to body tissues. Arteries are constructed of three coats of tissue called *tunics* and a hollow core, called a *lumen,* through which blood flows (Figure 18.1). The deep coat is called the *tunica interna* and consists of a lining of endothelium in contact with the blood and a layer of elastic tissue called the *internal elastic lamina.* The middle coat, or *tunica media,* is usually the thickest layer and consists of elastic fibers and smooth muscle fibers. This tunic is responsible for two major properties of arteries: *elasticity* and *contractility.* The superficial coat, or *tunica externa,* is composed principally of elastic and collagen fibers. An *external elastic lamina* may separate the tunica externa from the tunica media.

Obtain a prepared slide of a cross section of an artery and identify the tunics using Figure 18.1 as a guide.

As arteries approach various tissues of the body, they become smaller and are known as *arterioles* (ar-TER-rē-ōls; *arteriola* = small artery). Arterioles play a key role in regulating blood flow from arteries into capillaries. When arterioles enter a tissue, they branch into countless microscopic blood vessels called capillaries.

B. CAPILLARIES

Capillaries (KAP-i-lar'-ēs; *capillaris* = hairlike) are microscopic blood vessels that connect arterioles and venules. Their function is to permit the exchange of nutrients and wastes between blood and body tissues. This function is related to the fact that capillaries consist of only a single layer of endothelium.

C. VENULES AND VEINS

When several capillaries unite, they form small veins called *venules* (VEN-yools; *venula* = little vein). They collect blood from capillaries and drain it into veins.

Veins (VĀNS) are composed of the same three tunics as arteries, but there are variations in their relative thicknesses. The tunica interna of veins is thinner than that of their accompanying arteries. In addition, the tunica media of veins is much thinner than that of accompanying arteries with relatively little smooth muscle or elastic fibers. The tunica externa is the thickest layer consisting of collagen and elastic fibers (Figure 18.1). Functionally, veins return blood from tissues *to* the heart.

Obtain a prepared slide of a transverse section of an artery and its accompanying vein and compare them, using Figure 18.1 as a guide.

D. CIRCULATORY ROUTES

The two basic postnatal (after birth) circulatory routes are systemic and pulmonary circulation (Figure 18.2 on page 411). Some other circulatory routes, which are all subdivisions of systemic circulation, include hepatic portal circulation, coronary (cardiac) circulation, fetal circulation, and the cerebral arterial circle (circle of Willis). The latter is found at the base of the brain (see Table 18.2).

1. Systemic Circulation

The largest route is the *systemic circulation* (see Figures 18.3 through 18.11 and Tables 18.1 through 18.11). This route includes the flow of blood from the left ventricle to all parts of the body. The function of the systemic circulation is to carry oxygen and nutrients to body tissues and to remove carbon dioxide and other wastes from them. All systemic arteries branch from the *aorta.* As the aorta emerges from

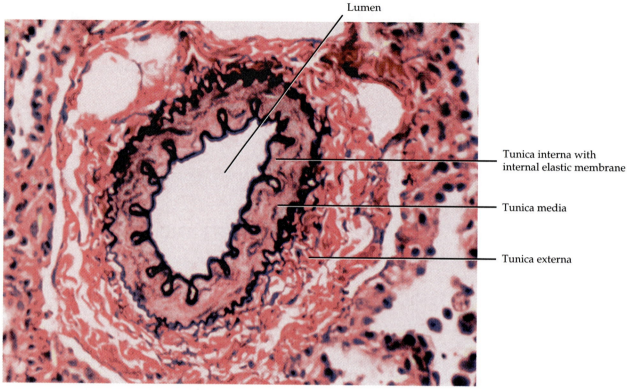

(a) Photomicrograph of an artery (25×)

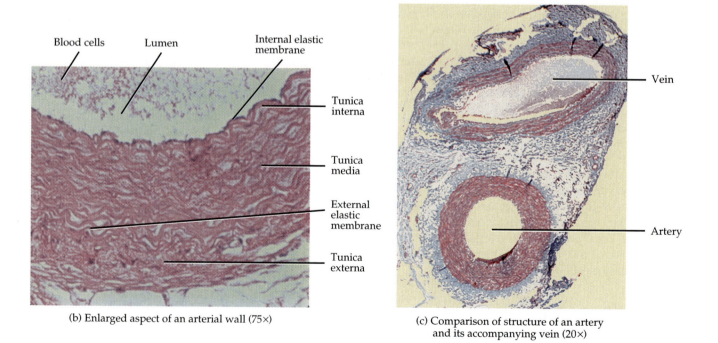

(b) Enlarged aspect of an arterial wall (75×)

(c) Comparison of structure of an artery and its accompanying vein (20×)

FIGURE 18.1 Histology of blood vessels.

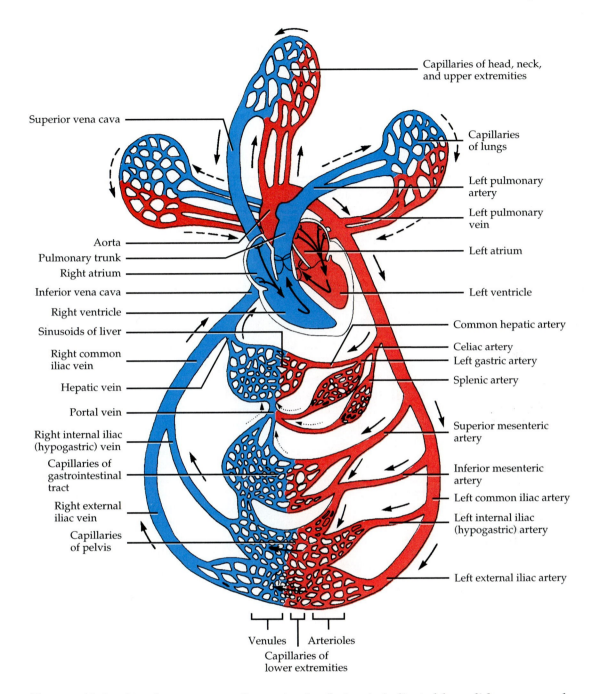

Superior vena cava

Capillaries of head, neck, and upper extremities

Capillaries of lungs

Left pulmonary artery

Left pulmonary vein

Aorta

Pulmonary trunk

Right atrium

Inferior vena cava

Right ventricle

Sinusoids of liver

Right common iliac vein

Hepatic vein

Portal vein

Right internal iliac (hypogastric) vein

Capillaries of gastrointestinal tract

Right external iliac vein

Capillaries of pelvis

Left atrium

Left ventricle

Common hepatic artery

Celiac artery

Left gastric artery

Splenic artery

Superior mesenteric artery

Inferior mesenteric artery

Left common iliac artery

Left internal iliac (hypogastric) artery

Left external iliac artery

Venules Arterioles

Capillaries of lower extremities

FIGURE 18.2 Circulatory routes. Systemic circulation is indicated by solid arrows; pulmonary circulation by broken arrows; and hepatic portal circulation by dotted arrows.

the left ventricle, it passes superiorly and posteriorly to the pulmonary trunk. At this point, it is called the *ascending aorta.* The ascending aorta gives off two coronary branches (right and left coronary arteries) to the heart muscle. Then it turns to the left, forming the *arch of the aorta* before descending to the level of the intervertebral disc between the fourth and the fifth thoracic vertebrae as the *descending aorta.* The descending aorta lies close to the vertebral bodies,

passes through the diaphragm, and divides at the level of the fourth lumbar vertebra into two *common iliac arteries,* which carry blood to the lower limbs. The section of the descending aorta between the arch of the aorta and the diaphragm is referred to as the *thoracic aorta.* The section between the diaphragm and the common iliac arteries is termed the *abdominal aorta.* Each section of the aorta gives off arteries that continue to branch into distributing arteries leading to organs and finally into the arterioles and capillaries that service the systemic tissues (except the air sacs of the lungs).

Deoxygenated blood is returned to the heart through the systemic veins. All the veins of the systemic circulation flow into either the *superior vena cava, inferior vena cava,* or *coronary sinus.* They in turn empty into the right atrium.

Refer to Tables 18.1 through 18.11 and Figures 18.3 through 18.12.

TABLE 18.1
Aorta and Its Branches (Figure 18.3)

OVERVIEW: The *aorta* (ā-OR-ta) is the largest artery of the body, about 2 to 3 cm (0.8 to 1.2 in.) in diameter. It begins at the left ventricle and contains a valve at its origin, called the aortic semilunar valve (see Figure 17.3a), which prevents backflow of blood into the left ventricle during its diastole (relaxation). The principal divisions of the aorta are the ascending aorta, arch of the aorta, thoracic aorta, and abdominal aorta.

Division of aorta	Arterial branch	Region supplied
Ascending aorta (ā-OR-ta)	Right and left coronary	Heart
Arch of aorta	Brachiocephalic (brā'-kē-ō-se FAL-ik) trunk → Right common carotid (ka-ROT-id) / Right subclavian (sub KLĀ-vē-an) Left common carotid Left subclavian	Right side of head and neck Right upper limb Left side of head and neck Left upper limb
Thoracic (thō-RAS-ik) *aorta*	Intercostals (in'-ter-KOS-tals) Superior phrenics (FREN-iks) Bronchials (BRONG-kē-als) Esophageals (e-sof'-a-JĒ-als)	Intercostal and chest muscles, pleurae Posterior and superior surfaces of diaphragm Bronchi of lungs Esophagus
Abdominal (ab-DOM-i-nal) *aorta*	Inferior phrenics (FREN-iks) Celiac → Common hepatic (he-PAT-ik) / Left gastric (GAS-trik) / Splenic (SPLĒN-ik) Superior mesenteric (MES-en-ter'-ik) Suprarenals (soo'-pra-RĒ-nals) Renals (RĒ-nals) Gonadals (gō-NAD-als) → Testiculars (tes-TIK-yoo-lars) or Ovarians (ō-VA-rē-ans) Inferior mesenteric (MES-en-ter'-ik) Common iliacs (IL-ē-aks) → External iliacs / Internal iliacs (hypogastrics)	Inferior surface of diaphragm Liver Stomach and esophagus Spleen, pancreas, stomach Small intestine, cecum, ascending and transverse colons, and pancreas Adrenal (suprarenal) glands Kidneys Testes Ovaries Transverse, descending, sigmoid colons and rectum Lower limbs Uterus, prostate gland, muscles of buttocks, and urinary bladder

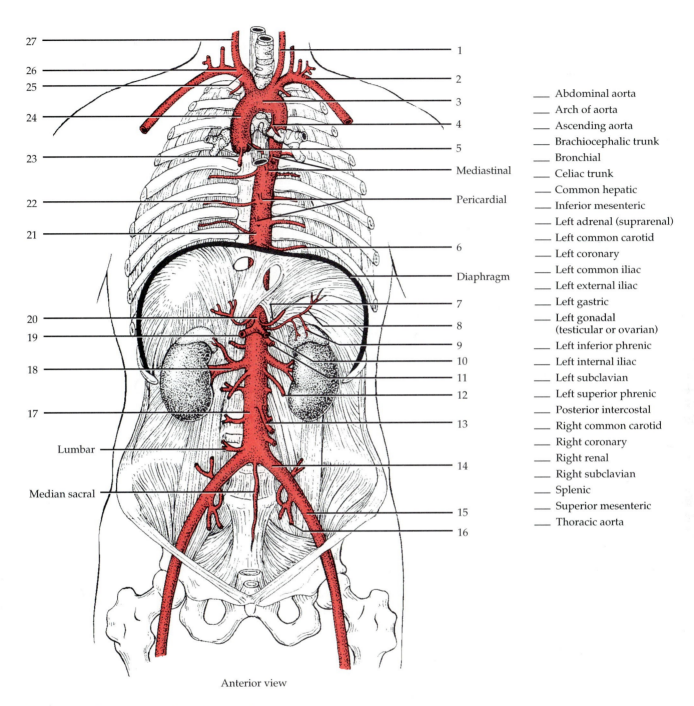

Anterior view

FIGURE 18.3   Principal branches of the aorta.

____ Abdominal aorta
____ Arch of aorta
____ Ascending aorta
____ Brachiocephalic trunk
____ Bronchial
____ Celiac trunk
____ Common hepatic
____ Inferior mesenteric
____ Left adrenal (suprarenal)
____ Left common carotid
____ Left coronary
____ Left common iliac
____ Left external iliac
____ Left gastric
____ Left gonadal
  (testicular or ovarian)
____ Left inferior phrenic
____ Left internal iliac
____ Left subclavian
____ Left superior phrenic
____ Posterior intercostal
____ Right common carotid
____ Right coronary
____ Right renal
____ Right subclavian
____ Splenic
____ Superior mesenteric
____ Thoracic aorta

TABLE 18.2
Arch of Aorta (Figures 18.4, 18.5, and 18.6)

OVERVIEW: The *arch of the aorta* is about 4.5 cm (1.8 in.) in length and is the continuation of the ascending aorta that emerges from the pericardium behind the sternum at the level of the sternal angle. Initially, the arch is directed upward, backward and to the left, and then downward on the left side of the body of the fourth thoracic vertebra. Actually, the arch is directed not only from right to left, but from anterior to posterior as well. The arch of the aorta terminates at the level of the invertebral disc between the fourth and fifth thoracic vertebrae, where it becomes the thoracic aorta. The thymus gland lies in front of the arch of the aorta, while the trachea lies behind it.
Three major arteries branch from the arch of the aorta. In order of their origination, they are the brachiocephalic trunk, left common carotid artery, and left subclavian artery.

Branch	Description and region
Brachiocephalic (brā'-kē-ō-se-FAL-ik)	The *brachiocephalic trunk,* which is found only on the right side, is the first and largest branch off the arch of the aorta. It bifurcates (divides) to form the right subclavian artery and right common carotid artery. The *right subclavian* (sub-KLĀ-vē-an) *artery* extends from the brachiocephalic to the first rib and then passes into the armpit (axilla) and supplies the arm, forearm, and hand. Continuation of the right subclavian into the axilla is called the *axillary* (AK-si-ler'-ē) *artery.* From here, it continues into the arm as the *brachial* (BRĀ-kē-al) *artery.* At the bend of the elbow, the brachial artery divides into the medial *ulnar* (UL-nar) and lateral *radial* (RĀ-dē-al) *arteries.* These vessels pass down to the palm, one on each side of forearm. In the palm, branches of the two arteries anastomose to form two palmar arches—the *superficial palmar* (PAL-mar) *arch* and the *deep palmar arch.* From these arches arise the *digital* (DIJ-i-tal) *arteries,* which supply the fingers and thumb (Figure 18.4). Before passing into the axilla, the right subclavian gives off a major branch to the brain called the *right vertebral* (VER-te-bral) *artery.* The right vertebral artery passes through the foramina of transverse processes of the cervical vertebrae and enters the skull through the foramen magnum to reach the inferior surface of the brain. Here it unites with the left vertebral artery to form the *basilar* (BAS-i-lar) *artery* (Figures 18.5 and 18.6). The *right common carotid artery* passes upward in the neck. At the upper level of the larynx, it divides into the *right external* and *right internal carotid* (ka-ROT-id) *arteries.* The external carotid supplies the right side of the thyroid gland, tongue, throat, face, ear, scalp, and dura mater. The internal carotid supplies the brain, right eye, and right sides of the forehead and nose (Figure 18.5). Inside the cranium, anastomoses of the left and right internal carotids along with the basilar artery form an arrangement of blood vessels at the base of the brain near the sella turcica called the *cerebral* (se-RĒ-bral) *arterial circle (circle of Willis).* From this circle arise arteries supplying most of the brain. Essentially the cerebral arterial circle is formed by union of the *anterior cerebral arteries* (branches of internal carotids) and *posterior cerebral arteries* (branches of basilar artery). Posterior cerebral arteries are connected with internal carotids by the *posterior communicating* (ko-MYOO-ni-kā'-ting) *arteries.* The anterior cerebral arteries are connected by the *anterior communicating arteries.* The *internal carotid* (ka-ROT-id) *arteries* are also considered part of the cerebral arterial circle. The function of the cerebral arterial circle is to equalize blood pressure to the brain and provide alternate routes for blood to the brain, should the arteries become damaged.
Left common carotid (ka-ROT-id)	The *left common carotid* is the the second branch off the the arch of the aorta. Corresponding to the right common carotid, it divides into basically the same branches with the same names, except that the arteries are now labeled "left" instead of "right."
Left subclavian (sub-KLĀ-vē-an)	The *left subclavian artery* is the third branch off the arch of the aorta. It distributes blood to the left vertebral artery and vessels of the left upper limb. Arteries branching from the left subclavian are named like those of the right subclavian.

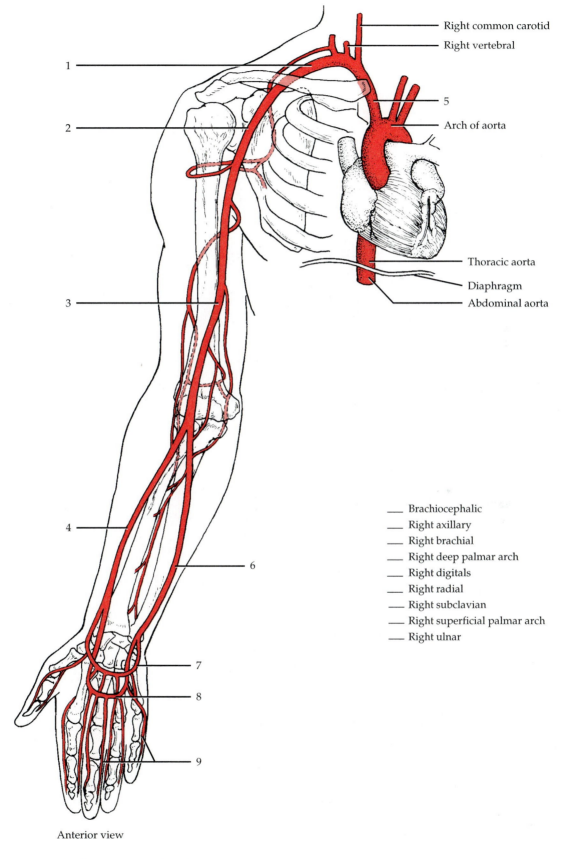

Right common carotid

Right vertebral

5

Arch of aorta

Thoracic aorta

Diaphragm

Abdominal aorta

1

2

3

4

6

7

8

9

___ Brachiocephalic
___ Right axillary
___ Right brachial
___ Right deep palmar arch
___ Right digitals
___ Right radial
___ Right subclavian
___ Right superficial palmar arch
___ Right ulnar

Anterior view

FIGURE 18.4 Arteries of right upper limb.

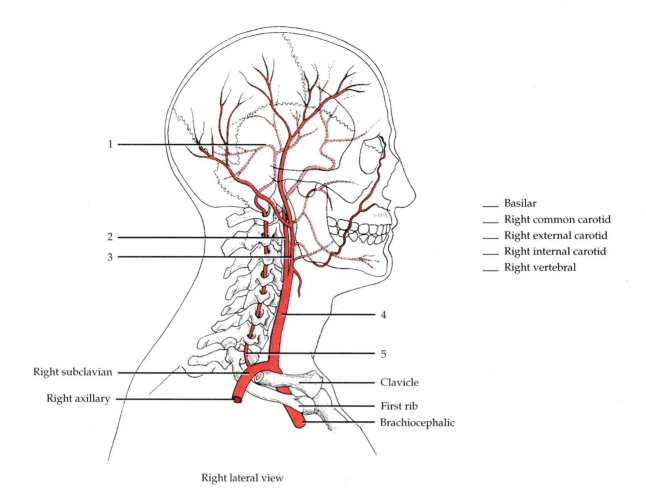

_____ Basilar
_____ Right common carotid
_____ Right external carotid
_____ Right internal carotid
_____ Right vertebral

1

2

3

4

5

Right subclavian

Right axillary

Clavicle

First rib

Brachiocephalic

Right lateral view

FIGURE 18.5   Arteries of neck and head.

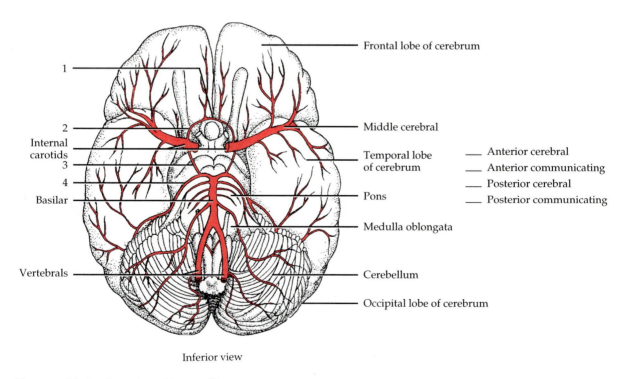

Frontal lobe of cerebrum

1

2

Internal
carotids

3

4

Basilar

Middle cerebral

Temporal lobe
of cerebrum

Pons

Medulla oblongata

Cerebellum

Occipital lobe of cerebrum

Vertebrals

_____ Anterior cerebral
_____ Anterior communicating
_____ Posterior cerebral
_____ Posterior communicating

Inferior view

FIGURE 18.6   Arteries of base of brain.

Write the names of the missing arteries in the following scheme of circulation. Be sure to indicate left or right where applicable.

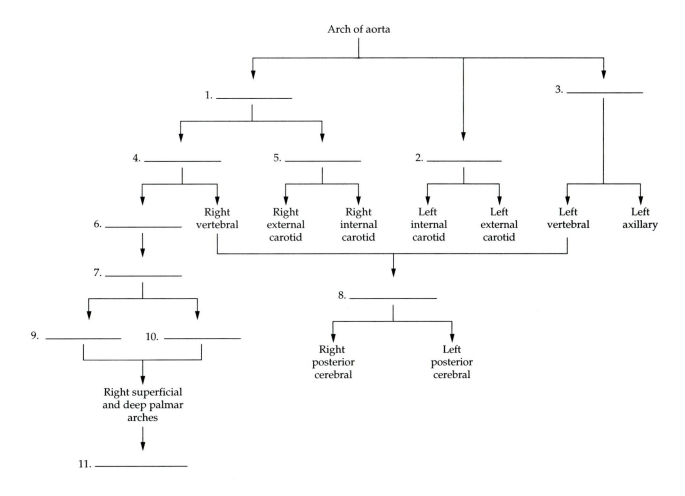

TABLE 18.3
Thoracic Aorta (Figure 18.3)

OVERVIEW: The ***thoracic aorta*** is about 20 cm (8 in.) long and is a continuation of the
   arch of the aorta. It begins at the level of the intervertebral disc between the fourth and
   fifth thoracic vertebrae, where it lies to the left of the vertebral column. As it descends,
   it moves closer to the midline and terminates at an opening in the diaphragm (aortic
   hiatus) in front of the vertebral column at the level of the intervertebral disc between
   the twelfth thoracic and first lumbar vertebra.
Along its course, the thoracic aorta sends off numerous small arteries to the viscera *(vis-*
   ceral branches) and body wall structures *(parietal branches).*

Branch	Description and region supplied
VISCERAL	
Pericardial (per'-i-KAR-dē-al)	Several minute *pericardial arteries* supply blood to the posterior aspect of the pericardium.
Bronchial (BRONG-kē-al)	One right and two left *bronchial arteries* supply the bronchial tubes, visceral pleurae, bronchial lymph nodes, and esophagus. (Whereas the right bronchial artery arises from the third posterior intercostal artery, the two left bronchial arteries arise from the thoracic aorta.)
Esophageal (e-sof'-a-JĒ-al)	Four or five *esophageal arteries* supply the esophagus.
Mediastinal (mē'-dē-as-TĪ-nal)	Numerous small *mediastinal arteries* supply blood to structures in the posterior mediastinum.
PARIETAL	
Posterior intercostal (in'-ter-KOS-tal)	Nine pairs of *posterior intercostal arteries* supply the intercostal, pectoral, and abdominal muscles; overlying subcutaneous tissue and skin; mammary glands; and vertebral canal and its contents.
Subcostal (SUB-kos-tal)	The left and right *subcostal arteries* have a distribution similar to that of the posterior intercostals.
Superior phrenic (FREN-ik)	Small *superior phrenic arteries* supply the posterior and superior surfaces of the diaphragm.

TABLE 18.4
Abdominal Aorta (Figure 18.7)

OVERVIEW: The *abdominal aorta* is the continuation of the thoracic aorta. It begins at the aortic hiatus in the diaphragm and ends at about the level of the fourth lumbar vertebra, where it divides into right and left common iliac arteries. The abdominal aorta lies in front of the vertebral column.

As with the thoracic aorta, the abdominal aorta gives off *visceral* and *parietal branches.* The unpaired visceral branches arise from the anterior surface of the aorta and include the celiac, superior mesenteric, and inferior mesenteric arteries. The paired visceral branches arise from the lateral surfaces of the aorta and include the adrenal (suprarenal), renal, and gonadal arteries. The paired parietal branches arise from the posterolateral surfaces of the aorta and include the inferior phrenic and lumbar arteries. The unpaired parietal artery is the median sacral.

Branch	Description and region supplied
VISCERAL	
Celiac (SĒ-lē-ak)	The *celiac artery (trunk)* is the first visceral aortic branch below the diaphragm. It has three branches: (1) *common hepatic* (he-PAT-ik) *artery,* (2) *left gastric* (GAS-trik) *artery,* and (3) *splenic* (SPLĒN-ik) *artery.* The common hepatic artery has three main branches: (1) *hepatic artery proper,* a continuation of the common hepatic artery, which supplies the liver and gallbladder; (2) *right gastric artery,* which supplies the stomach and duodenum; and (3) *gastroduodenal* (gas'-trō-doo'-ō-DE-nal) *artery,* which supplies the stomach, duodenum, and pancreas. The left gastric artery supplies the stomach and its *esophageal* (e-sof'-a-JĒ-al) *branch* supplies the esophagus. The splenic artery supplies the spleen and has three main branches: (1) *pancreatic* (pan'-krē-AT-ik) *arteries,* which supply the pancreas; (2) *left gastroepiploic* (gas'-trō-ep'-i-PLŌ-ik) *artery,* which supplies the stomach and greater omentum; and (3) *short gastric* (GAS-trik) *arteries,* which supply the stomach.
Superior mesenteric (MES-en-ter'-ik)	The *superior mesenteric artery* anastomoses extensively and has several principal branches: (1) *inferior pancreaticoduodenal* (pan'-krē-at'-i-kō-doo'-ō-DĒ-nal) *artery,* which supplies the pancreas and duodenum; (2) *jejunal* (je-JOO-nal) and *ileal* (IL-ē-al) *arteries,* which supply the jejunum and ileum, respectively; (3) *ileocolic* (il'-ē-ō-KŌL-ik) *artery,* which supplies the ileum and ascending colon; (4) *right colic* (KŌL-ik) *artery,* which supplies the ascending colon; and (5) *middle colic artery,* which supplies the transverse colon.
Suprarenals (soo'-pra-RĒ-nals)	Right and left *suprarenal arteries* supply blood to the adrenal (suprarenal) glands. The glands are also supplied by branches of the renal and inferior phrenic arteries.
Renals (RĒ-nals)	Right and left *renal arteries* carry blood to the kidneys and adrenal (suprarenal) glands
Gonadals (gō-NAD-als) [*testiculars* (tes-TIK-yoo-lars) or *ovarians* (ō-VA-rē-ans)]	Right and left *testicular arteries* extend into the scrotum and terminate in the testes; right and left *ovarian arteries* are distributed to the ovaries.
Inferior mesenteric (MES-en-ter'-ik)	The principal branches of the *inferior mesenteric artery,* which also anastomoses, are the (1) *left colic* (KŌL-ik) *artery,* which supplies the transverse and descending colons; (2) *sigmoid* (SIG-moyd) *arteries,* which supply the descending and sigmoid colons; and (3) *superior rectal* (REK-tal) *artery,* which supplies the rectum.
PARIETAL	
Inferior phrenics (FREN-iks)	The *inferior phrenic arteries* are distributed to the inferior surface of the diaphragm and adrenal (suprarenal) glands.
Lumbars (LUM-bars)	The *lumbar arteries* supply the spinal cord and its meninges and the muscles and skin of the lumbar region of the back.
Median sacral (SĀ-kral)	The *median sacral artery* supplies the sacrum, coccyx, and rectum.

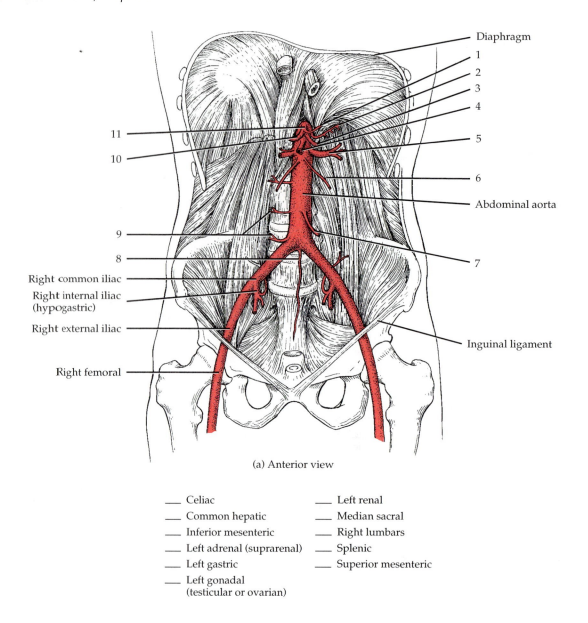

Diaphragm
1
2
3
4
5
6
Abdominal aorta
7
Inguinal ligament

11
10
9
8
Right common iliac
Right internal iliac (hypogastric)
Right external iliac
Right femoral

(a) Anterior view

____ Celiac   ____ Left renal
____ Common hepatic   ____ Median sacral
____ Inferior mesenteric   ____ Right lumbars
____ Left adrenal (suprarenal)   ____ Splenic
____ Left gastric   ____ Superior mesenteric
____ Left gonadal
      (testicular or ovarian)

FIGURE 18.7   Abdominal arteries. (a) Abdominal aorta and its principal branches.

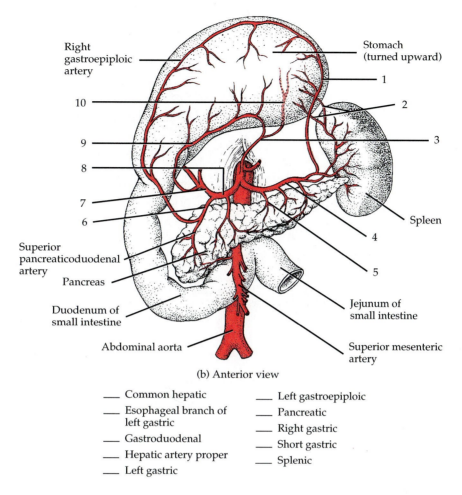

Right gastroepiploic artery

Stomach (turned upward)

1

10

2

9

3

8

7

6

Spleen

4

Superior pancreaticoduodenal artery

Pancreas

5

Duodenum of small intestine

Jejunum of small intestine

Abdominal aorta

Superior mesenteric artery

(b) Anterior view

___ Common hepatic

___ Esophageal branch of left gastric

___ Gastroduodenal

___ Hepatic artery proper

___ Left gastric

___ Left gastroepiploic

___ Pancreatic

___ Right gastric

___ Short gastric

___ Splenic

FIGURE 18.7 (*Continued*)   Abdominal arteries. (b) Branches of common hepatic, left gastric, and splenic arteries.

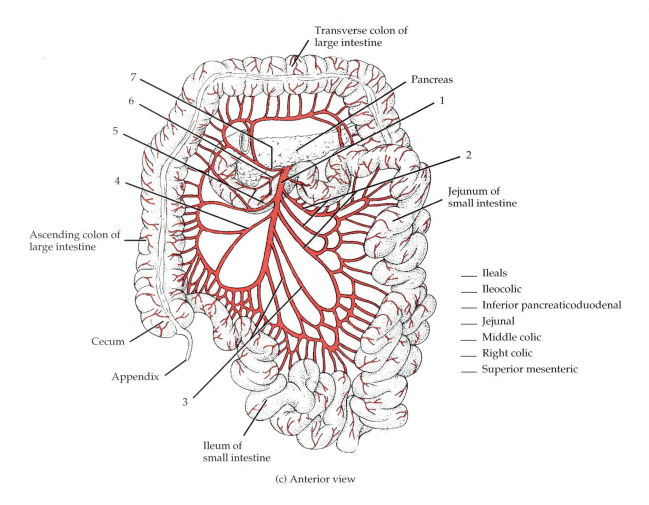

Transverse colon of
large intestine

Pancreas

7

6

5

1

4

2

Ascending colon of
large intestine

Jejunum of
small intestine

____ Ileals
____ Ileocolic
____ Inferior pancreaticoduodenal
____ Jejunal
____ Middle colic
____ Right colic
____ Superior mesenteric

Cecum

Appendix

3

Ileum of
small intestine

(c) Anterior view

FIGURE 18.7 (Continued)   Abdominal arteries. (c) Branches of superior mesenteric artery.

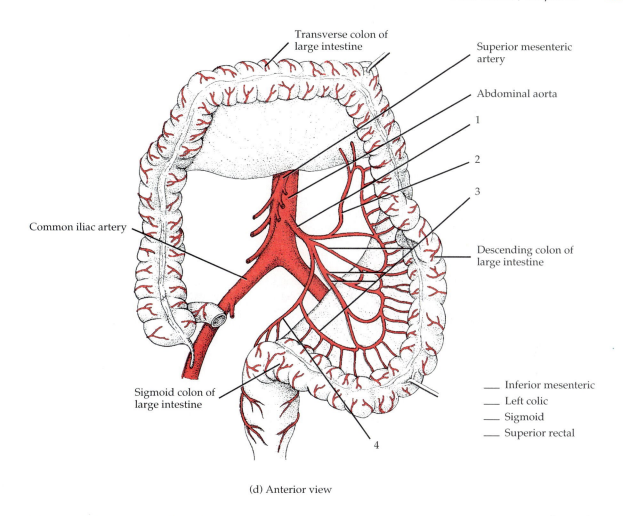

Transverse colon of
large intestine

Superior mesenteric
artery

Abdominal aorta

1

2

3

Common iliac artery

Descending colon of
large intestine

Sigmoid colon of
large intestine

____ Inferior mesenteric
____ Left colic
____ Sigmoid
____ Superior rectal

4

(d) Anterior view

FIGURE 18.7 *(Continued)*   Abdominal arteries. (d) Branches of inferior mesenteric artery.

TABLE 18.5
Arteries of Pelvis and Lower Limbs (Extremities) (Figure 18.8)

OVERVIEW: The *internal iliac arteries* enter the pelvic cavity in front of the sacroiliac joint and supply most of the blood to the pelvic viscera and wall. The *external iliacs* travel along the brim of the lesser (true) pelvis. Behind the midportion of the inguinal ligament, each external iliac artery enters the thigh where its name changes to the femoral artery.

Branch	Description and region supplied
Common iliacs (IL-ē-aks)	At about the level of the fourth lumbar vertebra, the abdominal aorta divides into the right and left *common iliac arteries.* Each passes inferiorly about 5 cm (2 in.) and gives rise to two branches: internal iliac and external iliac.
Internal iliacs	The *internal iliac (hypogastric) arteries* form branches that supply the psoas minor, gluteal muscles, quadratus lumborum, medial side of each thigh, urinary bladder, rectum, prostate gland, ductus (vas) deferens, uterus, and vagina.
External iliacs	The *external iliac arteries* diverge through the greater (false) pelvis and enter the thighs to become the right and left *femoral* (FEM-o-ral) *arteries.* Both femorals send branches back up to the genitals and the wall of the abdomen. Other branches run to the muscles of the thigh. The femoral continues down the medial and posterior side of the thigh posterior to the knee joint, where it becomes the *popliteal* (pop'-li-TĒ-al) *artery.* Between the knee and ankle, the popliteal runs down on the posterior aspect of the leg and is called the *posterior tibial* (TIB-ē-al) *artery.* Inferior to the knee, the *peroneal* (per'-ō-NĒ-al) *artery* branches off the posterior tibial to supply structures on the medial side of the fibula and calcaneus. In the calf, the *anterior tibial artery* branches off the popliteal and runs along the anterior surface of the leg. At the ankle, it becomes the *dorsalis pedis* (PED-is) *artery.* At the ankle, the posterior tibial divides into the *medial* and *lateral plantar* (PLAN-tar) *arteries.* The lateral plantar artery and the dorsalis pedis artery unite to form the *plantar arch.* From this arch, *digital arteries* supply the toes.

Write the names of the missing arteries in the following scheme of circulation. Be sure to indicate left or right where applicable.

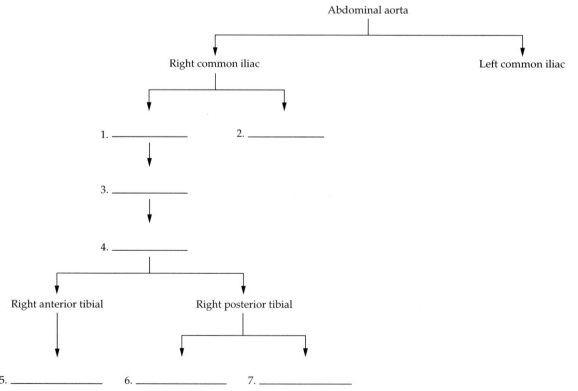

Abdominal aorta

Right common iliac   Left common iliac

1. _____   2. _____

3. _____

4. _____

Right anterior tibial   Right posterior tibial

5. _____   6. _____   7. _____

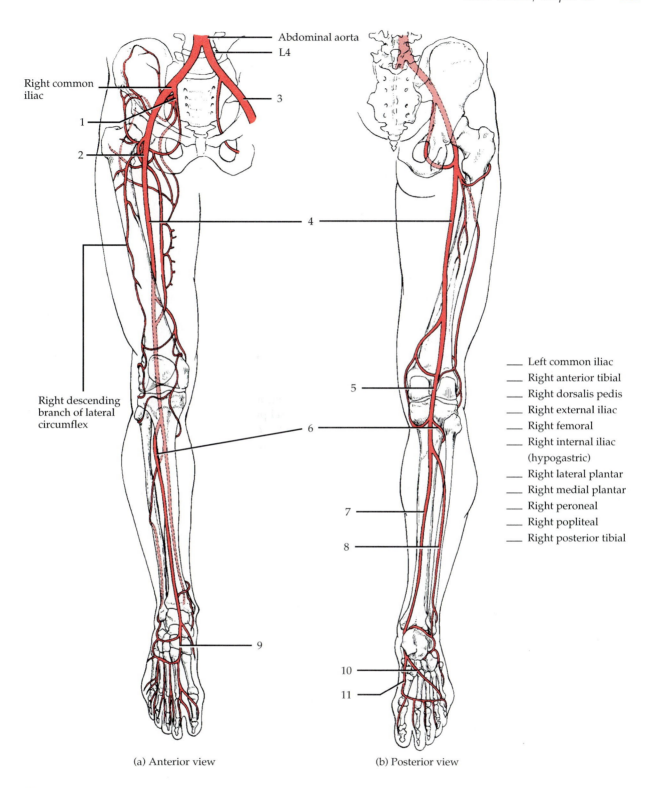

Abdominal aorta

L4

Right common iliac

3

1

2

4

Right descending branch of lateral circumflex

5

6

7

8

_____ Left common iliac
_____ Right anterior tibial
_____ Right dorsalis pedis
_____ Right external iliac
_____ Right femoral
_____ Right internal iliac (hypogastric)
_____ Right lateral plantar
_____ Right medial plantar
_____ Right peroneal
_____ Right popliteal
_____ Right posterior tibial

9

10

11

(a) Anterior view

(b) Posterior view

FIGURE 18.8 Arteries of pelvis and right lower limb.

TABLE 18.6
Veins of Systemic Circulation (Figure 18.10)

OVERVIEW: Deoxygenated blood returns to the right atrium from three veins: the *coronary sinus, superior vena cava,* and *inferior vena cava.* The coronary sinus receives blood from the cardiac veins; the superior vena cava receives blood from veins superior to the diaphragm, except the air sacs of the lungs. This includes the head, neck, upper limbs, and thoracic wall. The inferior vena cava receives blood from veins inferior to the diaphragm. This includes the lower limbs, most of the abdominal walls, and abdominal viscera.

Vein	Description and region drained
Coronary (KOR-o-nar-ē) *sinus*	The *coronary sinus* receives almost all venous blood from the myocardium. It is located in the coronary sulcus (see Fig. 17.4b) and opens into the right atrium between the orifice of the inferior vena cava and the tricuspid valve.
Superior vena cava (VĒ-na CĀ-va) *SVC*	The *SVC* is about 7.5 cm (3 in.) long and empties its blood into the upper part of the right atrium. It begins posterior to the right first costal cartilage by the union of the right and left brachiocephalic veins and ends at the level of the right third costal cartilage where it enters the right atrium.
Inferior vena cava (IVC)	The *IVC* is the largest vein the the body, about 3½ cm (1.4 in.) in diameter. It begins anterior to the fifth lumbar vertebra by the union of the common iliac veins, ascends behind the peritoneum to the right of the midline, pierces the costal tendon of the diaphragm at the level of the eighth thoracic vertebra, and enters the lower part of the right atrium. The inferior vena cava is commonly compressed during the later stages of pregnancy owing to the enlargement of the uterus. This produces edema of the ankles and feet and temporary varicose veins.

TABLE 18.7
Veins of Head and Neck (Figure 18.9)

OVERVIEW: The majority of blood draining from the head is pushed into three pairs of veins: *internal jugular, external jugular,* and *vertebral.* Within the cranium, all veins lead to the external jugular veins.

Vein	Description and region drained
Internal jugulars (JUG-yoo-lars)	Right and left *internal jugular veins* receive blood from the face and neck. They arise as a continuation of the *sigmoid* (SIG-moyd) *sinuses* at the base of the skull. Intracranial vascular sinuses are located between layers of the dura mater and receive blood from the brain. Other sinuses that drain into the internal jugulars include the *superior sagittal* (SAJ-i-tal) *sinus, inferior sagittal sinus, straight sinus,* and *transverse (lateral) sinuses.* Internal jugulars descend on either side of the neck and pass behind the clavicles, where they join with the right and left subclavian veins. Unions of the internal jugulars and subclavians form the right and left *brachiocephalic* (brā'-kē-ō-se-FAL-ik) *veins.* From here blood flows into the superior vena cava.
External jugulars	Right and left *external jugular veins* run down the neck along the outside of the internal jugulars. They drain blood from the parotid (salivary) glands, facial muscles, scalp, and other superficial structures into the subclavian veins.
Vertebrals (VER-te-brals)	Right and left *vertebral veins* descend through the transverse foramina of the cervical vertebrae and enter the subclavian veins. They drain deep structures of the neck such as vertebrae and muscles. In cases of heart failure, the venous pressure in the right atrium may rise. In such patients the pressure in the column of blood in the external jugular vein rises so that, even with the patient at rest and sitting in a chair, the external jugular vein will be visibly distended. Temporary distention of the vein is often seen in healthy adults when the intrathoracic pressure is raised in coughing and physical activity.

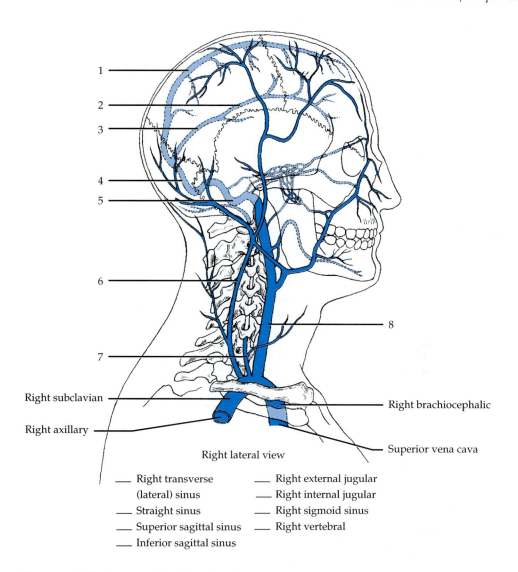

Right subclavian

Right axillary

Right brachiocephalic

Superior vena cava

Right lateral view

___ Right transverse
    (lateral) sinus
___ Straight sinus
___ Superior sagittal sinus
___ Inferior sagittal sinus

___ Right external jugular
___ Right internal jugular
___ Right sigmoid sinus
___ Right vertebral

FIGURE 18.9   Veins of head and neck.

TABLE 18.8
Veins of Upper Limbs (Extremities) (Figure 18.10)

OVERVIEW: Blood from each upper limb is returned to the heart by superficial and
deep veins. Both sets of veins contain valves. *Superficial veins* are located just below
the skin and are often visible. They anastomose extensively with each other and with
deep veins. *Deep veins* are located deep in the body. They usually accompany arteries,
and many have the same names as corresponding arteries. Most deep veins are paired
vessels.

Vein	Description and region drained
SUPERFICIAL	
Cephalics (se-FAL-iks)	The *cephalic vein* of each upper limb begins in the medial part of the *dorsal venous* (VĒ-nus) *arch* and winds upward around the radial border of the forearm. Anterior to the elbow, it is connected to the basilic vein by the *median cubital* (KYOO-bi-tal) *vein.* Just below the elbow, the cephalic vein unites with the *accessory cephalic vein* to form the cephalic vein of the upper limb. Ultimately, the cephalic vein empties into the axillary vein.
Basilics (ba-SIL-iks)	The *basilic vein* of each upper limb originates in the ulnar part of the *dorsal venous arch.* It extends along the posterior surface of the ulna to a point near the elbow where it receives the *median cubital vein.* If a vein must be punctured for an injection, transfusion, or removal of a blood sample, the median cubitals are preferred. After receiving the median cubital vein, the basilic continues ascending on the medial side until it reaches the middle of the arm. There it penetrates the tissues deeply and runs alongside the brachial artery until it joins the brachial vein. As the basilic and brachial veins merge in the axillary area, they form the axillary vein.
Median antebrachials (an'-tē-BRĀ-kē-als)	The *median antebrachial veins* drain the *palmar venous arch,* ascend on the ulnar side of the anterior forearm, and end in the median cubital veins.
DEEP	
Radials (RĀ-dē-als)	*Radial veins* receive the *dorsal metacarpal* (met'-a-KAR-pal) *veins.*
Ulnars (UL-nars)	*Ulnar veins* receive tributaries from the *palmar venous arch.* Radial and ulnar veins unite in the bend of the elbow to form the brachial veins.
Brachials (BRĀ-kē-als)	Located on either side of the brachial arteries, the *brachial veins* join into the axillary veins.
Axillaries (AK-si-ler'-ēs)	*Axillary veins* are a continuation of brachials and basilics. Axillaries end at the first rib, where they become subclavians.
Subclavians (sub-KLĀ-vē-ans)	Right and left *subclavian veins* unite with the internal jugulars to form brachiocephalic veins. The thoracic duct of the lymphatic system delivers lymph into the left subclavian vein at the junction with the internal jugular. The right lymphatic duct delivers lymph into the right subclavian vein at the corresponding junction.

Write the names of the missing veins in the following scheme of circulation. Be sure to indicate left or right where applicable.

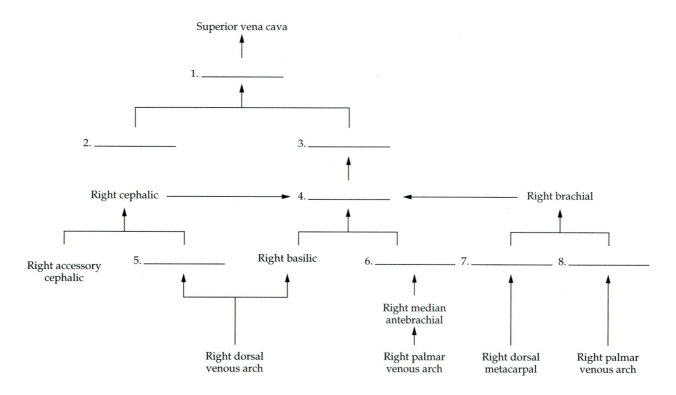

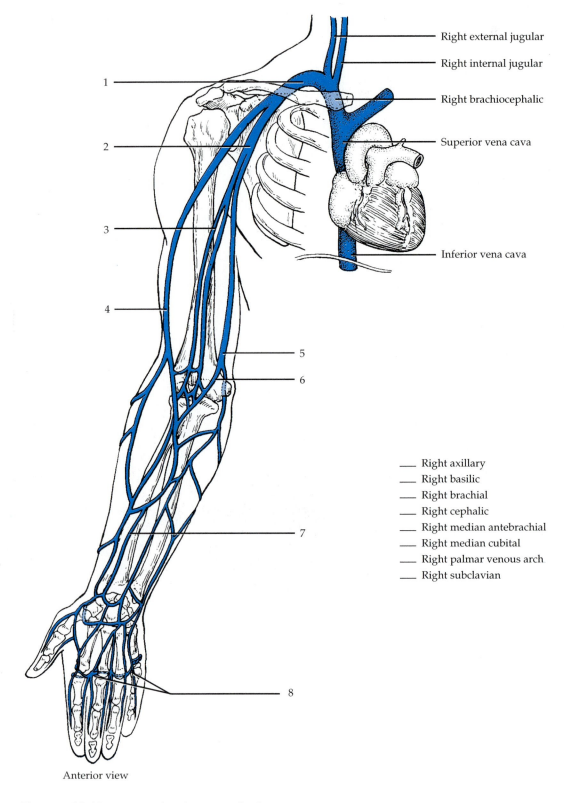

Right external jugular

Right internal jugular

Right brachiocephalic

Superior vena cava

Inferior vena cava

_____ Right axillary
_____ Right basilic
_____ Right brachial
_____ Right cephalic
_____ Right median antebrachial
_____ Right median cubital
_____ Right palmar venous arch
_____ Right subclavian

Anterior view

FIGURE 18.10   Veins of right upper limb.

TABLE 18.9
Veins of Thorax (Figure 18.11)

OVERVIEW: Although the brachiocephalic veins drain some portions of the thorax, most thoracic structures are drained by a network of veins called the *azygos system*. This is a network of veins on each side of the vertebral column: azygos, hemiazygos, and accessory hemiazygos. They show considerable variation in origin, course, tributaries, anastomoses, and termination. Ultimately, they empty into the superior vena cava.

Vein	Description and region drained
Brachiocephalic (brā-kē-ō-se-FAL-ik)	Right and left *brachiocephalic veins*, formed by the union of the subclavians and internal jugulars, drain blood from the head, neck, upper limbs, mammary glands, and upper thorax. Brachiocephalics unite to form the superior vena cava.
Azygos (az-Ī-gos) *system*	The *azygos system*, besides collecting blood from the thorax, may serve as a bypass for the inferior vena cava that drains blood from the lower body. Several small veins directly link the azygos system with the inferior vena cava. Large veins that drain the lower limbs and abdomen pass blood into the azygos system. If the inferior vena cava or hepatic portal vein becomes obstructed, the azygos system can return blood from the lower body to the superior vena cava.
Azygos	The *azygos vein* lies in front of the vertebral column and slightly right of midline. It begins as a continuation of the right ascending lumbar vein. It connects with the inferior vena cava, right common iliac, and lumbar veins. The azygos receives blood from the *left intercostal* (in'-ter-KOS-tal) *veins* that drain the chest muscles; from the hemiazygos and accessory hemiazygos veins; from several *esophageal* (e-sof'-a-JĒ-al), *mediastinal* (mē-dē-a-STĪ-nal), and *pericardial* (per'-i-KAR-dē-al) *veins*; and from the right *bronchial* (BRONG-kē-al) *vein*. The vein ascends to the fourth thoracic vertebra, arches over the right lung, and empties into the superior vena cava.
Hemiazygos (HEM-ē-az-ī-gos)	The *hemiazygos vein* is anterior to the vertebral column and slightly left of midline. It begins as a continuation of the left ascending lumbar vein. It receives blood from the lower four or five left *intercostal* (in'-ter-KOS-tal) *veins* and some *esophageal* (e-sof'-a-JĒ-al) and *mediastinal* (mē-dē-a-STI-nal) *veins*. At the level of the ninth thoracic vertebra, it joins the azygos vein.
Accessory hemiazygos	The *accessory hemizygos vein* is also anterior to the left of the vertebral column. It receives blood from three or four of the superior left *intercostal* (in'-ter-KOS-tal) *veins* and left *bronchial* (BRONG-kē-al) *vein*. It joins the azygos at the level of the eighth thoracic vertebra.

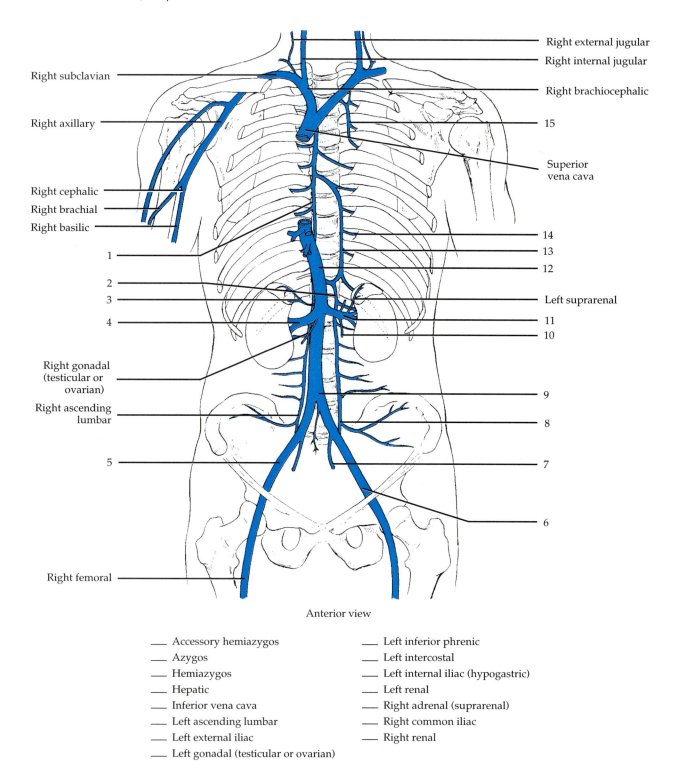

Right subclavian

Right axillary

Right cephalic

Right brachial

Right basilic

1

2

3

4

Right gonadal (testicular or ovarian)

Right ascending lumbar

5

Right femoral

Right external jugular

Right internal jugular

Right brachiocephalic

15

Superior vena cava

14

13

12

Left suprarenal

11

10

9

8

7

6

Anterior view

_____ Accessory hemiazygos
_____ Azygos
_____ Hemiazygos
_____ Hepatic
_____ Inferior vena cava
_____ Left ascending lumbar
_____ Left external iliac
_____ Left gonadal (testicular or ovarian)

_____ Left inferior phrenic
_____ Left intercostal
_____ Left internal iliac (hypogastric)
_____ Left renal
_____ Right adrenal (suprarenal)
_____ Right common iliac
_____ Right renal

FIGURE 18.11   Veins of thorax, abdomen, and pelvis.

TABLE 18.10
Veins of Abdomen and Pelvis (Figure 18.11)

OVERVIEW: Blood from the abdominopelvic viscera and abdominal wall returns to the heart via the *inferior vena cava*. Many small veins enter the inferior vena cava. Most carry return flow from parietal branches of the abdominal aorta and their names correspond to the names of the arteries. The inferior vena cava does not receive veins from the gastrointestinal tract, spleen, pancreas, and gallbladder. These organs pass their blood into a common vein, the hepatic portal vein, which delivers the blood to the liver. From here, the blood drains into the hepatic veins, which enter the inferior vena cava. This special flow of venous blood is called *hepatic portal circulation*, which is described shortly.

Vein	Description and region drained
Inferior vena cava (VĒ-na CA-va)	The *inferior vena cava* is formed by the union of two common iliac veins that drain the lower limbs and abdomen. The inferior vena cava extends superiorly through the abdomen and thorax to the right atrium.
Common iliacs (IL-ē-aks)	The *common iliac veins* are formed by the union of the internal (hypogastric) and external iliac veins and represent the distal continuation of the inferior vena cava at its bifurcation (branching).
Internal iliacs	Tributaries of the *internal iliac (hypogastric) veins* basically correspond to branches of internal iliac arteries. The internal iliacs drain the gluteal muscles, medial side of the thigh, urinary bladder, rectum, prostate gland, ductus (vas) deferens, uterus, and vagina.
External iliacs	The *external iliac veins* are a continuation of the femoral veins and receive blood from the lower limbs and inferior part of the anterior abdominal wall.
Renals (RĒ-nals)	The *renal veins* drain the kidneys.
Gonadals (gō-NAD-als) [*testicular* (tes-TIK-yoo-lor) or (ō-VA-rē-an) *ovarian*]	The *testicular veins* drain the testes (left testicular vein empties into the left renal vein and right testicular drains into the inferior vena cava); the *ovarian veins* drain the ovaries (left ovarian vein empties into the left renal vein and right ovarian drains into the inferior vena cava).
Suprarenals (soo'-pra-RĒ-nals)	The *suprarenal veins* drain the adrenal (suprarenal) glands (left suprarenal vein empties into the left renal vein and right suprarenal vein empties into the superior vena cava).
Inferior phrenics (FREN-iks)	The *inferior phrenic veins* drain the diaphragm (left interior phrenic vein sends a tributary to the left renal vein and right inferior phrenic vein empties into the superior vena cava).
Hepatics (he-PAT-iks)	The *hepatic veins* drain the liver.
Lumbars (LUM-bars)	A series of parallel *lumbar veins* drain blood from both sides of the posterior abdominal wall. The lumbars connect at right angles with the right and left *ascending lumbar veins,* which form the origin of the corresponding azygos or hemiazygos vein. The lumbars drain blood into the ascending lumbars and then run to the inferior vena cava, where they release the remainder of the flow.

TABLE 18.11
Veins of Lower Limbs (Extremities) (Figure 18.12)

OVERVIEW: Blood from each lower limb is drained by *superficial* and *deep veins*. The
   superficial veins often anastomose with each other and with deep veins along their
   length. Deep veins, for the most part, have the same names as their accompanying
   arteries. Both superficial and deep veins have valves.

Vein	Description and region drained
SUPERFICIAL VEINS	
Great saphenous (sa-FĒ-nus)	The *great saphenous vein,* the longest vein in the body, begins at the medial end of the *dorsal venous arch* of the foot. It passes anterior to the medial malleolus and then upward along the medial aspect of the leg and thigh just deep to the skin. It receives tributaries from superficial tissues and connects with the deep veins as well. It empties into the femoral vein in the groin. The great saphenous vein is frequently used for prolonged administration of intravenous fluids. This is particularly important in very young babies and in patients of any age who are in shock and whose veins are collapsed. It and the small saphenous vein are subject to varicosity.
Small saphenous	The *small saphenous vein* begins at the lateral end of the dorsal venous arch of the foot. It passes behind the lateral malleolus and ascends under the skin of the back of the leg. It receives blood from the foot and posterior portion of the leg. It empties into the popliteal vein behind the knee.
DEEP VEINS	
Posterior tibial (TIB-ē-al)	The *posterior tibial vein* is formed by the union of the *medial* and *lateral plantar* (PLAN-tar) *veins* behind the medial malleolus. It ascends deep in the muscle at the back of the leg, receives blood from the *peroneal* (per'-ō-NĒ-al) *vein,* and unites with the anterior tibial vein just below the knee.
Anterior tibial	The *anterior tibial vein* is the upward continuation of the *dorsalis pedis* (PED-is') *veins* in the foot. It runs between the tibia and fibula and unites with the posterior tibial to form the popliteal vein.
Popliteal (pop'-li-TĒ-al)	The *popliteal vein,* just behind the knee, receives blood from the anterior and posterior tibials and the small saphenous vein.
Femoral (FEM-o-ral)	The *femoral vein* is the upward continuation of the popliteal just above the knee. The femorals run up the posterior surface of the thighs and drain the deep structures of the thighs. After receiving the great saphenous veins in the groin, they continue as the right and left external iliac veins.

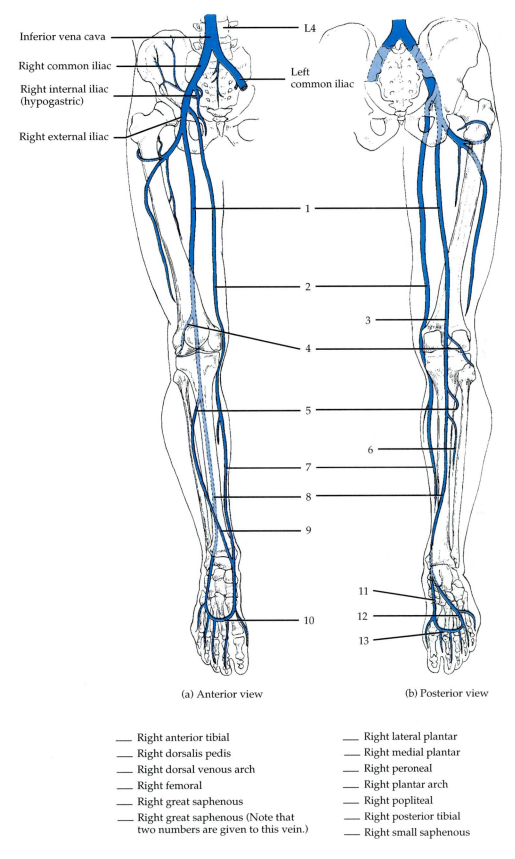

Inferior vena cava
Right common iliac
Right internal iliac (hypogastric)
Right external iliac
L4
Left common iliac

(a) Anterior view

(b) Posterior view

_____ Right anterior tibial
_____ Right dorsalis pedis
_____ Right dorsal venous arch
_____ Right femoral
_____ Right great saphenous
_____ Right great saphenous (Note that two numbers are given to this vein.)

_____ Right lateral plantar
_____ Right medial plantar
_____ Right peroneal
_____ Right plantar arch
_____ Right popliteal
_____ Right posterior tibial
_____ Right small saphenous

FIGURE 18.12 Veins of pelvis and right lower limb.

Write the names of the missing veins in the following scheme of circulation. Be sure to indicate left or right where applicable.

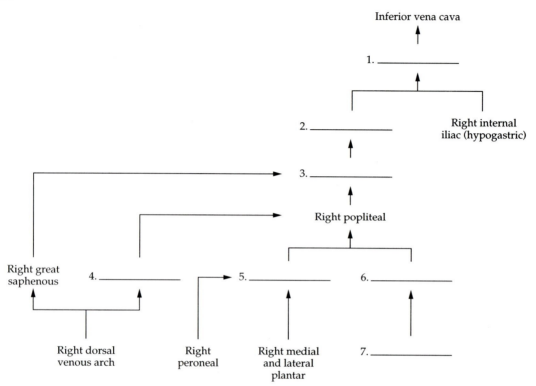

2. Hepatic Portal Circulation

The *hepatic* (*hepato* = liver) *portal circulation* detours venous blood from the gastrointestinal organs and spleen through the liver before it returns to the heart (Figure 18.13). A *portal system* carries blood between two capillary networks, from one location in the body to another without passing through the heart, in this case from capillaries of the gastrointestinal tract to sinusoids of the liver. After a meal, hepatic portal blood is rich with absorbed substances. The liver stores some and modifies others before they pass into the general circulation. For example, the liver converts glucose into glycogen for storage. It also modifies other digested substances so they may be used by cells, detoxifies harmful substances that have been absorbed by the gastrointestinal tract, and destroys bacteria by phagocytosis.

The *hepatic portal vein* is formed by the union of the (1) superior mesenteric and (2) splenic veins. The *superior mesenteric vein* drains blood from the small intestine, portions of the large intestine, stomach, and pancreas through the *jejunal, ileal, ileocolic, right colic, middle colic, pancreaticoduodenal,* and *right gastroepiploic veins.* The *splenic vein* drains blood from the stomach, pancreas, and portions of the large intestine through the *superior rectal, sigmoidal,* and *left colic veins.* The right and left gastric veins,

which open directly into the hepatic portal vein, drain the stomach. The *cystic vein,* which also opens into the hepatic portal vein, drains the gallbladder.

At the same time the liver receives deoxygenated blood via the hepatic portal system, it also receives oxygenated blood from the systemic circulation via the hepatic artery. Ultimately, all blood leaves the liver through the *hepatic veins,* which drain into the inferior vena cava.

Using your textbook, charts, or models for reference, label Figure 18.13.

3. Pulmonary Circulation

The *pulmonary* (*pulmo* = lung) *circulation* carries deoxygenated blood from the right ventricle to the air sacs within the lungs and returns oxygenated blood from the air sacs within the lungs to the left atrium (Figure 18.14). The *pulmonary trunk* emerges from the right ventricle and passes superiorly, posteriorly, and to the left. It then divides into two branches: the *right pulmonary artery* extends to the right lung; the *left pulmonary artery* goes to the left lung. The pulmonary arteries are the only postnatal (after birth) arteries that carry deoxygenated blood. On entering the lungs, the branches divide and subdivide until finally they form capillaries around the air sacs within the lungs. CO_2 passes from the blood into

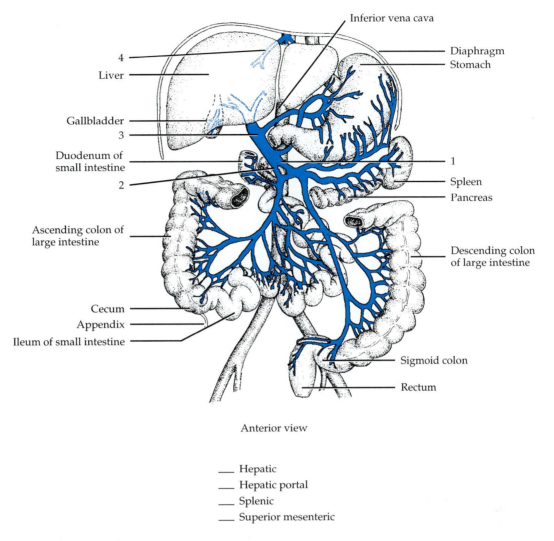

Anterior view

____ Hepatic
____ Hepatic portal
____ Splenic
____ Superior mesenteric

FIGURE 18.13 Hepatic portal circulation.

these air sacs and is exhaled. Inhaled O_2 passes from the air sacs into the blood. The pulmonary capillaries unite, form venules and veins, and eventually two *pulmonary veins* exit from each lung and transport the oxygenated blood to the left atrium. The pulmonary veins are the only postnatal veins that carry oxygenated blood. Contractions of the left ventricle then send the blood into the systemic circulation.

Study a chart or model of the pulmonary circulation, trace the path of blood through it, and label Figure 18.14.

4. Fetal Circulation

The circulatory system of a fetus, called *fetal circulation,* differs from the postnatal circulation because the lungs, kidneys, and gastrointestinal tract begin to function at birth. The fetus obtains its O_2 and nutrients by diffusion from the maternal blood and eliminates its CO_2 and wastes by diffusion into the maternal blood (Fig. 18.15).

The exchange of materials between fetal and maternal circulation occurs through a structure called the *placenta* (pla-SEN-ta). It is attached to the umbilicus (navel) of the fetus by the umbilical (um-BIL-i-kal) cord, and it communicates with the mother through countless small blood vessels that emerge from the uterine wall. The umbilical cord contains blood vessels that branch into capillaries in the placenta. Wastes from the fetal blood diffuse out of the capillaries, into spaces containing maternal blood (intervillous spaces) in the placenta, and finally into the mother's uterine blood vessels. Nutrients travel the opposite route—from the maternal blood vessels to the intervillous spaces to the fetal capillaries. Normally, there is no direct mixing of maternal and fetal blood since all exchanges occur by diffusion through capillary walls.

Blood passes from the fetus to the placenta via two *umbilical arteries.* These branches of the

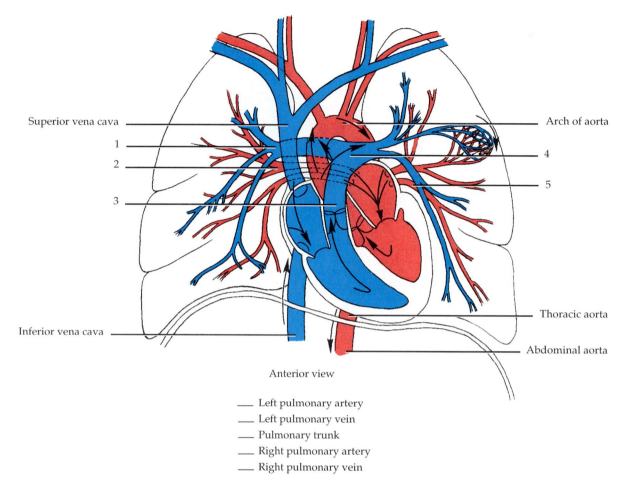

Superior vena cava

1

2

3

Inferior vena cava

Arch of aorta

4

5

Thoracic aorta

Abdominal aorta

Anterior view

___ Left pulmonary artery

___ Left pulmonary vein

___ Pulmonary trunk

___ Right pulmonary artery

___ Right pulmonary vein

FIGURE 18.14 Pulmonary circulation.

internal iliac (hypogastric) arteries are within the umbilical cord. At the placenta, fetal blood picks up O_2 and nutrients and eliminates CO_2 and wastes. The oxygenated blood returns from the placenta via a single ***umbilical vein.*** This vein ascends to the liver of the fetus, where it divides into two branches. Some blood flows through the branch that joins the hepatic portal vein and enters the liver. Most of the blood flows into the second branch, the ***ductus venosus*** (DUK-tus ve-NŌ-sus), which drains into the inferior vena cava. Thus a good portion of the O_2 and nutrients in the blood bypasses the fetal liver and is delivered to the developing fetal brain.

Circulation through other portions of the fetus is similar to postnatal circulation. Deoxygenated blood returning from the lower regions mingles with oxygenated blood from the ductus venosus in the inferior vena cava. This mixed blood then enters the right atrium. Deoxygenated blood returning from the upper regions of the fetus enters the superior vena cava and passes into the right atrium.

Most of the fetal blood does not pass from the right ventricle to the lungs, as it does in postnatal circulation, since the fetal lungs do not operate. In the fetus, an opening called the ***foramen ovale*** (fō-RĀ-men ō-VAL-ē) exists in the septum between the right and left atria. About one-third of the blood passes through the foramen ovale directly into the systemic circulation. The blood that does pass into the right ventricle is pumped into the pulmonary trunk, but little of this blood reaches the lungs. Most is sent through the ***ductus arteriosus*** (ar-tē-rē-Ō-sus). This vessel connects the pulmonary trunk with the aorta and allows most blood to bypass the fetal lungs. The blood in the aorta is carried to all parts of the fetus through the systemic circulation. When the common iliac arteries branch into the external and internal iliacs, part of the blood flows into the internal iliacs. It then goes to the umbilical arteries and back to the placenta for another exchange of materials. The only fetal vessel that carries fully oxygenated blood is the umbilical vein.

Label Figure 18.15.

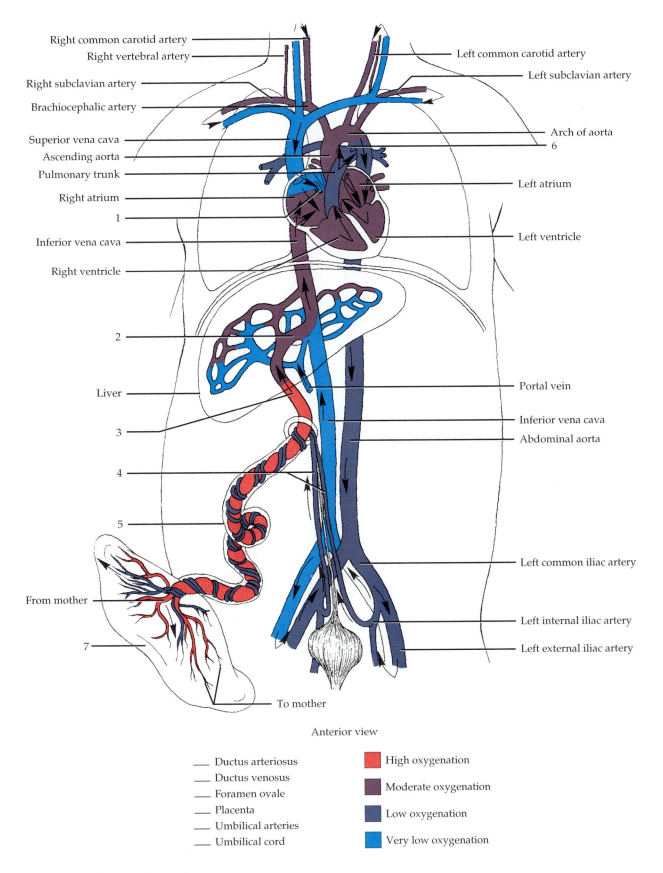

Right common carotid artery

Right vertebral artery

Right subclavian artery

Brachiocephalic artery

Superior vena cava

Ascending aorta

Pulmonary trunk

Right atrium

1

Inferior vena cava

Right ventricle

2

Liver

3

4

5

From mother

7

To mother

Left common carotid artery

Left subclavian artery

Arch of aorta

6

Left atrium

Left ventricle

Portal vein

Inferior vena cava

Abdominal aorta

Left common iliac artery

Left internal iliac artery

Left external iliac artery

Anterior view

___ Ductus arteriosus

___ Ductus venosus

___ Foramen ovale

___ Placenta

___ Umbilical arteries

___ Umbilical cord

High oxygenation

Moderate oxygenation

Low oxygenation

Very low oxygenation

FIGURE 18.15   Fetal circulation.

E. BLOOD VESSEL EXERCISE

For each vessel listed, indicate the region supplied (if an artery) or the region drained (if a vein):

1. *Coronary artery* _____

_____

2. *Internal iliac veins* _____

_____

3. *Lumbar arteries* _____

_____

4. *Renal artery* _____

_____

5. *Left gastric artery* _____

_____

6. *External jugular vein* _____

_____

7. *Left subclavian artery* _____

_____

8. *Axillary vein* _____

_____

9. *Brachiocephalic veins* _____

_____

10. *Transverse sinuses* _____

_____

11. *Hepatic artery* _____

_____

12. *Inferior mesenteric artery* _____

_____

13. *Suprarenal artery* _____

_____

14. *Inferior phrenic artery* _____

_____

15. *Great saphenous vein* _____

_____

16. *Popliteal vein* _____

_____

17. *Azygos vein* _____

_____

18. *Internal iliac (hypogastric) artery* _____

_____

19. *Internal carotid artery* _____

_____

20. *Cephalic vein* _____

_____

F. DISSECTION OF CAT CARDIOVASCULAR SYSTEM

CAUTION! *Please reread Section D, "Precautions Related to Dissection" at the beginning of the laboratory manual on page xiii before you begin dissection.*

In this section, you will dissect some of the principal blood vessels in the cat. Blood vessels should be studied in specimens that are at least doubly injected, meaning that the arteries have been injected with red latex and the veins have been injected with blue latex. Triply injected specimens are necessary to observe all of the hepatic portal system. This system is usually injected with yellow latex.

As you dissect blood vessels, bear in mind that they vary somewhat in position from one specimen to the next. Thus, the pattern of blood vessels in your specimen may not be exactly like those shown in the illustrations. Also remember that when you dissect blood vessels, you must free them from surrounding tissues so that they are clearly visible. When doing so, use forceps or a blunt probe, not a scalpel. Finally, you obtain best results if you dissect blood vessels on the side of the body opposite the side on which you dissected muscles.

1. Arteries

PROCEDURE

1. To examine the major blood vessels of the cat, extend the original thoracic longitudinal incisions posteriorly through the abdominal wall.

2. Using scissors, cut the wall from the diaphragm to the pubic region. Be careful not to damage viscera below.

3. On both sides of the abdomen make two additional cuts, laterally and dorsally, toward the spine about 4 in. below the diaphragm. Reflect the flaps of the abdominal wall.

4. Locate the *pulmonary trunk* exiting from the right ventricle (see Figure 17.5). The trunk branches into a right and a left pulmonary artery. The *left pulmonary artery* passes ventral to the aorta to reach the left lung whereas the *right pulmonary artery* passes between the arch of the aorta and the heart to reach the right lung.

5. Dorsal to the pulmonary trunk is the *ascending aorta,* which arises from the left ventricle. At the base of the ascending aorta, try to locate the origins of the coronary arteries. The *right coronary artery* passes in the *coronary sulcus,* a groove that separates the right atrium from the right ventricle. The *left coronary artery* passes in the *anterior longitudinal sulcus,* a groove that separates the right ventricle from the left ventricle.

6. The ascending aorta continues cranially and then turns to the left. This part of the aorta is called the *arch of the aorta.* Originating from the arch of the aorta are two large arteries—the brachiocephalic and left subclavian.

7. The *brachiocephalic artery* divides at about the level of the second rib into the *left common carotid artery* and then into the *right common carotid artery* and the *right subclavian artery* (Figure 18.16). The left common carotid artery ascends in the neck between the internal jugular vein and the trachea. Among the branches of the left common carotid artery are (1) the *superior thyroid artery,* which originates at the level of the thyroid cartilage and supplies the thyroid gland and muscles of the neck; (2) the *laryngeal artery,* which supplies the larynx; and (3) the *occipital artery,* which supplies the back of the neck. The right common carotid artery supplies the same structures on the right side of the body. The right and left subclavian arteries will be discussed shortly.

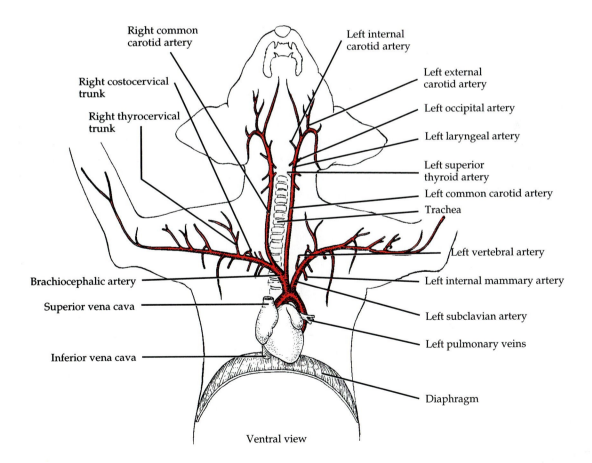

FIGURE 18.16 Arteries of forelimbs, neck, and head.

8. At the level of the cranial border of the larynx, the common carotid arteries divide into two arteries. The large *external carotid artery* primarily supplies the head structures outside the cranial cavity. The very small *internal carotid artery*, which may be absent, primarily supplies the structures within the cranial cavity. At the bifurcation of the common carotid into external and internal carotids, the dilated area is called the *carotid sinus.* This area contains stretch receptors that inform the brain of fluctuations in blood pressure so that necessary adjustments can be made. The *aortic sinus* in the aorta has a similar function.

9. The right and left subclavian arteries supply the head, neck, thoracic wall, and forelimbs. Among their important branches are (1) the *vertebral arteries,* which supply the brain; (2) the *internal mammary arteries,* which primarily supply the ventral thoracic wall; (3) the *costocervical trunk,* which supplies the neck, back, and intercostal muscles; and (4) the *thyrocervical trunk,* which supplies some neck and back muscles.

10. The continuation of the subclavian artery out of the thoracic cavity across the axillary space is called the *axillary artery* (Figure 18.17). Branches of the axillary artery include (1) the *ventral thoracic artery,* which supplies the pectoral muscles; (2) the *long thoracic artery,* which supplies the latissimus dorsi muscle; and (3) the *subscapular artery,* which supplies the shoulder muscles, dorsal muscles of the arm, and muscles of the scapula.

11. The continuation of the axillary artery into the arm is called the *brachial artery.* A branch of the brachial artery, the *anterior humeral circumflex artery,* arises just distal to the subscapular artery and supplies the biceps brachii muscle. Another branch of the brachial artery, the *deep brachial artery,* is distal to the anterior humeral circumflex. It supplies the triceps brachii, epitrochlearis, and latissimus dorsi muscles. Distal to the elbow, the brachial artery continues as the *radial artery.* In the forearm, the radial artery gives off a small branch, the *ulnar artery.* The radial and ulnar arteries supply the musculature of the forearm.

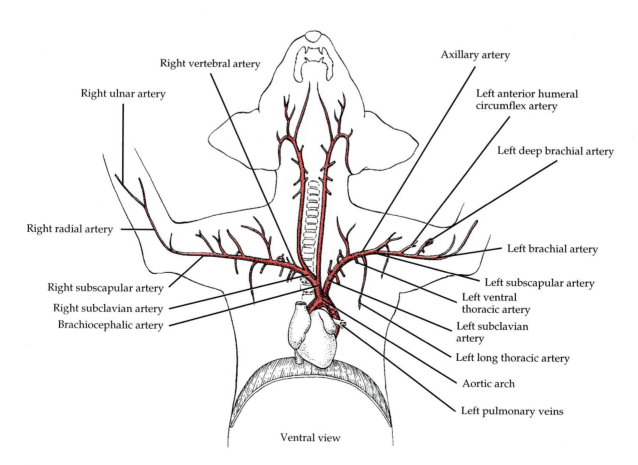

FIGURE 18.17 Arteries of forelimbs.

12. Continue your dissection of arteries by returning to the arch of the aorta. Follow it caudally as it passes through the thoracic cavity. At this point the aorta is known as the *thoracic aorta.* Arising from the thoracic aorta are (1) the *intercostal arteries* (10 pairs), which supply the intercostal muscles; (2) the *bronchial arteries,* which supply the bronchi and lungs; and (3) the *esophageal arteries,* which supply the esophagus.

13. The aorta passes through the diaphragm at the level of the second lumbar vertebra. The portion of the aorta between the diaphragm and pelvis is called the *abdominal aorta.*

14. The first branch of the abdominal aorta immediately caudal to the diaphragm is the *celiac trunk* (Figure 18.18). It consists of three branches. The *hepatic artery* supplies the liver, the *left gastric artery* along the lesser curvature of the stomach supplies the stomach, and the *splenic artery* supplies the spleen. See if you can also locate the *cystic artery,* a branch of the hepatic artery that supplies the gallbladder.

15. Caudal to the celiac trunk is the *superior mesenteric artery.* Its branches are the *inferior pancreatoduodenal artery* to the pancreas and duodenum, the *ileocolic artery* to the ileum and cecum, the *right colic artery* to the ascending colon, the *middle colic artery* to the transverse colon, and the *intestinal arteries* to the jejunum and ileum.

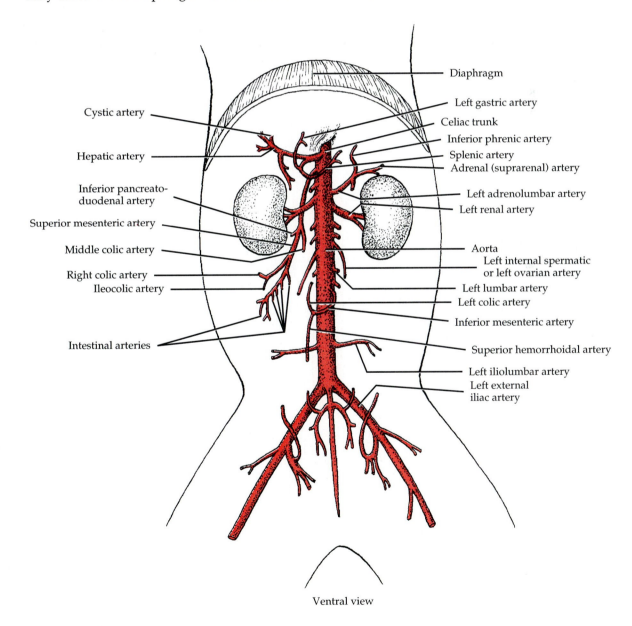

Ventral view

FIGURE 18.18 Arteries of abdomen.

16. Just caudal to the superior mesenteric artery, the abdominal aorta gives off a pair of *adrenolumbar arteries.* Each gives off an *inferior phrenic artery* to the diaphragm, an *adrenal (suprarenal) artery* to the adrenal gland, and a *parietal branch* to the muscles of the back.
17. The paired *renal arteries,* caudal to the adrenolumbar arteries, supply the kidneys.
18. In the male, the paired *internal spermatic arteries,* or *testicular arteries,* arise from the abdominal aorta caudal to the renal arteries. The internal spermatic arteries pass caudally to the internal inguinal ring and through the inguinal canal where they accompany the ductus deferens of each testis.
19. In the female, the paired *ovarian arteries* arise from the abdominal aorta caudal to the renal

arteries. They pass laterally in the broad ligament and supply the ovaries.
20. The unpaired *inferior mesenteric artery* arises caudal to the internal spermatic or ovarian arteries. Its branches include a *left colic artery* to the descending colon and a *superior hemorrhoidal artery* to the rectum.
21. Seven pairs of *lumbar arteries* arise from the dorsal surface of the abdominal aorta and supply the muscles of the dorsal abdominal wall. A pair of *iliolumbar arteries* arises near the inferior mesenteric artery and also supplies the muscles of the dorsal abdominal wall.
22. Near the sacrum, the abdominal aorta divides into right and left *external iliac arteries* (Figure 18.19). Also from this point of division, a short common artery extends caudally and

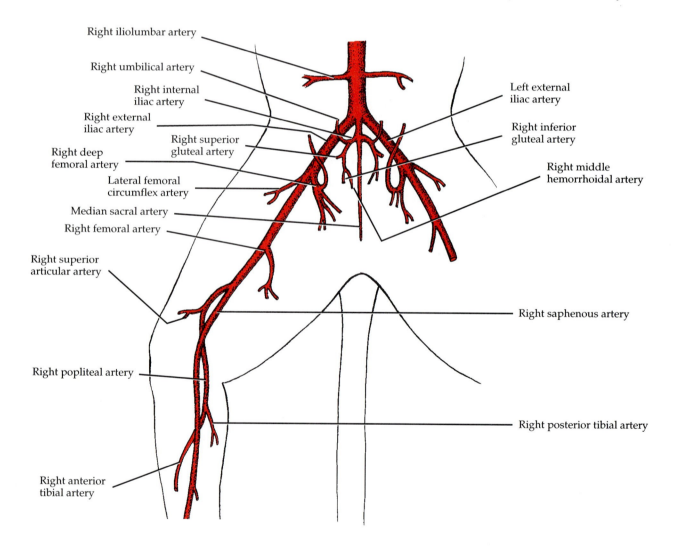

Ventral view

FIGURE 18.19   Arteries of right hindlimb.

gives rise to the paired *internal iliac (hypogastric) arteries* and unpaired *median sacral artery* that passes along the ventral aspect of the tail. The first branch off the internal iliac artery, the *umbilical artery*, supplies the urinary bladder. Another branch of the internal iliac artery, the *superior gluteal artery*, supplies the gluteal muscles and other muscles of the thigh. Still another branch of the internal iliac artery, the *middle hemorrhoidal artery*, supplies the rectum, anus, urethra, and adjacent structures. The terminal portion of the internal iliac artery, the *inferior gluteal artery*, supplies thigh muscles.

23. Now we can return to the external iliac arteries. Just before leaving the abdominal cavity, each external iliac artery gives off a branch, the *deep femoral artery*, which supplies the muscles of the thigh. Branches of the deep femoral artery supply the rectus abdominis muscle, the urinary bladder, and muscles on the medial aspect of the thigh.

24. The external iliac artery continues outside the abdominal cavity and into the thigh as the *femoral artery*. Passing through the thigh, it gives off one branch as the *lateral femoral circumflex artery* to the rectus femoris and vastus medialis. Just above the knee, the femoral artery gives off a branch as the *superior articular artery* and another as the *saphenous artery*. Both supply thigh muscles.

25. In the region of the knee, the femoral artery is known as the *popliteal artery*. It divides into an *anterior tibial artery* and a *posterior tibial artery*, which supply the leg and foot.

2. Veins

PROCEDURE

1. In learning the arteries of the cat, we began our study with the heart and then worked outward from the heart to various regions of the body. In learning the veins of the cat, we will begin our study with the various regions of the body and work toward the heart. Whereas arteries supply various parts of the body with blood, veins drain various parts of the body. That is, veins drain blood from capillary beds and take it back to the heart.

2. The veins of the pelvis and thigh generally correspond to the arteries in those regions. Whereas the artery supplies the region with blood, the corresponding vein drains the same region. An important difference between the arteries and veins of the pelvis is that the internal iliac vein joins the external iliac vein directly to form the common iliac vein. The cat has no common iliac artery.

3. The deep veins of the leg are the *anterior tibial vein* and the *posterior tibial vein* (Figure 18.20). At the knee the two unite to form the *popliteal vein*, another deep vein. In the thigh, the popliteal vein is known as the *femoral vein*. The superficial veins of the lower extremity are the *greater saphenous vein*, which runs along the medial surface and joins the femoral vein, and the *lesser saphenous vein*, which runs along the dorsal surface and joins a tributary of the internal iliac vein.

4. The *external iliac vein* is a continuation of the femoral vein. It has tributaries corresponding to the arteries and returns blood distributed by these arteries. The *internal iliac (hypogastric) vein*, with its tributaries, returns blood distributed by the internal iliac artery and its branches. The *common iliac vein* is formed by the union of the external and internal iliac veins.

5. The *inferior vena cava (postcava)*, formed by the union of the common iliac veins, enters the caudal portion of the right atrium of the heart. The tributaries of the inferior vena cava, exclusive of the hepatic portal system, include the *iliolumbar, lumbar, renal, adrenolumbar,* and *internal spermatic,* or *ovarian veins.* These correspond to the paired branches of the abdominal aorta. Cranial to the diaphragm, the inferior vena cava has no tributaries. After piercing the diaphragm, the inferior vena cava ascends and enters the right atrium.

6. The *hepatic portal system* consists of a series of veins that drain blood from the small and large intestines, stomach, spleen, and pancreas and convey the blood to the liver. The blood is delivered to the liver by the hepatic portal vein. Within the liver, the hepatic portal vein gives branches to each lobe. These branches terminate in capillaries called *liver sinusoids.* The sinusoids are drained by the hepatic veins, which leave the liver and join the inferior vena cava. Blood delivered to the liver by the hepatic artery also passes through the liver sinusoids.

7. Unless your specimen is triply injected, locating all the veins of the hepatic portal system will be very difficult. However, using Figure 18.21 as a guide, identify as many of the vessels as you can. Notice that the hepatic portal vein receives blood from the spleen and from

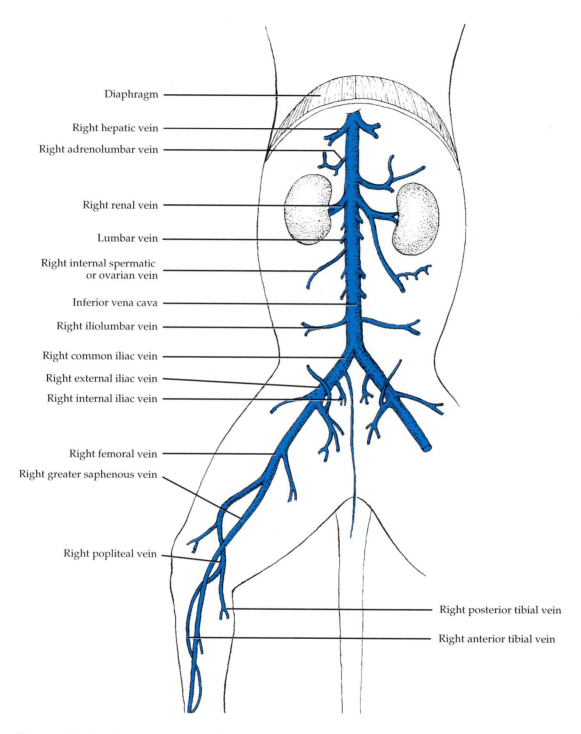

Diaphragm

Right hepatic vein

Right adrenolumbar vein

Right renal vein

Lumbar vein

Right internal spermatic
or ovarian vein

Inferior vena cava

Right iliolumbar vein

Right common iliac vein

Right external iliac vein

Right internal iliac vein

Right femoral vein

Right greater saphenous vein

Right popliteal vein

Right posterior tibial vein

Right anterior tibial vein

FIGURE 18.20   Veins of right hindlimb and abdomen.

all digestive organs except the liver. Blood is delivered to the hepatic portal vein by veins corresponding to the arteries that distribute blood to these viscera. In the cat, the **hepatic portal vein** is formed by the union of the **gastrosplenic vein** and the **superior mesenteric** **vein.** The superior mesenteric vein is formed by the **inferior mesenteric vein, intestinal veins,** and **ileocolic vein,** all of which drain the small intestine, and by the **anterior pancreato-duodenal vein** from the pancreas and small intestine (Figure 18.21).

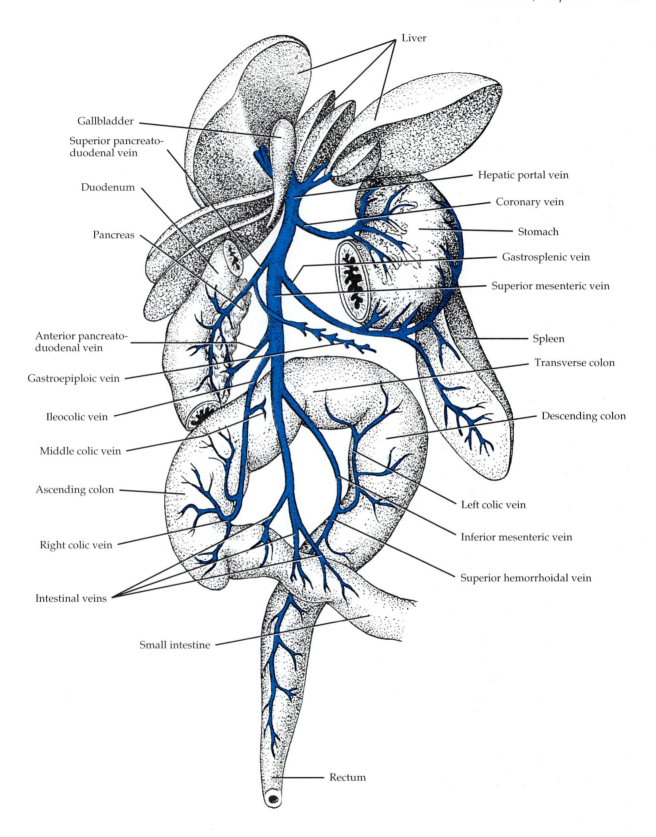

Ventral view

FIGURE 18.21 Hepatic portal system.

8. In the forearm identify the ulnar and radial veins. The *ulnar vein* joins the *radial vein* (Figure 18.22). Also find the *cephalic vein,* which joins the *transverse scapular vein,* and the *median cubital vein,* a communicating vein between the cephalic and brachial veins.

9. In the arm, identify the *brachial vein* and its tributaries. In general, this vein parallels the brachial artery and its branches.

10. The *axillary vein* is a continuation of the brachial vein in the axilla. The tributaries to the axillary vein generally correspond to the branches of the axillary artery.

11. The axillary vein is joined by the *subscapular vein* in the axillary region. In the shoulder, the axillary vein is known as the *subclavian vein.* Follow the subclavian medially and locate the superior vena cava.

12. The *superior vena cava (precava)* enters the anterior part of the right atrium and returns blood to the heart from all areas cranial to the diaphragm, except the heart itself. The superior vena cava is formed by the union of the two *brachiocephalic veins.*

13. Entering the right side of the superior vena cava immediately cranial to the heart is the single *azygos vein.* This vein collects blood from *intercostal veins* from the body wall, from the diaphragm, from the *esophageal veins,* and from the *bronchial veins.* Identify the azygos vein (Figure 18.22).

14. The superior vena cava also receives blood from the chest wall from the *internal mammary veins.* These veins unite as the *sternal vein* before entering the superior vena cava. The *right vertebral vein* may join the superior vena cava or the brachiocephalic vein.

15. From the superior vena cava, trace the *left brachiocephalic (innominate) vein* cranially (Figure 18.23). Identify the *left vertebral vein.* Note that the left brachiocephalic vein is formed by the union of the *left external jugular vein,* which drains the head outside the cranial cavity, and the *left subclavian vein,* which drains the arm. The *right brachiocephalic vein* is formed by the union of the right external jugular and right subclavian veins.

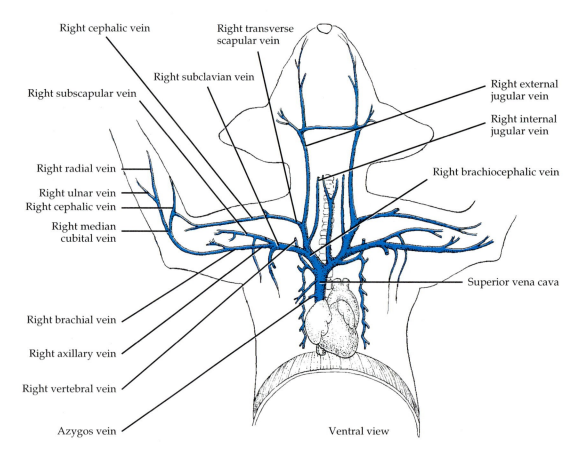

FIGURE 18.22   Veins of right forelimb.

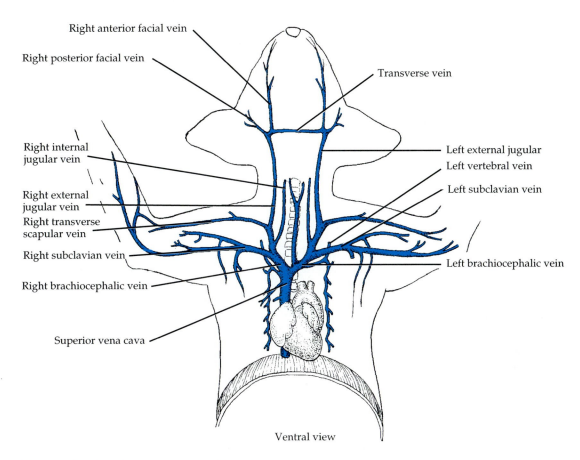

Right anterior facial vein

Right posterior facial vein

Transverse vein

Right internal jugular vein

Left external jugular

Left vertebral vein

Right external jugular vein

Left subclavian vein

Right transverse scapular vein

Right subclavian vein

Left brachiocephalic vein

Right brachiocephalic vein

Superior vena cava

Ventral view

FIGURE 18.23 Veins of neck and head.

16. If you trace the external jugular vein cranially, you will note that it receives the smaller *internal jugular vein* above the point of union of the external jugular and subclavian veins. The internal jugular vein drains the brain. Cranial to the internal jugular vein is the *transverse scapular vein,* which joins the external jugular.

17. The *external jugular vein* is formed by the union of the *anterior* and *posterior facial veins.*

These veins have tributaries that return blood distributed by the branches of the external carotid artery. The external jugular veins communicate near their points of formation by a large *transverse vein* that passes ventral to the neck.

ANSWER THE LABORATORY REPORT QUESTIONS AT THE END OF THE EXERCISE.

Blood Vessels 18

Student _____   Date _____

Laboratory Section _____   Score/Grade _____

PART 1. Multiple Choice

_____ 1. The largest of the circulatory routes is (a) systemic (b) pulmonary (c) coronary (d) hepatic portal

_____ 2. All arteries of systemic circulation branch from the (a) superior vena cava (b) aorta (c) pulmonary artery (d) coronary artery

_____ 3. The arterial system that supplies the brain with blood is the (a) hepatic portal system (b) pulmonary system (c) cerebral arterial circle (circle of Willis) (d) carotid system

_____ 4. An obstruction in the inferior vena cava would hamper the return of blood from the (a) head and neck (b) upper limbs (c) thorax (d) abdomen and pelvis

_____ 5. Which statement best describes arteries? (a) all carry oxygenated blood to the heart (b) all contain valves to prevent the backflow of blood (c) all carry blood away from the heart (d) only large arteries are lined with endothelium

_____ 6. Which statement is _not_ true of veins? (a) they have less elastic tissue and smooth muscle than arteries (b) their tunica externa is the thickest coat (c) most veins in the limbs have valves (d) they always carry deoxygenated blood

_____ 7. A thrombus in the first branch of the arch of the aorta would affect the flow of blood to the (a) left side of the head and neck (b) myocardium of the heart (c) right side of the head and neck and right upper limb (d) left upper limb

_____ 8. If a vein must be punctured for an injection, transfusion, or removal of a blood sample, the likely site would be the (a) median cubital (b) subclavian (c) hemiazygous (d) anterior tibial

_____ 9. In hepatic portal circulation, blood is eventually returned to the inferior vena cava through the (a) superior mesenteric vein (b) hepatic portal vein (c) hepatic artery (d) hepatic veins

_____ 10. Which of the following are involved in pulmonary circulation? (a) superior vena cava, right atrium, and left ventricle (b) inferior vena cava, right atrium, and left ventricle (c) right ventricle, pulmonary artery, and left atrium (d) left ventricle, aorta, and inferior vena cava

_____ 11. If a thrombus in the left common iliac vein dislodged, into which arteriole system would it first find its way? (a) brain (b) kidneys (c) lungs (d) left arm

_____ 12. In fetal circulation, the blood containing the highest amount of oxygen is found in the (a) umbilical arteries (b) ductus venosus (c) aorta (d) umbilical vein

_____ 13. The greatest amount of elastic tissue found in the arteries is located in which coat? (a) tunica interna (b) tunica media (c) tunica externa (d) tunica adventitia

_____ 14. Which coat of an artery contains endothelium? (a) tunica interna (b) tunica media (c) tunica externa (d) tunica adventitia

_____ 15. Permitting the exchange of nutrients and gases between the blood and tissue cells is the primary function of (a) capillaries (b) arteries (c) veins (d) arterioles

_____ 16. The circulatory route that runs from the gastrointestinal tract to the liver is called (a) coronary circulation (b) pulmonary circulation (c) hepatic portal circulation (d) cerebral circulation

_____ 17. Which of the following statements about systemic circulation is *not* correct? (a) its purpose is to carry oxygen and nutrients to body tissues and to remove carbon dioxide (b) all systemic arteries branch from the aorta (c) it involves the flow of blood from the left ventricle to all parts of the body except the lungs (d) it involves the flow of blood from the body to the left atrium

_____ 18. The opening in the septum between the right and left atria of a fetus is called the (a) foramen ovale (b) ductus venosus (c) foramen rotundum (d) foramen spinosum

_____ 19. The branch of the umbilical vein in the fetus that connects with the inferior vena cava, bypassing the liver, is the (a) foramen ovale (b) ductus venosus (c) ductus arteriosus (d) patent ductus

_____ 20. Which of the vessels does *not* belong with the others? (a) brachiocephalic artery (b) left common carotid artery (c) celiac artery (d) left subclavian artery

PART 2. Matching

_____ 21. Aortic branch that supplies the head and associated structures

_____ 22. Artery that distributes blood to the small intestine and part of the large intestine

_____ 23. Vessel into which veins of the head and neck, upper limbs, and thorax enter

_____ 24. Vessel into which veins of the abdomen, pelvis, and lower limbs enter

_____ 25. Vein that drains the head and associated structures

_____ 26. Longest vein in the body

_____ 27. Vein just behind the knee

_____ 28. Artery that supplies a major part of the large intestine and rectum

_____ 29. First branch off of the arch of the aorta

_____ 30. Arteries supplying the heart

A. Inferior vena cava

B. Superior vena cava

C. Superior mesenteric

D. Common carotid

E. Jugular

F. Brachiocephalic

G. Coronary

H. Inferior mesenteric

I. Popliteal

J. Great saphenous

Cardiovascular Physiology

19

A. CARDIAC CONDUCTION SYSTEM AND ELECTROCARDIOGRAM (ECG OR EKG)

The heart is innervated by the autonomic nervous system (ANS), which modifies, but does not initiate, the cardiac cycle. The heart can continue to contract if separated from the ANS. This is possible because the heart has an *intrinsic pacemaker* termed the *sinoatrial (SA) node.* The SA node is called an intrinsic pacemaker because of its ability to rhythmically, spontaneously depolarize, thereby reaching threshold value and initiating an action potential (nerve impulse). This ability to spontaneously depolarize is due to the SA node's permeability characteristics to sodium, potassium, and calcium ions. The SA node is connected to a series of specialized muscle cells that comprise the *cardiac conduction system.* This system distributes the action potentials that stimulate the cardiac muscle fibers to contract.

The SA node is located on the posterior aspect of the heart where the superior vena cava enters the right atrium. Once the SA node spontaneously depolarizes and reaches threshold value, the action potential is conducted through the remainder of the heart via the cardiac conduction system. From the SA node the action potential travels throughout the right atrium via a wavelike conduction process. The velocity for this is about 1 m/sec. In addition, the impulse is conducted to the left atrium and the *atrioventricular (AV) node*, which is located near the inferior portion of the interatrial septum, via the *anterior interatrial myocardial band (Bachmann's bundle).* The band is composed of *anterior, posterior,* and *middle internodal pathways*, which are three groups of specialized cardiac muscle cells. As the action potential spreads to the atrial musculature, it initiates atrial contraction. When the action potential reaches the AV node, it too, depolarizes, but in a manner that significantly slows the conduction velocity of the cardiac action potential. From the AV node, a tract of

conducting fibers called the *atrioventricular (AV) bundle (bundle of His)* extends to the superior portion of the interventricular septum and continues down the septum and divides into *right* and *left bundle branches.* The bundle branches, in turn, divide into a complex network of fibers called *conduction myofibers (Purkinje fibers)* that emerge from the bundle branches and pass into the cells of the ventricular myocardium. As the action potential passes from the AV node into the AV bundle the conduction velocity steadily and rapidly increases. Because of the large diameter of the conduction myofibers, the conduction velocity of the action potential reaches its highest speed within these fibers, thereby significantly increasing the speed with which the action potential is spread throughout the ventricles. As the action potentials reach the ventricular myocardial fibers, they are depolarized, thereby initiating ventricular contraction.

Label the components of the cardiac conduction system shown in Figure 19.1.

The electrocardiogram (ECG) provides a record of the electrical events happening within the heart. The ECG is obtained by placing recording electrodes on the surface of the body. These recording electrodes detect the changing potential differences of the heart on the body's surfaces, and a time-dependence plot of these differences is called an *electrocardiogram (ECG).* An *electrocardiograph* is an instrument used to record these changes. As the electrical impulses are transmitted throughout the cardiac conduction system and myocardial cells, a different electrical impulse is generated. These impulses are transmitted from the electrodes to a recording needle that graphs the impulses as a series of up-and-down (vertical) waves called *deflection waves* (Figure 19.2 on page 455).

1. Scalar Electrocardiographic Recordings

Electrocardiograms are usually recorded by *indirect leads* (recording electrodes placed some distance from the heart on the skin of the subject rather than

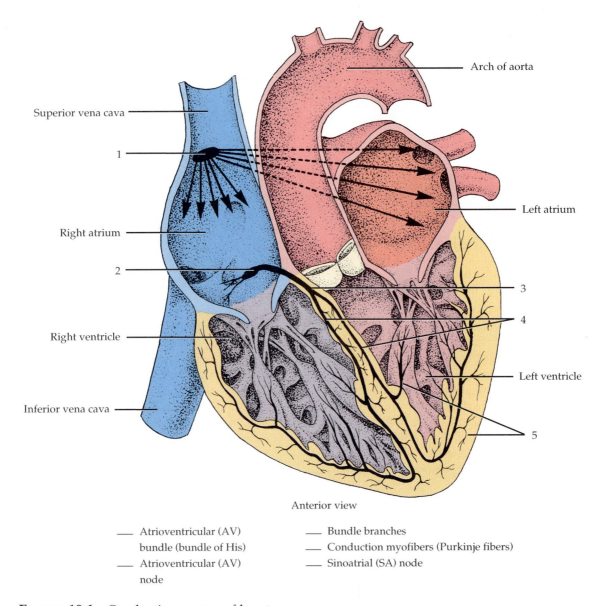

Superior vena cava

1

Right atrium

2

Right ventricle

Inferior vena cava

Arch of aorta

Left atrium

3

4

Left ventricle

5

Anterior view

—— Atrioventricular (AV)   —— Bundle branches
bundle (bundle of His)   —— Conduction myofibers (Purkinje fibers)
—— Atrioventricular (AV)   —— Sinoatrial (SA) node
node

FIGURE 19.1   Conduction system of heart.

directly on the heart). The scalar electrocardiogram represents a *two-dimensional* tracing of the moment-to-moment *three-dimensional* electrical changes that occur within the heart during the electrical processes of the cardiac cycle. Because of the differences in the size and shape of any one individual's heart and body dimensions, the electrocardiogram varies from individual to individual. In addition, within any one individual the ECG pattern varies with the placement of the leads.

The ECG recordings that will be obtained in the laboratory utilize the *three standard limb leads*. A **lead** is defined as two electrodes working in pairs. **Electrodes** are sensing devices of sensitive metal plates or small rods that can detect electrophysio-

logical phenomena such as changes in electrical potential of skin or nerves (see Exercise 9). For convenience, the three standard limb leads are typically attached to the right and left wrists and the left leg. Lead I records the potential difference between the left and right arms (LA and RA, respectively). Lead II records the potential difference between the right arm (RA) and left leg (LL). Lead III, therefore, would record the potential differences between the left arm (LA) and the left leg (LG).

A typical lead II ECG record (Figure 19.2) is composed of an orderly series of deflections and intervals. The first deflection, or wave, is termed the *P wave*, and represents *atrial depolarization*. After the P wave the ECG returns to baseline level because

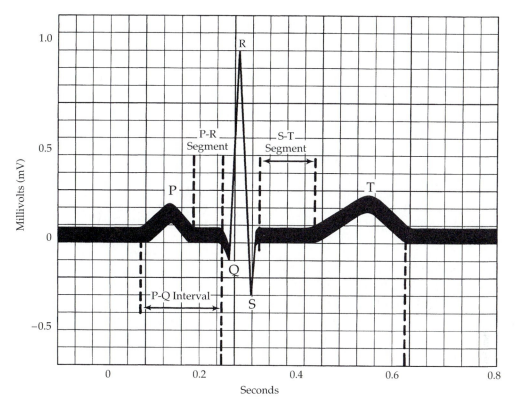

FIGURE 19.2   Recordings of a normal electrocardiogram (ECG), lead II.

the electrodes are no longer recording any potential differences. This time period is termed the ***P-Q interval.*** During this time period, however, the electrical activity of the heart is not quiescent, as the action potential is being propagated through the AV node, the AV bundle branches, the conduction myofibers, and ventricular muocardium. In response, ventricular muscle cells depolarize.

The ***QRS complex*** of the ECG represents *ventricular depolarization.* The QRS complex also represents *atrial repolarization,* since the atria repolarize at the same time the ventricles depolarize. No separate deflection for atrial repolarization is found because of the significantly smaller amount of atrial muscle mass as compared to the ventricular muscle mass.

In the time period between the QRS complex and the repolarization of the ventricles (during the T wave described below) the ECG again returns to, or very closely to its baseline level. This time period is termed the ***S-T segment.*** During the S-T segment of the ECG all of the ventricular muscle is depolarized. This segment of the ECG represents the plateau phase of the ventricular action potential.

The final wave is termed the ***T wave,*** and represents *ventricular repolarization.* The T wave is smaller

(but with a longer duration) than the QRS complex because the rate of repolarization is slower than the rate of depolarization within the ventricles.

In reading and interpreting an electrocardiogram, you must note the *lead* type of configuration from which the ECG was recorded, the size of the deflection waves, and the lengths of the P-Q interval and S-T segment. An attempt will not be made to interpret an electrocardiogram, but only to become familiar with a normal ECG. The ECG is valuable in diagnosing abnormal cardiac rhythms and conducting patterns, detecting the presence of fetal life, determining the presence of more than one fetus, and following the course of recovery from a heart attack.

2. Recording an ECG

Note: Depending on the type of laboratory equipment available, select from the following procedures related to electrocardiography.

PROCEDURE USING PHYSIOFINDER

PHYSIOFINDER is an interactive computer program that permits laboratory simulations that do not involve the use of animals or advanced or expensive laboratory equipment. It permits students to perform experiments, analyze data, draw conclusions, and form hypotheses on the basis of collected data. The program is available from HarperCollins Publishers (1-800-8HEART1).

For activities related to electrocardiography, select the appropriate experiments from PHYS-IOFINDER Module 4—Control of the Heart.

PROCEDURE USING CARDIOCOMP™

CARDIOCOMP™ is a computer hardware and software system designed to acquire data related to electrocardiography by permitting students to analyze their cardiac biopotentials. The program is available from INTELITOOL® (1-800-227-3805).

PROCEDURE USING A PHYSIOGRAPH-TYPE POLYGRAPH, ELECTROCARDIO-GRAPH, OR OSCILLOSCOPE

In the following procedure a physiograph-type polygraph is used, but an electrocardiograph or oscilloscope may also be used (Figure 19.3a).

In recording the ECG, we will utilize only the standard three limb leads. Usually only two electrodes will be in actual use at any one time. The third electrode will be automatically switched off by the *lead-selector switch* of the polygraph.

For most general purposes three leads are usually used. However, electrocardiologists use additional leads, including several chest wall electrodes (Figure 19.3b). These leads are positioned around the chest, and encircle the heart so that there are six intersecting lines on a horizontal plane through the atrioventricular node.

The procedure for recording electrocardiograms using the physiograph is as follows:

PROCEDURE

1. Either you or your laboratory partner should lie on a table or cot, roll down long stockings or socks to the ankles, and *remove all metal jewelry*.
2. Swab the skin with alcohol where electrodes will be applied. Electrode cream, jelly, or saline paste is applied to the skin only where the electrode will make contact, such as to the inner forearm just above the wrists and to the inside of the legs just above the ankles.
3. The cream is then applied on the concave surface of the electrode plates, which are then fastened securely to the limb area surfaces with rubber straps or adhesive tape (rings). Adjust the straps to fit snugly, but *not so tight as to impede blood flow*.

4. Firmly connect the proper end of the electrode cables to the electrode plates, using the standard limb leads previously described, and connect the other end to a cardiac preamplifier or to a self-contained electrocardiograph.
5. After you apply the electrodes and connect the instrument to the subject, ascertain the following before actually recording:
   a. Recording power switch is on.
   b. Paper is sufficient for the entire recording, and the pen is centered properly on the paper with ink flowing freely.
   c. Preamplifier sensitivity has been set at 10 mm per millivolt.
   d. Subject is lying quietly and is relaxed.
   Note: Usually five or six ECG complex recordings from each lead are sufficient. All electrocardiographers have chosen a standard paper speed of 25 mm per second.
6. Turn the instrument on and record for 30 seconds from each lead. Obtain directions from your laboratory instructor on the procedure to be followed when switching from lead to lead. When the recording leads are switched, note the change by marking it on the electrocardiogram. The subject must remain very still and relax completely during the recording period.
7. When the recording is finished, turn off the reading instrument and disconnect the leads from the subject and remove the electrodes, cleaning off the excess cream or paste. The skin can be washed with water to remove electrode cream.
8. Identify and letter the P, QRS, and T waves, and compare the waves with those shown in Figure 19.2.
9. Calculate the duration of the waves and the P-Q interval. Attach this recording to Section A of the LABORATORY REPORT RESULTS at the end of the exercise.

3. Determination of Mean Electrical Axis

As was previously outlined, depolarization of the entire ventricular myocardium does not occur simultaneously. The depolarization starts within the interventricular septum and then spreads to the ventricular walls. While the septum is depolarized and the ventricular walls are not, an electrical difference (voltage) exists between these two locations, thereby producing the R wave of the QRS complex. When both locations are

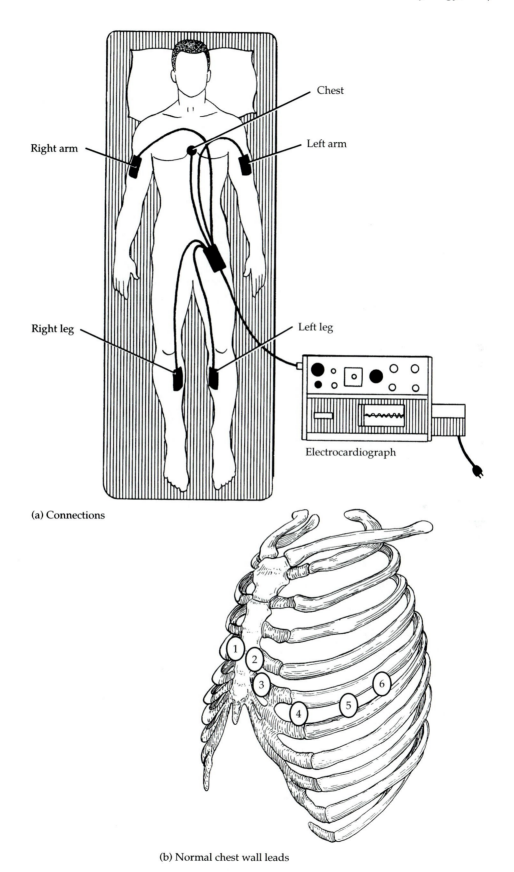

(a) Connections

(b) Normal chest wall leads

FIGURE 19.3 Electrocardiograph.

depolarized, no electrical difference occurs, thereby producing the S wave of the QRS complex and completing that phase of the ECG. The spread of the action potential through the ventricular myocardium depends upon several factors. These include the size of the ventricles, orientation of the heart during that particular instant of the cardiac cycle, and the ease with which action potentials spread throughout the various parts of the ventricular conducting system and myocardium. Because of this, it is clinically useful to determine the *mean electrical axis* (average direction of depolarization) of the cardiac cycle. This is done by observing the voltages of the QRS complex of the ECG from two different orientations using leads I and III of the ECG. Lead I crosses the heart from left arm to right arm, thereby giving a horizontal axis of observation, while lead III approximately parallels the heart, going from the left arm to the left leg, thereby providing a 120° axis of observation. When these two observations are taken together the average heart has a mean electrical axis of 59° (Figure 19.4).

PROCEDURE

The procedure for determining the mean electrical axis from the electrocardiogram is as follows:

1. Record an ECG utilizing lead I (with sensitivity setting at 1) from a subject that is reclining quietly.
2. Find a QRS complex and count the number of millimeters (small boxes on the paper) that it projects above the top of the *isoelectric point* (baseline) of the recording. Do this for several QRS complexes and record the average value in the space below.

  + _____

3. *Utilizing the same QRS complexes that were utilized for number 2 above* count the number of millimeters that the Q and S (or if the QRS complex is inverted utilize the R wave) waves project below the isoelectric point. Do this for several QRS complexes and record the average value in the space below.

  _ _____

4. Algebraically add the two values obtained above as demonstrated below and enter the value in the space, keeping the (–) sign if the sum is negative or the (+) sign if the sum is positive.
  (+59) + (–65) = (–6)

  Sum _____

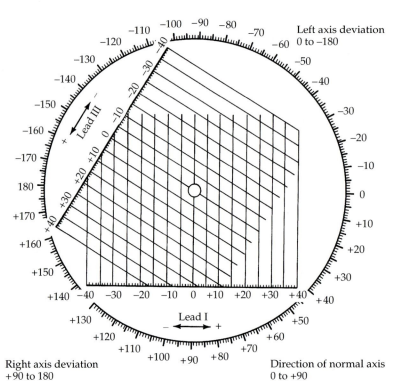

FIGURE 19.4 Grid chart showing the direction of the electrical axis of the heart as determined by a human electrocardiogram.

5. Record an ECG utilizing lead III (with the sensitivity at the same setting) from the same subject while he or she is reclining quietly.

6. Find several QRS complexes and repeat steps 3 through 5 above, recording the values in the spaces below.

+ _____

− _____

Sum _____

7. Utilizing a ruler or other straight edge mark a line on the grid chart in Section A of the LABORATORY REPORT RESULTS on the axis corresponding to the algebraic sum obtained from lead I.

8. Using a ruler or other straight edge mark a line on graph paper corresponding to the algebraic sum obtained from lead III.

9. Connect the two lines on the graph paper *making sure to pass through the center of the grid chart.* Record the mean electrical axis.

Mean electrical axis _____

B. CARDIAC CYCLE

In a normal *cardiac cycle* (heartbeat), the two atria contract while the two ventricles relax. Then, while the two ventricles contract, the two atria relax. The term *systole* (SIS-tō-lē; *systole* = contraction) refers to the phase of contraction; the phase of relaxation is *diastole* (dī-AS-tō-lē; *diastole* = expansion). A cardiac cycle consists of a systole and diastole of both atria plus a systole and diastole of both ventricles.

For the purposes of our discussion, we divide the cardiac cycle of a resting adult into three main phases (Figure 19.5).

1. *Relaxation period.* At the end of a heartbeat when the ventricles start to relax, all four chambers are in diastole. This is the beginning of the *relaxation* or *quiescent* (kwī-ES-ent) *period.* Repolarization of the ventricular muscle fibers (T wave in the ECG) initiates relaxation. As the ventricles relax, pressure within the chambers drops, and blood starts to flow from the pulmonary trunk and aorta back toward the ventricles. As this blood becomes trapped in the semilunar cusps, however, the valves close. Rebound of blood off the closed cusps produces a bump called the *dicrotic wave* on the aortic pressure curve.

With closing of the semilunar valves, there is a brief interval when ventricular blood volume does not change because both semilunar and AV valves are closed. This period is called *isovolumetric relaxation.* As the ventricles continue to relax, the space inside expands, and the pressure falls quickly. When ventricular pressure drops below atrial pressure, the AV valves open and ventricular filling begins.

2. *Ventricular filling.* The major part of ventricular filling occurs just after the AV valves open. Blood that has been flowing into the atria and building up while the ventricles were contracting now rushes into the ventricles. The first third of ventricular filling time thus is known as the period of *rapid ventricular filling.* During the middle third, called *diastasis,* a much smaller volume of blood flows into the ventricles.

Firing of the SA node results in atrial depolarization, noted as the P wave on the ECG. Atrial contraction follows the P wave, which also marks the end of the quiescent period. Atrial systole occurs in the last third of the ventricular filling period and accounts for the final 20–25 ml of the blood that fills the ventricles. At the end of ventricular diastole, there are about 130 ml in each ventricle. This volume of blood is called *end-diastolic volume (EDV).* Since atrial systole contributes only 20–30% of the total blood volume in the ventricles, atrial contraction is not absolutely necessary for adequate blood flow at normal heart rates. Throughout the period of ventricular filling, the AV valves are open and the semilunar valves are closed.

3. *Ventricular systole (contraction).* Near the end of atrial systole, the impulse from the SA node has passed through the AV node and into the ventricles, causing them to depolarize. This is represented by the QRS complex in the ECG. Then, ventricular contraction begins, and blood is pushed up against the AV valves, forcing them shut. For about 0.05 sec (50 msec), all four valves are closed again. This period is called *isovolumetric contraction.* During this time cardiac muscle fibers are contracting and exerting force, but are not yet shortening because it is very difficult to compress any liquid, including blood. Thus the muscle contraction is isometric (same length). Moreover, since

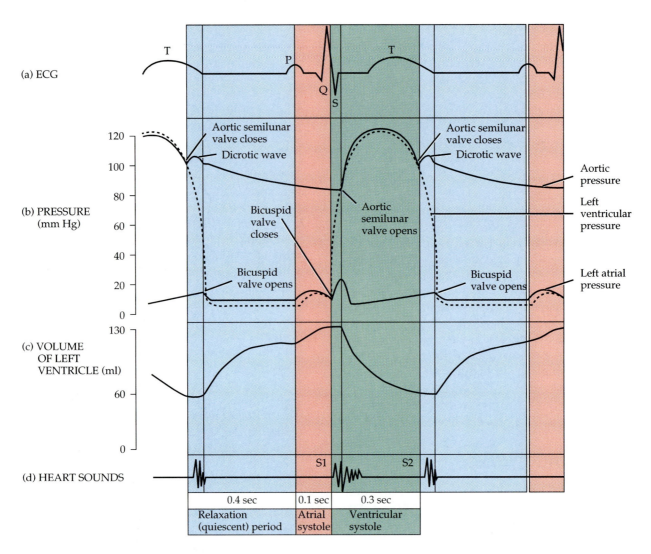

(a) ECG

(b) PRESSURE (mm Hg)

(c) VOLUME OF LEFT VENTRICLE (ml)

(d) HEART SOUNDS

FIGURE 19.5 Cardiac cycle. (a) ECG related to the cardiac cycle; (b) left atrial, left ventricular, and aortic pressure changes along with the opening and closing of the valves during cardiac cycle; (c) left ventricular volume during the cardiac cycle; (d) heart sounds related to the cardiac cycle.

there is no escape route for the blood, ventricular volume remains the same (isovolumetric).

As ventricular contraction continues, pressure inside the chambers rises sharply. When left ventricular pressure surpasses aortic pressure (about 80 mm Hg) and right ventricular pressure rises above the pressure in the pulmonary trunk (about 15–20 mm Hg), both semilunar valves open, and ejection of blood from the heart begins. This period is called *ventricular ejection* and lasts for about 0.25 sec (250 msec), until the ventricles start to relax. Then, the semilunar valves close and another relaxation period begins. The volume of blood still left in a ventricle following its systole is called *end-systolic volume (ESV)*. At rest, it is about 60 ml.

Since resting heart rate (HR) is about 75 beats/min, each cardiac cycle lasts about 0.8 sec (Figure 19.5). During the first 0.4 sec of the cycle, the relaxation period, all four chambers are in diastole. For the first part of the relaxation period, all valves are closed; during the latter part, the atrioventricular valves open and blood starts draining into the ventricles. During the next 0.1 sec, the atria contract and the atrioventricular valves are open, but the ventricles are still relaxed and the semilunar valves are closed. For the next 0.3 sec, the atria are relaxing and the ventricles are contracting. During the first part of this period, all valves are closed (isovolumetric contraction); during the second part, the semilunar valves are open (ventricular ejection). As the heart beats faster, the relaxation period (diastole) becomes shorter and shorter whereas the duration of the contraction period (systole) shortens only slightly.

C. CARDIAC CYCLE EXPERIMENTS

Note: Depending on the availability of live animals and the type of laboratory equipment, you may select from the following procedures related to heart experiments.

PROCEDURE USING PHYSIOFINDER

PHYSIOFINDER is an interactive computer program that permits laboratory simulations that do not involve the use of animals or advanced or expensive laboratory equipment. It permits students to perform experiments, analyze data, draw conclusions, and form hypotheses on the basis of collected data. The program is available from HarperCollins Publishers (1-800-8HEART1).

For activities related to the cardiac cycle, select the appropriate experiments from PHYSIOFINDER Module 4—Control of the Heart and Module 5—Cardiovascular Responses to Exercise.

PROCEDURE USING LIVE TURTLE (OR FROG)

Before beginning these experiments, review the procedure for pithing in Exercise 9, Section G.1 and proper use of the polygraph and stimulator in Exercise 9, Section G.2.

1. Observation of Cardiac Cycle

CAUTION! *Please reread Section D, "Precautions Related to Dissection" at the beginning of the laboratory manual on page xiii before you begin your dissection.*

PROCEDURE

Note: It is strongly suggested that either your instructor or a laboratory assistant perform the procedure for pithing a frog or turtle. The procedure requires some practice and may prove difficult for inexperienced individuals.

1. Obtain a double pithed turtle (or frog) from your instructor. The turtle heart is shown in Figure 19.6. Both the turtle and frog hearts

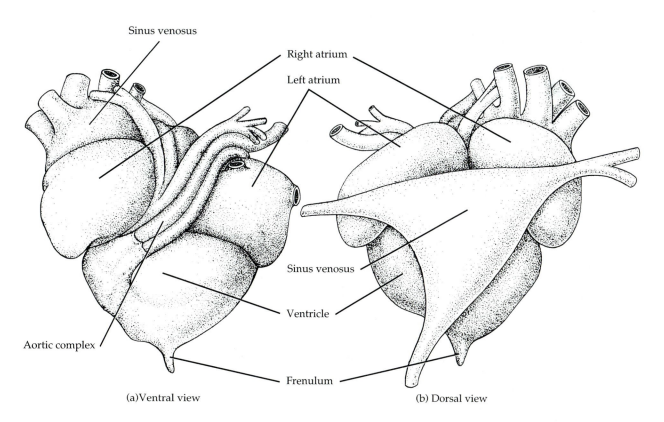

Sinus venosus

Right atrium

Left atrium

Sinus venosus

Ventricle

Aortic complex

Frenulum

(a) Ventral view

(b) Dorsal view

FIGURE 19.6 Turtle heart.

differ from the mammalian heart anatomically in that they have only a single ventricle from which all blood exits. The large sinus venosus on the dorsal surface serves as the pacemaker.

2. Fasten the specimen to a board ventral side up, and expose the heart by cutting through the body wall of the thoracic region.

Turtle	Frog
Using a saw and scalpel, remove the ventral shell and associated musculature. Carefully remove the mediastinum and pericardial sac.	Make a midline incision from the abdomen to the jaw and extend the incision laterally. Pin the body flaps to the board.

3. *Throughout this exercise keep the heart moist with Krebs-Ringer's solution.*
4. Observe the sequence of contraction of the atria and the ventricle and determine the rate of contraction for 30 sec. Calculate the heart rate per minute and record it.

Heart rate = _____ beats/minute.

   Estimate the relationship of atrial and ventricular cycles by shading in a portion of the bars below to indicate systole and leaving the portion representing diastole unshaded.

Atrial [ ]
Ventricular [ ]

2. Recording the Cardiac Cycle
PROCEDURE

1. Turtle	Frog
Tie a thread ligature tightly around the cardiac frenulum at the ventricular apex (Figure 19.6) and cut the frenulum *distal* to the thread.	Tie a thread to a small metal hook and carefully insert the hook into the tip of the ventricular myocardium. *Do not* penetrate the chamber.

2. Connect the thread to a force transducer.
3. Keep just enough tension on the thread to have a force tracing on the polygraph.
4. Connect an atrium to a second force transducer by either inserting a small metal hook or bent pin through the atrium and connect-

ing it to the transducer with a thread, or by tying a thread ligature to the atrium and connecting it to the transducer. Again regulate the tension in a manner similar to that for the ventricle.

5. Record several cardiac cycles and paste a strip of the recording paper in the space provided in Section C.2 of the LABORATORY REPORT RESULTS at the end of the exercise. Label the atrial (smaller) and ventricular (larger) systolic waves. Calculate and record the heart rate per minute.

Heart rate = _____ beats/minute
Why do the ventricular waves have greater

amplitude than the atrial waves? _____

3. Intrinsic Control of the Cardiac Cycle

Intrinsic control of the cardiac cycle refers to control mechanisms arising totally within the heart, devoid of any external influence from nerves or hormones, or both. The major intrinsic control mechanism is *Starling's Law of the Heart,* and is attributed to a length-tension interaction between the actin and myosin fibers of the myocardium. Simply stated, Starling's Law says that the heart will pump out all of the blood it receives, within physiological limits.

PROCEDURE

1. *Throughout this exercise keep the heart moist with Krebs-Ringer's solution.*
2. Utilizing the same preparation as that from the previous procedures, place minimal tension on the ventricle by regulating the tension of the ligature tied to the cardiac frenulum.
3. While continuously recording, increase the tension of the ligature tied to the cardiac frenulum in stepwise increments by raising the height of the transducer on its stand. Allow several beats at each tension setting.
4. Graph the relationship between resting tension and developed tension in Section C.3 of the LABORATORY REPORT RESULTS at the end of the exercise and explain your observations. What normal physiological event is mimicked by artificially increasing tension on the heart?

_____

_____

4. Extrinsic Control of the Cardiac Cycle

Extrinsic control of the cardiac cycle is exerted through the influences of the autonomic nervous system and hormones of the adrenal medulla. The parasympathetic nervous system innervates the heart through the vagus (X) nerves, while the sympathetic nervous system innervates the heart through branches of the stellate and cervical ganglia.

The sympathetic branch of the autonomic nervous system enhances atrial and ventricular contractility. The effects of increased sympathetic nervous system stimulation are asymmetrical in that they do not affect both sides of the heart evenly. The cardiac sympathetic nerves originating from the left stellate and cervical ganglia exert a significantly greater effect upon ventricular contractility than do those from the right stellate and cervical ganglia. Sympathetic nervous activity enhances myocardial performance, in that it increases both the strength and speed of ventricular myocardial contraction.

The parasympathetic branch of the ANS inhibits SA nodal pacemaker activity, slows AV nodal conduction velocity, and depresses ventricular myocardial performance. Increased parasympathetic nervous system activity decreases both the strength of contraction and the speed of contraction in ventricular myocardial cells.

PROCEDURE

1. Utilizing the same preparation as that from the previous procedures, *keep the heart moist with Krebs-Ringer's solution.*
2. Use a medicine dropper and apply a 1:1000 solution of epinephrine to the external surface of the heart. Note the changes, if any, in the cardiac cycle.
3. Monitor any changes in the cardiac cycle for 2–3 min.
4. Using a medicine dropper, rinse with Krebs-Ringer's solution and allow several minutes for recovery before proceeding to the next section. Using a medicine dropper, *gently* aspirate any accumulated fluid.
5. Attach the tracing in the space provided in Section C.4.a of LABORATORY REPORT RESULTS at the end of the exercise and explain your observations.
6. Expose the vagus (X) nerve in the neck and pass a ligature beneath it but *do not tie the ligature.*

Turtle	Frog
The vagus (X) nerve lies deep in the neck muscles beside the carotid artery. Use dissection scissors to expose the muscles of the neck and blunt dissection with a glass probe to isolate the left or right nerve.	The vagus (X) nerve lies beside the external jugular vein. Use blunt dissection with a glass probe to separate the left or right nerve.

7. Because different preparations will have varying thresholds, set the electrical stimulator at a frequency of 25 per second and a duration of 2 milliseconds (msec) and determine what is the threshold voltage necessary to effect a noticeable change in the heart rate or contraction strength while continuously recording.
8. Attach the tracing in the space provided in Section C.4.b of the LABORATORY REPORT RESULTS at the end of the exercise and explain your observations.
9. After allowing several minutes for recovery, use a medicine dropper and apply a 1:1000 solution of acetylcholine to the surface of the heart.
10. Monitor the changes, if any, in the cardiac cycle for 2–3 min and then, using a medicine dropper, rinse with Krebs-Ringer's solution. Allow several minutes for recovery. Using a medicine dropper, *gently* aspirate any accumulated fluid.
11. Attach the tracing in the space provided in Section C.4.c of the LABORATORY REPORT RESULTS at the end of the exercise and explain your observations.
12. After several minutes for recovery, use a medicine dropper and apply a 1:1000 solution of atropine sulfate to the surface of the heart. Note any changes in the cardiac cycle.
13. Monitor any changes in the cardiac cycle for 2–3 min and then, using a medicine dropper, rinse with Krebs-Ringer's solution. Allow several minutes for recovery. Using a medicine dropper, *gently* aspirate any accumulated fluid.
14. Attach the tracing in the space provided in Section C.4.d of the LABORATORY REPORT RESULTS at the end of the exercise and explain your observations.
15. Using medicine droppers, reapply atropine sulfate and immediately follow it with an external application of 1:1000 acetylcholine solution.

16. Note the effects, if any, of acetylcholine after application of atropine.
17. Attach the tracing in the space provided in Section C.4.e of the LABORATORY REPORT RESULTS at the end of the exercise and explain your observations.
18. After another application of atropine sulfate with a medicine dropper, restimulate the right and left vagi, as previously outlined. Note the changes, if any.
19. Using a medicine dropper, rinse the heart with Krebs-Ringer's solution.
20. Attach the tracing in the space provided in Section C.4.f of the LABORATORY REPORT RESULTS at the end of the exercise and explain your observations.

Which of the solution(s) mimics the effect of

the sympathetic stimulation? _____

Of parasympathetic stimulation? _____

5. Demonstration of Refractory Period

The concept of refractory period in cardiac muscle is very similar to that discussed in conjunction with nerve action potential propagation and skeletal muscle contraction. Compared to skeletal muscle, cardiac muscle has a long action potential duration and therefore a long refractory period. The normal refractory period of ventricular muscle is between 0.25 and 0.3 sec, which is approximately the duration of the action potential. In addition, there is a relative refractory period of about 0.05 sec. During this period the ventricular muscle is more difficult to stimulate than normal, but nonetheless can be excited.

The refractory period of atrial muscle is considerably shorter, being only about 0.15 sec, and a relative refractory period of an additional 0.03 sec. This is due to the shorter length of the atrial action potential. As a result of the shorter refractory period in atrial muscle the rate of contraction of the atria can be considerably faster than that of the ventricles.

PROCEDURE

1. Use the frog or turtle heart preparation from the preceding exercises, and *remember to keep the heart moist with Krebs-Ringer's solution.*
2. Set the paper drive on the polygraph to its highest speed.
3. While continuously recording, stimulate the ventricle with a single stimulus of about 50

volts (V) for 2 msec. Make sure that both electrodes touch the ventricle.
4. Allowing several contractions between stimuli, stimulate the ventricle at the following five points of ventricular systole.

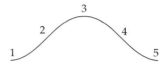

5. Note whether an extra systole occurs at any or all of the five points and, if so, whether it is followed by a compensatory pause.
6. Continuously stimulate the ventricle for 10 sec to see if you can elicit tetany.
7. Attach the tracing in the space provided in Section C.5 of the LABORATORY REPORT RESULTS at the end of the exercise and explain your observations.

D. HEART SOUNDS

The beating of a human heart usually produces four sounds, but only two may be detected with a *stethoscope.* (The remaining two sounds may be heard if adequately amplified electronically.) The first sound is created by blood turbulence associated with the closure of the atrioventricular valves soon after ventricular systole begins (See Figure 19.5d). This sound is the loudest and longest of the two sounds, and may be best heard over the apex of the heart. The sound produced by the tricuspid valve is best heard in the fifth intercostal space just lateral to the left border of the sternum, while the mitral valve sound is best heard in the fifth intercostal space at the apex of the heart (Figure 19.7).

The second heart sound is created by blood turbulence associated with closure of the semilunar valves at the beginning of ventricular diastole. This sound is of shorter duration and lower intensity, and has a more snapping sound as compared to the quality of the first heart sound. The second heart sound produced by the aortic semilunar valve closing is best heard in the second intercostal space to the right of the sternum. The second sound produced by the pulmonary semilunar valve is best heard in the second intercostal space just to the left of the sternum (Figure 19.7).

Heart sounds provide valuable information about the valves. Abnormal sounds made by the valves are termed *murmurs.* Some murmurs are caused by the noise made by blood flowing back

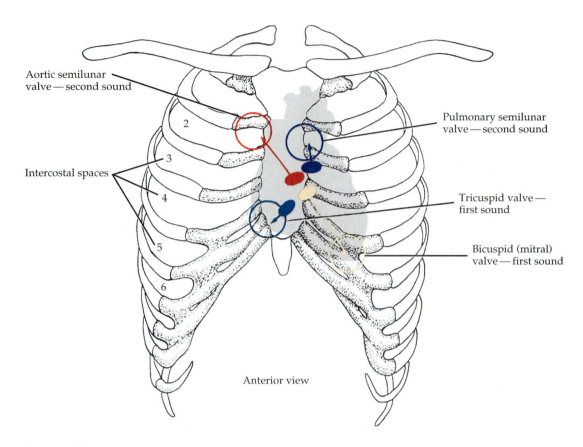

Aortic semilunar valve — second sound

Intercostal spaces

2

3

4

5

6

Pulmonary semilunar valve — second sound

Tricuspid valve — first sound

Bicuspid (mitral) valve — first sound

Anterior view

FIGURE 19.7   Surface areas where heart valve sounds are best heard.

into a chamber of the heart because of improper closure (incompetence) of a valve. Another murmur may be produced by the improper opening (incompetence) of an AV or semilunar valve, which restricts the flow of blood out of a cardiac chamber.

Murmurs do not always indicate that the valves are not functioning properly. Many individuals possess a *functional murmur* that has no clinical significance at all.

Listening to sounds of the body is called *auscultation* (aws-kul-TĀ-shun).

1. Use of Stethoscope

PROCEDURE

1. The stethoscope should be used in a quiet room.
2. The earpieces of the stethoscope *should be cleaned with alcohol* just before using and should also be pointed slightly forward when placed in the ears. They will be more comfortable in this position, and it will be easier to hear through them.

3. Listen to the heart sounds of your laboratory partner by placing the diaphragm of the stethoscope next to the skin at several positions on the chest wall illustrated in Figure 19.7.
4. The first sound is best heard at the apex of the heart (see above).
5. The second sound is best heard in the second intercostal space (see above).
6. **CAUTION!** *Assuming that your partner has no known or apparent cardiac or other health problems and is capable of such an activity, ask her/him to run in place about 25 steps. Listen to the heart sounds again.*
7. Answer all questions pertaining to Section D of the LABORATORY REPORT RESULTS at the end of the exercise.

E. PULSE RATE

During ventricular systole, a wave of pressure, called *pulse*, is produced in the arteries due to ventricular contraction. Pulse rate and heart rate are essentially the same. Pulse can be felt readily where an artery is near the surface of the skin and over the surface of a bone. Pulse rates vary considerably in individuals because of time of day, temperature, emotions, stress, and other factors. The normal adult pulse rate of the heart at rest is within a range of 72 to 80 beats per minute. With practice you can learn to take accurate pulse rates by counting the beats per 15 sec and multiplying by 4 to obtain beats per minute. The term *tachycardia* (tak'-e-KAR-dē-a) is applied to a rapid heart rate or pulse rate (over 100 beats/minute). *Bradycardia* (brād-e-KAR-dē-a) indicates a slow heart rate or pulse rate (below 50 beats/ minute).

Although the pulse may be detected in most surface arteries, the pulse rate is usually determined on the *radial artery* of the wrist.

1. Radial Pulse
PROCEDURE

1. Using your index and middle finger palpate your laboratory partner's radial artery.
2. The thumb should never be used because it has its own prominent pulse.
3. Palpate the area behind your partner's thumb just inside the bony prominence on the lateral aspect of the wrist.
4. **CAUTION!** *Do not apply too much pressure.*
5. Count the pulse, change positions, and record your results in Section E of the LABORATORY REPORT RESULTS at the end of the exercise.
6. Calculate the *median pulse rate* of the entire class.

2. Carotid Pulse
PROCEDURE

1. Using the same fingers that you used for the radial pulse, place them on either side of your partner's larynx (voice box).
2. *Gently* press downward and toward the back until you feel the pulse. You must feel the pulse clearly with at least two fingers, so adjust your hand accordingly.
3. The radial and carotid pulse can be compared under the following conditions: (a) sitting quietly, (b) standing quietly, (c) right after walking 60 steps, (d) right after running in place 60 steps, *assuming that your partner has no known or apparent cardiac or other health problems and is capable of such an activity.* Notice how long it takes the pulse to return to normal after the walking and running exercises.

4. Record your results in Section E of the LABORATORY REPORT RESULTS at the end of the exercise.
5. Compare your radial and carotid pulse in the table provided.

F. BLOOD PRESSURE (AUSCULTATION METHOD)

In physiological terms, the term *blood pressure* actually refers to the interaction of several different pressures (the pressure only within the arteries, or arterial pressure; the pressure only within the veins, or venous pressure; the pressure within the pulmonary system, or pulmonary pressure; and the pressure within all of the other vascular beds, termed systemic pressure). Clinically, however, the term *blood pressure* only refers to the pressure within the large arteries.

Arterial pressure can be measured either directly, by the insertion of a needle or catheter directly into the artery in such a way that the needle is pointing "upstream," or against the flow of blood within the vessel, or indirectly, by an instrument called a *sphygmomanometer* (sfig'-mō-ma-NOM-e-ter; *sphygmo* = pulse). Regardless of which method is utilized, blood pressure is recorded in millimeters of mercury (mm Hg) and is normally taken in the brachial artery.

A commonly used sphygmomanometer consists of an inflatable rubber cuff attached by a rubber tube to a compressible hand pump or bulb. Another tube attaches to a cuff and to a mercury column marked off in millimeters or an anaeroid gauge that measures the pressure in millimeters of mercury (mm Hg). The highest pressure in the artery, occurring during ventricular systole, is termed *systolic blood pressure.* The lowest pressure, occurring during ventricular diastole, is termed *diastolic blood pressure.* Blood pressures are usually expressed as a ratio of systolic to diastolic. The average blood pressure of a young adult is about 120 mm Hg systolic and 80 mm Hg diastolic, abbreviated to 120/80. The difference between systolic and diastolic pressure is called *pulse pressure.* Pulse pressure may be used clinically to indicate several physiological and pathological parameters, and usually averages 40 mm Hg in a healthy individual. The ratio of systolic pressure to diastolic pressure to pulse pressure may also be utilized clinically as a diagnostic tool, and is usually 3:2:1.

This exercise employs the indirect ***ausculatory method***, in which the sounds of blood flow are heard with a stethoscope. Blood flow in an artery is impeded by increasing pressure within a sphygmomanometer. When the cuff of the sphygmomanometer applies sufficient pressure to completely occlude blood flow, no sounds can be heard distal to the cuff because no blood can flow through the artery. When cuff pressure drops below the maximal (systolic) pressure in the artery, blood is heard passing through the vessel. When cuff pressure drops below the lowest (diastolic) pressure in the vessel, the sound becomes muffled and usually disappears. The sounds heard through the stethoscope via this procedure are termed *Korotkoff* (kō-ROT-kof) *sounds*.

Indirect blood pressure can be taken in any artery that can be occluded easily. The brachial artery has the advantage of being at approximately the same level as the heart, so brachial pressure closely reflects aortic pressure.

The procedure for determining blood pressure using the sphygmomanometer is as follows:

PROCEDURE

1. Either you or your laboratory partner should be comfortably seated, at ease, with your arm bared, slightly flexed, abducted, and perfectly relaxed. You may, for convenience, rest the forearm on a table in the supinated position.
2. Wrap the deflated cuff of the sphygmomanometer around the arm with the lower edge about 2.54 cm (1 in.) above the antecubital space. Close the valve on the neck of the rubber bulb.
3. Clean the earpieces of the stethoscope with alcohol before using it. Using the diaphragm of the stethoscope, find the pulse in the brachial artery just above the bend of the elbow, on the inner margin of the biceps brachii muscle.
4. Inflate the cuff by squeezing the bulb until the air pressure within it just exceeds 170 mm Hg. At this point the wall of the brachial artery is compressed tightly, and no blood should be able to flow through.
5. Place the diaphragm of the stethoscope firmly over the brachial artery and while watching the pressure gauge, slowly turn the valve, releasing air from the cuff. Listen carefully for Korotkoff's sounds as you watch the pressure fall. The first loud, rapping sound you hear will be the *systolic pressure*.
6. Continue listening as the pressure falls. The pressure recorded on the mercury column

when the sounds become faint or disappear is the *diastolic pressure* reading. It measures the force of blood in arteries during ventricular relaxation and specifically reflects the peripheral resistance of the arteries.

7. Repeat this procedure for both readings two or three times to see if you get consistent results. *Allow a few minutes between readings.* Record all results in Section F of the LABORATORY REPORT RESULTS at the end of the exercise.
8. Have a partner stand and record his or her blood pressure several times for each arm. Record all results in the table provided in Section F of the LABORATORY REPORT RESULTS at the end of the exercise.
9. *Now, assuming that your partner has no known or apparent cardiac or other health problems, and is capable of such an activity* have your partner do some exercise, such as running in place 40 or 50 steps, and measure the blood pressure again *immediately after the completion of the exercise.* Record the pulse pressure in the table provided in Section F of the LABORATORY REPORT RESULTS at the end of the exercise.

G. OBSERVING BLOOD FLOW

Note: Depending on the availability of live animals and the type of laboratory equipment, you may select from the following procedures related to observing blood flow.

PROCEDURE USING PHYSIOFINDER

PHYSIOFINDER is an interactive computer program that permits laboratory simulations that do not involve the use of animals or advanced or expensive laboratory equipment. It permits students to perform experiments, analyze data, draw conclusions, and form hypotheses on the basis of collected data.

For activities related to observing blood flow, select the appropriate experiments from PHYSIOFINDER Module 6—Control of Microcirculation.

PROCEDURE USING LIVE FROG

This experiment may be performed on either an intact or pithed frog. Review the proper use of a microscope in Exercise 1, Sections A.1 through A.4, and if the experiment is to be performed on a pithed frog review the procedures for pithing in Exercise 9, Section G.1.

Note: It is strongly suggested that either your instructor or a laboratory assistant perform the procedure for pithing a frog. The procedure requires some practice and may prove difficult for inexperienced individuals.

PROCEDURE

1. Strap a frog to a frog board in the prone position by wrapping a piece of moist cloth around the body, head, and legs and securing with several rubber bands. Be sure not to secure the animal too tightly, thereby causing internal injury.
2. Pin the webbing of one foot over the hole in the frog board and secure the frog board to the stage of the microscope with rubber bands in such a way that the hole in the board is located directly over the hole in the microscope stage.
3. Continuously moisten the restraining cloth and foot webbing with water throughout the entire experiment.

1. Control Observations

PROCEDURE

1. Using an adequate amount of light, observe the blood flow through the vasculature of the frog's foot with either the 10× or 20× objective.
2. Identify an arteriole by its characteristic pulsating, rapid flow of blood.
3. A venule can be characterized by a slower, more constant blood flow, while a capillary may be recognized by its small diameter and slow blood flow. Red blood cells usually pass through a capillary in single file, thereby aiding in the vessel's identification.
4. Attempt to locate the junction of an arteriole and capillary (metarterioles cannot be distinguished readily in such a preparation) in the vascular bed. Although the precapillary sphincter probably will not be visible, one should be able to observe the irregular blood flow from the arteriole into the capillary due to the activity of the precapillary sphincter.
5. Make a drawing of the vascular bed visible in your microscope and label the various vessels in the space provided in Section G.1 of the LABORATORY REPORT RESULTS at the end of the exercise.

2. Effect of Histamine

PROCEDURE

1. Dry the webbing of the foot by gently blotting with paper towel.
2. *While observing the peripheral blood flow through the microscope,* use a medicine dropper to add several drops of 1:10,000 histamine solution and note the changes, if any, in peripheral blood flow.
3. Record your observations in Section G.2 of the LABORATORY REPORT RESULTS at the end of the exercise.

3. Effect of Antihistamines

PROCEDURE

1. After rinsing the foot several times with plain water, gently blot dry and add several drops of antihistamine solution to the webbing of the foot *while observing the peripheral blood flow through the microscope.*
2. *Without rinsing the foot,* use a medicine dropper and add several drops of 1:10,000 histamine solution and note the changes, if any, in peripheral blood flow.
3. Record and explain your observations in Section G.3 of the LABORATORY REPORT RESULTS at the end of the exercise.

4. Effect of Epinephrine

PROCEDURE

1. After rinsing the foot several times with plain water, gently blot dry and use a medicine dropper to add several drops of 1:1000 epinephrine solution to the webbing of the foot *while observing the peripheral blood flow through the microscope.*
2. Record and explain your observations in Section G.4 of the LABORATORY REPORT RESULTS at the end of the exercise.

ANSWER THE LABORATORY REPORT QUESTIONS AT THE END OF THE EXERCISE.

Cardiovascular Physiology 19

Student _____   **Date** _____

Laboratory Section _____   **Score/Grade** _____

SECTION A. CARDIAC CONDUCTION SYSTEM
AND ELECTROCARDIOGRAM (ECG OR EKG)

Attach examples of the electrocardiogram strips you obtained.

LEAD I
_____

Electrocardiogram Strip

LEAD II
_____

Electrocardiogram Strip

LEAD III
_____

Electrocardiogram Strip
_____

SECTION C. CARDIAC CYCLE EXPERIMENTS

2. Recording the cardiac cycle:

3. Intrinsic control of the cardiac cycle:

4. Extrinsic control of the cardiac cycle:
   a. Application of epinephrine:

   b. Vagal stimulation:

   c. Application of acetylcholine:

d. Application of atropine sulfate:

e. Application of atropine sulfate immediately followed by acetylcholine:

f. Application of atropine sulfate and vagal stimulation:

5. Refractory period:

SECTION D. HEART SOUNDS

1. Which heart sound is the loudest? _____

2. Did you hear a third sound? _____

3. Where does the first sound originate? _____

4. Where does the second sound originate? _____

5. After you exercised, how did the heart sounds differ from before? _____

6. Did they differ in rate and intensity? _____

7. Did the first or second increase in loudness? _____

SECTION E. PULSE RATE

1. Radial pulse rate count results: _____ pulses per minute.
2. Have all radial pulse rates put on the blackboard, arranging them from the highest to the lowest. The median pulse rate is found exactly half-way down from the top.

What is the median radial pulse rate of the class? _____

What was the highest rate? _____

What was the lowest rate? _____
3. Compare your radial and carotid pulse rates by filling in the following table.

	Radial pulse rate	Carotid pulse rate
Sitting quietly		
Standing quietly		
After walking		
After running in place		

SECTION F. BLOOD PRESSURE (AUSCULTATION METHOD)

Record your systolic and diastolic blood pressures in the following table.

	Systolic pressure		Diastolic pressure		Pulse pressure
	Left arm	Right arm	Left arm	Right arm	
Sitting					
Standing					
After running					

SECTION G. OBSERVING BLOOD FLOW

1. Draw a diagram indicating the vascular bed observed during your control observations.

2. Indicate whether blood flow increased, decreased, or remained unchanged following addition of histamine. Explain the physiological mechanism for your observations. _____

_____

3. a. Indicate how blood flow changed, if at all, following addition of antihistamine solution to the webbing of the foot. Explain the physiological mechanism for your observation. _____

_____

   b. Indicate how blood flow changed following the addition of the histamine solution while the antihistamine solution was still in place and indicate a possible physiological mechanism for antihistamine

_____

4. Indicate how blood flow changed, if at all, following the addition of epinephrine solution to the webbing of the frog's foot and give a physiological mechanism for your observation. _____

_____

   Also indicate how the autonomic nervous system could give rise to a physiological condition opposite to that observed in your experiment. _____

_____

Cardiovascular Physiology 19

Student _____   Date _____

Laboratory Section _____   Score/Grade _____

PART 1. Multiple Choice

_____ 1. When the semilunar valves are open during a cardiac cycle, which of the following occur?

I—atrioventricular valves are closed;
II—ventricles are in systole;
III—ventricles are in diastole;
IV—blood enters the aorta;
V—blood enters the pulmonary trunk;
VI—atrial contraction.
(a) I, II, IV, and V (b) I, II, and VI (c) II, IV, and V (d) I, III, IV, and VI

_____ 2. When ventricular pressure drops below atrial pressure, which phase of the cardiac cycle occurs? (a) ventricular filling (b) ventricular systole (c) isovolumetric contraction (d) isovolumetric relaxation

_____ 3. During which phase of the cardiac cycle are all four chambers in diastole? (a) ventricular filling (b) relaxation (c) isovolumetric contraction (d) isovolumetric relaxation

_____ 4. Which of the following statements is *not true?* (a) Pulse pressure is the difference between systolic and diastolic blood pressures. (b) Both systolic and diastolic blood pressures can be obtained via the pulse method. (c) Systolic blood pressure is obtained at the first loud, rapping sound you hear when you are measuring blood pressure via the ausculatory method. (d) Diastolic blood pressure is obtained when the sound heard through the stethoscope when you are measuring blood pressure via the ausculatory method becomes muffled and usually disappears.

_____ 5. The two distinct heart sounds, described phonetically as lubb and dupp, represent (a) contraction of the ventricles and relaxation of the atria (b) contraction of the atria and relaxation of the ventricles (c) blood turbulence associated with closing of the atrioventricular and semilunar valves (d) surging of blood into the pulmonary artery and aorta.

PART 2. Completion

6. Systole and diastole of both atria plus systole and diastole of both ventricles is called

_____.

7. Blood flow through the heart is controlled by speed of the cardiac cycle, venous return to the heart, opening and closing of the valves, and _____.

8. Heart sounds provide valuable information about the _____.

9. Abnormal or peculiar heart sounds are called _____.

10. Although the pulse may be detected in most surface arteries, pulse rate is usually determined on the

_____.

11. The heart has an intrinsic regulating system called the cardiac _____ system.

12. Electrical impulses accompanying the cardiac cycle are recorded by the _____.

13. The typical ECG produces three clearly recognizable waves. The first wave, which indicates depolar-

ization of the atria, is called the _____.

14. Various up-and-down impulses produced by an ECG are called _____.

15. The instrument normally used to measure blood pressure is called a(n) _____.

16. The artery that is normally used to evaluate blood pressure is the _____.

17. Rapping or thumping sounds heard clinically when blood pressure is being taken are called

_____ sounds.

18. The difference between systolic and diastolic pressure is called _____.

19. An average blood pressure value for an adult is _____.

20. An average pulse pressure is _____.

Lymphatic System 20

The *lymphatic* (lim-FAT-ik) *system* is composed of a pale yellow fluid called lymph, vessels that transport lymph called lymphatic vessels (lymphatics), a number of structures and organs that contain lymphatic tissue, and red bone marrow, which is the site of lymphocyte production. Lymphatic tissue is a specialized form of reticular connective tissue that contains large numbers of lymphocytes.

A. LYMPHATIC VESSELS

Lymphatic vessels originate as microscopic *lymph capillaries* in spaces between cells. They are found in most parts of the body; they are absent in avas-cular tissue, the central nervous system, splenic pulp, and bone marrow. They are slightly larger than blood capillaries and have a unique structure that permits interstitial fluid to flow into them but not out. Lymph capillaries also differ from blood capillaries in that they end blindly; blood capillaries have an arterial and a venous end. In addition, lymph capillaries are structurally adapted to ensure the return of proteins to the cardiovascular system when they leak out of blood capillaries.

Just as blood capillaries converge to form venules and veins, lymph capillaries unite to form larger and larger lymph vessels called *lymphatic vessels* (Figure 20.1). Lymphatic vessels resemble veins in structure, but have thinner walls and more

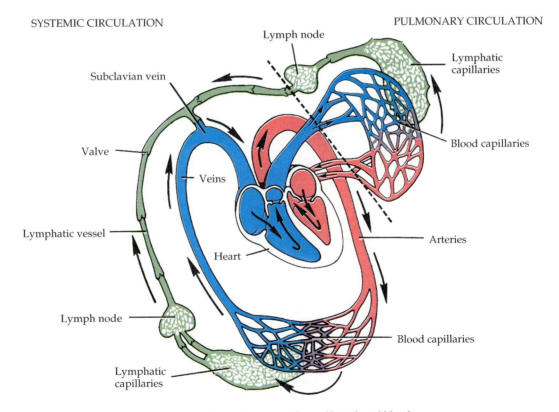

Arrows show direction of flow of lymph and blood

FIGURE 20.1 Relationship of lymphatic system to cardiovascular system.

477

valves, and contain lymph nodes at various intervals along their length (Figure 20.1). Ultimately, lymphatic vessels deliver lymph into two main channels—the thoracic duct and the right lymphatic duct. These will be described shortly.

Lymphangiography (lim-fan'-jē-OG-ra-fē) is the x-ray examination of lymphatic vessels and lymph organs after they are filled with a radiopaque substance. Such an x-ray is called a *lymphangiogram* (lim-FAN-jē-ō-gram). Lymphangiograms are useful in detecting edema and carcinomas, and in locating lymph nodes for surgical and radiotherapeutic treatment.

B. LYMPHATIC TISSUE

1. Lymph Nodes

The oval or bean-shaped structures located along the length of lymphatic vessels are called *lymph nodes*. A lymph node contains a slight depression on one side called a *hilus* (HĪ-lus), where blood vessels and efferent lymphatic vessels leave the node. Each node is covered by a *capsule* of dense connective tissue that extends into the node. The capsular extensions are called *trabeculae* (tra-BEK-yoo;-lē; *trabecula* = little beam). Internal to the capsule is a supporting network of reticular fibers and fibroblasts. The capsule, trabeculae, and reticular fibers and fibroblasts constitute the stroma (framework) of a lymph node. The parenchyma (functioning part) of a lymph node is specialized into two regions: cortex and medulla. The outer *cortex* contains densely packed lymphocytes arranged in masses called *lymphatic follicles*. The outer rim of each follicle contains T cells. The central area of each follicle, called the *germinal center*, is the site where B cells proliferate into antibody-secreting plasma cells. The inner region of a lymph node is called the *medulla*. In the medulla, the lymphocytes are arranged in strands called *medullary cords*. These cords also contain macrophages and plasma cells.

Lymph flows through a node in one direction. It enters through *afferent* (*ad* = to; *ferre* = to carry) *lymphatic vessels* that enter the convex surface of the node at several points. They contain valves that open toward the node so that the lymph is directed *inward*. Once inside the node, the lymph enters the sinuses, which are a series of irregular channels. Lymph from the afferent lymphatic vessels enters the *cortical sinuses* just inside the capsule. From here it circulates to the *medullary sinuses* between the medullary cords. From these sinuses the lymph usually circulates into one or

two *efferent* (*ex* = away) *lymphatic vessels*, located at the hilus of the lymph node. Efferent lymphatic vessels are wider than afferent vessels and contain valves that open away from the node to convey lymph *out* of the node.

Lymph passing from tissue spaces through lymphatic vessels on its way back to the cardiovascular system is filtered through lymph nodes. As lymph passes through the nodes it is filtered of foreign substances. These substances are trapped by the reticular fibers within the node. Then, macrophages destroy some foreign substances by phagocytosis and lymphocytes bring about destruction of others by immune responses. Plasma cells and T cells that proliferate within lymph nodes can circulate to other parts of the body.

Label the parts of a lymph node in Figure 20.2 and the various groups of lymph nodes in Figure 20.3.

2. Tonsils

Tonsils are multiple aggregations of large lymphatic nodules embedded in a mucous membrane. The tonsils are arranged in a ring at the junction of the oral cavity and pharynx. The tonsils are situated strategically to protect against invasion of foreign substances (Figure 20.3a). The *pharyngeal* (fa-RIN-jē-al) *tonsil* is embedded in the posterior wall of the nasopharynx. The paired *palatine* (PAL-a-tīn) *tonsils* are situated in the space between the pharyngopalatine and palatoglossal arches. They are the ones commonly removed by a tonsillectomy. The *lingual* (LIN-gwal) *tonsil* is located at the base of the tongue and may also have to be removed by a tonsillectomy. Functionally, the tonsils participate in immune responses by producing lymphocytes and antibodies.

3. Spleen

The oval *spleen* is the largest mass of lymphatic tissue in the body (Figure 20.3a). It is situated in the left hypochondriac region between the fundus of the stomach and diaphragm.

The splenic artery and vein and the efferent lymphatic vessels pass through the hilus. Since the spleen has no afferent lymphatic vessels or lymph sinuses, it does not filter lymph. One key splenic function related to immunity is the production of B cells, which develop into anti-body-producing plasma cells. The spleen also phagocytizes bacteria and worn-out and damaged red blood cells and platelets. During early fetal development, the spleen participates in blood cell formation.

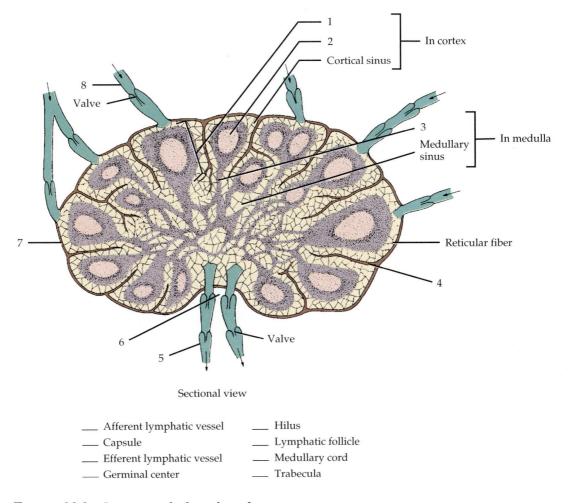

1

2

Cortical sinus

In cortex

8

Valve

3

Medullary
sinus

In medulla

7

Reticular fiber

4

6

Valve

5

Sectional view

___ Afferent lymphatic vessel   ___ Hilus
___ Capsule   ___ Lymphatic follicle
___ Efferent lymphatic vessel   ___ Medullary cord
___ Germinal center   ___ Trabecula

FIGURE 20.2   Structure of a lymph node.

4. Thymus Gland

Usually a bilobed lymphatic organ, the ***thymus
gland*** is located in the mediastinum, posterior to the
sternum and between the lungs (Figure 20.3a). Its
role in immunity is to synthesize hormones that
help produce T cells that destroy invading microbes,
including the AIDS (acquired immune deficiency
syndrome) virus, directly or indirectly by producing
various substances.

C. LYMPH CIRCULATION

When plasma is filtered by blood capillaries, it
passes into the interstitial spaces; it is then known
as interstitial fluid. When this fluid passes from
interstitial spaces into lymph capillaries, it is
called ***lymph*** (*lympha* = clear water). Lymph from
lymph capillaries flows into lymphatic vessels that
run toward lymph nodes. At the nodes, afferent
vessels penetrate the capsules at numerous points,

and the lymph passes through the sinuses of the
nodes. Efferent vessels from the nodes unite to
form ***lymph trunks***.

The principal trunks pass their lymph into two
main channels, the thoracic duct and the right lymphatic duct. The ***thoracic (left lymphatic) duct*** begins
as a dilation in front of the second lumbar vertebra
called the ***cisterna chyli*** (sis-TER-na KĪ-lē). The thoracic duct is the main collecting duct of the lymphatic
system and receives lymph from the left side of the
head, neck, and chest, the left upper limb, and the
entire body inferior to the ribs (Figure 20.3b).

The ***right lymphatic duct*** drains lymph from
the upper right side of the body (Figure 20.3b).
Ultimately, the thoracic duct empties all of its
lymph into the junction of the left internal jugular
vein and left subclavian vein, and the right lymphatic duct empties all of its lymph into the junction of the right internal jugular vein and right
subclavian vein. Thus, lymph is drained back into
the blood and the cycle repeats itself continuously.

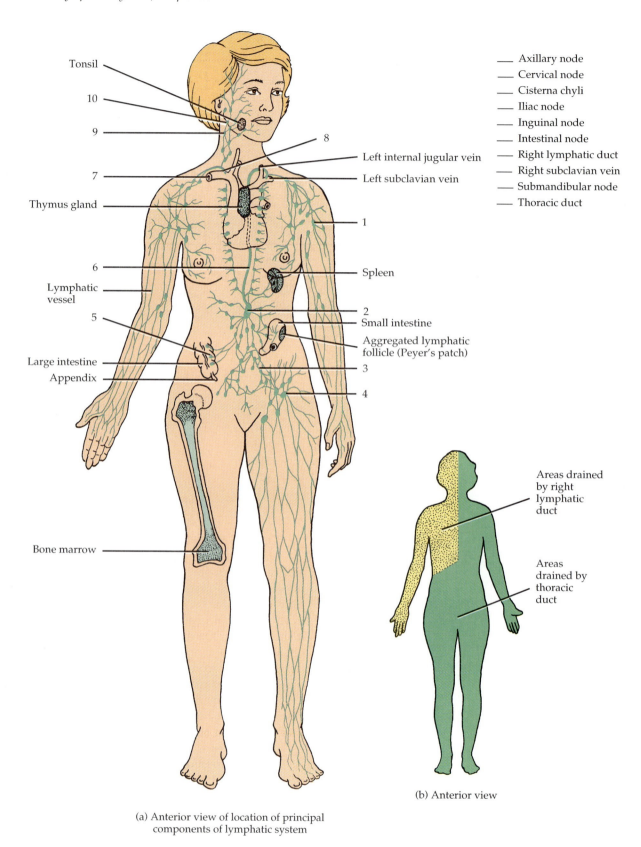

Tonsil

10

9

7

Thymus gland

6

Lymphatic vessel

5

Large intestine

Appendix

Bone marrow

8

Left internal jugular vein

Left subclavian vein

1

Spleen

2
Small intestine
Aggregated lymphatic follicle (Peyer's patch)

3

4

_____ Axillary node
_____ Cervical node
_____ Cisterna chyli
_____ Iliac node
_____ Inguinal node
_____ Intestinal node
_____ Right lymphatic duct
_____ Right subclavian vein
_____ Submandibular node
_____ Thoracic duct

Areas drained by right lymphatic duct

Areas drained by thoracic duct

(b) Anterior view

(a) Anterior view of location of principal components of lymphatic system

FIGURE 20.3   Lymphatic system.

Edema, an excessive accumulation of interstitial fluid in tissue spaces, may be caused by an obstruction, such as an infected node or a blockage of vessels, in the pathway between the lymphatic capillaries and the subclavian veins. Another cause is excessive lymph formation and increased permeability of blood capillary walls. A rise in capillary blood pressure, in which interstitial fluid is formed faster than it is passed into lymphatic vessels, also may result in edema.

D. DISSECTION OF CAT LYMPHATIC SYSTEM

CAUTION! *Please reread Section D, "Precautions Related to Dissection" at the beginning of the laboratory manual on page xiii before you begin your dissection.*

Even if your cat has not been injected for examination of the lymphatic system, certain parts can be observed.

1. *Lymphatic vessels* The major lymphatic vessel is the *thoracic duct*, a brownish vessel in the left pleural cavity dorsal to the aorta. It passes deep to most of the blood vessels at the base of the neck and enters the left external jugular vein near the entrance of the internal jugular vein (Figure 20.4). Using the diagram as a guide, see if you can find the *right lymphatic duct*.
2. *Lymph nodes* These are small ovoid or round masses of lymphatic tissue connected by lymph vessels. You may be able to identify lymph nodes in the axillary and inguinal regions, at the base of the neck, at the angle of the jaws, and in the mesentery of the small intestine (Figure 20.4).
3. *Lymphatic organs* Lymphatic organs include the *thymus gland, tonsils,* and *spleen*.

ANSWER THE LABORATORY REPORT QUESTIONS AT THE END OF THE EXERCISE.

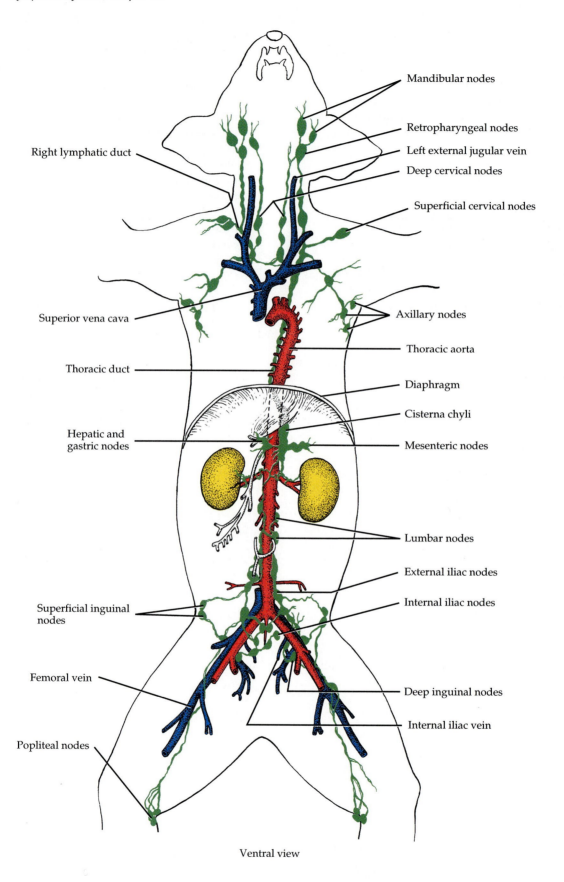

Mandibular nodes

Retropharyngeal nodes

Left external jugular vein

Deep cervical nodes

Superficial cervical nodes

Right lymphatic duct

Superior vena cava

Axillary nodes

Thoracic aorta

Thoracic duct

Diaphragm

Cisterna chyli

Hepatic and gastric nodes

Mesenteric nodes

Lumbar nodes

External iliac nodes

Internal iliac nodes

Superficial inguinal nodes

Femoral vein

Deep inguinal nodes

Internal iliac vein

Popliteal nodes

Ventral view

FIGURE 20.4 Lymphatic system of cat.

Lymphatic System 20

Student _____   **Date** _____

Laboratory Section _____   **Score/Grade** _____

PART 1. Completion

1. Small masses of lymphatic tissue located along the length of the lymphatic vessels are called

   _____.

2. Lymphatic vessels have thinner walls than veins, but resemble veins in that they also have

   _____.

3. All lymphatic vessels converge, get larger, and eventually merge into two main channels, the thoracic

   duct and the _____.

4. Cells in the lymph nodes that carry on phagocytosis are _____.

5. Lymph is conveyed out of lymph nodes in _____ vessels.

6. B cells in lymph nodes produce certain cells that are responsible for the production of antibodies.

   These cells are called _____.

7. The x-ray examination of lymphatic vessels and lymph organs after they are filled with a radiopaque

   substance is called _____.

8. This x-ray examination is useful in detecting edema and _____.

9. The largest mass of lymphatic tissue is the _____.

10. The main collecting duct of the lymphatic system is the _____ duct.

11. _____ are the areas within a lymph node that contain T and B cells.

12. The _____ tonsils are commonly removed by a tonsillectomy.

Respiratory System *21*

Cells continually use oxygen (O_2) for the metabolic reactions that release energy from nutrient molecules and produce ATP. At the same time, these reactions release carbon dioxide (CO_2). Since an excessive amount of CO_2 produces acidity that is toxic to cells, the excess CO_2 must be eliminated quickly and efficiently. The two systems that cooperate to supply O_2 and eliminate CO_2 are the cardiovascular system and the respiratory system. The respiratory system provides for gas exchange, intake of O_2, and elimination of CO_2, whereas the cardiovascular system transports the gases in the blood between the lungs and the cells. Failure of either system has the same effect on the body: disruption of homeostasis and rapid death of cells from oxygen starvation and buildup of waste products. In addition to functioning in gas exchange, the respiratory system also contains receptors for the sense of smell, filters inspired air, produces sounds, and helps eliminate wastes.

The *respiratory system* consists of the nose, pharynx (throat), larynx (voice box), trachea (windpipe), bronchi, and lungs (Fig. 21.1). Structurally, the respiratory system consists of two portions. (1) The term *upper respiratory system* refers to the nose, pharynx, and associated structures. (2) The *lower respiratory system* refers to the larynx, trachea, bronchi, and lungs. Functionally, the respiratory system also consists of two portions. (1) The *conducting portion* consists of a series of interconnecting cavities and tubes—nose, pharynx, larynx, trachea, bronchi, and terminal bronchioles—that conduct air into the lungs. (2) The *respiratory portion* consists of those portions of the respiratory system where the exchange of gases occurs—respiratory bronchioles, alveolar ducts, alveolar sacs, and alveoli.

Respiration is the exchange of gases between the atmosphere, blood, and cells. It takes place in three basic steps:

1. *Pulmonary ventilation* The first process, *pulmonary* (*pulmo* = lung) *ventilation,* or breathing, is the inspiration (inflow) and expiration (outflow) of air between the atmosphere and the lungs.
2. *External (pulmonary) respiration* This is the exchange of gases between the air spaces of the lungs and blood in pulmonary capillaries. The blood gains O_2 and loses CO_2.
3. *Internal (tissue) respiration* The exchange of gases between blood in systemic capillaries and tissue cells is known as internal (tissue) respiration. The blood loses O_2 and gains CO_2. Within cells, the metabolic reactions that consume O_2 and produce CO_2 during production of ATP are termed *cellular respiration.*

Using your textbook, charts, or models for reference, label Figure 21.1

A. ORGANS OF THE RESPIRATORY SYSTEM

1. Nose

The *nose* has an external portion and an internal portion inside the skull. The external portion consists of a supporting framework of bone and hyaline cartilage covered with muscle and skin and lined by mucous membrane. The bridge of the nose is formed by the nasal bones, which hold it in a fixed position. Because it has a framework of pliable hyaline cartilage, the rest of the external nose is somewhat flexible. On the undersurface of the external nose are two openings called the *external nares* (NA-rēz; singular is *naris*), or *nostrils.* The interior structures of the nose are specialized for three functions: (1) incoming air is warmed, moistened, and filtered; (2) olfactory stimuli are received; and (3) large, hollow resonating chambers modify speech sounds.

The internal portion of the nose is a large cavity in the skull that lies inferior to the anterior cranium and superior to the mouth. Anteriorly, the

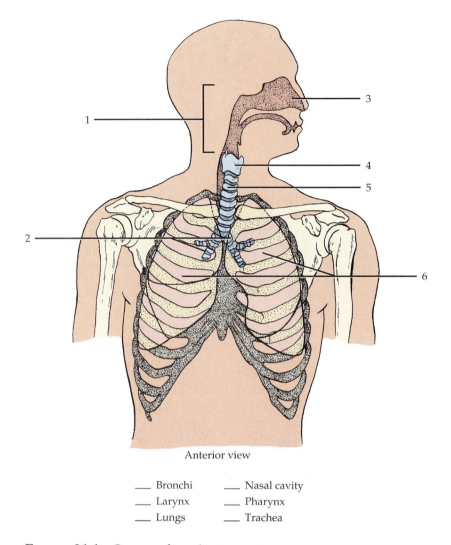

Anterior view

___ Bronchi   ___ Nasal cavity
___ Larynx   ___ Pharynx
___ Lungs   ___ Trachea

FIGURE 21.1   Organs of respiratory system.

internal nose merges with the external nose, and posteriorly it communicates with the pharynx through two openings called the ***internal nares (choanae).*** Ducts from the paranasal sinuses (frontal, sphenoidal, maxillary, and ethmoidal) and the nasolacrimal ducts also open into the internal nose. The lateral walls of the internal nose are formed by the ethmoid, maxillae, lacrimal, palatine, and inferior nasal conchae bones. The ethmoid also forms the roof. The floor of the internal nose is formed mostly by the palatine bones and palatine processes of the maxillae, which together comprise the hard palate.

The inside of both the external and internal nose is called the ***nasal cavity.*** It is divided into right and left sides by a vertical partition called the ***nasal septum.*** The anterior portion of the septum consists primarily of hyaline cartilage. The remainder is formed by the vomer, perpendicular plate of the

ethmoid, maxillae, and palatine bones (see Figure 21.2). The anterior portion of the nasal cavity, just inside the nostrils, is called the ***vestibule*** and is surrounded by cartilage. The superior nasal cavity is surrounded by bone.

When air enters the nostrils, it passes first through the vestibule. The vestibule is lined by skin containing coarse hairs that filter out large dust particles. The air then passes into the superior nasal cavity. Three shelves formed by projections of the superior, middle, and inferior nasal conchae extend out of each lateral wall of the cavity. The conchae, almost reaching the septum, subdivide each side of the nasal cavity into a series of grovelike passageways—the ***superior, middle,*** and ***inferior meatuses*** (mē-Ā-tes-ez; *meatus* = passage; singular is ***meatus***). Mucous membrane lines the cavity and its shelves.

The olfactory receptors lie in the membrane lining the superior nasal conchae and adjacent septum.

This region is called the **olfactory epithelium.** Inferior to the olfactory epithelium, the mucous membrane contains capillaries and pseudostratified ciliated columnar epithelium with many goblet cells. As the air whirls around the conchae and meatuses, it is warmed by blood in the capillaries. Mucus secreted by the goblet cells moistens the air and traps dust particles. Drainage from the nasolacrimal ducts and perhaps secretions from the paranasal sinuses also help moisten the air. The

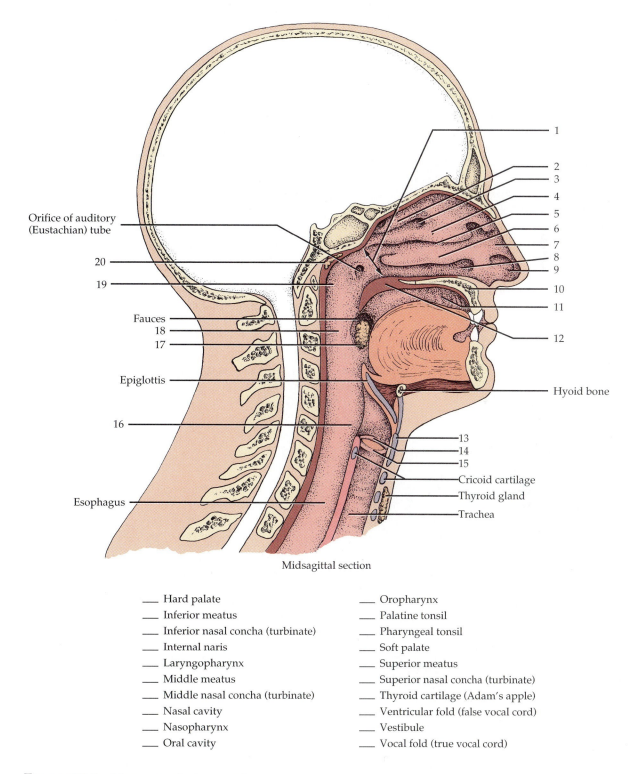

Midsagittal section

FIGURE 21.2   Upper respiratory system.

____ Hard palate

____ Inferior meatus

____ Inferior nasal concha (turbinate)

____ Internal naris

____ Laryngopharynx

____ Middle meatus

____ Middle nasal concha (turbinate)

____ Nasal cavity

____ Nasopharynx

____ Oral cavity

____ Oropharynx

____ Palatine tonsil

____ Pharyngeal tonsil

____ Soft palate

____ Superior meatus

____ Superior nasal concha (turbinate)

____ Thyroid cartilage (Adam's apple)

____ Ventricular fold (false vocal cord)

____ Vestibule

____ Vocal fold (true vocal cord)

cilia move the mucus–dust packages toward the pharynx so that they can be eliminated from the respiratory tract by swallowing or expectoration (spitting). Substances in cigarette smoke inhibit movement of cilia. When this happens, only coughing can remove mucus–dust particles from the airways. This is one reason that smokers cough often.

Label the hard palate, inferior meatus, inferior nasal concha, internal naris, middle meatus, middle nasal concha, nasal cavity, oral cavity, soft palate, superior meatus, superior nasal concha, and vestibule in Figure 21.2.

The surface anatomy of the nose is shown in Figure 21.3.

2. Pharynx

The *pharynx* (FAR-inks) (throat) is a somewhat funnel-shaped tube about 13 cm (5 in.) long that starts at the internal nares and extends to the level of the cricoid cartilage, the most inferior cartilage of the larynx (voice box). Lying posterior to the nasal and oral cavities and just anterior to the cervical vertebrae, the pharynx is a passageway for air and food, a resonating chamber for speech sounds, and a housing for tonsils.

The pharynx is composed of a superior portion, called the *nasopharynx,* an intermediate portion, the *oropharynx,* and an inferior portion, the *laryngopharynx* (la-rin'-gō-FAR-inks) or *hypopharynx.*

The nasopharynx consists of *pseudostratified ciliated columnar epithelium* and has four openings in its wall: two *internal nares* plus two openings into the *auditory (Eustachian) tubes.* The nasopharynx also contains the *pharyngeal tonsil (adenoid).* The oropharynx is lined by *nonkeratinized stratified squamous epithelium* and receives one opening: the *fauces* (FAW-sēz). The oropharynx contains the *palatine* and *lingual tonsils.* The laryngopharynx is also lined by *nonkeratinized stratified squamous epithelium* and becomes continuous with the esophagus posteriorly and the larynx anteriorly.

Label the laryngopharynx, nasopharynx, oropharynx, palatine tonsil, and pharyngeal tonsil in Figure 21.2.

3. Larynx

The *larynx* (LAIR-inks), or voice box, is a short passageway connecting the laryngopharynx with the trachea. Its wall is composed of nine pieces of cartilage.

a. *Thyroid cartilage (Adam's apple)* Large anterior piece that gives larynx its triangular shape.
b. *Epiglottis* (*epi* = above; *glotta* = tongue) Leaf-shaped cartilage on top of larynx that closes off the larynx so that foods and liquids are routed into the esophagus and kept out of the respiratory system.

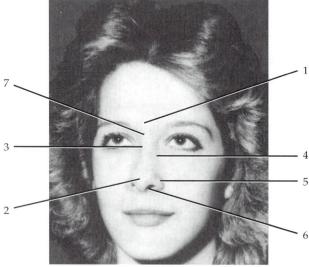

Anterior view

1. **Root.** Superior attachment of nose at forehead located between eyes.
2. **Apex.** Tip of nose.
3. **Dorsum nasi.** Rounded anterior border connecting root and apex; in profile, may be straight, convex, concave, or wavy.
4. **Nasofacial angle.** Point at which side of nose blends with tissues of face.
5. **Ala.** Convex flared portion of inferior lateral surface.
6. **External nares.** External openings into nose (nostrils).
7. **Bridge.** Superior portion of dorsum nasi, superficial to nasal bones.

FIGURE 21.3 Surface anatomy of nose.

c. *Cricoid* (KRĪ-koyd; *krikos* = ring) *cartilage* Ring of cartilage forming the inferior portion of the larynx that is attached to the first ring of tracheal cartilage.

d. *Arytenoid* (ar-i-TĒ-noyd; *arytaina* = ladle) *cartilages* Paired, pyramid-shaped cartilages at superior border of cricoid cartilage that attach vocal folds to the intrinsic pharyngeal muscles.

e. *Corniculate* (kor-NIK-yoo-lāt; *corniculate* = shaped like a small horn) *cartilages* Paired, horn-shaped cartilages at apex of arytenoid cartilages.

f. *Cuneiform* (kyoo-NĒ-i-form; *cuneus* = wedge) *cartilages* Paired, club-shaped cartilages anterior to the corniculate cartilages.

With the aid of your textbook, label the laryngeal cartilages shown in Figure 21.4. Also label the thyroid cartilage in Figure 21.2.

The mucous membrane of the larynx is arranged into two pairs of folds, a superior pair called the *ventricular folds (false vocal cords)* and an inferior pair called the *vocal folds (true vocal cords)*. The space between the vocal folds when they are apart is called the *rima glottidis*. Together, the vocal folds and rima glottidis are referred to as the *glottis*. Movement of the vocal folds produces sounds; variations in pitch result from (1) varying degrees of tension and (2) varying lengths in males and females.

With the aid of your textbook, label the ventricular folds, vocal folds, and rima glottidis in Figure 21.5. Also label the ventricular and vocal folds in Figure 21.2.

4. Trachea

The *trachea* (TRĀ-kē-a) (windpipe) is a tubular air passageway about 12 cm (4½ in.) in length and 2.5 cm (1 in.) in diameter. It lies anterior to the esophagus and, at its inferior end (fifth thoracic vertebra), divides into right and left primary bronchi

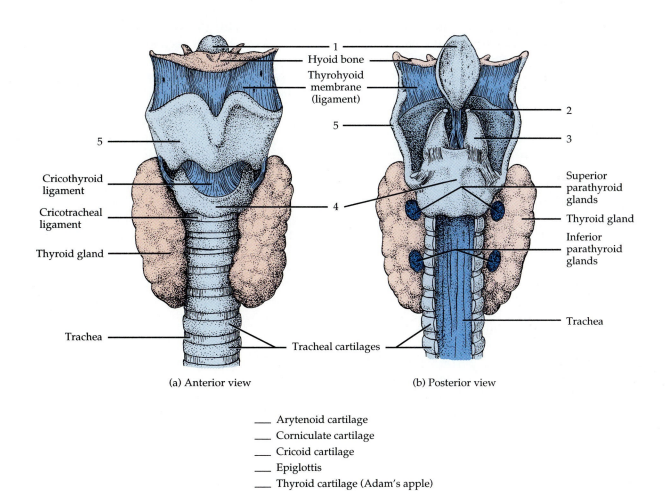

(a) Anterior view   (b) Posterior view

___ Arytenoid cartilage
___ Corniculate cartilage
___ Cricoid cartilage
___ Epiglottis
___ Thyroid cartilage (Adam's apple)

FIGURE 21.4   Larynx.

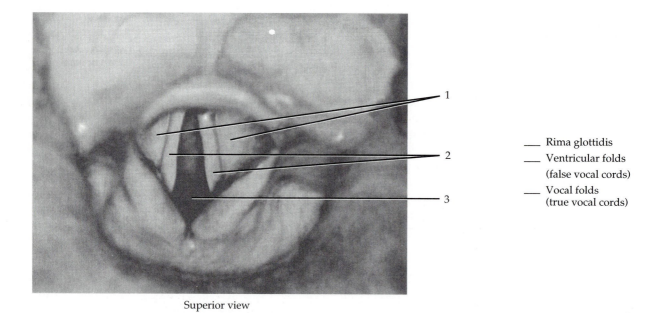

_____ Rima glottidis
_____ Ventricular folds
        (false vocal cords)
_____ Vocal folds
        (true vocal cords)

Superior view

FIGURE 21.5 Photograph of larynx.

(Figure 21.6). The epithelium of the trachea consists of *pseudostratified ciliated columnar epithelium.* This epithelium consists of ciliated columnar cells, goblet cells, and basal cells. The epithelium offers the same protection against dust as the membrane lining the nasal cavity and larynx.

Obtain a prepared slide of pseudostratified ciliated columnar epithelium from the trachea, and, with the aid of your textbook, label the ciliated columnar cells, cilia, goblet cells, and basal cells in Figure 21.7.

The trachea consists of smooth muscle, elastic connective tissue, and incomplete *rings of cartilage* (hyaline) shaped like a series of letter Cs. The open ends of the Cs are held together by the *trachealis muscle.* The cartilage provides a rigid support so that the tracheal wall does not collapse inward and obstruct the air passageway, and, because the open parts of the Cs face the esophagus, the latter can expand into the trachea during swallowing. If the trachea should become obstructed, a *tracheostomy* (trā-kē-OS-tō-mē) may be performed. Another method of opening the air passageway is called *intubation,* in which a tube is passed into the mouth and down through the larynx and the trachea.

5. Bronchi

The trachea terminates by dividing into a *right primary bronchus* (BRONG-kus), going to the right lung, and a *left primary bronchus,* going to the left lung. They continue dividing in the lungs into smaller bronchi, the *secondary (lobar) bronchi* (BRONG-kē), one for each lobe of the lung. These bronchi, in turn, continue dividing into still smaller bronchi called *tertiary (segmental) bronchi,* which divide into *bronchioles.* The next division is into even smaller tubes called *terminal bronchioles.* This entire branching structure of the trachea is commonly referred to as the *bronchial tree.*

Label Figure 21.6.

Bronchography (brong-KOG-ra-fē) is a technique for examining the bronchial tree. With this procedure, an intratracheal catheter is passed transorally or transnasally through the rima glottidis into the trachea. Then an opaque contrast medium is introduced into the trachea and distributed through the bronchial branches. Radiographs of the chest in various positions are taken and the developed film, called *bronchogram* (BRONG-kō-gram), provides a picture of the bronchial tree.

6. Lungs

The *lungs* (*lunge* = light, since the lungs float) are paired, cone-shaped organs lying in the thoracic cavity (see Figure 21.1). The *pleural* (*pleura* = side) *membrane* encloses and protects each lung. Whereas the *superficial parietal pleura* lines the wall of the thoracic cavity, the *deep visceral pleura* covers the lungs; the potential space between parietal and visceral pleurae, the *pleural cavity,* contains a lubricating fluid to reduce friction as the lungs expand and recoil.

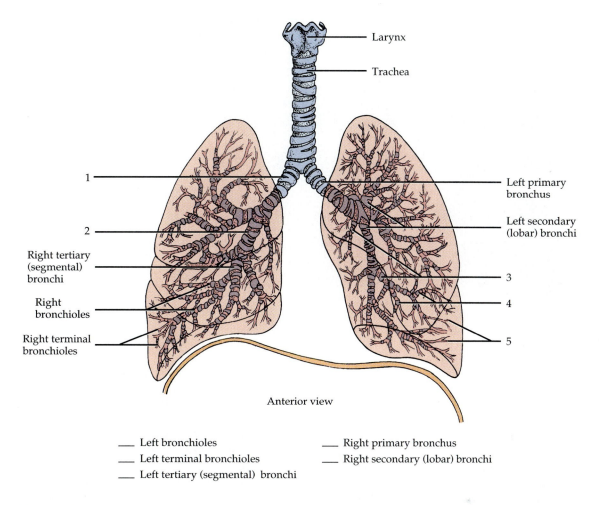

Larynx

Trachea

1

2

Right tertiary
(segmental)
bronchi

Right
bronchioles

Right terminal
bronchioles

Left primary
bronchus

Left secondary
(lobar) bronchi

3

4

5

Anterior view

____ Left bronchioles

____ Left terminal bronchioles

____ Left tertiary (segmental) bronchi

____ Right primary bronchus

____ Right secondary (lobar) bronchi

FIGURE 21.6 Air passageways to the lungs. Shown is the bronchial tree in relationship to lungs.

Major surface features of the lungs include

a. *Base* Broad inferior portion resting on diaphragm.

b. *Apex* Narrow superior portion just above clavicles.

c. *Costal surface* Surface lying against ribs.

d. *Mediastinal surface* Medial surface.

e. *Hilus* Region in mediastinal surface through which bronchial tubes, blood vessels, lymphatic vessels, and nerves enter and exit the lung.

f. *Cardiac notch* Medial concavity in left lung in which heart lies.

Each lung is divided into *lobes* by one or more *fissures*. The right lung has three lobes, *superior*, *middle*, and *inferior*; the left lung has two lobes, *superior* and *inferior*. The *horizontal fissure* separates the superior lobe from the middle lobe in the right lung; an *oblique fissure* separates the middle lobe from the inferior lobe in the right

lung and the superior lobe from the inferior lobe in the left lung. Each lobe receives its own secondary bronchus.

Using your textbook as a reference, label Figure 21.8a on page 493.

Each lobe of a lung is divided into regions called *bronchopulmonary segments,* each supplied by a tertiary bronchus. Each bronchopulmonary segment is composed of many smaller compartments called *lobules.* Each lobule is wrapped in elastic connective tissue and contains a lymphatic vessel, arteriole, venule, and branch from a terminal bronchiole. Terminal bronchioles divide into *respiratory bronchioles,* which, in turn, divide into several *alveolar* (al-VĒ-ō-lar) *ducts.* Around the circumference of alveolar ducts are numerous alveoli and alveolar sacs. *Alveoli* (al-VĒ-ō-lī) are cup-shaped outpouchings lined by epithelium and supported by a thin elastic membrane. The singular is *alveolus. Alveolar sacs* are two or more alveoli that share a common opening. Over

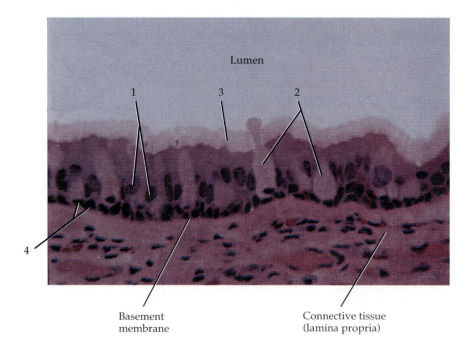

Lumen

1   3   2

_____ Basal cells
_____ Cilia
_____ Ciliated columnar cells
_____ Goblet cells

4

Basement
membrane

Connective tissue
(lamina propria)

FIGURE 21.7   Histology of trachea.

the alveoli, an arteriole and venule disperse into a network of capillaries. Gas is exchanged between the lungs and blood by diffusion across the alveolar and the capillary walls.

Using your textbook as a reference, label Figure 21.8b.

Each alveolus consists of

a. *Type I alveolar (squamous pulmonary epithelial) cells* Cells that form a continuous lining of the alveolar wall, except for occasional type II alveolar (septal) cells.

b. *Type II alveolar (septal) cells* Cuboidal cells dispersed among type I alveolar cells that secrete a phospholipid substance called *surfactant* (sur-FAK-tant), a surface tension-lowering agent.

c. *Alveolar macrophages (dust cells)* Phagocytic cells that remove fine dust particles and other debris from the alveolar spaces.

Obtain a slide of normal lung tissue and examine it under high power. Using your textbook as a reference, see if you can identify a terminal bronchiole, respiratory bronchiole, alveolar duct, alveolar sac, and alveoli.

If available, examine several pathological slides of lung tissue, such as slides that show emphysema and lung cancer. Compare your observations to the normal lung tissue.

The exchange of respiratory gases between the lungs and blood takes place by diffusion across the

alveolar and capillary walls. This membrane, through which the respiratory gases move, is collectively known as the *alveolar-capillary (respiratory) membrane* (Figure 21.9 on page 494). It consists of

1. A layer of type I alveolar (squamous pulmonary epithelial) cells with type II alveolar (septal) cells and alveolar macrophages (dust cells) that constitute the alveolar (epithelial) wall.
2. An epithelial basement membrane underneath the alveolar wall.
3. A capillary basement membrane that is often fused to the epithelial basement membrane.
4. The endothelial cells of the capillary.

Label the components of the alveolar-capillary (respiratory) membrane in Figure 21.9.

Examine the scanning electron micrograph of the cells of the alveolar wall in Figure 21.10 on page 494.

B. DISSECTION OF CAT RESPIRATORY SYSTEM

CAUTION! *Please reread Section D, "Precautions Related to Dissection" at the beginning of the laboratory manual on page xiii before you begin your dissection.*

Air enters the respiratory system by one of two routes, the nose or the mouth. Air entering the

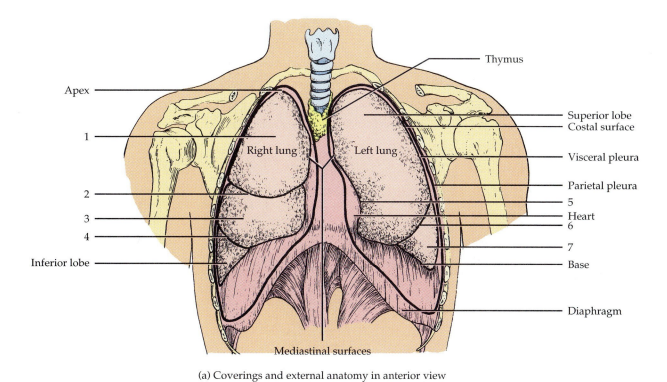

Apex

Thymus

Right lung   Left lung

Superior lobe
Costal surface

Visceral pleura

Parietal pleura

1

2

3

4

Inferior lobe

5

Heart

6

7

Base

Diaphragm

Mediastinal surfaces

(a) Coverings and external anatomy in anterior view

15

8

9

10

Elastic connective
tissue

11

12

Pulmonary
(alveolar)
capillary

13

Visceral
pleura

14

(b) Portion of a lobule of the lung

___ Alveolar ducts
___ Alveolar sac
___ Alveoli
___ Cardiac notch
___ Horizontal fissure
___ Inferior lobe
___ Lymphatic vessel
___ Middle lobe
___ Oblique fissure
     of right lung
___ Oblique fissure
     of left lung
___ Pulmonary
     arteriole
___ Pulmonary
     venule
___ Respiratory
     bronchiole
___ Superior lobe
___ Terminal
     bronchiole

FIGURE 21.8   Lungs.

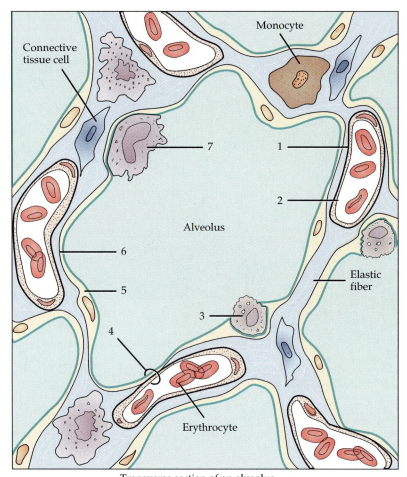

Transverse section of an alveolus

FIGURE 21.9 Alveolar-capillary (respiratory) membrane.

___ Alveolar-capillary (respiratory) membrane

___ Alveolar macrophage (dust cell)

___ Capillary basement membrane

___ Capillary endothelium

___ Epithelial basement membrane

___ Type I alveolar (squamous pulmonary epithelial) cell

___ Type II alveolar (septal) cell

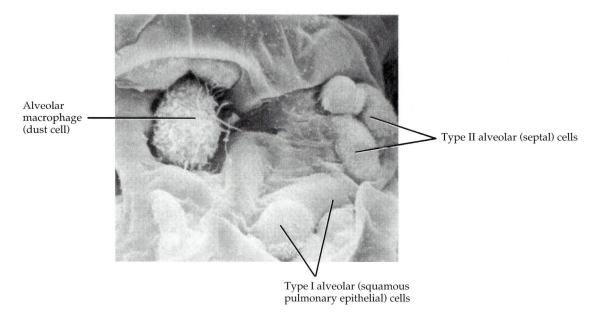

FIGURE 21.10 Scanning electron micrograph of alveolar wall at magnification of 3420×. (Reproduced by permission from R. G. Kessel and R. H. Kardon, *Tissues and Organs: A Text Atlas of Scanning Electron Microscopy*, W. H. Freeman 1979.)

external nares (nostrils) passes into the nasal cavity, through the internal nares (choanae), and into the nasopharynx, oropharynx, and laryngopharynx. From here it enters the larynx. Air entering the mouth passes through the vestibule, oral cavity, fauces, subdivisions of the pharynx, and into the larynx. The mouth and pharynx of the cat will be studied in detail in Exercise 22, Digestive System.

PROCEDURE

1. Larynx

1. The *larynx* (voice box) consists of five cartilages (Figure 21.11).
2. Expose the laryngeal cartilages by stripping off the surrounding tissues on the dorsal, ventral, and left lateral surfaces.
3. Identify the following parts of the larynx:
   a. *Thyroid cartilage* Forms most of the ventral and lateral walls.
   b. *Cricoid cartilage* Caudal to the thyroid cartilage; its expanded dorsal portion forms most of the dorsal wall of the larynx.
   c. *Arytenoid cartilages* Paired, small, triangular cartilages cranial to dorsal part of cricoid cartilage. These are best seen if the larynx is separated from the esophagus.
   d. *Epiglottis* Projects cranially from the thyroid cartilage to which it is attached; guards the opening into the larynx. To examine the cranial portion of the larynx, make a mid-

ventral incision through the larynx and cranial part of the trachea. Be careful not to damage the blood vessels on either side of the trachea. Spread the cut edges back and identify the following structures.
   e. *False vocal cords* Cranial pair of mucous membranes extending across the larynx from the arytenoid cartilages to the base of the epiglottis.
   f. *True vocal cords* Caudal, larger pair of mucous membranes extending across the larynx from the arytenoid cartilages to the thyroid cartilage; important in sound production.
   g. *Glottis* Space between the free margins of the true vocal cords.
   h. *Tracheal cartilages* C-shaped, dorsally incomplete rings of cartilage that support the trachea and prevent it from collapsing.

2. Trachea

1. The *trachea* (windpipe) is a tube that extends from the larynx and bifurcates into the bronchial tubes (Figure 21.12a). Note the tracheal cartilages.
2. On either side of the trachea near the larynx you can locate the dark lobes of the *thyroid gland.* This gland belongs to the endocrine system. Its lobes are connected by a ventral strip of thyroid tissue called the *isthmus.*

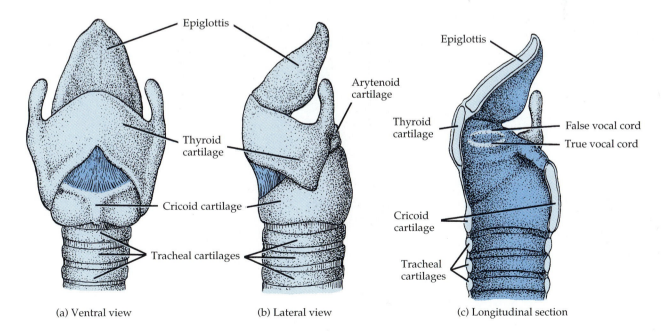

(a) Ventral view   (b) Lateral view   (c) Longitudinal section

FIGURE 21.11   Larynx of cat.

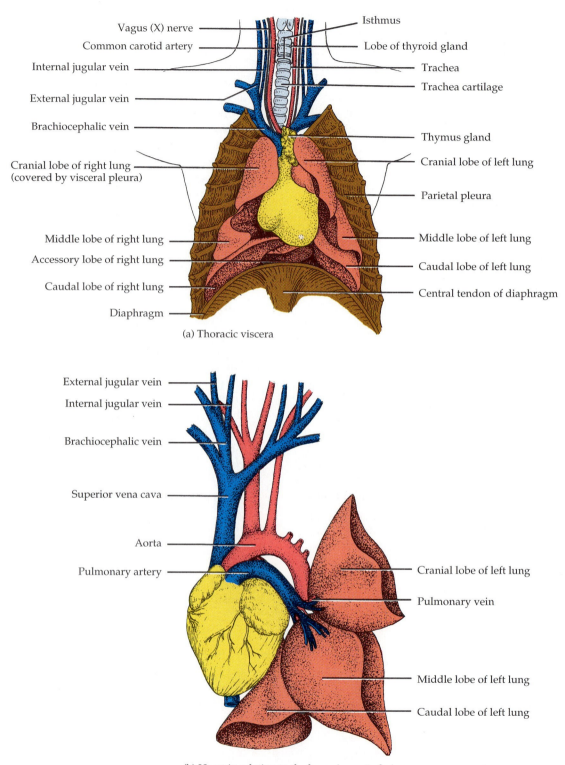

Vagus (X) nerve

Common carotid artery

Internal jugular vein

External jugular vein

Brachiocephalic vein

Cranial lobe of right lung
(covered by visceral pleura)

Middle lobe of right lung

Accessory lobe of right lung

Caudal lobe of right lung

Diaphragm

Isthmus

Lobe of thyroid gland

Trachea

Trachea cartilage

Thymus gland

Cranial lobe of left lung

Parietal pleura

Middle lobe of left lung

Caudal lobe of left lung

Central tendon of diaphragm

(a) Thoracic viscera

External jugular vein

Internal jugular vein

Brachiocephalic vein

Superior vena cava

Aorta

Pulmonary artery

Cranial lobe of left lung

Pulmonary vein

Middle lobe of left lung

Caudal lobe of left lung

(b) Heart in relation to the lungs in ventral view

FIGURE 21.12 Respiratory system of cat.

3. Also on either side of the trachea you can identify the large *common carotid arteries*, the smaller *internal jugular veins*, and the white, threadlike *vagus nerve* adjacent to the common carotid artery.

3. Bronchi

1. The *bronchi* are two short tubes formed by the bifurcation of the trachea at the level of the sixth rib. One bronchus enters each lung. The bronchi, like the trachea, consist of C-shaped cartilages.
2. Within the lungs, the bronchi branch repeatedly into *secondary bronchi*. These branch into *tertiary bronchi*, which, in turn, branch into *bronchioles*.
3. Ultimately, the bronchioles terminate in alveoli, or air sacs, where gases are exchanged.
4. After you examine the parts of the lungs, you can trace the bronchial tree into the substance of the lungs. Use Figure 21.6 for reference.

4. Lungs

1. The *lungs* are paired, lobed, spongy organs in which respiratory gases are exchanged (Figure 21.12).
2. Each lung is covered by *visceral pleura*, whereas *parietal pleura* lines the thoracic wall.
3. The potential space between the visceral pleurae of the lungs is called the *mediastinum.* This contains the thymus gland of the endocrine system, heart, trachea, esophagus, nerves, and blood vessels. In the midventral thorax, visceral and parietal pleurae meet to form a ventral partition called the *mediastinal septum.* This structure is caudal to the heart.
4. Note that the right lung has four lobes and the left lung has three lobes. Try to identify the *pulmonary arteries* entering the lungs. They are usually injected with blue latex.
5. Now try to find the *pulmonary veins* leaving the lungs. They are usually injected with red latex.
6. If you cut through a lobe of the lung, you should be able to identify branches of the pulmonary arteries, pulmonary veins, and bronchi. The lobes of the lung are labeled in Figure 21.12a.

C. DISSECTION OF SHEEP PLUCK

CAUTION! *Please reread Section D, "Precautions Related to Dissection" at the beginning of the* laboratory manual on page xiii before you begin your dissection.

PROCEDURE

1. Preserved sheep pluck may or may not be available for dissection.
2. Pluck consists mainly of a sheep *trachea, bronchi, lungs, heart,* and *great vessels,* and a small portion of the *diaphragm.* It is a good demonstration because it is large and shows the close anatomical correlation between these structures and the systems to which they belong, namely, the respiratory and the cardiovascular systems.
3. The heart and its great blood vessels have been described in detail in Exercise 17.
4. Pluck can also be used to examine in great detail the trachea and its relationship to the development of the bronchi until they branch into each lung. In addition, this specimen is sufficiently large that the bronchial tree can be exposed by careful dissection.
5. This dissection is done by starting at the primary and secondary bronchi and slowly and carefully removing lung tissue as the trees form even smaller branches into lungs.
6. These specimens should be used primarily by the instructor for demonstration, but you can help expose the bronchial tree.

D. LABORATORY TESTS ON RESPIRATION

1. Observation of Living Lung Tissue

CAUTION! *Please reread Section D, "Precautions Related to Dissection" at the beginning of the laboratory manual on page xiii before you begin your dissection.*

PROCEDURE

1. *Note: It is strongly suggested that either your instructor or a laboratory assistant perform the procedure for pithing a frog. The procedure requires some practice and may prove difficult for inexperienced individuals.*
2. Pin the freshly pithed frog to the dissecting tray so the ventral surface is upward.
3. As illustrated in Figure 2.5, make a midsagittal cut through the skin from the abdomen to the level of the forelimbs. Make two lateral cuts to create flaps of skin which can be reflected and pinned to the tray.
4. *Taking care not to cut too deeply,* carefully make an incision parallel to the midline over

either the right or left lung. Observe the inflated lung.

5. With a probe, gently lift the lung out through the incision and keep the lung moist with Ringer's solution that has been maintained at room temperature.

6. Place the tray under the dissecting microscope and locate the capillary network that surrounds the alveoli. Observe the blood flow through the capillaries of the lungs.

Describe your observation._____

_____

2. Mechanics of Pulmonary Ventilation (Breathing)

Pulmonary ventilation (breathing) is the process by which gases are exchanged between the atmosphere and the alveoli. Air moves throughout the respiratory system as a result of pressure gradients (differences) between the external environment and the respiratory system. Likewise, oxygen and carbon dioxide then move between the respiratory alveoli and the pulmonary capillaries of the cardiovascular system, and the peripheral capillaries of the cardiovascular system and the tissues of the body as a result of diffusion gradients. We breathe in (inhale) when the pressure inside the thoracic cavity and lungs is less than the air pressure in the atmosphere; similarly we breathe out (exhale) when the pressure inside the lungs and thoracic cavity is greater than the pressure in the atmosphere.

a. INSPIRATION

Breathing in is called *inspiration (inhalation)*. When the thoracic cavity is at rest, and no air movement is occurring, the pressure inside the lungs equals the pressure of the atmosphere, which is approximately 760 mm Hg, or 1 atmosphere (atm), at sea level. For air to flow into the lungs, something must happen to reduce the pressure within the lungs to a value lower than atmospheric pressure. This condition is achieved by increasing the volume of the thoracic cavity.

In order for inspiration to occur, the thoracic cavity must be expanded. This increases lung volume and thus decreases pressure in the lungs. The first step toward increasing lung volume (size) involves contraction of the principal inspiratory muscles—the diaphragm and external intercostals. Contraction of the diaphragm causes it to flatten. This increases the vertical length of the thoracic

cavity and accounts for about 75% of the air that enters the lungs during inspiration. At the same time the diaphragm contracts, the external intercostals contract. As a result, the ribs are pulled superiorly and outward, increasing the anterior-posterior diameter of the thoracic cavity. This movement of the ribs by the external intercostals is much like the movement of a bucket handle when a bucket is placed on its side and the handle is moved from a vertical to a more horizontal position. Because the thoracic cavity is a closed system, this increase in the vertical length and anterior-posterior diameter of the thoracic cavity causes the volume of the thoracic cavity to increase, and the pressure within the cavity to decrease. As a result of this decreased pressure within the thoracic cavity, pressure within the lungs decreases, thereby causing air to move into the lungs.

During normal breathing, the pressure between two pleural layers, called *intrapleural (intrathoracic) pressure,* is always subatmospheric. (It may become temporarily positive, but only during modified respiratory movements such as coughing or straining during child birth or defecation.) When the respiratory system is at rest no air movement in or out of the respiratory tree is occurring, and intrapleural pressure is approximately 4 mm Hg lower than atmospheric pressure, or approximately 756 mm Hg. The overall increase in the size of the thoracic cavity causes intrapleural pressure to fall to approximately 754 mm Hg. The parietal and visceral pleural membranes are normally strongly attached to each other due to surface tension created by a very thin layer of water between the membranes. Therefore, as the walls of the thoracic cavity expand, the parietal pleural lining the thoracic cavity is pulled in all directions, and the visceral pleura is pulled along with it. Consequently the size of the lungs increase, thereby decreasing the pressure within the lungs (called *alveolar (intrapulmonic) pressure.* This decrease in alveolar pressure causes a pressure gradient between the alveoli of the lungs and the external environment. Because the respiratory tree is open to the external environment via the oral and nasal cavities, air moves down the pressure gradient from the atmosphere into the pulmonary alveoli. Air continues to move into the lungs until alveolar pressure equals atmospheric pressure.

b. EXPIRATION

Breathing out, called *expiration (exhalation),* is also achieved by a pressure gradient, but in this case the gradient is reversed so that the pressure in the lungs is greater than the pressure of the atmosphere. Normal quiet expiration, unlike inspiration, is a passive process, in that it does not involve the active contraction of muscles. When inspiration is completed, the diaphragm and external intercostal muscles relax. As the diaphragm returns to its domelike position, the vertical length of the thoracic cavity decreases, returning to its original dimension. Likewise, when the external intercostals relax, the ribs move posteriorly and inferiorly, decreasing the anterior-posterior diameter of the thoracic cavity. These movements return intrapleural pressure to its normal resting value of 756 mm Hg, or 4 mm Hg below atmospheric pressure.

As intrapleural pressure returns to its preinspiration level, the walls of the lungs are no longer pulled outward by the parietal pleura. The elastic recoil of the connective tissue within the lungs and respiratory tree allows the lungs to return to their resting shape and volume, thereby causing alveolar pressure to become slightly greater than atmospheric pressure (or approximately 762 mm Hg). Now air again moves down its pressure gradient, moving from the alveoli through the respiratory tree and out the oral and nasal openings.

In order to demonstrate pulmonary ventilation, a model lung (a bell-jar demonstrator) will be used. This apparatus is basically an artificial thoracic cavity that mimics the organs of the respiratory system and allows the pressure/volume ratios to be manipulated. A diagram of such a device is shown in Figure 21.13. Some of the models will not have a stopcock (rima glottidis); some will not have a tube opening into the pleural space.

PROCEDURE

1. Label the diagram in Figure 21.13 by writing the name of the anatomical structure that corresponds to the following parts of the apparatus: rima glottidis, trachea, primary bronchus, lungs, alveolar pressure, pleural space, and diaphragm.

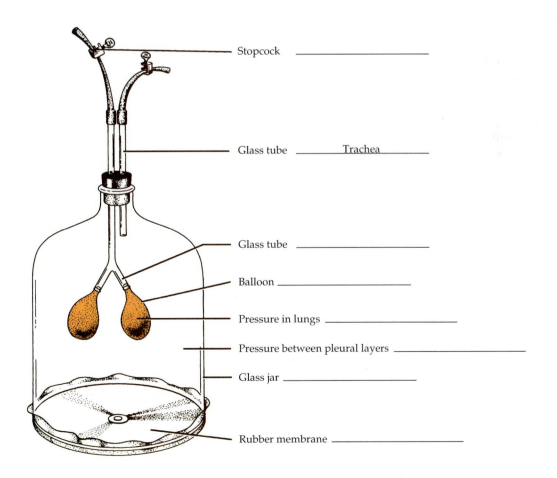

Stopcock _____

Glass tube ____Trachea____

Glass tube _____

Balloon _____

Pressure in lungs _____

Pressure between pleural layers _____

Glass jar _____

Rubber membrane _____

FIGURE 21.13   Model lung.

2. Pull on the rubber membrane (diaphragm) to simulate inspiration. What changes occur in intrapleural pressure?_____

_____

In alveolar pressure?_____

_____

3. Push the rubber membrane (diaphragm) upward to simulate expiration. What changes occur in intrapleural pressure?_____

_____

In alveolar pressure?_____

_____

Under what normal processes would such changes in intrapleural and alveolar pressures be observed?_____

_____

4. Open the clamp on the tube leading to the intrapleural space. This simulates pneumothorax (air in the pleural cavity). Try to cause inspiration and expiration.

Explain your results. _____

_____

5. If present, close the stopcock (rima glottidis) and simulate expiration. This procedure mimics the Valsalva maneuver (forced expiration against a closed rima glottidis as during periods of straining). What changes occur in intrapleural and alveolar pressure?_____

_____

3. Measurement of Chest and Abdomen in Respiration

When the diaphragm contracts, the dome shape of this muscle flattens; the flattening pushes against the abdominal viscera and decreases thoracic volume, which, in turn, decreases alveolar pressure. The volume changes in the abdomen and chest can easily be measured.

PROCEDURE

1. Place a tape measure at the level of the fifth rib (about armpit level) and determine the size of the chest immediately after a
   a. Normal inspiration
   b. Normal expiration

c. Maximal inspiration
d. Maximal expiration

2. Repeat step 1 with the tape measure at the level of the waist to determine abdominal size.
3. Record your results in Section D.1 of the LABORATORY REPORT RESULTS at the end of the exercise.
   Which measurement increased during inspiration?_____

   _____

   Which increased during expiration?_____

   _____

4. Respiratory Sounds

Air flowing through the respiratory tree creates characteristic sounds that can be detected through the use of a stethoscope. Normal breathing sounds include a soft, breezy sound caused by air filling the lungs during inspiration. As the air exits the lungs during expiration a short, lower-pitched sound may be heard. Perform the following exercises.

PROCEDURE

1. *Clean the earplugs of a stethoscope with alcohol.*
2. Place the diaphragm of the stethoscope just below the larynx and listen for bronchial sounds during both inspiration and expiration.
3. Move the stethoscope slowly downward toward the bronchial tubes until the sounds are no longer heard.
4. Place the stethoscope under the scapula, under the clavicle, over different intercostal spaces (the spaces between the ribs) on the chest, and listen for any sound during inspiration and expiration.

5. Use of a Pneumograph

The *pneumograph* is an instrument that measures variations in breathing patterns caused by various physical or chemical factors. The chest pneumograph, which is attached to a polygraph recorder via electrical leads, consists of a rubber bellows that fits around the chest just below the rib cage. As the subject breathes, chest movements cause changes in the air pressure within the pneumograph that are transmitted to the recorder. Normal inspiration and expiration can thus be recorded, and the effects of a wide range of physical and chemical factors on these movements can be studied.

PROCEDURE

Perform the following exercises and record your values in Section D.2 of the LABORATORY REPORT

RESULTS at the end of the exercise. Label and save all recordings and attach them to Section D.2 of the LABORATORY REPORT RESULTS. In addition, label the inspiratory and expiratory phases of each recording, determine their duration, and then calculate the respiratory rate.

1. Place a respiratory pneumograph around the chest at the level of the sixth rib and attach it at the back. Connect the electrical lead to the recorder and adjust the instrument's centering and sensitivity so that the needle deflects as the subject breathes. If it does not, adjust the pneumograph bellows up or down on the chest, or loosen the degree of the tightness around the subject's chest.

2. Seat the subject so that the pneumograph recording is not visible to the subject.

3. Set the polygraph at a slow speed (approximately 1 cm/sec) and record normal, quiet breathing (eupnea) for approximately 30 sec.

4. Have the subject inhale deeply and hold his or her breath for as long as possible. Record the breathing pattern during the breath holding and 30 to 60 sec after the resumption of breathing at the end of breath holding.

5. Record the respiratory movements of a subject in Section D.2 of the LABORATORY REPORT RESULTS at the end of the exercise during the following activities:
    a. Reading
    b. Swallowing water
    c. Laughing
    d. Yawning
    e. Coughing
    f. Sniffing
    g. Doing a rather difficult long-division calculation

6. Measurement of Respiratory Volumes

In clinical practice, *respiration* refers to one complete respiratory cycle, that is, one inspiration and one expiration. A normal adult has 14 to 18 respirations in a minute, during which the lungs exchange specific volumes of air with the atmosphere.

As the following respiratory volumes and capacities are discussed, keep in mind that the values given vary with age, height, sex, and physiological state. The volume of air expired under normal, quiet inspiration is approximately 500 ml. The volume of air expired under normal, quiet breathing conditions is equal to that of inspiration, and this volume of air is called *tidal volume* (Figure 21.14). Of this volume of air, only approximately 350 ml reaches the alveoli. The other 150 ml of air does not reach the alveoli because it remains in the spaces not designed for air exchange *(anatomical dead space)* (nose, pharynx, larynx, trachea, and bronchi) as well as those areas designed for air exchange but not currently being utilized by the lungs *(physiological dead space).*

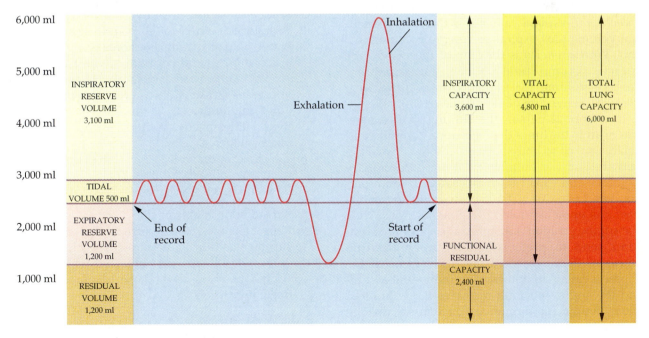

FIGURE 21.14   Spirogram of lung volumes and capacities.

If we take a very deep breath, we can inspire much more than the 500-ml tidal volume taken in during normal, quiet respiration. The additional inhaled air, called the *inspiratory reserve volume*, averages 3100 ml above the tidal volume of quiet respiration. If we inspire *normally* and then expire *as forcibly as possible*, a normal individual would be able to exhale the 500 ml of air taken in during a normal, quiet tidal volume, as well as an additional 1200 ml of air, which is termed *expiratory reserve volume.* Even after a forceful expiration there is still some air remaining in the lungs, which is termed *residual volume* (1200 ml). This volume of air ensures gas exchange between the lungs and the cardiovascular system during brief time periods between respiratory cycles, or during extended periods of no respiratory activity (apnea).

Opening the thoracic cavity allows the intrapleural pressure to equal the atmospheric pressure, forcing out some of the residual volume. The air remaining is called the *minimal volume.* Minimal volume provides a medical and legal tool for determining whether a baby was born dead or died after birth. The presence of minimal volume can be demonstrated by placing a piece of lung in water and watching it float. Fetal lungs contain no air, and so the lung of a stillborn will not float in water.

Lung capacities are combinations of specific lung volumes. *Inspiratory capacity,* the total inspiratory ability of the lungs, is the sum of tidal volume plus inspiratory reserve volume (3600 ml). *Functional residual capacity* is the sum of residual volume plus expiratory reserve volume (2400 ml). *Vital capacity* is the sum of the inspiratory reserve volume, tidal volume, and expiratory reserve volume (4800 ml). Finally, *total lung capacity* is the sum of all volumes (6000 ml).

In a normal person the volumes and capacities of air in the lungs depends upon body size and build, as well as body position and the physiological state of the pulmonary system. Pulmonary volumes and capacities may change when a person lies down due to the movement of abdominal contents upon assuming a supine position. When the supine position is assumed the abdominal contents tend to press superiorly upon the diaphragm, thereby decreasing the volume of the thoracic cavity. In addition, an individual assuming the supine position increases the volume of blood within the pulmonary circulation, thereby decreasing the space for pulmonary air.

Note: Depending on the type of laboratory equipment available, select from the following procedures related to measurement of respiratory volumes.

PROCEDURE USING PHYSIOFINDER

PHYSIOFINDER is an interactive computer program that permits laboratory simulations that do not involve the use of animals or advanced or expensive laboratory equipment. It permits students to perform experiments, analyze data, draw conclusions, and form hypotheses on the basis of collected data. The program is available from HarperCollins Publishers (1-800-8HEART1).

For activities related to respiratory volumes, select the appropriate experiments from PHYSIOFINDER module 10—Control of Breathing.

PROCEDURE USING SPIROCOMP™

SPIROCOMP™ is a computerized spirometry system consisting of hardware and software designed to allow for quick and easy measurement of standard lung volumes. A stacked bar graph display of the data clearly depicts the relationship between the various volumes. The built-in group analysis features also allows you to view gender-specific data. The program is available from INTELITOOL® (1-800-227-3805).

PROCEDURE USING VITAL CAPACITY APPARATUS

Using the vital capacity apparatus (Figure 21.15), determine your own particular vital capacity value.

The procedure for determining forced vital capacity is as follows:

PROCEDURE

1. Place a *sterile* air shield over the mouthpiece.
2. Push black pointer to red pointer at zero.
3. Stand, hold apparatus horizontally at eye level, and completely fill your lungs with air.
4. Apply mouth to air shield and blow air into apparatus as forcefully as possible, completely emptying lungs.
5. Read vital capacity, in liters, at upper edge of black pointer.
6. Compare value with printed tables on apparatus using your height in centimeters as a guide.
7. Discard the air shield.
8. Record your vital capacity: _____
9. Repeat the above procedure while lying in the supine position. Compare the various lung volumes and capacities obtained while in the supine position with those obtained in the upright position.

PROCEDURE USING HANDHELD RESPIROMETER

A *respirometer (spirometer)* is an instrument used to measure volumes of air exchanged in breathing.

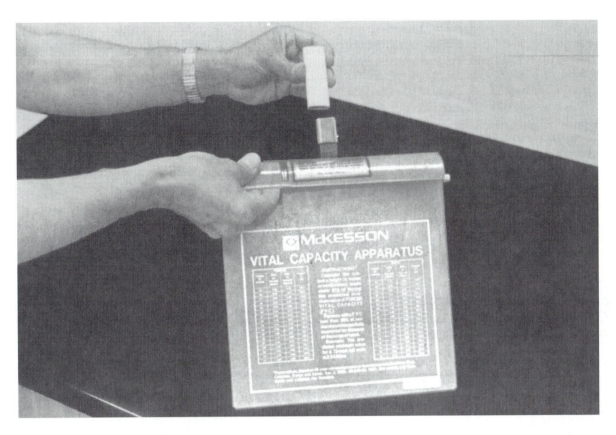

FIGURE 21.15 Placing a disposable air shield on a vital capacity apparatus.

Of the several different respirometers available, we will make use of two: a handheld respirometer and the Collins respirometer. One type of handheld respirometer is the Pulmometer (Figure 21.16). It measures and provides a direct digital display of certain respiratory volumes and capacities.

PROCEDURE

1. Set the needle or the digital indicator to zero.
2. Place a *clean* mouthpiece in the spirometer tube and hold the tube in your hand while you breathe normally for a few respirations. (***Note:*** *Inhale through the nose and exhale through the mouth.*)
3. Place the mouthpiece in your mouth and inhale; exhale three normal breaths through the mouthpiece.
4. Divide the total volume of air expired by three to determine your average tidal volume. Record in Section D.3 of the LABORATORY REPORT RESULTS at the end of the exercise.
5. Return the scale indicator to zero.
6. Breathe normally for a few respirations. Following a normal expiration of tidal volume, forcibly exhale as much air as possible into the mouthpiece.

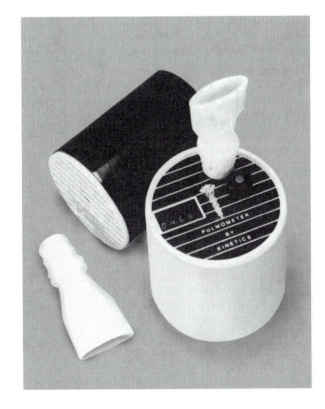

FIGURE 21.16 Handheld respirometer.

7. Repeat step 6 twice. Divide the total amount of air expired by three to determine your average expiratory reserve volume. Record this value in Section D.3 of the LABORATORY REPORT RESULTS at the end of the exercise.

8. To determine vital capacity, breathe deeply a few times and inhale as much air as possible. Exhale as fully and as steadily as possible into the spirometer tube.

9. Repeat step 8 twice. Divide the total volume of expired air by three to determine your average vital capacity. Record this value in Section D.3 of the LABORATORY REPORT RESULTS at the end of the exercise.

10. Calculate your inspiratory reserve volume by substituting the known volumes in this equation: Vital capacity = Inspiratory reserve volume + tidal volume + expiratory reserve volume. Record this volume in Section D.3 of the LABORATORY REPORT RESULTS at the end of the exercise.

11. Compare your vital capacity with the normal values shown in Tables 21.1 and 21.2.

12. Sit quietly and count your respirations per minute. Calculate your minute volume of respirations (MVR) by multiplying the breaths per minute by your tidal volume. Record this value in Section D.3 of the LAB-

TABLE 21.1
Predicted Vital Capacities for Females

| Age | cm | 152 | 154 | 156 | 158 | 160 | 162 | 164 | 166 | 168 | 170 | 172 | 174 | 176 | 178 | 180 | 182 | 184 | 186 | 188 |
	in.	59.8	60.6	61.4	62.2	63.0	63.7	64.6	65.4	66.1	66.9	67.7	68.5	69.3	70.1	70.9	71.7	72.4	73.2	74.0
16		3070	3110	3150	3190	3230	3270	3310	3350	3390	3430	3470	3510	3550	3590	3630	3670	3715	3755	3800
17		3055	3095	3135	3175	3215	3255	3295	3335	3375	3415	3455	3495	3535	3575	3615	3655	3695	3740	3780
18		3040	3080	3120	3160	3200	3240	3280	3320	3360	3400	3440	3480	3520	3560	3600	3640	3680	3720	3760
20		3010	3050	3090	3130	3170	3210	3250	3290	3330	3370	3410	3450	3490	3525	3565	3605	3645	3695	3720
22		2980	3020	3060	3095	3135	3175	3215	3255	3290	3330	3370	3410	3450	3490	3530	3570	3610	3650	3685
24		2950	2985	3025	3065	3100	3140	3180	3200	3260	3300	3335	3375	3415	3455	3490	3530	3570	3610	3650
26		2920	2960	3000	3035	3070	3110	3150	3190	3230	3265	3300	3340	3380	3420	3455	3495	3530	3570	3610
28		2890	2930	2965	3000	3040	3070	3115	3155	3190	3230	3270	3305	3345	3380	3420	3460	3495	3535	3570
30		2860	2895	2935	2970	3010	3045	3085	3120	3160	3195	3235	3270	3310	3345	3385	3420	3460	3495	3535
32		2825	2865	2900	2940	2975	3015	3050	3090	3125	3160	3200	3235	3275	3310	3350	3385	3425	3460	3495
34		2795	2835	2870	2910	2945	2980	3020	3055	3090	3130	3165	3200	3240	3275	3310	3350	3385	3425	3460
36		2765	2805	2840	2875	2910	2950	2985	3020	3060	3095	3130	3165	3205	3240	3275	3310	3350	3385	3420
38		2735	2770	2810	2845	2880	2915	2950	2990	3025	3060	3095	3130	3170	3205	3240	3275	3310	3350	3385
40		2705	2740	2775	2810	2850	2885	2920	2955	2990	3025	3060	3095	3135	3170	3205	3240	3275	3310	3345
42		2675	2710	2745	2780	2815	2850	2885	2920	2955	2990	3025	3060	3100	3135	3170	3205	3240	3275	3310
44		2645	2680	2715	2750	2785	2820	2855	2890	2925	2960	2995	3030	3060	3095	3130	3165	3200	3235	3270
46		2615	2650	2685	2715	2750	2785	2820	2855	2890	2925	2960	2995	3030	3060	3095	3130	3165	3200	3235
48		2585	2620	2650	2685	2715	2750	2785	2820	2855	2890	2925	2960	2995	3030	3060	3095	3130	3160	3195
50		2555	2590	2625	2655	2690	2720	2755	2785	2820	2855	2890	2925	2955	2990	3025	3060	3090	3125	3155
52		2525	2555	2590	2625	2655	2690	2720	2755	2790	2820	2855	2890	2925	2955	2990	3020	3055	3090	3125
54		2495	2530	2560	2590	2625	2655	2690	2720	2755	2790	2820	2855	2885	2920	2950	2985	3020	3050	3085
56		2460	2495	2525	2560	2590	2625	2655	2690	2720	2755	2790	2820	2855	2885	2920	2950	2980	3015	3045
58		2430	2460	2495	2525	2560	2590	2625	2655	2690	2720	2750	2785	2815	2850	2880	2920	2945	2975	3010
60		2400	2430	2460	2495	2525	2560	2590	2625	2655	2685	2720	2750	2780	2810	2845	2875	2915	2940	2970
62		2370	2405	2435	2465	2495	2525	2560	2590	2620	2655	2685	2715	2745	2775	2810	2840	2870	2900	2935
64		2340	2370	2400	2430	2465	2495	2525	2555	2585	2620	2650	2680	2710	2740	2770	2805	2835	2865	2895
66		2310	2340	2370	2400	2430	2460	2495	2525	2555	2585	2615	2645	2675	2705	2735	2765	2800	2825	2860
68		2280	2310	2340	2370	2400	2430	2460	2490	2520	2550	2580	2610	2640	2670	2700	2730	2760	2795	2820
70		2250	2280	2310	2340	2370	2400	2425	2455	2485	2515	2545	2575	2605	2635	2665	2695	2725	2755	2780
72		2220	2250	2280	2310	2335	2365	2395	2425	2455	2480	2510	2540	2570	2600	2630	2660	2685	2715	2745
74		2190	2220	2245	2275	2305	2335	2360	2390	2420	2450	2475	2505	2535	2565	2590	2620	2650	2680	2710

Source: E. A. Gaensler and G. W. Wright, *Archives of Environmental Health,* 12 (Feb.): 146–189 (1966).

TABLE 21.2
Predicted Vital Capacities for Males

		Height in centimeters and inches																		
	cm	152	154	156	158	160	162	164	166	168	170	172	174	176	178	180	182	184	186	188
Age	in.	59.8	60.6	61.4	62.2	63.0	63.7	64.6	65.4	66.1	66.9	67.7	68.5	69.3	70.1	70.9	71.7	72.4	73.2	74.0
16		3920	3975	4025	4075	4130	4180	4230	4285	4335	4385	4440	4490	4540	4590	4645	4695	4745	4800	4850
18		3890	3940	3995	4045	4095	4145	4200	4250	4300	4350	4405	4455	4505	4555	4610	4660	4710	4760	4815
20		3860	3910	3960	4015	4065	4115	4165	4215	4265	4320	4370	4420	4470	4520	4570	4625	4675	4725	4775
22		3830	3880	3930	3980	4030	4080	4135	4185	4235	4285	4335	4385	4435	4485	4535	4585	4635	4685	4735
24		3785	3835	3885	3935	3985	4035	4085	4135	4185	4235	4285	4330	4380	4430	4480	4530	4580	4630	4680
26		3755	3805	3855	3905	3955	4000	4050	4100	4150	4200	4250	4300	4350	4395	4445	4495	4545	4595	4645
28		3725	3775	3820	3870	3920	3970	4020	4070	4115	4165	4215	4265	4310	4360	4410	4460	4510	4555	4605
30		3695	3740	3790	3840	3890	3935	3985	4035	4080	4130	4180	4230	4275	4325	4375	4425	4470	4520	4570
32		3665	3710	3760	3810	3855	3905	3950	4000	4050	4095	4145	4195	4240	4290	4340	4385	4435	4485	4530
34		3620	3665	3715	3760	3810	3855	3905	3950	4000	4045	4095	4140	4190	4225	4285	4330	4380	4425	4475
36		3585	3635	3680	3730	3775	3825	3870	3920	3965	4010	4060	4105	4155	4200	4250	4295	4340	4390	4435
38		3555	3605	3650	3695	3745	3790	3840	3885	3930	3980	4025	4070	4120	4165	4210	4260	4305	4350	4400
40		3525	3575	3620	3665	3710	3760	3805	3850	3900	3945	3990	4035	4085	4130	4175	4220	4270	4315	4360
42		3495	3540	3590	3635	3680	3725	3770	3820	3865	3910	3955	4000	4050	4095	4140	4185	4230	4280	4325
44		3450	3495	3540	3585	3630	3675	3725	3770	3815	3860	3905	3950	3995	4040	4085	4130	4175	4220	4270
46		3420	3465	3510	3555	3600	3645	3690	3735	3780	3825	3870	3915	3960	4005	4050	4095	4140	4185	4230
48		3390	3435	3480	3525	3570	3615	3655	3700	3745	3790	3835	3880	3925	3970	4015	4060	4105	4150	4190
50		3345	3390	3430	3475	3520	3565	3610	3650	3695	3740	3785	3830	3870	3915	3960	4005	4050	4090	4135
52		3315	3353	3400	3445	3490	3530	3575	3620	3660	3705	3750	3795	3835	3880	3925	3970	4010	4055	4100
54		3285	3325	3370	3415	3455	3500	3540	3585	3630	3670	3715	3760	3800	3845	3890	3930	3975	4020	4060
56		3255	3295	3340	3380	3425	3465	3510	3550	3595	3640	3680	3725	3765	3810	3850	3895	3940	3980	4025
58		3210	3250	3290	3335	3375	3420	3460	3500	3545	3585	3630	3670	3715	3755	3800	3840	3880	3925	3965
60		3175	3220	3260	3300	3345	3385	3430	3470	3500	3555	3595	3635	3680	3720	3760	3805	3845	3885	3930
62		3150	3190	3230	3270	3310	3350	3390	3440	3480	3520	3560	3600	3640	3680	3730	3770	3810	3850	3890
64		3120	3160	3200	3240	3280	3320	3360	3400	3440	3490	3530	3570	3610	3650	3690	3730	3770	3810	3850
66		3070	3110	3150	3190	3230	3270	3310	3350	3390	3430	3470	3510	3550	3600	3640	3680	3720	3760	3800
68		3040	3080	3120	3160	3200	3240	3280	3320	3360	3400	3440	3480	3520	3560	3600	3640	3680	3720	3760
70		3010	3050	3090	3130	3170	3210	3250	3290	3330	3370	3410	3450	3480	3520	3560	3600	3640	3680	3720
72		2980	3020	3060	3100	3140	3180	3210	3250	3290	3330	3370	3410	3450	3490	3530	3570	3610	3650	3680
74		2930	2970	3010	3050	3090	3130	3170	3200	3240	3280	3320	3360	3400	3440	3470	3510	3550	3590	3630

Source: E. A. Gaensler and G. W. Wright, *Archives of Environmental Health,* 12 (Feb.): 146–189 (1966).

ORATORY REPORT RESULTS at the end of the exercise.

$$\frac{\text{\# breaths}}{\text{minute}} \times \frac{\text{\# ml}}{\text{breath}} = \frac{\text{\# ml}}{\text{minute}}$$

13. Repeat the above procedures while lying in the supine position. Compare the various lung volumes and capacities obtained while in the supine position with those obtained in the upright position.

PROCEDURE USING COLLINS RESPIROMETER

The Collins respirometer, shown in Figure 21.17, is a closed system that allows the measurement of volumes of air inhaled and exhaled. The instru-ment consists of a weighted drum, containing air, inverted over a chamber of water. The air-filled chamber is connected to the subject's mouth by a tube. When the subject inspires, air is removed from the chamber, causing the drum to sink and producing an upward deflection. This deflection is recorded by the stylus on the graph paper on the kymograph (rotating drum). When the subject expires, air is added, causing the drum to rise and producing a downward deflection. These deflec-tions are recorded as a *spirogram* (see Figure 21.14). The horizontal (x) axis of a spirogram is graduated in millimeters (mm) and records elapsed time; the vertical (y) axis is graduated in milliliters (ml) and records air volumes. Spiro-metric studies measure lung capacities and rates

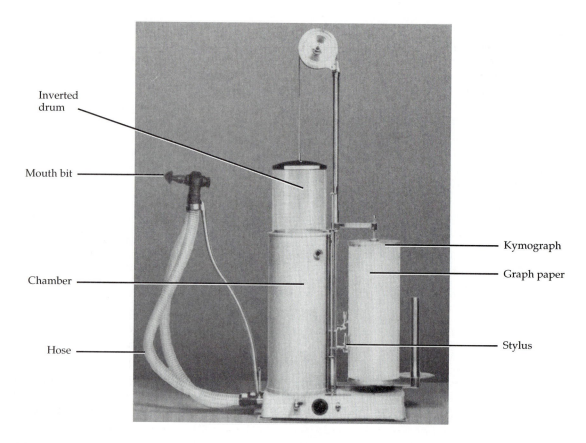

FIGURE 21.17   Collins respirometer. This type is commonly used in college biology laboratories.

and depths of ventilation for diagnostic purposes. Spirometry is indicated for individuals with labored breathing, and is used to diagnose respiratory disorders such as bronchial asthma and emphysema.

The air in the body is at a different temperature than the air contained in the respirometer; body air is also saturated with water vapor. To make the correction for these differences, one measures the temperature of the respirometer and uses Table 21.3 to obtain a conversion factor: *body temperature, pressure (atmospheric), saturated* (with water vapor) *(BTPS).* All measured volumes are then multiplied by this factor.

PROCEDURE

1. *When using the Collins respirometer, the disposable mouthpiece is discarded after use by each subject, and the hose is then detached and rinsed with 70% alcohol.* **CAUTION!** *This procedure must be repeated with every student using the equipment.*
2. Students should work in pairs, with one student operating the respirometer while the other student is tested.

TABLE 21.3
Temperature Variation Conversion Factors

Temperature °C (°F)	Conversion factor
20 (68.0)	1.102
21 (69.8)	1.096
22 (71.6)	1.091
23 (73.4)	1.085
24 (75.2)	1.080
25 (77.0)	1.075
26 (78.8)	1.068
27 (80.6)	1.063
28 (82.4)	1.057
29 (84.2)	1.051
30 (86.0)	1.045
31 (87.8)	1.039
32 (89.6)	1.032
33 (91.4)	1.026
34 (93.2)	1.020
35 (95.0)	1.014
36 (96.8)	1.007
37 (98.6)	1.000

3. Before starting the recording, a little practice may be necessary to learn to inhale and exhale only through your mouth and into the hose. A nose clip may be used to prevent leakage from the nose.

4. Make sure that the respirometer is functioning. Your instructor may have already prepared the instrument for use.
   a. The leveling screws must be raised or lowered so that the bell does not rub on the metal body.
   b. Paper must be fed from a roll onto the kymograph *or* taped onto the kymograph in single sheets.
   c. The soda-lime that absorbs carbon dioxide must be pink (fresh) and not purple (saturated with CO_2).
   d. Pen reservoirs must be full and primed.

5. Raise and lower the drum several times to flush the respirometer of stale air and position the ventilometer pen (the upper pen, usually with black ink) so that it will begin writing in the center of the spirogram paper.

6. Set the free-breathing valve so that the opening is seen through the side of the valve.

7. Place a *sterile or disposable mouthpiece in your mouth* and close your nose with a clamp or with the fingers.

8. Breathe several times to become accustomed to the use of this instrument.

9. Have your lab partner close the free-breathing valve (turn it completely to the opposite direction so that no opening is seen) and turn on the respirometer to the slow speed. At this setting, the kymograph moves 32 mm (the distance between two vertical lines) each minute.

10. Perform the following exercises and record your values in Section D.4 of the LABORATORY REPORT RESULTS at the end of the exercise. In each case, * indicates the placement of the mouth on the mouthpiece.

11. Inspire normally, then * exhale normally three times. This volume is your tidal volume. Repeat two more times and average and record the values.

12. Expire normally, then * exhale as much air as possible three times, recording this volume. This value is your expiratory reserve volume. Repeat two more times and record the values.

13. After taking a deep breath, * exhale as much air as possible three times. This volume is your vital capacity. Repeat two more times and record the values. Compare your vital capacity to the normal value shown in Tables 21.1 and 21.2.

14. Because your vital capacity consists of tidal volume, inspiratory reserve volume, and expiratory reserve volume, and because you have already measured tidal volume in step 11 and expiratory reserve volume in step 12, you can calculate your inspiratory reserve volume by subtracting tidal volume and expiratory reserve volume from vital capacity. Inspiratory reserve volume = (vital capacity) – (tidal volume + expiratory reserve volume). Record this value.

In some cases, an individual with a pulmonary disorder has a nearly normal vital capacity. If the rate of expiration is timed, however, the extent of the pulmonary disorder becomes apparent. In order to do this, an individual expels air into a Collins respirometer as fast as possible and the expired volume is measured per unit of time. Such a test is called *forced expiratory volume* (FEV_T). The T indicates that the volume of air is timed. FEV_1 is the volume of air forcefully expired in 1 sec, FEV_2 is the volume expired in 2 sec, and so on. A normal individual should be able to expel 83% of the total capacity during the first second, 94% in 2 sec, and 97% in 3 sec. For individuals with disorders such as emphysema and asthma, the percentage can be considerably lower, depending on the extent of the problem.

FEV_1 is determined according to the following procedure.

PROCEDURE

1. Apply a noseclip to prevent leakage of air through the nose.
2. Turn on the kymograph.
3. Before placing the mouthpiece in your mouth, inhale as deeply as possible.
4. Expel all the air you can into the mouthpiece.
5. Turn off the kymograph.
6. Draw a vertical line on the spirogram at the starting point of exhalation. Mark this A.
7. Using the Collins VC timed interval ruler, draw a vertical line to the left of A and label it line B. The time between lines A and B is 1 sec.
8. The FEV_1 is the point where the spirogram tracing crosses line B.
9. Record your value here _____
10. Now read your vital capacity from the spirogram and record the value here _____
11. In order to adjust for differences in temperature in the respirometer, use Table 21.3 as a guide. Determine the temperature in the respirometer, find the appropriate conversion factor and multiply the conversion factor by your vital capacity.

12. If, for example, the temperature of the respirometer is 75.2°F (24°C), the conversion factor is 1.080. And, if your vital capacity is 5600 ml, then

$$1.080 \times 5600 \text{ ml} = 6048 \text{ ml}$$

13. Use the same conversion factor and multiply it by your FEV_1. If your FEV_1 is 4000 ml, then

$$1.080 \times 4000 \text{ ml} = 4320 \text{ ml}$$

14. To calculate FEV_1, divide 4320 by 6048.

$$FEV_1 = \frac{4320}{6048} = 71\%$$

15. Repeat the procedure three times and record your FEV_1 in Section D.4 of the LABORATORY REPORT RESULTS at the end of the exercise.

Compare your results using the Collins respirometer with those using the handheld respirometer.

7. Chemical Regulation of Respiration

The size of the thorax is affected by the action of the respiratory muscles. These muscles contract and relax as a result of nerve impulses transmitted to them from centers in the brain. The respiratory center is composed of widely dispersed groups of neurons and is functionally divided into three areas: (1) the medullary rhythmicity area, contained within the medulla oblongata, which appears to control the basic rhythm of respiration; (2) the pneumotaxic (noo-mō-TAK-sik) area in the pons, which facilitates expiration; and (3) the apneustic (ap-NOO-stik) center in the pons, which inhibits expiration.

Although the basic rhythm of respiration is set and coordinated by the respiratory center, the rhythm can be modified in response to the demands of the body by neural input to the center. Among the factors that can alter the rate of respiration are the levels of CO_2 and O_2 within the blood.

Under normal circumstances, the *partial pressure of carbon dioxide within the arteries* (arterial pCO_2) is 40 mm Hg. If there is even a slight increase in arterial pCO_2—a condition known as *hypercapnia*—the central chemoreceptors in the medulla oblongata and peripheral chemoreceptors in the carotid and aortic bodies are stimulated. This causes the inspiratory area to become highly active, and the rate of respiration increases, a condition called *hyperventilation.* This increased rate allows the body to expel more CO_2 until the arterial pCO_2 returns to 40 mm Hg. If arterial pCO_2 is lower than 40 mm Hg *(hypocapnia)*, the central peripheral chemoreceptors are not stimulated, and no stimulatory impulses are sent to the inspiratory center. Consequently, respiratory rate decreases *(hypoventilation)*, and pCO_2 returns to 40 mm Hg.

The oxygen chemoreceptors found within the medulla oblongata and the aortic and carotid bodies are sensitive only to large decreases in the pO_2 because hemoglobin remains about 85% or more saturated at pO_2 values all the way down to 50 mm Hg. If arterial pO_2 falls from a normal of 100 mm Hg to approximately 50 mm Hg, the oxygen chemoreceptors become stimulated and send impulses to the inspiratory center and respirations increase. But if the pO_2 falls much below 50 mm Hg, the cells of the inspiratory area will suffer oxygen starvation and will not respond well to any chemical receptors. They would therefore send fewer impulses to the inspiratory muscles, and the respiration rate would decrease or breathing would cease altogether.

8. Measuring CO_2 During Exhalation

We can measure the effect of various factors on respiration by measuring (1) the CO_2 content of exhaled air or (2) the rate of respiration.

In order to measure CO_2 content, a gas is bubbled through lime water and the time is measured until the liquid becomes cloudy. Lime water is a solution of calcium hydroxide [$Ca(OH)_2$]. The cloudiness occurs because carbon dioxide combines with calcium hydroxide to form calcium carbonate, a white precipitate, in the following reaction.

$$CO_2 + Ca(OH)_2 \rightarrow CaCO_3 + H_2O$$

Carbon dioxide + Calcium hydroxide → Calcium carbonate + Water

PROCEDURE

CAUTION! *Assuming that your partner has no known or apparent cardiac or other health problems and is capable of such an activity, ask her/him to perform the following activities.*

1. Label three test tubes N, H, and E (Normal, Hyper-ventilation, Exercise). Fill each about half full of calcium hydroxide [$Ca(OH)_2$] solution.

2. Start the stopwatch and *exhale* normal expirations through the straw into tube N. Note the time elapsed until the solution becomes turbid.
3. Hyperventilate for about 30 sec. **CAUTION!** *Stop hyperventilation at the first sign of dizziness or lightheadedness.* Repeat step 2, *exhaling* normally into tube H.
4. After a 3-min rest, exercise briskly for 2 min.
5. Repeat step 2, *exhaling* into tube E.
6. Record all times in Section D.5 of the LABORATORY REPORT RESULTS at the end of the exercise.

In which situation was the CO_2 content of

exhaled air the greatest? _____

Why? _____

_____

E. LABORATORY TESTS COMBINING RESPIRATORY AND CARDIOVASCULAR INTERACTIONS

1. *Experimental setup* Review earlier explanations on recording the following:
   a. Respiratory movements with a pneumograph (Section D.5 in Exercise 21).
   b. Blood pressure with sphygmomanometer and stethoscope (Section F in Exercise 19).
   c. Radial pulse (Section E.1 in Exercise 19).

CAUTION! *Assuming that your partner has no known or apparent cardiac or other health problems and is capable of such an activity, ask her/him to perform the following activities after attaching the various pieces of apparatus in order to record respiratory movements and blood pressure.*

2. *Experimental procedure* Some form of regulated exercise is necessary for this experiment. Choose one of the following forms of exercise to have your subject participate in:
   a. *Riding exercise cycle* If this form is chosen, set the resistance to be felt while riding the cycle, but *not so high* that the subject cannot complete the 4-min exercise period without difficulty.
   b. *Harvard Step Test* In this form of exercise the subject is to step up onto a 20-in. plat-

form (a chair will substitute quite well) with one foot at a time. The subject is to bring *both* feet up onto the platform before stepping back down to the floor. The subject is also to remain erect at all times, and to do 30 complete cycles (up onto the platform and back down) per minute.

3. *Experimental protocol* Record respiratory movements, blood pressure, and pulse during each of the procedures listed. Record your data in the table provided in Section E of the LABORATORY REPORT RESULTS at the end of the exercise. All data should be recorded *simultaneously*, thereby requiring participation of all members of the experimental group.
   a. *Basal readings* Have the subject sit erect and quiet for 3 min. Obtain readings for

   Respiratory rate and depth

   Systolic and diastolic blood pressure

   Pulse rate

   When the basal readings have been recorded, obtain additional readings after (1) sitting quietly on the exercise cycle for 3 min, if this form of exercise is to be utilized, or (2) standing quietly for 3 min in front of the platform that is to be utilized for the Harvard Step Test.
   b. *Readings after 1 min of exercise* After the subject has exercised for 1 min, obtain additional readings.
   c. *Readings after 2 and 3 min of exercise* Again, obtain additional readings after the subject has completed 2 and 3 min of exercise.
   d. *Readings upon completion of exercise* When the subject has completed 4 min of exercise, obtain readings for respiratory rate, respiratory depth, heart rate, and systolic and diastolic blood pressure immediately upon completion *while the subject remains seated on the exercise cycle or stands erect on the floor*, depending upon the type of exercise utilized.
   e. *Readings 1, 2, 3, and 5 min after completion of exercise* With the subject *still sitting on the exercise cycle or still standing* erect on the floor obtain additional readings at the above time intervals after completion of the exercise period.

ANSWER THE LABORATORY REPORT QUESTIONS AT THE END OF THE EXERCISE.

Respiratory System 21

Student _____   Date _____

Laboratory Section _____   Score/Grade _____

SECTION D. LABORATORY TESTS ON RESPIRATION

1. Measurement of Chest and Abdomen in Respiration

Record the results of your chest measurements in inches, in the following table.

Size after a	Chest	Abdomen
Normal inspiration		
Normal expiration		
Maximal inspiration		
Maximal expiration		

2. Use of Pneumograph

Attach a sample of any one of the following activities.

Reading
Swallowing water
Laughing
Yawning
Coughing
Sniffing

3. Use of Handheld Respirometer

Tidal volume _____

Expiratory reserve volume _____

Vital capacity _____

Inspiratory reserve volume _____

Minute volume of respiration (MVR) _____

4. Use of Collins Respirometer

Record the results of your exercises using the Collins respirometer in the following table.

	Tidal volume (1)	Expiratory reserve (2)	Vital capacity (3)	Inspiratory reserve (4)	FEV$_1$ (5)
First time	ml	ml	ml	ml	ml
Second time	ml	ml	ml	ml	ml
Third time	ml	ml	ml	ml	ml
Your average	ml	ml	ml	ml	ml
Normal value	ml	ml	ml	ml	ml

5. Measuring CO$_2$ During Exhalation

	Time until turbidity
Tidal respirations	
After hyperventilation	
After exercise	

SECTION E. COMBINED RESPIRATORY AND CARDIOVASCULAR INTERACTIONS

Record the results of your exercises in the following table.

	Respiratory rate and depth	Systolic and diastolic pressure	Pulse rate
Basal readings while sitting erect and quiet for 3 min			
Basal readings while standing on cycle or in front of platform for 3 min			
Readings after 1 min of exercise			
Readings after 2 min of exercise			
Readings after 3 min of exercise			
Readings after 4 min of exercise			
Readings after 1 min following completion of exercise			
Readings after 2 min following completion of exercise			
Readings after 3 min following completion of exercise			
Readings after 5 min following completion of exercise			

Respiratory System 21

Student _____   Date _____

Laboratory Section _____   Score/Grade _____

PART 1. Multiple Choice

_____ 1. The overall exchange of gases between the atmosphere, blood, and cells is called (a) inspiration (b) respiration (c) expiration (d) none of these

_____ 2. The lateral walls of the internal nose are formed by the ethmoid bone, maxillae, lacrimal, inferior conchae, and the (a) hyoid bone (b) nasal bone (c) palatine bone (d) occipital bone

_____ 3. The portion of the pharynx that contains the pharyngeal tonsils is the (a) oropharynx (b) laryngopharynx (c) nasopharynx (d) pharyngeal orifice

_____ 4. The Adam's apple is a common term for the (a) thyroid cartilage (b) cricoid cartilage (c) epiglottis (d) none of these

_____ 5. The C-shaped rings of cartilage of the trachea not only prevent the trachea from collapsing but also aid in the process of (a) lubrication (b) removing foreign particles (c) gas exchange (d) swallowing

_____ 6. Of the following structures, the smallest in diameter is the (a) left primary bronchus (b) bronchioles (c) secondary bronchi (d) alveolar ducts

_____ 7. The structures of the lung that actually contain the alveoli are the (a) respiratory bronchioles (b) fissures (c) lobules (d) terminal bronchioles

_____ 8. From superficial to deep, the structure(s) that you would encounter first among the following is (are) the (a) bronchi (b) parietal pleura (c) pleural cavity (d) secondary bronchi

PART 2. Completion

9. An advantage of nasal breathing is that the air is warmed, moistened, and _____.

10. Improper fusion of the palatine and maxillary bones results in a condition called

_____.

11. The protective lid of cartilage that prevents food from entering the trachea is the

_____.

12. After removal of the _____ an individual would be unable to speak.

13. The upper respiratory tract is able to trap and remove dust because of its lining of

_____.

14. The passage of a tube into the mouth and down through the larynx and trachea to bypass an

    obstruction is called _____.

15. A radiograph of the bronchial tree after administration of an iodinated medium is called a

    _____.

16. The sequence of respiratory tubes from largest to smallest is trachea, primary bronchi, secondary

    bronchi, bronchioles, _____, respiratory bronchioles, and alveolar ducts.

17. Both the external and internal nose are divided internally by a vertical partition called the

    _____.

18. The undersurface of the external nose contains two openings called the nostrils or

    _____.

19. The functions of the pharynx are to serve as a passageway for air and food and to provide a resonat-

    ing chamber for _____.

20. An inflammation of the membrane that encloses and protects the lungs is called

    _____.

21. The anterior portion of the nasal cavity just inside the nostrils is called the _____.

22. Groovelike passageways in the nasal cavity formed by the conchae are called

    _____.

23. The portion of the pharynx that contains the palatine and lingual tonsils is the

    _____.

24. Each bronchopulmonary of a lung is subdivided into many compartments called

    _____.

25. A(n) _____ is an outpouching lined by squamous epithelium and supported by a
    thin elastic membrane.

26. The _____ cartilage attaches the larynx to the trachea.

27. The portion of a lung that rests on the diaphragm is the _____.

28. Phagocytic cells in the alveolar wall are called _____.

29. The surface of a lung lying against the ribs is called the _____ surface.

30. The _____ is a structure in the medial surface of a lung through which bronchi,
    blood vessels, lymphatic vessels, and nerves pass.

31. The nose, pharynx, and associated structures comprise the _____ respiratory sys-
    tem.

32. The _____ is the rounded, anterior border of the nose that connects the root and
    apex.

33. The _____ cartilages of the larynx attach the vocal folds to the intrinsic pharyngeal
    muscles.

34. Each _____ of a lung is supplied by a tertiary bronchus.

PART 3. Matching

_____	**35.** Tidal volume	A.	1200 ml of air
_____	**36.** Inspiratory reserve volume	B.	3600 ml of air
_____	**37.** Inspiratory capacity	C.	4800 ml of air
_____	**38.** Expiratory reserve volume	D.	500 ml of air
_____	**39.** Vital capacity	E.	3100 ml of air
_____	**40.** Total lung capacity	F.	6000 ml of air

Digestive System

22

Digestion occurs basically as two events—mechanical digestion and chemical digestion. *Mechanical digestion* consists of various movements of the gastrointestinal tract that help chemical digestion. These movements include physical breakdown of food by the teeth and complete churning and mixing of this food with enzymes by the smooth muscles of the stomach and small intestine. *Chemical digestion* consists of a series of catabolic (hydrolysis) reactions that break down the large nutrient molecules that we eat, such as carbohydrates, lipids, and proteins, into much smaller molecules that can be absorbed and used by body cells.

A. GENERAL ORGANIZATION OF DIGESTIVE SYSTEM

Digestive organs are usually divided into two main groups. The first is the *gastrointestinal (GI) tract*, or *alimentary* (*alimentum* = nourishment) *canal*, a continuous tube running from the mouth to the anus, and measuring about 9 m (30 ft) in length in a cadaver. This tract is composed of the mouth, pharynx, esophagus, stomach, small intestine, and large intestine. The small intestine has three regions: duodenum, jejunum, and ileum. The large intestine has four regions: cecum, colon, rectum, and anal canal. The colon is divided into ascending colon, transverse colon, descending colon, and sigmoid colon.

The second group of organs composing the digestive system consists of the *accessory structures* such as the teeth, tongue, salivary glands, liver, gallbladder, and pancreas (see Figure 22.1).

Using your textbook, charts, or models for reference, label Figure 22.1.

The wall of the gastrointestinal tract, especially from the stomach to the anal canal, has the same basic arrangement of tissues. The four layers (tunics) of the tract, from the deep to superficial, are the *mucosa, submucosa, muscularis*, and *serosa* (see Figure 22.9).

Inferior to the diaphragm, the serosa is also called the *peritoneum* (per'-i-tō-NĒ-um; *peri* = around; *tonos* = tension). The peritoneum is composed of a layer of simple squamous epithelium (called mesothelium) and an underlying layer of connective tissue. The *parietal peritoneum* lines the wall of the abdominopelvic cavity, and the *visceral peritoneum* covers some of the organs in the cavity. The potential space between the parietal and visceral portions of the peritoneum is called the *peritoneal cavity*. Unlike the two other serous membranes of the body, the pericardium and the pleura, which smoothly cover the heart and lungs, the peritoneum contains large folds that weave in between the viscera. The important extensions of the peritoneum are the *mesentery* (MEZ-en-ter'-ē; *meso* = middle; *enteron* = intestine) *mesocolon, falciform* (FAL-si-form) *ligament, lesser omentum* (ō-MENT-um), and *greater omentum.*

Inflammation of the peritoneum, called *peritonitis,* is a serious condition because the peritoneal membranes are continuous with one another, enabling the infection to spread to all the organs in the cavity.

B. ORGANS OF DIGESTIVE SYSTEM

1. Mouth (Oral Cavity)

The *mouth*, also called the *oral*, or *buccal* (BUK-al; *bucca* = cheeks), *cavity*, is formed by the cheeks, hard and soft palates, and tongue. The *hard palate* forms the anterior portion of the roof of the mouth and the *soft palate* forms the posterior portion. The *tongue* forms the floor of the oral cavity and is composed of skeletal muscle covered by mucous membrane. Partial digestion of carbohydrates and triglycerides occurs in the mouth.

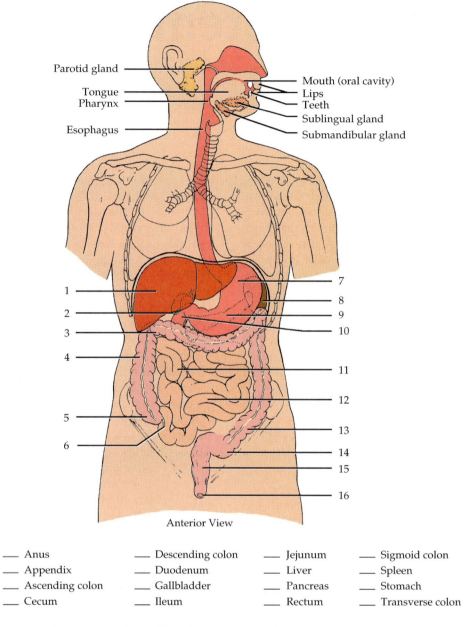

Parotid gland

Tongue
Pharynx

Esophagus

Mouth (oral cavity)
Lips
Teeth
Sublingual gland
Submandibular gland

1
2
3
4
5
6

7
8
9
10
11
12
13
14
15
16

Anterior View

___ Anus	___ Descending colon	___ Jejunum	___ Sigmoid colon
___ Appendix	___ Duodenum	___ Liver	___ Spleen
___ Ascending colon	___ Gallbladder	___ Pancreas	___ Stomach
___ Cecum	___ Ileum	___ Rectum	___ Transverse colon

FIGURE 22.1   Organs of the digestive system and related structures.

a. **Cheeks** Lateral walls of oral cavity. Muscular structures covered by skin and lined by nonkeratinized stratified squamous epithelium; anterior portions terminate in the *superior* and *inferior labia* (lips).

b. **Vermilion** (ver-MIL-yon) Transition zone of lips where outer skin and inner mucous membranes meet.

c. **Labial frenulum** (LĀ-bē-al FREN-yoo-lum; *labium* = fleshy border; *frenulum* = small bridle) Midline fold of mucous membrane that attaches the inner surface of each lip to its corresponding gum.

d. **Vestibule** (= entrance to a canal) Space bounded externally by cheeks and lips and internally by gums and teeth.

e. **Oral cavity proper** Space extending from the gums and teeth to the *fauces* (FAW-sēz; *fauces* = passages), opening of oral cavity proper with pharynx. Area is enclosed by the dental arches.

f. **Hard palate** Formed by maxillae and palatine bones and covered by mucous membrane.

g. *Soft palate* Arch-shaped muscular partition between oropharynx and nasopharynx lined by mucous membrane. Hanging from free border of soft palate is a muscular projection, the *uvula* (YOU-vyoo-la = little grape).

h. *Palatoglossal arch (anterior pillar)* Muscular fold that extends inferiorly, laterally, and anteriorly to the side of the base of tongue.

i. *Palatopharyngeal* (PAL-a-tō-fa-rin'-jē-al) *arch (posterior pillar)* Muscular fold that extends inferiorly, laterally, and posteriorly to the side of pharynx. *Palatine tonsils* are between arches and *lingual tonsil* is at base of tongue.

j. *Tongue* Movable, muscular organ on floor of oral cavity. *Extrinsic muscles* originate outside tongue (to bones in the area), insert into connective tissues, and move tongue from side to side and in and out to maneuver food for chewing and swallowing; *intrinsic muscles* originate and insert into connective tissues within the tongue and alter shape and size of tongue for speech and swallowing.

k. *Lingual* (*lingua* = tongue) *frenulum* Midline fold of mucous membrane on undersurface of tongue that helps restrict its movement posteriorly.

l. *Papillae* (pa-PIL-ē = nipple-shaped projections) Projections of lamina propria on surface of tongue covered with epithelium; *filiform* (= threadlike) *papillae* are conical projections in parallel rows over anterior two-thirds of tongue; *fungiform* (= shaped like a mushroom) *papillae* are mushroomlike elevations distributed among filiform papillae and more numerous near tip of tongue (appear as red dots and most contain taste buds); *circumvallate* (*circum* = around; *vallare* = to wall) *papillae* are arranged in the form of an inverted V on the posterior surface of tongue (all contain taste buds).

Using a mirror, examine your mouth and locate as many of the structures (a through l) as you can. Label Figure 22.2.

2. Salivary Glands

Most saliva is secreted by the *salivary glands*, which lie outside the mouth and pour their contents into ducts that empty into the oral cavity. The carbohydrate-digesting enzyme in saliva is salivary amylase. The three pairs of salivary glands are the *parotid* (*para* = near; *otia* = ear) *glands* (anterior and inferior to the ears), which secrete into the oral cavity vestibule through *parotid (Stensen's) ducts; submandibular glands* (deep to the base of the tongue in the posterior part of the floor of the mouth), which secrete on either side of the lingual frenulum in the floor of the oral cavity through *submandibular (Wharton's) ducts;* and *sublingual glands* (superior to the submandibular glands), which secrete into the floor of the oral cavity through *lesser sublingual (Rivinus') ducts.*

Label Figure 22.3 on page 523.

The parotid glands are compound tubuloacinar glands, whereas the submandibulars and sublinguals are compound acinar glands (see Figure 22.4).

Examine prepared slides of the three different types of salivary glands and compare your observations to Figure 22.4 on page 523.

3. Teeth

Teeth (dentes) are located in the sockets of the alveolar processes of the mandible and maxillae. The alveolar processes are covered by *gingivae* (jin-JĪ-vē) or gums, which extend slightly into each socket. The sockets are lined by a dense fibrous connective tissue called a *periodontal* (*peri* = around; *odous* = tooth) *ligament*, which anchors the teeth in position and acts as a shock absorber during chewing.

Following are the parts of a tooth:

a. *Crown* Exposed portion above level of gums.

b. *Root* One to three projections embedded in socket.

c. *Neck* Constricted junction line of the crown and root near the gum line.

d. *Dentin* Calcified connective tissue that gives teeth their basic shape and rigidity.

e. *Pulp cavity* Enlarged part of cavity in crown within dentin.

f. *Pulp* Connective tissue containing blood vessels, lymphatic vessels, and nerves.

g. *Root canal* Narrow extension of pulp cavity in root.

h. *Apical foramen* Opening in base of root canal through which blood vessels, lymphatic vessels, and nerves enter tooth.

i. *Enamel* Covering of crown that consists primarily of calcium phosphate and calcium carbonate.

j. *Cementum* Bonelike substance that covers and attaches root to periodontal ligament.

With the aid of your textbook, label the parts of a tooth shown in Figure 22.5 on page 524.

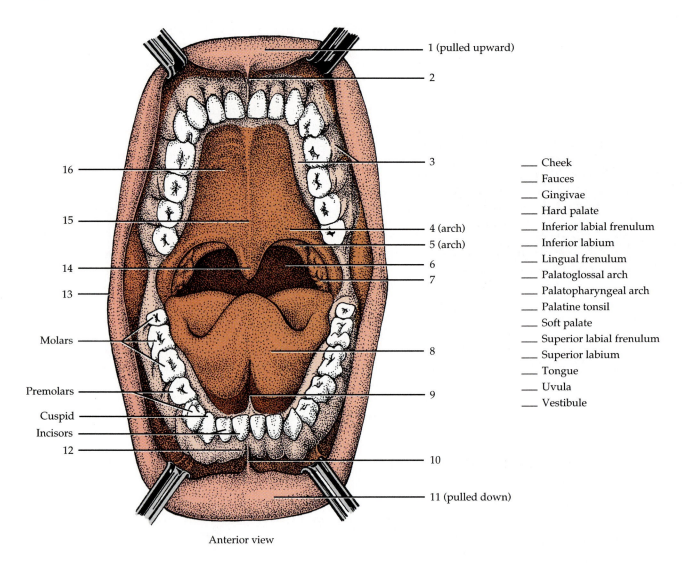

1 (pulled upward)

2

___ Cheek
___ Fauces
___ Gingivae
3
___ Hard palate
4 (arch)
___ Inferior labial frenulum
5 (arch)
___ Inferior labium
6
___ Lingual frenulum
7
___ Palatoglossal arch
___ Palatopharyngeal arch
___ Palatine tonsil
___ Soft palate
8
___ Superior labial frenulum
___ Superior labium
9
___ Tongue
___ Uvula
___ Vestibule

16

15

14

13

Molars

Premolars

Cuspid

Incisors

12

10

11 (pulled down)

Anterior view

FIGURE 22.2 Mouth (oral cavity).

4. Dentitions

Dentitions (sets of teeth) are of two types: *deciduous* (baby) and *permanent*. Deciduous teeth begin to erupt at about 6 months of age, and one pair appears at about each month thereafter until all 20 are present. The deciduous teeth are as follows:

a. *Incisors* Central incisors closest to midline, with lateral incisors on either side. Incisors are chisel-shaped, adapted for cutting into food, have only one root.
b. *Cuspids (canines)* Posterior to incisors. Cuspids have pointed surfaces (cusps) for tearing and shredding food, have only one root.
c. *Molars* First and second molars posterior to canines. Molars crush and grind food. Upper molars have four cusps and three roots, lower molars have four cusps and two roots.

All deciduous teeth are usually lost between 6 and 12 years of age and replaced by permanent dentition consisting of 32 teeth that appear between age 6 and adulthood. The permanent teeth are:

a. *Incisors* Central incisors and lateral incisors replace those of deciduous dentition.
b. *Cuspids (canines)* These replace those of deciduous dentition.
c. *Premolars (bicuspids)* First and second premolars replace deciduous molars. Premolars crush and grind food, have two cusps and one root (upper first premolars have two roots).
d. *Molars* These erupt behind premolars as jaw grows to accommodate them and do not replace any deciduous teeth. First molars erupt at age 6, second at age 12, and third (wisdom teeth) after age 18.

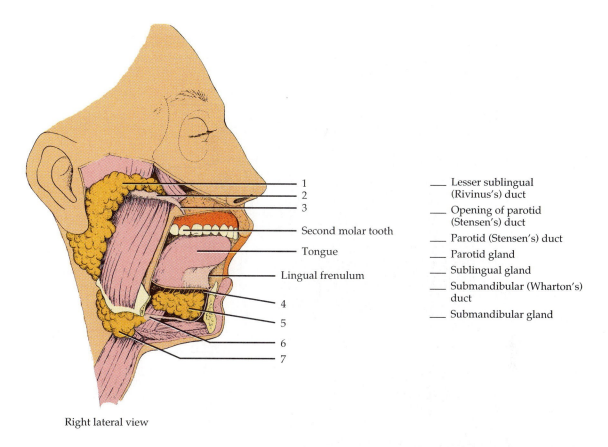

1
2
3
___ Second molar tooth
___ Tongue
___ Lingual frenulum
4
5
6
7

___ Lesser sublingual (Rivinus's) duct
___ Opening of parotid (Stensen's) duct
___ Parotid (Stensen's) duct
___ Parotid gland
___ Sublingual gland
___ Submandibular (Wharton's) duct
___ Submandibular gland

Right lateral view

FIGURE 22.3   Location of salivary glands.

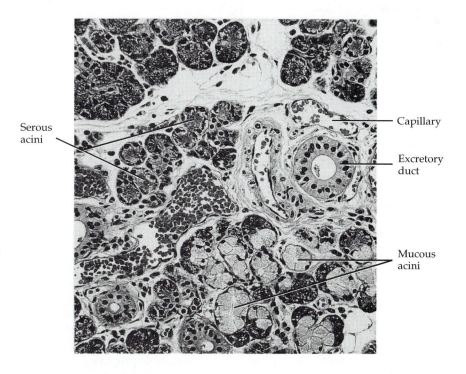

Serous acini

Capillary

Excretory duct

Mucous acini

FIGURE 22.4   Histology of salivary glands.

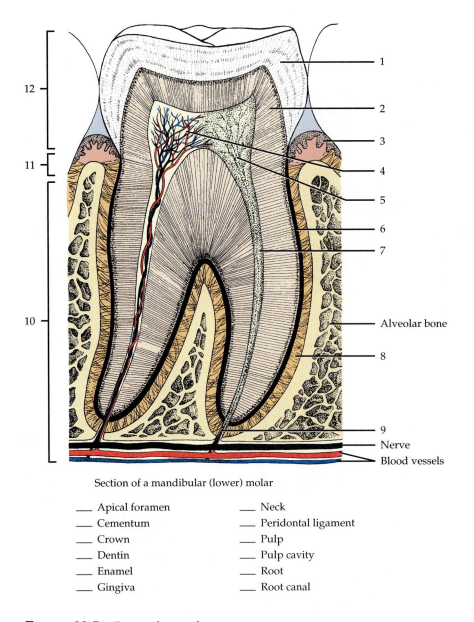

12

11

10

1

2

3

4

5

6

7

Alveolar bone

8

9

Nerve

Blood vessels

Section of a mandibular (lower) molar

____ Apical foramen

____ Cementum

____ Crown

____ Dentin

____ Enamel

____ Gingiva

____ Neck

____ Peridontal ligament

____ Pulp

____ Pulp cavity

____ Root

____ Root canal

FIGURE 22.5   Parts of a tooth.

Using a mirror, examine your mouth and locate as many teeth of the permanent dentition as you can.

With the aid of your textbook, label the deciduous and permanent dentitions in Figure 22.6.

5. Esophagus

The **esophagus** (e-SOF-a-gus; *oisein* = to carry; *phagema* = food) is a muscular, collapsible tube posterior to the trachea. The structure is 23 to 25 cm (10 in.) long and extends from the laryngopharynx through the mediastinum and esophageal hiatus in the diaphragm and terminates in the superior portion of the stomach. The esophagus conveys food from the pharynx to the stomach by peristalsis.

Histologically, the esophagus consists of a **mucosa** (nonkeratinized stratified squamous epithelium, lamina propria, muscularis mucosae), **submucosa** (areolar connective tissue, blood vessels, mucous glands), **muscularis** (superior third striated, middle third striated and smooth, inferior third smooth), and **adventitia** (ad-ven-TISH-ya). The esophagus is not covered by a serosa.

Examine a prepared slide of a cross section of the esophagus that shows its various coats. With the aid of your textbook, label Figure 22.7.

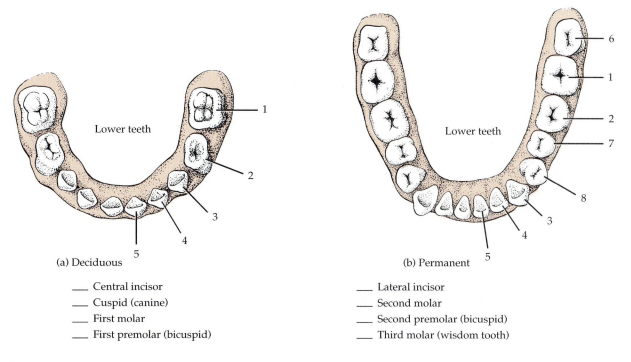

FIGURE 22.6   Dentitions.

(a) Deciduous

Lower teeth

___ Central incisor
___ Cuspid (canine)
___ First molar
___ First premolar (bicuspid)

(b) Permanent

Lower teeth

___ Lateral incisor
___ Second molar
___ Second premolar (bicuspid)
___ Third molar (wisdom tooth)

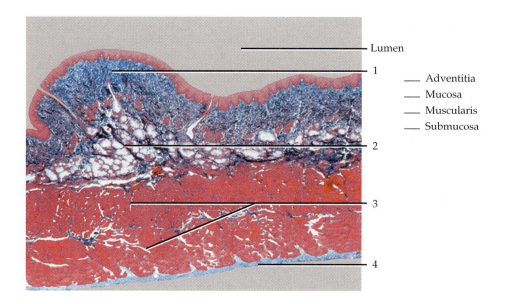

Lumen

___ Adventitia
___ Mucosa
___ Muscularis
___ Submucosa

FIGURE 22.7   Histology of esophagus.

6. Stomach

The **stomach** is a J-shaped enlargement of the gastrointestinal tract inferior to the diaphragm (see Figure 22.1). It is in the epigastric, umbilical, and left hypochondriac regions of the abdomen. The superior part is connected to the esophagus; the inferior part empties into the duodenum, the first portion of the small intestine. The stomach is divided into four main areas: cardia, fundus, body, and pylorus. The **cardia** (CAR-dē-a) surrounds the lower esophageal sphincter, a physiological sphincter in the esophagus just superior to the diaphragm. The rounded portion superior to and to the left of the cardia is the *fundus* (FUN-dus). Inferior to the fundus, the large central portion of the stomach is called the *body*. The narrow, inferior region is the *pylorus* (pī-LOR-us; *pyle* = gate; *ouros* = guard). The pylorus consists of a *pyloric antrum* (AN-trum = cave), which is closer to the body of the stomach, and a *pyloric canal*, which is closer to the duodenum. The concave medial border of the stomach is called the *lesser curvature*, and the convex lateral border is the *greater curvature*. The pylorus communicates with the duodenum of the small intestine via a sphincter called the *pyloric sphincter (valve)*. The main chemical activity of the stomach is to begin the digestion of proteins.

Label Figure 22.8.

The *mucosa* of the stomach consists of simple columnar epithelium, lamia propria, and muscularis mucosae. The mucosa is arranged in large folds called *rugae* (ROO-jē = wrinkles). The columnar epithelium of the mucosa contains many narrow channels that extend down into the lamina propria and are referred to as *gastric glands (pits)*. The glands are lined with four kinds of cells: (1) *chief (zymogenic) cells* that secrete inactive pepsinogen, which is converted to active pepsin, a protein-digesting enzyme; (2) *parietal (oxyntic) cells* that secrete hydrochloric acid and intrinsic factor; (3) *mucous neck cells* that secrete mucus and (4) *enteroendocrine cells* called *G cells* that secrete the hormone gastrin. The *submucosa* consists of areolar connective tissue. The *muscularis* has three layers of smooth muscle—superficial longitudinal, middle circular, and deep oblique. The oblique layer is limited chiefly to the body of the stomach. The *serosa* is part of the visceral peritoneum.

Examine a prepared slide of a section of the stomach that shows its various layers. With the aid of your textbook, label Figure 22.9.

7. Pancreas

The *pancreas* (*pan* = all; *kreas* = flesh) is a retroperitoneal gland posterior to the greater curvature of the stomach (see Figure 22.1). The gland consists of a *head* (expanded portion near duodenum), *body* (central portion), and *tail* (terminal tapering portion).

Histologically, the pancreas consists of *pancreatic islets (islets of Langerhans)* that contain (1) glucagon-producing *alpha cells*, (2) insulin-producing *beta cells*, (3) somatostatin-producing *delta cells*, and pancreatic polypeptide-producing *F cells* (see Figure 15.5). The pancreas also consists of *acini* that produce pancreatic juice (see Figure

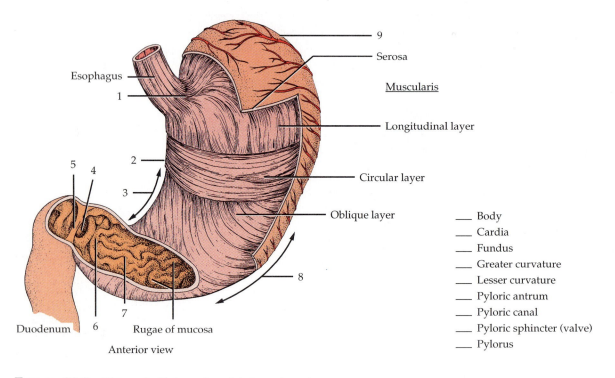

Esophagus

1

Serosa

Muscularis

Longitudinal layer

2

Circular layer

3

5 4

Oblique layer

Duodenum 6

7

Rugae of mucosa

8

9

___ Body
___ Cardia
___ Fundus
___ Greater curvature
___ Lesser curvature
___ Pyloric antrum
___ Pyloric canal
___ Pyloric sphincter (valve)
___ Pylorus

Anterior view

FIGURE 22.8 Stomach. External and internal anatomy.

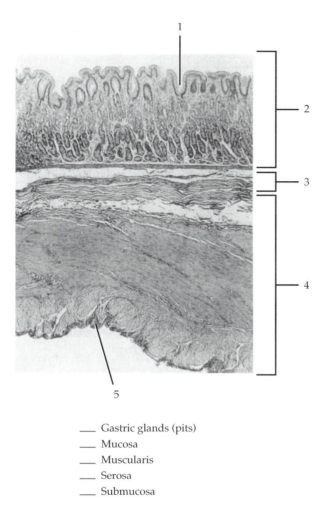

_____ Gastric glands (pits)

_____ Mucosa

_____ Muscularis

_____ Serosa

_____ Submucosa

FIGURE 22.9 Histology of stomach.

15.5). Pancreatic juice contains enzymes that assist in the chemical breakdown of carbohydrates, proteins, triglycerides, and nucleic acids.

Pancreatic juice is delivered from the pancreas to the duodenum by a large main tube, the *pancreatic duct (duct of Wirsung)*. This duct unites with the common bile duct from the liver and pancreas and enters the duodenum in a common duct called the *hepatopancreatic ampulla (ampulla of Vater)*. The ampulla opens on an elevation of the duodenal mucosa, the *duodenal papilla*. An *accessory pancreatic duct (duct of Santorini)* may also lead from the pancreas and empty into the duodenum about 2.5 cm (1 in.) superior to the hepato-pancreatic ampulla. With the aid of your textbook, label the structures associated with the pancreas in Figure 22.10.

8. Liver

The *liver* is located inferior to the diaphragm (see Figure 22.1). It occupies most of the right hypochon-driac and part of the epigastric regions of the abdomen. The gland is divided into two principal lobes, the *right lobe* and *left lobe*, separated by the *falciform ligament*. The falciform ligament attaches the liver to the anterior abdominal wall and diaphragm. The right lobe consists of an inferior *quadrate lobe* and a posterior *caudate lobe*.

Each lobe is composed of microscopic functional units called *lobules*. Among the structures in a lobule are cords of *hepatocytes (liver cells)* arranged in a radial pattern around a *central vein; sinusoids*, endothelial lined spaces between hepatocytes through which blood flows; and *stellate reticuloendothelial (Kupffer) cells* that destroy bacteria and worn-out blood cells by phagocytosis.

Examine a prepared slide of several liver lobules. Compare your observations with Figure 22.11.

Bile is manufactured by hepatocytes and functions in the emulsification of triglycerides in the small intestine. The liquid is passed to the small intestine as follows: Hepatocytes secrete bile into *bile canaliculi* (kan'-a-LIK-yoo-lī = small canals) that empty into small ducts. The small ducts merge into larger *right* and *left hepatic ducts*, one in each principal lobe of the liver. The right and left hepatic ducts unite outside the liver to form a single *common hepatic duct*. This duct joins the *cystic (kystis =* bladder) *duct* from the gallbladder to become the *common bile duct*, which empties into the duodenum at the hepatopancreatic ampulla (ampulla of Vater). When triglycerides are not being digested, a valve around the hepatopancreatic ampulla, the *sphincter of the hepatopancreatic ampulla (sphincter of Oddi)*, closes, and bile backs up into the gallbladder via the cystic duct. In the gallbladder, bile is stored and concentrated.

With the aid of your textbook, label the structures associated with the liver in Figure 22.10.

9. Gallbladder

The *gallbladder* (*galla* = bile) is a pear-shaped sac in a fossa along the posterior surface of the liver (see Figure 22.1). The gallbladder stores and concentrates bile. The cystic duct of the gallbladder and common hepatic duct of the liver merge to form the common bile duct. The *mucosa* of the gallbladder consists of simple columnar epithelium that contains rugae. The *muscularis* consists of smooth muscle and the outer coat consists of visceral peritoneum.

With the aid of your textbook, label the structures associated with the gallbladder in Figure 22.10.

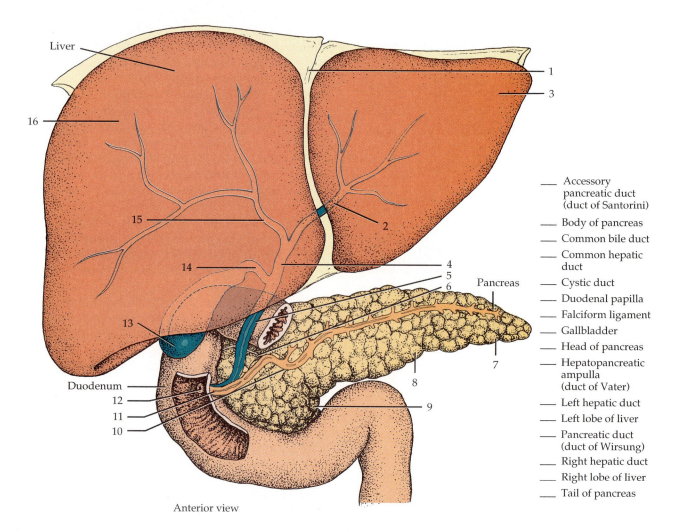

Liver

16

15

14

13

Duodenum
12
11
10

1

3

2

4
5
6

Pancreas

7

8

9

____ Accessory
        pancreatic duct
        (duct of Santorini)
____ Body of pancreas
____ Common bile duct
____ Common hepatic
        duct
____ Cystic duct
____ Duodenal papilla
____ Falciform ligament
____ Gallbladder
____ Head of pancreas
____ Hepatopancreatic
        ampulla
        (duct of Vater)
____ Left hepatic duct
____ Left lobe of liver
____ Pancreatic duct
        (duct of Wirsung)
____ Right hepatic duct
____ Right lobe of liver
____ Tail of pancreas

Anterior view

FIGURE 22.10 Relations of the liver, gallbladder, duodenum, and pancreas.

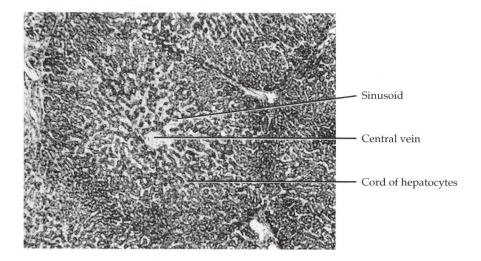

Sinusoid

Central vein

Cord of hepatocytes

FIGURE 22.11 Photomicrograph of liver lobule.

10. Small Intestine

The bulk of digestion and absorption occurs in the **small intestine**, which begins at the pyloric sphincter (valve) of the stomach, coils through the central and inferior part of the abdomen, and joins the large intestine at the ileocecal sphincter (see Figure 22.1). The mesentery attaches the small intestine to the posterior abdominal wall. The small intestine is about 6.35 m (21 ft) long and is divided into three segments: **duodenum** (doo'-ō-DĒ-num), which begins at the stomach; **jejunum** (jē-JOO-num), the middle segment; and **ileum** (IL-ē-um), which terminates at the large intestine (Figure 22.12).

Histologically, the **mucosa** contains many pits lined with glandular epithelium called **intestinal glands (crypts of Lieberkühn)**; they secrete enzymes that digest carbohydrates, proteins, and nucleic acids (see Figure 22.13). Some of the simple columnar cells of the mucosa are **goblet cells** that secrete

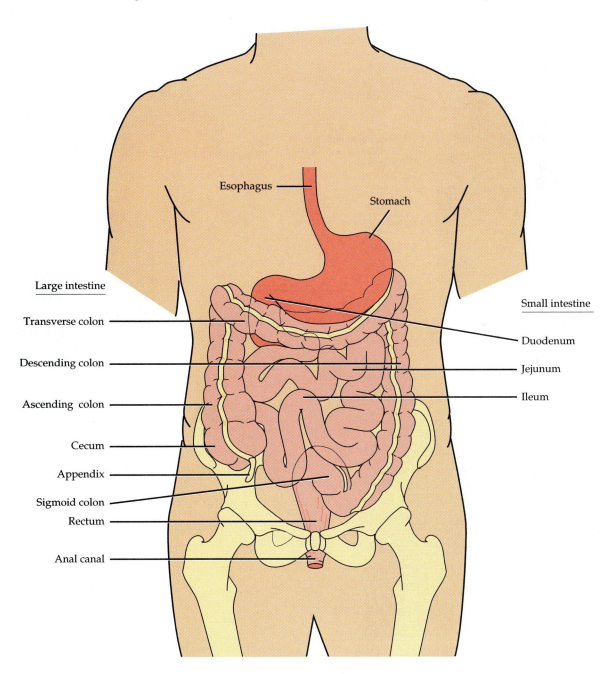

Esophagus

Stomach

Large intestine

Transverse colon

Descending colon

Ascending colon

Cecum

Appendix

Sigmoid colon

Rectum

Anal canal

Small intestine

Duodenum

Jejunum

Ileum

FIGURE 22.12   Intestines. Note the parts of the small intestine on the right side of the illustration and the parts of the large intestine on the left side.

mucus (see Figure 22.13); others contain *microvilli* to increase the surface area for absorption (see Figure 3.1). The mucosa contains a series of finger-like projections, the *villi* (see Figure 22.13). Each villus contains a blood capillary and lymphatic vessel called a *lacteal*; these absorb digested nutrients. The 4 million to 5 million villi in the small intestine greatly increase the surface area for absorption. The mucosa and *submucosa* also contain deep permanent folds, the *circular folds,* or *plicae circulares* (PLĪ-kē SER-kyoo-lar-es), which also help to increase the surface area for absorption. In the sub-mucosa of the duodenum are *duodenal (Brunner's) glands,* which secrete an alkaline mucus to protect the mucosa from excess acid and the action of digestive enzymes. The *muscularis* of the small intestine consists of an outer longitudinal layer and an inner circular layer of smooth muscle. Except for a major portion of the duodenum, the *serosa* (visceral peritoneum) completely covers the small intestine.

Obtain prepared slides of the small intestine (section through its tunics and villi) and identify as many structures as you can, using Figure 22.13 and your textbook as references.

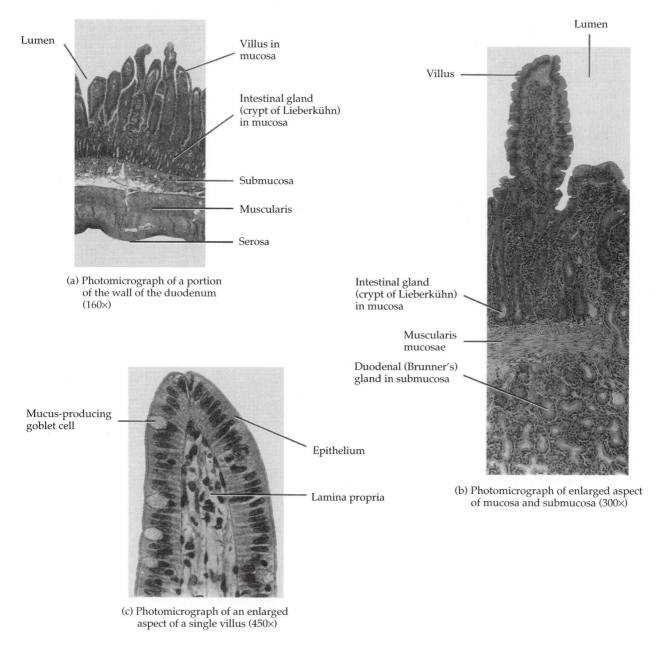

(a) Photomicrograph of a portion of the wall of the duodenum (160×)

(b) Photomicrograph of enlarged aspect of mucosa and submucosa (300×)

(c) Photomicrograph of an enlarged aspect of a single villus (450×)

FIGURE 22.13   Histology of small intestine.

11. Large Intestine

The *large intestine* functions in the completion of absorption of water that leads to the formation of feces, and in the expulsion of feces from the body. Bacteria residing in the large intestine manufacture certain vitamins (some B vitamins and vitamin K). The large intestine is about 1½ m (5 ft) long and extends from the ileum to the anus (see Figure 22.12). It is attached to the posterior abdominal wall by an extension of visceral peritoneum called mesocolon. The large intestine is divided into four principal regions: cecum, colon, rectum, and anal canal.

The opening from the ileum into the large intestine is guarded by a fold of mucous membrane, the *ileocecal sphincter* (*valve*). Hanging below the valve is a blind pouch, the *cecum*, to which is attached the *vermiform appendix* (*vermis* = worm; *appendix* = appendage) by an extension of visceral peritoneum called the *mesoappendix*. Inflammation of the vermiform appendix is called *appendicitis*. The open end of the cecum merges with the *colon* (*kolon* = food passage). The first division of the colon is the *ascending colon*, which ascends on the right side of the abdomen and turns abruptly to the left at the inferior surface of the liver *(right colic [hepatic] flexure)*. The *transverse colon* continues across the abdomen, curves at the inferior surface of the spleen *(left colic [splenic] flexure)*, and passes down the left side of the abdomen as the *descending colon*. The *sigmoid colon* begins near the iliac crest, projects medially toward the midline, and terminates at the rectum at the level of the third sacral vertebra. The *rectum* is the last 20 cm (7 to 8 in.) of the gastrointestinal tract. Its terminal 2 to 3 cm (1 in.) is known as the *anal canal*. The opening of the anal canal to the exterior is the *anus*.

With the aid of your textbook, label the parts of the large intestine in Figure 22.14.

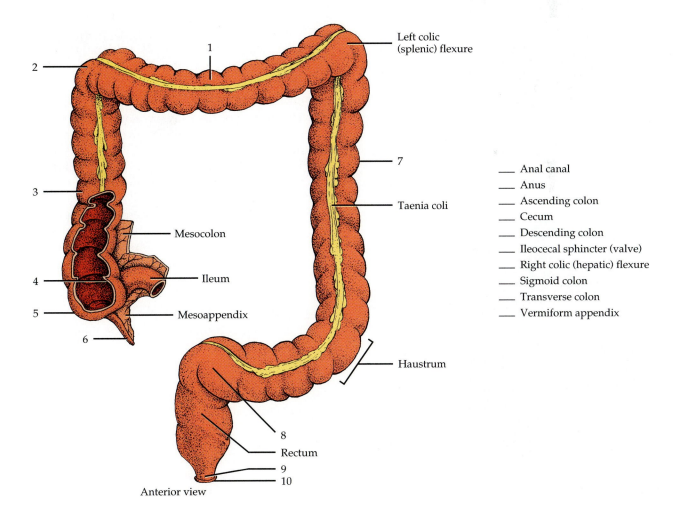

Left colic (splenic) flexure

Taenia coli

Haustrum

Mesocolon

Ileum

Mesoappendix

Rectum

Anterior view

___ Anal canal
___ Anus
___ Ascending colon
___ Cecum
___ Descending colon
___ Ileocecal sphincter (valve)
___ Right colic (hepatic) flexure
___ Sigmoid colon
___ Transverse colon
___ Vermiform appendix

FIGURE 22.14   Large intestine.

The *mucosa* of the large intestine consists of simple columnar epithelium with numerous *goblet cells* (Figure 22.15). The *submucosa* is similar to that in the rest of the gastrointestinal tract. The *muscularis* consists of an outer longitudinal layer and an inner circular layer of smooth muscle. However, the longitudinal layer is not continuous; it is broken up into three flat bands, the *taeniae coli* (TĒ-nē-a KŌ-lī; *taenia* = flat), which gather the colon into a series of pouches called *haustra* (HAWS-tra). Singular is *haustrum* (= shaped like a pouch) (see Figure 22.14). The *serosa* of the large intestine is visceral peritoneum.

Examine a prepared slide of the large intestine showing its tunics. Compare your observations to Figure 22.15.

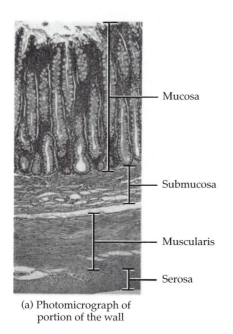

— Mucosa

— Submucosa

— Muscularis

— Serosa

(a) Photomicrograph of portion of the wall

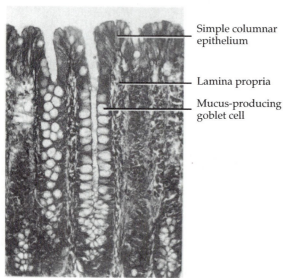

Simple columnar epithelium

Lamina propria

Mucus-producing goblet cell

(b) Photomicrograph of enlarged aspect of mucosa

FIGURE 22.15 Histology of large intestine.

C. DISSECTION OF CAT DIGESTIVE SYSTEM

CAUTION! *Please reread Section D, "Precautions Related to Dissection" at the beginning of the laboratory manual on page xiii **before** you begin your dissection.*

PROCEDURE

1. Salivary Glands

1. The head of the cat contains five pairs of *salivary glands*.
2. Working on the right side of the head of your specimen, locate the following glands (Figure 22.16).

a. PAROTID GLAND

1. The largest of the salivary glands, the parotid is found in front of the ear over the masseter muscle.
2. The *parotid duct* passes over the masseter muscle and pierces the cheek opposite the last upper premolar tooth.
3. Look inside the cheek to identify the opening of the duct. Do not confuse the parotid duct with the branches of the anterior facial nerve. One branch is dorsal and one is ventral to the parotid duct.

b. SUBMANDIBULAR GLAND

1. This gland is below the parotid at the angle of the jaw. The posterior facial vein passes over it.
2. Below the submandibular gland is a lymph node that you should remove. Lymph nodes have smooth surfaces, whereas salivary glands have lobulated surfaces.
3. Lift the anterior margin of the submandibular gland to find the *submandibular duct*.
4. To trace the duct forward, dissect, transect, and reflect the digastric and mylohyoid muscles. The submandibular duct is joined by the duct from the sublingual gland, which is usually difficult to identify.

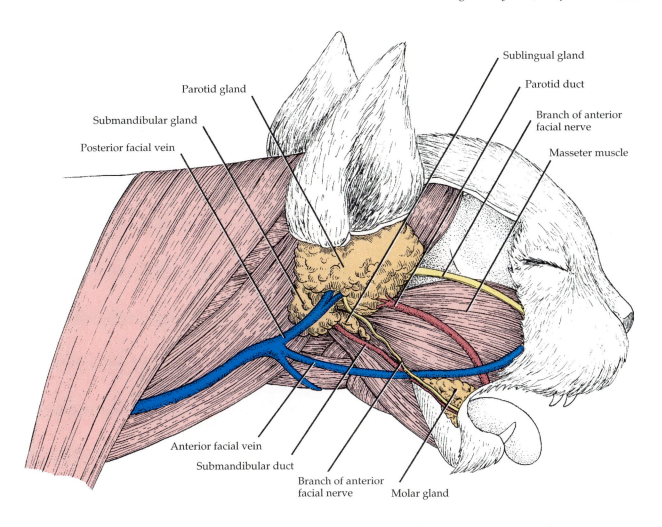

Right lateral view

FIGURE 22.16   Salivary glands of cat.

5. The ducts of the submandibular and the sublingual glands open into the floor of the mouth near the lower incisors.

c. SUBLINGUAL GLAND
This gland is cranial to the submandibular gland. Its duct joins that of the submandibular gland.

d. MOLAR GLAND
This is a very small mass located near the angle of the mouth between the masseter muscle and mandible. Dissection is not necessary.

e. INFRAORBITAL GLAND
This gland is within the floor of the orbit (it is not shown in Figure 22.16).

2. Mouth
1. Open the *mouth* by cutting through the muscles at each angle of the jaw with bone shears.
2. Press down on both sides of the mandible to expose the mouth.
3. Identify the following parts (Figure 22.17).

a. LIPS
The *lips* are the folds of skin around the orifice of the mouth, lined internally by mucous membrane. They are attached to the mandible by a fold of mucous membrane at the midline.

b. CHEEKS
The *cheeks* are the lateral borders of the oral cavity.

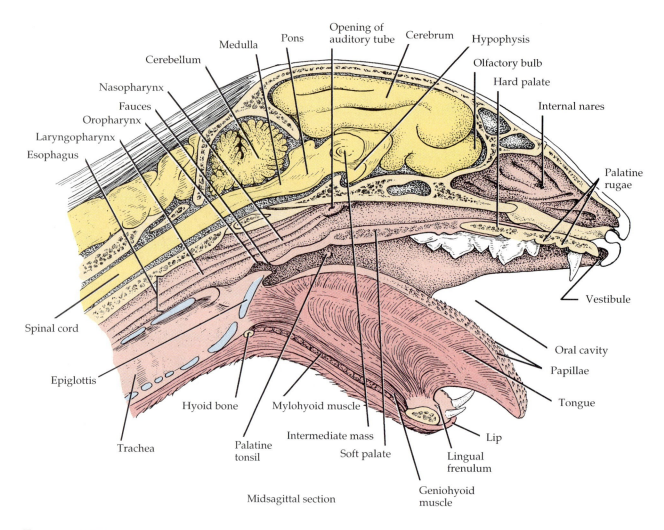

Medulla
Pons
Opening of auditory tube
Cerebrum
Hypophysis
Cerebellum
Olfactory bulb
Nasopharynx
Hard palate
Fauces
Internal nares
Oropharynx
Laryngopharynx
Esophagus
Palatine rugae
Spinal cord
Vestibule
Oral cavity
Epiglottis
Papillae
Tongue
Hyoid bone
Mylohyoid muscle
Trachea
Intermediate mass
Lip
Palatine tonsil
Soft palate
Lingual frenulum
Geniohyoid muscle
Midsagittal section

FIGURE 22.17 Head and neck of cat.

c. VESTIBULE

The *vestibule* is the space between lips and teeth.

d. ORAL CAVITY

The *oral cavity* is the part of the mouth behind the teeth.

e. TEETH

1. The *teeth* of a cat may be represented by the dental formula $\frac{3-1-3-1}{3-1-2-1}$. The numbers represent, from left to right, the number of incisors, canines, premolars, and molars on each side. The upper row of numbers represents teeth in the upper jaw and the lower row represents teeth in the lower jaw.
2. Observe the teeth. How many teeth are in a complete permanent set?

f. TONGUE

1. The *tongue* forms the floor of the oral cavity.
2. Observe the *lingual frenulum*, a fold of mucous membrane that anchors the tongue to the floor of the mouth.
3. Now examine the dorsum (free surface) of the tongue and, using a hand lens, find the papillae.
4. The *filiform papillae* are located mostly in the front and middle of the tongue. They are pointed (spinelike) and are the most numerous of the papillae.
5. The *fungiform papillae* are small, mushroom-shaped structures between and behind the filiform papillae.
6. The *circumvallate (vallate) papillae*, about 12 in number, are large and rounded and are surrounded by a circular groove. They are found near the back of the tongue.

7. The *foliate papillae*, leaf-shaped and relatively few in number, are lateral to the vallate papillae. Taste buds are located between papillae.

g. HARD PALATE

1. This forms the cranial part of the palate. The bones of the hard palate are the palatine process of the premaxilla, the palatine process of the maxilla, and the palatine bone.
2. The mucosa of the hard palate contains transverse ridges called *palatine rugae*.

h. SOFT PALATE

This forms the caudal portion of the palate and has no bony support.

i. PALATINE TONSILS

1. These are small rounded masses of lymphatic tissue in the lateral wall of the soft palate near the base of the tongue.
2. The tonsils are between lateral folds of tissue called the *glossopalatine arches*. The arches are best seen by pulling the tongue ventrally.

3. Pharynx

1. The *pharynx* (throat) is a common passageway for the digestive and respiratory systems.
2. Although the pharynx will not be dissected at this time, you can locate some of its important parts in Figure 22.17. These include the *nasopharynx*, the part of the pharynx above the soft palate; the *auditory (Eustachian) tubes*, a pair of slitlike openings in the laterodorsal walls of the nasopharynx; the *internal nares (choanae)*, which enter the cranial portion of the nasopharynx; the *oropharynx*, the portion of the pharynx between the glossopalatine arches and free posterior margin of the soft palate (the *fauces* is the opening between the oral cavity and oropharynx); and the *laryngopharynx*, the part of the pharynx behind the larynx.
3. The laryngopharynx opens into the esophagus and larynx.

4. Esophagus

1. The *esophagus* is a muscular tube that begins at the termination of the laryngopharynx (Figure 22.17). It extends through the thoracic cavity, pierces the diaphragm, and terminates at the stomach.

2. The esophagus is dorsal to and a little left of the trachea, running posteriorly toward the diaphragm.
3. Just above the diaphragm, it lies dorsal to the heart and ventral to the aorta.

5. Abdominal Structures

a. PERITONEUM

1. The interior of the abdominal wall is lined by the *parietal peritoneum.*
2. The various abdominal organs are covered by *visceral peritoneum*.
3. Extensions of the peritoneum between the abdominal wall and viscera are termed *mesenteries, ligaments*, and *omenta*. Within them are blood vessels, lymphatics, and nerves that supply the viscera.
4. Other peritoneal extensions are the greater omentum, lesser omentum, and mesentery (Figure 22.18).

b. GREATER OMENTUM

1. This is a double sheet of peritoneum that attaches the greater curvature of the stomach to the dorsal body wall.
2. The greater omentum encloses the spleen and part of the pancreas and covers the transverse colon and a large part of the small intestine.
3. The cavity between the layers of the greater omentum is called the *omental bursa*. It contains considerable fat and is often entwined with the intestine.
4. The portion of the greater omentum between the stomach and spleen is called the *gastrosplenic ligament*.
5. Remove the greater omentum.

c. LESSER OMENTUM

1. This extends from the left lateral lobe of the liver to the stomach and the duodenum.
2. The portion between the stomach and the liver is called the *hepatogastric ligament*, and the portion between the duodenum and the liver is called the *hepatoduodenal ligament*.
3. Within the right lateral border of the lesser omentum are the common bile duct, hepatic artery, and portal vein.

d. MESENTERY

1. The *mesentery proper* suspends the small intestine from the dorsal body wall.
2. The *mesocolon* suspends the colon from the dorsal body wall.

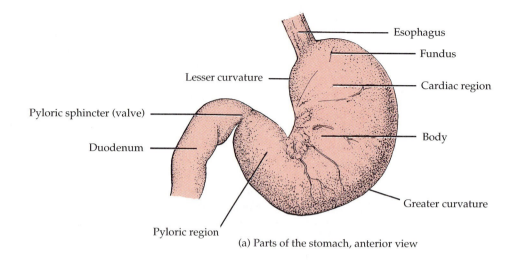

Esophagus

Fundus

Lesser curvature

Cardiac region

Pyloric sphincter (valve)

Duodenum

Body

Greater curvature

Pyloric region

(a) Parts of the stomach, anterior view

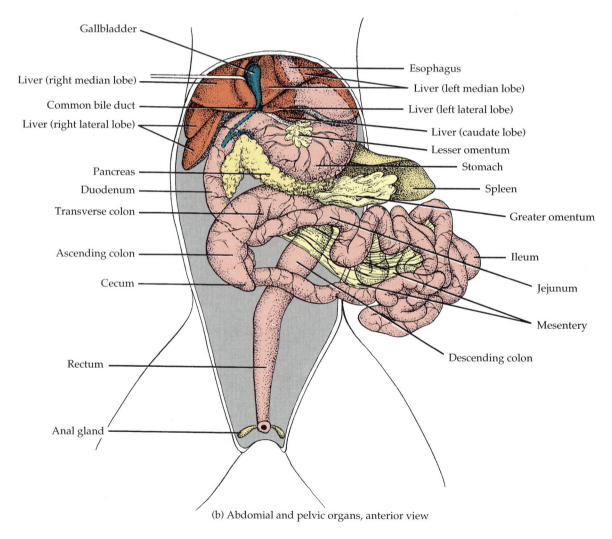

Gallbladder

Liver (right median lobe)

Common bile duct

Liver (right lateral lobe)

Pancreas

Duodenum

Transverse colon

Ascending colon

Cecum

Rectum

Anal gland

Esophagus

Liver (left median lobe)

Liver (left lateral lobe)

Liver (caudate lobe)

Lesser omentum

Stomach

Spleen

Greater omentum

Ileum

Jejunum

Mesentery

Descending colon

(b) Abdomial and pelvic organs, anterior view

FIGURE 22.18   Digestive system of cat.

3. Note the blood vessels and lymph nodes in the mesenteries.

e. LIVER

1. The *liver* is the largest abdominal organ. It is reddish brown and is located immediately caudal to the diaphragm (Figure 22.18).
2. If you pull the liver and diaphragm apart, you can see the central portion of the diaphragm is formed by a tendon into which its muscle fibers insert. This is the *central tendon* of the diaphragm.
3. The liver is divided into *right* and *left lobes*. Between these two main lobes is the *falciform ligament.*
4. The falciform ligament is continuous with the *coronary ligament*, which attaches the liver to the central tendon of the diaphragm.
5. At the free edge of the falciform ligament is a fibrous strand that represents the vestige of the umbilical vein. This strand is known as the *round ligament.*
6. Each main lobe of the liver is further subdivided into two lobes called the lateral and the medial lobes. Thus, there is a *left lateral lobe*, *left median lobe*, *right lateral lobe*, and *right median lobe*.
7. Caudal to the left lateral lobe is another small lobe, the *caudate lobe*.

f. GALLBLADDER

The *gallbladder* is a saclike structure located in a depression on the dorsal surface of the right median lobe of the liver (Figure 22.18).

g. STOMACH

1. The *stomach* lies to the left of the abdominal cavity (Figure 22.18). Its position can be noted by lifting the lobes of the liver.
2. Cut through the diaphragm to note the point at which the esophagus enters the stomach.
3. Identify the following parts of the stomach.
   a. *Cardiac region* Portion of the stomach adjacent to the esophagus.
   b. *Fundus* Dome-shaped portion extending cranially to the left of the cardiac region.
   c. *Body* Large section between the fundus and the pyloric region.
   d. *Pyloric region* Narrow caudal portion that empties into the duodenum. Between the pyloric region and the duodenum is a valve called the *pyloric sphincter (valve)*, which can be seen if the stomach is cut open. To see the pyloric sphincter, make a longitudinal

incision from the body to the duodenum and then wash the contents out of the stomach. You can also see the longitudinal *rugae* if the stomach is cut open.
   i. *Greater curvature* The long, left, and caudal margin of the stomach.
   ii. *Lesser curvature* The short, right, and cranial margin of the stomach.

h. SPLEEN

The *spleen* (which is lymphatic tissue rather than part of the digestive system) is a reddish brown structure on the left side of the abdominal cavity near the greater curvature of the stomach (Figure 22.18).

i. SMALL INTESTINE

1. To examine the small intestine, reflect the greater omentum, which covers the transverse colon and most of the *small intestine*.
2. The small intestine is divided into three major regions (Figure 22.18). The first region is the *duodenum*, which is U-shaped and about 6 in. long. The common bile duct and the pancreatic duct empty into the duodenum near the pyloric sphincter.
3. The enlargement of the duodenum where the ducts unite and enter the duodenum is called the *hepatopancreatic ampulla* (Figure 22.19).
4. The *common bile duct* is formed by the union of the *hepatic ducts* from the lobes of the liver and the *cystic duct* from the gallbladder.
5. The second region of the small intestine is the *jejunum*, which composes about the first half of its remaining length (see Figure 22.18).
6. The *ileum*, the third region of the small intestine, composes about the second half of its remaining length. There is no definite demarcation between the jejunum and ileum. Trace the ileum to its junction with the large intestine.
7. If you cut open the jejunum or ileum, you can note the velvety appearance of the mucosa. This appearance is due to villi, microscopic fingerlike projections that carry on absorption.
8. Remove any parasitic roundworms or tapeworms.

j. LARGE INTESTINE

1. The beginning of the *large intestine* consists of a short, blind pouch called the *cecum* (see Figure 22.18). In humans, the appendix is attached to the cecum. Cats have no appendix.
2. Cut open the wall of the cecum and colon opposite the entrance of the ileum.

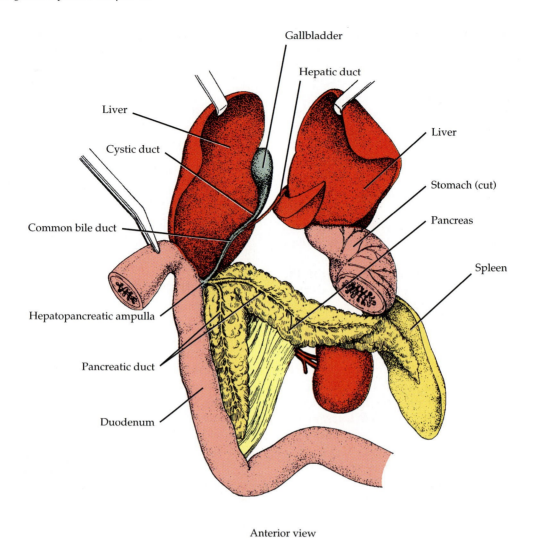

Anterior view

FIGURE 22.19   Relations of the liver, gallbladder, duodenum, and pancreas in cat.

3. Identify the *ileocecal sphincter (valve)* at the junction of the ileum and colon.
4. Cranial to the cecum is the *ascending colon* on the right side of the abdominal cavity. It extends from the cecum to the liver. Here, it passes across the abdominal cavity as the *transverse colon*.
5. The transverse colon continues as the *descending colon*, which runs caudally toward the midline. The cat has no sigmoid colon (which is the next part of the large intestine in humans).
6. Following the descending colon in the cat is the *rectum*, which descends into the pelvic cavity and opens to the exterior as the *anus*.
7. Lateral to the anus are the *anal (scent) glands*. Their ducts empty into the rectum.

K. PANCREAS
1. The *pancreas* is a lobulated organ lying against the descending portion of the duodenum and across the body of the spleen (see Figure 22.18).
2. If you carefully tease away the pancreatic tissue, you can identify the *pancreatic duct* inside the pancreas. This duct is a white, threadlike structure that conveys pancreatic secretions to the duodenum.
3. The pancreatic duct unites with the common bile duct at the hepatopancreatic ampulla.

D. DEGLUTITION

Swallowing, or *deglutition* (dē-gloo-TISH-un), is the mechanism that moves food from the mouth through the esophagus to the stomach. It is facilitated by saliva and mucus and involves a complex series of actions by the mouth, pharynx, and esophagus. Swallowing can be initiated voluntarily, but the remainder of the activity is almost completely under reflex control. Swallowing is a very orderly sequence of events involving the movement of food from the mouth to the stomach, while simultaneously involving the inhibition of respiratory activity and the prevention of entrance of foreign particles into the trachea.

Swallowing is divided into three phases: (1) the *voluntary phase*, when a *bolus* (soft flexible mass of food) is moved into the oropharynx; (2) the *pharyngeal phase*, involving the movement of the bolus through the pharynx into the esophagus; and (3) the *esophageal phase*, which involves the movement of the bolus from the esophagus, through the lower esophageal (gastroesophageal) sphincter, and into the stomach. Both the pharyngeal and esophageal phases are involuntary. The passage of solid or semisolid food from the mouth to the stomach takes 4 to 8 sec. Very soft foods and liquids pass through in about 1 sec.

PROCEDURE

1. Chew a cracker and note the movements of your tongue as you begin to swallow. Chew another cracker, lean over your chair, placing your mouth below the level of your pharynx, and attempt to swallow again.

2. Note the movements of your laboratory partner's larynx during the pharyngeal stage of deglutition.

3. *Clean the earpieces of a stethoscope with alcohol.* Using the stethoscope, listen to the sounds that occur when your partner swallows water. Place the stethoscope just to the left of the midline at the level of the sixth rib. You should hear the sound of water as it makes contact with the lower esophageal sphincter (valve), followed by the sound of water passing through the sphincter after the sphincter relaxes. The sphincter is at the junction of the esophagus and stomach.

4. Record your observations in Section D of the LABORATORY REPORT RESULTS at the end of the exercise. At what point can deglutition

be stopped voluntarily? _____

_____

E. OBSERVATION OF MOVEMENTS OF THE GASTROINTESTINAL TRACT

Substances are moved throughout the gastrointestinal tract by involuntary muscular contractions called *peristalsis* (per'-i-STAL-sis; *peri* = around; *stalsis* = contraction). As a result of these contractions, substances are digested and absorbed, and wastes are eliminated. Several different types of movements occur in various regions of the gastrointestinal tract as a result of rhythmic contractions of the smooth muscle. The motor functions of the hollow organs of the gastrointestinal tract are performed by the different bundles of smooth muscle arranged in the wall of the tract. Unlike skeletal muscle, it is the bundles of smooth muscle that conduct the action potentials throughout the organ, as the individual smooth muscle cells cannot generate sufficient electrical activity to initiate or propagate an action potential. Smooth muscle cells within each bundle are connected via low-electrical-resistance gap junctions and therefore function as a *syncytium.* Muscle that functions in this manner propagates any electrical activity that is initiated within it throughout all of the muscles in the layer, regardless of direction from the initial electrical event. Even though the individual muscle layers are separated by connective tissue, there is a sufficient amount of branching and interconnecting between cells by gap junctions to allow the smooth muscle to function as a syncytium.

Two basic types of movements occur within the gastrointestinal tract: mixing movements and propulsive movements. Mixing movements keep the intestinal contents thoroughly mixed, while propulsive movements move the contents within the gastrointestinal tract.

Mixing movements may be caused by either local constrictive movements within a small region of the wall of the gastrointestinal tract or by peristaltic contractions. Mixing movements differ from region to region within the gastrointestinal system. These contractions macerate food, mix it with the secretions of the gastric glands, and reduce it to a thick liquid called *chyme* (kīm). As digestion proceeds in the stomach, more vigorous mixing contractions begin at the body of the stomach and intensify as they reach the pylorus. Each mixing contraction forces a small amount of gastric contents into the duodenum through the pyloric sphincter. However, most of the food is forced back into the body of the stomach by the mixing contractions, where it is subjected to further mixing. This forward and backward movement of the gastric contents is responsible for almost all of the mixing in the stomach.

In the small intestine mixing contractions are termed *segmental contractions.* These contractions constrict the intestine into small localized segments. Circular fibers in adjacent segments contract and relax alternately, spilling the contents back and forth. As one set of contractions relaxes, the next set occurs at a point midway between the previous contractions, thus "chopping" the intestinal contents.

Pendular movements involve alternate contraction and relaxation of the longitudinal layer of smooth muscle within the small intestine. These contractions alternately shorten and lengthen the small intestine, thereby moving the intestinal contents back and forth.

In the colon, or large intestine, *haustral churning,* a type of segmental contraction, churns the contents of one haustrum (pouch), thereby allowing the contents of the colon to come in contact with the mucosa so that absorption can occur.

In addition to mixing contractions, propulsive movements are the basic type of contraction which results in movement of the gastrointestinal contents distally towards the rectum, and these contractions are peristaltic in nature. In a peristaltic contraction a ring of contracted muscle appears around the hollow organ, and moves progressively forward. Peristalsis is typically stimulated by distention of a gastrointestinal organ. Distention is usually caused by food collecting at a point within the gastrointestinal tract. This distention stimulates the initiation of a peristaltic contraction approximately 3 cm proximal to the distention, and the peristaltic wave is propagated distally along the organ. Other stimuli that may initiate a peristaltic contraction include an irritation of the mucosa of the organ, nervous and hormonal signals.

CAUTION! *Please reread Section D, "Precautions Related to Dissection" at the beginning of the laboratory manual on page xiii before you begin your dissection.*

PROCEDURE

1. Obtain a rat from your laboratory instructor. (The rats will have been euthanized just prior to the beginning of the laboratory period by either cervical dislocation or injection of an air embolism.)

2. Using surgical scissors with a blunt point, quickly make a midline incision through the skin and muscle of the abdomen just below the xiphoid process to the pubis (see Figure 2.5). Hold your scissors so that the points of the scissors cause the abdominal wall to bulge slightly outward as you cut, so to avoid injury to the viscera.

3. Locate the inferiormost projection of the ribs on the lateral walls of the animal. Make two lateral incisions just beneath this level, thereby avoiding damage to the diaphragm and entry into the thoracic cavity. Now make two additional lateral incisions at the level of the pubis. Reflect the abdominal skin to expose the viscera.

4. Note the parietal peritoneum lining the abdominal wall.

5. Identify the stomach. Describe any movements of the stomach musculature. _____

_____

6. Using a blunt forceps, *gently* squeeze the pyloric sphincter (valve). *Gently* squeeze a loop of the small intestine. Describe the reaction to these stimuli. _____

_____

8. Cover the stomach with a small sponge soaked in Locke's solution that has been maintained at 37°C. Reflect the greater omentum backward onto the sponge.

9. Now examine the mesentery. Note the distribution of blood vessels and the presence of whitish lacteals.

10. Observe the movements of the small intestine. Note any pendular, segmental, and peristaltic movements. Describe any movements observed. _____

_____

ALTERNATE PROCEDURE

If Exercise E is to be done, have your instructor or a laboratory assistant remove a section of jejunum from the rat at this time and immediately immerse it in Locke's solution that has been maintained at 37°C.

11. Describe the movements of the intestinal tract as the tissues become ischemic. _____

_____

12. Open the stomach along the greater curvature. Test the pH of the gastric contents with pH paper. Slit open the ileum and test the pH of the ileal contents. Record the pH.

  a. Gastric pH _____

  b. Ileal pH _____

13. Dispose of the rat according to the procedure given by your laboratory instructor. *Clean your equipment thoroughly.*

F. PHYSIOLOGY OF INTESTINAL SMOOTH MUSCLE

Contraction of smooth muscle within the wall of the intestine is responsible for propelling foods through the organ and mixing the food with digestive enzymes. In these experiments you will have an opportunity to observe some of the factors that affect intestinal smooth muscle contraction. You will also be able to compare the physiology of intestinal smooth muscle to that of skeletal muscle (Exercise 9) and cardiac muscle (Exercise 19).

CAUTION! *Please reread Section A, "General Safety Precautions and Procedures" on page xi and Section C, "Precautions Related to Working with Reagents" on page xii at the beginning of the laboratory manual before you begin any of the following experiments. You should also read the experiments before you perform them to be sure that you understand all the procedures and safety precautions.*

1. Isolation of the Intestinal Segment

PROCEDURE

1. Your instructor or a laboratory assistant will provide you with a 2-in. segment of jejunum from the small intestine of a rat. *Do not stretch it.*
2. Immediately immerse the jejunum in 37°C Locke's solution.
3. Fill a medicine dropper with 37°C Locke's solution and *gently* flush out the luminal contents of the jejunal segment. *Be careful not to stretch the intestinal segment.* Cut off any excess mesentery and cut the segment into two pieces approximately 2.54 cm (1 in.) in length.
4. Place the segments in a beaker of clean 37°C Locke's solution.

5. Using a sharp probe, make a small hole in each end of the segment. Using blunt forceps or hemostat, force a long segment of thread through each hole and *gently* tie one end of each thread to the intestinal segment.
6. Tie a loop in a thread at one end so that the loop is approximately 1 in. in length and loop it over one end of the L tube connected to the air supply. Tie the thread to the L tube to prevent any movement of this segment of thread (Figure 22.20).
7. Trim the second thread to approximately 20 cm (8 in.) in length and tie a small loop into the free end. Slip this loop over the hook of a muscle transducer.
8. Adjust the tension on the threads and the intestinal segment so that the string is taut but such that the tissue is not overly stretched. (**Note:** *Depending on the model of the myograph utilized, either turn up the sensitivity or attach the string to the most sensitive arm of the transducer.*)
9. Adjust the air flow so that a continuous, light stream of air passes across the intestinal segment.
10. Adjust the transducer so that the amplitude of the contractions is from 1.5 to 2.5 cm (0.5 to 1 in.) If contractions are not observed after oxygen has bubbled through the solution, raise the heat to 39°C until contractions commence. If contractions are still not observed after 5 min at 39°C, use a medicine dropper to add 1 drop of 1:10,000 acetylcholine to the bath.

2. Experimental Procedures

a. INHERENT CONTRACTILE ACTIVITY
After allowing an initial 5 to 10 min for the intestinal segment to stabilize following the initiation of contractions, observe its basal activity while recording with the polygraph.

1. Is the segment quiescent, or does it exhibit intrinsic contractile activity? _____ _____
2. Is the extent of relaxation consistent, or does the preparation exhibit changes? _____ _____
3. Does the preparation resemble skeletal or cardiac muscle with respect to its contractile activity? _____ _____

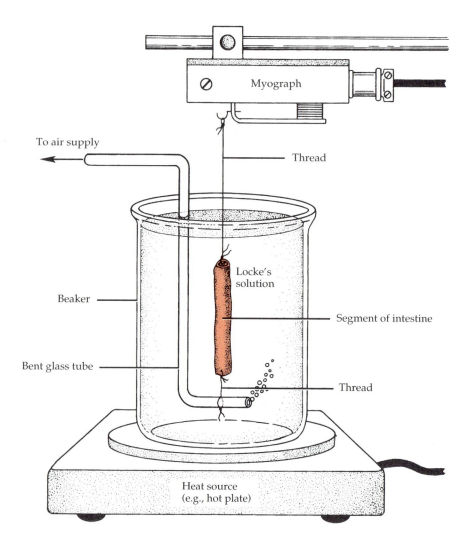

To air supply

Thread

Locke's solution

Beaker

Segment of intestine

Bent glass tube

Thread

Heat source (e.g., hot plate)

Myograph

FIGURE 22.20   Experimental setup for the physiology of intestinal smooth muscle.

4. On the basis of your knowledge of the physiology of skeletal and cardiac muscle, explain your observation. _____

_____

b. EFFECT OF OXYGEN

PROCEDURE

1. Shut off the oxygen supply for 10 min, or until a change in contractile activity is noticed.

   Describe your observations. _____

_____

2. Resupply oxygen and describe your observations. _____

_____

_____

c. EFFECT OF TEMPERATURE

PROCEDURE

1. Lower the temperature of Locke's solution to 20°C by adding ice to it.
2. Record the contractile activity for approximately 2 min with the polygraph. Record the

temperature on the polygraph paper above the recording.

3. Rewarm the Locke's solution *slowly.* Turn the polygraph on and control the heat so that the temperature of the solution rises at a rate of 2°C every 2 min. Continue to warm the solution until the temperature reaches 40°C.

4. Mark the temperature on the paper every 2 min until the bath has reached 40°C. Record for 2 min at 40°C. Turn off the physiograph. Describe the changes that occurred in the intestinal segment with regard to strength and frequency of contraction as temperature

   increased. _____

   _____

   _____

   _____

   At what temperature are the contractions

   strongest? _____

   Explain your observations. _____

   _____

   _____

   _____

   _____

d. EFFECT OF NOREPINEPHRINE

PROCEDURE

1. Place 5 to 8 drops of a 1:50,000 norepinephrine solution into the Locke's solution.

2. Observe any changes in the following: basal tone, strength of contraction, frequency of contraction, and speed of contraction.

3. How do each of these responses compare with those seen when norepinephrine was added to cardiac muscle in Exercise 19? Consider that norepinephrine is released from the sympathetic branch of the autonomic nervous system; did your results meet your expectations? Provide a physiological explanation in the space provided.

   Basal tone _____

   _____

   _____

Strength of contraction _____

_____

_____

Frequency of contraction _____

_____

_____

Speed of contraction _____

_____

_____

e. EFFECT OF ACETYLCHOLINE

PROCEDURE

1. Before performing this experiment, drain the Locke's solution containing norepinephrine and add fresh solution. Wait for 5 min before proceeding.

2. Using a medicine dropper, now place 1 to 5 drops of a 1:1,000 acetylcholine solution, one drop at a time, into the Locke's solution.

3. Observe any changes in the following: basal tone, strength of contraction, frequency of contraction, and speed of contraction.

4. How do each of these responses compare with those seen when acetylcholine was added to cardiac muscle in Exercise 19? Consider that acetylcholine is released from the parasympathetic branch of the autonomic nervous system; did your results meet your expectations? Provide a physiological explanation in the space provided.

   Basal tone _____

   _____

   _____

   Strength of contraction _____

   _____

   _____

   Frequency of contraction _____

   _____

   _____

Speed of contraction _____

_____

Does acetylcholine modify the contractile

properties of skeletal muscle? _____

_____

Does the autonomic nervous system play a

role in skeletal muscle contraction? _____

_____

_____

f. EFFECT OF ATROPINE

PROCEDURE

1. Before performing this experiment, drain Locke's solution containing acetylcholine and add a fresh solution. Wait 5 min before proceeding.
2. Using a medicine dropper, place 5 to 10 drops of a 1:50,000 atropine solution into Locke's solution.
3. Explain the effects of atropine on the following:

Basal tone _____

_____

_____

Strength of contraction _____

_____

_____

Frequency of contraction _____

_____

_____

Speed of contraction _____

_____

_____

g. EFFECT OF ATROPINE AND ACETYLCHOLINE

PROCEDURE

1. Using a medicine dropper, add 1 to 5 drops of a 1:1000 solution of acetylcholine to Locke's solution already containing atropine.

2. Wait 3 to 5 min and compare the effect of acetylcholine in the presence of atropine on the following:

Basal tone _____

_____

_____

Strength of contraction _____

_____

_____

Frequency of contraction _____

_____

_____

Speed of contraction _____

_____

_____

Based on your observations, what type of acetylcholine receptors does the smooth-

muscle segment possess? _____

_____

_____

h. EFFECT OF ATROPINE, ACETYLCHOLINE, AND NOREPINEPHRINE

PROCEDURE

1. Using a medicine dropper, add 5 to 8 drops of a 1:50,000 solution of norepinephrine to Locke's solution already containing atropine and acetylcholine.
2. Wait 3 to 5 min and compare the effect of norepinephrine in the presence of atropine and acetylcholine on the following:

Basal tone _____

_____

_____

Strength of contraction _____

_____

Frequency of contraction _____

_____

_____

Speed of contraction _____

_____

_____

i. EFFECT OF CALCIUM REMOVAL

PROCEDURE

1. Before performing this experiment, drain Locke's solution containing atropine, acetylcholine, and norepinephrine and fill the beaker with fresh Locke's solution. Allow 5 min before proceeding.
2. To determine whether the intestinal segment has completely recovered, record about 3 min of activity and then compare your observations to those recorded in Section 3.a.
3. Now replace Locke's solution with Locke's solution containing no calcium $O[Ca]_e$.
4. Explain the effects of Locke's solution containing $O[Ca]_e$ on the following:

Basal tone _____

_____

_____

Strength of contraction _____

_____

_____

Frequency of contraction _____

_____

_____

Speed of contraction _____

_____

_____

Would you expect similar results if you were to expose cardiac muscle to Locke's solution containing $O[Ca]_e$? _____

_____

_____

Would you expect similar results if you were to expose skeletal muscle to Locke's solution containing $O[Ca]_e$? _____

_____

What substance(s) used in this exercise stimulate (depolarize) intestinal smooth muscle?

_____

_____

What substances used in this exercise inhibit (hyperpolarize) intestinal smooth muscle?

_____

_____

G. CHEMISTRY OF DIGESTION

In order to understand how food is digested, let us examine the various processes that occur between absorption and final utilization of nutrients by the body's cells.

Nutrients are chemical substances in food that provide energy, act as building blocks to form new body components, or assist body processes. The six major classes of nutrients are carbohydrates, lipids, proteins, minerals, vitamins, and water. Cells break down carbohydrates, lipids, and proteins to release energy, or use them to build new structures and new regulatory substances, such as hormones and enzymes.

Enzymes are produced by living cells to catalyze or speed up many of the reactions in the body. Enzymes are proteins and act as **catalysts** to speed up reactions without being permanently altered by the reaction. Digestive enzymes function as catalysts to speed chemical reactions in the gastrointestinal tract.

Because digestive enzymes function outside the cells that produce them, they are capable of also reacting within a test tube and therefore provide an excellent means of studying enzyme activity.

CAUTION! *Please reread Section A, "General Safety Precautions and Procedures" on page xi and Section C, "Precautions Related to Working with Reagents" on page xii at the beginning of the laboratory manual before you begin any of the following experiments. You should also read the experiments before you perform them to be sure that you understand all the procedures and safety precautions.*

1. Positive Tests for Sugar and Starch

Salivary amylase is an enzyme produced by the salivary glands. This enzyme starts starch digestion and *hydrolyzes* (splits using water) it into maltose (a disaccharide), maltotriose (a trisaccharide) and α-dextrins. We measure the amount of starch and sugar present before and after enzymatic activity. It is expected that the amount of starch should *decrease* and the sugar level should *increase* as a result of salivary amylase activity.

a. TEST FOR SUGAR

Benedict's test is commonly used to detect sugars. Glucose (monosaccharide), maltose (disaccharide), or any other reducing sugars react with *Benedict's solution,* forming insoluble red cuprous oxide. The precipitate of cuprous oxide can usually be seen in the bottom of the tube when standing. Benedict's solution turns green, yellow, orange, or red depending on the amount of reducing sugar present according to the following scale:

blue (–)
green (+)
yellow (++)
orange (+++)
red (++++)

Test for the presence of sugar (maltose) as follows:

PROCEDURE

1. Using separate medicine droppers, place 2 ml of maltose and 2 ml of Benedict's solution in a Pyrex test tube.
2. *Using a test tube holder*, place the test tube in a boiling water bath and heat for 5 minutes. **CAUTION!** *Make sure that the mouth of the test tube is pointed away from you and all other persons in the area.* Note the color change.
3. *Using a test tube holder*, remove the test tube from the water bath.
4. Repeat the same procedure using a starch solution instead of maltose. Notice that the color does not change, because the solution contains no sugar. Now you have a method for detecting sugar.

b. TEST FOR STARCH

Lugol's solution is a brown-colored iodine solution used to test certain polysaccharides, especially starch. Starch, for example, gives a *deep blue* to *black* color with Lugol's solution (the black is really a concentrated blue color). Cellulose, monosaccharides, and disaccharides do not react. A negative test is indicated by a yellow to brown color of the solution itself, or possibly some other color (other than blue to black) resulting from pigments present in the substance being tested.

PROCEDURE

1. Using separate medicine droppers, place a drop of starch solution on a spot plate and then add a drop of Lugol's solution to it. Notice the black color that forms as the starch-iodine complex develops.
2. Repeat the test using a maltose solution in place of the starch solution. Note that there is no color change, because Lugol's solution and maltose do not combine. Now you have a method for detecting starch.

c. DIGESTION OF STARCH

PROCEDURE

1. Using separate medicine droppers, transfer 3 ml of a starch solution to a small beaker, add a fresh enzyme solution consisting of a pinch of amylase powder in 3 ml of water, and mix thoroughly with a glass rod.
2. Wait for 1 min, record the time, remove 1 drop of the mixture with a glass rod to the depression of a spot plate, and then test for starch with Lugol's solution.
3. At 1-min intervals, test 1-drop samples of the mixture until you no longer note a positive test for starch. Keep the glass rod in the mixture, stirring it from time to time.
4. After the starch test is seen to be negative, test the remaining mixture for the presence of glucose. Do this by adding 2 ml of the mixture with a medicine dropper to 2 ml of the Benedict's solution in a Pyrex test tube and heat in a boiling water bath as per the procedure outlined in a.2, "Test for Sugar." Answer questions a through c in Section G.1 of the LABORATORY REPORT RESULTS at the end of the exercise.

2. Effect of Temperature on Starch Digestion

In the following procedure, you will test starch digestion at five different temperatures to determine how temperature influences enzyme activity. Lugol's solution is again used for presence or absence of starch.

A fresh enzyme solution consisting of a pinch of amylase powder in 3 ml of water should be used as before. Five constant-temperature water baths should be available. Starting with the lowest temperature, these are: 0°C or cooler, 10°C, 40°C, 60°C, and boiling.

PROCEDURE

1. Prepare 10 test Pyrex tubes, 5 containing 1 ml each of enzyme solution and 5 containing 1 ml each of starch solution.

2. Using rubber bands, pair the tubes (i.e., a tube containing enzyme with one containing starch) and use a test tube holder to place one pair into each water bath. **CAUTION!** *Make sure that the mouth of the test tube is pointed away from you and everyone else in the area.*

3. Permit the tubes to adapt to the bath temperatures for about 5 min, then mix the enzyme and starch solutions of each pair together, and *using a test tube holder* place the single test tube in its respective water bath.

4. After 30 sec, *using a test tube holder*, remove the test tubes from the water baths and test all five tubes for starch on a spot plate using Lugol's solution.

5. Repeat every 30 sec until you have determined the time required for the starch to disappear (that is, to be digested).

6. Record and graph your results in Section G.2 of the LABORATORY REPORT RESULTS at the end of the exercise.

3. Effect of pH on Starch Digestion

You can demonstrate the effectiveness of salivary amylase digestion at different pH readings.

PROCEDURE

1. Prepare three buffer solutions as follows:

   Solution A   pH 4.0
   Solution B   pH 7.0
   Solution C   pH 9.0

2. Once again a fresh enzyme solution consisting of a pinch of amylase powder in 3 ml of water should be used.

3. Using medicine droppers, mix 4 ml of a starch solution with 2 ml of buffer solution A in a Pyrex test tube.

4. Repeat this procedure with buffer solutions B and C.

5. You now have three test tubes of a starch-buffer solution, each at a different pH (4.0, 7.0, and 9.0).

6. Using separate medicine droppers, place one drop of starch-buffer solution A on a spot plate and immediately add one drop of the saliva.

7. Test for starch disappearance using Lugol's solution, and record the time when starch first disappears completely.

8. Repeat this test for the other two starch buffers (solutions B and C) and record the time when starch is no longer present at each pH.

9. Record and explain your results in Section G.3 of the LABORATORY REPORT RESULTS at the end of the exercise.

4. Action of Bile on Triglycerides

Bile is important in the process of lipid digestion because of its emulsifying effects (breaking down of large globules to smaller, uniformly distributed particles) on triglycerides (fats) and oils. *Bile does not contain any enzymes.* Emulsification of triglycerides by means of bile salts serves to increase the surface area of the triglyceride that will be exposed to the action of the lipase.

PROCEDURE

1. Place 5 ml of water into one Pyrex test tube and 5 ml of bile solution into a second.

2. Using a medicine dropper, add one drop of vegetable oil that has been colored with a fat-soluble dye, such as Sudan B, into each tube.

3. Place a stopper in both tubes. Shake them *vigorously,* and then let them stand in a test tube rack undisturbed for 10 min. Triglycerides or oils that are broken into sufficiently small droplets will remain suspended in water in the form of an *emulsion.* If emulsification has not occurred, the triglyceride or oil will lie on the surface of the water.

4. Answer questions in Section G.4 of the LABORATORY REPORT RESULTS at the end of the exercise.

5. Digestion of Triglycerides

You can demonstrate the effect of pancreatic juice on triglycerides by the use of pancreatin, which contains all the enzymes present in pancreatic juice. Because the optimum pH of the pancreatic enzymes ranges from 7.0 to 8.8, the pancreatin is prepared in sodium carbonate. The enzyme used in this test is **pancreatic lipase**, which digests triglycerides to fatty acids and glycerol. The fatty acid produced changes the color of **blue** litmus to **red.**

PROCEDURE

1. Using a medicine dropper, place 5 ml of litmus cream (heavy cream to which powdered litmus has been added to give it a blue color) in a Pyrex test tube, and, *using a test tube holder,* place the test tube in a 40°C water bath.
2. Repeat the procedure with another 5-ml portion in a second tube, but put it in an ice bath.
3. When the tubes have adapted to their respective temperatures (in about 5 min), use a medicine dropper to add 5 ml of pancreatin to each tube and, *using a test tube holder,* replace them in their water baths until a color change occurs in one tube.
4. Summarize your results and your explanation in Section G.5 of the LABORATORY REPORT RESULTS at the end of the exercise.

6. Digestion of Protein

Here you demonstrate the effect of pepsin on protein and the factors affecting the rate of action of pepsin. *Pepsin,* a proteolytic enzyme, is secreted in inactive form (pepsinogen) by chief (zymogenic) cells in the lining of the stomach. Pepsin digests proteins (fibrin in this experiment) to peptides. The efficiency of pepsin activity depends on the pH of the solution, the optimum being 1.5 to 2.5. Pepsin is almost completely inactive in neutral or alkaline solutions.

PROCEDURE

1. Prepare and number the following five Pyrex test tubes. For this test, the quantity of each solution must be *measured carefully.*

   Tube 1—5 ml of 0.5% pepsin; 5 ml of 0.8% HCl
   Tube 2—5 ml of pepsin; 5 ml of water

   Tube 3—5 ml of pepsin, boiled for 10 min in a water bath; 5 ml of 0.8% HCl

CAUTION! *Make sure that the mouth of the test tube is pointed away from you and all other persons in the area. Using a test tube holder, remove the test tube from the water bath.*

   Tube 4—5 ml of pepsin; 5 ml of 0.5% NaOH
   Tube 5—5 ml of water; 5 ml of 0.8% HCl

2. First determine the approximate pH of each test tube using Hydrion paper (range 1 to 11) and put the values in the table provided in Section G.6 of the LABORATORY REPORT RESULTS at the end of the exercise.
3. Using forceps, place a small amount of fibrin (the protein) in each test tube. An amount near the size of a pea will be sufficient. *Using a test tube holder,* put the tubes in a 40°C water bath and *carefully* shake occasionally. Maintain 40°C temperature closely. The tubes should remain in the water bath for *at least 1½ hr.*
4. Watch the changes that the fibrin undergoes. The swelling that occurs in some tubes should not be confused with digestion. Digested fibrin becomes transparent and disappears (dissolves) as the protein is digested to soluble peptides.
5. Finish the experiment when the fibrin is digested in one of the five tubes.
6. Record all your observations in the table provided in Section G.6 of the LABORATORY REPORT RESULTS at the end of the exercise.

ANSWER THE LABORATORY REPORT QUESTIONS AT THE END OF THE EXERCISE.

Digestive System 22

Student _____   Date _____

Laboratory Section _____   Score/Grade _____

SECTION D. DEGLUTITION

1. Tongue movement at beginning of deglutition: _____

_____

2. Description of swallowing with head below pharynx: _____

_____

3. Laryngeal movements during pharyngeal deglutition: _____

_____

4. Time lapse between two sounds of deglutition: _____

_____

SECTION G. CHEMISTRY OF DIGESTION

1. Digestion of Starch

a. How long did it take for the starch to be digested? _____

b. Did your observation indicate the presence of maltose? _____

c. What is the meaning of this result? _____

2. Effect of Temperature on Starch Digestion

d. Record the time required for starch to disappear at the temperatures tested.

_____0°C   _____40°C   _____Boiling

_____10°C   _____60°C

e. What is the optimum temperature for starch digestion? _____

f. What happens to amylase when it is boiled? _____

g. Is the effect of boiling reversible or irreversible? _____

h. Is the effect of freezing reversible or irreversible? _____

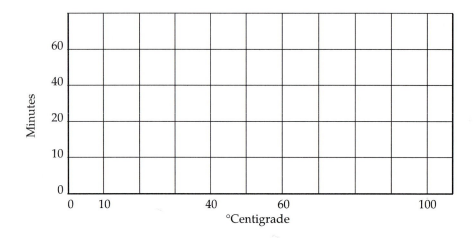

3. Effect of pH on Starch Digestion

i. Record the time required for starch to disappear at the pH readings tested.

Solution A: pH 4.0 _____

Solution B: pH 7.0 _____

Solution C: pH 9.0 _____

j. Explain your results

A: pH 4.0 _____

B: pH 7.0 _____

C: pH 9.0 _____

k. What is the optimum pH for the action of salivary amylase? _____

_____

4. Action of Bile on Triglycerides

l. What difference can you detect in the appearance of the mixtures in the two test tubes? _____

_____

m. How does emulsification of lipids by means of bile aid in the digestion of triglycerides? _____

_____

5. Digestion of Triglycerides

Record the results of the digestion of triglycerides by pancreatic lipase in pancreatic juice in the following table.

Tube no.	Temperature of water bath	Change in pH (color)	Explanation of results
1			
2			

6. Digestion of Protein

Record the results of the digestion of protein by pepsin in the following table.

Tube no.	Tube contents	pH	Digestion observed (yes or no)	Explanation of results
1				
2				
3				
4				
5				

Digestive System 22

Student _____   Date _____

Laboratory Section _____   Score/Grade _____

PART 1. Multiple Choice

_____ 1. If an incision has to be made in the small intestine to remove an obstruction, the first layer of tissue to be cut is the (a) muscularis (b) mucosa (c) serosa (d) submucosa

_____ 2. Mesentery, lesser omentum, and greater omentum are all directly associated with the (a) peritoneum (b) liver (c) esophagus (d) mucosa of the gastrointestinal tract

_____ 3. Chemical digestion of carbohydrates is initiated in the (a) stomach (b) small intestine (c) mouth (d) large intestine

_____ 4. A tumor of the villi and circular folds would interfere most directly with the body's ability to carry on (a) absorption (b) deglutition (c) mastication (d) peristalsis

_____ 5. The main chemical activity of the stomach is to begin the digestion of (a) triglycerides (b) proteins (c) carbohydrates (d) all of the above

_____ 6. Surgical cutting of the lingual frenulum would occur in which part of the body? (a) salivary glands (b) esophagus (c) nasal cavity (d) tongue

_____ 7. To free the small intestine from the posterior abdominal wall, which of the following would have to be cut? (a) mesocolon (b) mesentery (c) lesser omentum (d) falciform ligament

_____ 8. The cells of gastric glands that produce secretions directly involved in chemical digestion are the (a) mucous (b) parietal (c) chief (d) pancreatic islets (islets of Langerhans)

_____ 9. An obstruction in the hepatopancreatic ampulla (ampulla of Vater) would affect the ability to transport (a) bile and pancreatic juice (b) gastric juice (c) salivary amylase (d) intestinal juice

_____ 10. The terminal portion of the small intestine is known as the (a) duodenum (b) ileum (c) jejunum (d) pyloric sphincter (valve)

_____ 11. The portion of the large intestine closest to the liver is the (a) right colic flexure (b) rectum (c) sigmoid colon (d) left colic flexure

_____ 12. The lamina propria is found in which coat? (a) serosa (b) muscularis (c) submucosa (d) mucosa

_____ 13. Which structure attaches the liver to the anterior abdominal wall and diaphragm? (a) lesser omentum (b) greater omentum (c) mesocolon (d) falciform ligament

_____ 14. The opening between the oral cavity and pharynx is called (a) vermilion border (b) fauces (c) vestibule (d) lingual frenulum

_____ 15. All of the following are parts of a tooth *except* the (a) crown (b) root (c) cervix (d) papilla

_____ **16.** Cells of the liver that destroy worn-out white and red blood cells and bacteria are termed (a) hepatocytes (b) stellate reticuloendothelial (Kupffer) cells (c) alpha cells (d) beta cells

_____ **17.** Bile is manufactured by which cells? (a) alpha (b) beta (c) hepatocytes (d) stellate reticuloendothelial (Kupffer)

_____ **18.** Which part of the small intestine secretes the intestinal digestive enzymes? (a) intestinal glands (b) duodenal (Brunner's) glands (c) lacteals (d) microvilli

_____ **19.** Structures that give the colon a puckered appearance are called (a) taenia coli (b) villi (c) rugae (d) haustra

PART 2. Completion

20. An acute inflammation of the serous membrane lining the abdominal cavity and covering the abdominal viscera is referred to as _____.

21. The _____ is a sphincter (valve) between the ileum and large intestine.

22. The portion of the small intestine that is attached to the stomach is the _____.

23. The portion of the stomach closest to the esophagus is the _____.

24. The _____ forms the floor of the oral cavity and is composed of skeletal muscle covered with mucous membrane.

25. The convex lateral border of the stomach is called the _____.

26. The three special structures found in the wall of the small intestine that increase its efficiency in absorbing nutrients are the villi, circular folds, and _____.

27. The three pairs of salivary glands are the parotids, submandibulars, and _____.

28. The small intestine is divided into three segments: duodenum, ileum, and _____.

29. The large intestine is divided into four main regions: the cecum, colon, rectum, and

_____.

30. The enzyme that is present in saliva is called _____.

31. The transition of the lips where the outer skin and inner mucous membrane meet is called the

_____.

32. The _____ papillae are arranged in the form of an inverted V on the posterior surface of the tongue.

33. The portion of a tooth containing blood vessels, lymphatic vessels, and nerves is the

_____.

34. The teeth present in permanent dentition, but not in deciduous dentition, that replace the deciduous molars are the _____.

35. The portion of the gastrointestinal tract that conveys food from the pharynx to the stomach is the

_____.

36. The inferior region of the stomach connected to the small intestine is the _____.

37. The clusters of cells in the pancreas that secrete digestive enzyme are called _____.

38. The caudate lobe, quadrate lobe, and central vein are all associated with the _____.

39. The common bile duct is formed by the union of the common hepatic duct and

_____ duct.

40. The pear-shaped sac that stores bile is the _____.

41. _____ glands of the small intestine secrete an alkaline substance to protect the mucosa from excess acid.

42. The _____ attaches the large intestine to the posterior abdominal wall.

43. The _____ is the last 20 cm (7 to 8 in.) of the gastrointestinal tract.

44. A midline fold of mucous membrane that attaches the inner surface of each lip to its corresponding

gum is the _____.

45. The bonelike substance that gives teeth their basic shape is called _____.

46. The _____ anchors teeth in position and helps to dissipate chewing forces.

47. The portion of the colon that terminates at the rectum is the _____ colon.

48. The palatine tonsils are between the palatoglossal and _____ arches.

49. The teeth closest to the midline are the _____.

50. The vermiform appendix is attached to the _____.

PART 3. Matching

_____ **51.** Pancreatic lipase	A. Commonly used solution in the test for starch
_____ **52.** Benedict's solution	B. Capable of digesting starch
_____ **53.** Lugol's solution	C. Capable of digesting protein
_____ **54.** Salivary amylase	D. Commonly used solution for detecting reducing sugars
_____ **55.** Pepsin	E. Digests triglycerides into fatty acids and glycerol and changes the color of blue litmus to red

Urinary System

The *urinary system* functions to keep the body in homeostasis by controlling the composition and volume of the blood. The system accomplishes these functions by removing and restoring selected amounts of water and various solutes. The kidneys also excrete selected amounts of various wastes, assume a role in erythropoiesis by secreting erythropoietin, help control blood pH, help regulate blood pressure by secreting renin (which activates the renin-angiotensin pathway), participate in the synthesis of vitamin D, and perform gluconeogenesis (synthesis of glucose molecules) during periods of fasting or starvation.

The urinary system consists of two kidneys, two ureters, one urinary bladder, and a single urethra (see Figure 23.1). Other systems that help in waste elimination are the respiratory, integumentary, and digestive systems.

Using your textbook, charts, or models for reference, label Figure 23.1.

A. ORGANS OF URINARY SYSTEM

1. Kidneys

The paired *kidneys,* which resemble kidney beans in shape, are found just superior to the waist between the parietal peritoneum and the posterior wall of the abdomen. Because they are behind the peritoneal lining of the abdominal cavity, they are referred to as *retroperitoneal* (re'-tro-per-i-to-NE-al; *retro* = behind). The kidneys are located between the levels of the last thoracic and third lumbar vertebrae, with the right kidney slightly lower than the left because of the position of the liver.

Three layers of tissue surround each kidney: the *renal* (*renalis* = kidney) *capsule, adipose capsule,* and *renal fascia.* They function to protect the kidney and hold it firmly in place.

Near the center of the kidney's concave border, which faces the vertebral column, is a vertical fissure called the *renal hilus,* through which the

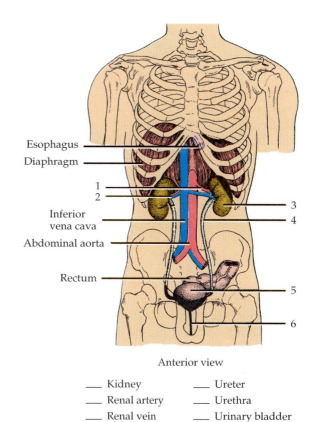

Esophagus
Diaphragm
1
2
Inferior vena cava
Abdominal aorta
Rectum
3
4
5
6

Anterior view

___ Kidney   ___ Ureter
___ Renal artery   ___ Urethra
___ Renal vein   ___ Urinary bladder

FIGURE 23.1 Organs of male urinary system and associated structures.

ureter leaves the kidney and through which blood vessels, lymphatics, and nerves enter and exit the kidney. The hilus is the entrance to a cavity in the kidney called the *renal sinus.*

If a frontal section is made through a kidney, the following structures can be seen:

a. *Renal cortex* (*cortex* = rind or bark) Superficial, narrow, reddish area.
b. *Renal medulla* (*medulla* = inner portion) Deep, wide, reddish-brown area.
c. *Renal (medullary) pyramids* Striated triangular structures, 8 to 18 in number, in the renal

medulla. The bases of the renal pyramids face the renal cortex, and the apices, called *renal papillae,* are directed toward the center of the kidney.

d. *Renal columns* Cortical substance between renal pyramids.

e. *Renal pelvis* Large cavity in the renal sinus, enlarged proximal portion of ureter.

f. *Major calyces* (KĀ-li-sēz) Consist of 2 or 3 cuplike extensions of the renal pelvis.

g. *Minor calyces* Consist of 8 to 18 cuplike extensions of the major calyces.

Within the renal cortex and renal pyramids of each kidney are more than 1,000,000 microscopic units called nephrons, the functional units of the kidneys (described shortly). As a result of their activity in regulating the volume and chemistry of the blood, they produce urine. Urine passes from the nephrons to the minor calyces, major calyces, renal pelvis, ureter, urinary bladder, and urethra.

Examine a specimen, model, or chart of the kidney and with the aid of your textbook, label Figure 23.2.

2. Nephrons

Basically, a *nephron* (NEF-ron) consists of (1) a renal corpuscle (KOR-pus-sul; *corpus* = body; *cle* = tiny) where fluid is filtered and (2) a renal tubule into which the filtered fluid (filtrate) passes. A *renal corpuscle* has two components: a tuft (knot) of capillaries called a *glomerulus* (glō-MER-yoo-lus; *glomus* = ball; *ulus* = small) surrounded by a double-walled epithelial cup, called a *glomerular (Bowman's) capsule,* lying in the renal cortex of the kidney. The inner wall of the capsule, the *visceral layer,* consists of epithelial cells called *podocytes* and surrounds the glomerulus. A space called the *capsular (Bowman's) space* separates the visceral layer from the outer wall of the capsule, the *parietal layer,* which is composed of simple squamous epithelium.

The visceral layer of the glomerular (Bowman's) capsule and endothelium of the glomerulus form an *endothelial-capsular (filtration) membrane,* a very effective filter. Electron microscopy has determined that the membrane consists of the following components, given in the order in which substances are filtered (Figure 23.3 on page 560).

a. *Endothelial fenestrations (pores) of the glomerulus* The single layer of endothelial cells has large fenestrations (pores) that pre-

vent filtration of blood cells but allow all components of blood plasma to pass through.

b. *Basement membrane of the glomerulus* This layer of extracellular material lies between the endothelium and the visceral later of the glomerular capsule. It consists of fibrils in a glycoprotein matrix and prevents filtration of larger proteins.

c. *Slit membranes between pedicels.* The specialized epithelial cells that cover the glomerular capillaries are called *podocytes* (*podos* = foot). Extending from each podocyte are thousands of footlike structures called *pedicels* (PED-i-sels; *pediculus* = little foot). The pedicels cover the basement membrane, except for spaces between them, which are called *filtration slits.* A thin membrane, the *slit membrane,* extends across filtration slits and prevents filtration of medium-sized proteins.

The endothelial-capsular membrane filters blood passing through the kidney. Blood cells and large molecules, such as proteins, are retained by the filter and eventually are recycled into the blood. The filtered substances pass through the membrane and into the space between the parietal and visceral layers of the glomerular (Bowman's) capsule, and then enter the renal tubule (described shortly).

As the filtered fluid (filtrate) passes through the remaining parts of a nephron, substances are selectively added and removed. The end product of these activities is urine. Nephrons are frequently classified into two kinds. A *cortical nephron* usually has its glomerulus in the superficial renal cortex, and the remainder of the nephron penetrates only into the superficial renal medulla. A *juxtamedullary nephron* usually has its glomerulus close to the corticomedullary junction, and other parts of the nephron penetrate deeply into the renal medulla (see Figure 23.4 on page 561). The following description of the remaining components of a nephron applies to juxtamedullary nephrons.

After the filtrate leaves the glomerular (Bowman's) capsule, it passes through the following parts of a renal tubule:

a. *Proximal convoluted tubule (PCT)* Coiled tubule in the renal cortex that originates at the glomerular (Bowman's) capsule; consists of simple cuboidal epithelium with microvilli.

b. *Loop of Henle (nephron loop)* U-shaped tubule that connects the proximal and distal

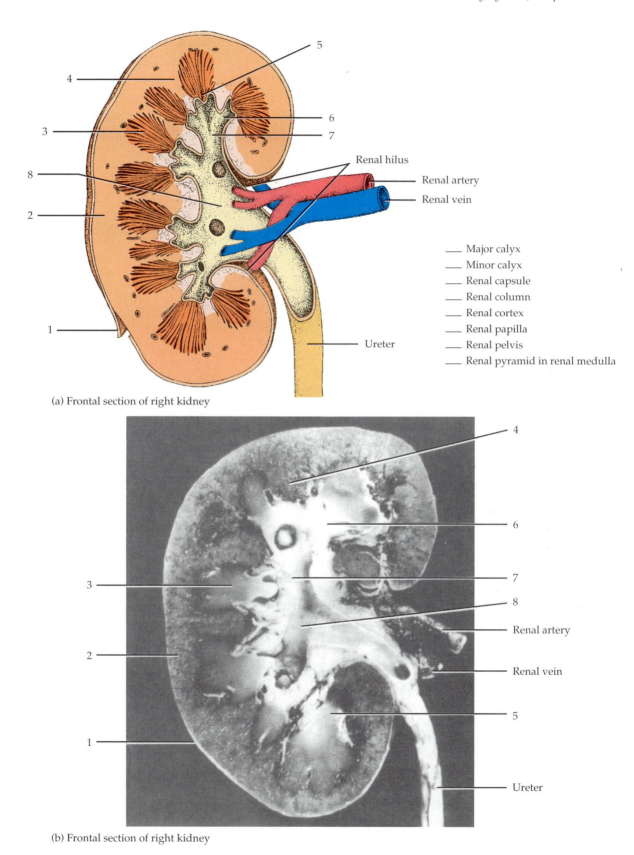

5

4

6

7

Renal hilus

Renal artery

Renal vein

3

8

2

___ Major calyx
___ Minor calyx
___ Renal capsule
___ Renal column
___ Renal cortex
___ Renal papilla
___ Renal pelvis
___ Renal pyramid in renal medulla

1

Ureter

(a) Frontal section of right kidney

4

6

7

8

Renal artery

Renal vein

3

2

5

1

Ureter

(b) Frontal section of right kidney

FIGURE 23.2   Kidney. (a) Diagram. (b) Photograph.

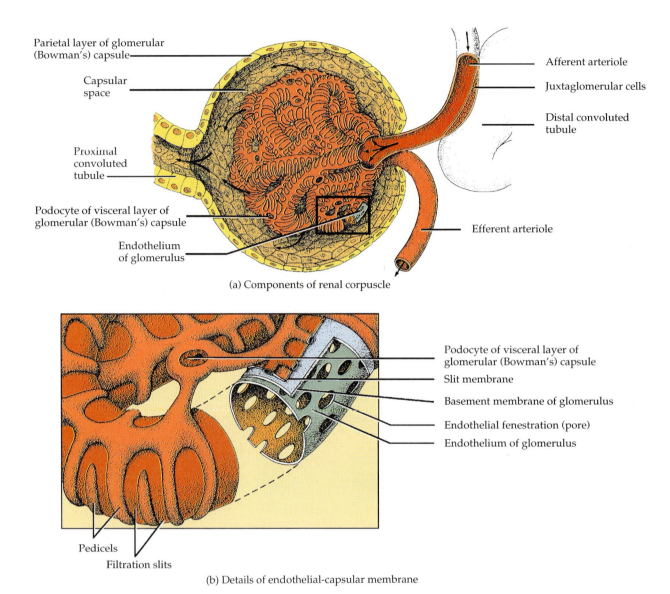

Parietal layer of glomerular (Bowman's) capsule

Capsular space

Proximal convoluted tubule

Podocyte of visceral layer of glomerular (Bowman's) capsule

Endothelium of glomerulus

Afferent arteriole

Juxtaglomerular cells

Distal convoluted tubule

Efferent arteriole

(a) Components of renal corpuscle

Podocyte of visceral layer of glomerular (Bowman's) capsule

Slit membrane

Basement membrane of glomerulus

Endothelial fenestration (pore)

Endothelium of glomerulus

Pedicels

Filtration slits

(b) Details of endothelial-capsular membrane

FIGURE 23.3 Endothelial-capsular membrane.

convoluted tubules. It consists of a ***descending limb of the loop of Henle,*** an extension of the proximal convoluted tubule that dips down into the renal medulla and consists of simple squamous epithelium, and an ***ascending limb of the loop of Henle*** that ascends in the renal medulla and approaches the renal cortex and consists of simple squamous, cuboidal, and columnar epithelium; the ascending limb is wider in diameter than the descending limb.

c. ***Distal convoluted tubule (DCT)*** Coiled extension of ascending limb in the renal cortex; consists of simple cuboidal epithelium with fewer microvilli than in the proximal convoluted tubule.

Distal convoluted tubules terminate by merging with straight ***collecting ducts.*** In the renal medulla, collecting ducts receive distal convoluted tubules from several nephrons, pass through the renal pyramids, and open at the renal papillae into minor calyces through about 30 large ***papillary ducts.*** The processed filtrate, called urine, passes from the collecting ducts to papillary ducts, minor calyces, major calyces, renal pelvis, ureter, urinary bladder, and urethra.

With the aid of your textbook, label the parts of a nephron and associated structures in Figure 23.4.

Examine prepared slides of various components of nephrons and compare your observations to Figure 23.5 on page 562.

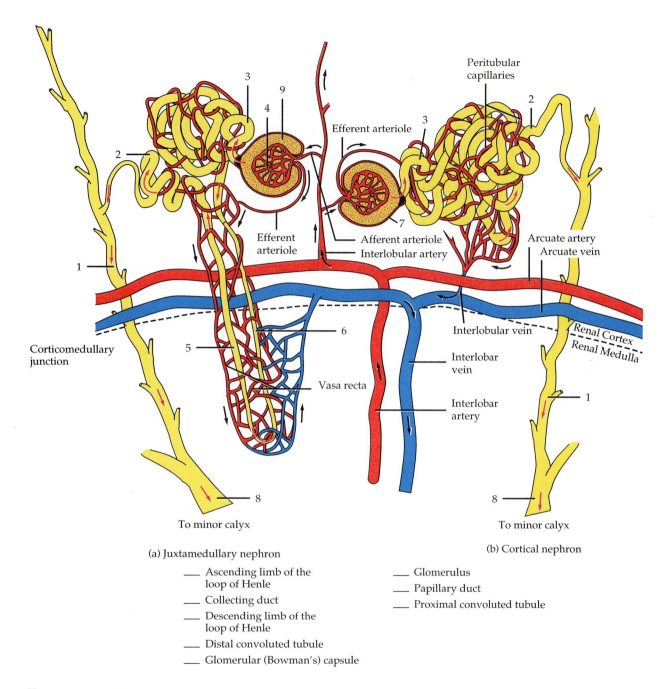

Peritubular capillaries

Efferent arteriole

Efferent arteriole

Afferent arteriole
Interlobular artery

Arcuate artery
Arcuate vein

Interlobular vein

Renal Cortex
Renal Medulla

Interlobar vein

Interlobar artery

Efferent arteriole

Corticomedullary junction

Vasa recta

To minor calyx

To minor calyx

(a) Juxtamedullary nephron

(b) Cortical nephron

___ Ascending limb of the loop of Henle

___ Collecting duct

___ Descending limb of the loop of Henle

___ Distal convoluted tubule

___ Glomerular (Bowman's) capsule

___ Glomerulus

___ Papillary duct

___ Proximal convoluted tubule

FIGURE 23.4 Nephrons. The colored arrows indicate the direction in which the filtrate flows.

3. Blood and Nerve Supply

Nephrons are abundantly supplied with blood vessels, and the kidneys actually receive around 20 to 25% of the total cardiac output, or approximately 1200 ml, every minute. The blood supply originates in each kidney with the ***renal artery,*** which divides into many branches, eventually supplying the nephron and its complete tubule.

Before or immediately after entering the renal hilus, the renal artery divides into a larger anterior branch and a smaller posterior branch. From these branches, five ***segmental arteries*** originate. Each gives off several branches, the ***interlobar arteries,*** which pass between the renal pyramids in the renal columns. At the bases of the pyramids, the interlobar arteries arch between the renal medulla and renal cortex and here are known as ***arcuate*** (*arcuatus* = shaped like a bow) ***arteries.*** Branches of the arcuate arteries, called ***interlobular arteries,*** enter the renal cortex. ***Afferent*** (*ad* = toward; *ferre* = to carry) ***arterioles,*** branches of the interlobular

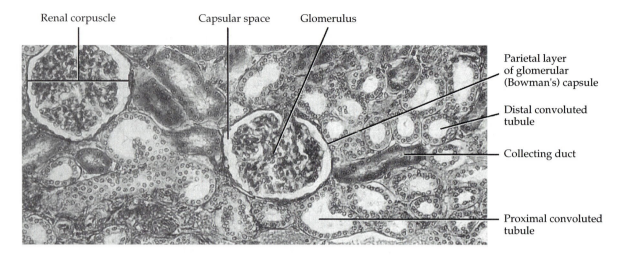

FIGURE 23.5 Histology of nephrons.

arteries, are distributed to the *glomeruli.* Blood leaves the glomeruli via *efferent* (*efferers* = to bring out) *arterioles.*

The next sequence of blood vessels depends on the type of nephron. Around convoluted tubules, efferent arterioles of cortical nephrons divide to form capillary networks called *peritubular* (*peri* = around) *capillaries.* Efferent arterioles of juxtamedullary nephrons also form peritubular capillaries and, in addition, form long loops of blood vessels around medullary structures called *vasa recta* (VĀ-sa REK-ta; *vasa* = vessels; *recta* = straight). Peritubular capillaries eventually reunite to form *interlobular veins.* Blood then drains into *arcuate veins, interlobar veins,* and *segmental veins.* Blood leaves the kidneys through the *renal vein* that exits at the hilus. (The vasa recta pass blood into the interlobular veins, arcuate veins, interlobar veins, and renal veins.)

In each nephron, the final portion of the ascending limb of the loop of Henle makes contact with the afferent arteriole serving its own renal corpuscle. The cells of the renal tubule in this region are tall and crowded together. Collectively, they are known as the *macula densa* (*macula* = spot; *densa* = dense). These cells monitor the Na$^+$ and Cl$^-$ concentration of fluid in the tubule lumen. Next to the macula densa, the wall of the afferent arteriole (and sometimes efferent arteriole) contains modified smooth muscle fibers called *juxtaglomerular (JG) cells.* Together with the macula densa, they constitute the *juxtaglomerular apparatus,* or *JGA.* The JGA helps regulate arterial blood pressure and the rate of blood filtration by the kidneys. The dis-

tal convoluted tubule begins a short distance past the macula densa.

Using Figures 23.4 and 23.6 as guides, trace a drop of blood from its entrance into the renal artery to its exit through the renal vein. As you do so, name in sequence each blood vessel through which blood passes for both cortical and juxtaglomerular nephrons.

Label Figure 23.7 on page 564.

The nerve supply to the kidneys comes from the *renal plexus* of the autonomic system. The nerves are vasomotor because they regulate the circulation of blood in the kidney by regulating the diameters of the arterioles.

4. Ureters

The body has two retroperitoneal *ureters* (YOO-re-ters), one for each kidney; each ureter is a continuation of the renal pelvis and runs to the urinary bladder (see Figure 23.1). Urine is carried through the ureters mostly by peristaltic contractions of the muscular layer of the ureters. Each ureter extends 25 to 30 cm (10 to 12 in.). At the base of the urinary bladder, the ureters turn medially and enter the posterior aspect of the urinary bladder.

Histologically, the ureters consist of an inner *mucosa* of transitional epithelium and an underlying lamina propria (connective tissue), a middle *muscularis* (inner longitudinal and outer circular smooth muscle), and an outer *adventitia* (areolar connective tissue).

Examine a prepared slide of the wall of the ureter showing its various layers. With the aid of your textbook, label Figure 23.8 on page 565.

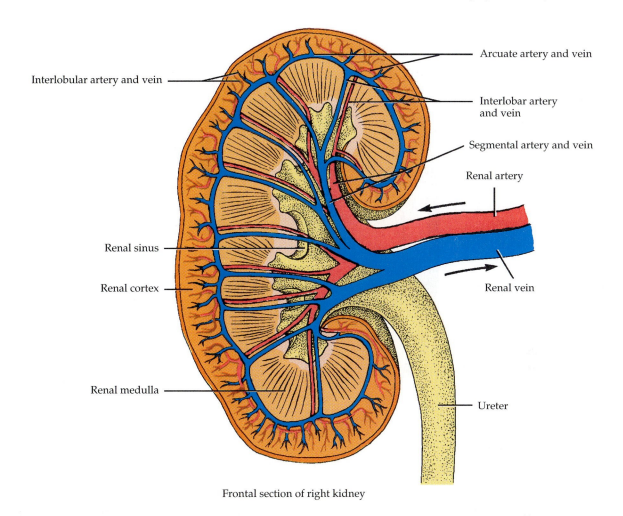

Interlobular artery and vein

Arcuate artery and vein

Interlobar artery and vein

Segmental artery and vein

Renal artery

Renal sinus

Renal cortex

Renal vein

Renal medulla

Ureter

Frontal section of right kidney

FIGURE 23.6 Macroscopic blood vessels of the kidney. Macroscopic and microscopic blood vessels are shown in Figure 23.4.

5. Urinary Bladder

The *urinary bladder* is a hollow muscular organ located in the pelvic cavity posterior to the pubic symphysis (see Figure 23.1). In the male, the bladder is directly anterior to the rectum; in the female, it is anterior to the vagina and inferior to the uterus.

At the base of the interior of the urinary bladder is the *trigone* (TRĪ-gōn; *trigonium* = triangle), a triangular area bounded by the opening into the urethra (internal urethral orifice) and ureteral openings into the bladder. The *mucosa* of the urinary bladder consists of transitional epithelium and lamina propria (connective tissue). Rugae (folds in the mucosa) are also present. The *muscularis,* also called the *detrusor* (de-TROO-ser; *detrudere* = to push down) *muscle,* consists of three layers of smooth muscle: inner longitudinal, middle circular, and outer longitudinal. In the region around the opening to the ure-

thra, the circular muscle fibers form an *internal urethral sphincter.* Inferior to this is the *external urethral sphincter* which is composed of skeletal muscle and is a modification of the urogenital diaphragm. The superficial coat of the urinary bladder is the *adventitia,* a layer of areolar connective tissue that is continuous with that of the ureters. Over the superior surface of the urinary bladder, the adventitia is also covered with peritoneum and the two together constitute the *serosa.*

Urine is expelled from the bladder by an act called *micturition* (mik'-too-RISH-un; *micturire* = to urinate), commonly known as urination, or voiding. The average capacity of the urinary bladder is 700 to 800 ml.

Using your textbook as a guide, label the external urethral sphincter, internal urethral sphincter, ureteral openings, and ureters in Figure 23.9 on page 566.

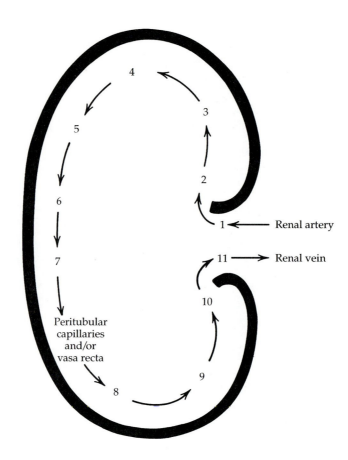

FIGURE 23.7 Blood supply of the right kidney. This view is designed to show the *sequence* of blood flow, not the anatomical location of blood vessels, which is shown in Figure 23.6.

Examine a prepared slide of the wall of the urinary bladder. With the aid of your textbook, label Figure 23.10 on page 567.

6. Urethra

The *urethra* is a small tube leading from the internal urethral orifice in the floor of the urinary bladder to the exterior of the body. In females, this tube is posterior to the pubic symphysis and is embedded in the anterior wall of the vagina; its length is approximately 3.8 cm (1½ in.). The opening of the urethra to the exterior, the *external urethral orifice,* is between the clitoris and vaginal orifice. In males, its length is around 20 cm (8 in.), and it follows a route different from that of the female. Immediately inferior to the urinary bladder, the urethra passes through the prostate gland (prostatic urethra), pierces the urogenital diaphragm (membranous urethra), and traverses the penis (spongy urethra). The urethra is the terminal portion of the urinary system, and serves as the pas-

sageway for discharging urine from the body. In addition, in the male, the urethra serves as the duct through which reproductive secretions are discharged from the body.

Label the urethra and urethral orifice in Figure 23.9 on page 566.

B. DISSECTION OF CAT URINARY SYSTEM

PROCEDURE

CAUTION! *Please reread Section D, "Precautions Related to Dissection" at the beginning of the laboratory manual on page xiii before you begin your dissection.*

1. Kidneys

1. The *kidneys* are situated on either side of the vertebral column at about the level of the third to fifth lumbar vertebrae. The right kidney is

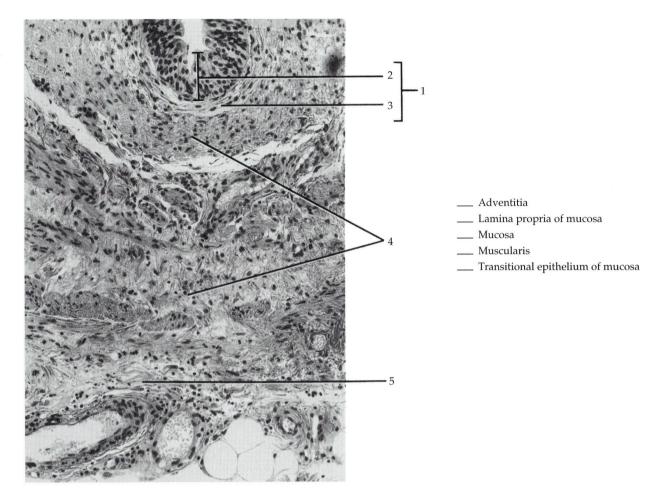

Adventitia
Lamina propria of mucosa
Mucosa
Muscularis
Transitional epithelium of mucosa

FIGURE 23.8   Histology of ureter.

slightly higher than the left (Figure 23.11 on page 568).

2. Each is surrounded by a mass of fat called the *adipose capsule,* which should be removed.

3. Unlike other abdominal organs that are suspended by an extension of the peritoneum, the kidneys are covered by parietal peritoneum only on their ventral surfaces. Because of this, the kidneys are said to be *retroperitoneal.*

4. Identify the *adrenal (suprarenal) glands* located cranial and medial to the kidneys. They are part of the endocrine system.

5. Note that the medial surface of each kidney contains a concave opening, the *renal hilus,* through which blood vessels and the ureter enter or leave the kidney.

6. Identify the *renal artery* branching off the aorta and entering the renal hilus and the *renal vein* leaving the renal hilus and joining the inferior vena cava. Also, identify the ureter caudal to the renal vein.

7. If a detailed dissection of the cat kidney is to be done, follow the directions of the sheep kidney dissection and identify the following structures of the cat that correlate with the sheep kidney.
   a. Renal hilus
   b. Renal pelvis
   c. Calyces
   d. Renal cortex
   e. Renal (medullary) pyramids
   f. Renal column
   g. Renal papillae

2. Ureters

1. The *ureters,* like the kidneys, are retroperitoneal.

2. Each begins as the renal pelvis in the renal sinus and passes caudally to the urinary bladder.

3. The ureters pass posterior to the urinary bladder and open into its floor.

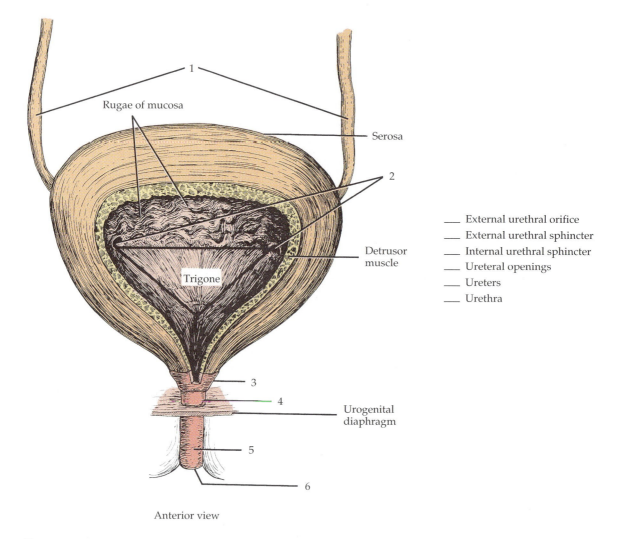

Rugae of mucosa

Serosa

Detrusor muscle

Trigone

1

2

____ External urethral orifice
____ External urethral sphincter
____ Internal urethral sphincter
____ Ureteral openings
____ Ureters
____ Urethra

3

4

5

6

Urogenital diaphragm

Anterior view

FIGURE 23.9   Urinary bladder and female urethra.

3. Urinary Bladder

1. The *urinary bladder,* a pear-shaped musculo-membranous sac located just cranial to the pubic symphysis (Figure 23.11 on page 568), is also retro-peritoneal.
2. The urinary bladder is attached to the abdominal wall by peritoneal folds termed *ligaments.* The *ventral suspensory ligament* extends from the ventral side of the bladder to the *linea alba*. The *lateral ligaments,* one on either side, connect the sides of the bladder to the dorsal body wall. They contain considerable fat.
3. The broad, rounded, cranial portion of the bladder is called the *fundus.*
4. The narrow, caudal, attached portion is referred to as the *neck.*

5. If you have a male cat, identify the *rectovesical pouch,* the space between the urinary bladder and the rectum.
6. If you have a female cat, identify the *vesicouterine pouch,* the space between the urinary bladder and the uterus.
7. If you cut open the urinary bladder, you might be able to see the openings of the ureters into the bladder.

4. Urethra

1. The *urethra* is a duct that conducts urine from the neck of the urinary bladder to the exterior.
2. Dissection of the urethra will be delayed until the reproductive system is studied.

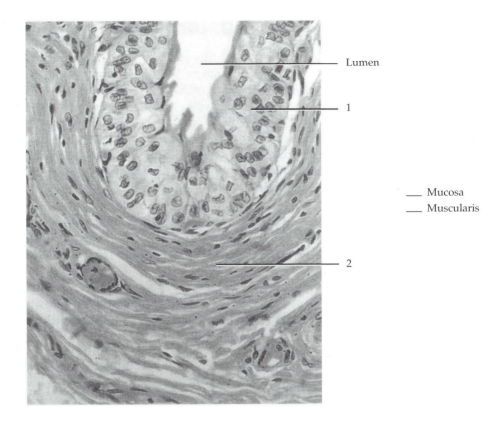

Lumen

1

___ Mucosa
___ Muscularis

2

FIGURE 23.10   Histology of urinary bladder.

C. DISSECTION OF SHEEP (OR PIG) KIDNEY

The sheep kidney is very similar to both human and cat kidney. You may use Figure 23.2 as a reference for this dissection.

CAUTION! *Please reread Section D, "Precautions Related to Dissection" at the beginning of the laboratory manual on page xiii before you begin your dissection.*

PROCEDURE

1. Examine the intact kidney and notice the renal hilus and the fatty tissue that normally surrounds the kidney. Strip away the fat.
2. As you peel the fat off, look carefully for the *adrenal (suprarenal) gland.* This gland is usually found attached to the superior surface of the kidney, as it is in the human. Most preserved kidneys do not have this gland. If it is present, remove it, cut it in half, and note its distinct outer *cortex* and inner *medulla.*
3. Look at the *renal hilus,* which is the concave area of the kidney. From here the *ureter, renal artery,* and *renal vein* enter and exit.

4. Differentiate these blood vessels by examining the thickness of their walls. Which vessel has the thicker wall?
5. With a sharp scalpel *carefully* make a frontal section through the kidney.
6. Identify the *renal capsule* as a thin, tough layer of connective tissue completely surrounding the kidney.
7. Immediately beneath this capsule is an outer light-colored area called the *renal cortex.* The inner dark-colored area is the *renal medulla.*
8. The *renal pelvis* is the large chamber formed by the expansion of the ureter inside the kidney. This renal pelvis divides into many smaller areas called *renal calyces,* each of which has a dark tuft of kidney tissue called a *renal pyramid.*
9. The bases of these pyramids face the cortical area. Their apices, called *renal papillae,* are directed toward the center of the kidney.
10. The calyces collect urine from collecting ducts and drain it into the renal pelvis and out through the ureter.
11. The renal artery divides into several branches that pass between the renal pyramids. These vessels are small and delicate and may be too difficult to dissect and trace through the renal medulla.

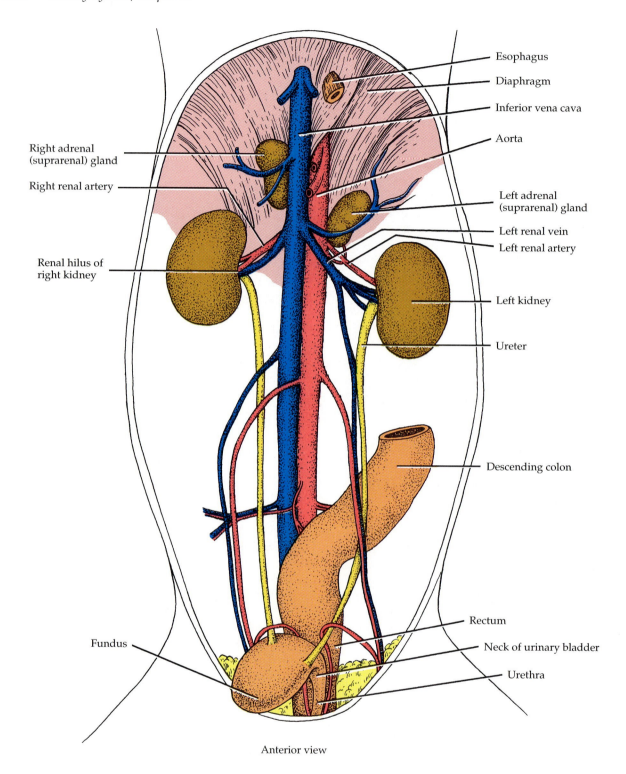

Right adrenal (suprarenal) gland

Right renal artery

Renal hilus of right kidney

Fundus

Esophagus

Diaphragm

Inferior vena cava

Aorta

Left adrenal (suprarenal) gland

Left renal vein

Left renal artery

Left kidney

Ureter

Descending colon

Rectum

Neck of urinary bladder

Urethra

Anterior view

FIGURE 23.11   Urinary system of male cat.

D. RENAL PHYSIOLOGY EXPERIMENTS

Note: Depending on the availability of live animals and the type of laboratory equipment, you may select from the following procedures related to renal physiology.

PROCEDURE USING PHYSIOFINDER

PHYSIOFINDER is an interactive computer program that permits laboratory simulations which do not involve the use of animals or of advanced or expensive laboratory equipment. It permits students to perform experiments, analyze data, draw conclusions, and form hypotheses on the basis of collected data. The program is available from HarperCollins Publishers (1-800-8HEART1).

For activities related to renal physiology, select the appropriate experiments from PHYSIOFINDER Module 7—Renal Water Regulation, Module 8—Renal Sodium Regulation, and Module 9—Renal Hydrogen Ion Regulation.

PROCEDURE USING GOLDFISH TO DEMONSTRATE TUBULAR SECRETION

Through filtration, the glomeruli of the kidneys form a filtrate. The filtrate contains everything in the blood except for high-molecular-weight substances and blood cells. The filtrate thus formed passes through the tubules of the nephron and is modified by several processes involving both active transport and diffusion. During this modification process, the movement of a substance from the tubular lumen into the peritubular capillaries is termed **tubular reabsorption.** This process can occur via either diffusion or active transport, or some combination of both processes (one process occurring at the luminal surface, with the other occurring at the opposite surface). Movement of a substance from the peritubular capillaries through the tubular cells and into the tubular lumen is termed **tubular secretion.** Like reabsorption, this process can occur via either diffusion or active transport, or some combination of both processes. Typically, materials secreted by the nephron are compounds that are incompletely metabolized or not metabolized at all, weak acids or bases, or substances foreign to the body. Secretion may occur in both the proximal and distal convoluted tubules.

A very simple technique to demonstrate tubular secretion utilizes isolated goldfish renal tubules. The fish kidney is ideally suited because the lack of supportive connective tissue between the tubules allows easy and rapid dissection of single tubules. Tubules suspended in Ringer's solu-

tion remain viable for several hours and readily demonstrate the secretion of several substances.

1. Isolation of Renal Tubules

CAUTION! *Please reread Section C, "Precautions Related to Working with Reagents," and Section D, "Precautions Related to Dissection" on page xiii at the beginning of the laboratory manual before you begin any of the following experiments. You should also read the experiments before you perform them to be sure that you understand all the procedures and safety precautions.*

PROCEDURE

1. After decapitating the fish, make a midline ventral incision and expose the kidneys. Identification of the kidneys is aided by their location lateral to the vertebral column and by their reddish-brown color.
2. Gently remove the kidneys and place them in a dish containing Ringer's solution.
3. Utilizing dissection needles, break the kidneys into progressively smaller pieces until the renal tubules are separated. If additional separation is necessary, gently pass some of the tubules into and out of a wide-bore medicine dropper.

2. Experimental Procedure

a. SECRETION OF PHENOL RED

PROCEDURE

1. Place several drops of Ringer's solution onto a depression slide and place one or two renal tubules into the depression via the dropper.
2. Observe the tubules under a microscope and become familiar with their anatomy. Be able to locate the lumen of the tubules.
3. Now place several drops of Ringer's solution containing phenol red (0.05 M) into the depression of another slide, and add one or two tubules to this solution.
4. Immediately place the slide under the microscope and determine (a) time required for the *first* appearance of phenol red within the lumen of the tubule and (b) time required for the *maximum concentration* of dye within the lumen of the tubule.
5. Record your results in Section D.1 of the LABORATORY REPORT RESULTS at the end of the exercise.

b. INHIBITION OF SECRETION

Several substances are secreted by similar or identical tubular carrier mechanisms. Therefore, if two substances are competing for the same carrier, the substance that has a higher concentration or greater affinity for the carrier will inhibit the secretion of the second substance. Such is the case with phenol red and penicillin G.

PROCEDURE

1. Place several tubules in a depression slide containing 0.05 M phenol red and 1 M penicillin G.
2. Determine the values for the time required for phenol red to first appear in the lumen of the tubule, and the time required for maximum concentration of the dye within the tubular lumen.
3. Record your results in Section D.2 of the LABORATORY REPORT RESULTS at the end of the exercise.
4. What type of inhibitory mechanism does this

   experiment demonstrate? _____

   _____

   _____

   _____

c. METABOLIC REQUIREMENTS FOR SECRETION

Phenol red is secreted via active transport, that is, movement of a substance against a concentration gradient, utilizing a carrier and involving expenditure of energy. The energy in this process is obtained by splitting ATP into ADP and P.

PROCEDURE

1. To demonstrate that the secretion of phenol red requires energy, suspend several tubules in a solution containing 0.05 M phenol red and either 1 mM trichlorophenol or 0.2 mM 2,4 dinitrophenol.
2. Record the times required for dye transport in Section D.3 of the LABORATORY REPORT RESULTS at the end of the exercise.
3. Look up the actions of the chemical you utilized, and discuss how this would affect the transport process in the secretion of phenol

   red. _____

   _____

   _____

   _____

E. URINE

The kidneys function to maintain bodily homeostasis. This is accomplished by three processes: (1) filtration of the blood by the glomeruli, (2) tubular reabsorption, and (3) tubular secretion. As a result of these three functions, *urine* is formed and eliminated from the body. Urine contains a high concentration of solutes, and in a healthy person, the volume, pH, and solute concentration of urine will vary with the needs of the internal environment. In certain pathological conditions, the characteristics of urine may change drastically. An analysis of the volume and the physical and chemical properties of urine tells us much about the state of the body.

1. Physical Characteristics

Normal urine usually varies between a straw yellow and an amber transparent color, and possesses a characteristic odor. Urine color varies considerably according to the ratio of solutes to water and according to an individual's diet.

Cloudy urine sometimes reflects the secretion of mucin from the urinary tract lining and is not necessarily an indicator of a pathological condition. The normal pH of urine ranges between 4.6 and 8.0 and averages 6.0. The pH of urine is also strongly affected by diet, with a high-protein diet lowering the pH, and a mostly vegetable diet increasing the pH of the urine.

Specific gravity is the ratio of the weight of a volume of a substance to the weight of an equal volume of distilled water. Water has a specific gravity of 1.000. The specific gravity of urine depends on the amount of solids in solution, and normally ranges from 1.001 to 1.035. The greater the concentration of solutes, the higher the specific gravity. In certain conditions, such as diabetes mellitus, specific gravity is high because of the high glucose content.

2. Abnormal Constituents

When the body's metabolism becomes abnormal, many substances not normally found in urine may appear in varying amounts, while normal constituents may appear in abnormal amounts. *Urinalysis* is the analysis of the physical and chemical properties of urine, and is a vital tool in diagnosing pathological conditions.

  a. *Albumin* Normally absent from a urine sample because the molecules are too large to be filtered out of the blood via the endothelial-capsular membrane. When albumin is found in the urine, the condition is called *albuminuria*.

   b. *Glucose* Urine normally contains such small amounts of *glucose* that clinically glucose is considered to be absent from urine samples. Its presence in significant amounts is called *glucosuria*, and the most common cause is a high blood sugar (glucose) level seen in certain diseases such as diabetes mellitus.

   c. *Erythrocytes Hematuria* is the term utilized to describe the presence of red blood cells in a urine sample. Hematuria usually indicates the presence of a pathological condition within the kidneys.

   d. *Leucocytes Pyuria* (pī-YOO-rē-a) is the condition that occurs when white blood cells and other components of pus are found in the urine, and this usually indicates a pathological condition.

   e. *Ketone bodies* Normal urine contains a small amount of *ketone (acetone) bodies.* Their appearance in large quantities in the urine produces the condition called *ketosis (acetonuria)* and may indicate physiological abnormalities.

   f. *Casts* Tiny masses of various substances that have hardened and assumed the shape of the lumens of the nephron tubules. They are microscopic in size and are composed of many different substances.

   g. *Calculi* Insoluble *calculi* (stones) are various salts that have solidified in the urinary tract. They may be found anywhere from the kidney tubules to the external opening, and their presence causes considerable pain as they attempt to pass through the various lumens of the urinary system.

F. URINALYSIS

In this exercise you will determine some of the characteristics of urine and perform tests for some abnormal constituents that may be present in urine. Some of these tests may be used in determining unknowns in urine specimens.

CAUTION! *Please reread Section A, "General Safety Precautions and Procedures" on page xi, and Section C, "Precautions Related to Working with Reagents" on page xii at the beginning of the laboratory manual before you begin any of the following experiments. Read the experiments before you perform them, to be sure that you understand all the procedures and safety precautions.*

When working with urine, avoid any kind of contact with an open sore, cut, or wound. Wear tight-fitting surgical gloves and safety goggles.

Work with your own urine only. After you have completed your experiments, place all glassware in a fresh household bleach solution or other comparable disinfectant, wash the laboratory tabletop with a fresh household bleach solution or comparable disinfectant, and dispose of the glove and Chemstrips® in the appropriate biohazard container provided by your instructor.

1. Urine Collection

PROCEDURE

1. A specimen of urine may be collected at any time for routine tests; urine voided within 3 hr after meals, however, may contain abnormal constituents. For this reason the first voiding in the morning is preferred.
2. Both males and females should collect a midstream sample of urine in a sterile container. A midstream sample is essential to avoid contamination from the external genitalia, and to avoid the presence of pus cells and bacteria that are normally found in the urethra.
3. If not examined immediately, the specimen should be refrigerated to prevent unnecessary bacterial growth.
4. Before testing *always* mix urine by swirling, inverting the container, or stirring with a wooden swab stick.
5. *Keep all containers clean!*
6. Wrap all papers and sticks and put them in the garbage pail.
7. Rinse all test tubes and glass containers carefully with *cold water* after they have cooled.
8. Flush sinks well with cold water.
9. Obtain either your own freshly voided urine sample or a provided sample if one is available.

2. Chemstrip® Testing

Alternate methods can be employed for several of the tests you are about to perform. One alternate method is the use of plastic strips to which are attached paper squares impregnated with various reagents. These strips display a color reaction when dipped into urine with any abnormal constituents.

PROCEDURE

1. At this point, you should take a Chemstrip® and test your urine sample for the following:

pH, protein, glucose, ketones, bilirubin, and blood (hemoglobin). Record your results below.

2. Determine which component(s) of urine are measured by the Chemstrip® you are using. If you are using a strip that tests for multiple substances, determine which squares measure which substances. Locate the color chart used to read the results. Note the appropriate time for reading each test.

3. Remove a test strip from the vial and *replace the cap*. Dip the test strip into your urine sample for no longer than 1 sec, being sure that all reagents on the strip are immersed.

4. Remove any excess urine from the strip by drawing the edge of the strip along the rim of the container holding your urine sample.

5. After the appropriate time, as indicated on the Chemstrip® vial, hold the strip close to the color blocks on the vial.

6. Make sure that the strip blocks are properly lined up with the color chart on the vial.

Test	Chemstrip® result
pH	
Protein	
Glucose	
Ketones	
Bilirubin	
Blood	

3. Physical Analysis

a. COLOR

Normal urine varies in color from straw yellow to amber because of the pigment *urochrome*, a by-product of hemoglobin destruction. Observe the color of your urine sample. Some abnormal colors are as follows:

Color	Possible Cause
Silvery, milky	Pus, bacteria, epithelial cells
Smoky brown, rust	Blood
Orange, green, blue, red	Medications or liver disease

Record the color of your urine in Section F.3.a of the LABORATORY REPORT RESULTS at the end of the exercise.

How would sickle-cell anemia affect urine color?

_____

_____

b. TRANSPARENCY

A fresh urine sample should be clear. Cloudy urine may be due to substances such as mucin, phosphates, urates, fat, pus, mucus, microbes, crystals, and epithelial cells.

PROCEDURE

1. To determine transparency, cover the container and shake your urine sample and observe the degree of cloudiness.

2. Record your observations in Section F.3.b of the LABORATORY REPORT RESULTS at the end of the exercise.

c. pH

The pH of urine varies with several factors already indicated. You can test the pH of your urine by using either a Chemstrip® or pH paper. Because you have already determined the pH of your urine by using a Chemstrip®, you might want to verify the results using pH paper.

PROCEDURE

1. Place a strip of pH paper into your urine sample three consecutive times.

2. Shake off any excess urine.

3. Let the pH paper sit for 1 min and then compare it to the color chart provided.

4. Record your observations in Section F.3.c of the LABORATORY REPORT RESULTS at the end of the exercise.

How would the consumption of antacid

(sodium bicarbonate) affect urine pH? _____

_____

_____

d. SPECIFIC GRAVITY

The specific gravity is easily determined using a urinometer (hydrometer). The urinometer is a float with a numbered scale near the top that indicates specific gravity directly.

PROCEDURE

1. Familiarize yourself with the scale on the urinometer neck. Determine the change in specific gravity represented by each calibration.
2. Allow your urine to reach room temperature. Urinometers are calibrated to read the specific gravity at 15°C (69°F). If the temperature differs, add or subtract 0.001 for each 3°C above or below 15°C.
3. Fill the cylinder ¾ full of urine and insert the urinometer. Make sure that it is free-floating; if not, spin the neck gently.
4. Read the scale at the bottom of the meniscus when the urinometer is at rest.
5. Record the specific gravity in Section F.3.d of the LABORATORY REPORT RESULTS at the end of the exercise.
6. Rinse the urinometer and cylinder. *Follow your instructor's directions for cleaning them.*
   How would dehydration affect the specific

   gravity of urine? _____

   _____

   _____

4. Chemical Analysis

a. CHLORIDE AND SODIUM CHLORIDE

Most of the sodium chloride (NaCl) present in the renal filtrate is reabsorbed; a small amount remains as a normal component of urine. The normal value for chloride (Cl⁻) is 476 mg/100 ml; for sodium (Na⁺) it is 294 mg/100 ml.

PROCEDURE

1. Using a medicine dropper, place 10 drops of urine into a Pyrex test tube.
2. Using a medicine dropper, add 1 drop of 20% potassium chromate and *gently* agitate the tube. It should be a yellow color.
3. Using a medicine dropper, add 2.9% silver nitrate solution *one drop at a time*, counting the drops and *gently* agitating the test tube during the time the solution is being added.
4. Count the number of drops needed to change the color of the solution from bright yellow to brown.
5. Determine the chloride and sodium chloride concentrations of the sample and record your results in Section F.4.a of the LABORATORY REPORT RESULTS at the end of the exercise. Each drop of silver nitrate added in step 4 is

equivalent to (1) 61 mg of Cl⁻ per 100 ml of urine, and (2) 100 mg of NaCl per 100 ml of urine.

Thus, to determine the chloride concentration of the sample, multiply the number of drops times 61 to obtain the number of mg of Cl⁻ per 100 ml of urine.

To determine the sodium chloride concentration of the sample, multiply the number of drops times 100 to obtain the number of mg of NaCl per 100 ml of urine.

What would a high NaCl content in the urine indi-

cate? _____

_____

_____

b. GLUCOSE

Glucosuria (the presence of glucose in the urine) occurs in patients with diabetes mellitus or other disorders. Traces of glucose may occur in normal urine, but detection of these small amounts requires special tests.

(1) Benedict's Test   Benedict's solution is commonly used to detect reducing sugars in urine and is not specific for just glucose.

PROCEDURE

1. In a Pyrex test tube, combine 10 drops of urine with 5 ml of Benedict's solution. Mix the solution.
2. Using a test tube holder, place the test tube in a boiling water bath for 5 min.
   CAUTION! *Make sure that the mouth of the test tube is pointed away from you and all other persons in the area.*
3. *Using a test tube holder*, remove the test tube from the water bath and read the results according to the following chart.
4. Record your results in Section F.4.b of the LABORATORY REPORT RESULTS at the end of the exercise.

Color	Results
Blue	Negative
Greenish yellow	1 + (0.5 g/100 ml)
Olive green	2 + (1 g/100 ml)
Orange-yellow	3 + (1.5 g/100 ml)
Brick red (with precipitate)	4 + (> 2 g/100 ml)

(2) Clinitest® Reagent Method

PROCEDURE

1. Using medicine droppers, place 10 drops of water and 5 drops of urine in a Pyrex test tube.

CAUTION! *Place the test tube in a test tube rack because it will become too hot to handle.*

2. With forceps, add one Clinitest® tablet.

CAUTION! *The concentrated sodium hydroxide in the tablet generates enough heat to make the liquid in the test tube boil. Make sure that the mouth of the test tube is pointed away from you and all other persons in the area.*

3. The color of the solution is graded as in Benedict's test.
4. Fifteen seconds after boiling has stopped, shake the test tube *gently* and evaluate the color according to the Benedict's test table above. Disregard any color change that occurs after 15 sec.

CAUTION! *Make sure the mouth of the test tube is pointed away from you and others.*

5. Record your results in Section F.4.b of the LABORATORY REPORT RESULTS at the end of the exercise.
   How would a high-sugar diet affect the urine?

   Explain. _____

   _____

   _____

(3) Chemstrip® Method
Record your results in Section F.4.b of the LABORATORY REPORT RESULTS at the end of the exercise.

c. PROTEIN
Normal urine contains traces of proteins that are hard to detect through regular laboratory procedures. Albumin is the most abundant serum protein and is the one usually detected. Because tests for albumin are determined by precipitating the protein either by heat (coagulation) or by adding a reagent, the urine sample should either be filtered or centrifuged (Figure 23.12). The test for protein will be done by the Albutest® reagent method and the Chemstrip® method.

FIGURE 23.12 Tabletop centrifuge used for spinning urine samples to obtain sediment for microscopic analysis.

(1) Albutest® Reagent Method

PROCEDURE

1. Using forceps, place an Albutest® tablet on a clean, dry paper towel and add one drop of urine.
2. After the drop has been absorbed, add two drops of water and allow these to penetrate before reading.
3. Compare the color (in daylight or fluorescent light) on top of the tablet with the color chart provided in lab.
4. If albumin is present in the urine, a *blue-green* spot will remain on the surface of the tablet after the water is added. The amount of protein is indicated by the intensity of the blue-green color.
5. If the test is negative, the original color of the tablet will not be changed at the completion of the test.
6. Record your results in Section F.4.c of the LABORATORY REPORT RESULTS at the end of the exercise.
   Why is protein not normally found in urine?

   _____

   _____

(2) Chemstrip® Method
Record your results in Section F.4.c of the LABORATORY REPORT RESULTS at the end of the exercise.

d. KETONE (ACETONE) BODIES
The presence of ketone (acetone) bodies in urine is a result of abnormal fat catabolism. Reagents such

as sodium nitroprusside, ammonium sulfate, and ammonium hydroxide are available in the form of tablets. Ketones turn purple when added to these chemicals.

(1) Acetest® Tablet Method

PROCEDURE

1. Using forceps, place an Acetest® tablet on a clean, dry paper towel, and, using a medicine dropper, place one drop of urine on the tablet.
2. If acetone or ketone is present, a *lavender-purple* color develops within 30 sec. If the tablet becomes cream-colored from wetting, the results are negative. Compare results with the color chart that comes with the reagent.
3. Record your results in Section F.4.d of the LAB-ORATORY REPORT RESULTS at the end of the exercise.

   Why would starvation cause ketones? _____

   _____

   _____

(2) Chemstrip® Method   Record your results in Section F.4.d of the LABORATORY REPORT RESULTS at the end of the exercise.

e. BILE PIGMENTS
Bile pigments, biliverdin and bilirubin, are not normally present in urine. The presence of large quantities of bilirubin in the extracellular fluids produces jaundice, a yellowish tint to the body tissues, including yellowness of the skin and deep tissues.

(1) Shaken Tube Test for Bile Pigments

PROCEDURE

1. Fill a Pyrex test tube halfway with urine, stopper the tube, and shake it vigorously, being careful to keep the stopper in the tube.
2. A yellow color of foam indicates the presence of bile pigments.
3. Record your results in Section F.2.e of the LAB-ORATORY REPORT RESULTS at the end of the exercise.

(2) Ictotest® for Bilirubin

PROCEDURE

1. Using a medicine dropper, place a drop of urine on one square of the special mat provided in the Ictotest® kit.

2. Using forceps, place one Ictotest® reagent tablet in the center of the moistened area.

CAUTION! *Do not touch the tablets with your fingers. Recap the bottle.*

3. Add one drop of water directly to the tablet; after 5 sec, add another drop of water to the tablet so that the water runs off onto the mat. Observe the color of the mat around the tablet at 60 sec. The presence of bilirubin will turn the mat *blue* or *purple*. A slight *pink* or *red* color is negative for bilirubin.
4. Record your results in Section F.2.e of the LAB-ORATORY REPORT RESULTS at the end of the exercise.

(3) Chemstrip® Method   Record your results in Section F.2.e of the LABORATORY REPORT RESULTS at the end of the exercise.

f. HEMOGLOBIN
Hemoglobin is not normally found in urine.

(1) Chemstrip® Method   Record your results in Section F.4.f of the LABORATORY REPORT RESULTS at the end of the exercise.

5. Microscopic Analysis

If you allow a urine specimen to stand undisturbed for a few hours, many suspended materials will settle to the bottom. A much faster method is to centrifuge a urine sample.

PROCEDURE

1. Place 5 ml of fresh urine in a centrifuge tube (Figure 23.12). Follow the directions of your instructor to centrifuge the tube for 5 min at a slow speed (1500 revolutions per minute [rpm]).
2. Dispose of the supernatant (the clear urine) and mix the sediment by shaking the test tube.
3. Using a long medicine dropper or a Pasteur pipette with a bulb, place a small drop of sediment on a clean glass slide, add one drop of Sedi-stain® or methylene blue, and place a cover glass over the specimen.
4. Using low power and reduced light, examine the sediment for any of the microscopic elements pictured in Figure 23.13.
5. Increase the light and use high power or oil immersion to look for crystals.

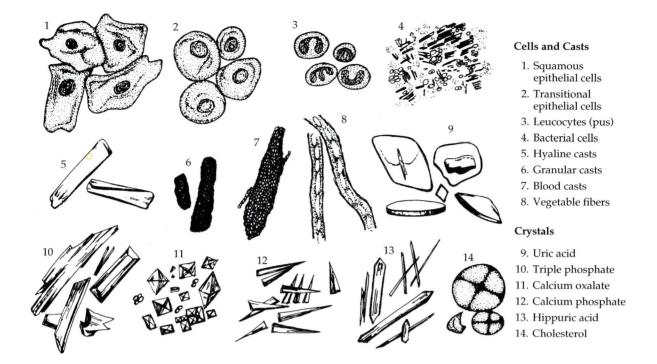

Cells and Casts

1. Squamous epithelial cells
2. Transitional epithelial cells
3. Leucocytes (pus)
4. Bacterial cells
5. Hyaline casts
6. Granular casts
7. Blood casts
8. Vegetable fibers

Crystals

9. Uric acid
10. Triple phosphate
11. Calcium oxalate
12. Calcium phosphate
13. Hippuric acid
14. Cholesterol

Figure 23.13   Diagram of microscopic elements in urine.

6. Crystals can be identified as follows (Figure 23.13):
   a. *Calcium oxalate* Dumbbell and octahedral shapes.
   b. *Calcium phosphate* Very pointed, wedge-shaped formations that may occur as individual crystals or grouped together to form rosettes.
   c. *Cholesterol* Spherical crystals that have a crosslike configuration inside.
   d. *Hippuric acid* Long, needlelike crystals.
   e. *Triple phosphates* Prisms or feathery forms.
   f. *Uric acid* Rhombic prisms, wedges, dumbbells, rosettes, irregular crystals. These are pigmented in sediment, and the color varies from yellow to dark reddish brown.

Draw the results of your observation in Section F.5 of the LABORATORY REPORT RESULTS at the end of the exercise.

Why might some red blood cells in urinary sediment be crenated? _____

_____

Why might yeast be seen in the urine of a person with diabetes mellitus? _____

_____

6. Unknown Specimens

a. UNKNOWNS PREPARED BY INSTRUCTOR

When the composition of a substance has not been defined, it is called an *unknown.* In this exercise, the unknowns will be urine specimens to which the instructor has added glucose, albumin, or any other detectable substance. Each unknown contains only one added substance. Perform the previously outlined tests until you identify the substance in your unknown.

Record your results in Section F.6.a of the LABORATORY REPORT RESULTS at the end of the exercise.

b. UNKNOWNS PREPARED BY CLASS (OPTIONAL)

The class should be divided into two groups. Each group adds certain substances, such as glucose, protein, starch, or fat, to normal, freshly voided urine, keeping accurate records as to what was added to each sample. The two groups then exchange samples, and each group does the basic chemical tests on urine to detect which substances were added. Each student should add one substance to one urine sample and see if another student can detect what was added. Record your results in Section F.6.b of the LABORATORY REPORT RESULTS at the end of the exercise.

ANSWER THE LABORATORY REPORT QUESTIONS AT THE END OF THE EXERCISE.

Urinary System 23

Student _____   Date _____

Laboratory Section _____   Score/Grade _____

SECTION D. TUBULAR SECRETION

1. Secretion of Phenol Red

Time required for first appearance _____

Time required for maximum concentration _____

2. Inhibition of Secretion

Time required for first appearance _____

Time required for maximum concentration _____

3. Metabolic Requirements for Secretion

Time required for first appearance _____

Time required for maximum concentration _____

SECTION F. URINALYSIS

3. Physical Analysis

Characteristic	Normal	Your sample
a. Color	Straw yellow to amber	_____
b. Sediment	None	_____
c. pH	5.0–7.8	_____
d. Specific gravity	1.008–1.030	_____

4. Chemical Analysis

Record your results in the spaces provided.

a. CHLORIDE AND SODIUM CHLORIDE

Chloride concentration _____ mg/100 ml

Sodium chloride concentration _____ mg/100 ml

b. GLUCOSE

Benedict's test _____

Clinitest® tablet _____

Chemstrip® _____

c. PROTEIN

Albutest® reagent tablets _____

Chemstrip® _____

d. KETONE (ACETONE) BODIES

Acetest® tablet _____

Chemstrip® _____

e. BILE PIGMENTS

Shaken tube (bile pigments) _____

Ictotest® (bilirubin) _____

Chemstrip® _____

f. HEMOGLOBIN

Chemstrip® _____

5. Microscopic Analysis

Draw some of the substances (types of cells, types of crystals, or other elements) that you found in the microscopic examination of urinary sediment.

6. Unknown Specimens

a. What substance did you find in the unknown specimen that was prepared by your instructor?

_____

b. What substance did you find in the unknown specimen that was prepared by other students?

_____

Urinary System 23

Student _____   Date _____

Laboratory Section _____   Score/Grade _____

PART 1. Multiple Choice

_____ 1. Beginning at the deepest layer and moving toward the superficial layer, identify the order of tissue layers surrounding the kidney. (a) renal capsule, renal fascia, adipose capsule (b) renal fascia, adipose capsule, renal capsule (c) adipose capsule, renal capsule, renal fascia (d) renal capsule, adipose capsule, renal fascia

_____ 2. The functional unit of the kidney is the (a) nephron (b) ureter (c) urethra (d) hilus

_____ 3. Substances filtered by the kidney must pass through the endothelial-capsular membrane, which is composed of several parts. Which of the following choices lists the correct order of the parts as substances pass through the membrane? (a) epithelium of the visceral layer of glomerular (Bowman's) capsule, endothelium of the glomerulus, basement membrane of the glomerulus (b) endothelium of the glomerulus, basement membrane of the glomerulus, epithelium of the visceral layer of glomerular (Bowman's) capsule (c) basement membrane of the glomerulus, endothelium of the glomerulus, epithelium of the visceral layer of the glomerular (Bowman's) capsule (d) epithelium of the visceral layer of the glomerular (Bowman's) capsule, basement membrane of the glomerulus, endothelium of the glomerulus

_____ 4. In the glomerular (Bowman's) capsule, the afferent arteriole divides into a capillary network called a(n) (a) glomerulus (b) interlobular artery (c) peritubular capillary (d) efferent arteriole

_____ 5. Transport of urine from the renal pelvis into the urinary bladder is the function of the (a) urethra (b) calculi (c) casts (d) ureters

_____ 6. The terminal portion of the urinary system is the (a) urethra (b) urinary bladder (c) ureter (d) nephron

_____ 7. Damage to the renal medulla would interfere first with the functioning of which parts of a juxtamedullary nephron? (a) glomerular (Bowman's) capsule (b) distal convoluted tubule (c) collecting ducts (d) proximal convoluted tubules

_____ 8. An obstruction in the glomerulus would affect the flow of blood into the (a) renal artery (b) efferent arteriole (c) afferent arteriole (d) intralobular artery

_____ 9. Urine that leaves the distal convoluted tubule passes through the following structures in which sequence? (a) collecting duct, hilus, calyces, ureter (b) collecting duct, calyces, pelvis, ureter (c) calyces, collecting duct, pelvis, ureter (d) calyces, hilus, pelvis, ureter

_____ 10. The position of the kidneys posterior to the peritoneal lining of the abdominal cavity is described by the term (a) retroperitoneal (b) anteroperitoneal (c) ptosis (d) inferoperitoneal

_____ 11. Of the following structures, the one to receive filtrate *last* as it passes through the nephron is the (a) proximal convoluted tubule (b) ascending limb of the loop of Henle (c) glomerulus (d) collecting duct

_____ 12. Peristalsis of the ureter is a function of the (a) serosa (b) mucosa (c) submucosa (d) muscularis

_____ 13. The trigone and the detrusor muscle are associated with the (a) kidney (b) urinary bladder (c) urethra (d) ureters

_____ 14. The notch on the medial surface of the kidney through which blood vessels enter and exit is called the (a) renal medulla (b) major calyx (c) renal hilus (d) renal column

_____ 15. Blood is drained from the kidneys by the (a) renal arteries (b) interlobar arteries (c) interlobular veins (d) renal veins

_____ 16. The epithelium of the urinary bladder that permits distension is (a) stratified squamous (b) transitional (c) simple squamous (d) pseudostratified

_____ 17. How many times a day is the entire volume of blood in the body filtered by the kidneys? (a) 100 times (b) 5 times (c) 30 times (d) 60 times

_____ 18. The average urine capacity of the urinary bladder is (a) 1000 to 1200 ml (b) 50 to 100 ml (c) 700 to 800 ml (d) 200 to 300 ml

_____ 19. The normal pH of urine is between (a) 4.6 and 8.0 (b) 2.0 and 4.8 (c) 10.0 and 12.0 (d) none of the above

_____ 20. Normal urine has a specific gravity of approximately (a) 1.001 to 1.035 (b) 1.030 to 1.080 (c) 1.100 to 1.200 (d) none of the above

_____ 21. The special chemical that may be used to detect glucose in the urine is (a) sulfosalicylic acid (b) Benedict's solution (c) Lugol's solution (d) nitric acid

PART 2. Completion

22. In addition to the urinary system, other systems that help eliminate wastes are the respiratory,

integumentary, and _____ systems.

23. The double-walled cup found in a nephron is called a(n) _____.

24. The special capillary network found inside of this double-walled cup is the _____.

25. The major blood vessel that enters each kidney is the _____.

26. The nerve supply to the kidneys comes from the autonomic nervous system and is called the

_____.

27. Urine is expelled from the urinary bladder by an act called urination, voiding, or

_____.

28. The small tube in the urinary system that leads from the floor of the urinary bladder to the outside is

the _____.

29. The apices of renal pyramids are referred to as renal _____.

30. The cortical substance between renal pyramids is called a renal _____.

31. Cuplike extensions of the renal pelvis, usually two or three in number, are referred to as

_____.

32. Epithelial cells of the visceral layer of the glomerular (Bowman's) capsule are called

_____.

33. Distal convoluted tubules terminate by merging with _____.

34. Long loops of blood vessels around the medullary structures of juxtamedullary nephrons are called

_____.

35. Which blood vessel comes next in this sequence? Interlobar artery, arcuate artery, interlobular artery,

_____.

36. The abnormal condition in which red blood cells are found in the urine in appreciable amounts is

called _____.

37. Various salts that solidify in the urinary tract are called _____.

38. Various substances that have hardened and assumed the shape of the lumens of the nephron tubules

are the _____.

39. The pH of urine in individuals on high-protein diets tends to be _____ than normal.

40. The greater the concentration of solutes in urine, the greater will be its _____.

pH and Acid-Base Balance

24

A. THE CONCEPT OF pH

When molecules of inorganic acids, bases, or salts dissolve in water, they undergo *ionization* (ī'-on-i-ZĀ-shun), or *dissociation* (dis'-sō-sē-Ā-shun); that is, they separate into ions.

An *acid* can be defined as a substance that dissociates into one or more *hydrogen ions (H+)* and one or more *anions* (negative ions). Because H^+ is a single proton with a charge of +1, an acid can also be defined as a proton donor. A *base,* by contrast, dissociates into one or more *hydroxide ions (OH−)* and one or more *cations* (positive ions). A base can also be viewed as a proton acceptor. Hydroxide ions have a strong attraction for protons. A *salt,* when dissolved in water, dissociates into cations and anions, neither of which is H^+ or OH^-. Acids and bases react with one another to form salts. Body fluids must constantly contain balanced quantities of acids and bases. In solutions such as those found inside or outside body cells, acids dissociate into hydrogen ions (H^+) and anions. Bases, on the other hand, dissociate into hydroxide ions (OH^-) cations. The more hydrogen ions that exist in a solution, the more acidic the solution; conversely, the more hydroxide ions, the more basic (alkaline) the solution.

Biochemical reactions—those that occur in living systems—are very sensitive to even small changes in acidity or alkalinity. Any departure from the narrow limits of normal H^+ and OH^- concentrations may greatly modify cell functions and disrupt homeostasis. For this reason, the acids and bases that are constantly formed in the body must be kept in balance.

A solution's acidity or alkalinity is expressed on the *pH scale,* which runs from 0 to 14. This scale is based on the concentration of H^+ in a solution. The midpoint of the scale is 7, where the concentrations of H^+ and OH^- are equal. A substance with a pH of 7, such as distilled (pure) water, is neutral. A solution that has more H^+ than OH^- is an *acidic solution* and has a pH below 7. A solution that has more OH^- than H^+ is a *basic (alkaline) solution* and has a pH above 7. A change of one whole number on the pH scale represents a 10-fold change from the previous concentration. A pH of 1 denotes 10 times more H^+ than a pH of 2. A pH of 3 indicates 10 times fewer H^+ than a pH of 2 and 100 times fewer H^+ than a pH of 1.

B. MEASURING pH

1. Using Litmus Paper

PROCEDURE

CAUTION! *Please reread Section A, "General Safety Precautions and Procedures," on page xi, and Section C, "Precautions Related to Working with Reagents," on page xii at the beginning of the laboratory manual, before you begin any of the following experiments. Read the experiments before you perform them, to be sure that you understand all the procedures and safety precautions.*

1. Before you begin, it is important to know that *an acid solution will turn blue litmus paper red* and *a basic (alkaline) solution will turn red litmus paper blue.*
2. Using forceps, dip a strip of red litmus paper into each of the solutions to be tested. Use a *new strip* for each solution. Record you observations in the Table in Section B.1 of the LABORATORY REPORT RESULTS at the end of the exercise.
3. Using forceps, dip a strip of blue litmus paper into each of the solutions to be tested. Use a *new strip* for each solution. Record your observations in the Table in Section B.1 of the LABORATORY REPORT RESULTS at the end of the exercise.
4. Using your textbook as a guide, determine the pH of the following body fluids:

Body Fluid	pH
Bile	
Saliva	
Gastric juice	
Blood	
Pancreatic juice	
Semen	
Urine	

2. Using pH Paper

PROCEDURE

CAUTION! *Please reread Section A, "General Safety Precautions and Procedures," on page xi, and Section C, "Precautions Related to Working with Reagents," on page xii, at the beginning of the laboratory manual, before you begin any of the following experiments. Read the experiments before you perform them, to be sure that you understand all the procedures and safety precautions.*

1. Using forceps, dip a strip of pH paper into each of the solutions to be tested. Use a *new strip* for each solution. Use wide-range pH paper to determine to approximate pH and narrow-range pH paper to determine a more accurate pH.
2. Compare the color of the strip of pH paper to the color chart on the pH paper container.
3. Record your observations in the Table in Section B.2 of the LABORATORY REPORT RESULTS at the end of the exercise.

3. Using a pH Meter

Because there are different types of pH meters, your instructor will demonstrate how to use the pH meter in your laboratory.

PROCEDURE

CAUTION! *Please reread Section A, "General Safety Precautions and Procedures," on page xi, and Section C, "Precautions Related to Working with Reagents," on page xii, at the beginning of the laboratory manual, before you begin any of the following experiments. Read the experiments before you perform them, to be sure that you understand all the procedures and safety precautions.*

1. Examine the pH meter that has been made available to you and identify the following parts: (1) electrodes, (2) pH dial or digital display, (3) temperature control, and (4) calibration control.
2. Plug in the pH meter and turn it on. (*NOTE: Some models take up to one-half hour to warm up.*)
3. Using the temperature control knob, adjust the pH meter for the temperature of the solutions to be tested.
4. To calibrate the pH meter, place the electrode(s) in a beaker that contains a pH 7 buffer solution. The electrode(s) should be immersed at least 1 in. into the solution. Adjust the pH meter with the appropriate controls so that the meter will show a pH value of 7. Now the instrument is calibrated.
5. Depress the standby button and remove the electrode(s) from the buffer solution. *The electrode(s) should not touch anything.* Rinse the electrode(s) with distilled water, using a wash bottle. The rinse water can be collected in an empty beaker.
6. Immerse the electrode(s) into the first solution to be tested. Release the standby button and note the pH. Record the pH in the Table in Section B.3 of the LABORATORY REPORT RESULTS at the end of the exercise.
7. Depress the standby button, remove the electrode(s) from the solution, and rinse with distilled water. Test the pH of the remaining solutions and record each pH in the Table in Section B.3 of the LABORATORY REPORT RESULTS at the end of the exercise.

C. ACID-BASE BALANCE

A very important electrolyte in terms of the body's acid-base balance is the hydrogen ion (H^+). Although some hydrogen ions enter the body in ingested foods, most are produced as a result of the cellular metabolism of substances such as glucose, fatty acids, and amino acids. One of the major challenges to homeostasis is keeping the H^+ concentration at an appropriate level to maintain proper acid-base balance.

The balance of acids and bases is maintained by controlling the H^+ concentration of body fluids, particularly extracellular fluid. In a healthy person, the pH of the extracellular fluid remains between 7.35 and 7.45. Metabolism typically produces a significant excess of H^+. If there were no mechanisms for disposal of acids, the rising concentration of H^+

in body fluids would quickly lead to death. Homeostasis of H^+ concentration within a narrow pH range is essential to survival and depends on three major mechanisms.

1. *Buffer systems* Buffers act quickly to bind H^+ temporarily, which removes excess H^+ from solution but not from the body.
2. *Exhalation of carbon dioxide* By increasing the rate and depth of breathing, more carbon dioxide can be exhaled. This reduces the level of carbonic acid and is effective within minutes.
3. *Kidney excretion* The slowest mechanism, taking hours or days, but the only way to eliminate acids other than carbonic acid is through their passage into urine and their excretion by the kidneys.

In the following experiments, you will note the relationship between buffers and the exhalation of carbon dioxide to pH.

1. Buffers and pH

Most *buffer systems* of the body consist of a weak acid and the salt of that acid, which functions as a weak base. Buffers function to prevent rapid, drastic changes in the pH of a body fluid by changing strong acids and bases into weak acids and bases. Buffers work within fractions of a second. A strong acid dissociates into H^+ more easily than does a weak acid. Strong acids therefore lower pH more than weak ones because strong acids contribute more H^+. Similarly, strong bases dissociate more easily into hydroxide ions (OH^-). The principal buffer systems of the body fluids are the carbonic acid–bicarbonate system, the phosphate system, and the protein buffer system.

a. CARBONIC ACID–BICARBONATE BUFFER SYSTEM

The *carbonic acid–bicarbonate buffer system* is based on the bicarbonate ion (HCO_3^-), which can act as a weak base, and carbonic acid (H_2CO_3), which can act as a weak acid. Thus, the buffer system can compensate for either an excess or a shortage of H^+. For example, if these is an excess of H^+ (an acid condition), HCO_3^- can function as a weak base and remove the excess H^+ as follows:

$$H^+ + HCO_3^- \rightarrow H_2CO_3 \rightarrow H_2O + CO_2$$

Hydrogen ion Bicarbonate ion (weak base) Carbonic acid Water Carbon dioxide

On the other hand, if there is a shortage of H^+ ions (an alkaline condition), H_2CO_3 can function as a weak acid and provide H^+ as follows:

$$H_2CO_3 \rightarrow H^+ + HCO_3^-$$

Carbonic acid (weak acid) Hydrogen ion Bicarbonate ion

A typical bicarbonate buffer system consists of a mixture of carbonic acid (H_2CO_3) and its salt, sodium bicarbonate ($NaHCO_3$). The carbonic acid–bicarbonate buffer system is an important regulator of blood pH. When a strong acid, such as hydrochloric acid (HCl) is added to a buffer solution containing sodium bicarbonate, which behaves like a weak base, the following reaction occurs:

$$HCl + NaHCO_3 \rightarrow NaCl + H_2CO_3$$

Hydrochloric acid (strong acid) Sodium bicarbonate (weak base) Sodium chloride Carbonic acid (weak acid)

If a strong base, such as sodium hydroxide ($NaOH$), is added to a buffer solution containing a weak acid, such as carbonic acid, the following reaction occurs:

$$NaOH + H_2CO_3 \rightarrow H_2O + NaHCO_3$$

Sodium hydroxide (strong base) Carbonic acid (weak acid) Water Sodium bicarbonate (weak base)

Normal metabolism produces more acids than bases and thus tends to acidify the blood rather than make it more alkaline. Accordingly, the body needs more bicarbonate salt than it needs carbonic acid. Bicarbonate molecules outnumber carbonic acid molecules 20:1.

b. PHOSPHATE BUFFER SYSTEM

The *phosphate buffer system* acts in the same manner as the carbonic acid–bicarbonate buffer system. The components of the phosphate buffer system are the sodium salts of dihydrogen phosphate and sodium monohydrogen phosphate ions. The dihydrogen phosphate ion acts as the weak acid and is capable of buffering strong bases.

$$NaOH + NaH_2PO_4 \rightarrow H_2O + Na_2HPO_4$$

Sodium hydroxide (strong base) Sodium dihydrogen phosphate (weak acid) Water Sodium monohydrogen phosphate (weak base)

The monohydrogen phosphate ion acts as the weak base and is capable of buffering strong acids.

$$HCl + Na_2HPO_4 \rightarrow NaCl + NaH_2PO_4$$

| Hydrochloric acid (strong acid) | Sodium monohydrogen phosphate (weak base) | Sodium chloride (salt) | Sodium dihydrogen phosphate (weak acid) |

Because the phosphate concentration is highest in intracellular fluid, the phosphate buffer system is an important regulator of pH in the cytosol. It also is present at a lower level in extracellular fluids and acts to buffer acids in urine. NaH_2PO_4 is formed when excess H^+ in the kidney tubules combines with Na_2HPO_4. In this reaction, $Na+$ released from Na_2HPO_4 forms sodium bicarbonate ($NaHCO_3$) and passes into the blood. The H^+ that replaces Na^+ becomes part of the NaH_2PO_4 that passes into the urine. This reaction is one of the mechanisms by which the kidneys help maintain pH by the acidification of urine.

C. PROTEIN BUFFER SYSTEM

The *protein buffer system* is the most abundant buffer in body cells and plasma. Inside red blood cells the protein hemoglobin is an especially good buffer. Proteins are composed of amino acids. An amino acid is an organic compound that contains at least one carboxyl group (COOH) and at least one amine group (NH_2). The free carboxyl group at one end of a protein acts like an acid by releasing hydrogen ions (H^+) ions when pH rises and can dissociate in this way:

$$\underset{\underset{H}{|}}{\overset{\overset{R}{|}}{NH_2-C-COOH}} \rightarrow \underset{\underset{H}{|}}{\overset{\overset{R}{|}}{NH_2-C-COO}} + H^+$$

The H^+ is then able react with any excess hydroxide ion (OH^-) in the solution to form water.

The free amine group at the other end of a protein can act as a base by combining with hydrogen ions when pH falls as follows:

$$\underset{\underset{H}{|}}{\overset{\overset{R}{|}}{COOH-C-NH_2}} + H^+ \rightarrow \underset{\underset{H}{|}}{\overset{\overset{R}{|}}{COOH-C-NH_3^+}}$$

Thus, proteins act as both acidic and basic buffers.

The following exercise will demonstrate how buffers resist changes in pH.

PROCEDURE

CAUTION! *Please reread Section A, "General Safety Precautions and Procedures," on page xi, and Section C, "Precautions Related to Working with Reagents," on page xii, at the beginning of the laboratory manual, before you begin any of the following experiments. Read the experiments before you perform them, to be sure that you understand all the procedures and safety precautions.*

1. Using a pH meter, immerse the electrode(s) in a beaker containing distilled water and determine the pH. _____

2. Drop by drop, slowly add 0.05 M hydrochloric acid (HCl) to the distilled water. Gently swirl the beaker after each drop is added. Note how many drops it takes for the pH of the solution to change one whole number. _____

   What is the pH of the solution? _____

3. Remove the electrode(s) from the solution and rinse with distilled water.

4. Now immerse the electrode(s) in a pH 7 buffer solution. Drop by drop, slowly add 0.05 M HCl to the solution. Gently swirl the beaker after each drop is added. Note how many drops it takes for the pH of the solution to change one whole number. _____

   What conclusion can you draw from this observation? _____

   _____

   _____

5. Remove the electrode(s) from the solution and rinse with distilled water.

6. Immerse the electrode(s) in a beaker of fresh distilled water and determine the pH. _____

   _____

7. Drop by drop, slowly add 0.05 M sodium hydroxide (NaOH) to the distilled water. Gently swirl the beaker after each drop is added. Note how many drops it takes for the pH of the solution to change one whole number.

   _____

   What is the pH of the solution? _____

8. Remove the electrode(s) from the solution and rinse with distilled water.

9. Now immerse the electrode(s) in a pH 7 buffer solution. Drop by drop, slowly add 0.05 M NaOH to the solution. Gently swirl the beaker after each drop is added. Note how many drops it takes for the pH of the solution to change one whole number. _____

   What conclusion can you draw from this observation? _____

2. Respirations and pH

Breathing also plays a role in maintaining the pH of the body. An increase in the carbon dioxide (CO_2) concentration in body fluids increases H^+ concentration and thus lowers the pH (makes it more acidic). This is illustrated by the following reactions:

$$CO_2 + H_2O \rightleftharpoons H_2CO_3 \rightleftharpoons H^+ + HCO_3^-$$

Conversely, a decrease in the CO_2 concentration of body fluids raises the pH (makes it more basic).

The pH of body fluids can be adjusted, usually in 1 to 3 mins, by a change in the rate and depth of breathing. If the rate and depth of breathing increase, more CO_2 is exhaled, the reaction just given is driven to the left, H^+ concentration falls, and the blood pH rises. Because carbonic acid can be eliminated by exhaling CO_2, it is called a *volatile acid.* If the rate of respiration slows down, less carbon dioxide is exhaled, and the blood pH falls. Doubling the breathing rate increases the pH by about 0.23, from 7.4 to 7.63. Reducing the breathing rate to one-quarter its normal rate lowers the pH by 0.4, from 7.4 to 7.0. These examples show the powerful effect of alterations in breathing on pH of body fluids.

The pH of body fluids, in turn, affects the rate of breathing. If, for example, the blood becomes more acidic, the increase in hydrogen ions is detected by chemoreceptors that stimulate the inspiratory center in the medulla. As a result, the diaphragm and other muscles of respiration contract more forcefully and frequently—the rate and depth of breathing increase.

The same effect is achieved if the blood level of CO_2 increases. The increased rate and depth of respiration remove more CO_2 from blood to reduce the H^+ concentration, and blood pH increases. On the other hand, if the pH of the blood increases, the respiratory center is inhibited and respirations decrease. A decrease in the CO_2 concentration of blood has the same effect. The decreased rate and depth of respirations cause CO_2 to accumulate in blood and the H^+ concentration increases. The respiratory mechanism normally can eliminate more acid or base than can all the buffers combined, but it is limited to eliminating only the single volatile acid, carbonic acid.

The following exercise will demonstrate the relationship of exhalation of carbon dioxide to pH.

PROCEDURE

CAUTION! *Please reread Section A, "General Safety Precautions and Procedures," on page xi, and Section C, "Precautions Related to Working with Reagents," on page xii, at the beginning of the laboratory manual, before you begin any of the following experiments. Read the experiments before you perform them, to be sure that you understand all the procedures and safety precautions.*

1. Fill a large beaker with 100 ml of distilled water.
2. Add 5 ml of 0.10 normal sodium hydroxide (NaOH) solution and 5 drops of phenol red.
3. Phenol red is a pH indicator. It remains red in a basic solution, changes to orange in a neutral solution, and changes to yellow in an acidic solution.
4. While at rest, exhale through a straw into the solution. Your partner should determine how long it takes for the solution to change from orange to yellow. _____

CAUTION! *Perform step 5 only if you have no known or apparent cardiac or other health problems and are capable of such an activity.*

5. *Run in place for about 100 steps.*
6. Exhale through a straw into a fresh solution as prepared in steps 1 and 2. Your partner should determine how long it takes for the solution to change from orange to yellow.
7. Change places with your partner, and repeat the experiment. Explain the difference in time it took for the solution to change from orange to yellow at rest and following exercise. _____

   _____

   _____

   _____

D. ACID-BASE IMBALANCES

The normal blood pH range is 7.35 to 7.45. *Acidosis* (or *acidemia*) is a condition in which blood pH is below 7.35. *Alkalosis* (or *alkalemia*) is a condition in which blood pH is higher than 7.45.

A change in blood pH that leads to acidosis or alkalosis can be compensated to return pH to normal. *Compensation* refers to the physiological response to an acid–base imbalance. If a person has an altered pH due to metabolic causes, respiratory mechanisms (hyperventilation or hypoventilation) can help compensate for the alteration. Respiratory compensation occurs within minutes and is maximized within hours. On the other hand, if a person has a altered pH due to respiratory causes, metabolic mechanisms (kidney excretion) can compensate for the alteration. Metabolic compensation may begin in minutes but takes days to reach a maximum.

1. Physiological Effects

The principal physiological effect of acidosis is depression of the CNS through depression of synaptic transmission. If the blood pH falls below 7, depression of the nervous system is so severe that the individual becomes disoriented and comatose and dies. Patients with severe acidosis usually die in a state of coma. On the other hand, the major physiological effect of alkalosis is overexcitability in both the CNS and peripheral nerves. Nerves conduct impulses repetitively, even when not stimulated by normal stimuli, resulting in nervousness, muscle spasms, and even convulsions and death.

In the discussion that follows, note that both respiratory acidosis and alkalosis are primary disorders of blood pCO_2 (normal range 35 to 45 mm Hg). On the other hand, both metabolic acidosis and alkalosis are primary disorders of bicarbonate (HCO_3^-) concentration (normal range 22 to 26 mEq/liter).

2. Respiratory Acidosis

The hallmark of *respiratory acidosis* is an elevated pCO_2 of arterial blood (above 45 mm Hg). Inadequate exhalation of CO_2 decreases the blood pH. It occurs as a result of any condition that decreases the movement of CO_2 from the blood to the alveoli of the lungs to the atmosphere and therefore causes a buildup of carbon dioxide, carbonic acid, and hydrogen ions. Such conditions include emphysema, pulmonary edema, injury to the respiratory center of the medulla, airway obstruction, or disorders of the muscles involved in breathing. Metabolic compensation involves increased excretion of H^+ and increased reabsorption of HCO_3^- by the kidneys. Treatment of respiratory acidosis aims to increase the exhalation of CO_2. Excessive secretions can be suctioned out of the respiratory tract, and artificial respiration can be given. In addition, intravenous administration of bicarbonate and ventilation therapy to remove excessive carbon dioxide can be used.

3. Respiratory Alkalosis

In *respiratory alkalosis* arterial blood pCO_2 is decreased (below 35 mm Hg). Hyperventilation causes the pH to increase. It occurs in conditions that stimulate the respiratory center. Such conditions include oxygen deficiency due to high altitude or pulmonary disease, cerebrovascular accident (CVA), severe anxiety, and aspirin overdose. The kidneys attempt to compensate by decreasing excretion of H^+ and decreasing reabsorption of HCO_3^-. Treatment of respiratory alkalosis is aimed at increasing the level of CO_2 in the body. One simple measure is to have the person breathe into a paper bag and then rebreathe the exhaled mixture of CO_2 and oxygen from the bag.

4. Metabolic Acidosis

In *metabolic acidosis* there is a decrease in HCO_3^- concentration (below 22 mEq/liter). The decrease in pH is caused by loss of bicarbonate, such as may occur with severe diarrhea or renal dysfunction; accumulation of an acid, other than carbonic acid, as may occur in ketosis; or failure of the kidneys to excrete H^+ derived from metabolism of dietary proteins. Compensation is respiratory by hyperventilation. Treatment of metabolic acidosis consists of intravenous solutions of sodium bicarbonate and correcting the cause of acidosis.

5. Metabolic Alkalosis

In *metabolic alkalosis* HCO_3^- concentration is elevated (above 26 mEq/liter). A nonrespiratory loss of acid by the body or excessive intake of alkaline drugs causes the pH to increase. Excessive vomiting of gastric contents results in a substantial loss of hydrochloric acid and is probably the most frequent cause of metabolic alkalosis. Other causes of metabolic alkalosis include gastric suctioning, use of certain diuretics, endocrine disorders, and administration of alkali. Compensation is respiratory by hypoventilation. Treatment of metabolic alkalosis consists of fluid therapy to replace chloride, potassium, and other electrolyte deficiencies and correcting the cause of alkalosis.

A summary of acidosis and alkalosis is presented in Table 24.1.

Based on an analysis of respiratory gases, you can determine if a person has acidosis or alkalosis and whether the acidosis or alkalosis is respiratory or metabolic, as reflected by the change in pH.

In the following table (bottom of page), note the normal ranges for pH, pCO_2, and HCO_3^-. Also note how values above and below normal relate to acidosis and alkalosis.

If a change in pH is a result of an abnormal pCO_2 value, then the condition is respiratory in nature. If, instead, a change in pH is a result of an abnormal HCO_3^- value, the condition is metabolic in nature.

Problems related to acid-base imbalance are reflected in each of the following conditions. Determine whether each is (1) acidosis or alkalosis and (2) metabolic or respiratory:

a. pH = 7.32
HCO_3^- = 10 mEq/liter

b. pH = 7.48
pCO_2 = 32 mmHg
c. pH = 7.52
HCO_3^- = 28 mEq/liter
d. pH = 7.30
pCO_2 = 48 mmHg

E. RENAL REGULATION OF HYDROGEN ION CONCENTRATION

PHYSIOFINDER is an interactive computer program that permits laboratory simulations that do not involve the use of animals or expensive laboratory equipment. It permits students to perform experiments, analyze data, draw conclusions, and form hypotheses on the basis of collected data. The program is available from HarperCollins Publishers (1-800-8HEART1).

For activities related to renal regulation of hydrogen ion concentration, select the appropriate experiments from PHYSIOFINDER Module 9—Renal Hydrogen Ion Regulation.

ANSWER THE LABORATORY REPORT QUESTIONS AT THE END OF THE EXERCISE

	pH	pCO_2	HCO_3^-
Normal range	7.35–7.45	35–45 mmHg	22–26 mEq/liter
Acidosis	Below 7.35	Above 45 mmHg	Below 22 mEq/liter
Alkalosis	Above 7.45	Below 35 mmHg	Above 26 Eq/liter

TABLE **24.1**
Summary of Acidosis and Alkalosis

Condition	Definition	Common cause	Compensatory mechanism
Respiratory acidosis	Increased pCO$_2$ (above 45 mm Hg) and decreased pH (below 7.35) if there is no compensation.	Hypoventilation due to emphysema, pulmonary edema, trauma to respiratory center, airway obstructions, dysfunction of muscles of respiration.	Renal: increased excretion of H$^+$; increased reabsorption of HCO$_3^-$. If compensation is complete, pH will be within normal range, but pCO$_2$ will be high.
Respiratory alkalosis	Decreased pCO$_2$ (below 35 mm Hg) and increased pH (above 7.45) if there is no compensation.	Hyperventilation due to oxygen deficiency, pulmonary disease, cerebrovascular accident (CVA), anxiety, or aspirin overdose.	Renal: decreased excretion of H$^+$; decreased reabsorption of HCO$_3^-$. If compensation is complete, pH will be within norman range, but pCO$_2$ will be low.
Metabolic acidosis	Decreased bicarbonate (below 22 mEq/liter) and decreased pH (below 7.35) if there is no compensation.	Loss of bicarbonate due to diarrhea, accumulation of acid (ketosis), renal dysfunction.	Respiratory: hyperventilation, which increases loss of CO$_2$. If compensation is complete, pH will be within normal range but HCO$_3^-$ will be low.
Metabolic alkalosis	Increased bicarbonate (above 26 mEq/liter) and increased pH (above 7.45) if there is no compensation.	Loss of acid or excessive intake of alkaline drugs; due to vomiting, gastric suctioning, use of certain diuretics, and administration of alkali.	Respiratory: hypoventilation, which slows loss of CO$_2$. If compensation is complete, pH will be within normal range, but HCO$_3^-$ will be high.

pH and Acid-Base Balance **24**

Student _____   **Date** _____

Laboratory Section _____   **Score/Grade** _____

SECTION B. MEASURING pH

1. Using Litmus Paper

Solution	Color of red litmus paper	Color of blue litmus paper	Is the solution acidic or basic?
Milk of magnesia			
Vinegar			
Coffee			
Carbonated soft drink			
Orange juice			
Distilled water			
Baking soda			
Lemon juice			

2. Using pH Paper

Solution	pH
Milk of magnesia	
Vinegar	
Coffee	
Carbonated soft drink	
Orange juice	
Distilled water	
Baking soda	
Lemon juice	

3. Using a pH Meter

Solution	pH
Milk of magnesia	
Vinegar	
Coffee	
Carbonated soft drink	
Orange juice	
Distilled water	
Baking soda	
Lemon juice	

pH and Acid-Base Balance 24

Student _____   **Date** _____

Laboratory Section _____   **Score/Grade** _____

PART 1. Multiple Choice

_____ 1. An acid is a substance that dissociates into (a) OH^- (b) H^+ (c) HCO_3^- (d) Na^+

_____ 2. Which of the following pHs is more acidic? (a) 6.89 (b) 6.91 (c) 7.00 (d) 6.83

_____ 3. The pH of bile is (a) 4.2 (b) 6.35 to 6.85 (c) 7.6 to 8.6 (d) 3.0 to 3.5

_____ 4. Which mechanism is quickest to restore pH? (a) exhalation of CO_2 (b) buffers (c) kidney excretion (d) inhalation of oxygen

_____ 5. In the carbonic acid–bicarbonate buffer system, which substance functions to buffer a strong base? (a) NaOH (b) $NaHCO_3$ (c) H_2CO_3 (d) NaCl

_____ 6. The most abundant buffer in body cells and plasma is the (a) protein buffer (b) phosphate buffer (c) hemoglobin buffer (d) carbonic acid–bicarbonate buffer

_____ 7. Doubling the breathing rate increases pH by about (a) 0.75 (b) 2.21 (c) 1.86 (d) 0.23

PART 2. Completion

8. Bases dissociate into _____ ions and cations.

9. A change of one whole number on the pH scale represents a _____-fold change from the previous concentration.

10. A(n) _____ solution will turn red litmus paper blue.

11. The pH of blood is _____ .

12. Most H^+ in the body is produced as a result of _____ .

13. In the protein buffer system, the carboxyl group acts as a(n) _____ .

14. If the rate and depth of respiration increase, pH will _____ .

15. If blood becomes more basic, the rate and depth of respiration will _____ .

16. A pH higher than 7.45 is referred to as _____ .

17. Metabolic acidosis and alkalosis are disorders of _____ concentration.

18. The principal physiological effect of _____ is depression of the CNS.

19. _____ acidosis is characterized by a pH below 7.35 and an elevated pCO_2.

20. Compensation for metabolic acidosis is _____.

Reproductive Systems

25

Reproduction is the process by which new individuals of a species are produced and the genetic material is passed from generation to generation. This maintains continuation of the species. Cell division in a multicellular organism is necessary for growth as well as repair and it involves passing of genetic material from parent cells to daughter cells. In somatic cell division a parent cell produces two identical daughter cells. This process is involved in replacing cells and growth. In reproductive cell division, sperm and egg cells are produced for continuity of the species.

The organs of the male and female reproductive systems may be grouped by function. (1) The testes and ovaries, also called *gonads* (*gonos* = seed), function in the production of gametes—sperm cells and ova, respectively. The gonads also secrete hormones. (2) The *ducts* of the reproductive systems transport, receive, and store gametes. (3) Still other reproductive organs, called *accessory sex glands,* produce materials that support gametes. (4) Finally, several *supporting structures,* including the penis, have various roles in reproduction. In this exercise, you will study the structure of the male and female reproductive organs and associated structures.

A. ORGANS OF MALE REPRODUCTIVE SYSTEM

The *male reproductive system* includes (1) the testes, or male gonads, which produce sperm and secrete hormones; (2) a system of ducts that transport, receive, or store sperm; (3) accessory glands, whose secretions contribute to semen; and (4) several supporting structures, including the penis.

1. Testes

The *testes,* or *testicles,* are paired oval glands that lie in the pelvic cavity for most of fetal life. They usually begin to enter the scrotum during the latter half of the seventh month of fetal development; full descent is not complete until just before birth. If the testes do not descend, the condition is called *cryptorchidism* (krip-TOR-ki-dizm; *kryptos* = hidden; *orchis* = testis). Cryptorchidism results in sterility, because the cells that stimulate the initial development of sperm cells are destroyed by the higher temperature of the pelvic cavity. The chance of testicular cancer is 30 to 50% greater in cryptorchid testes.

Each testis is partially covered by a serous membrane called the *tunica* (*tunica* = sheath) *vaginalis,* which is derived from the peritoneum. Internal to the tunica vaginalis is a dense white fibrous capsule, the *tunica albuginea* (al'-byoo-JIN-ē-a; *albus* = white), which extends inward and divides the testis into a series of 200 to 300 internal compartments called *lobules.* Each lobule contains one to three tightly coiled *seminiferous* (*semen* = seed; *ferre* = to carry) *tubules* where sperm production *(spermatogenesis)* occurs.

Label the structures associated with the testes in Figure 25.1.

Spermatogenic cells are sperm-forming cells in various stages that undergo mitosis and differentiation to eventually produce sperm. Together with supporting cells, they line the seminiferous tubules. The most immature spermatogenic cells are called *spermatogonia* (sper'-ma-tō-GŌ-nē-a; *sperm* = seed; *gonium* = generation or offspring; singular is *spermatogonium*). They lie next to the basement membrane. Toward the lumen of the tubule are layers of progressively more mature cells. In order of advancing maturity, these are *primary spermatocytes* (SPER-ma-tō-sīts), *secondary spermatocytes,* and *spermatids.* By the time a *sperm cell,* or *spermatozoon* (sper'-ma-tō-ZŌ-on; *zoon* = life; plural is *sperm,* or *spermatozoa*), has nearly reached maturity, it is released into the lumen of the tubule and begins to move out of the rete testis.

Embedded among the spermatogenic cells in the tubules are large *sustentacular* (sus'-ten-TAK-yoo-lar;

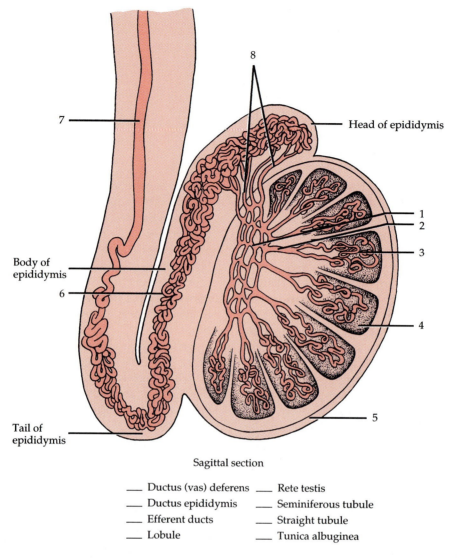

7

8

Head of epididymis

1
2

3

Body of
epididymis

6

4

Tail of
epididymis

5

Sagittal section

___ Ductus (vas) deferens   ___ Rete testis
___ Ductus epididymis   ___ Seminiferous tubule
___ Efferent ducts   ___ Straight tubule
___ Lobule   ___ Tunica albuginea

FIGURE 25.1   Testis showing its system of ducts.

sustentare = to support), or **Sertoli, cells** that extend from the basement membrane to the lumen of the tubule. Sustentacular cells support and protect developing spermatogenic cells; nourish spermatocytes, spermatids, and sperm; phagocytize excess spermatid cytoplasm as development proceeds; and regulate the effects of testosterone and follicle-stimulating hormone (FSH). Sustentacular cells also control movements of spermatogenic cells and the release of sperm into the lumen of the seminiferous tubules. They produce fluid for sperm transport and secrete the hormone inhibin, which helps regulate sperm production by inhibiting the secretion of FSH. In the spaces between adjacent seminiferous tubules are clusters of cells called **interstitial endocrinocytes,** or **Leydig cells.** These cells secrete testosterone, the most important androgen (male sex hormone). Because they produce both sperm

and hormones, the testes are both exocrine and endocrine glands.

Using your textbook, charts, or models as reference, label Figure 25.2.

Sperm are produced at the rate of about 300 million per day. Once ejaculated, they usually live about 48 hr in the female reproductive tract. The parts of a sperm are as follows:

a. *Head* Contains the *nucleus* and *acrosome* (produces hyaluronic acid and proteinases to bring about penetration of secondary oocyte).
b. *Midpiece* Contains numerous mitochondria in which the energy for locomotion is generated.
c. *Tail* Typical flagellum used for locomotion.

With the aid of your textbook, label Figure 25.3 on page 598.

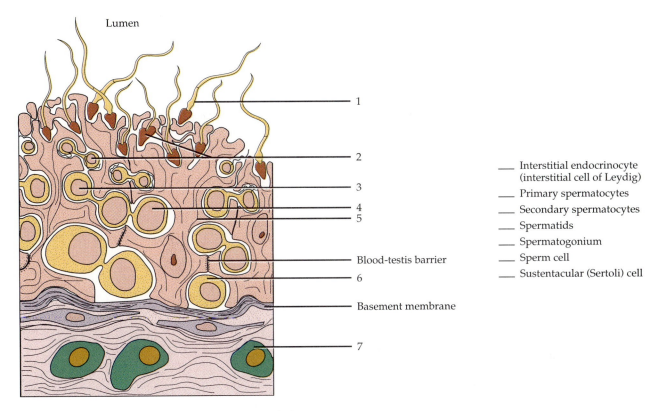

Lumen

_____ 1

_____ 2

_____ 3

_____ 4
_____ 5

Blood-testis barrier

_____ 6

Basement membrane

_____ 7

_____ Interstitial endocrinocyte
        (interstitial cell of Leydig)
_____ Primary spermatocytes
_____ Secondary spermatocytes
_____ Spermatids
_____ Spermatogonium
_____ Sperm cell
_____ Sustentacular (Sertoli) cell

(a) Transverse section of a portion of a seminiferous tubule

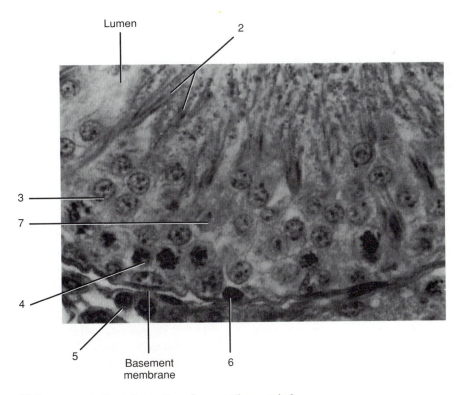

Lumen   2

3 —

7 —

4 —

5   Basement   6
         membrane

(b) Transverse section of a portion of a seminiferous tubule

FIGURE 25.2   Seminiferous tubules showing various stages of spermatogenesis. (a) Diagram.
(b) Photomicrograph.

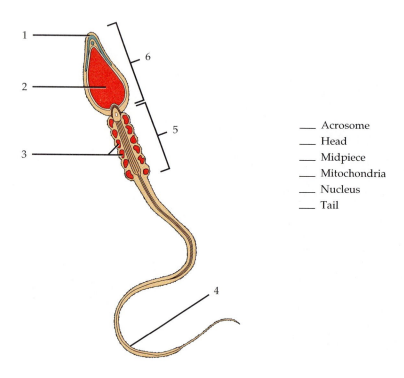

_____ Acrosome
_____ Head
_____ Midpiece
_____ Mitochondria
_____ Nucleus
_____ Tail

FIGURE 25.3   Parts of a sperm cell.

2. Ducts

As sperm cells mature, they are moved through seminiferous tubules into tubes called *straight tubules,* from which they are transported into a network of ducts, the *rete* (RĒ-tē; *rete* = network) *testis.* The sperm cells are next transported out of the testes through a series of coiled *efferent ducts* that empty into a single *ductus epididymis* (ep'-i-DID-i-mis; *epi* = above; *didymos* = testis). From there, they are passed into the *ductus (vas) deferens,* which ascends along the posterior border of the testis, penetrates the inguinal canal, enters the pelvic cavity, and loops over the side and down the posterior surface of the urinary bladder. The ductus (vas) deferens and duct from the seminal vesicle (gland) together form the *ejaculatory* (e-JAK-yoo-la-tō'-rē; *ejectus* = to throw out) *duct,* which propels the sperm cells into the *urethra,* the terminal duct of the system. The male urethra is divisible into (1) a *prostatic portion,* which passes through the prostate gland; (2) a *membranous portion,* which passes through the urogenital diaphragm; and (3) a *spongy (penile) portion,* which passes through the corpus spongiosum of the penis (see Figure 25.6 on page 600). The *epididymis* is a comma-shaped organ that is divisible into a head, body, and tail. The head is the superior portion that contains the efferent ducts; the body is the middle portion that contains the duc-

tus epididymis; and the tail is the inferior portion in which the ductus epididymis continues as the ductus (vas) deferens.

Label the various ducts of the male reproductive system in Figure 25.1.

The ductus epididymis is lined with *pseudostratified columnar epithelium.* The free surfaces of the cells contain long, branching microvilli called *stereocilia.* The muscularis deep to the epithelium consists of smooth muscle. Functionally, the ductus epididymis is the site of sperm maturation (increased motility and fertility potential). They require between 10 and 14 days to complete their maturation—that is, to become capable of fertilizing a secondary oocyte. The ductus epididymis also stores sperm cells and propels them toward the urethra during emission by peristaltic contraction of its smooth muscle. Sperm cells may remain in storage in the ductus epididymis up to a month or more. After that, they are expelled from the epididymis or reabsorbed in the epididymis.

Obtain a prepared slide of the ductus epididymis showing its mucosa and muscularis. Compare your observations to Figure 25.4.

Histologically, the *ductus (vas) deferens (seminal duct)* is also lined with *pseudostratified columnar epithelium* and its muscularis consists of three layers of smooth muscle. Peristaltic contractions of the muscularis propel sperm cells toward the urethra during ejaculation. One method of sterilization in

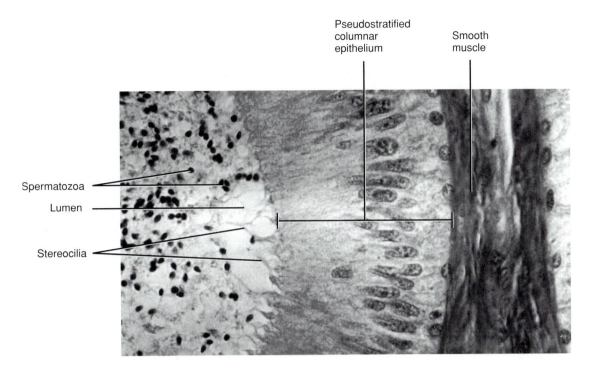

FIGURE 25.4   Histology of ductus epididymis.

males, *vasectomy,* involves removal of a portion of each ductus (vas) deferens.

Obtain a prepared slide of the ductus (vas) deferens showing its mucosa and muscularis. Compare your observations to Figure 25.5.

3. Accessory Sex Glands

Whereas the ducts of the male reproductive system store or transport sperm, a series of *accessory sex glands* secrete most of the liquid portion of

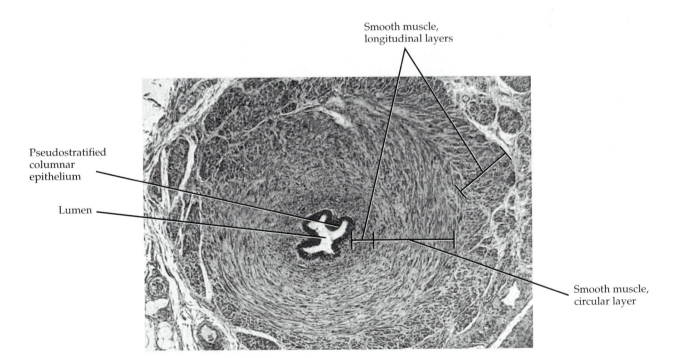

FIGURE 25.5   Histology of ductus (vas) deferens.

semen. Semen is a mixture of sperm cells and the secretions of the seminal vesicles, prostate gland, and bulbourethral glands.

The *seminal* (*seminalis* = pertaining to seed) *vesicles* (VES-i-kuls) are paired, convoluted, pouchlike structures posterior to and at the base of the urinary bladder anterior to the rectum. The glands secrete the alkaline viscous component of semen into the ejaculatory duct. The seminal vesicles contribute about 60% of the volume of semen.

The *prostate* (PROS-tāt) *gland,* a single doughnut-shaped gland inferior to the urinary bladder, surrounds the prostatic urethra. The prostate secretes a slightly acidic fluid into the prostatic urethra. The prostatic secretion constitutes about 25% of the total semen produced.

The paired *bulbourethral* (bul'-bō-yoo-RĒ-thral), or *Cowper's, glands,* located inferior to the prostate on either side of the membranous urethra, are about the size of peas. They secrete an alkaline substance through ducts that open into the spongy (penile) urethra.

With the aid of your textbook, label the accessory glands and associated structures in Figure 25.6.

4. Penis

The *penis* conveys urine to the exterior and introduces sperm cells into the vagina during copulation. Its principal parts are

a. *Glans* (*glandes* = acorn) *penis* Slightly enlarged distal end.
b. *Corona* Margin of glans penis.
c. *Prepuce* (PRĒ-pyoos) Foreskin; loosely fitting skin covering glans penis.
d. *Corpora cavernosa* (*corpus* = body; *caverna* = hollow) *penis* Two dorsolateral masses of erectile tissue.
e. *Corpus spongiosum penis* Midventral mass of erectile tissue that contains spongy (penile) urethra.
f. *External urethral orifice* Opening of spongy (penile) urethra to exterior.

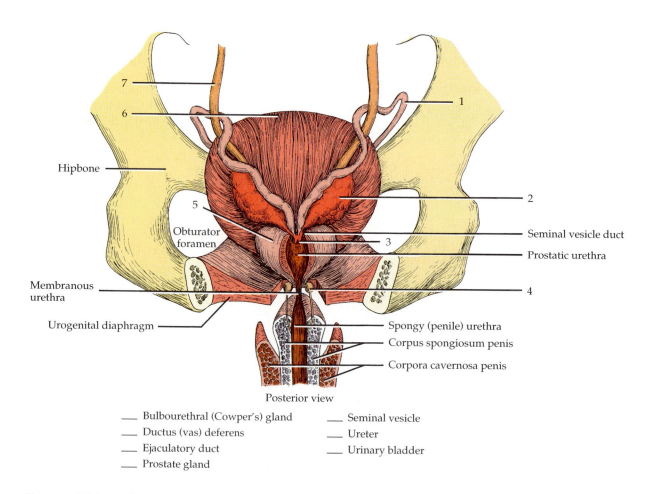

Posterior view

___ Bulbourethral (Cowper's) gland   ___ Seminal vesicle
___ Ductus (vas) deferens   ___ Ureter
___ Ejaculatory duct   ___ Urinary bladder
___ Prostate gland

FIGURE 25.6 Relationships of some male reproductive organs.

With the aid of your textbook, label the parts of the penis in Figure 25.7.

Now that you have completed your study of the organs of the male reproductive system, label Figure 25.8.

B. ORGANS OF FEMALE REPRODUCTIVE SYSTEM

The *female reproductive system* includes the female gonads (ovaries), which produce secondary oocytes; uterine (Fallopian) tubes, or oviducts, which transport secondary oocytes and fertilized ova to the uterus; vagina; external organs that compose the vulva; and the mammary glands.

1. Ovaries

The *ovaries* (*ovarium* = egg receptacle) are paired glands that resemble almonds in size and shape. Functionally, the ovaries produce secondary oocytes, discharge them about once a month by a process called ovulation, and secrete female sex hormones (estrogens, progesterone, relaxin, and inhibin). The point of entrance for blood vessels and nerves is the *hilus*. The ovaries are positioned in the superior pelvic cavity, one on each side of the uterus, by a series of ligaments:

a. *Mesovarium* Double-layered fold of peritoneum that attaches ovaries to broad ligaments of uterus.
b. *Ovarian ligament* Anchors ovary to uterus.
c. *Suspensory ligament* Attaches ovary to pelvic wall.

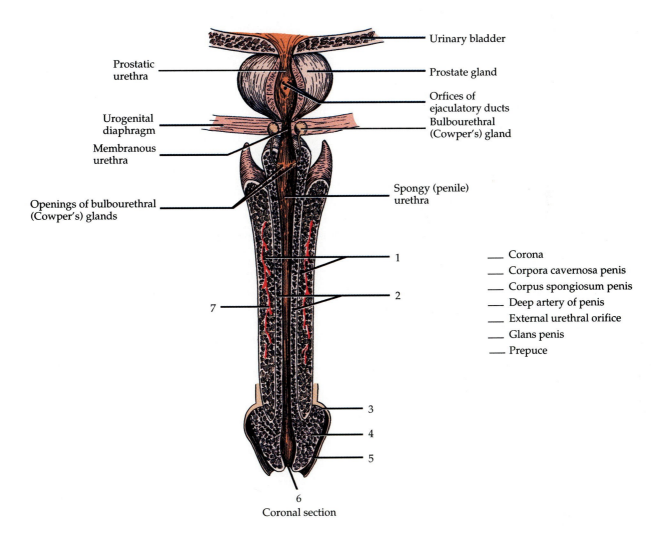

Prostatic urethra

Urogenital diaphragm

Membranous urethra

Openings of bulbourethral (Cowper's) glands

Urinary bladder

Prostate gland

Orfices of ejaculatory ducts

Bulbourethral (Cowper's) gland

Spongy (penile) urethra

1

2

7

3

4

5

6

Coronal section

___ Corona
___ Corpora cavernosa penis
___ Corpus spongiosum penis
___ Deep artery of penis
___ External urethral orifice
___ Glans penis
___ Prepuce

FIGURE 25.7 Internal structure of penis viewed from floor of penis.

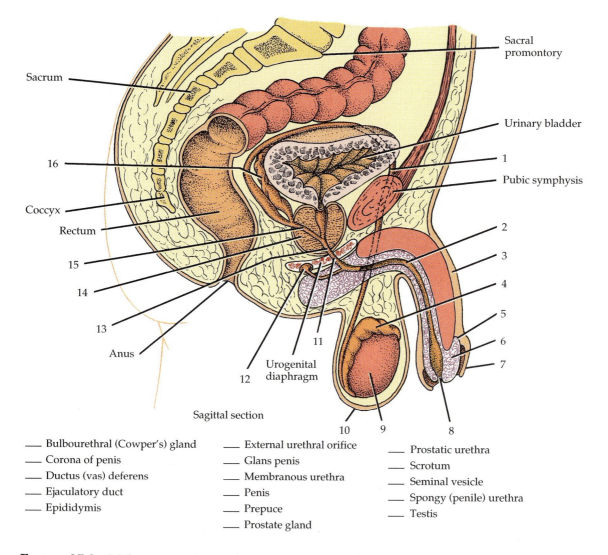

Sacral promontory

Sacrum

Urinary bladder

16

1

Pubic symphysis

Coccyx

2

Rectum

3

15

4

14

5

13

6

Anus

7

11

Urogenital diaphragm

12

10   9   8

Sagittal section

FIGURE 25.8   Male organs of reproduction and surrounding structures.

___ Bulbourethral (Cowper's) gland
___ Corona of penis
___ Ductus (vas) deferens
___ Ejaculatory duct
___ Epididymis

___ External urethral orifice
___ Glans penis
___ Membranous urethra
___ Penis
___ Prepuce
___ Prostate gland

___ Prostatic urethra
___ Scrotum
___ Seminal vesicle
___ Spongy (penile) urethra
___ Testis

With the aid of your textbook, label the ovarian ligaments in Figure 25.9.

Histologically, the ovaries consist of the following parts:

1. *Germinal epithelium* A layer of simple epithelium (low cuboidal or squamous) that covers the surface of the ovary and is continuous with the mesothelium that covers the mesovarium. The term *germinal epithelium* is a misnomer since it does not give rise to oocytes, although at one time it was believed that it did.
2. *Tunica albuginea* A whitish capsule of dense, irregular connective tissue immediately deep to the germinal epithelium.
3. *Stroma* A region of connective tissue deep to the tunica albuginea and composed of a superficial, dense layer called the *cortex* and a deep, loose layer known as the *medulla.*

4. *Ovarian follicles* (*folliculus* = little bag) Lie in the cortex and consist of *oocytes* (immature ova) in various stages of development and their surrounding cells. When the surrounding cells form a single layer, they are called *follicular cells.* Later in development, when they form several layers, they are referred to as *granulosa cells.* The surrounding cells nourish the developing oocyte and begin to secrete estrogens as the follicle grows larger. Ovarian follicles undergo a series of changes prior to ovulation, progressing through several distinct stages. The most numerous and peripherally arranged follicles are termed *primordial follicles.* If a primordial follicle progresses to ovulation (release of a mature ovum), it will sequentially transform into a *primary (preantral) follicle,* then a *secondary (antral) follicle,* and finally a *mature (Graafian) follicle.*

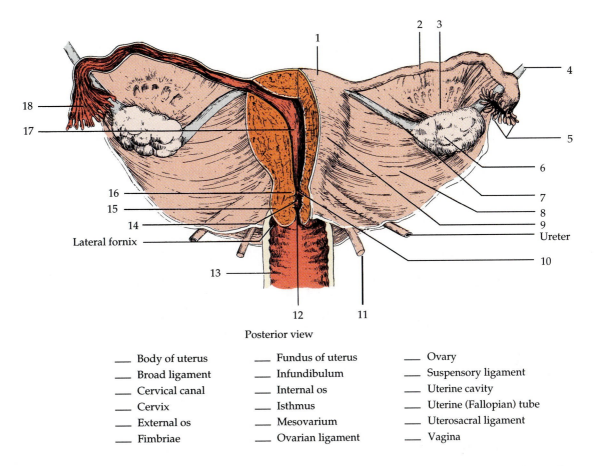

FIGURE 25.9   Uterus and associated female reproductive structures. The left side of figure has been sectioned to show internal structures.

___ Body of uterus	___ Fundus of uterus	___ Ovary
___ Broad ligament	___ Infundibulum	___ Suspensory ligament
___ Cervical canal	___ Internal os	___ Uterine cavity
___ Cervix	___ Isthmus	___ Uterine (Fallopian) tube
___ External os	___ Mesovarium	___ Uterosacral ligament
___ Fimbriae	___ Ovarian ligament	___ Vagina

5. *Mature (Graafian) follicle*   A large, fluid-filled follicle that soon will rupture and expel a secondary oocyte, a process called *ovulation.*
6. *Corpus luteum* (= yellow body)   Contains the remnants of an ovulated mature follicle. The corpus luteum produces progesterone, estrogens, relaxin, and inhibin until it degenerates and turns into fibrous tissue called a *corpus albicans* (= white body).

With the aid of your textbook, label the parts of an ovary in Figure 25.10.

Obtain prepared slides of the ovary, examine them, and compare your observations to Figure 25.11 on page 605.

2. Uterine (Fallopian) Tubes

The *uterine (Fallopian) tubes,* or *oviducts,* extend laterally from the uterus and transport secondary oocytes from the ovaries to the uterus. Fertilization normally occurs in the uterine tubes. The tubes are positioned between folds of the broad ligaments of the uterus. The funnel-shaped, open distal end of each uterine tube, called the *infundibulum,* is surrounded by a fringe of fingerlike projections called *fimbriae* (FIM-bre-ē; *fimbrae* = fringe). The *ampulla* (am-POOL-la) of the uterine tube is the widest, longest portion, constituting about two-thirds of its length. The *isthmus* (IS-mus) is the short, narrow, thick-walled portion that joins the uterus.

With the aid of your textbook, label the parts of the uterine tubes in Figure 25.9.

Histologically, the mucosa of the uterine tubes consists of ciliated columnar cells and secretory cells. The muscularis is composed of inner circular and outer longitudinal layers of smooth muscle. Wavelike contractions of the muscularis help move the ovum down into the uterus. The serosa is the outer covering.

Examine a prepared slide of the wall of the uterine tube, and compare your observations to Figure 25.12 on page 606.

3. Uterus

The *uterus (womb)* is the site of menstruation, implantation of a fertilized ovum, development of

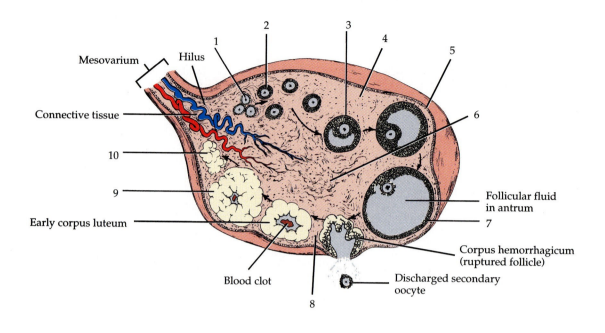

Mesovarium

Hilus

Connective tissue

10

9

Early corpus luteum

Blood clot

1 2 3 4 5

6

Follicular fluid
in antrum

7

Corpus hemorrhagicum
(ruptured follicle)

Discharged secondary
oocyte

8

—— Corpus albicans
—— Corpus luteum (mature)
—— Cortex of stroma
—— Germinal epithelium
—— Mature (Graafian) follicle
—— Medulla of stroma

—— Primary (preantral) follicle
—— Primordial follicle
—— Secondary (antral) follicle
—— Tunica albuginea

FIGURE 25.10   Histology of ovary. Arrows indicate sequence of developmental stages that occur as part of ovarian cycle.

the fetus during pregnancy, and labor. Located between the urinary bladder and the rectum, the organ is shaped like an inverted pear. The uterus is subdivided into the following regions:

a. Fundus Dome-shaped portion superior to the uterine tubes.
b. Body Major, tapering portion.
c. Cervix Inferior narrow opening into vagina.
d. Isthmus (IS-mus) Constricted region between body and cervix.
e. Uterine cavity Interior of the body.
f. Cervical canal Interior of the cervix.
g. Internal os Site where cervical canal opens into uterine cavity.
h. External os Site where cervical canal opens into vagina.

Label these structures in Figure 25.9.
The uterus is maintained in position by the following ligaments:

a. Broad ligaments Double folds of parietal peritoneum that anchor the uterus to either side of the pelvic cavity.
b. Uterosacral ligaments Parietal peritoneal extensions that connect the uterus to the sacrum.
c. Cardinal (lateral cervical) ligaments Tissues containing smooth muscle, uterine blood vessels, and nerves. These ligaments extend below the bases of the broad ligaments between the pelvic wall and the cervix and vagina, and are the chief ligaments that maintain the position of the uterus, helping to keep it from dropping into the vagina.
d. Round ligaments Extend from uterus to external genitals (labia majora) between folds of broad ligaments.

Label the uterine ligaments in Figure 25.9.
Histologically, the uterus consists of three principal layers: endometrium, myometrium, and perimetrium (serosa). The inner *endometrium*

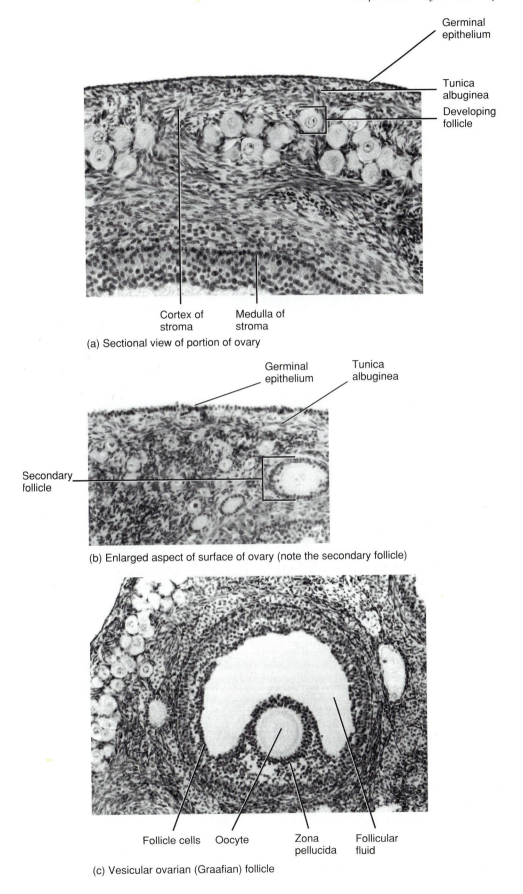

(a) Sectional view of portion of ovary

(b) Enlarged aspect of surface of ovary (note the secondary follicle)

(c) Vesicular ovarian (Graafian) follicle

FIGURE 25.11   Histology of ovary.

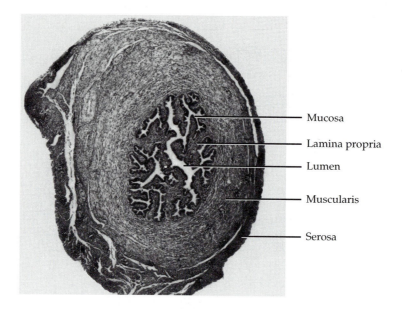

FIGURE 25.12   Histology of uterine (Fallopian) tube.

(*endo* = within) is a mucous membrane that consists of simple columnar epithelium, an underlying endometrial stroma composed of connective tissue and endometrial glands. The endometrium consists of two layers: (1) *stratum functionalis,* the layer closer to the uterine cavity that is shed during menstruation; and (2) *stratum basalis* (ba-SAL-is), the permanent layer that produces a new stratum functionalis after menstruation. The middle *myometrium* (*myo* = muscle) forms the bulk of the uterine wall and consists of three layers of smooth muscle. During labor its coordinated contractions help to expel the fetus. The outer layer is the *perimetrium* (*peri* = around; *metron* = uterus), or *serosa,* part of the visceral peritoneum.

4. Vagina

A muscular, tubular organ lined with a mucous membrane, the *vagina* (*vagina* = sheath) is the passageway for menstrual flow, the receptacle for the penis during copulation, and the inferior portion of the birth canal. The vagina is situated between the urinary bladder and rectum and extends from the cervix of the uterus to the vestibule of the vulva. Recesses called *fornices* (FOR-ni-sēz'; *fornix* = arch or vault) surround the vaginal attachment to the cervix (see Figure 25.9) and make possible the use of contraceptive diaphragms. The opening of the vagina to the exterior, the *vaginal orifice,* may be bordered by a thin fold of vascularized membrane, the *hymen* (*hymen* = membrane).

Label the vagina in Figure 25.9.

Histologically, the mucosa of the vagina consists of nonkeratinized stratified squamous epithelium and connective tissue that lies in a series of transverse folds, the *rugae.* The muscularis is composed of an outer circular and an inner longitudinal layer of smooth muscle.

5. Vulva

The *vulva* (VUL-va; *volvere* = to wrap around), or *pudendum* (pyoo-DEN-dum), is a collective term for the external genitals of the female. It consists of the following parts:

a. *Mons pubis* (MONZ PŪ-bis) Elevation of adipose tissue over the pubic symphysis covered by skin and pubic hair.

b. *Labia majora* (LĀ-bē-a ma-JŌ-ra; *labium* = lip) Two longitudinal folds of skin that extend inferiorly and posteriorly from the mons pubis. Singular is *labium majus.* The folds, covered by pubic hair on their superior lateral surfaces, contain abundant adipose tissue and sebaceous (oil) and sudoriferous (sweat) glands.

c. *Labia minora* (MĪ-nō-ra) Two folds of mucous membrane medial to labia majora. Singular is *labium minus.* The folds have numerous sebaceous glands but few sudoriferous glands and no fat or pubic hair.

d. *Clitoris* (KLI-to-ris) Small cylindrical mass of erectile tissue and nerves at anterior junction of labia minora. The exposed portion is called the *glans,* the covering is called the *prepuce* (foreskin).

e. *Vestibule* Cleft between labia minora containing vaginal orifice, hymen (if present), external urethral orifice, and openings of several ducts.

f. *Vaginal orifice* Opening of vagina to exterior.

g. *Hymen* Thin fold of vascularized membrane that borders vaginal orifice.

h. *External urethral orifice* Opening of urethra to exterior.

i. *Orifices of paraurethral (Skene's) glands* Located on either side of external urethral orifice. The glands secrete mucus.

j. *Orifices of ducts of greater vestibular* (ves-TIB-yoo-lar), or *Bartholin's, glands* Located in a groove between hymen and labia minora. These glands produce a mucoid secretion that supplements lubrication during intercourse.

k. *Orifices of ducts of lesser vestibular glands* Microscopic orifices opening into vestibule.

Using your textbook as an aid, label the parts of the vulva in Figure 25.13.

6. Mammary Glands

The *mammary* (*mamma* = breast) *glands* are modified sweat glands that lie over the pectoralis major muscles and are attached to them by a layer of deep fascia. They consist of the following structures:

a. *Lobes* Around 15 to 20 compartments separated by adipose tissue.

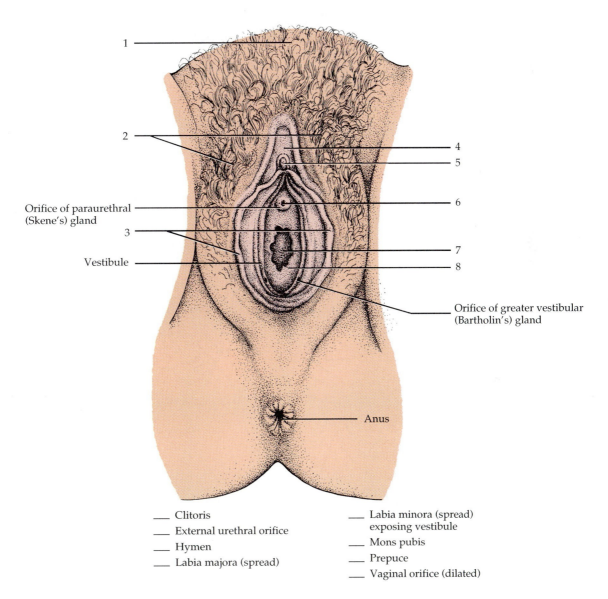

FIGURE 25.13 Vulva.

___ Clitoris
___ External urethral orifice
___ Hymen
___ Labia majora (spread)

___ Labia minora (spread) exposing vestibule
___ Mons pubis
___ Prepuce
___ Vaginal orifice (dilated)

b. *Lobules* Smaller compartments in lobes that contain clusters of milk-secreting glands called *alveoli* (*alveolus* = small cavity).

c. *Secondary tubules* Receive milk from alveoli.

d. *Mammary ducts* Receive milk from secondary tubules.

e. *Lactiferous* (*lact* = milk; *ferre* = to carry) *sinuses* Expanded distal portions of mammary ducts that store milk.

f. *Lactiferous ducts* Receive milk from lactiferous sinuses.

g. *Nipple* Projection on anterior surface of mammary gland that contains lactiferous ducts.

h. *Areola* (a-RĒ-ō-la; *areola* = small space) Circular pigmented skin around nipple.

With the aid of your textbook, label the parts of the mammary gland in Figure 25.14.

Examine a prepared slide of alveoli of the mammary gland, and compare your observations to Figure 25.15.

Now that you have completed your study of the organs of the female reproductive system, label Figure 25.16.

7. Female Reproductive Cycle

During their reproductive years, nonpregnant females normally experience a cyclical sequence of changes in the ovaries and uterus. Each cycle takes about a month and involves both oogenesis and preparation of the uterus to receive a fertilized ovum. Hormones secreted by the hypothalamus, anterior pituitary gland, and ovaries control the principal events. The *ovarian cycle* is a series of events associated with the maturation of an oocyte. The *uterine (menstrual) cycle* is a series of changes in the endometrium of the uterus. Each month, the endometrium is prepared for the arrival of a fertilized ovum that will develop in the uterus until birth. If fertilization does not occur, the stratum functionalis portion of the endometrium is shed. The general term *female reproductive cycle* encompasses the ovarian and uterine cycles, the hormonal changes that regulate them, and cyclical changes in the breasts and cervix.

a. HORMONAL REGULATION

The uterine cycle and ovarian cycle are controlled by gonadotropin releasing hormone (GnRH) from the hypothalamus. See Figure 25.17 on page 610. GnRH

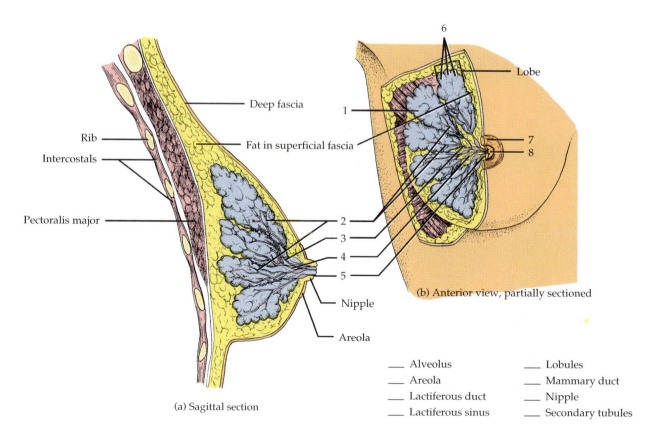

(a) Sagittal section

(b) Anterior view, partially sectioned

____ Alveolus	____ Lobules
____ Areola	____ Mammary duct
____ Lactiferous duct	____ Nipple
____ Lactiferous sinus	____ Secondary tubules

FIGURE 25.14 Mammary glands.

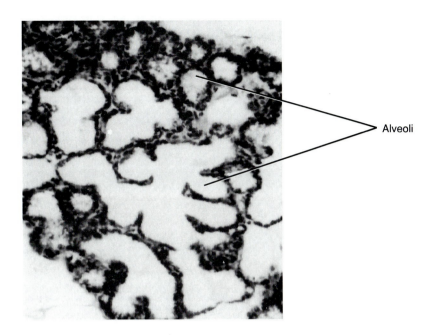

FIGURE 25.15   Histology of mammary gland showing alveoli.

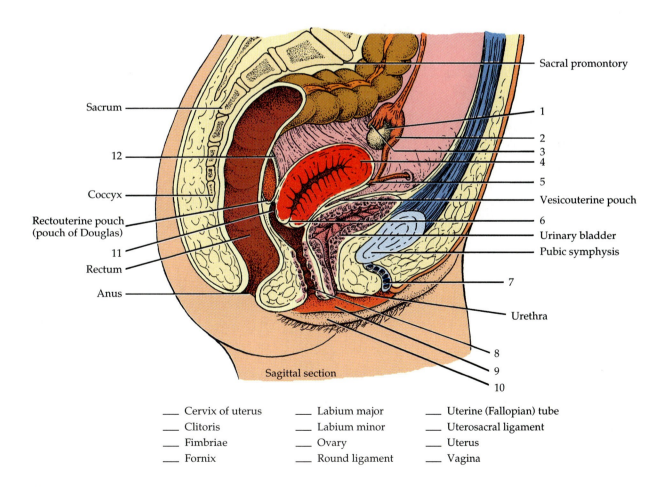

Sacral promontory

Sacrum

1

2

3

4

12

5

Coccyx

Vesicouterine pouch

Rectouterine pouch
(pouch of Douglas)

6

11

Urinary bladder

Rectum

Pubic symphysis

Anus

7

Urethra

8

9

Sagittal section

10

___ Cervix of uterus   ___ Labium major   ___ Uterine (Fallopian) tube
___ Clitoris   ___ Labium minor   ___ Uterosacral ligament
___ Fimbriae   ___ Ovary   ___ Uterus
___ Fornix   ___ Round ligament   ___ Vagina

FIGURE 25.16   Female organs of reproduction and surrounding structures.

stimulates the release of follicle-stimulating hormone (FSH) and luteinizing hormone (LH) from the anterior pituitary gland. FSH stimulates the initial secretion of estrogens by the follicles. LH stimulates the further development of ovarian follicles and their full secretion of estrogens, brings about ovulation, and stimulates the production of estrogens, progesterone, relaxin, and inhibin by the corpus luteum.

b. PHASES OF THE FEMALE REPRODUCTIVE CYCLE

The duration of the female reproductive cycle typically is 24–35 days. For this discussion, we shall assume a duration of 28 days, divided into three phases: the menstrual phase, preovulatory phase, and postovulatory phase (Figure 25.17).

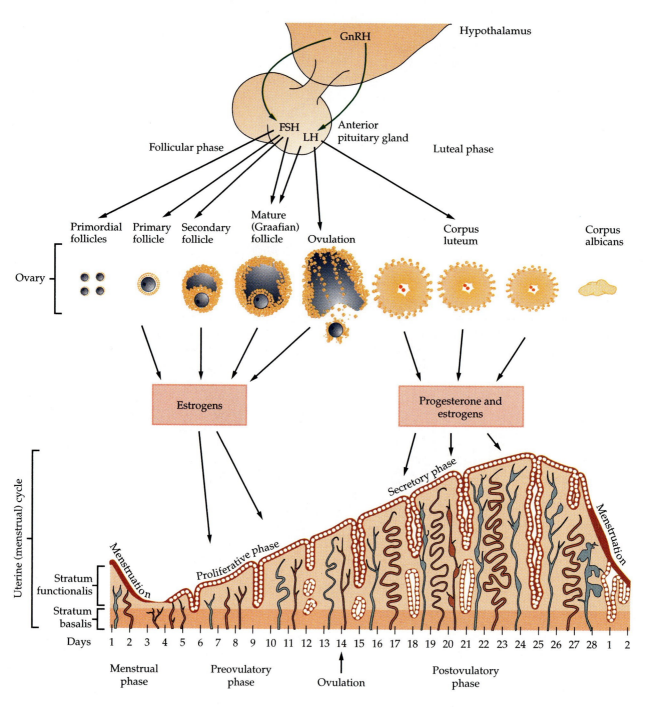

FIGURE 25.17   Menstrual cycle.

(1) Menstrual Phase (Menstruation) The *menstrual* (MEN-stroo-al) *phase,* also called *menstruation* (men'-stroo-Ā-shun) or **menses** (*mensis* = month), lasts for roughly the first five days of the cycle. (By convention, the first day of menstruation marks the first day of a new cycle.)

Events in the Ovaries During the menstrual phase, 20 or so small secondary (antral) follicles, some in each ovary, begin to enlarge. Follicular fluid, secreted by the granulosa cells and oozing from blood capillaries, accumulates in the enlarging antrum while the oocyte remains near the edge of the follicle (see Figure 25.10).

Events in the Uterus Menstrual flow from the uterus consists of 50–150 ml of blood, tissue fluid, mucus, and epithelial cells derived from the endometrium. This discharge occurs because the declining level of estrogens and progesterone causes the uterine spiral arteries to constrict. As a result, the cells they supply become ischemic (deficient in blood) and start to die. Eventually, the entire stratum functionalis sloughs off. At this time the endometrium is very thin because only the stratum basalis remains. The menstrual flow passes from the uterine cavity to the cervix and through the vagina to the exterior.

(2) Preovulatory Phase The *preovulatory phase,* the second phase of the female reproductive cycle, is the time between menstruation and ovulation. The preovulatory phase of the cycle is more variable in length than the other phases and accounts for most of the difference when cycles are shorter or longer than 28 days. It lasts from days 6 to 13 in a 28-day cycle.

Events in the Ovaries Under the influence of FSH, the group of about 20 secondary follicles continues to grow and begins to secrete estrogens and inhibin. By about day 6, one follicle in one ovary has outgrown all the others and is called the *dominant follicle.* Estrogens and inhibin secreted by the dominant follicle decrease the secretion of FSH, which causes the other less well-developed follicles to stop growing and undergo atresia.

The one dominant follicle becomes the *mature (Graafian) follicle* that continues to enlarge until it is more than 20 mm in diameter and ready for ovulation (see Figure 25.10). This follicle forms a blisterlike bulge on the surface of the ovary. Fraternal (nonidentical) twins may result if two secondary follicles achieve dominance and both ovulate. During the final maturation process, the dominant follicle continues to increase its production of estrogens under the influence of an increasing level of LH. Estrogens are the primary ovarian hormones before ovulation, but small amounts of progesterone are produced by the mature follicle a day or two before ovulation.

With reference to the ovaries, the menstrual phase and preovulatory phase together are termed the *follicular* (fō-LIK-yoo-lar) *phase* because ovarian follicles are growing and developing.

Events in the Uterus Estrogens being liberated into the blood by growing follicles stimulate the repair of the endometrium. Cells of the stratum basalis undergo mitosis and produce a new stratum functionalis. As the endometrium thickens, the short, straight endometrial glands develop and the arterioles coil and lengthen as they penetrate the stratum functionalis. The thickness of the endometrium approximately doubles to about 4–6 mm. With reference to the uterus, the preovulatory phase is also termed the *proliferative phase* because the endometrium is proliferating.

(3) Ovulation The rupture of the mature (Graafian) follicle with release of the secondary oocyte into the pelvic cavity, called *ovulation,* usually occurs on day 14 in a 28-day cycle. During ovulation, the secondary oocyte remains surrounded by its zona pellucida and corona radiata. It generally takes a total of about 20 days (spanning the last 6 days of the previous cycle and the first 14 days of the current cycle) for a secondary follicle to develop into a fully mature follicle. During this time the developing ovum completes reduction division (meiosis I) and reaches metaphase of equatorial division (meiosis II). The *high* levels of estrogens during the last part of the preovulatory phase exert a *positive feedback* effect on both LH and GnRH and cause ovulation.

An over-the-counter home test that detects the LH surge associated with ovulation is now available. The test predicts ovulation a day in advance. FSH also increases at this time, but not as dramatically as LH because FSH is stimulated only by the increase in GnRH. The positive feedback effect of estrogens on the hypothalamus and anterior pituitary gland does not occur if progesterone is present at the same time.

After ovulation, the mature follicle collapses and blood within it forms a clot due to minor bleeding during rupture and collapse of the follicle to become the *corpus hemorrhagicum* (*hemo* = blood; *rhegnynai* = to burst forth). (See Figure 25.10.) The clot is eventually absorbed by the remaining follicular cells. In time, the follicular cells enlarge, change character, and form the corpus luteum under the influence of LH. Stimulated by LH, the corpus luteum secretes progesterone, estrogens, relaxin, and inhibin.

(4) Postovulatory Phase The *postovulatory phase* of the female reproductive cycle is the most constant in duration and lasts for 14 days, from days 15 to 28 in a 28-day cycle. It represents the time between ovulation and the onset of the next menses. After ovulation, LH secretion stimulates the remnants of the mature follicle to develop into the corpus luteum. During its 2-week lifespan, the corpus luteum secretes increasing quantities of progesterone and some estrogens.

Events in One Ovary If the egg is fertilized and begins to divide, the corpus luteum persists past its normal 2-week lifespan. It is maintained by *human chorionic* (kō-rē-ON-ik) *gonadotropin (hCG),* a hormone produced by the chorion of the embryo as early as 8–12 days after fertilization. The chorion eventually develops into the placenta and the presence of hCG in maternal blood or urine is an indication of pregnancy. As the pregnancy progresses, the placenta itself begins to secrete estrogens to support pregnancy and progesterone to support pregnancy and breast development for lactation. Once the placenta begins its secretion, the role of the corpus luteum becomes minor. With reference to the ovaries, this phase of the cycle is also called the *luteal phase.*

   If hCG does not rescue the corpus luteum, after 2 weeks its secretions decline and it degenerates into a corpus albicans (see Figure 25.10). The lack of progesterone and estrogens due to degeneration of the corpus luteum then causes menstruation. In addition, the decreased levels of progesterone, estrogens, and inhibin promote the release of GnRH, FSH, and LH, which stimulate follicular growth, and a new ovarian cycle begins.

Events in the Uterus Progesterone produced by the corpus luteum is responsible for preparing the endometrium to receive a fertilized ovum. Preparatory activities include growth and coiling of the endometrial glands, which begin to secrete glycogen, vascularization of the superficial endometrium, thickening of the endometrium, and an increase in the amount of tissue fluid. These preparatory changes are maximal about 1 week after ovulation, corresponding to the time of possible arrival of a fertilized ovum. With reference to the uterus, this phase of the cycle is called the *secretory phase* because of the secretory activity of the endometrial glands.

   Obtain microscope slides of the endometrium showing the menstrual, preovulatory, and postovulatory phases of the menstrual cycle. See if you can note the differences in thickness of the endometrium, distribution of blood vessels, and distribution and size of endometrial glands.

C. DISSECTION OF CAT REPRODUCTIVE SYSTEMS

CAUTION! *Please reread Section D, "Precautions Related to Dissection" at the beginning of the laboratory manual on page xiii* **before** *you begin your dissection.*

1. Male Reproductive System

The male reproductive system of the cat is shown in Figure 25.18.

PROCEDURE
a. SCROTUM
1. The *scrotum,* a double sac ventral to the anus that contains the testes, consists of skin, muscle, and connective tissue.
2. Internally, the scrotum is divided into two chambers by a *median septum.* A testis is found in each chamber.

b. TESTES
1. Carefully cut open the scrotum to examine the *testes.*
2. The glistening covering of peritoneum around the testes is the *tunica vaginalis.*

c. EPIDIDYMIDES
1. Cut away the tunica vaginalis.
2. Each *epididymis* is a long, coiled tube that receives spermatozoa from a testis. It is located anterolateral to its testis.
3. Each epididymis consists of an enlarged cranial end *(head),* a narrow middle portion *(body),* and an expanded caudal portion *(tail).*

d. DUCTUS (VAS) DEFERENS
1. Each *ductus (vas) deferens* is a continuation from the tail of the epididymis. It carries sperm cells from the epididymis through the inguinal canal to the urethra.
2. The *inguinal canal* is a short passageway through the abdominal musculature.

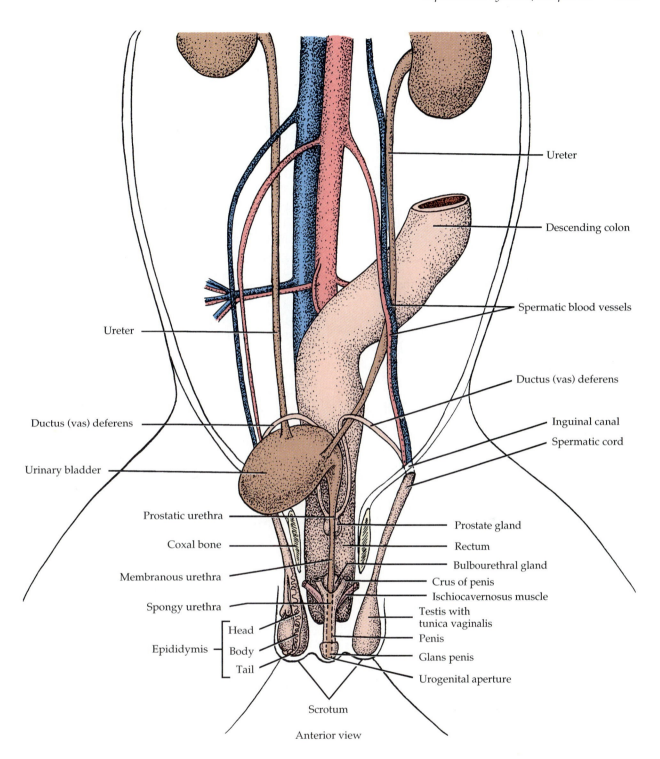

Ureter

Descending colon

Spermatic blood vessels

Ductus (vas) deferens

Inguinal canal

Spermatic cord

Ureter

Ductus (vas) deferens

Urinary bladder

Prostatic urethra

Coxal bone

Membranous urethra

Spongy urethra

Epididymis { Head Body Tail }

Prostate gland

Rectum

Bulbourethral gland

Crus of penis

Ischiocavernosus muscle

Testis with tunica vaginalis

Penis

Glans penis

Urogenital aperture

Scrotum

Anterior view

FIGURE 25.18   Male reproductive and urinary systems of cat.

3. Passing through the inguinal canal with the ductus deferens are the testicular artery, vein, lymphatic vessel, and nerve. Collectively, these structures constitute the *spermatic cord,* which is covered by connective tissue.

4. In the abdomen, the ductus deferens separates from the spermatic cord, passes ventral to the ureter, and then medially to the dorsal surface of the urinary bladder. From there it enters the *prostatic urethra,* the portion of the urethra that passes through the prostate gland.

5. The pelvic cavity must be exposed by cutting through the pubic symphysis at the midline and opening the pelvic girdle.

e. PROSTATE GLAND

1. At the junction of the ductus (vas) deferens and the neck of the urinary bladder, locate the *prostate gland.*
2. The ducts of the prostate gland empty into the prostatic urethra.

f. BULBOURETHRAL GLANDS

1. These two glands are located on either side of the membranous urethra, the portion of the urethra between the prostatic urethra and the spongy urethra (the part passing through the penis).
2. The bulbourethral (Cowper's) glands are just dorsal to the penis.

g. PENIS

1. This is the external organ of copulation ventral to the scrotum.
2. The free (distal) end of the penis is enlarged and is referred to as the *glans penis.* The loose skin covering the penis is the *foreskin,* or *prepuce.*
3. Cut open the prepuce to examine the glans penis. The opening of the penis to the exterior is the *urogenital aperture.*
4. Internally the penis consists of two dorsally located cylinders and one ventrally located cylinder of erectile tissue. The dorsal masses are side by side and are known as the *corpora cavernosa penis.* Their proximal ends are known as the *crura.* Each crus is attached to the ischium on its own side by the *ischiocavernosus muscle.* The ventral mass of erectile tissue is called the *corpus spongiosum penis.* Through it passes the *spongy (cavernosus) urethra.*

h. OS PENIS

1. This is the small bone embedded on one surface of the urethra in the glans penis.
2. The bone helps to stiffen this part of the penis.

i. ANAL GLANDS

1. These are a pair of round glands beneath the skin on either side of the rectum near the anus.
2. They produce an odoriferous secretion into the anus that is believed to be used for sexual attraction.

2. Female Reproductive System

The female reproductive system of the cat is shown in Figure 25.19.

PROCEDURE

a. OVARIES

1. The *ovaries* are a pair of small, oval organs located slightly caudal to the kidneys.
2. The lumpy appearance of the ovaries is due to the presence, within the ovaries, of *vesicular ovarian (Graafian) follicles.* These are small vesicles that contain ova and protrude to the ovarian surface.
3. Each ovary is suspended by the *mesovarium,* a peritoneal fold extending from the dorsal body wall to the ovary.

b. UTERINE (FALLOPIAN) TUBES

1. The paired *uterine (Fallopian) tubes* lie on the cranial surfaces of the ovaries.
2. The expanded end of the uterine tube, the *infundibulum,* has a fringed border of small fingerlike projections called *fimbriae* and also has an opening termed the *ostium,* which receives the ovum.
3. The peritoneum of the uterine tubes is called the *mesosalpinx.* It is continuous with the mesovarium.

c. UTERUS

1. The uterus is a Y-shaped structure consisting of two principal parts—uterine horns and body.
2. The *uterine horns* are the enlarged caudal communications of the uterine tubes. The embryos develop in the horns.
3. The *body* of the uterus is the caudal, median portion of the uterus. It is located between the urinary bladder and the rectum.

d. BROAD LIGAMENT

1. The uterine horns and body of the uterus are supported by a fold of peritoneum called the *mesometrium.* It extends from these structures to the lateral body walls.
2. Together, the mesometrium, mesosalpinx, and mesovarium are referred to as the *broad ligament.*

e. ROUND LIGAMENT

1. The *round ligament* is a thin, fibrous band that extends from the dorsal body wall to the middle of each uterine horn.
2. The remainder of the female reproductive system can be seen only if the pelvic cavity is exposed. You will have to cut through the pelvic muscles and pubic symphysis at the midline.
3. Spread the thighs back.

f. VAGINA

1. Before examining the vagina, locate the *urethra,* a tube caudal to the urinary bladder.
2. Dorsal to the urethra is the *vagina,* a tube leading from the body of the uterus to the *urogenital sinus (vestibule).* This is a common pas-

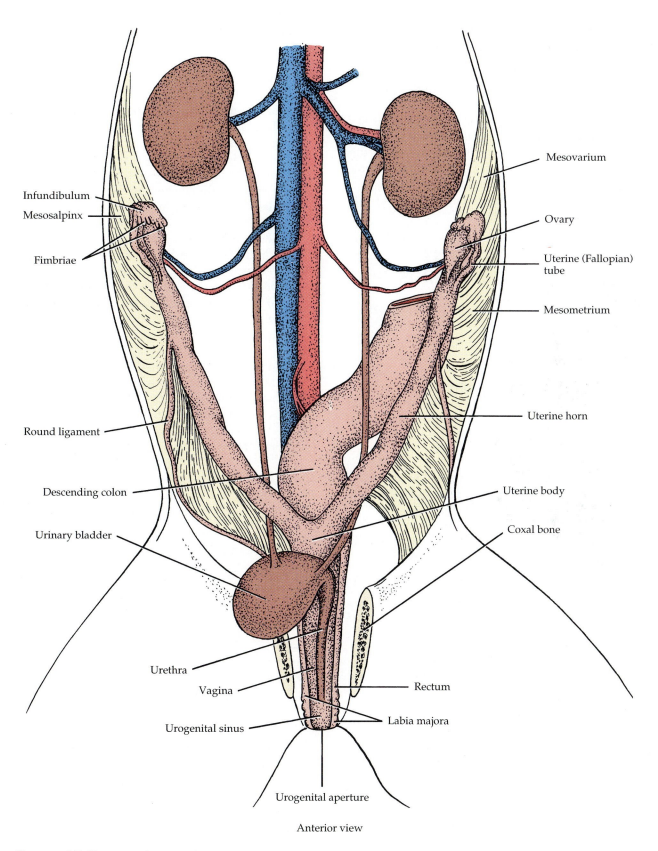

Mesovarium

Ovary

Uterine (Fallopian) tube

Mesometrium

Uterine horn

Uterine body

Coxal bone

Infundibulum

Mesosalpinx

Fimbriae

Round ligament

Descending colon

Urinary bladder

Urethra

Vagina

Urogenital sinus

Rectum

Labia majora

Urogenital aperture

Anterior view

FIGURE 25.19   Female reproductive and urinary systems of cat.

sageway formed by the union of the urethra and vagina.

g. VULVA

1. The urogenital sinus opens to the exterior as the *urogenital aperture,* just ventral to the anus.
2. On either side of the urogenital aperture are folds of skin called the *labia majora.*
3. In the ventral wall of the urogenital sinus is the *clitoris,* a mass of erectile tissue that is homologous to portions of the penis.
4. The urogenital aperture, labia majora, and clitoris constitute the *vulva.*

D. DISSECTION OF FETUS-CONTAINING PIG UTERUS

Examination of the uterus of a pregnant pig reveals that the fetuses are equally spaced in the two uterine horns. Each fetus produces a local enlargement of the horn. The litter size normally ranges from 6 to 12. Your instructor may have you dissect the fetus-containing uterus of a pregnant pig or have one available as a demonstration (Figure 25.20). If you do a dissection, use the following directions. Also examine a chart or model of a human fetus and pregnant uterus if they are available.

CAUTION! *Please reread Section D, "Precautions Related to Dissection" at the beginning of the* laboratory manual on page xiii before you begin your dissection.

PROCEDURE

1. Using a sharp scissors, cut open one of the enlargements of the horn and you will see that each fetus is enclosed together with an elongated, sausage-shaped *chorionic vesicle.*
2. You will also notice many round bumps called *areolae* located over the chorionic surface.
3. The lining of the uterus together with the wall of the chorionic vesicle forms the *placenta.*
4. Carefully cut open the chorionic vesicle, avoiding cutting or breaking the second sac lying within that surrounds the fetus itself.
5. The vesicle wall is the fusion of two extra-embryonic membranes, the outer *chorion* (KOR-ē-on) and the inner *allantois* (a-LAN-tō-is). The allantois is the large sac growing out from the fetus, and the umbilical cord contains its stalk.
6. The *umbilical blood vessels* are seen in the allantoic wall spreading out in all directions and are also seen entering the *umbilical cord.*
7. A thin-walled nonvascular *amnion* surrounds the fetus. This membrane is filled with *amniotic fluid,* which acts as a protective water cushion and prevents adherence of the fetus and membranes.

ANSWER THE LABORATORY REPORT QUESTIONS AT THE END OF THE EXERCISE.

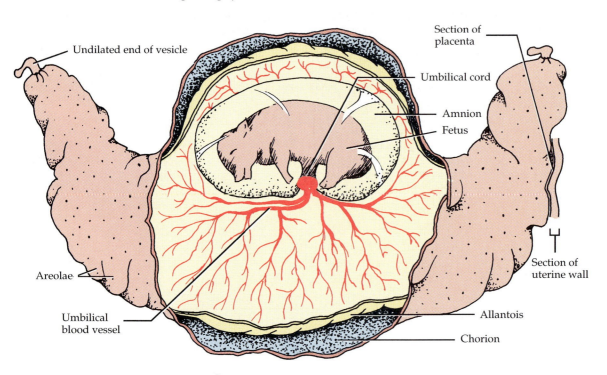

Section of placenta
Umbilical cord
Amnion
Fetus
Section of uterine wall
Allantois
Chorion
Undilated end of vesicle
Areolae
Umbilical blood vessel

FIGURE 25.20   Fetal pig in opened chorionic vesicle.

Reproductive Systems 25

Student _____   Date _____

Laboratory Section _____   Score/Grade _____

PART 1. Multiple Choice

_____ 1. Structures of the male reproductive system responsible for the production of sperm cells are the (a) efferent ducts (b) seminiferous tubules (c) seminal vesicles (d) rete testis

_____ 2. The superior portion of the male urethra is encircled by the (a) epididymis (b) testes (c) prostate gland (d) seminal vesicles

_____ 3. Cryptorchidism is a condition associated with the (a) prostate gland (b) testes (c) seminal vesicles (d) bulbourethral (Cowper's) glands

_____ 4. Weakening of the suspensory ligament would directly affect the position of the (a) mammary glands (b) uterus (c) uterine (Fallopian) tubes (d) ovaries

_____ 5. Organs in the female reproductive system responsible for transporting secondary oocytes and ova from the ovaries to the uterus are the (a) uterine (Fallopian) tubes (b) seminal vesicles (c) inguinal canals (d) none of the above

_____ 6. The name of the process that is responsible for the actual production of sperm cells is called (a) cryptorchidism (b) oogenesis (c) spermatogenesis (d) spermatogonia

_____ 7. Fertilization normally occurs in the (a) uterine (Fallopian) tubes (b) vagina (c) uterus (d) ovaries

_____ 8. The portion of the uterus that assumes an active role during labor is the (a) serosa (b) endometrium (c) peritoneum (d) myometrium

_____ 9. Stereocilia are associated with the (a) ductus (vas) deferens (b) oviduct (c) epididymis (d) rete testis

_____ 10. The major portion of the volume of semen is contributed by the (a) bulbourethral (Cowper's) glands (b) testes (c) prostate gland (d) seminal vesicles

_____ 11. The chief ligament supporting the uterus and keeping it from dropping into the vagina is the (a) cardinal ligament (b) round ligament (c) broad ligament (d) ovarian ligament

_____ 12. Which sequence, from inside to outside, best represents the histology of the uterus? (a) stratum basalis, stratum functionalis, myometrium, perimetrium (b) myometrium, perimetrium, stratum functionalis, stratum basalis (c) stratum functionalis, stratum basalis, myometrium, perimetrium (d) stratum basalis, stratum functionalis, perimetrium, myometrium

_____ 13. The white fibrous capsule that divides the testis into lobules is called the (a) dartos (b) raphe (c) tunica albuginea (d) germinal epithelium

_____ **14.** Which sequence best represents the course taken by sperm cells from their site of origin to the exterior? (a) seminiferous tubules, efferent ducts, epididymis, ductus (vas) deferens, ejaculatory duct, urethra (b) seminiferous tubules, efferent ducts, epididymis, ductus (vas) deferens, urethra, ejaculatory duct (c) seminiferous tubules, efferent ducts, ductus (vas) deferens, epididymis, ejaculatory duct, urethra (d) seminiferous tubules, epididymis, efferent ducts, ductus (vas) deferens, ejaculatory duct, urethra

_____ **15.** The ovaries are anchored to the uterus by the (a) ovarian ligament (b) broad ligament (c) suspensory ligament (d) mesovarium

_____ **16.** The terminal duct for the male reproductive system is the (a) urethra (b) ductus (vas) deferens (c) inguinal canal (d) ejaculatory duct

_____ **17.** The site of sperm cell maturation is the (a) ductus (vas) deferens (b) spermatic cord (c) epididymis (d) testes

_____ **18.** Which of the following is the site of menstruation, implantation of a fertilized ovum, development of the fetus during pregnancy, and labor? (a) uterus (b) uterine (Fallopian) tubes (c) vagina (d) cervix

_____ **19.** Glands lying over the pectoralis major muscles are (a) lesser vestibular glands (b) adrenal (suprarenal) glands (c) mammary glands (d) greater vestibular (Bartholin's) glands

PART 2. Completion

20. Discharge of a secondary oocyte from the ovary about once each month is a process referred to as

_____.

21. The inferior, narrow portion of the uterus that opens into the vagina is the _____.

22. The clusters of milk-secreting cells of the mammary glands are referred to as

_____.

23. The distal end of the penis is a slightly enlarged region called the _____.

24. Covering the slightly enlarged region of the penis is a loosely fitting skin called the

_____.

25. The circular pigmented area surrounding each nipple of the mammary glands is the

_____.

26. After a secondary oocyte leaves the ovary, it enters the open, funnel-shaped distal end of the uterine

(Fallopian) tube called the _____.

27. The portion of a sperm cell that contains the nucleus and acrosome is the

_____.

28. Vasectomy refers to removal of a portion of the _____.

29. The mass of erectile tissue in the penis that contains the spongy urethra is the

_____.

30. Both the mature (Graafian) follicle and _____ of the ovary secrete hormones.

31. The superior dome-shaped portion of the uterus is called the _____.

32. The _____ anchor the uterus to either side of the pelvic cavity.

33. The passageway for menstrual flow and inferior portion of the birth canal is the

_____.

34. Two longitudinal folds of skin that extend inferiorly and posteriorly from the mons pubis and are covered with pubic hair are the _____.

35. The _____ is a small mass of erectile tissue at the anterior junction of the labia minora.

36. The thin fold of vascularized membrane that borders the vaginal orifice is the

_____.

37. Complete the following sequence for the passage of milk: alveoli, secondary tubules, lactiferous

sinuses, _____, lactiferous ducts, nipple.

38. The layer of simple epithelium covering the free surface of the ovary is the _____.

39. The phase of the menstrual cycle between days 6 and 13 during which endometrial repair occurs is

the _____ phase.

40. During menstruation, the stratum _____ of the endometrium is sloughed off.

41. The most immature spermatogenic cells are called _____.

42. The _____ contains the remnants of an ovulated mature follicle.

43. The hypothalamic hormone that controls the uterine and ovarian cycles is _____.

44. High levels of estrogens exert a positive feedback on LH and GnRH that cause

_____.

Development

26

Development refers to the sequence of events starting with fertilization of a secondary oocyte and ending with the formation of a complete organism. Consideration will be given to how reproductive cells are produced and to a few developmental events associated with pregnancy.

A. SPERMATOGENESIS

The process by which the testes produce haploid (*n*) sperm cells involves several phases, including meiosis, and is called *spermatogenesis* (sper'-ma-tō-JEN-e-sis; *spermato* = sperm; *genesis* = to produce). In order to understand spermatogenesis, review the following concepts.

1. In sexual reproduction, a new organism is produced by the union and fusion of sex cells called *gametes* (*gameto* = to marry). Male gametes, produced in the testes, are called sperm cells, and female gametes, produced in the ovaries, are called oocytes.
2. The cell resulting from the union and fusion of gametes, called a *zygote* (*zygosis* = a joining), contains two full sets of chromosomes (DNA), one set from each parent. Through repeated mitotic cell divisions, a zygote develops into a new organism.
3. Gametes differ from all other body cells (somatic cells) in that they contain the *haploid* (one-half) *chromosome number,* symbolized as *n*. In humans, this number is 23, which composes a single set of chromosomes. The nucleus of a somatic cell contains the *diploid chromosome number,* symbolized as 2*n*. In humans, this number is 46, which composes two sets of paired chromosomes. One set of 23 chromosomes comes from the mother and the other set comes from the father.
4. In a diploid cell, two chromosomes that belong to a pair are called *homologous* (*homo* = same) *chromosomes (homologues).* In human diploid

cells, the members of 22 of the 23 pairs of chromosomes are morphologically similar and are called *autosomes.* The other pair, termed X and Y chromosomes, are called the *sex chromosomes* because they determine one's gender. In the female, the homologous pair of sex chromosomes are two similar X chromosomes; in the male, the pair consists of an X and a Y chromosome.
5. If gametes were diploid (2*n*), like somatic cells, the zygote would contain twice the diploid number (4*n*), and with every succeeding generation the chromosome number would continue to double and normal development could not occur.
6. This continual doubling of the chromosome number does not occur because of *meiosis* (*meio* = less), a process by which gametes produced in the testes and ovaries receive the haploid chromosome number. Thus, when haploid (*n*) gametes fuse, the zygote contains the diploid chromosome number (2*n*) and can undergo normal development.

In humans, spermatogenesis takes about 74 days. The seminiferous tubules are lined with immature cells called *spermatogonia* (sper'-ma-tō-GŌ-nē-a; *sperm* = seed; *gonium* = generation or offspring), or sperm mother cells (see Figure 26.2 on page 623). Singular is *spermatogonium.* These cells develop from *primordial* (*primordialis* = primitive or early form) *germ cells* that arise from yolk sac endoderm and enter the testes early in development. In the embryonic testes, the primordial germ cells differentiate into spermatogonia but remain dormant until they begin to undergo mitotic proliferation at puberty. Spermatogonia contain the diploid (2*n*) chromosome number. Some spermatogonia remain relatively undifferentiated and capable of extensive mitotic division. Following division, some of the daughter cells remain undifferentiated and serve as a reservoir of precursor cells to prevent depletion of the stem cell population. Such cells remain near the basement membrane. The remainder of the daughter cells differentiate into spermatogonia that lose contact with the basement membrane of the seminiferous tubule, undergo certain developmental changes, and become known as *primary spermatocytes* (SPER-ma-tō-sītz'). Primary spermatocytes, like spermatogonia, are diploid (2*n*); that is, they have 46 chromosomes.

1. Reduction Division (Meiosis I)

Each primary spermatocyte enlarges before dividing. Then two nuclear divisions take place as part of meiosis. In the first, DNA is replicated and 46 chromosomes (each made up of two identical chromatids from the replicated DNA) form and move toward the equatorial plane of the cell. There they line up in homologous pairs so that there are 23 pairs of duplicated chromosomes in the center of the cell. This pairing of homologous chromosomes is called *synapsis*. The four chromatids of each homologous pair then become associated with each other to form a *tetrad*. In a tetrad, portions of one chromatid may be exchanged with portions of another. This process, called *crossing over,* permits an exchange of genes among maternal and paternal chromosomes (Figure 26.1) that results in the *recombination* of genes. Thus, the sperm cells eventually produced are genetically unlike each other and unlike the cell that produced them—one reason for the great variation among humans. Next, the meiotic spindle forms and the kinetochore microtubules organized by the centromeres extend toward the poles of the cell. As the pairs separate, one member of each pair migrates to opposite poles of the dividing cell. The random arrangement of chromosome pairs on the spindle is another reason for variation among humans. The cells formed by the first nuclear division (reduction division) are called *secondary spermatocytes.* Each cell has 23 chromosomes—the haploid number. Each chromosome of the secondary spermatocytes, however, is made up of two chromatids (two copies of the DNA) still attached by a centromere. Moreover, the genes of the chromosomes of secondary spermatocytes may be rearranged as a result of crossing-over.

2. Equatorial Division (Meiosis II)

The second nuclear division of meiosis is *equatorial division.* There is no replication of DNA. The chromosomes (each composed of two chromatids) line up in single file along the equatorial plane, and the chromatids of each chromosome separate from each other. The cells formed from the equatorial division are called *spermatids.* Each contains half the original chromosome number, or 23 chromosomes, and is haploid. Each primary spermatocyte therefore produces four spermatids by meiosis (reduction division and equatorial division). Spermatids lie close to the lumen of the seminiferous tubule.

3. Spermiogenesis

The final stage of spermatogenesis, called *spermiogenesis* (sper'-mē-ō-JEN-e-sis), involves the maturation of spermatids into sperm cells. Each spermatid embeds in a sustentacular (Sertoli) cell and develops a head with an acrosome (enzyme-containing granule) and a flagellum (tail). Sustentacular cells extend from the basement membrane to the lumen of the seminiferous tubule, where they nourish the developing spermatids. Since there is no cell division in spermiogenesis, each spermatid develops into a single *sperm cell (spermatozoon).* The release of a sperm cell from a sustentacular cell is known as *spermiation.*

Sperm cells enter the lumen of the seminiferous tubule and migrate to the ductus epididymis,

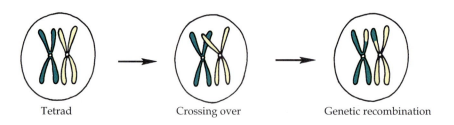

Tetrad   Crossing over   Genetic recombination

FIGURE 26.1   Crossing over within a tetrad resulting in genetic recombination.

where in 10 to 14 days they complete their maturation and become capable of fertilizing a secondary oocyte. Sperm cells are also stored in the ductus (vas) deferens. Here, they can retain their fertility for up to several months.

With the aid of your textbook, label Figure 26.2.

B. OOGENESIS

The formation of haploid (*n*) secondary oocytes in the ovary involves several phases, including meiosis, and is referred to as ***oogenesis*** (ō'-ō-JEN-e-sis; *oo* = egg; *genesis* = to produce). With some important

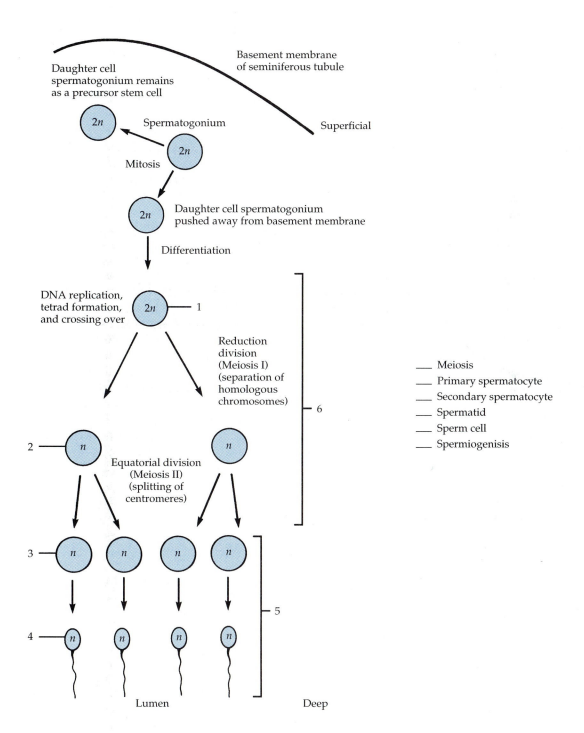

___ Meiosis
___ Primary spermatocyte
___ Secondary spermatocyte
___ Spermatid
___ Sperm cell
___ Spermiogenisis

FIGURE 26.2 Spermatogenesis.

exceptions, oogenesis occurs in essentially the same manner as spermatogenesis.

1. Reduction Division (Meiosis I)

During early fetal development, primordial (primitive) germ cells migrate from the endoderm of the yolk sac to the ovaries. There, germ cells differentiate within the ovaries into *oogonia* ($\bar{o}'$-$\bar{o}$-GŌ-nē-a; singular is *oogonium* ($\bar{o}'$-$\bar{o}$-GŌ-nē-um). Oogonia are diploid ($2n$) cells that divide mitotically to produce millions of germ cells. Even before birth, many of these germ cells degenerate, a process known as *atresia.* A few develop into larger cells called *primary (primus = first) oocytes* (Ō'-$\bar{o}$-sītz) that enter prophase of reduction division (meiosis I) during fetal development but do not complete it until puberty. At birth 200,000–2,000,000 oogonia and primary oocytes remain in each ovary. Of these, about 400 will mature and ovulate during a woman's reproductive lifetime; the remaining 99.98% undergo atresia.

Each primary oocyte is surrounded by a single layer of follicular cells, and the entire structure is called *primordial follicle* (see Figure 25.10). Although the stimulating mechanism is unclear, a few primordial follicles start to grow, even during childhood. They become *primary (preantral) follicles,* which are surrounded first by one layer of cuboidal-shaped follicular cells and then by six to seven layers of cuboidal and low-columnar cells called *granulosa cells.* As a follicle grows, it forms a clear glycoprotein layer, called the *zona pellucida* (pe-LOO-si-da) between the oocyte and the granulosa cells. The innermost layer of granulosa cells becomes firmly attached to the zona pellucida and is called the *corona radiata* (*corona* = crown; *radiata* = radiation). The outermost granulosa cells rest on a basement membrane that separates them from the surrounding ovarian stroma. This outer region is called the *theca folliculi.* As the primary follicle continues to grow, the theca differentiates into two layers: (1) the *theca interna,* a vascularized internal layer of secretory cells, and (2) the *theca externa,* an outer layer of connective tissue cells. The granulosa cells begin to secrete follicular fluid, which builds up in a cavity called the *antrum* in the center of the follicle. The follicle is now termed a *secondary (antral) follicle.* During early childhood, primordial and developing follicles continue to undergo atresia.

After puberty, under the influence of the gonadotropin hormones secreted by the anterior pituitary gland, each month meiosis resumes in one secondary follicle. The diploid primary oocyte completes reduction division (meiosis I) and two haploid cells of unequal size, both with 23 chromosomes (n) of two chromatids each, are produced. The follicle in which these events are taking place, termed the *mature (Graafian) follicle* (also called a *vesicular ovarian follicle*) will soon rupture and release its oocyte.

The smaller cell produced by meiosis I, called the *first polar body,* is essentially a packet of discarded nuclear material. The larger cell, known as the *secondary oocyte,* receives most of the cytoplasm. Once a secondary oocyte is formed, it proceeds to the metaphase of equatorial division (meiosis II) and then stops at this stage.

2. Equatorial Division (Meiosis II)

At ovulation, usually one secondary oocyte (with the first polar body and corona radiata) is expelled into the pelvic cavity. Normally, the cells are swept into the uterine (Fallopian) tube. If fertilization does not occur, the oocyte and other cells degenerate. If sperm cells are present in the uterine tube and one penetrates the secondary oocyte (fertilization), however, equatorial division (meiosis II) resumes. The secondary oocyte splits into two haploid (n) cells of unequal size. The larger cell is the *ovum,* or mature egg; the smaller one is the *second polar body.* The nuclei of the sperm cell and the ovum then unite, forming a diploid ($2n$) *zygote.* The first polar body may also undergo another division to produce two polar bodies. If it does, the primary oocyte ultimately gives rise to a single haploid (n) ovum and three haploid (n) polar bodies, which all degenerate. Thus an oogonium gives rise to a single gamete (ovum), whereas a spermatogonium produces four gametes (sperm).

With the aid of your textbook, label Figure 26.3.

C. EMBRYONIC PERIOD

The *embryonic period* is the first two months of development, and the developing human is called an *embryo.*

1. Fertilization

During *fertilization* (fer-til-i-ZĀ-shun; *fertilis* = reproductive) the genetic material from the sperm cell and secondary oocyte merges into a single nucleus (Figure 26.4a). Of the 300–500 million sperm introduced into the vagina, less than 1% reach the

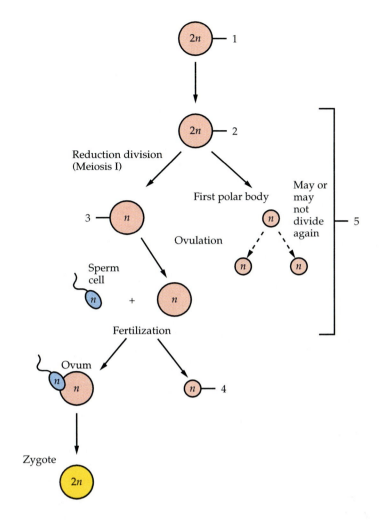

FIGURE 26.3 Oogenesis.

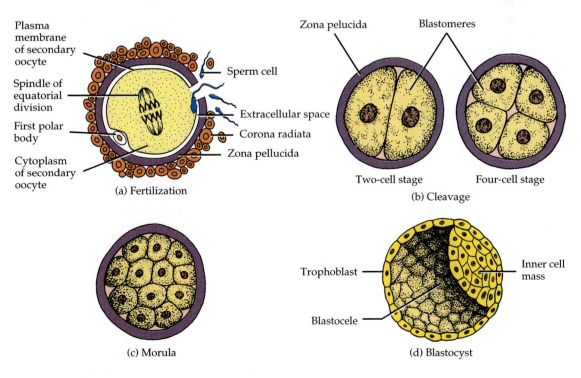

FIGURE 26.4 Fertilization.

secondary oocyte. Fertilization normally occurs in the uterine (Fallopian) tube about 12–24 hours after ovulation. Since ejaculated sperm cells remain viable for about 48 hours and an oocyte is viable for about 24 hours after ovulation, there typically is a 3-day window during which pregnancy can occur—from 2 days before to 1 day after ovulation. Peristaltic contractions and action of cilia transport the oocyte through the uterine tube. Sperm cells swim up the female tract by whiplike movements of their tail (flagellum). The acrosome of sperm cells produces an enzyme called *acrosin* that stimulates sperm cell motility and migration within the female reproductive tract. Also, muscular contractions of the uterus, stimulated by prostaglandins in semen, probably aid sperm cell movement toward the uterine tube. Finally, the oocyte is thought to secrete a chemical substance that attracts sperm cells.

Besides contributing to sperm cell movement, the female reproductive tract also confers on sperm cells the capacity to fertilize a secondary oocyte. Although sperm cells undergo maturation in the epididymus, they are still not able to fertilize an oocyte until they have been in the female reproductive tract for several hours.

Capacitation (ka-pas'-i-TĀ-shun) refers to the functional changes that sperm cells undergo in the female reproductive tract that allow them to fertilize a secondary oocyte. During this process, the membrane around the acrosome becomes fragile so that several destructive enzymes—hyaluronidase, acrosin, and neuraminidase—are secreted by the acrosomes. It requires the collective action of many sperm cells to have just one penetrate the secondary oocyte. The enzymes help penetrate the corona radiata and zona pellucida around the oocyte. Sperm cells bind to receptors in the zona pellucida. Normally only one sperm cell penetrates and enters a secondary oocyte. Sperm cells bind to receptors in the zona pellucida. Normally only one sperm cell penetrates and enters a secondary oocyte. This event is called *syngamy* (*syn* = together; *gamos* = marriage). Syngamy causes depolarization, which triggers the release of calcium ions inside the cell. Calcium ions stimulate the release of granules by the oocyte that, in turn, promote changes in the zona pellucida to block entry of other sperm cells . This prevents *polyspermy,* fertilization by more than one sperm cell. Once a sperm cell has entered a secondary oocyte, the oocyte completes equatorial division (meiosis II). It divides into a larger ovum (mature egg) and a smaller second polar body that fragments and disintegrates (see Figure 26.3).

When a sperm cell has entered a secondary oocyte, the tail is shed and the nucleus in the head develops into a structure called the *male pronucleus.* The nucleus of the secondary oocyte develops into a *female pronucleus.* After the pronuclei are formed, they fuse to produce a *segmentation nucleus.* The segmentation nucleus is diploid since it contains 23 chromosomes (*n*) from the male pronucleus and 23 chromosomes (*n*) from the female pronucleus. Thus the fusion of the haploid (*n*) pronuclei restores the diploid number (2*n*). The fertilized ovum, consisting of a segmentation nucleus, cytoplasm, and zona pellucida, is called a *zygote* (ZĪ-gōt; *zygosis* = a joining).

2. Formation of the Morula

After fertilization, rapid mitotic cell divisions of the zygote take place. These early divisions of the zygote are called *cleavage* (Figure 26.4b). Although cleavage increases the number of cells, it does not increase the size of the embryo, which is still contained within the zona pellucida.

The first cleavage begins about 24 hours after fertilization and is completed about 30 hours after fertilization, and each succeeding division takes slightly less time. By the second day after fertilization, the second cleavage is completed. By the end of the third day, there are 16 cells. The progressively smaller cells produced by cleavage are called *blastomeres* (BLAS-tō-mērz; *blast* = germ, sprout; *meros* = part). Successive cleavages produce a solid sphere of cells, still surrounded by the zona pellucida, called the *morula* (MOR-yoo-la; *morula* = mulberry). (Figure 26.4c). A few days after fertilization, the morula is about the same size as the original zygote.

3. Development of the Blastocyst

By the end of the fourth day, the number of cells in the morula increases and it continues to move through the uterine (Fallopian) tube toward the uterine cavity. At 4½–5 days, the dense cluster of cells has developed into a hollow ball of cells and enters the uterine cavity; it is now called a *blastocyst* (*kystis* = bag) (Figure 26.4d).

The blastocyst has an outer covering of cells called the *trophoblast* (TRŌ-fō-blast; *troph* = nourish), an *inner cell mass (embryoblast),* and an internal fluid-filled cavity called the *blastocele* (BLAS-tō-sēl; *koilos* = hollow). The trophoblast ultimately forms part of the membranes composing the fetal portion of the placenta; part of the inner cell mass develops into the embryo.

PROCEDURE

1. Obtain prepared slides of the embryonic development of the sea urchin. First try to find a zygote. This will appear as a single cell surrounded by an inner fertilization membrane and an outer, jellylike membrane. Draw a zygote in the spaces provided.
2. Now find several cleavage stages. See if you can isolate two-cell, four-cell, eight-cell, and sixteen-cell stages. Draw the various stages in the spaces provided.
3. Try to find a blastula (called a blastocyst in humans), a hollow ball of cells with a lighter center due to the presence of the blastocele. Draw a blastula in the space provided.

Zygote

Two-cell stage

Four-cell stage

Eight-cell stage

Sixteen-cell stage

Blastula

4. Implantation

The blastocyst remains free within the cavity of the uterus for a short period of time before it attaches to the uterine wall. During this time, the zona pellucida disintegrates and the blastocyte enlarges. The blastocyst receives nourishment from glycogen-rich secretions of endometrial (uterine) glands, sometimes called uterine milk. About 6 days after fertilization the blastocyst attaches to the endometrium, a process called *implantation* (Figure 26.5). At this time, the endometrium is in its secretory phase.

As the blastocyst implants, usually on the posterior wall of the fundus or body of the uterus, it is oriented so that the inner cell mass is toward the endometrium. The trophoblast develops two layers in the region of contact between the blastocyst and endometrium. These layers are a **syncytiotrophoblast** (sin-sīt'-ē-ō-TROF-ō-blast; *syn* = joined; *cyto* = cell) that contains no cell boundaries and a **cytotrophoblast** (sī-tō-TROF-ō-blast) between the inner cell mass and syncytiotrophoblast that is composed of distinct cells (Figure 26.5c). These two layers of trophoblast become part of the chorion (one of the fetal membranes) as they undergo further growth. During implantation, the syncytiotrophoblast secretes enzymes that enable the blastocyst to penetrate the uterine lining. The enzymes digest and liquefy the endometrial cells. The fluid and nutrients further nourish the burrowing blastocyst for about a week after implantation. Eventually, the blastocyst becomes buried in the endometrium. The trophoblast also secretes human chorionic gonadotropin (hCG) that rescues the corpus luteum from degeneration and sustains its secretion of progesterone and estrogens.

5. Primary Germ Layers

After implantation, the first major event of the embryonic period occurs. The inner cell mass of the blastocyst begins to differentiate into the three *primary germ layers:* ectoderm, endoderm, and mesoderm. These are the major embryonic tissues from which all tissues and organs of the body will develop. The process by which the two-layered inner cell mass is converted into a structure composed of the primary germ layers is called *gastrulation* (gas'-troo-LĀ-shun; *gastrula* = little belly).

Within eight days after fertilization, the cells of the inner cytotrophoblast proliferate and form the amnion (a fetal membrane) and a space, the

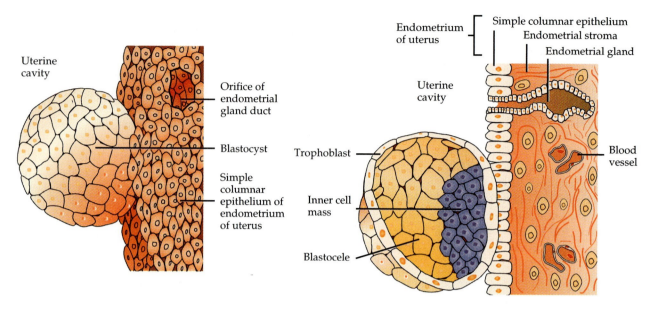

Uterine cavity

Orifice of endometrial gland duct

Blastocyst

Simple columnar epithelium of endometrium of uterus

(a) External view, about 5 days after fertilization

Endometrium of uterus

Simple columnar epithelium
Endometrial stroma
Endometrial gland

Uterine cavity

Trophoblast

Inner cell mass

Blastocele

Blood vessel

(b) Internal view, about 6 days after fertilization

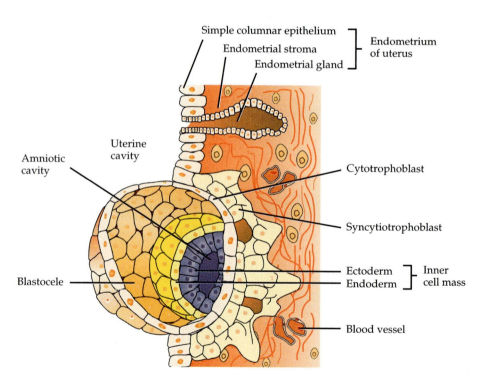

Simple columnar epithelium
Endometrial stroma
Endometrial gland

Endometrium of uterus

Amniotic cavity

Uterine cavity

Cytotrophoblast

Syncytiotrophoblast

Blastocele

Ectoderm
Endoderm

Inner cell mass

Blood vessel

(c) Internal view, about 7 days after fertilization

FIGURE 26.5 Implantation.

amniotic (am-nē-OT-ik; *amnion* = lamb) or *amniotic cavity,* over the inner cell mass. The layer of cells of the inner cell mass that is closer to the amniotic cavity develops into the *ectoderm* (*ecto* = outside; *derm* = skin). The layer of the inner cell mass that borders the blastocele develops into the *endoderm* (*endo* = inside). As the amniotic cavity

forms, the inner cell mass at this stage is called the *embryonic disc.* It will form the embryo. At this stage, the embryonic disc contains ectodermal and endodermal cells; the mesodermal cells are scattered external to the disc.

About the 12th day after fertilization, striking changes appear (Figure 26.6a). The cells of the

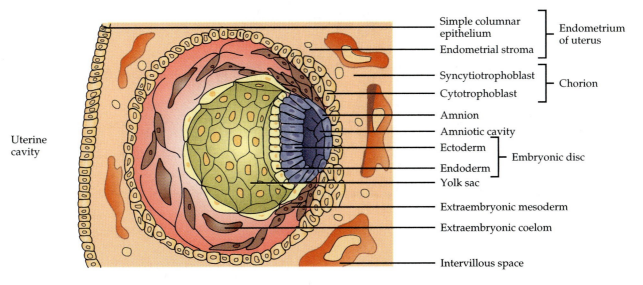

(a) Internal view, about 12 days after fertilization

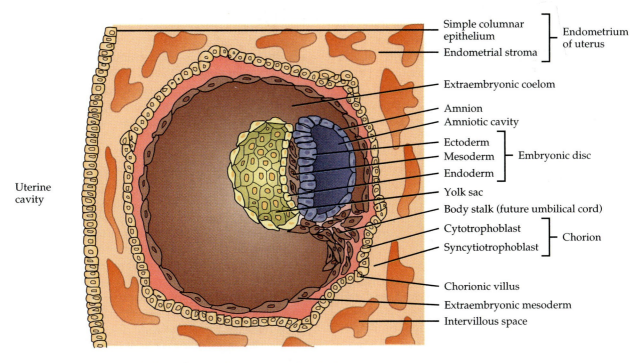

(b) Internal view, about 14 days after fertilization

FIGURE 26.6 Formation of the primary germ layers and associated structures.

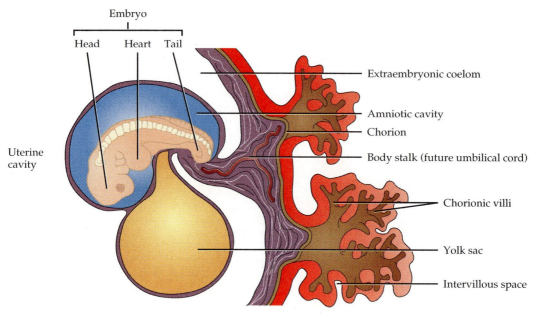

(c) External view, about 25 days after fertilization

FIGURE 26.6 *(Continued)*   Formation of the primary germ layers and associated structures.

endodermal layer have been dividing rapidly, so that groups of them now extend around in a circle, forming the yolk sac, another fetal membrane (described shortly). The cells of the *mesoderm* (*meso* = middle), which develop between the ectodermal and endodermal layers, also have been dividing, and many have left the area of the embryonic disc and can be seen around the structures that are becoming fetal membranes.

About the 14th day, the cells of the embryonic disc differentiate into three distinct layers: the ectoderm, the mesoderm, and the endoderm (Figure 26.6b). The mesoderm in the disc soon splits into two layers, and the space between the layers becomes the *extraembryonic coelom* (SĒ-lōm; *koiloma* = cavity), the future ventral body cavity.

As the embryo develops (Figure 26.6c), the endoderm becomes the epithelial lining of most of the gastrointestinal tract, urinary bladder, gallbladder, liver, pharynx, larynx, trachea, bronchi, lungs, vagina, urethra, and thyroid, parathyroid, and thymus glands, among other structures. The mesoderm develops into muscle; cartilage, bone and other connective tissues; red bone marrow, lymphoid tissue, endothelium of blood and lymphatic vessels, gonads, dermis of the skin, and other structures. The ectoderm develops into the entire nervous system, epidermis of skin, epidermal derivatives of the skin, and portions of the eye and other sense organs.

6. Embryonic Membranes

A second major event that occurs during the embryonic period is the formation of the *embryonic (extraembryonic) membranes* (Figure 26.7). These membranes lie outside the embryo and protect and nourish the embryo and, later, the fetus. The membranes are the yolk sac, amnion, chorion, and allantois.

In many species, the **yolk sac** is a membrane that is the primary source of nourishment for the embryo. A human embryo receives nutrients from the endometrium, however; the yolk sac remains small and functions as an early site of blood formation. The yolk sac also contains cells that migrate into the gonads and differentiate into the primitive germ cells (spermatogonia and oogonia).

The **amnion** is a thin, protective membrane that forms by the eighth day after fertilization and initially overlies the embryonic disc. As the embryo grows, the amnion comes to entirely surround the embryo, creating a cavity that becomes filled with **amniotic fluid.** Most amniotic fluid is initially derived from a filtrate of maternal blood. Later, the fetus makes daily contributions to the fluid by

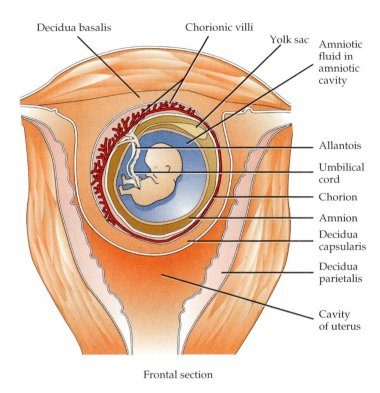

Decidua basalis   Chorionic villi   Yolk sac   Amniotic fluid in amniotic cavity

Allantois

Umbilical cord

Chorion

Amnion

Decidua capsularis

Decidua parietalis

Cavity of uterus

Frontal section

FIGURE 26.7   Embryonic membranes.

excreting urine into the amniotic cavity. Amniotic fluid serves as a shock absorber for the fetus, helps regulate fetal body temperature, and prevents adhesions between the skin of the fetus and surrounding tissues. Embryonic cells are sloughed off into amniotic fluid; they can be examined in the procedure called **amniocentesis** (am'-nē-ō-sen-TĒ-sis). The amnion usually ruptures just before birth and with its fluid constitutes the "bag of waters."

The **chorion** (KŌ-rē-on) is derived from the trophoblast of the blastocyst and the mesoderm that lines the trophoblast. It surrounds the embryo and, later, the fetus. Eventually, the chorion becomes the principal embryonic part of the placenta, the structure for exchange of materials between the mother and fetus. It also produces human chorionic gonadotropin (hCG). The amnion, which also surrounds the fetus, eventually fuses to the inner layer of the chorion.

The **allantois** (a-LAN-tō-is; *allas* = sausage) is a small, vascularized outpouching of the yolk sac. It serves as an early site of blood formation. Later its blood vessels serve as the umbilical connection in the placenta between mother and fetus. This connection is the umbilical cord.

7. Placenta and Umbilical Cord

Development of the *placenta* (pla-SEN-ta; *placenta* = flat cake), the third major event of the embryonic period, is accomplished by the third month of pregnancy. The placenta has the shape of a flat cake when fully developed and is formed by the chorion of the embryo and a portion of the endometrium (decidua basalis) of the mother (see Figure 26.7). Functionally, the placenta allows oxygen and nutrients to diffuse into fetal blood from maternal blood. Simultaneously, carbon dioxide and wastes diffuse from fetal blood into maternal blood at the placenta.

The placenta also is a protective barrier since most microorganisms cannot cross it. However, certain viruses, such as those that cause AIDS, German measles, chickenpox, measles, encephalitis, and poliomyelitis, may pass through the placenta. The placenta also stores nutrients such as carbohydrates, proteins, calcium, and iron, which are released into fetal circulation as required. Finally, the placenta produces several hormones that are necessary to maintain pregnancy. Almost all drugs, including alcohol, and many substances that can cause birth defects pass freely through the placenta.

If implantation occurs, a portion of the endometrium becomes modified and is known as the *decidua* (dē-SID-yoo-a; *deciduus* = falling off). The decidua includes all but the stratum basalis layer of the endometrium and separates from the endometrium after the fetus is delivered. Different regions of the decidua, all areas of the stratum functionalis, are named based on their positions relative to the site of the implanted, fertilized ovum (see Figure 26.7). The *decidua basalis* is the portion of the endometrium between the chorion and the stratum basalis of the uterus. It becomes the maternal part of the placenta. The *decidua capsularis* is the portion of the endometrium that covers the embryo and is located between the embryo and the uterine cavity. The *decidua parietalis* (par-rī-e-TAL-is) is the remaining modified endometrium that lines the noninvolved areas of the entire pregnant uterus. As the embryo and later the fetus enlarge, the decidua capsularis bulges into the uterine cavity and initially fuses with the decidua parietalis, thus obliterating the uterine cavity. By about 27 weeks, the decidua capsularis degenerates and disappears.

During embryonic life, fingerlike projections of the chorion, called *chorionic villi* (kō'-rē-ON-ik VIL-ī), grow into the decidua basalis of the endometrium (see Figure 26.7). These will contain fetal blood vessels of the allantois. They continue growing until they are bathed in maternal blood sinuses called *intervillous* (in-ter-VIL-us) *spaces*. Thus maternal and fetal blood vessels are brought into proximity. It should be noted, however, that maternal and fetal blood do not normally mix. Rather, oxygen and nutrients in the blood of the mother's intervillous spaces diffuse across the cell membranes into the capillaries of the villi while waste products diffuse in the opposite direction. From the capillaries of the villi, nutrients and oxy-

gen enter the fetus through the umbilical vein. Wastes leave the fetus through the umbilical arteries, pass into the capillaries of the villi, and diffuse into the maternal blood. A few materials, such as IgG antibodies, pass from the blood of the mother into the capillaries of the villi.

The *umbilical* (um-BIL-i-kul) *cord* is a vascular connection between mother and fetus. It consists of two umbilical arteries that carry deoxygenated fetal blood to the placenta, one umbilical vein that carries oxygenated blood into the fetus, and supporting mucous connective tissue called Wharton's jelly from the allantois. The entire umbilical cord is surrounded by a layer of amnion (see Figure 26.7).

At delivery, the placenta detaches from the uterus and is termed the *afterbirth*. At this time, the umbilical cord is severed, leaving the baby on its own. The small portion (about an inch) of the cord that remains still attached to the infant begins to wither and falls off, usually within 12–15 days after birth. The area where the cord was attached becomes covered by a thin layer of skin and scar tissue forms. The scar is the *umbilicus (navel)*.

D. FETAL PERIOD

During the *fetal period,* the months of development after the second month, all the organs of the body grow rapidly from the original primary germ layers, and the organism takes on a human appearance. During this time, the developing human is called a *fetus.* Some of the principal changes associated with fetal growth are summarized in Table 26.1.

ANSWER THE LABORATORY REPORT QUESTIONS AT THE END OF THE EXERCISE.

TABLE 26.1
Changes Associated with Embryonic and Fetal Growth

End of Month	Approximate size and weight	Representative changes
1	0.6 cm (⁶⁄₁₆ in.)	Eyes, nose, and ears not yet visible. Vertebral column and vertebral canal form. Small buds that will develop into limbs form. Heart forms and starts beating. Body systems begin to form. The central nervous system appears at the start of the third week.
2	3 cm (1¼ in.) 1 g (¹⁄₃₀ oz)	Eyes far apart, eyelids fused, nose flat. Ossification begins. Limbs become distinct and digits are well formed. Major blood vessels form. Many internal organs continue to develop.
3	7 ½ cm (3 in.) 30 g (1 oz)	Eyes almost fully developed but eyelids still fused, nose develops bridge, and external ears are present. Ossification continues. Limbs are fully formed and nails develop. Heartbeat can be detected. Urine starts to form. Fetus begins to move, but it cannot be felt by mother. Body systems continue to develop.
4	18 cm (6½–7 in.) 100 g (4 oz)	Head large in proportion to rest of body. Face takes on human features and hair appears on head. Skin bright pink. Many bones ossified, and joints begin to form. Rapid development of body systems.
5	25–30 cm (10–12 in.) 200–450 g (½–1 lb)	Head less disproportionate to rest of body. Fine hair (lanugo) covers body. Skin still bright pink. Brown fat forms and is the site of heat production. Fetal movements commonly felt by mother (quickening). Rapid development of body systems.
6	27–35 cm (11–14 in.) 550–800 g (1¼–1½ lb)	Head becomes even less disproportionate to rest of body. Eyelids separate and eyelashes form. Substantial weight gain. Skin wrinkled and pink. Type II alveolar cells begin to produce surfactant.
7	32–42 cm (13–17 in.) 1100–1350 g (2½–3 lb)	Head and body more proportionate. Skin wrinkled and pink. Seven-month fetus (premature baby) is capable of survival. Fetus assumes an upside-down position. Testes descend into scrotum.
8	41–45 cm (16½–18 in.) 2000–2300 g (4½–5 lb)	Subcutaneous fat deposited. Skin less wrinkled. Chances of survival much greater at end of eighth month.
9	50 cm (20 in.) 3200–3400 g (7–7½ lb)	Additional subcutaneous fat accumulates. Lanugo shed. Nails extend to tips of fingers and maybe even beyond.

Development 26

Student _____   **Date** _____

Laboratory Section _____   **Score/Grade** _____

PART 1. Multiple Choice

_____ 1. The basic difference between spermatogenesis and oogenesis is that (a) two more polar bodies are produced in spermatogenesis (b) the secondary oocyte contains the haploid chromosome number, whereas the mature sperm cell contains the diploid number (c) in oogenesis, one secondary oocyte is produced, and in spermatogenesis four mature sperm cells are produced (d) both mitosis and meiosis occur in spermatogenesis, but only meiosis occurs in oogenesis

_____ 2. The union of a sperm cell nucleus and a secondary oocyte nucleus resulting in formation of a zygote is referred to as (a) implantation (b) fertilization (c) gestation (d) parturition

_____ 3. The most advanced stage of development for these stages is the (a) morula (b) zygote (c) ovum (d) blastocyst

_____ 4. Damage to the mesoderm during embryological development would directly affect the formation of (a) muscle tissue (b) the nervous system (c) the epidermis of the skin (d) hair, nails, and skin glands

_____ 5. The placenta, the organ of exchange between mother and fetus, is formed by union of the endometrium with the (a) yolk sac (b) amnion (c) chorion (d) umbilicus

_____ 6. One oogonium produces (a) one ovum and three polar bodies (b) two ova and two polar bodies (c) three ova and one polar body (d) four ova

_____ 7. Implantation is defined as (a) attachment of the blastocyst to the uterine (Fallopian) tube (b) attachment of the blastocyst to the endometrium (c) attachment of the embryo to the endometrium (d) attachment of the morula to the endometrium

_____ 8. Epithelium lining most of the gastrointestinal tract and a number of other organs is derived from (a) ectoderm (b) mesoderm (c) endoderm (d) mesophyll

_____ 9. The nervous system is derived from the (a) ectoderm (b) mesoderm (c) endoderm (d) mesophyll

_____ 10. Which of the following is _not_ an embryonic membrane? (a) amnion (b) placenta (c) chorion (d) allantois

PART 2. Completion

11. A normal human sperm cell, as a result of meiosis, contains _____ chromosomes.
12. The process that permits an exchange of genes resulting in their recombination and a part of the variation among humans is called _____.

13. The result of meiosis in spermatogenesis is that each primary spermatocyte produces four

    _____.

14. The stage of spermatogenesis that results in maturation of spermatids into sperm cells is called

    _____.

15. The afterbirth expelled in the final stage of delivery is the _____.

16. After the second month, the developing human is referred to as a(n) _____.

17. Embryonic tissues from which all tissues and organs of the body develop are called the

    _____.

18. The cells of the inner cell mass divide to form two cavities: amniotic cavity and

    _____.

19. Somatic cells that contain two sets of chromosomes are referred to as _____.

20. At the end of the _____ month of development, a heartbeat can be detected.

21. In oogenesis, primordial follicles develop into _____ follicles.

22. The clear, glycoprotein layer between the oocyte and granulosa cells is called the

    _____.

23. _____ refers to the functional changes that sperm cells undergo in the female
    reproductive tract that allow them to fertilize a secondary oocyte.

24. The _____ of a blastocyst develops into an embryo.

25. The decidua _____ is the portion of the endometrium between the chorion and
    stratum basalis of the uterus.

26. At the end of the _____ month of development, the testes descend into the scrotum.

Genetics

27

Genetics (je-NET-iks) is the branch of biology that studies inheritance. *Inheritance* is the passage of hereditary traits from one generation to another. It is through the passage of hereditary traits that you acquired your characteristics from your parents and will transmit your characteristics to your children. If all individuals were brown-eyed, we could learn nothing of the hereditary basis of eye color. However, because some people are blue-eyed and marry brown-eyed people, we can gain some knowledge of how hereditary traits are transmitted. We constantly analyze the genetic bases of the *differences* between individuals. Some of these differences occur normally, such as differences in eye color, blood groups, or ability to taste PTC (phenylthiocarbamide). Other differences are abnormal, such as physical abnormalities and abnormalities in the processes of metabolism.

A. GENOTYPE AND PHENOTYPE

The vast majority of human cells, except gametes, contain 23 pairs of chromosomes (diploid number) in their nuclei. One chromosome from each pair comes from the mother, and the other comes from the father. The two chromosomes that belong to a pair are called *homologous* (hō-MOL-ō-gus) *chromosomes,* and these homologues contain genes that control the same traits. The homologue of a chromosome that contains a gene for height also contains a gene for height.

The relationship of genes to heredity can be illustrated by the disorder called *phenylketonuria,* or *PKU* (see Figure 27.1). People with PKU are unable to manufacture the enzyme phenylalanine hydroxylase. Current belief is that PKU results from the presence of an abnormal gene symbolized as *p*. The normal gene is symbolized as *P*. *P* and *p* are said to be alleles. An *allele* is one of many alternative forms of a gene, occupying the same *locus* (position of a gene on a chromosome) in homologous chromosomes. The chromosome that has the gene that directs phenylalanine hydroxylase production will have either *p* or *P* on it. Its homologue will also have either *p* or *P*. Thus every individual will have one of the following genetic makeups, or *genotypes* (JĒ-nō-tīps): *PP, Pp,* or *pp*. Although people with genotypes of *Pp* have the abnormal gene, only those with genotype *pp* suffer from the disorder because the normal gene masks the abnormal one. A gene that masks the expression of its allele is called the *dominant gene,* and the trait expressed is said to be a dominant trait. The homologous gene that is masked is called the *recessive gene.* The trait expressed when two recessive genes are present is called the recessive trait. Several dominant and recessive traits inherited in human beings are listed in Table 27.1.

Traditionally, the dominant gene is symbolized with a capital letter and the recessive one with a lowercase letter. When the same genes appear on homologous chromosomes, as in *PP* or *pp,* the person is said to be *homozygous* for a trait. When the genes on homologous chromosomes are different, however, as in *Pp,* the person is said to be *heterozygous* for the trait. *Phenotype* (FĒ-nō-tīp; *pheno* = showing) refers to how the genetic composition is expressed in the body. An individual with *Pp* has a different genotype from one with *PP,* but both have the same phenotype—which in this case is normal production of phenylalanine hydroxylase.

B. PUNNETT SQUARES

To determine how gametes containing haploid chromosomes unite to form diploid fertilized eggs, special charts called *Punnett squares* are used. The Punnett square is merely a device that helps one visualize all the possible combinations of male and female gametes, and is invaluable as a learning exercise in genetics. Usually, the possible paternal alleles in sperm cells are placed at the side of the chart and the possible maternal alleles in secondary

TABLE 27.1
Selected Hereditary Traits in Humans

Dominant	Recessive
Coarse body hair	Fine body hair
Male pattern baldness	Baldness
Normal skin pigmentation	Albinism
Freckles	Absence of freckles
Astigmatism	Normal vision
Near- or farsightedness	Normal vision
Normal hearing	Deafness
Broad lips	Thin lips
Tongue roller	Inability to roll tongue into a U shape
PTC taster	PTC nontaster
Large eyes	Small eyes
Polydactylism (extra digits)	Normal digits
Brachydactylism (short digits)	Normal digits
Syndactylism (webbed digits)	Normal digits
Feet with normal arches	Flat feet
Hypertension	Normal blood pressure
Diabetes insipidus	Normal excretion
Huntington's chorea	Normal nervous system
Normal mentality	Schizophrenia
Migraine headaches	Normal
Widow's peak	Straight hairline
Curved (hyperextended) thumb	Straight thumb
Normal Cl⁻ transport	Cystic fibrosis
Hypercholesterolemia (familial)	Normal cholesterol level

oocytes are placed at the top (Figure 27.1). The spaces in the chart represent the possible genotypes for that trait in fertilized ova formed by the union of the male and female gametes. Possible combinations are determined simply by dropping the female gamete on the left into the two boxes below it and dropping the female gamete on the right into the two spaces under it. The upper male gamete is then moved across to the two spaces in line with it, and the lower male gamete is moved across to the two spaces in line with it.

C. SEX INHERITANCE

Lining up human chromosomes in pairs reveals that the last pair (the twenty-third pair) differs in males and in females (Figure 27.2a). In females, the pair consists of two rod-shaped chromosomes designated as X chromosomes. One X chromosome is also present in males, but its mate is hook-shaped and called a Y chromosome. The XX pair in the female and the XY pair in the male are called the *sex chromosomes,* and all other pairs of chromosomes are called *autosomes.*

The sex of an individual is determined by the sex chromosomes (Figure 27.2b). When a spermatocyte undergoes meiosis to reduce its chromosome number from diploid to haploid, one daughter cell will contain the X chromosome and the other will contain the Y chromosome. When the secondary oocyte is fertilized by an X-bearing sperm, the offspring normally will be a female (XX). Fertilization by a Y sperm cell normally produces a male (XY).

Sometimes chromosomes fail to move toward opposite poles of a cell in meiotic anaphase. This is called *nondisjunction* and results in one sex cell having two members of a chromosome pair while the other receives none. Thus, eggs can contain two X's or no X (symbolized as 0), and sperm cells may contain both an X and a Y chromosome, two X's or two Y's, or no sex chromosomes at all.

Because the X chromosome contains so many genes unrelated to sex that are necessary for development, a zygote must contain at least one X chromosome to survive. Thus, Y0 and YY zygotes do not develop. However, other zygotes with sex chromosome anomalies do develop. Examples are Turner's syndrome (the presence of only one X chromosome, X0) and Kleinfelter's syndrome (an extra Y chromosome, XYY).

"Extra" X chromosomes (more than two in the female and more than one in the male) have a surprisingly minor effect on the individual, compared with the significant effect of other additional chromosomes. Studies show that only one X chromosome is active in any cell. Any additional X chromosomes are randomly inactivated early in development and do not express the genes contained on them.

These inactivated X chromosomes remain tightly coiled against the cell membrane and can be seen as what are called *Barr bodies* (Figure 27.3 on page 640). Since XY males do not have inactivated X chromosomes, no Barr bodies will be seen in normal male cells.

PROCEDURE

1. Make a buccal smear by *gently* scraping the inside of your cheek with the flat end of a toothpick. Discard the toothpick and *gently* scrape the same area again. This will produce more live cells from a deeper layer of the epithelium of the mucous membrane.

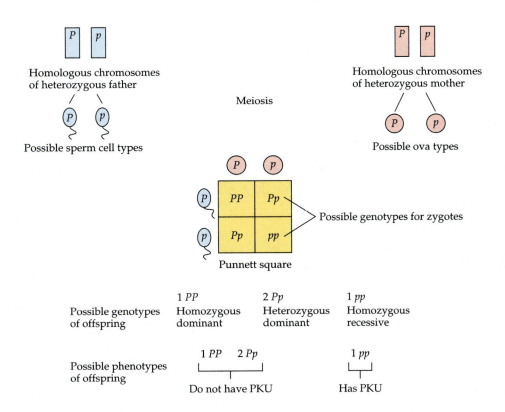

FIGURE 27.1 Inheritance of phenylketonuria (PKU).

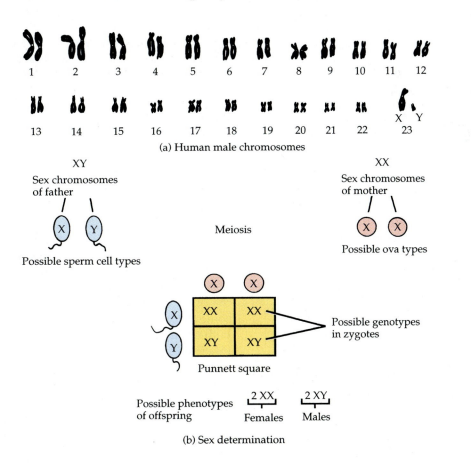

FIGURE 27.2 Inheritance of sex. In (a), note the sex chromosomes, X and Y.

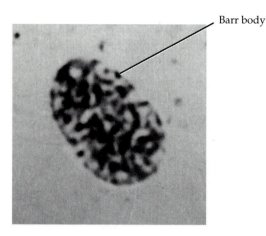

Barr body

FIGURE 27.3   Barr body in cell from buccal mucosa of human female. Feulgen stain; magnification ×300.

2. Spread the material over a clean glass slide.
3. Promptly place the slide in a Coplin jar filled with fixative for 1 min.
4. Remove the slide and wash gently under running tap water.
5. Place the slide in a Coplin jar filled with Giemsa stain for 10 to 20 min. Fresh solutions of stain prepared within 2 hr require 10 min; older stains require a longer time.
6. Wash the slide *gently* under running water and air-dry.
7. Examine the slide under high power and look for interphase nuclei. Identify Barr bodies, small disc-shaped chromatin bodies lying against the nuclear membrane (see Figure 27.3). Depending on the position of the nuclei and the staining technique, Barr bodies should be seen in 30 to 70% of the cells of a normal female.
8. Examine a slide prepared from the buccal epithelium of a class member not of your sex.
9. Draw a cell containing a Barr body in the space provided.

Barr body

D. SEX-LINKED INHERITANCE

As do the other 22 pairs of chromosomes, the sex chromosomes contain genes that are responsible for the transmission of a number of nonsexual traits. Genes for these traits appear on X chromosomes, but many of these genes are absent from Y chromosomes. Traits transmitted by genes on the X chromosome are called **sex-linked traits.** This pattern of heredity is different from the pattern described earlier. About 150 sex-linked traits are known in humans. Examples of sex-linked traits are red–green color blindness and hemophilia.

1. Red–Green Color Blindness

Let us consider the most common type of color blindness, called red–green color blindness. In this condition, there is a deficiency in either red or green cones and red and green are seen as the same color, either red or green, depending on which cone is present. The gene for **red–green color blindness** is a recessive one designated *c*. Normal color vision, designated *C*, dominates. The *C/c* genes are located on the X chromosome. The Y chromosome does not contain these genes. Thus the ability to see colors depends entirely on the X chromosomes. The possible combinations are:

Genotype	Phenotype
$X^C X^C$	Normal female
$X^C X^c$	Normal female (carrying the recessive gene)
$X^c X^c$	Red–green color-blind female
$X^C Y$	Normal male
$X^c Y$	Red–green color-blind male

Only females who have two X^c genes are red–green color-blind. This rare situation can result only from the mating of a color-blind male and a color-blind or carrier female. In $X^C X^c$ females, the trait is masked by the normal, dominant gene. Males, on the other hand, do not have a second X chromosome that would mask the trait. Therefore all males with an X^c gene will be red–green color blind. The inheritance of red–green color blindness is illustrated in Figure 27.4.

2. Hemophilia

Hemophilia is a condition in which the blood fails to clot or clots very slowly after an injury. Hemophilia is a much more serious defect than color blindness because individuals with severe hemophilia can bleed to death from even a small cut.

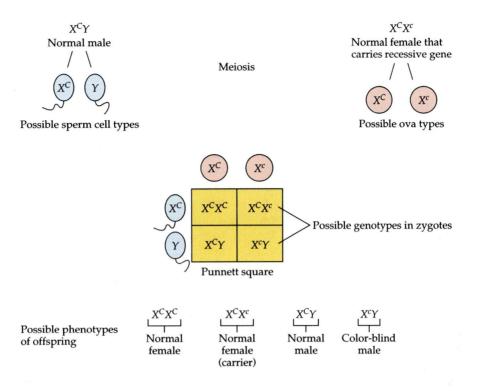

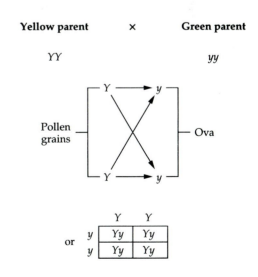

FIGURE 27.4   Inheritance of red–green color blindness.

Hemophilia is caused by a recessive gene as is color blindness. If H represents normal clotting and h represents abnormal clotting, X^hX^h females will be hemophiliacs. Males with X^HY will be normal and males with X^hY will be hemophiliacs. Other sex-linked traits in humans are fragile X syndrome, nonfunctional sweat glands, certain forms of diabetes, some types of deafness, uncontrollable rolling of the eyeballs, absence of central incisors, night blindness, one form of cataract, juvenile glaucoma, and juvenile muscular dystrophy.

E. MENDELIAN LAWS

In any genetic cross, all the offspring in the first (that is, the parental, or P_1) generation are symbolized as F_1. The F is from the Latin word *filial*, which means progeny. The second generation is symbolized as F_2, the third as F_3, and continues that way. The recognized "father" of genetics is Gregor Mendel, whose basic experiments were performed on garden peas. As a result of his tests, Gregor Mendel postulated what are now called **Mendelian Laws**, or **Mendelian Principles**. The **First Mendelian Law**, or the **Law of Segregation**, asserts that, in cells of individuals, genes occur in pairs, and that when those individuals produce germ cells, each germ cell receives only one member of the pair.

This law applies equally to pollen grains (or sperm) and to ova. The genetic cross is represented as follows:

All possible combinations of pollen grains and ova are indicated by the arrows. Notice that all combinations yield the genotype Yy. All these F_1 seeds were yellow, yellow being dominant to green, or, in genetic terms, Y being dominant to y.

These F$_1$ individuals resembled the yellow parent in phenotype (being yellow) but not in genotype (Yy as opposed to YY). Both parents were homozygous. Both members of that pair of alleles were the same. The yellow parent was homozygous for Y and the green parent for y. The F$_1$ individuals were heterozygous, having one Y and one y.

When the F$_1$ plants were self-fertilized, the F$_2$ seeds appeared in the ratio of 3 yellow/1 green. Mendel found similar 3:1 ratios for the other traits he studied, and this type of result has been reported in many species of animals and plants for a variety of traits. Not only does the recessive trait reappear in the F$_2$, but also in a definite proportion of the individuals, one-fourth of the total. If the sample is small, the ratio may deviate considerably from 3:1, but as the progeny or sampling numbers get larger, the ratio usually comes closer and closer to an exact 3:1 ratio. The reason is that the ratio depends on the random union of gametes. The result is a 3:1 phenotypic ratio, or a 1:2:1 genotypic ratio.

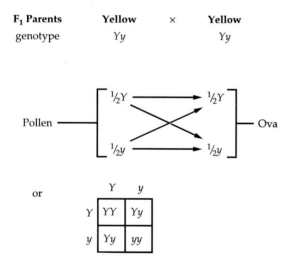

F$_1$ Parents   Yellow   ×   Yellow

genotype Yy Yy

Pollen ——   Ova

or

	Y	y
Y	YY	Yy
y	Yy	yy

Thus, the four combinations of pollen and ova are expected to occur as follows:

$\frac{1}{4}YY$ = yellow ⎤

$\frac{1}{4}Yy$ = yellow ⎬ ¾

$\frac{1}{4}Yy$ = yellow ⎦

$\frac{1}{4}yy$ = green ⎤ ¼

It is important to realize that these fractions depend on the operation of the laws of probability.

A model using coins will emphasize the point. This model consists of two coins, a nickel and a penny, tossed at the same time. The penny may represent the pollen (male parent). At any given toss, the chances are equal that the penny will come up "heads" or that it will come up "tails." Similarly, at any given fertilization the chances are equal that a Y-bearing pollen grain or that a y-bearing one will be transmitted. The nickel represents the ovum. Again, the chances are equal for "heads" or "tails," just as the chances are equal that in any fertilization a Y-bearing or a y-bearing ovum will take part. If we toss the two coins together and do it many times, we will obtain approximately the following:

¼ nickel heads; penny heads	(= YY)
¼ nickel heads; penny tails	(= Yy)
¼ nickel tails; penny heads	(= yY)
¼ nickel tails; penny tails	(= yy)

If we assume "heads" as dominant, we find that three-fourths of the time there is at least one "head" and one-fourth of the time no "heads" (both coins "tails"). Hence this gives us a model of the 3:1 ratio dependent on the laws of probability.

The genetic cross just demonstrated considers only one pair of alleles (yellow versus green or "heads" versus "tails") and is therefore called a ***monohybrid cross***. Mendel's second principle applies to genetic crosses in which two traits or two pairs of alleles are considered. These ***dihybrid crosses*** enabled him to postulate his second principle, the ***Principle of Independent Assortment.*** This principle states that the segregation of one pair of traits occurrs independently of the segregation of a second pair of traits. This is the case only if the traits are caused by genes located on nonhomologous chromosomes.

When Mendel crossed garden peas with round yellow seeds with garden peas with wrinkled green seeds, his F$_1$ generation showed that yellow and round were dominant. If self-fertilization then occurred, the F$_2$ generation resulted as follows:

Round yellow	¾ × ¾ =	$\frac{9}{16}$
Round green	¾ × ¼ =	$\frac{3}{16}$
Wrinkled yellow	¼ × ¾ =	$\frac{3}{16}$
Wrinkled green	¼ × ¼ =	$\frac{1}{16}$

Therefore, in a dihybrid cross, the expected phenotypic ratio was 9:3:3:1, with $\frac{9}{16}$ of the F$_2$ being doubly dominant, and only $\frac{1}{16}$ being doubly recessive.

F. MULTIPLE ALLELES

In the genetics examples we have considered to this point, we have discussed only two alleles of each gene. However, many, and possibly all genes, have *multiple alleles;* that is, they exist in more than two allelic forms even though a diploid cell cannot carry more than two alleles.

One example of multiple alleles in humans involves ABO blood groups (Exercise 16). The four basic blood types (phenotypes) of the ABO system are determined by three alleles: I^A, I^B, and i. Alleles I^A and I^B are not dominant over each other. Rather, they are **codominant;** that is, both genes are expressed equally. Both I^A and I^B alleles, however, are dominant over allele i. These three alleles can give rise to six genotypes, as follows:

Genotype	Phenotype (blood type)
$I^A I^A$ or $I^A i$	A
$I^B I^B$ or $I^B i$	B
$I^A I^B$	AB
ii	O

Given this information, is it possible for a child with type O blood to have a mother with type O blood and a father with type AB blood?

Explain _____

_____

_____

If two children in a family have type O blood, the mother has type B blood and the father has type A blood, what is the genotype of the father?

_____

What is the genotype of the mother? _____

G. GENETICS EXERCISES

1. Karyotyping

A group of cytogeneticists meeting in Denver, Colorado, in 1960 adopted a system for classifying and identifying human chromosomes. Chromosome *length* and *centromere position* were the bases for classification. The Denver classification has become a standard for human chromosome studies. By the early 1970s, most human chromosomes could be identified microscopically.

Every chromosome pair could not be identified consistently until chromosome *banding techniques*

finally distinguished all 46 human chromosomes. Bands are defined as parts of chromosomes that appear lighter or darker than adjacent regions with particular staining methods.

A *karyotype* is a chart made from a photograph of the chromosomes in metaphase. The chromosomes are cut out and arranged in matched pairs according to length (see Figure 27.2a). Their comparative size, shape, and morphology are then examined to determine if they are normal.

Karyotyping helps scientists to visualize chromosomal abnormalities. For example, individuals with Down syndrome typically have 47 chromosomes, instead of the usual 46, with chromosome 21 being represented three times rather than only twice. The syndrome is characterized by mental retardation, retarded physical development, and distinctive facial features (round head, broad skull, slanting eyes, and large tongue). With chronic myelogenous leukemia, part of the long arm of a chromosome 22 is missing, resulting in the blood disease. The chromosome is referred to as the Philadelphia chromosome, named for the city where it was first detected.

2. PKU Screening

Phenylketonuria (PKU), an inherited metabolic disorder that occurs in approximately 1 in 16,000 births, is transmitted by an autosomal recessive gene (see Figure 27.1). Individuals with this condition do not have the enzyme phenylalanine hydroxylase, which converts the amino acid phenylalanine to tyrosine. As a result, phenylalanine and phenylpyruvic acid accumulate in the blood and urine. These substances are toxic to the central nervous system and can produce irreversible brain damage. Most states in the United States require routine screening for this disorder at birth. The test is accomplished by a simple color change in treated urine.

The procedure for testing for PKU is as follows.

PROCEDURE

1. A Phenistix® test strip is made specifically for testing urine for phenylpyruvic acid. Dip this test strip in freshly voided urine.
2. Compare the color change with the color chart on the Phenistix® bottle. The test is based on the reaction of ferric ions with phenylpyruvic acid to produce a gray-green color.
3. Record your results in Section G.1 of the LABORATORY REPORT RESULTS at the end of the exercise.

3. PTC Inheritance

The ability to taste the chemical compound known as phenylthiocarbamide, commonly called PTC, is inherited. On the average, 7 out of 10 people, on chewing a small piece of paper treated with PTC, detect a definite bitter or sweet taste. Others do not taste anything.

Individuals who can taste something (bitter or sweet) are called "tasters" and have the dominant allele T, either as TT or Tt. A nontaster is a homozygous recessive and is designated as tt.

Determine your phenotype for tasting PTC and record your results in Section G.2 of the LABORATORY REPORT RESULTS at the end of the exercise.

Note: If PTC paper is not available, a 0.5% solution of phenylthiourea (PTT) can be substituted because the capacity to taste PTT is also inherited as a dominant.

4. Corn Genetics

Genetic corn may be purchased and used in this exercise. Each ear of corn represents a family of offspring. Mark a starting row with a pin to avoid repetition. Count the kernels (individuals) for each trait (color, wrinkled, or smooth). Record your results in Section G.3 of the LABORATORY REPORT RESULTS at the end of the exercise.

Develop a ratio by using your lowest number as "1" and dividing it into the others to determine what multiples of it they are. See how close you come to Mendel's ratios. Figure out the probable genotype and phenotype of the parent plants if you can. Monohybrid crosses, test crosses, dihybrid crosses, and trihybrid crosses are available.

5. Color Blindness

Using either Stilling or Ishihara test charts, test the entire class for red–green color blindness. Tests for color blindness depend on the person's ability to distinguish various colors from one another and also on his or her ability to judge correctly the degree of contrast between colors.

Of all men, 2% are color-blind to red and 6% to green, so 8% of all men are red–green color-blind. Red–green color blindness is rare in the female, occurring in only 1 of every 250 women. Record your results in Section G.4 of the LABORATORY REPORT RESULTS at the end of the exercise.

6. Mendelian Laws of Inheritance

Follow the procedure outlined in the explanation of the Mendelian Law of Segregation, tossing a nickel and a penny simultaneously to prove the law and determine ratios.

PROCEDURE

1. Toss the nickel and the penny together 10 times to get the genotypes of a family of 10. Repeat this procedure for a total of five times to obtain five families of 10 offspring. Record all of the results on the chart in Section G.5 of the LABORATORY REPORT RESULTS at the end of the exercise.
   Note: Use the following symbols for the following exercises.

   G = gene for yellow
   g = gene for green
   GG = the genotype of an individual pure (homozygous) for yellow
   gg = the genotype of an individual pure (homozygous) for green
   Gg = the genotype of the hybrid (heterozygous) individual, phenotypically yellow
   ♀ = symbol for female
   ♂ = symbol for male

2. Obeying the Mendelian Law of Segregation and using the Punnett square shown in Section G.6 of the LABORATORY REPORT RESULTS at the end of the exercise, cross yellow garden peas with green garden peas (a monohybrid cross). Show the P_1, F_1, and F_2 generations and all the different phenotypes and genotypes.

3. Obeying the Mendelian Law of Independent Assortment and using the Punnett square shown in Section G.8 of the LABORATORY REPORT RESULTS at the end of the exercise, cross the round yellow seeds with the wrinkled green seeds (a dihybrid cross). Show the P_1, F_1, and F_2 generations and all the different phenotypes and genotypes. The F_2 generation can be generated from a Punnett square comparable to that for the monohybrid cross, but with 16 rather than 4 squares.

7. Observing Phenotypes

The pattern of inheritance of many human traits is complex and involves many genes; the inheritance of other traits is controlled by single genes. You are asked to record your phenotype and your genotype, if it can be determined, for several traits controlled by single genes. For example, if you have the phenotypically dominant trait A, your genotype will be $A-$, indicating that the allele symbolized as "–" is not known, since you could be homozygous dominant (*AA*) or heterozygous (*Aa*) for the trait. If you have the recessive trait *a−*, your genotype will be *aa*. Record your phenotype and genotype for the following traits in Section G.10 of the LABORATORY REPORT RESULTS at the end of the exercise.

1. *Attached earlobes* The dominant gene *E* causes earlobes that develop free from the neck; *ee* results in adherent earlobes connected to the cervical skin.
2. *Tongue rolling* The dominant gene *R* causes the development of muscles that allow the tongue to be rolled into a U shape. The *rr* genotype prohibits such rolling.
3. *Hair whorl direction* The dominant gene *W* causes the hair whorl on the cranial surface of the scalp to turn in a clockwise direction; the genotype *ww* determines a counterclockwise whorl.
4. *Little-finger bending* The dominant gene *B* causes the distal segment of the little finger to bend laterally. The genotype *bb* results in a straight distal segment.

5. *Double-jointed thumbs* The dominant gene *J* results in loose ligaments that allow the thumb to be bent out of the constricted orientation caused by the recessive genotype *jj*.
6. *Widow's peak* The dominant gene *W* causes the hairline to extend caudally in the midline of the forehead. The recessive genotype *ww* results in a straight hairline.
7. *Rh factor* A dominant gene *Rh* results in the presence of the Rh antigen on red blood cells. This antigen is not present with the recessive genotype *rhrh*. Use the results obtained in Exercise 16.I.4 to determine your phenotype.

   What additional information would you need to determine your complete genotype if you have the dominant phenotype for these traits?

   _____

   _____

   _____

ANSWER THE LABORATORY REPORT QUESTIONS AT THE END OF THE EXERCISE.

Genetics 27

Student _____   **Date** _____

Laboratory Section _____   **Score/Grade** _____

SECTION G. GENETICS EXERCISES

1. PKU Screening

1. _____ negative—cream color

   _____ 15 mg%—light green

   _____ 40 mg%—medium green

   _____ 100 mg%—dark green

2. PTC Inheritance

2. _____ bitter taste

   _____ sweet taste

   _____ negative (no taste)

3. Corn Genetics

3. Ratio_____ monohybrid cross

   Ratio_____ test cross

   Ratio_____ dihybrid cross

   Ratio_____ trihybrid cross

4. Color Blindness

	Male students	Female students
4. Red color-blind	_____	_____
Green color-blind	_____	_____

5. Mendelian Laws of Inheritance

5. Record the results of tossing a nickel and a penny together 10 times.

	Female (nickel)	Male (penny)	1	2	3	4	5	Total	Class total
Dominant offspring	A Heads	A Heads							
	A Heads	a Tails							
	a Tails	A Heads							
Recessive offspring	a Tails	a Tails							
Ratio dominant to recessive									

6. Complete the following monohybrid cross. Fill in genotypes (within circles) and phenotypes (under circles).

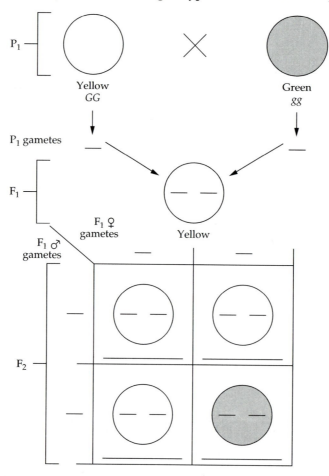

Monohybrid cross in the garden pea (*Pisum sativum*). G = allele for yellow, g = allele for green, P_1 = parental generation, F_1 = first filial generation, F_2 = second filial generation.

7. What is the phenotype ratio of the F$_2$ generation? _____ yellow/ _____ green.

8. Complete the following dihybrid cross. Fill in genotypes (within circles) and phenotypes (under circles).

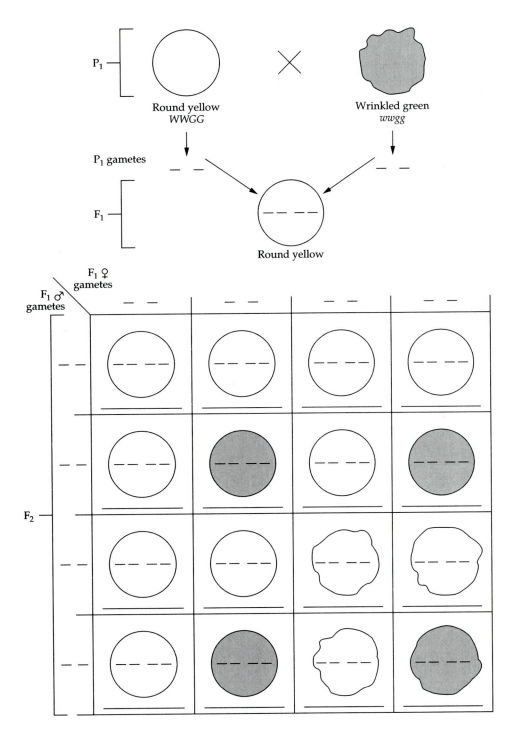

Dihybrid cross in the garden pea *(Pisum sativum)*. *W* = allele for round, *w* = allele for wrinkled, *G* = allele for yellow, *g* = allele for green.

9. What is the phenotype ratio of the F$_2$ generation?

_____round yellow/_____ wrinkled yellow/_____ round green/

_____wrinkled green.

10. Observe phenotypes.

Trait	Phenotype	Genotype
Earlobes		
Tongue rolling		
Hair whorl		
Double-jointed thumbs		
Widow's peak		
Rh factor		

Genetics 27

Student _____   Date _____

Laboratory Section _____   Score/Grade _____

PART 1. Multiple Choice

_____ 1. Using the symbols *Aa* to represent genes, which of the following is true? (a) the trait is homozygous for the dominant characteristic (b) the trait is homozygous for the recessive characteristic (c) the trait is heterozygous (d) sex-linked inheritance is in operation

_____ 2. Which statement concerning the normal inheritance of sex is correct? (a) all zygotes contain a Y chromosome (b) some ova contain a Y chromosome (c) all ova and all sperm cells contain an X chromosome (d) all ova have an X chromosome, some sperm cells have an X chromosome, and some sperm cells have a Y chromosome

_____ 3. The genotype that will express characteristics associated with hemophilia (assume that *H* represents the gene for normal blood) is (a) $X^H X^h$ (b) $X^h Y$ (c) $X^H X^H$ (d) $X^H Y$

_____ 4. The exact position of a gene on a chromosome is called the (a) homologue (b) locus (c) triad (d) allele

PART 2. Completion

5. When the same genes appear on homologous chromosomes, as in *PP* or *pp*, the individual is said to

   be _____ for the trait.

6. If different genes appear on homologous chromosomes, as in *Pp*, the individual is

   _____ for the trait.

7. Genetic composition expressed in the body or morphologically is called the body's

   _____.

8. The device that helps one visualize all the possible combinations of male and female gametes is

   called the _____.

9. The twenty-third pair of human chromosomes are the sex chromosomes. All of the other pairs of

   chromosomes are called _____.

10. Red–green color blindness is an inherited trait that is specifically called a(n) _____ trait.

11. The recognized "father" of genetics is _____.

12. A genetic cross that involves only one pair of alleles (or traits) is called a(n) _____ cross.

13. Passage of hereditary traits from one generation to another is called _____.

14. The genetic makeup of an individual is called the person's _____.

15. One of the many alternative forms of a gene is called its _____.

16. Two chromosomes that belong to a pair are called _____ chromosomes.

17. A gene that masks the expression of its allele is called a(n) _____ gene.

18. Failure of chromosomes to move to opposite poles of a cell during meiotic prophase is called

_____.

19. A normal female who carries the recessive gene for red–green color blindness would have the follow-

ing genotype: _____.

20. Genes that are expressed equally are said to be _____.

Some Important Units of Measurement

English Units of Measurement

Fundamental or Derived Unit	Units and Equivalents
Length	12 inches (in.) = 1 foot (ft) = 0.333 yard (yd)
	3 ft = 1 yd
	1760 yd = 1 mile (mi)
	5280 ft = 1 mi
Mass	1 ounce (oz) = 28.35 grams (g); 1 g = 0.0353 oz
	1 pound (lb) = 453 g = 16 oz; 1 kilogram (kg) = 2.205 lb
	1 ton = 2000 lb = 907 kg
Time	1 second (sec) = 1/86 400 of a mean solar day
	1 minute (min) = 60 sec
	1 hour (hr) = 60 min = 3600 sec
	1 day = 24 hr = 1440 min = 86 400 sec
Volume	1 fluid dram (fl dr) = 0.125 fluid ounce (fl oz)
	1 fl oz = 8 fl dr = 0.0625 quart (qt) = 0.008 gallon (gal)
	1 qt = 256 fl dr = 32 fl oz = 2 pints (pt) = 0.25 gal
	1 gal = 4 qt = 128 fl oz = 1024 fl dr

Metric Units of Length and Some English Equivalents

Metric Unit	Meaning of Prefix	Metric Equivalent	English Equivalent
1 kilometer (km)	kilo = 1000	1000 m	3280.84 ft or 0.62 mi; 1 mi = 1.61 km
1 hectometer (hm)	hecto = 100	100 m	328 ft
1 dekameter (dam)	deka = 10	10 m	32.8 ft
1 meter (m)	Standard unit of length		39.37 in. or 3.28 ft or 1.09 yd
1 decimeter (dm)	deci = $\frac{1}{10}$	0.1 m	3.94 in.
1 centimeter (cm)	centi = $\frac{1}{100}$	0.01 m	0.394 in.; 1 in. = 2.54 cm
1 millimeter (mm)	milli = $\frac{1}{1000}$	0.001 m = $\frac{1}{10}$ cm	0.0394 in.
1 micrometer (μm) [formerly micron (μ)]	micro = $\frac{1}{1,000,000}$	0.0000001 m = $\frac{1}{10,000}$ cm	3.94×10^{-5} in.
1 nanometer (nm) [formerly millimicrons (mμ)]	nano = $\frac{1}{1,000,000,000}$	0.000000001 m = $\frac{1}{10,000,000}$ cm	3.94×10^{-8} in.

Temperature

Unit	K	ºF	ºC
1 degree Kelvin (K)	1	$\frac{9}{5}$(K) − 459.7	K + 273.16*
1 degree Fahrenheit (ºF)	$\frac{5}{9}$(ºF) + 255.4	1	$\frac{5}{9}$(ºF − 32)
1 degree Celsius (ºC)	ºC − 273	$\frac{9}{5}$(ºC) + 32	1

* Absolute zero (K) = −273.16ºC

Volume

Unit	ml	cm³	qt	oz
1 milliliter (ml)	1	1	1.06×10^{-3}	3.392×10^{-2}
1 cubic centimeter (cm³)	1	1	1.06×10^{-3}	3.392×10^{-2}
1 quart (qt)	943	943	1	32
1 fluid ounce (fl oz)	29.5	29.5	3.125×10^{-2}	1

Periodic Table of the Elements

KEY

6	Atomic Number
C	Symbol
12.01	Atomic Weight
Carbon	Name

Periodic Table

1	2	3	4	5	6	7	8	9	10	11	12	13	14	15	16	17	18
1 H 1.0080 Hydrogen																	2 He 4.003 Helium
3 Li 6.940 Lithium	4 Be 9.013 Berilium											5 B 10.82 Boron	6 C 12.011 Carbon	7 N 14.008 Nitrogen	8 O 16.000 Oxygen	9 F 19.00 Fluorine	10 Ne 20.183 Neon
11 Na 22.991 Sodium	12 Mg 24.32 Magnesium											13 Al 26.98 Alminum	14 Si 28.09 Silicon	15 P 30.975 Phosphorus	16 S 32.066 Sulfur	17 Cl 35.457 Chlorine	18 Ar 39.944 Argon
19 K 39.100 Potassium	20 Ca 40.08 Calcium	21 Sc 44.96 Scandium	22 Ti 47.90 Titanium	23 V 50.95 Vanadium	24 Cr 52.01 Chromium	25 Mn 54.94 Manganese	26 Fe 55.85 Iron	27 Co 58.94 Cobalt	28 Ni 58.71 Nickel	29 Cu 63.54 Copper	30 Zn 65.38 Zinc	31 Ga 69.72 Gallium	32 Ge 72.60 Germanium	33 As 74.91 Arsenic	34 Se 78.96 Selenium	35 Br 79.916 Bromine	36 Kr 83.80 Krypton
37 Rb 85.48 Rubidium	38 Sr 87.63 Strontium	39 Y 88.92 Yttrium	40 Zr 91.22 Zirconium	41 Nb 92.91 Niobium	42 Mo 95.95 Molybdenum	43 Tc (99) Technetium	44 Ru 101.1 Ruthenium	45 Rh 102.91 Rhodium	46 Pd 106.4 Palladium	47 Ag 107.880 Silver	48 Cd 112.41 Cadmium	49 In 114.82 Indium	50 Sn 118.70 Tin	51 Sb 121.76 Antimony	52 Te 127.61 Tellurium	53 I 126.91 Iodine	54 Xe 131.30 Xenon
55 Cs 132.91 Cesium	56 Ba 137.36 Barium	57 La 138.92 Lanthanum	72 Hf 178.50 Hafnium	73 Ta 180.95 Tantalum	74 W 183.86 Wolfram	75 Re 186.22 Rhenium	76 Os 190.2 Osmium	77 Ir 192.2 Iridium	78 Pt 195.09 Platinum	79 Au 197.0 Gold	80 Hg 200.61 Mercury	81 Tl 204.39 Thallium	82 Pb 207.21 Lead	83 Bi 209.00 Bismuth	84 Po (210) Polonium	85 At (210) Astatine	86 Rn (222) Radon
87 Fr (223) Francium	88 Ra (226) Radium	89 Ac (227) Actinium	104 Unq (261) Unnilquadium	105 Unp (262) Unnilpentium	106 Unh (263) Unnilhexium	107 Uns (262) Unnilseptium	108 Uno (265) Unniloctium	109 Une (267) Unnilennium									

Lanthanide series

58 Ce 140.13 Cerium	59 Pr 140.92 Praseodymium	60 Nd 144.27 Neodymium	61 Pm (147) Promethium	62 Sm 150.35 Samarium	63 Eu 152.0 Europium	64 Gd 157.26 Gadolinium	65 Tb 158.93 Terbium	66 Dy 162.51 Dysprosium	67 Ho 164.94 Holmium	68 Er 167.27 Erbium	69 Tm 168.94 Thulium	70 Yb 173.04 Ytterbium	71 Lu 174.99 Lutetium

Actinide series

90 Th (232) Thorium	91 Pa (231) Protactinium	92 U 238.07 Uranium	93 Np (237) Neptunium	94 Pu (242) Plutonium	95 Am (243) Americium	96 Cm (247) Curium	97 Bk (249) Berkelium	98 Cf (251) Californium	99 Es (254) Einsteinium	100 Fm (253) Fermium	101 Md (256) Mendelevium	102 No (253) Nobelium	103 Lw 257 Lawrencium

Eponyms Used in This Laboratory Manual

Eponym	Current Terminology
Achilles tendon	calcaneal tendon
Adam's apple	thyroid cartilage
ampulla of Vater (VA-ter)	hepatopancreatic ampulla
Bartholin's (BAR-tō-linz) gland	greater vestibular gland
Billroth's (BIL-rōtz) cord	splenic cord
Bowman's (BŌ-manz) capsule	glomerular capsule
Bowman's (BŌ-manz) gland	olfactory gland
Broca's (BRŌ-kaz) area	motor speech area
Brunner's (BRUN-erz) gland	duodenal gland
bundle of His (HISS)	atrioventricular (AV) bundle
canal of Schlemm (SHLEM)	scleral venous sinus
circle of Willis (WIL-is)	cerebral arterial circle
Cooper's (KOO-perz) ligament	suspensory ligament of the breast
Cowper's (KOW-perz) gland	bulbourethral gland
crypt of Lieberkühn (LĒ-ber-kyoon)	intestinal gland
duct of Rivinus (ri-VĒ-nus)	lesser sublingual duct
duct of Santorini (san'-tō-RĒ-nē)	accessory duct
duct of Wirsung (VĒR-sung)	pancreatic duct
end organ of Ruffini (roo-FĒ-nē)	type-II cutaneous mechanoreceptor
Eustachian (yoo-STĀ-kē-an) tube	auditory tube
Fallopian (fal-LŌ-pē-an) tube	uterine tube
gland of Zeis (ZĪS)	sebaceous ciliary gland
Golgi (GOL-jē) tendon organ	tendon organ
Graafian (GRAF-ē-an) follicle	mature follicle
Hassall's (HAS-alz) corpuscle	thymic corpuscle
Haversian (ha-VĒR-shun) canal	central canal
Haversian (ha-VĒR-shun) system	osteon
interstitial cell of Leydig (LĪ-dig)	interstitial endocrinocyte
islet of Langerhans (LANG-er-hanz)	pancreatic islet
Kupffer's (KOOP-ferz) cells	stellate reticuloendothelial cell
loop of Henle (HEN-lē)	loop of the nephron
Malpighian (mal-PIG-ē-an) corpuscle	splenic nodule
Meibomian (mī-BŌ-mē-an) gland	tarsal gland
Meissner's (MĪS-nerz) corpuscle	corpuscle of touch
Merkel's (MER-kelz) disc	tactile disc
Müller's (MIL-erz) duct	paramesonephric duct
Nissl (NISS-l) bodies	chromatophilic substance
node of Ranvier (ron-VĒ-ā)	neurofibral node
organ of Corti (KOR-tē)	spiral organ
Pacinian (pa-SIN-ē-an) corpuscle	lamellated corpuscle
Peyer's (PI-erz) patches	aggregated lymphatic follicles
plexus of Auerbach (OW-er-bak)	myenteric plexus
plexus of Meissner (MĪS-ner)	submucous plexus
pouch of Douglas	rectouterine pouch
Purkinje (pur-KIN-jē) fiber	conduction myofiber

Eponym	Current Terminology
Rathke's (RATH-kēz) pouch	hypophyseal pouch
Schwann (SCHVON) cell	neurolemmocyte
Sertoli (ser-TŌ-lē) cell	sustentacular cell
Skene's (SKĒNZ) gland	paraurethral gland
sphincter of Oddi (OD-dē)	sphincter of the hepatopancreatic ampulla
Stensen's (STEN-senz) duct	parotid duct
Volkmann's (FŌLK-manz) canal	perforating canal
Wharton's (HWAR-tunz) duct	submandibular duct
Wharton's (HWAR-tunz) jelly	mucous connective tissue
Wormian (WER-mē-an) bone	sutural bone

Figure Credits

1.1 Courtesy of Olympus America, Inc.

3.5 (a)–(f) © Carolina Biological Supply/Photo-Take NYC.

4.1 (a) Biophoto Associates/Photo Researchers, Inc. (b) M. I. Walker/Photo Researchers, Inc. (c) Biophoto Associates/Photo Researchers, Inc. (d) G.W. Willis/Biological Photo Service (e) © Biophoto Associates/Science Source/Photo Researchers, Inc. (f), (g) Ed Reschke.

4.3 (a) (b) Biophoto Associates/Photo Researchers, Inc. (c) Robert Brons/Biological Photo Service (d) Biophoto Associates/Photo Researchers, Inc. (e) Ed Reschke (f) Bruce Iverson/Visuals Unlimited (g) Fred Hossler/Visuals Unlimited (h) Frederick C. Skvara (i) Chuck Brown/Photo Researchers, Inc.

5.2 Reproduced by permission from R. G. Kessel and R. H. Kardon, *Tissues and Organs: A Text-Atlas of Scanning Electron Microscopy*, W. H. Freeman, 1979.

6.3 Biophoto Associates/Photo Researchers, Inc.

8.2 (a), (b), (c), (e) Copyright © 1983 by Gerard Tortora, Courtesy of Matt Iacobino and Lynne Borghesi (d) Courtesy of Evan J. Colella (f), (g), (h) Courtesy of Matt Iacobino (i), (j) © 1991 Evan J. Colella.

8.4 © by John Eads.

9.1 James R. Smail and Russell A. Whitehead, Macalester College.

9.3 (a) Alfred Owczarzak/Biological Photo Service (b) G. W. Willis/Biological Photo Service.

9.4 (a) Johanes Rhodin, Don Fawcett/Visuals Unlimited (b) Don Fawcett/Visuals Unlimited.

9.7 Physiograph is a registered trademark of Narco BioSystems Division of International Biomedical, Inc.

11.2–11.8 Biomedical Graphics Department, University of Minnesota Hospitals.

11.9 (a) Biomedical Graphics Department, University of Minnesota Hospitals (b) O. Richard Johnson.

11.10–11.12 Biomedical Graphics Department, University of Minnesota Hospitals.

12.1 © Carolina Biological Supply/PhotoTake NYC.

13.18 By permission of Tektronix, Inc.

14.6 By Andrew Kuntzman, Wright State University.

14.7 (b) © Carolina Biological Supply/PhotoTake NYC.

14.10 Biomedical Graphics Department, University of Minnesota Hospitals.

14.18 (a) John Cunningham/Visuals Unlimited. (b) Biophoto Associates/Photo Researchers, Inc.

14.19 Copyright © 1983 by Gerard Tortora, Courtesy of James Borghesi.

15.2 By James R. Smail and Russell A. Whitehead, Macalester College.

15.3 By James R. Smail and Russell A. Whitehead, Macalester College.

15.4 © G. W. Willis/Biological Photo Service.

15.5 © Don W. Fawcett/Visuals Unlimited.

15.6 By Andrew Kuntzman, Wright State University.

16.3 Becton-Dickinson, Division of Becton, Dickinson and Company.

16.4 Modified from *RBC Determination for Manual Methods and WBC Determination for Manual Methods*. Becton-Dickinson, Division of Becton, Dickinson and Company.

16.6, 16.7 Photographs courtesy of Fisher Scientific.

16.8 Photograph courtesy of Lenni Patti.

16.9 Courtesy of Leica, Inc. Deerfield, IL.

17.2 Courtesy of Chihiro Yokochi, MD and Johannes W. Rohen, MD from the book *Photographic Anatomy of the Human Body*, Igaku-Shoin Medical Publishers, New York, NY; 1978.

17.3 (b) John Eads.

18.1 (a) © Martin Rotker/PhotoTake NYC (b) © Carolina Biological Supply/PhotoTake NYC (c) © CNRI / PhotoTake NYC.

Index

A (anisotropic) band, 136
Abdominal aorta, 412, 413
Abdominal cavity, 19
Abdominopelvic cavity, 19
Abdominopelvic quadrants, 21
Abdominopelvic regions, 19–21
ABO blood grouping, 386–388
Accessory sex gland, 599–600
Accessory structures, of digestion, 519
Accommodation, 328
Acetest® reagent test for ketone bodies, 575
Acetylcholine (ACh), 139
Achilles tendon. See Calcaneal tendon
Acid, 583
Acid-base balance, 584–587
Acid-base imbalance, 588–589
Acidosis
    metabolic, 588
    physiological effects, 588
    respiratory, 588
Acini, 526
Acrosome, 596
Active transport, 33, 38–39
    primary, 38
    secondary, 39
Adam's apple. See Thyroid cartilage
Adaptation, 308
Adenohypophysis. See Anterior pituitary gland
Adipocyte, 58, 63
Adipose tissue, 63
Adrenal cortex, 356
Adrenal medulla, 356–357
Adrenal (suprarenal) gland, 355–357
Adrenergic fiber, 297
Adventitia, 524, 562–563
Afferent neuron. See Sensory neuron
Afterimage, 308
Agglutination, 386
Albino, 73
Albumin, 570
Albuminuria, 571
Alimentary canal. See Gastrointestinal (GI) tract
Alkalosis
    metabolic, 588–589
    physiological effects, 588
    respiratory, 588
Allantois, 631
Albutest® reagent test for protein, 574
Allele, 637
All-or-none principle, 140
Alpha cell, 357, 526
Alveolar–capillary (respiratory) membrane, 492
Alveolar duct, 491
Alveolar macrophage, 492
Alveolar pressure, 498

Alveolar sac, 491
Alveolus, 491, 608
Amnion, 627, 630
Amniotic cavity, 629
Amphiarthrosis, 123–124
Ampulla of Vater. See Hepatopancreatic ampulla
Anal canal, 531
Anal triangle, 176
Anaphase, 43
Antagonist, 157
Anatomical dead space, 501
Anatomy
    subdivisions of, 9
Anterior cavity of eyeball, 324
Anterior root of spinal nerve, 260
Anterior triangle, of neck, 170, 226
Anterolateral (spinothalamic) pathways, 269, 314
Anus, 531
Aorta, 396, 409
Aortic semilunar valve, 405
Apocrine (sweat) gland, 58, 75
Aponeurosis, 160
Appendix. See Vermiform appendix
Aqueous humor, 324
Arachnoid, 259
Arbor vitae, 280
Arch of aorta, 411
Areola, 608
Areolar connective tissue, 59, 63
Arrector pili muscle, 74
Arteries
    abdominal aorta, 412, 413
    afferent arteriole, 561
    anterior cerebral, 414
    anterior communicating, 414
    anterior interventricular, 399
    anterior tibial, 424
    arch of aorta, 411
    arcuate, 561
    ascending aorta, 411, 412
    axillary, 414
    basilar, 414
    brachial, 414
    brachiocephalic, 414
    bronchial, 418
    celiac, 419
    cerebral arterial circle, 414
    of cerebral circulation, 414
    circumflex, 399
    coats of, 409
    colic, 419
    common carotid, 414
    common hepatic, 419
    common iliac, 424
    coronary, 396, 399, 412
    deep palmar arch, 414
    definition of, 409
    digital, 414
    dorsalis pedis, 424
    ductus arteriosus, 438
    efferent arteriole, 562

esophageal, 418
external carotid, 414
external iliac, 424
femoral, 424
of fetal circulation, 437–439
gastric, 419
gastroduodenal, 419
gastroepiploic, 419
gonadals, 419
hepatic, 419
histology of, 409
ileal, 419
ileocolic, 419
inferior mesenteric, 419
inferior pancreaticoduodenal, 419
inferior phrenic, 419
interlobar, 561
interlobular, 561
internal carotid, 414
internal iliac, 424
jejunal, 419
lateral plantar, 424
lumbar, 419
marginal, 396
medial plantar, 424
median sacral, 419, 445
mediastinal, 418
ovarian, 419
pancreatic, 419
pericardial, 418
peroneal, 424
plantar arch, 424
popliteal, 424
posterior cerebral, 414
posterior communicating, 414
posterior intercostal, 418
posterior interventricular, 396
posterior tibial, 424
pulmonary, 436
of pulmonary circulation, 436–437
pulmonary trunk, 436
radial, 414
renal, 419, 561
short gastric, 419
sigmoid, 419
splenic, 419
subclavian, 414
subcostal, 418
superficial palmar arch, 414
superior mesenteric, 419
superior phrenic, 418
superior rectal, 419
suprarenal, 419
of systemic circulation, 409, 411–412
testicular, 419
thoracic aorta, 412, 413, 418
ulnar, 414
umbilical, 437
vertebral, 414
Arteriole, 409
Arthrology, 123
Articular capsule, 124

Articular cartilage, 84, 124
Articulations
    amphiarthrosis, 123
    ball-and-socket, 127
    cartilaginous, 123
    condyloid, 119
    definition of, 123
    diarthrosis, 123, 124–127
    fibrous, 124
    functional classification of, 123
    gliding, 125
    hinge, 125
    knee, 127–128
    pivot, 125, 127
    saddle, 127
    structural classification of, 123
    suture, 100, 123
    symphysis, 124
    synarthrosis, 123
    synchondrosis, 123
    syndesmosis, 124
    synovial, 124–127
Ascending aorta, 411
Ascending colon, 531
Ascending limb of loop of nephron, 560
Ascending tract, 266–269
Association fibers, 279
Association neuron, 246
Astigmatism, 327–328
Astrocyte, 248
Atresia, 624
Atrioventricular (AV) bundle, 453
Atrioventricular (AV) node, 453
Atrium, 396
Audiometer, 336
Auditory ossicle, 331
Auditory sensation, 331–339
Auditory tube, 331, 488
Auricle, of heart, 396
Autonomic nervous system
    activities of, 296
    parasympathetic division, 297
    sympathetic division, 296
Autosome, 621, 638
Axolemma, 246
Axon, 137, 246
Axon collateral, 246
Axon hillock, 246
Axon terminal, 137, 246
Axoplasm, 246

B cell, 361
Backbone. See Vertebral column
Ball-and-socket joint, 127
Barr body, 638
Bartholin's gland. See Greater vestibular gland
Basal ganglia, 279–280
Base, 583
Basement membrane, 51
Basilar membrane, 332
Basophil, 382
Benedict's test, 573
Beta cell, 357, 526

Bicuspid (mitral) valve, 396
Bile, 547
Bile canaliculi, 527
Blastocoel, 626
Blastocyst, 626
Blastomere, 626
Blind spot. *See* Optic disc
Blood
    formed elements, 369
    groupings of, 386–389
    origin, 369
    plasma, 369
    tests, 370–389
Blood flow in frog, 467–468
Blood grouping (typing)
    ABO, 386–388
    Rh, 388–389
Blood plasma, 369–371
Blood pressure, 466–467
Blood vessels
    arteries, 409, 414–424
    arterioles, 409
    capillaries, 409
    cat, 440–449
    definition of, 409
    of fetal circulation, 437–439
    of hepatic portal circulation, 436
    human, 466
    pressure of, 466
    of pulmonary circulation, 436–437
    of pulse, 466
    of systemic circulation, 409–435
    veins, 409, 426–436
    venules, 409
Body cavity, 19
Body temperature, homeostasis of, 76–78
Bolus, 539
Bone growth, 85–86
    in length, 85–86
    in thickness, 86
Bone marrow
    red, 81
    yellow, 81
Bones
    atlas, 102
    auditory ossicles, 93
    axis, 103
    calcaneus, 114
    capitate, 111
    carpus, 110
    cervical vertebrae, 102–103
    chemistry of, 84
    clavicle, 108
    coccyx, 106
    cranial, 93–100
    cuboid, 114
    ethmoid, 93
    facial, 93
    femur, 114
    fibula, 114
    fontanels of, 100
    foramina of, 100
    formation of, 85
    fractures of, 87, 88
    frontal, 93
    growth of, 85–86
    hamate, 111
    hip, 113–114
    histology of, 83
    humerus, 110
    hyoid, 93
    incus, 331
    intermediate cuneiform, 114
    lacrimal, 93
    lateral cuneiform, 114
    lower limb, 114, 118
    lumbar vertebrae, 103
    lunate, 111
    malleus, 331

mandible, 95
    markings of, 89–90
    maxillae, 93
    medial cuneiform, 114
    metacarpus, 111
    metatarsus, 118
    nasal, 93
    navicular, of foot, 114
    occipital, 93
    organization into skeleton, 93
    palatine, 93
    parietal, 93
    patella, 114
    pectoral (shoulder) girdle, 108
    pelvic (hip) girdle, 113–114
    phalanges, of fingers, 111
    phalanges, of toes, 118
    pisiform, 111
    radius, 110
    ribs, 106–108
    sacrum, 106
    scaphoid, 111
    scapula, 108
    sesamoid, 89
    skull, 93–100
    sphenoid, 93
    stapes, 331
    sternum, 106–108
    sutural, 89
    sutures of, 100
    talus, 114
    tarsus, 114
    temporal, 93
    thoracic vertebrae, 103
    thorax, 106
    tibia, 114
    trapezium, 111
    trapezoid, 111
    triquetral, 111
    types of, 87, 89
    ulna, 110
    upper limb, 110–112
    vertebral column, 102
    vomer, 93
    zygomatic, 93
Bone surface markings, 89–90
Bone tissue, 81–90
Bowman's capsule. *See* Glomerular capsule
Brain
    basal ganglia, 279–280
    brain stem, 273
    cat, 288–292
    cerebellum, 280–281
    cerebrospinal fluid, 273–274
    cerebrum, 277–279
    cranial meninges, 273
    cranial nerves and, 284–288
    diencephalon, 273
    human, 273–281
    hypothalamus, 277
    lobes of, 277
    medulla oblongata, 274, 276
    midbrain, 276–277
    parts of, 273
    pons, 276
    sheep, 295–296
    thalamus, 277
    ventricles of, 273
Brain macrophage. *See* Microglial cell
Brain sand, 360
Brain stem, 273
Breast. *See* Mammary gland
Broad ligament, 359, 604
Bronchi
    alveolar duct, 491
    bronchioles, 490
    primary, 490
    secondary (lobar), 490

terminal, 490
    tertiary (segmental), 490
Bronchial tree, 490, 491
Bronchiole, 490
Bronchogram, 490
Bronchopulmonary segment, 491
Brownian movement, 34
Brunner's gland. *See* Duodenal gland
Buffer
    carbonic acid–bicarbonate, 585
    phosphate, 585–586
    protein, 586
Bulb, of hair follicle, 74
Bulbourethral gland, 600
Bundle branch, 453
Bundle of His. *See* Atrioventricular (AV) bundle
Bursae, 125

Calcaneal tendon, 201
Calculus, renal, 571
Calyx, of kidney, 558
Canaliculi, 81
Canal of Schlemm. *See* Scleral venous sinus
Capacitation, 626
Capillaries, 409
Cardiac cycle, 459–464
Cardiac muscle tissue, 136–137
Cardiac notch, 491
Cardinal ligament, 604
Cardiovascular system
    arteries, 414–424
    blood, 369–389
    blood pressure, 466–467
    capillaries, 409
    cat, 401–402, 440–449
    circulatory routes, 409–440
    definition of, 369
    dissection of, 401–405, 440–449
    heart, 395–405
    physiology of, 453–468
    sheep, 402, 405
    turtle, 461–464
    veins, 426–436
Carotene, 73
Cartilage, 64–65
Cartilaginous joint, 123
Cast, 571
Cat
    cardiovascular system of, 440–449
    digestive system of, 532–538
    lymphatic system of, 481–482
    muscular system of, 202–218
    nervous system of, 288–295
    reproductive systems of, 612–616
    respiratory system of, 492, 495–497
    skeletal system of, 118, 120
    skinning of, 202, 205
    urinary system of, 564–566
Cauda equina, 260
Cecum, 531
Cell
    definition of, 31
    diversity of, 33
    division of, 40–43
    extracellular materials of, 40
    generalized animal, 32
    inclusions of, 39
    movement of substances into, 33–38
    parts of, 31–33
Cell body, of neuron, 246
Cell division, 40–43
Cell inclusions, 39

Cell membrane. *See* Plasma membrane
Cell physiology, 10
Cementum, 521
Central canal, 81, 260
Central fovea, 324
Central nervous system (CNS), 245, 259–296
Central sulcus, 277
Centromere, 41
Centrosome, 32
Cerebellar peduncle, 280
Cerebellum
    determining functions of, 281
    structure of, 280
Cerebral arterial circle, 414
Cerebral circulation, 414
Cerebral cortex, 277, 279
Cerebral nuclei. *See* Basal ganglia
Cerebral peduncle, 277
Cerebrospinal fluid (CSF), 273–274
Cerebrum, 268, 277
Ceruminous gland, 76, 331
Cervical enlargement, 260
Cervical vertebrae, 102–102
Cervix, of uterus, 604
Chemical digestion, 519, 545–548
Chemistry of bone, 84
Chemstrip® testing of urine, 571–572
Cholinergic fiber, 297
Chondrocyte, 64
Chordae tendineae, 396
Chorion, 631
Chorionic villi, 632
Choroid, 322
Choroid plexus, 274
Chromaffin cell, 356
Chromatid, 41
Chromatin, 41
Chromatophilic substance, 246
Chromosome, 43
Chromosome number, 621
Chyme, 539
Cilia, 33
Ciliary body, 322
Circle of Willis. *See* Cerebral arterial circle
Circular fold, 530
Circulatory routes, 409–440
Cleavage, 626
Cleavage furrow, 43
Clinitest® reagent test for glucose, 574
Clitoris, 606
Coccyx, 106
Cochlea, 332
Cochlear duct, 332
Codominance, 643
Collagen fiber, 59
Collecting duct, 560
Collins respirometer, 505–508
Column, of spinal cord, 260
Commissural fibers, 279
Common bile duct, 527
Common hepatic duct, 527
Compact (dense) bone, 81
Compound light microscope, 1–6
Conduction myofiber, 453
Conduction system, of heart, 453–454
Condyloid joint, 119
Cone, 324
Conjunctiva, 321
    bulbar, 321
    palpebral, 321
Connective tissue, 58–65
Contractility, 135
Contraction, muscular, 137–149
Conus medullaris, 260

Convergence, 249, 329
Convolution. *See* Gyrus
Cornea, 322
Corona, 600
Corona radiata, 624
Coronary sinus, 396, 399
Coronary sulcus, 396
Corpora cavernosa penis, 600
Corpora quadrigemina, 277
Corpus hemorrhagicum, 604, 611
Corpuscle of touch, 72, 309
Corpus albicans, 359, 603
Corpus luteum, 359, 603
Corpus spongiosum penis, 600
Covering and lining epithelium,
    51–56
Cowper's gland. *See*
    Bulbourethral gland
Cranial cavity, 19
Cranial nerves
    cat, 292
    human, 284–288
    tests for function of,
        284–288
Crenation, 36
Crossing over, 622
Crypt of Lieberkühn. *See* Intestinal
    gland
Crystals, in urine, 576
Cupula, 337
Cystic duct, 527
Cytokinesis, 41, 43
Cytology, 9, 31
Cytoplasm, 31
Cytoskeleton, 32

Deafness
    conduction, 335
    sensorineural, 335
Decidua
    basalis, 632
    capsularis, 632
    parietalis, 632
Deciduous teeth, 522
Decussation of pyramids,
    276
Deflection waves, 453
Deglutition, 538–539
Delta cell, 357, 526
Dendrite, 246
Dense connective tissue, 63–64
Dense irregular connective tissue,
    64
Dense regular connective tissue,
    63
Dentin, 521
Dentition, 522, 524, 525
Deoxyribonucleic acid (DNA), 41
Dermis, 71–73
Descending colon, 531
Descending limb of loop of Henle,
    560
Descending tract, 266, 269–270
Detrusor muscle, 563
Development
    definition of, 621
    embryonic period, 624–632
    fertilization, 624–626
    fetal period, 632, 633
    implantation, 627
Developmental anatomy, 9,
    621–633
Dialysis, 38
Diaphragm
    pelvic, 174
    respiratory, 173
    urogenital, 176
Diaphysis, 84
Diarthrosis, 123, 124–127
Diastasis, 459
Diastole, 459

Diastolic blood pressure, 466
Diencephalon, 273
Differential white blood cell
    count, 383–385
Digestion
    chemistry of, 545–548
    definition of, 519
    in mouth, 519, 546
    in small intestine, 529–530
    in stomach, 526
    of triglycerides, 547
    of protein, 548
    of starch, 546
Digestive system
    cat, 532–538
    chemistry of, 545–548
    definition of, 519
    dissection of, 532–558
    general histology, 519
    human, 519–532
    organization of GI tract, 519
    organs of, 519–532
    peritoneum of, 519
    physiology of, 541–545
Dihybrid cross, 642
Diploid cell, 621, 624
Directional terms, 15, 18
Direct (pyramidal) pathways,
    340
Dissection
    cat blood vessels, 440–449
    cat brain, 288–292
    cat cranial nerves, 292
    cat digestive system, 532–538
    cat heart, 401–402
    cat lymphatic system, 481
    cat muscular system, 202–218
    cat reproductive systems,
        612–616
    cat respiratory system, 492,
        495–497
    cat spinal cord, 293, 295
    cat spinal nerves, 295
    cat urinary system, 564–566
    precautions related to, xi
    pregnant pig uterus, 616
    sheep brain, 295–296
    sheep heart, 402, 405
    sheep kidney, 567
    sheep pluck, 497
    vertebrate eye, 324–326
    white rat, 21–24
Distal convoluted tubule, 560
Divergence, 249
Dominant gene, 637
Dorsal body cavity, 19
Duct of Wirsung. *See* Pancreatic
    duct
Ductus arteriosus, 438
Ductus (vas) deferens, 598–599
Ductus epididymis, 598, 599
Ductus venosus, 438
Duodenal gland, 530
Duodenum, 529
Dura mater, 259, 273

Ear
    and equilibrium, 337–339
    external, 331
    and hearing, 331–337
    internal, 331–334
    middle, 331
    otoscopy, 334
    surface anatomy of, 336
    tests involving, 334–339
Eardrum, 331
Eccrine (sweat) glands, 75
ECG. *See* Electrocardiogram
Ectoderm, 629
Edema, 481
EDV. *See* End-diastolic volume

EEG. *See* Electroencephalogram
Effector, 251
Efferent duct, 598
Efferent neuron. *See* Motor neuron
Ejaculatory duct, 598
Elastic connective tissue, 64
Elastic fiber, 59
Elasticity, 135
Electrocardiogram
    definition of, 453
    deflection waves of, 453
    recording of, 455–456
Electrocardiograph, 453, 457
Electroencephalogram
    preparation of subject, 282, 284
    recording, 284
    waves of, 281–282
Electromyography, 148–149
Electron micrograph, 6
Electron microscope, 6
Embryo, 624
Embryonic membranes, 630, 631
Embryonic period, 624–632
Emmetropic eye, 327
Enamel, 521
End-diastolic volume (EDV), 459
Endocardium, 395
Endochondral ossification, 85
Endocrine glands, 56, 351
Endocrine system
    definition of, 351
    glands of, 339–349
    physiology of, 361–363
Endocrinology, 10
Endocytosis, 39
Endoderm, 629
Endolymph, 331
Endometrium, 605–606
Endomysium, 136, 160
Endoneurium, 262
Endoplasmic reticulum (ER), 31
End organ of Ruffini. *See* Type II
    cutaneous mechanoreceptor
Endosteum, 82
Endothelial-capsular membrane,
    558
Endothelium, 52
End-plate potential, 139
End-systolic volume, 460
Enteroendocrine cell, 526
Enzymes
    definition of, 545
    pancreatic lipase, 547
    pepsin, 548
    salivary amylase, 546
Eosinophil, 382
Ependymal cells, 248
Epicardium, 395
Epidermis
    definition of, 71
    layers of, 71
    pigments of, 73
Epididymis, 598
Epiglottis, 488
Epimysium, 135, 159
Epineurium, 262
Epiphyseal line, 86
Epiphyseal plate, 85–86
Epiphysis, 84
Epithelial tissue, 51–58
Equatorial division (meiosis II),
    622, 624
Equilibrium, 337–339
Erythroblastosis fetalis. *See*
    Hemolytic disease of new-
    born
Erythrocyte, 371
Erythropoiesis, 369
Esophagus, 524, 525
Eustachian tube. *See* Auditory
    tube

Excitability, 135
Excitation-contraction coupling,
    137
Exercise physiology, 10
Exocrine gland
    definition of, 57, 351
    functional classification of, 58
    structural classification of, 57
Expiration, 498–499
Expiratory reserve volume, 502
Extensibility, 135
Extensor retinaculum, 183
External auditory canal, 331
External ear, 331
External nares, 485
External respiration, 485
External root sheath, 74
Extracellular materials, 40
Extraembryonic coelom, 630
Eyeball
    accessory structures of, 321
    dissection of, 324–326
    extrinsic muscles of, 164
    image formation, 321–330
    ophthalmoscopic examination
        of, 326
    sensory pathway, 330
    structure of, 321–324
    surface anatomy of, 325
    and vision, 321–330

Facilitated diffusion, 34–35
Falciform ligament, 527
Fallopian tube. *See* Uterine tube
Fascia lata, 192
Fascicle, 158, 262
Fat. *See* Triglyceride
Fatty acid, 547
Fauces, 520
Feedback system, 13
    negative, 13
    positive, 13–14
Female pronucleus, 626
Female reproductive cycle,
    608–612
Female reproductive system,
    601–612
Fertilization, 624–626
Fetal circulation, 437–439
Fetal period, 632, 633
Fetus, 632
Fibroblast, 58
Fibrous joints, 123
Fibrous tunic, 322
Filtrate, glomerular, 558
Filtration, 37–38
Filtration slit, 558
Filum terminale, 260
Fimbriae, 603
Final common pathway, 270
First-order neuron, 307
Fissure
    of brain, 277
    of lung, 491
Fixator, 157
Flagellum, 33
Flexor retinaculum, 183
Follicular cell, 354, 602
Fontanel
    anterior, 100
    anterolateral, 100
    definition of, 100
    posterior, 100
    posterolateral, 100
Foramen ovale, of heart, 396, 438
Foramina, of skull, 100
Forced expiratory volume (FEV_T),
    507
Foreskin. *See* Prepuce
Formed elements, of blood, 369
Fossa ovalis, 396

Fracture, 87, 88
Free nerve ending, 310
Frog
  isolated muscle preparation of, 143–144
  muscle experiments using, 142–148
  nerve muscle preparation of, 144–145
  observing lung tissue of, 497–498
  peripheral blood flow in, 467–468
  pithing of, 142–143
  spinal reflexes of, 251–254
Functional residual capacity, 502
Fundus
  of stomach, 526
  of uterus, 604

Gallbladder, 527, 528
Gamete, 621
Gastric gland, 526
Gastrointestinal (GI) tract
  general histology of, 519
  organs of, 519–532
  peritoneum of, 519
  physiology of, 519–532
Gastrulation, 627
Generalized animal cell, 32
General senses
  pain, 310
  proprioceptive, 310–311
  tactile, 309–310
  tests for determining, 316–317, 319, 321, 324–330, 334–339
  thermoreceptive, 310
Genetics, 637–650
Genotype, 637
Germinal epithelium, 359, 602
Gingivae, 521
Gland of Zeis. *See* Sebaceous ciliary gland
Glandular epithelium, 56–58
Glans penis, 600–601
Gliding joint, 125
Glomerular capsule, 558
Glomerular filtration, 558
Glomerulus, 558
Glottis, 489
Glucose, in urine, 571
Glucosuria, 571
Goblet cell, 55, 529
Goldfish, renal tubules of, 569
Golgi complex, 31
Golgi tendon organ. *See* Tendon organ
Gomphosis, 123
Gonad, 595
Graafian follicle. *See* Mature follicle
Granular layer, 74
Granulosa cells, 359, 602, 624
Gray matter, 260
Greater vestibular gland, 607
Gross anatomy, 9
Ground substance, 58
Group actions of skeletal muscles, 157–158
Gums. *See* Gingivae
Gustatory receptor, 317–319
Gustatory sensation, 317–321
  tests for, 319, 321
Gyrus, 277

H zone, 136
Hair
  root, 73
  shaft, 73
Hair cell, 337
Hair follicle, 73

Hair root plexus, 74, 309
Haploid cell, 621, 624
Hard palate, 519, 520
Haustra, 532
Haustral churning, 540
Haversian canal. *See* Central canal
Haversian system. *See* Osteon
Heart
  atria of, 396
  blood supply of, 396, 399
  cardiac cycle of, 459–464
  cat, 401–402
  chambers, 396
  conduction system of, 453–454
  electrocardiogram of, 453–459
  great vessels of, 396
  human, 395–400
  pericardium of, 395
  sounds of, 464–465
  turtle, 461–464
  valves of, 396
  ventricles of, 396
Heartbeat, 459–461
Heart murmur, 464–465
Heart sounds, listening to, 464–465
Helper T cell, 71
Hematocrit, 376–377
Hematopoietic stem cell, 369
Hematuria, 571
Hemisphere
  cerebellar, 280
  cerebral, 277
Hemocytoblast. *See* Hematopoietic stem cell
Hemocytometer, filling of, 372
Hemoglobin, 371
  determination of, 378–381
Hemoglobinometer, 379
Hemolysis, 36, 386
Hemolytic disease of newborn, 389
Hemophilia, inheritance of, 640–641
Hemopoiesis, 81, 369
Henle's layer. *See* Pallid layer
Hepatic duct, 527
Hepatic portal circulation, 436
Hepatocyte, 527
Hepatopancreatic ampulla, 527
Heterozygous, 637
Hilus, 359, 465, 478, 491, 601
Hinge joint, 125
Histology, 9, 51
Holocrine gland, 58
Homeostasis, 13
Homologous chromosome, 621, 637
Homozygous, 637
Hormones
  adrenocorticotropic (ACTH), 353
  aldosterone, 356
  androgens, 356
  antidiuretic (ADH), 353
  atrial natriuretic peptide (ANP), 361
  calcitonin (CT), 355
  calcitriol, 361
  cholecystokinin (CCK), 361
  corticosterone, 356
  cortisol, 356
  cortisone, 356
  definition of, 351–361
  epinephrine, 357
  erythropoietin, 361
  estrogens, 359
  gastric inhibitory peptide (GIP), 361
  gastrin, 361
  glucagon, 357
  glucocorticoids, 356

gonadotropic, 353
human chorionic gonadotropin (HCG), 361
human chorionic somatomam-motropin (HCS), 361
human growth hormone (hGH), 353
inhibin, 359
insulin, 358
intestinal gastrin, 361
luteinizing (LH), 353
melanocyte-stimulating (MSH), 353
melatonin, 360
mineralocorticoids, 356
norepinephrine (NE), 357
oxytocin (OT), 353
pancreatic polypeptide, 358
parathyroid (PTH), 355
progesterone, 359
prolactin (PRL), 353
relaxin, 359
secretin, 361
somatostatin, 357
testosterone, 359
thymic humoral factor (THF), 361
thymosin, 361
thyroid, 353
thyroid-stimulating (TSH), 353
thyroxine (T_4), 354
triiodothyronine (T_3), 354
tropic, 353
vitamin D, 361
Horn, of spinal cord, 260
Huxley's layer. *See* Granular layer
Hymen, 606, 607
Hypertonic solution, 36
Hypophysis. *See* Pituitary gland
Hypothalamus, 277
Hypotonic solution, 36

I (isotropic) band, 136
Ictotest® for bilirubin, 575
Ileocecal sphincter (valve), 531
Ileum, 529
Iliotibial tract, 192
Immunology, 10
Implantation, of fertilized egg, 627, 628
Independent assortment, 642
Indirect (extrapyramidal) pathways, 340
Inferior colliculi, 277
Inferior extensor retinaculum, 199
Inferior vena cava, 396
Infundibulum
  of pituitary, 351
  of uterine tube, 603
Inguinal canal, 171
Inguinal ligament, 171
Inheritance
  of red-green color blindness, 640
  definition of, 637
  genotype, 637
  of hemophilia, 640–641
  phenotype, 637
  of PKU, 637
  PTC, 644
  of sex, 638–640
  sex-linked, 640
Initial segment, 246
Inner cell mass, 626
Inspiration, 498
Inspiratory capacity, 502
Inspiratory reserve volume, 502
Insula, 277
Integrating center, 251
Integumentary system
  definition of, 71
  glands, 75–76

hair, 73–74
nails, 76
skin, 71–73
Intercalated disc, 136–137
Intermediate mass, 277
Internal ear, 331–334
Internal nares, 486, 488
Internal respiration, 485
Internal root sheath, 74
Interphase, 41
Interstitial cell of Leydig. *See* Interstitial endocrinocyte
Interstitial endocrinocyte, 358, 596
Intestinal gland, 529
Intestinal smooth muscle, physiology of, 541–545
Intramembranous ossification, 85
Intrapleural pressure, 498
Ionization, 583
Iris, 322
Island of Reil. *See* Insula
Islets of Langerhans. *See* Pancreatic islets
Isoantibody, 386
Isoantigen, 386
Isolated muscle preparation, 143–144
Isotonic solution, 36
Isovolumetric contraction, 459
Isthmus
  of thyroid gland, 354
  of uterine tube, 603
  of uterus, 603

Jejunum, 529
Joint kinesthetic receptor, 310
Joints. *See* Articulations
Juxtaglomerular apparatus, 562
Juxtaglomerular cell, 562

Karyotype, 643
Keratin, 55, 71
Keratinocyte, 71
Ketone bodies, 571
Ketosis, 571
Kidney, 557–558, 559
Kidney stone. *See* Renal calculus
Kinesthetic sense. *See* Proprioception
Kinetochore, 41
Knee jerk. *See* Patellar reflex
Knee joint, 127–128
Korotkoff sound, 467
Kupffer cell. *See* Stellate reticu-loendothelial cell

Labial frenulum, 520
Labia majora, 606
Labia minora, 606
Laboratory safety precautions, xi–xiii
Labyrinth. *See* Internal ear
Lacrimal apparatus, 321
Lacteal, 530
Lactiferous duct, 608
Lactiferous sinus, 608
Lacuna, 64, 81
Lamella, 81
Lamellated corpuscle, 72, 310
Langerhan cell, 71
Large intestine, 531–532
Laryngopharynx, 488
Larynx, 488–489, 490
Law of segregation, 641
Lens, 324
Lesser vestibular gland, 607
Leukocyte, 382
Levels of structural organization, 10
Life processes, 12–13

Ligaments
    accessory, 124
    anterior cruciate, 127
    extracapsular, 124
    falciform, 527
    fibular collateral, 127
    intracapsular, 125
    oblique popliteal, 127
    patellar, 127
    posterior cruciate, 127
    tibial collateral, 127
Linea alba, 171, 229
Lingual frenulum, 521
Lipase, 547
Litmus paper, 583
Liver, 527
Locus, 637
Longitudinal fissure, 277
Loop of Henle. *See* Loop of the nephron
Loop of the nephron, 558, 560
Loose connective tissue, 59, 63
Lumbar enlargement, 260
Lumbar vertebrae, 102
Lungs, 490, 492–493
Lymphangiogram, 478
Lymphangiography, 478
Lymphatic follicle, 478
Lymphatic system
    cat, 481, 482
    definition of, 477
    dissection of, 481
    human, 477–481
Lymphatic vessel, 478
Lymph capillary, 477
Lymph circulation, 479–480
Lymph node, 478
Lymphocyte, 360, 382
Lysosome, 31

M line, 136
Macrophage, 58
Macula, 337
Macula densa, 562
Macula lutea, 324
Male pronuculus, 626
Male reproductive system, 595–601, 602
Mammary duct, 608
Mammary gland, 607–608, 609
Mast cell, 58
Matrix
    of connective tissue, 58
    of hair follicle, 74
    of nail, 76
Mature connective tissue, 59
Mature follicle, 359, 602–603
Mean electrical axis, 456, 458–459
Meatuses, of nasal cavity, 486
Mechanical digestion, 519
Mechanoreceptor, 308
Medial lemniscus, 269
Mediastinum, 19
Medulla oblongata, 274, 276
Medullary (marrow) cavity, 82
Meibomian gland. *See* Tarsal gland
Meiosis, 41, 621
Meissner's corpuscle. *See* Corpuscle of touch
Melanin, 73
Melanocyte, 71, 73
Membrane
    mucous, 65
    pleural, 490
    serous, 65–66
    synovial, 66
Mendelian laws of inheritance, 641–642
Meninges, 259, 273

Meniscus, 125, 128
Menstrual cycle, 608–612
    menstruation, 611
    postovulatory phase, 610, 612
    preovulatory phase, 610, 611
Merkel cell, 71
Merkel disc. *See* Tactile disc
Merocrine gland, 58
Mesoderm, 630
Mesothelium, 52
Mesovarium, 359, 601
Metaphase, 41, 43
Metaphase plate, 43
Metaphysis, 84
Microglial cell, 248
Microscope
    compound light, 1–6
    electron, 6
    magnification of, 1, 6
    parfocal, 5
    parts of, 1–3
    procedure for using, 4–6
    rules when using, 3–4
    setting up, 4
Microscopic examination of urine, 575–576
Microtubule, 41
Microvillus, 530
Micturition, 563
Midbrain, 276–277
Middle ear, 331
Mitochondrion, 31
Mitosis, 41–43
Mitotic spindle, 41
Mitral valve. *See* Bicuspid valve
Modality, 308
Modiolus, 332
Monocyte, 382
Monohybrid cross, 642
Mons pubis, 606
Morula, 626
Motor end plate, 138
Motor neuron, 137, 251
Motor pathways, 340
Mouth. *See* Oral cavity
Mucous cell, 526
Mucous membrane, 65
Muller's Doctrine of Specific Energizes, 308
Multicellular exocrine gland, 57
Multiple allele, 643
Muscarinic receptor, 299
Muscle contraction
    all-or-none principle, 140
    biochemistry of, 148
    laboratory tests on, 140–148
Muscle fatigue, 146
Muscles. *See* Skeletal muscles
Muscle spindle, 310
Muscle tissue
    all-or-none principle of, 140
    cardiac, 136–137
    characteristics of, 135
    connective tissue components of, 158–160
    contraction of, 137–149
    fatigue of, 146
    functions of, 135
    histology of, 135–140
    kinds of, 135
    physiology of, 135–140
    polygraph and, 141–142
    skeletal, 135–149
    sliding-filament theory of, 140
    smooth (visceral), 135
    types of, 135
Muscle twitch, 146
Muscular system
    cat, 202–218
    connective tissue components of, 158–160

contraction and, 137–149
    dissection of, 202–218
    functions of, 135
    histology of, 135–140
    human, 157–218
    naming skeletal muscles, 158
    physiology of, 135–140
Myelin sheath, 246
Myocardium, 395
Myofibril, 135
Myofilament, 135
Myogram, 141
Myograph transducer, 141
Myometrium, 606
Myoneural junction. *See* Neuromuscular junction

Nails, 76
Naming skeletal muscles, 158
Nasal cavity, 486
Nasal conchae, 486
Nasopharynx, 488
Near-point accommodation, 328
Nephron, 558–560, 561, 562
Nerve impulse, 249–250
Nerve-muscle preparation, 144–145
Nervous system
    autonomic, 296–299
    brain, 273–281
    cat, 288–295
    cranial nerves, 284–288
    dissection of, 288–295
    electroencephalogram and, 281–284
    histology, 246
    human, 259–288
    plexuses, 263, 265
    reaction times, 272–273
    reflexes and, 272
    sensory receptors and, 307–340
    spinal cord, 259–272
    spinal nerves, 259–272
Nervous tissue
    histology of, 246
    neuroglia, 248–249
    reflex arc and, 250–251
Nervous tunic, 322
Neurofibral node, 246
Neuroglia, 246
Neurohypophysis. *See* Posterior pituitary gland
Neurolemma, 246
Neurolemmocyte, 248–249
Neurology, 245
Neuromuscular junction, 138
Neuronal circuit, 249–250
Neurons
    association, 246
    classification of, 246
    first-order, 266
    lower motor, 270
    motor, 246
    second-order, 269
    sensory, 246
    structure of, 246
    third-order, 269
    upper-motor, 270
Neurophysiology, 10
Neurosecretory cell, 353
Neutrophil, 382
Nicotinic receptor, 297
Nipple, 608
Nissl body. *See* Chromatophilic substance
Nociceptor, 310
Node of Ranvier. *See* Neurofibral node
Nondisjunction, 638
Nose, 485–488
Nostril. *See* External nares

Nuclei, 137
Nucleus, of cell, 31
Nutrient, 545
Nystagmus, 338

Oil gland. *See* Sebaceous gland
Olfactory adaptation, 316–317
Olfactory receptor, 314–316
Olfactory sensation, 314–317
Oligodendrocyte, 248
Oogenesis, 623–624, 625
Oogonium, 624
Ophthalmoscope, 326
Ophthalmoscopic examination of eyeball, 326
Optic chiasm, 330
Optic disc, 324
Oral cavity, 519–521
Organ of Corti. *See* Spiral organ
Oropharynx, 488
Oscilloscope, use of, 282
Osmosis, 35
Osseous tissue. *See* Bone Tissue
Ossification, 85
    endochondral, 85
    intramembranous, 85
Osteoblast, 84
Osteocyte, 82
Osteology, 81
Osteon, 82
Osteoprogenitor cell, 84
Otolith, 337
Otolithic membrane, 337
Oval window, 331
Ovarian cycle, 608
Ovarian follicle, 602, 611
Ovarian ligament, 359, 601
Ovary, 359, 601–603, 604, 605
Ovulation, 611
Ovum, 624
Oxyhemoglobin saturation, 381
Oxyphil cell, 355

P wave, 454
Pacinian corpuscle. *See* Lamellated corpuscle
Pain sensation, 310
Pallid layer, 74
Pancreas, 357–358, 526–527, 528
Pancreatic duct, 527
Pancreatic islet, 357, 526
Pancreatic lipase, 547
Papilla
    dermal, 72
    duodenal, 527
    of hair, 74
    of kidney, 558
    of tongue, 319, 521
Papillary duct, 560
Papillary layer of dermis, 72
Papillary muscle, 396
Parafollicular cell, 354
Parallel after-discharge circuit, 250
Paranasal sinus, 100, 102
Parathyroid gland, 355
Paraurethral gland, 607
Parietal cell, 526
Parotid gland, 521
Pars intermedia, 353
Passive transport, 33–38
Pathological anatomy, 9
Pathophysiology, 10
Patellar ligament, 196
Patellar reflex, 272
Pedicel, 558
Pelvic cavity, 19
Pendular movement, 540
Penis, 600, 601

Pepsin, 548
Perforating canal, 82
Pericardial cavity, 19, 395
Pericardium, 395
Perichondrium, 64
Periodontal ligament, 521
Perilymph, 331
Perimetrium, 606
Perimysium, 135, 159
Perineum, 176
Periosteum, 84
Peripheral nervous system, 245, 260–265, 284–288
Peristalsis, 539–540
Peritoneum, 519
Peritubular capillary, 562
Permanent teeth, 522
Peroxisome, 31
Perspiration, 76
pH
    measuring, 583
    meter, 584
    paper, 584
    scale, 583
Phagocytosis, 39
Pharynx, 488
Phenotype, 637, 644–645
Phenylketonuria (PKU), inheritance of, 637, 643
Phenylthiocarbamide (PTC) and taste, 321, 644
Photomicrograph, 1
Physiology
    subdivisions of, 10
Pia mater, 259
Pig, pregnant uterus of, 616
Pineal gland, 360
Pinealocyte, 360
Pinocytosis, 39
Pithing procedure, 142–143
Pituicyte, 353
Pituitary gland, 351–354
Pivot joint, 125, 127
Placenta, 437, 631–632
Plane, of body, 18
Plasma membrane, 31
Plasma cell, 58
Platelet, 385
Pleural cavity, 19, 490
Pleural membrane, 490
Plexus
    brachial, 263
    cervical, 263
    lumbar, 263
    sacral, 265
Pneumograph, use of, 500–501
Podocyte, 558
Polar body, 624
Polycythemia, 373
Polygraph, use of, 141–142
Polysaccharide, 546
Pons, 276
Posterior column–medial lemniscus pathway, 269, 314
Posterior pituitary gland, 351
Posterior root ganglion, 260
Posterior root of spinal nerve, 260
Posterior triangle, of neck, 170, 226
Prepuce, 600, 606
Pressure sensation, 310
Primary bronchus, 490
Primary follicle, 359, 602
Primary germ layers, 627–630
Primary oocyte, 624
Primary spermatocyte, 358, 595, 622
Prime mover (agonist), 157
Primordial follicle, 359, 602

Principal cell, 355
Projection fibers, 279, 308
Pronucleus
    female, 626
    male, 626
Prophase, 41
Proprioceptive sensation, 310–311
Prostate gland, 600
Protein, 548
Proximal convoluted tubule, 558
Pseudopod, 39
Pseudostratified columnar epithelium, 56
PTC, inheritance of, 644
Pubic symphysis, 124
Pulmonary artery, 396
Pulmonary circulation, 436–437
Pulmonary ventilation, 498–499
Pulmonary semilunar valve, 396
Pulmonary trunk, 396
Pulmonary vein, 396
Pulp cavity, 521
Pulse, determination of, 465–466
Pulse pressure, 466
Punctate distribution, 309
Punnett square, 637–638
Pupil, 322
Purkinje fiber. See Conduction myofiber
Pyloric sphincter (valve), 526
Pylorus, of stomach, 526
Pyramid, of medulla oblongata, 276
Pyuria, 571

QRS wave (complex), 455
Quadrants, of abdominopelvic cavity, 21
Quadriceps tendon, 196

Radiographic anatomy, 9
Rapid ventricular filling, 459
Rat
    dissection of, 21–24
    endocrine physiology experiments, 361–363
    experiments on small intestine, 540–545
    observing gastrointestinal movements, 540–541
    standard metabolic determination in, 361–363
Reaction times, 272–273
Receptor
    alpha, 299
    auditory, 322
    beta, 299
    classifications, 308–309
    definition of, 251
    for equilibrium, 337
    for gustation, 317–319
    for olfaction, 314–316
    for pain, 310
    for pressure, 310
    for proprioception, 310–311
    for touch, 309–310
    for vibration, 310
    for vision, 321–330
    muscarinic, 299
    nicotinic, 297
    thermoreceptive, 310
Recessive gene, 637
Rectum, 531
Red blood cell count, 372–376
Red blood cells. See Erythrocytes
Red-green color blindness
    inheritance of, 640, 641
    testing for, 330
Reduction division (meiosis I), 622

Reflex
    Achilles, 272
    definition of, 250–251
    experiments, 251–254, 272
    patellar, 272
    plantar, 272
Reflex arc, 250–251
Reflex experiments, 251–254, 272
Refraction, 328
Regional anatomy, 9
Relaxation period, 459
Renal column, 558
Renal corpuscle, 558
Renal cortex, 557
Renal hilus, 557
Renal medulla, 557
Renal papilla, 558
Renal pelvis, 558
Renal plexus, 562
Renal pyramid, 558
Renal sinus, 557
Renal tubule, experiments with, 569–570
Reproductive cell division, 41, 621–624
Reproductive systems
    cat, 612–616
    definition of, 595
    dissection of, 612–616
    female organs of, 601–612
    gamete formation and, 621–624
    human, 595–612
    male organs of, 595–601, 602
    menstrual cycle and, 608–612
    pig, 616
Residual volume, 502
Resolution, 1
Resorption of bone, 84
Respiration
    air volumes exchanged, 501–508
    definition of, 485
    and pH, 587
Respiratory bronchiole, 491
Respiratory-cardiovascular interactions, 509
Respiratory center, 508
Respiratory experiments, 499–509
Respiratory sounds, 500
Respiratory system
    air volumes exchanged, 501–508
    cat, 492, 495–497
    definition of, 485
    dissection of, 492, 495–497
    human, 485–492
    organs of, 485–492
    physiology of, 10, 485–497
    sheep, 497
Respirometer (spirometer), use of, 502–508
Rete testis, 598
Reticular connective tissue, 63
Reticular fiber, 59
Reticular layer of dermis, 72
Reticulocyte count, 372
Reticuloendothelial cell, 369
Retina. See Nervous tunic
Retroperitoneal, 557
Reverberating circuit, 249–250
Rh blood grouping, 388–389
Ribosome, 31
Right lymphatic duct, 479
Rima glottidis, 489
Rod, 323
Root canal, 521
Rotator (musculotendinous) cuff, 179
Round ligament, 604

Round window, 331
Rugae
    of gallbladder, 527
    of stomach, 526
    of vagina, 606

Saccule, 331
Sacrum, 106
Saddle joint, 127
Salivary amylase, 546
Salivary gland, 521
Salt, 583
Sarcolemma, 135, 136, 137
Sarcomere, 136
Sarcoplasm, 135
Sarcoplasmic reticulum (SR), 135
Satellite cell, 249
Scala tympani, 332
Scala vestibuli, 332
Schwann cell. See Neurolemmocyte
Sclera, 322
Scleral venous sinus, 322
Sebaceous ciliary gland, 321
Sebaceous (oil) gland, 75
Secondary (lobar) bronchus, 490
Secondary follicle, 359, 602
Secondary oocyte, 624
Secondary spermatocyte, 358, 595, 622
Secondary tubule, 608
Sedimentation rate, 378
Segmental contractions, 540
Segmentation nucleus, 626
Semen, 600
Semicircular canal, 331
Semicircular duct, 331
Seminal vesicle, 600
Seminiferous tubule, 358, 595
Sensations
    auditory, 331–337
    characteristics of, 307–308
    definition of, 307
    equilibrium, 337–339
    general, 309
    gustatory, 317–321
    olfactory, 314–317
    pain, 310
    prerequisites for, 307–308
    pressure, 310
    proprioceptive, 310–311
    special, 314–340
    tactile, 309–310
    tests for, 326–330, 335–336
    thermoreceptive, 310
    touch, 309–310
    vibration, 310
    visual, 321–330
Sensory modality, 307
Sensory-motor integration, 340
Sensory neuron, 251
Sensory pathways, 314
Sensory unit, 307
Septal cell, 492
Serosa, 519, 526, 532, 563, 606
Serous membrane, 65–66
Sertoli cell. See Sustentacular cell
Sesamoid bone, 89
Sex chromosome, 621, 638
Sex inheritance, 638–640
Sex-linked trait, 640
Sheath of Schwann. See Neurolemma
Sheep
    brain, 295–296
    cranial nerves, 295–296
    eye, 324–326
    heart, 402, 405
    kidney, 567
    pluck, dissection of, 497

Sigmoid colon, 531
Simple columnar epithelium, 55
Simple cuboidal epithelium, 52, 55
Simple diffusion, 34
Simple gland, 57
Simple squamous epithelium, 52
Sinoatrial (SA) node, 453
Skeletal muscle
  abductor pollicis longus, 184
  acromiodeltoid, 208
  acromiotrapezius, 211
  actions of, 157–158
  adductor brevis, 192, 196
  adductor femoris, 214
  adductor longus, 192, 196, 214
  adductor magnus, 196
  anconeus, 181, 211
  anterior scalene, 189
  arytenoid, 168
  belly, 157
  biceps brachii, 181, 211
  biceps femoris, 197, 214
  brachialis, 181, 211
  brachioradialis, 181, 214
  buccinator, 161
  bulbospongiosus, 176
  caudofemoralis, 214
  clavobrachialis, 208
  clavotrapezius, 208
  cleidomastoid, 206
  coccygeus, 174
  connective tissue components
    of, 158–160
  coracobrachialis, 179
  corrugator supercilii, 161
  cricothyroid, 168
  criteria for naming, 158
  deep transverse perineus,176
  deltoid, 179
  depressor labii inferioris, 161
  diaphragm, 173
  digastric, 166, 208
  epicranius, 161
  epitrochlearis, 211
  erector spinae, 188
  extensor carpi radialis brevis,
    184, 214
  extensor carpi radialis longus,
    184, 214
  extensor carpi ulnaris,
    184, 214
  extensor digiti minimi, 184
  extensor digitorum, 184
  extensor digitorium communis,
    214
  extensor digitorum lateralis,
    214
  extensor digitorum longus, 199,
    218
  extensor hallucis longus, 199
  extensor indicis, 184
  extensor pollicis brevis, 214
  extensor pollicis longus, 184
  external anal sphincter, 176
  external intercostals, 173, 208
  external oblique, 171, 206
  flexor carpi radialis, 183, 211
  flexor carpi ulnaris, 183, 211
  flexor digitorum longus, 200,
    214
  flexor digitorum profundus,
    184, 211
  flexor digitorum sublimis, 211
  flexor digitorum superficialis,
    183
  flexor hallucis longus, 200, 218
  flexor pollicis longus, 184
  frontalis, 161
  gastrocnemius, 200, 218
  genioglossus, 165, 208
  geniohyoid, 166, 208

gluteus maximus, 192, 214
gluteus medius, 192, 214
gluteus minimus, 192
gracilis, 196, 214
group actions, 157–158
hamstrings, 197
hyoglossus, 165, 208
iliacus, 192
iliococcygeus, 174
iliocostalis, 211
iliocostalis cervicis, 188
iliocostalis lumborum, 188
iliocostalis thoracis, 188
iliopsoas, 210
inferior constrictor, 160, 168
inferior gemellus, 192
inferior oblique, 164
inferior rectus, 164
infrahyoid (strap), 168
infraspinatus, 179, 211
internal intercostals, 173, 208
internal oblique, 171, 208
interspinales, 189
intertransversarii, 189
ischiocavernosus, 176
lateral cricoarytenoid, 168
lateral pterygoid, 163
lateral rectus, 164
latissimus dorsi, 179, 211
levator ani, 174
levator labii superioris, 161
levator palpebrae superioris,
  161
levator scapulae, 177, 207
levator scapulae ventralis,
  208
longissimus capitis, 170, 188
longissimus cervicis, 188
longissimus dorsi, 211
longissimus thoracis, 188
masseter, 163, 208
medial pterygoid, 163
medial rectus, 164
mentalis, 161
middle constrictor, 168
middle scalene, 188
multifidus, 189
multifidus spinae, 211
mylohyoid, 166, 208
naming, 158
obturator externus, 193
obturator internus, 193
occipitalis, 161
omohyoid, 168
orbicularis oculi, 161
orbicularis oris, 161
origin and insertion, 157
palatoglossus, 165
palatopharyngeus, 168
palmaris longus, 183, 214
pectineus, 192, 196, 214
pectoantebrachialis, 206
pectoralis major, 179, 206
pectoralis minor, 177, 206
peroneus brevis, 199, 218
peroneus longus, 199, 218
peroneus tertius, 199, 218
piriformis, 193
plantaris, 200, 218
platysma, 161
popliteus, 200
posterior cricoarytenoid, 168
posterior scalene, 189
posterior triangle, 170
pronator quadratus, 181, 214
pronator teres, 181, 211
psoas major, 192
pubococcygeus, 174
quadratus lumborum, 171
quadriceps femoris, 196
rectus abdominis, 171

rectus femoris, 197, 214
rhomboideus, 211
rhomboideus capitis, 211
rhomboideus major, 177
rhomboideus minor, 177
risorius, 161
rotatores, 189
sacrospinalis. *See* Erector spinal
salpingopharyngeus, 167
sartorius, 197, 214
scalene, 189
scalenus, 207
segmental, 189
semimembranosus, 197, 214
semispinalis capitis, 170, 189
semispinalis cervicus, 189
semispinalis thoracis, 189
semitendinosus, 197, 214
serratus anterior, 177
serratus dorsalis caudalis, 211
serratus dorsalis cranialis, 211
serratus ventralis, 208
soleus, 200, 218
spinalis capitis, 189
spinalis cervicis, 189
spinalis dorsi, 211
spinalis thoracis, 189
spinodeltoid, 208
spinotrapezius, 211
splenius, 188
splenius capitis, 170, 188
splenius cervicis, 188
sternocleidomastoid, 170
sternohyoid, 168, 208
sternomastoid, 206
sternothyroid, 168, 208
styloglossus, 165, 211
stylohyoid, 166
stylopharyngeus, 168
subclavius, 177
subscapularis, 179, 211
superficial inguinal ring, 171
superficial transverse perineus,
  176
superior gemellus, 192
superior oblique, 164
superior rectus, 164
supinator, 181
suprahyoid, 166
supraspinatus, 179, 211
temporalis, 163, 208
tensor fasciae latae, 192, 214
tenuissimus, 214
teres major, 179, 211
teres minor, 179, 211
thyroarytenoid, 168
thyrohyoid, 168, 208
tibialis anterior, 199, 218
tibialis posterior, 200
transversospinalis, 189
transversus abdominis, 171
transversus costarum, 207
trapezius, 177
triceps brachii, 181, 206
urethral sphincter, 176
vastus intermedius, 197, 214
vastus lateralis, 197, 214
vastus medialis, 197, 214
xiphihumeralis, 207
zygomaticus major, 161
Skeletal muscle tissue
all-or-none principle of, 140
connective tissue components
  of, 158–160
contraction of, 137–149
definition of, 135
electromyography and, 148–149
histology of, 135–140
physiology of, 135–140
polygraph and, 141–142
sliding-filament theory of, 140

Skeletal system
  appendicular skeleton of,
    110–118
  articulations of, 123–128
  axial skeleton of, 93–108
  bones of, 93–120
  cat, 118, 120
  definition of, 81
  divisions of, 93
  fontanels, 100
  functions of, 81–84
  histology of, 83
  human, 93–120
  paranasal sinuses of, 100, 102
  skull, 93–100
  sutures, 100
Skene's gland. *See* Paraurethral
  gland
Skin
  color of, 73
  dermis of, 71–73
  epidermis of, 71
  functions of, 71
  glands of, 75–76
  hair of, 73–74
  nails of, 76
  structure of, 71–73
Skull
  bones of, 93–100
  fontanels of, 100
  foramina of, 100
  paranasal sinuses of, 100, 102
  sutures of, 100
Sliding-filament theory, 140
Slit membrane, 558
Small intestine, 529–530
Smooth (visceral) muscle tissue
  contraction of, 539–545
  effect of acetylcholine on,
    543–544
  effect of atropine on, 544
  effect of calcium removal on,
    545
  effect of norepinephrine on,
    543
  effect of oxygen on, 542
  effect of temperature on,
    542–543
  histology of, 137
  inherent contractile activity of,
    541–542
  isolation of, 541
  observation of, 539–541
Snellen chart, 326
Soft palate, 487, 519, 521
Somatic cell division, 41–43
Somatic motor pathways, 269–270,
  340
  direct, 269–270, 340
  indirect, 270, 340
Somatic nervous system, 245
Somatic sensory pathways, 266,
  269, 314
  anterolateral (spinothalamic) to
    the cortex, 269
  to the cerebellum, 269
  posterior column–medial
    lemniscus to the cortex,
    269, 314
Special senses
  auditory, 331–337
  equilibrium, 337–339
  gustatory, 317–321
  olfactory, 314–317
Specific gravity, of urine, 572–573
Sperm cell, 358, 595, 622
Spermatid, 358, 595, 622
Spermatogenesis, 358, 621–623
Spermatogenic cell, 358, 595
Spermatogonium, 358, 595, 621
Spermatozoon. *See* Sperm cell

Spermiogenesis, 622–623
Sphincter of hepatopancreatic
   ampulla, 527
Sphincter of Oddi. *See* Sphincter
   of hepatopancreatic ampulla
Sphygmomanometer, use of,
   466, 467
Spinal cord
   columns of, 260
   external anatomy, 259–260
   horns of, 260
   meninges, 259
   nerves of, 259–272
   tracts of, 266–270
   transverse-sectional structure,
      260
Spinal nerves, 259–273
Spinal reflexes of frog, 251–254
Spiral organ, 332
Spirogram, 501, 505
Spleen, 478
Spongy (cancellous) bone, 81
Squamous pulmonary epithelial
   cell, 492
Standard metabolic rate, determi-
   nation of in rats, 361–363
Starch, test for, 546
Starling's law of the heart, 462
Stellate reticuloendothelial cell, 527
Stereocilia, 337, 598
Stethoscope, use of, 465
Stimulus, 307
Stomach, 525–526, 527
Straight tubule, 598
Stratified squamous epithelium,
   55–56
Stratum basale, 71
Stratum basalis, 606
Stratum corneum, 71
Stratum functionalis, 606
Stratum granulosum, 71
Stratum lucidum, 71
Stratum spinosum, 71
Striations, 136
Stroma, 359, 602
Subarachnoid space, 259
Subcutaneous layer, 71
Sublingual gland, 521
Submandibular gland, 521
Sudoriferous (sweat) gland, 75
Sugar, test for, 546
Sulcus
   of brain, 277
Superficial fascia. *See*
   Subcutaneous layer
Superior colliculi, 277
Superior extensor retinaculum,
   199
Superior vena cava, 396
Surface anatomy
   abdomen and pelvis, 229
   arm, 229–233
   back, 226–227
   buttocks, 233
   chest, 227–229
   cranium, 225
   definition of, 225
   ear, 336
   eye, 325
   face, 225
   foot, 237
   forearm, 231, 233
   hand, 233
   head, 225
   knee, 236–237
   leg, 237
   lower limb, 233–237
   neck, 225–226
   nose, 488
   shoulder, 229–231
   thigh, 233, 236

trunk, 226–229
   upper limb, 229–233
Suspensory ligament
   of lens, 324
   of ovary, 359, 601
Sustentacular cell, 358, 595–596
Sutural bone, 89
Suture, 100, 123
Sweat gland. *See* Sudoriferous
   gland
Symphysis joint, 124
Synapse, 249
Synapsis, 622
Synaptic cleft, 138
Synaptic end bulb, 138, 246
Synaptic gutter, 138
Synaptic vesicle, 246
Synarthrosis, 123
Synchondrosis, 123
Syndesmosis, 123–124
Synergist, 157
Syngamy, 626
Synovial joint, 124–127
Synovial (joint) cavity, 123
Synovial membrane, 66, 124
System
   cardiovascular, 11, 369–468
   definition of, 71
   digestive, 12, 519–548
   endocrine, 11, 351–364
   integumentary, 10, 71
   lymphatic, 11, 477–479, 480
   muscular, 11, 157
   nervous, 11, 245–299
   origin and insertion, 157
   reproductive, 12, 595–616
   respiratory, 12, 485–509
   skeletal, 11, 81–90, 93–120
   urinary, 12, 557–576
Systemic anatomy, 9
Systemic circulation, 409–435
Systole, 459
Systolic blood pressure, 466

T cell, 361
T wave, 455
Tactile disc, 309
Tactile sensation, 309–310
Taenia coli, 532
Target cell, 351
Tarsal gland, 321
Taste bud, 317
Tectorial membrane, 334
Teeth, 521
Telophase, 43
Tendinous intersection, 171
Tendon, 160
Tendon organ, 310
Terminal bronchiole, 490, 491
Tertiary (segmental) bronchus, 490
Testes, 358–359, 595–596
Testing for
   ABO blood, 386–388
   action of bile on fats, 547
   afterimages, 329–330
   astigmatism, 327–328
   auditory acuity, 335–337
   bile pigments in urine, 575
   binocular vision, 329
   blind spot, 329
   blood flow in frog, 467–468
   blood pressure, 467
   chemical constituents of blood,
      371
   chemical regulation of respira-
      tion, 508–509
   chloride in urine, 573
   color blindness, 330, 644
   color of urine, 572
   constriction of pupil, 329
   convergence, 329

cranial nerve function, 284–288
   crystals in urine, 575–576
   deglutition, 538–539
   determining hyperinsulinemia
      in fish, 363–364
   determining mean electrical
      axis, 456, 458–459
   determining standard metabolic
      rate in rats, 361–363
   differential white blood cell
      count, 383–385
   effect of pH on starch digestion,
      547
   effects of estrogens and testos-
      terone in rats, 361–363
   electrocardiogram, 453–456
   equilibrium, 337–339
   extrinsic control of cardiac
      cycle, 463–464
   fat digestion, 547
   glucose, 573–574
   gustation, 319, 321
   hearing impairment, 335–336
   heart sounds, 464–465
   hematocrit, 377–378
   hemoglobin determination,
      378–381
   hemoglobin in urine, 575
   intrinsic control of cardiac
      cycle, 462
   ketone bodies in urine, 574–575
   measuring pH, 583–584
   muscle contraction, 140–149
   near-point accommodation, 328
   observing cardiac cycle,
      461–462
   olfactory adaptation, 316–317
   pain receptors, 312–313
   peripheral blood flow in frog,
      467–468
   phenylketonuria, 637, 643
   pH of urine, 572
   physical characteristics of
      blood, 370–371
   pressure receptors, 312
   proprioceptors, 313–314
   protein digestion, 548
   PTC inheritance, 644
   pulse rate, 466
   recording cardiac cycle, 462
   red blood cell count, 373–376
   refractory period, 464
   respirations and pH, 587
   respiratory-cardiovascular
      interactions, 509
   respiratory sounds, 500
   respiratory volumes, 501–508
   Rh factor, 388–389
   sedimentation rate, 378
   sodium chloride in urine, 573
   specific gravity of urine,
      572–573
   starch digestion, 546
   starch, presence of, 546
   sugar, detection of, 546
   thermoreceptors, 312
   touch receptors, 311–312
   transparency of urine, 572
   tubular secretion, 569–570
   two-point discrimination, 311
   visual acuity, 326–327
   white blood cell count,
      383–385
Tetanus, 147
Tetrad, 622
Thalamus, 277
Theca folliculi, 624
Thermoreceptive sensation,
   310
Thoracic aorta, 411
Thoracic cavity, 19

Thoracic duct, 479
Thoracic vertebrae, 103
Threshold stimulus, 137
Thrombocyte. *See* Platelet
Thymus gland, 360–361, 479
Thyroid cartilage, 225
Thyroid colloid, 354
Thyroid follicle, 354
Thyroid gland, 354–355
Tidal volume, 501
Tissue
   bone, 81–90
   connective, 58–65
   definition of, 51
   epithelial, 51–58
   muscle, 135–149
   nervous, 246
Tongue, 519, 521
Tonsils
   lingual, 488
   palatine, 488
   pharyngeal, 488
Total lung capacity, 502
Touch sensation, 309
Trachea, 489–490, 492
Tract, of spinal cord
   anterior corticospinal, 270
   anterior spinocerebellar, 269
   anterior spinothalamic, 269,
      314
   corticobulbar, 270
   lateral corticospinal, 270
   lateral reticulospinal, 270
   lateral spinothalamic, 269, 314
   medial reticulospinal, 270
   motor (descending), 270, 271
   posterior column, 269
   posterior spinocerebellar, 269
   sensory (ascending), 266, 269,
      271
   tectospinal, 270
   vestibulospinal, 270
Transduction, 307
Transitional epithelium, 56
Transverse colon, 531
Transverse (T) tubule, 135
Triad, 135
Tricuspid valve, 396
Trigger zone, 246
Triglyceride, 547
Trigone, 563
Trophoblast, 626
Tubular reabsorption, 569
Tubular secretion, 569
Tunica albuginea, 358, 359, 595,
   602
Tunica externa, 409
Tunica interna, 409
Tunica media, 409
Tunica vaginalis, 358, 595
Turtle, heart, 461–464
Twitch contraction, 146
Two-point discrimination test, 311
Tympanic membrane, *See*
   Eardrum
Type II cutaneous mechanorecep-
   tor, 310

Umbilical cord, 437
Umbilicus, 229, 632
Unicellular exocrine gland, 57
Unit summation, 145–146
Universal donor, 386
Universal recipient, 386
Ureter, 562, 565
Urethra, 564, 566, 598
Urinalysis, 571–576
Urinary bladder, 563, 566, 567
Urinary system
   cat, 564–566
   definition of, 557